Lecture Notes in Computer Science 5467

Commenced Publication in 1973
Founding and Former Series Editors:
Gerhard Goos, Juris Hartmanis, and Jan van Leeuwen

Matthias Ehrgott Carlos M. Fonseca
Xavier Gandibleux Jin-Kao Hao
Marc Sevaux (Eds.)

Evolutionary Multi-Criterion Optimization

5th International Conference, EMO 2009
Nantes, France, April 7-10, 2009
Proceedings

Volume Editors

Matthias Ehrgott
The University of Auckland, Department of Engineering Science
70 Symonds Street, Room 415, Auckland 1001, New Zealand
E-mail: m.ehrgott@auckland.ac.nz

Carlos M. Fonseca
Universidade do Algarve, Faculty of Science and Technology
Department of Electronic Engineering and Informatics
Campus de Gambelas, 8005-139 Faro, Portugal
E-mail: cmfonsec@ualg.pt

Xavier Gandibleux
Université de Nantes, Faculty of Sciences and Technology
Laboratoire d' Informatique de Nantes-Atlantique, UMR CNRS 6241
2, Rue de la Houssinière, BP 92208, 44322 Nantes Cedex 03, France
E-mail: Xavier.Gandibleux@univ-nantes.fr

Jin-Kao Hao
Université d'Angers, LERIA, Faculty of Sciences
2 Boulevard Lavoisier, 49045, Angers Cedex 01, France
E-mail: Jin-Kao.Hao@univ-angers.fr

Marc Sevaux
Université de Bretagne-Sud - UEB, Lab-STICC - UMR CNRS 3192
Centre de Recherche BP 92116, 56321 Lorient Cedex, France
E-mail: marc.sevaux@univ-ubs.fr

Library of Congress Control Number: Applied for

CR Subject Classification (1998): F.2, G.1.6, G.1.2, I.2.8

LNCS Sublibrary: SL 1 – Theoretical Computer Science and General Issues

ISSN 0302-9743
ISBN-10 3-642-01019-9 Springer Berlin Heidelberg New York
ISBN-13 978-3-642-01019-4 Springer Berlin Heidelberg New York

springer.com

Printed in Germany

Typesetting: Camera-ready by author, data conversion by Scientific Publishing Services, Chennai, India
Printed on acid-free paper SPIN: 12647408 06/3180 5 4 3 2 1 0

Preface

Multi-criterion optimization refers to optimization problems with two or more objectives expressing conflicting goals that are formulated within a mathematical programming framework. The problems addressed may involve linear or nonlinear objective functions and/or constraints, continuous or discrete variables, and may or may not be affected by uncertainty in the data. This branch of multiple criteria decision making (MCDM) finds application in numerous domains: engineering design, health, transportation, telecommunications, bioinformatics, etc.

The concept of a unique optimal solution does not apply as soon as multiple objectives are optimized simultaneously. The models and methods introduced in multi-criterion optimization deal with the concept of a set of efficient (also called Pareto optimal) solutions. Efficient solutions imply trade-offs between the different criteria. The computation of the efficient solution set may be hard when the size of the problem is large, when the problem is computationally complex, when the data are not crisp. It is then often impossible to guarantee the computation of exact solutions. In that case, approximate solutions, i.e., sub-optimal solutions computed with limited and controlled resources, such as available time, are of interest. This is the domain of multi-objective metaheuristics, of which evolutionary multi-criterion optimization (EMO) is definitely the most prominent representative. The success of EMO is due to the simplicity of its concepts and the generality of its methods, and is clearly expressed by the many impressive success stories reported in the literature.

Research activities in EMO have boomed since the mid-1990s. Three generations of work are identifiable throughout the years. In the first generation, research focused on the design of efficient algorithmic methods for the approximation of efficient solutions. The second generation dealt with the problem of measuring the quality of the approximations generated by the algorithms. The current generation stresses the hybridization with other currents of optimization. "Hot" questions concern the integration of a decision maker (interactivity, preferences), the robustness of the generated solutions, and the coupling with other optimization approaches (operations research, constraint programming, other metaheuristics). Research in the area of EMO is evolving very fast, and continuously investigates new challenging open questions in order to enlarge and refine its position of useful technology for multi-criterion optimization. Today EMO algorithms are recognized as being among the most valuable and promising methods for tackling complex and diverse multi-criterion optimization problems.

To capitalize on, and promote exchanges within, the growing community of researchers involved in evolutionary multi-criterion optimization, an international conference series devoted to EMO was launched in 2001. This conference brings together researchers and practitioners from different disciplines of computer science, operations research, engineering optimization, mathematical programming and multi-criteria analysis. Theoretical results and algorithmic developments in the field of EMO are covered, including practice and applications of EMO in real-life situations. After Zürich (Switzerland) in 2001, Faro (Portugal) in 2003, Guanajuato (Mexico) in 2005 and Matsushima-Sendai (Japan) in 2007, Nantes (France) hosted the 5th International Conference on Evolutionary Multi-Criterion Optimization (EMO 2009). The conference took place at the Faculty of Science of the University of Nantes during April 7-10, 2009.

To emphasize the current generation of EMO research, the subtitle of the conference was "Where Optimization Technologies Meet Evolutionary Multi-Criterion Optimization". The conference was structured around invited speakers (tutorial sessions, keynote sessions, industrial session) and selected presentations (oral sessions and poster session). The EMO 2009 scientific program included five invited talks given by Denis Bouyssou (France) on "Choice and Preferences," Kathrin Klamroth (Germany) on "Discrete Multiobjective Optimization," Manual Laguna (USA) on "Scatter Search and Path-Relinking," Thomas Stützle (Belgium) on "Ant Colony Optimization," and Pascal Van Hentenryck (USA) on "Constraint Programming." The selection of contributed papers was based on full-paper submissions, rigorously refereed by at least three members of the International Program Committee. The EMO 2009 International Program Committee was composed of 118 international well-known researchers from 25 countries. This volume includes the 39 research papers that were selected for presentation at the conference, from the 72 submissions received. The papers published in the volume represent the most recent developments on evolutionary multi-criterion optimization, and cover a large spectrum of current research topics: applications, algorithm development, theoretical analysis, performance analysis and comparison, alternative methods, MCDM, many objectives, uncertainty and noise, and the interface between EMO and MCDA.

We would like to express our appreciation to the keynote, tutorial and industrial speakers for accepting our invitation. We also thank all the authors who submitted their work to EMO 2009. Our sincerest gratitude goes to the members of the International Program Committee for the considerable work they have invested in the reviewing process and for their contribution to making this volume an up-to-date reference in the field of evolutionary multi-criterion optimization. The organizers are especially grateful to all the sponsors for financial support, and to the members of the Local Organizing Committee for their investment in the preparation of the conference, in particular, Valérie Coutand, the conference secretary.

April 2009

Matthias Ehrgott
Carlos M. Fonseca
Xavier Gandibleux
Jin-Kao Hao
Marc Sevaux

Organization

EMO 2009 was organized by the Faculté des Sciences of the Université de Nantes in cooperation with Ecole Centrale de Nantes, Ecole des Mines de Nantes, Université d'Angers and Université de Bretagne-Sud.

Committees and Chairs

Conference Chairs
: Carlos M. Fonseca (Universidade do Algarve, Portugal)
: Xavier Gandibleux (Université de Nantes, France)

Steering Committee
: David Corne (Heriot-Watt University, UK)
: Kalyanmoy Deb (IIT Kanpur, India)
: Peter J. Fleming (University of Sheffield, UK)
: Carlos M. Fonseca (Universidade do Algarve, Portugal)
: J. David Schaffer (Phillips Research, USA)
: Lothar Thiele (ETH Zürich, Switzerland)
: Eckart Zitzler (ETH Zürich, Switzerland)

Local Organizing Committee
: Vincent Barichard (Université d'Angers, France)
: Philippe Dépincé (Ecole Centrale de Nantes, France)
: Xavier Gandibleux (Université de Nantes, France)
: Jin Kao Hao (Université d'Angers, France)
: Fabien Le Huédé (Ecole des Mines de Nantes, France)
: Anthony Przybylski (Université de Nantes, France)
: Marc Sevaux (Université de Bretagne-Sud, France)

International Program Committee

Program Chairs

Matthias Ehrgott (University of Auckland, New Zealand)
Jin-Kao Hao (Université d'Angers, France)
Marc Sevaux (Université de Bretagne-Sud, France)

MCDM Track Chair

José Figueira (Instituto Superior Técnico, Portugal)

Program Committee

H. Abbass (Australia)
H.E. Aguirre (Japan)
R.T.F. Ah King (Mauritius)
E. Alba (Spain)
S. Azarm (USA)
V. Barichard (France)
M. Basseur (France)
V. Belton (UK)
N. Beume (Germany)
J. Branke (Germany)
C.A. Brizuela (Mexico)
R. Caballero (Spain)
C.A. Coello Coello (Mexico)
D.W. Corne (UK)
L. Costa (Portugal)
K. Deb (India)
X. Delorme (France)
P. Dépincé (France)
K. Doerner (Austria)
R. Drechsler (Germany)
M. Ehrgott (New Zealand)
M. Emmerich (The Netherlands)
R. Everson (UK)
J. Fieldsend (UK)
J. Figueira (Portugal)
P. Fleming (UK)
J. Fodor (Hungary)
C.M. Fonseca (Portugal)
X. Gandibleux (France)
A. Gaspar-Cunha (Portugal)
G. Salvatore (Italy)
D. Greiner (Spain)
C. Grimme (Germany)
T. Hanne (Switzerland)
J-K. Hao (France)
C. Haubelt (Germany)
C. Henggeler (Portugal)
A. Hernandez (Mexico)
M. I. Heywood (Canada)
P. Hingston (Australia)
E. J Hughes (UK)
C. Igel (Germany)
H. Ishubuchi (Japan)
A.j Jaszkiewicz (Poland)
S. Jeong (Japan)
Y. Jin (Germany)
M. João Alves (Portugal)
D. Jones (UK)
L. Jourdan (France)
N. Katoh (Japan)
J. Knowles (UK)
M. Koeppen (Japan)
R. Kumar (India)
M. E. Kurz (USA)
G. B. Lamont (USA)
D. Landa Silva (UK)
F. Le Huédé (France)
X. Li (Australia)
S. Louis (USA)
J. Antonio Lozano (Spain)
C. Mariano (Mexico)
E. Mezura-Montes (Mexico)
M. Middendorf (Germany)
K. Miettinen (Finland)
J. Molina (Spain)
S. Mostaghim (Germany)
T. Murata (Japan)
H. Nakayama (Japan)
B. Naujoks (Germany)
A. J. Nebro (Spain)
S. Obayashi (Japan)
T. Okabe (Japan)
J. Olvander (Sweden)
I. Ono (Japan)
A. Oyama (Japan)
L. Paquete (Portugal)
G. Parks (UK)
V. Pediroda (Italy)
S. Poles (Italy)
R. Purshouse (UK)
R. S. Ranjithan (USA)
P. Reed (USA)
M. Reyes-Sierra (Mexico)
P. Rockett (UK)
K. Rodrguez (Mexico)
T. Philip Runarsson (Ireland)
J. Sakuma (Japan)
D. Savic (UK)

M. Savill (UK)
J. D. Schaffer (USA)
H. Schmeck (Germany)
M. Schoenauer (France)
D. Seese (Germany)
B. Sendhoff (Germany)
M. Sevaux (France)
P. Siarry (France)
K. Sörensen (Belgium)
R. Steuer (USA)
T. Stützle (Belgium)
R. Takahashi (Brazil)
E-G. Talbi (France)
K. Chen Tan (Singapore)
K. Tanaka (Japan)
J. Teghem (Belgium)
J. Teich (Germany)
J. Antonio Tenreiro (Portugal)
L. Thiele (Switzerland)
A. Tiwari (USA)
V. Tkind't (France)
A. Toffolo (Italy)
D. Tuyttens (Belgium)
D. Vanderpooten (France)
S. Watanabe (Japan)
L. While (Australia)
G. G. Yen (USA)
Y. Yun (Japan)
A. Zell (Germany)
E. Zitzler (Switzerland)

Sponsoring Institutions

Université de Nantes
Université de Bretagne-Sud
Université d'Angers
Ecole Centrale de Nantes
Ecole des Mines de Nantes
Laboratoire d'Informatique de Nantes Atlantique (LINA)
Laboratoire d'Electronique des Systèmes TEmps Réel (LABSTICC)
Laboratoire d'Etude et de Recherche en Informatique d'Angers (LERIA)
Fédération des laboratoires STIC nantais (AtlanSTIC)
EURO working group on metaheuristics (EU/ME)
Esteco (Italy)
Sirehna (France)

Support Institution

Institut de Recherche en Communications et en Cyberntique de Nantes

Scientific Sponsorships

Centre National de la Recherche Scientifique (CNRS)
Société française de recherche opérationnelle et d'aide à la décision (ROADEF)
Groupe de travail français "programmation mathématique multiobjectif" (PM2O)

Table of Contents

Invited Talks

Theoretical Analysis

Uncertainty and Noise

Algorithm Development

Performance Analysis and Comparison

Applications

MCDM Track

Many Objectives

Alternative Methods

EMO and MCDA

Scatter Search and Path Relinking

Manuel Laguna

Leeds School of Business
University of Colorada
Boulder, CO 80309-0419 – USA
laguna@colorado.edu

This presentation explores the evolutionary approach called scatter search, which originated from strategies for creating composite decision rules and surrogate constraints. Recent studies have demonstrated the practical advantages of this approach for solving a diverse collection of optimization problems from both classical and real world settings. Scatter search contrasts with other evolutionary procedures, such as genetic algorithms, by providing unifying principles for joining solutions based on generalized path constructions in Euclidean space and by utilizing strategic designs where other approaches resort to randomization. Additional advantages are provided by intensification and diversification mechanisms that exploit adaptive memory, drawing on foundations that link scatter search to tabu search.

We also address the scatter search generalization called path relinking. Features that have been added to scatter search by extension of its basic philosophy, are captured in the path relinking framework. From a spatial orientation, the process of generating linear combinations of a set of reference solutions (as typically done in scatter search) may be characterized as generating paths between and beyond these solutions, where solutions on such paths also serve as sources for generating additional paths. This leads to a broader conception of the meaning of creating combinations of solutions. By natural extension, such combinations may be conceived to arise by generating paths between and beyond selected solutions in neighborhood space, rather than in Euclidean space.

The presentation also includes applications of scatter search/path relinking to multi-objective optimization and directions for future research.

M. Ehrgott et al. (Eds.): EMO 2009, LNCS 5467, p. 1, 2009.

Ant Colony Optimization

Thomas Stützle

IRIDIA-CoDE, Université Libre de Bruxelles (ULB), Brussels, Belgium
stuetzle@ulb.ac.be

Ant Colony Optimization (ACO) is a stochastic local search method that has been inspired by the pheromone trail laying and following behavior of some ant species [1]. Artificial ants in ACO essentially are randomized construction procedures that generate solutions based on (artificial) pheromone trails and heuristic information that are associated to solution components. Since the first ACO algorithm has been proposed in 1991, this algorithmic method has attracted a large number of researchers and in the meantime it has reached a significant level of maturity. In fact, ACO is now a well-established search technique for tackling a wide variety of computationally hard problems.

The vast majority of the available ACO applications concern NP-hard combinatorial optimization problems and, among these, mainly those with only one single objective. However, many realistic problems involve two or more, typically competing objective. Therefore, it is not surprising that several researches have investigated the extension of ACO algorithm to handle multiple objective functions. These approaches range from applications to problems with lexicographicically ordered objectives to problems that are tackled in the Pareto sense.

In this tutorial we will first give an overview of ACO, highlighting its inspiring source, the main algorithmic variants, and the main application areas. The core part of this tutorial then reviews ways of how ACO algorithms can be used to tackle multiobjective *combinatorial* optimization problems (MCOPs). We will review the main approaches that have been proposed so far with a special emphasis on the application to MCOPs that are tackled in the Pareto sense. In fact, there exists a large number of degrees of freedom for the algorithm designer when applying ACO algorithms to MCOPs. These range from the use of one or several pheromone matrices, the usage of one or several ant colonies, variations on the pheromone update schemes, the usage or not of local search procedures and so on. We will support our discussion with results obtained from some experimental analyses for conceptually simple multiobjective problems such as the multiobjective quadratic assignment problem and the multiobjective traveling salesman problem.

Reference

1. Dorigo, M., Stützle, T.: Ant Colony Optimization. MIT Press, USA (2004)

M. Ehrgott et al. (Eds.): EMO 2009, LNCS 5467, p. 2, 2009.

Constraint Programming

Pascal Van Hentenryck

Brown University, Providence RI 02912, USA

Constraint programming is a remarkable success story. It quickly moved from research laboratories to industrial applications in the late 1980s and is in daily use to solve complex optimization throughout the world. At the same time, constraint programming has continued to evolve, addressing new needs and opportunities. This talk reviews some recent progress in constraint programming. The first part of the talk starts with its fundamental contribution, the ability to express and exploit combinatorial substructures to prune infeasible solutions and find feasible solutions. It also reviews some of the benefits of its architecture on a variety of applications in scheduling, rostering, and combinatorial matching, emphasizing the underlying modeling and computation techniques. The second part of the talk argues that constraint programming is an integration technology and reviews some hybridizations of constraint programming, including local search, mathematical programming, and global optimization. The final part of the talk gives a brief overview of some promising, novel applications of constraint programming.

Acknowledgments. This work was supported in part by the U.S. Department of Homeland Security's National Infrastructure Analysis and Simulation Center (NISAC) Program, by NSF award DMI-0600384, and ONR Award N000140610607.

M. Ehrgott et al. (Eds.): EMO 2009, LNCS 5467, p. 3, 2009.

Discrete Multiobjective Optimization

Kathrin Klamroth

University of Wuppertal, 42097 Wuppertal, Germany
klamroth@math.uni-wuppertal.de
http://www.math.uni-wuppertal.de/~klamroth

1 Problems and Applications

Multiple objective combinatorial optimization (MOCO) has become a quickly growing field in multiple objective optimization, and has recently attracted the attention of researchers both from the fields of multiple objective optimization and from single objective integer programming. Typical examples of MOCO problems include multiple objective knapsack problems with applications, among others, in capital budgeting, multiple objective assignment problems, the multiple objective traveling salesman problem, and other network problems like multiple objective minimum spanning tree, shortest path, and minimum cost flow problems. Formally, a general MOCO problem can be stated as $\min\{f(x) = (f_1(x), \ldots, f_p(x))^T : x \in X\}$, where the decision space X is a given *discrete* feasible set that usually has some additional combinatorial structure.

Since the set of feasible solutions of a MOCO problem is discrete and usually finite, it can at least theoretically be enumerated to identify all Pareto optimal solutions. This is, however, generally impractical due to the exponentially growing number of feasible (and sometimes also Pareto optimal) solutions.

2 Neighborhood Search and Connectedness

Structural properties of the efficient set of MOCO problems play a crucial role for the development of efficient solution methods. A central question in the context of metaheuristics relates to the connectedness of the efficient set with respect to combinatorially or topologically motivated neighborhood structures. A positive answer to this question would provide a theoretical justification for the application of fast neighborhood search techniques, not only for multiple objective but also for appropriate formulations of single objective problems.

Unfortunately, most of the classical MOCO problems do in general not possess a connectedness property with respect to reasonable neighborhoods. This includes, among others, knapsack problems (and even several special cases of knapsack problems) and linear assignment problems, shortest path problems as well as spanning tree and minimum cost flow problems. Numerical tests performed for different variants of the knapsack problem suggest, however, that the likelihood with which non-connected adjacency graphs occur in randomly generated problem instances is very small.

M. Ehrgott et al. (Eds.): EMO 2009, LNCS 5467, p. 4, 2009.

Preference Models Used in Multiple Criteria Decision Making: Foundations and Assessment

Denis Bouyssou

CNRS–LAMSADE, Université Paris Dauphine,
F-75775 Paris Cedex 16, France
bouyssou@lamsade.dauphine.fr

Abstract. This purpose of this talk is to offer a nontechnical introduction to the main preference models used in multiple criteria decision making. The emphasis is on the, central, additive value function model. We outline its axiomatic foundations and present various possible assessment techniques to implement it. Some extensions of this model, e.g., nonadditive models or models tolerating intransitive preferences are then briefly reviewed.

M. Ehrgott et al. (Eds.): EMO 2009, LNCS 5467, p. 5, 2009.

Approximating the Least Hypervolume Contributor: NP-Hard in General, But Fast in Practice

Karl Bringmann[1] and Tobias Friedrich[2]

[1] Universität des Saarlandes, Saarbrücken, Germany
[2] International Computer Science Institute, Berkeley, CA, USA

Abstract. The hypervolume indicator is an increasingly popular set measure to compare the quality of two Pareto sets. The basic ingredient of most hypervolume indicator based optimization algorithms is the calculation of the hypervolume contribution of single solutions regarding a Pareto set. We show that exact calculation of the hypervolume contribution is #**P**-hard while its approximation is **NP**-hard. The same holds for the calculation of the minimal contribution. We also prove that it is **NP**-hard to decide whether a solution has the least hypervolume contribution. Even deciding whether the contribution of a solution is at most $(1+\varepsilon)$ times the minimal contribution is **NP**-hard. This implies that it is neither possible to efficiently find the least contributing solution (unless $\mathbf{P} = \mathbf{NP}$) nor to approximate it (unless $\mathbf{NP} = \mathbf{BPP}$).

Nevertheless, in the second part of the paper we present a very fast approximation algorithm for this problem. We prove that for arbitrarily given $\varepsilon, \delta > 0$ it calculates a solution with contribution at most $(1 + \varepsilon)$ times the minimal contribution with probability at least $(1-\delta)$. Though it cannot run in polynomial time for all instances, it performs extremely fast on various benchmark datasets. The algorithm solves very large problem instances which are intractable for exact algorithms (e.g., 10000 solutions in 100 dimensions) within a few seconds.

1 Introduction

Multi-objective optimization deals with the task of optimizing several objective functions at the same time. As these functions are often conflicting, we cannot aim for a single optimal solution but for a set of Pareto optimal solutions. Unfortunately, the Pareto set frequently grows exponentially in the problem size. In this case, it is not possible to compute the whole front efficiently and the goal is to compute a good approximation of the Pareto front.

There are many indicators to measure the quality of a Pareto set, but there is only one widely used that is strictly Pareto compliant [22], namely the hypervolume indicator. Strictly Pareto compliant means that given two Pareto sets A and B the indicator values A higher than B if the Pareto set A dominates the Pareto set B. The hypervolume (HYP) measures the volume of the dominated portion of the objective space. It was first proposed and employed for multi-objective optimization by Zitzler and Thiele [20].

M. Ehrgott et al. (Eds.): EMO 2009, LNCS 5467, pp. 6–20, 2009.

It has become very popular recently and several algorithms have been developed to calculate it. The first one was the Hypervolume by Slicing Objectives (HSO) algorithm which was suggested independently by Zitzler [19] and Knowles [9]. To improve its runtime on practical instances, various speed up heuristics of HSO have been suggested [16, 18]. The currently best asymptotic runtime of $O(n \log n + n^{d/2})$ is obtained by Beume and Rudolph [2] by an adaption of Overmars and Yap's algorithm [11] for Klee's Measure Problem [8].

From a geometric perspective, the hypervolume indicator is just measuring the volume of the union of a certain kind of boxes in $\mathbb{R}^d_{\geq 0}$, namely of boxes which share the reference point[1] as a common point. We will use the terms point and box interchangeably for solutions as the dominated volume of a point defines a box and vice versa. Given a set M of n points in $\mathbb{R}^d$, we define the hypervolume of M to be

$$\text{HYP}(M) := \text{VOL}\left(\bigcup_{(x_1,\ldots,x_d)\in M} [0,x_1] \times \ldots \times [0,x_d] \right)$$

In [4] the authors have proven that it is **#P**-hard[2] in the number of dimension to calculate HYP precisely. Therefore, all hypervolume algorithms must have an exponential runtime in the number of objectives (unless $\mathbf{P} = \mathbf{NP}$). Without the widely accepted assumption $\mathbf{P} \neq \mathbf{NP}$, the only known lower bound for any d is $\Omega(n \log n)$ [3]. Note that the worst-case combinatorial complexity (i.e., the number of faces of all dimensions on the boundary of the union) of $\Theta(n^d)$ does not imply any bounds on the computational complexity.

Though the **#P**-hardness of HYP dashes the hope for an exact subexponential algorithm, there are a few estimation algorithms [1, 4] for approximating the hypervolume based on Monte Carlo sampling. However, the only approximation algorithm with proven bounds is presented in [4]. There, the authors describe an FPRAS for HYP which gives an ε-approximation of the hypervolume with probability $(1 - \delta)$ in time $\mathcal{O}(\log(1/\delta)\, nd/\varepsilon^2)$.

New Complexity Results

We will now describe a few problems related to the calculation of the hypervolume indicator and state our results. For this, observe that calculating the hypervolume itself is actually not necessary in a hypervolume-based evolutionary multi-objective optimizer as the algorithm actually only has to find a box with the minimal contribution to the hypervolume.

The *contribution* of a box $x \in M$ to the hypervolume of a set M of boxes is the volume dominated by x and no other element of M. We define the contribution $\text{CON}(M, x)$ of x to be

$$\text{CON}(M, x) := \text{HYP}(M) - \text{HYP}(M \setminus x).$$

[1] Without loss of generality we assume the reference point to be 0^d.

[2] **#P** is the analog of **NP** for counting problems. For details see either the original paper by Valiant [15] or the standard textbook on computational complexity by Papadimitriou [12].

In Section 2 we show that this problem is **#P**-hard to solve exactly. Furthermore, approximating CON by a factor of $2^{d^{1-\varepsilon}}$ is **NP**-hard for any $\varepsilon > 0$. Hence, CON is not approximable. Note that this is no contradiction to the above-mentioned FPRAS for HYPas an approximation of HYP yields no approximation of CON.

As a hypervolume-based optimizer is only interested in the box with the *minimal contribution*, we also consider the following problem. Given a set M of n boxes in $\mathbb{R}^d$, find the least contribution of any box in M, that is,

$$\mathrm{MINCON}(M) := \min_{x \in M} \mathrm{CON}(M, x).$$

The reduction in Section 2 shows that MINCON is **#P**-hard and not approximable, even if we know the box which is the least contributor.

Both mentioned problems can be used to find the box contributing the least hypervolume, but their hardness does not imply hardness of the problem itself, which we are trying to solve, namely calculating *which box has the least contribution.* Therefore we also examine the following problem. Given a set M of n boxes in $\mathbb{R}^d$, we want to find a box with the least contribution in M, that is,

$$\mathrm{LC}(M) := \operatorname*{argmin}_{x \in M} \mathrm{CON}(M, x).$$

If there are multiple boxes with the same (minimal) contribution, we are, of course, satisfied with any of them. In Section 2 we prove that this problem is **NP**-hard to decide.

However, for practical purposes it most often suffices to solve a relaxed version of the above problem. That is, we just need to find a box which contributes not much more than the minimal contribution, meaning that is is only a $(1 + \varepsilon)$ factor away. If we then throw out such a box, we have an error of at most ε. We will call this ε-LC(M) as it is an "approximation" of the problem LC. Given a set M of n boxes in $\mathbb{R}^d$ and $\varepsilon > 0$, we want to find a box with contribution at most $(1 + \varepsilon)$ times the minimal contribution of any box in M, that is,

$$\mathrm{CON}(M, \varepsilon\text{-}\mathrm{LC}(M)) \leq (1 + \varepsilon)\,\mathrm{MINCON}(M).$$

The final result of Section 2 is the **NP**-hardness of ε-LC. This shows, that there is no way of computing the least contributor efficiently, and even no way to approximate it.

New Approximation Algorithm

In Section 3 we will give a "practical" algorithm for determining a small contributor. Technically speaking, it solves the following problem we call ε-δ-LC(M): Given a set M of n boxes in $\mathbb{R}^d$, $\varepsilon > 0$ and $\delta > 0$, with probability at least $1 - \delta$ find a box with contribution at most $(1 + \varepsilon)\,\mathrm{MINCON}(M)$.

$$\Pr[\mathrm{CON}(M, \varepsilon\text{-}\delta\text{-}\mathrm{LC}(M)) \leq (1 + \varepsilon)\,\mathrm{MINCON}(M)] \geq 1 - \delta.$$

As we will be able to choose δ arbitrarily, solving this problem is of high practical interest. By the **NP**-hardness of ε-LC there is no way of solving ε-δ-LC efficiently, unless **NP** = **BPP**. This means, our algorithm cannot run in polynomial time for all instances. Its runtime depends on some hardness measure H (cf. Section 3.2), which is an intrinsic property of the given input, but generally unbounded, i.e., not bounded by some function in n and d.

However, in Section 4 we show that our algorithm is practically very fast on various benchmark datasets, even for dimensions completely intractable for exact algorithms like $d = 100$ for which we can solve instances with $n = 10000$ points within seconds. This implies a huge shift in the practical usability of the hypervolume indicator.

2 Hardness of Approximation

In this section we first show hardness of approximating MINCON, which we will use afterwards to show hardness of LC and ε-LC. We will reduce #MON-CNF to MINCON, which is the problem of counting the number of satisfying assignments of a Boolean formula in conjunctive normal form in which all variables are unnegated. While the problem of deciding satisfiability of such formula is trivial, counting the number of satisfying assignments is #**P**-hard and even approximating it by a factor of $2^{d^{1-\varepsilon}}$ for any $\varepsilon > 0$ is **NP**-hard, where d is the number of variables (see Roth [14] for a proof).

Theorem 1. *MINCON is #**P**-hard and approximating it by a factor of* $2^{d^{1-\varepsilon}}$ *is **NP**-hard for any* $\varepsilon > 0$.

Proof. To show the theorem, we reduce #MON-CNF to MINCON. Let $\square(a_1, \dots, a_d)$ denote a box $[0, a_1] \times \dots \times [0, a_d]$. Let $f = \bigwedge_{k=1}^{n} \bigvee_{i \in C_k} x_i$ be a monotone Boolean formula given in CNF with $C_k \subseteq [d] := \{1, \dots, d\}$, for $k \in [n]$, d the number of variables, n the number of clauses. First, we construct a box $A_k = \square(a_1^k, \dots, a_d^k, 2^d + 2) \subseteq \mathbb{R}^{d+1}$ for each clause C_k with one vertex at the origin and the opposite vertex at $(a_1^k, \dots, a_d^k, 2^d + 2)$, where we set

$$a_i^k = \begin{cases} 1, & \text{if } \ i \in C_k \\ 2, & \text{otherwise} \end{cases}, \quad i \in [d].$$

Additionally, we need a box $B = \square(2, \dots, 2, 1) \subseteq \mathbb{R}^{d+1}$ and set $M = \{A_1, \dots, A_n, B\}$. Since we can assume without loss of generality that no clause is dominated by another, meaning $C_i \not\subseteq C_j$ for every $i \neq j$, every box A_k uniquely overlaps a region $[x_1, x_1+1] \times \dots \times [x_d, x_d+1] \times [1, 2^d+2]$ with $x_i \in \{0, 1\}, i \in [d]$, so that the contribution of every box A_k is greater than 2^d and the contribution of B is at most 2^d, so that B is indeed the least contributor.

Observe that the contribution of B to HYP(M) can be written as a union of boxes of the form $B_{x_1,\dots,x_d} = [x_1, x_1 + 1] \times \dots \times [x_d, x_d + 1] \times [0, 1]$ with $x_i \in \{0, 1\}, i \in [d]$. Moreover, $B_{x_1,\dots,x_d}$ is not a subset of the contribution of B to HYP(M) iff it is a subset of $\bigcup_{k=1}^{n} A_k$ iff it is a subset of some A_k iff we have $a_i^k \geq x_i + 1$ for $i \in [d]$ iff $a_i^k = 2$ for all i with $x_i = 1$ iff $i \notin C_k$ for all i with $x_i = 1$ iff $(x_1, \dots, x_d)$ satisfies $\bigwedge_{i \in C_k} \neg x_i$ for some k iff $(x_1, \dots, x_d)$

satisfies the negated formula $\bar{f} = \bigvee_{k=1}^{n} \bigwedge_{i \in C_k} \neg x_i$. This implies that $B_{x_1,\dots,x_d}$ is a subset of the contribution of B iff $(x_1,\dots,x_d)$ satisfies f. Hence, since $\text{VOL}(B_{x_1,\dots,x_d}) = 1$, we have $\text{MINCON}(M) = \text{CON}(M,B) = |\{(x_1,\dots,x_d) \in \{0,1\}^d \mid (x_1,\dots,x_d) \text{ satisfies } f\}|$. Thus a polynomial time algorithm solving MINCON(M) would result in a polynomial time algorithm for #MON-CNF, which proves the claim. □

Note that the reduction from above implies that MINCON is **#P**-hard and **NP**-hard to approximate even if the least contributor is known. Moreover, since we constructed boxes with integer coordinates in $[0, 2^d + 2]$ a number of $b = \mathcal{O}(d^2 n)$ bits suffices to represent all $d+1$ coordinates of the $n+1$ constructed points. Hence, MINCON is hard even if all coordinates are integral. We define as input size $b + n + d$, where b is the number of bits in the input. We will use this result in the next proof. Also note that the same hardness for CON follows immediately, as it is hard to compute CON(M, B) as constructed above.

By reducing MINCON to LC, one can now show **NP**-hardness of LC. We skip this proof and directly prove **NP**-hardness of ε-LC by using the hardness of approximating MINCON in the following theorem.

Theorem 2. *ε-LC is* **NP**-*hard for any constant ε. More precisely, it is* **NP**-*hard for $(1+\varepsilon)$ bounded from above by $2^{d^{1-c}-1}$ for some $c > 0$.*

Proof. We reduce MINCON to ε-LC. Let M be a set of n boxes in $\mathbb{R}^d$, i.e., a problem instance of MINCON represented by a number of b bits, so that the input size is $b + n + d$.

As discussed above, we can assume that the coordinates are integral. We can further assume that $d \geq 2$ as MINCON is trivial for $d = 1$. The minimal contribution of M might be 0, but this occurs if and only if one box in M dominates another. As the latter can be checked in polynomial time, we can without loss of generality also assume that MINCON(M) > 0.

Now, let V be the volume of the bounding box of all the boxes in M, i.e., the product of all maximal coordinates in the d dimensions. We know that V is an integer with $1 \leq V \leq 2^b$, as there are only b bits in the input.

We now define a slightly modified set of boxes:

$$\begin{aligned}
A &= \{\square(a_1 + 2V, a_2, \dots, a_d) \mid \square(a_1, \dots, a_d) \in M\}, \\
B &= \square(2V, \dots, 2V), \\
C_\lambda &= \square(1, \dots, 1, 2V + \lambda), \\
M_\lambda &= A \cup \{B\} \cup \{C_\lambda\}.
\end{aligned}$$

The boxes in A are the boxes of M, but shifted along the x_1-axis. By definition, $a_i \leq V$, $i \in [d]$ for all $\square(a_1, \dots, a_d) \in M$. The contribution to HYP(M_λ) of a box in A is the same as the contribution to HYP(M) of the corresponding box in M as the additional part is overlapped by the "blocking" box B. Also note that the contribution of a box in A is less or equal than V.

The box B uniquely overlaps at least the space $[V, 2V] \times \ldots \times [V, 2V]$ (as every coordinate of a point in M is less than equal to V) which has volume at least V. Hence, B is never the least contributor of M_λ. The box C_λ then has a contribution of $\text{VOL}([0,1] \times \ldots \times [0,1] \times [2V, 2V + \lambda]) = \lambda$, so that C_λ is a least contributor iff λ is less than or equal to the minimal contribution of any box in A to $\text{HYP}(M_\lambda)$ which holds iff we have $\lambda \leq \text{MINCON}(M)$.

Since we can decide, whether C_λ is the least contributor, by one call to $\text{LC}(M_\lambda)$, we can do kind of a binary search on λ. As we are interested in a multiplicative approximation, we search for $\kappa := \log_2(\lambda)$ to be the largest value less than equal to $\log_2(\text{MINCON}(M))$, where κ now is an integer in the range $[0, b]$. As we can only answer ε-LC-queries we cannot do exact binary search. But we can still follow its lines, recurring on the left half of the current interval, if for the median value κ_{m} we get $\varepsilon\text{-LC}(M_{\lambda_{\text{m}}}) = C_{\lambda_{\text{m}}}$, where $\lambda_{\text{m}} = 2^{\kappa_{\text{m}}}$, and on the right half, if we get any other result.

The incorrectness of ε-LC may misguide our search, but since we have $\text{CON}(M, \varepsilon\text{-LC}(M)) \leq (1+\varepsilon)\,\text{MINCON}(M)$ it can give a wrong answer (i.e., not the least contributor) only if we have $(1+\varepsilon)^{-1}\text{MINCON}(M) \leq 2^\kappa \leq (1+\varepsilon)\,\text{MINCON}(M)$. Outside of this interval our search goes perfectly well. Thus, after the binary search, i.e, after at most $\lceil \log_2(b) \rceil$ many calls to ε-LC, we end up at a value κ which is either inside the above interval (in which case we are satisfied) or the largest integer smaller than $\log_2((1+\varepsilon)^{-1}\text{MINCON}(M))$ or the smallest integer greater than $\log_2((1+\varepsilon)\,\text{MINCON}(M))$. Hence, we have $\kappa \leq \log_2((1+\varepsilon)\,\text{MINCON}(M)) + 1$ implying $\lambda = 2^\kappa \leq 2(1+\varepsilon)\,\text{MINCON}(M)$. Analogously, we get $\lambda = 2^\kappa \geq \text{MINCON}(M)/(2(1+\varepsilon))$. Therefore after $\mathcal{O}(\log(b))$ many calls to ε-LC we get a $2\,(1+\varepsilon)$ approximation of $\text{MINCON}(M)$. Since this is **NP**-hard for $2\,(1+\varepsilon)$ bounded from above by $2^{d^{1-c}}$ for some $c > 0$, we showed **NP**-hardness of ε-LC in this case. Note that this includes any constant ε. □

The **NP**-hardness of ε-LC not only implies **NP**-hardness of LC, but also the non-existence of an efficient algorithm for ε-δ-LC unless **NP** = **BPP**. The above proof also gives a very good intuition about the problem ε-LC: As we can approximate the minimal contribution by a small number of calls to ε-LC, there cannot be a much faster way to solve ε-LC but to approximate the contributions – approximating at least the least contribution can be only a factor of $\mathcal{O}(\log(b))$ slower than solving ε-LC. This motivates the algorithm we present in the next section, which tries to approximate the contributions of the various boxes.

3 Practical Approximation Algorithm

The last section ruled out the possibility of a worst case efficient algorithm for computing or approximating the least contributor. Nevertheless, we are now presenting an algorithm $\mathcal{A}$ that is "safe" and has a good practical runtime, but no polynomial worst case runtime (as this is not possible). By "safe" we mean that it provably solves ε-δ-LC, i.e., it holds that

$$\Pr[\text{CON}(M, \mathcal{A}(M, \varepsilon, \delta)) \leq (1+\varepsilon)\,\text{MINCON}(M)] \geq 1 - \delta.$$

As the algorithm is going to approximate the contributions, we cannot avoid ε and solve LC directly, as with no $(1+\Delta)$-approximation, for any $\Delta > 0$ we can decide whether two contributions are equal or just nearly equal (and in the latter case which one is greater). We consider an ε around 10^{-2} or 10^{-3} as sufficient for typical instances. This implies for most instances that we return the correct result as there are no two small contributions which are only a $(1+\varepsilon)$-factor apart. For the remaining cases we return at least a box which has contribution at most $(1+\varepsilon)$ times the minimal contribution, which means we make an "error" of ε.

Additionally, the algorithm is going to be a randomized Monte Carlo algorithm, which is why we need the δ and do not always return the correct result. However, we will be able to set $\delta = 10^{-6}$ or even $\delta = 10^{-12}$ without increasing the runtime overly. In the following we will describe algorithm $\mathcal{A}$, prove its correctness and describe its runtime.

3.1 The Algorithm $\mathcal{A}$

Our algorithm works as follows. For each box A it determines the minimal bounding box of the space that is uniquely overlapped by the box. To do so we start with the box A itself. Then we iterate over all other boxes B. If B dominates A in all but one dimension, then we can cut the bounding box in the non-dominated dimension. This can be realized in $\mathcal{O}(dn^2)$.

Having the bounding box BB_A of the contribution of A we start to sample randomly in it. For each random point we determine if it is uniquely dominated by A. If we checked $\textsc{noSamples}(A)$ random points and $\textsc{noSuccSamples}(A)$ of them were uniquely dominated by A, then the contribution of A is about $\widetilde{V}_A := \frac{\textsc{noSuccSamples}(A)}{\textsc{noSamples}(A)} \textsc{vol}(\mathrm{BB}_A)$, where $\textsc{vol}(\mathrm{BB}_A)$ denotes the volume of the bounding box of the contribution of A. Additionally, we can give an estimate of the deviation of $\widetilde{V}_A$ from V_A, the correct contribution of A (i.e., $V_A = \mathrm{CON}(M, A)$): Using Chernoff's inequality we get that for

$$\Delta(A) := \sqrt{\frac{\log(2n/\delta)}{2\textsc{noSamples}(A)}} \textsc{vol}(\mathrm{BB}_A) \tag{1}$$

the probability that V_A deviates from $\widetilde{V}_A$ by more than $\Delta(A)$ is small enough.

We would like to sample in the bounding boxes in parallel such that every $\widetilde{V}_A$ deviates about the same Δ. We do this by initializing Δ arbitrarily (e.g., $\Delta = 1$) and then in every iteration decrease Δ by some factor (e.g., $\frac{1}{2}$) and sample in each bounding box until we have $\Delta(A) \leq \Delta$. If we then have at any point two boxes A and B with

$$\widetilde{V}_A - \Delta(A) > \widetilde{V}_B + \Delta(B) \tag{2}$$

we can with good probability assume that A is not a least contributor as we would need to have $\widetilde{V}_A - V_A > \Delta(A)$ or $V_B - \widetilde{V}_B > \Delta(B)$ for A having a less contribution than B (which is necessary for A being the least contributor). Hence, whenever such a situation occurs we can delete A from our race, meaning that we do not have to sample in its bounding box anymore. Note that we never have to compare two arbitrary boxes, but only a box A to the currently smallest box $\widetilde{LC}$, i.e., the box with $\widetilde{V}_{\widetilde{LC}}$ minimal.

We can run this race, deleting boxes if their contribution is clearly too much by the above selection equation until either there is just one box left, in which case we have found the least contributor, or until we have reached a point where we have approximated all contributions well enough. Given an abortion criterion ε we can just return $\widetilde{LC}$ (the box with currently smallest approximated contribution) when we have

$$0 < \frac{\widetilde{V}_{\widetilde{LC}} + \Delta(\widetilde{LC})}{\widetilde{V}_A - \Delta(A)} \leq 1 + \varepsilon,$$

for any box $A \neq \widetilde{LC}$ still in the race. If this equation holds, then we can be quite sure that any box has contribution at least $\frac{1}{1+\varepsilon} V_{\widetilde{LC}}$, and, similarly, all other boxes that are still in the race, too. So, after all, we have solved ε-δ-LC.

Due to space limitations, the proof that the described algorithm is indeed correct has been removed. It can be found in Section 3.3 of [5].

3.2 Runtime

As discussed above, our algorithm needs a runtime of at least $\Omega(dn^2)$. This seems to be the true runtime on many practical instances (cf. Section 4). However, by Theorem 2 we cannot hope for a matching upper bound. In this section we present an upper bound on the runtime depending on some characteristics of the input.

For an upper bound, observe that we have to approximate each box A up to $\Delta = (V_A - \text{MINCON}(M))/4$ to be able to delete it with high probability: At this point, $\widetilde{V}_A \geq V_A - \Delta$ and $\widetilde{V}_B \leq V_B + \Delta$, for B a least contributor, so that $\widetilde{V}_A - \widetilde{V}_B \geq 2\Delta$ with probability at least $1 - \delta/n$. Similarly, we can show that the expected value of Δ where we delete box A is $\Omega(V_A - \text{MINCON}(M))$. By equation (1) we observe that we need a number of

$$\frac{\log(2n/\delta)\text{VOL}(\text{BB}_A)^2}{2\Omega(V_A - \text{MINCON}(M))^2} = \mathcal{O}\left(\frac{\log(n/\delta)\text{VOL}(\text{BB}_A)^2}{(V_A - \text{MINCON}(M))^2}\right)$$

samples to delete box A on average. For the least contributor LC, we need $\mathcal{O}\left(\frac{\log(n/\delta)\text{VOL}(\text{BB}_{LC})^2}{(\text{sec-min}(V) - \text{MINCON}(M))^2}\right)$ many samples until we have finally deleted all other boxes, where sec-min(V) denotes the second smallest contribution of any box in M. Since each sample takes runtime $\mathcal{O}(dn)$ and everything besides the sampling takes much less runtime, we get an overall runtime of $\mathcal{O}(dn\,(n + \log(n/\delta)\,\text{H}))$, where

$$\text{H} = \frac{\text{VOL}(\text{BB}_{LC})^2}{(\text{sec-min}(V) - \text{MINCON}(M))^2} + \sum_{LC \neq A \in S} \frac{\text{VOL}(\text{BB}_A)^2}{(V_A - \text{MINCON}(M))^2}$$

Algorithm 1. $\mathcal{A}(M, \varepsilon, \delta)$ solves ε-δ-LC(M) for a set M of n boxes in $\mathbb{R}^d$ and $\varepsilon, \delta > 0$, i.e., it determines a box $x \in M$ s.t. $\Pr[\mathrm{CON}(M,x) \leq (1+\varepsilon)\,\mathrm{MINCON}(M)] \geq 1-\delta$.

determine the bounding boxes BB_A for all $A \in M$
initialize $\textsc{noSamples}(A) = \textsc{noSuccSamples}(A) = 0$ for all $A \in M$
initialize Δ
set $S := M$
repeat
  set $\Delta := \Delta/2$
  for all $A \in S$ **do**
    repeat
      sample a random point in BB_A
      increase $\textsc{noSamples}(A)$ and possibly $\textsc{noSuccSamples}(A)$
      update $\widetilde{V}_A$ and $\Delta(A)$
    until $\Delta(A) \leq \Delta$
  od
  set $\widetilde{LC} := \mathrm{argmin}\{\widetilde{V}_A \mid A \in S\}$
  for all $A \in S$ **do**
    if $\widetilde{V}_A - \Delta(A) > \widetilde{V}_{\widetilde{LC}} + \Delta(\widetilde{LC})$ **then**
      $S := S \backslash \{A\}$
    od
  od
until $|S| = 1$ or $0 < \frac{\widetilde{V}_{\widetilde{LC}} + \Delta(\widetilde{LC})}{\widetilde{V}_A - \Delta(A)} \leq 1+\varepsilon$ for any $\widetilde{LC} \neq A \in S$
return $\widetilde{LC}$

is a certain measure of hardness of the input. This value is unbounded and can even be undefined if there is no unique least contributor. In this case our abortion criterion comes into play: With probability $(1-\delta)$ after approximating every contribution up to $\Delta = \frac{\varepsilon}{4+2\varepsilon}\mathrm{MINCON}(M)$ we have $\widetilde{V}_{LC} \leq V_{LC} + \Delta$, thus $\widetilde{V}_{\widetilde{LC}} \leq V_{LC} + \Delta$, and $\widetilde{V}_A \geq V_{LC} - \Delta$ for every other box A still in the race. Then we conclude

$$\frac{\widetilde{V}_{\widetilde{LC}} + \Delta(\widetilde{LC})}{\widetilde{V}_A - \Delta(A)} \leq \frac{V_{LC} + 2\Delta}{V_{LC} - 2\Delta} = \frac{1 + 2\frac{\varepsilon}{4+2\varepsilon}}{1 - 2\frac{\varepsilon}{4+2\varepsilon}} = 1 + \varepsilon$$

for every box $\widetilde{LC} \neq A \in S$. Hence, the above defined value for Δ suffices to enforce abortion. Since we get this Δ after $\textsc{noSamples}(A) = \frac{\log(2n/\delta)\textsc{vol}(\mathrm{BB}_A)^2}{2(\frac{\varepsilon}{4+2\varepsilon}\mathrm{MINCON}(M))^2}$ samples, this yields another upper bound for the overall number of samples, a still unbounded but always finite value:

$$\mathcal{O}\left(\frac{\log(n/\delta)}{\varepsilon^2\mathrm{MINCON}(M)^2} \sum_{A \in M} \textsc{vol}(\mathrm{BB}_A)^2\right)$$

However, for the random testcases that we consider in Section 4 the above defined hardness H is a more realistic measure of runtime as there are never two identical contributions and not too many equally small contributions. There one observes values for H that roughly lie in the interval $[n, 10n]$.

4 Experimental Analysis

To demonstrate the performance of the described approximation algorithm for the hypervolume contribution, we have implemented it and measured its performance on different datasets. To yield a practically relevant algorithm, we have implemented several heuristical improvements which are described in detail in Section 3.4 of [5]. The most important for the correct interpretation of the experiments is that we use a classical exact algorithm for small n and d. We now first describe the used benchmark datasets and then our results.

4.1 Datasets

We used five different fronts similar to the DTLZ test suite [7]. As we do not want to compare the hypervolume algorithms for point distributions specific to different optimizers like NSGA-II [6] or SPEA2 [21], we have sampled the points from different surfaces randomly. This allows full scalability of the datasets in the number of points and the number of dimensions.

To define the datasets, we use random variables with two different distributions. Simple uniformly distributed random variables are provided by the build-in random number generator rand() of C++. To get random variables with a Gaussian distribution, we used the polar form of the Box-Muller transformation as described in [13].

Linear Dataset: The first dataset consists of points $(x_1, x_2, \dots, x_d) \in [0,1]^d$ with $\sum_{i=1}^{d} x_i = 1$. They are obtained by generating d Gaussian random variables y_1, y_2, ..., y_d and then using the normalized points

$$(x_1, x_2, \dots, x_n) := \frac{(|y_1|, |y_2|, \dots, |y_n|)}{|y_1| + |y_2| + \dots + |y_d|}.$$

Spherical Dataset: To obtain uniformly distributed points $(x_1, x_2, \dots, x_d) \in [0,1]^d$ with $\sum_{i=1}^{d} x_i^2 = 1$ we follow the method of Muller [10]. That is, we generate d Gaussian random variables y_1, y_2, ..., y_d and take the points

$$(x_1, x_2, \dots, x_n) := \frac{(|y_1|, |y_2|, \dots, |y_n|)}{\sqrt{y_1^2 + y_2^2 + \dots + y_d^2}}.$$

Concave Dataset: Analogously to the spherical dataset we choose points $(x_1, x_2, \dots, x_d) \in [0,1]^d$ with $\sum_{i=1}^{d} \sqrt{x_i} = 1$. For this, we generate again d Gaussian random variables y_1, y_2, ..., y_d and use the points

$$(x_1, x_2, \dots, x_n) := \frac{(|y_1|, |y_2|, \dots, |y_n|)}{(\sqrt{|y_1|} + \sqrt{|y_2|} + \dots + \sqrt{|y_d|})^2}.$$

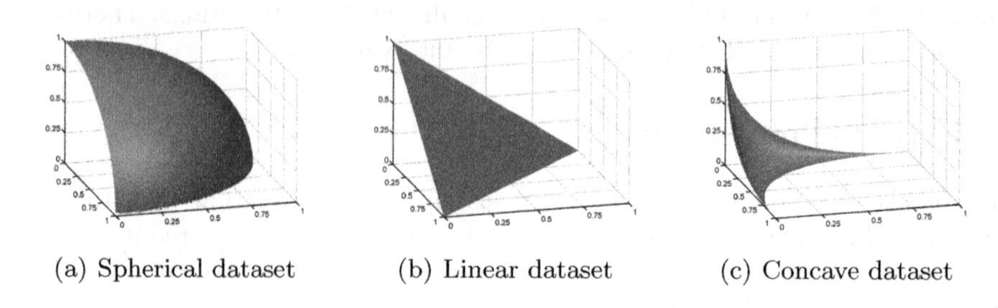

(a) Spherical dataset (b) Linear dataset (c) Concave dataset

Fig. 1. Visualization of the first three datasets

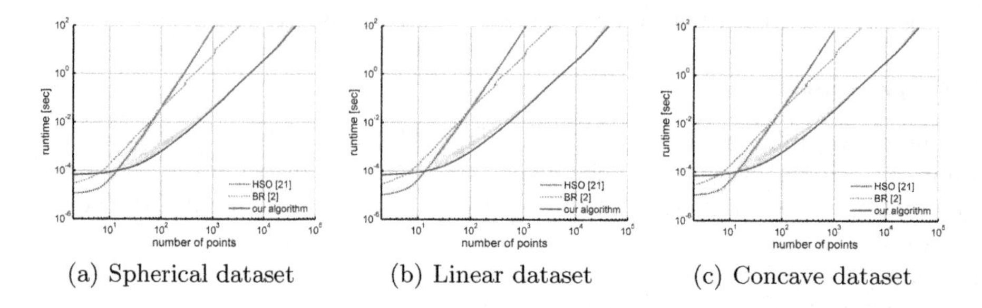

(a) Spherical dataset (b) Linear dataset (c) Concave dataset

Fig. 2. Experimental results for $d = 3$

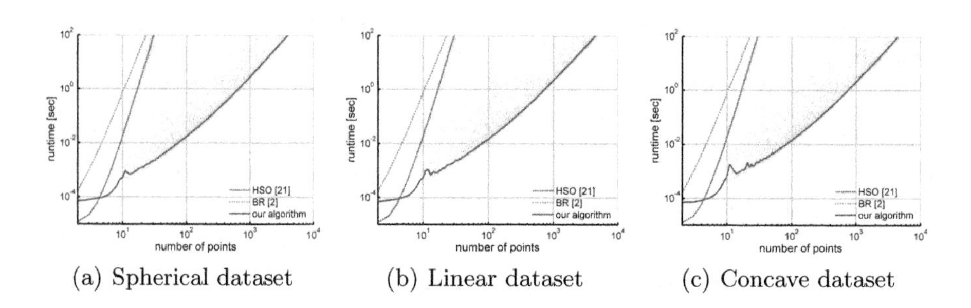

(a) Spherical dataset (b) Linear dataset (c) Concave dataset

Fig. 3. Experimental results for $d = 10$

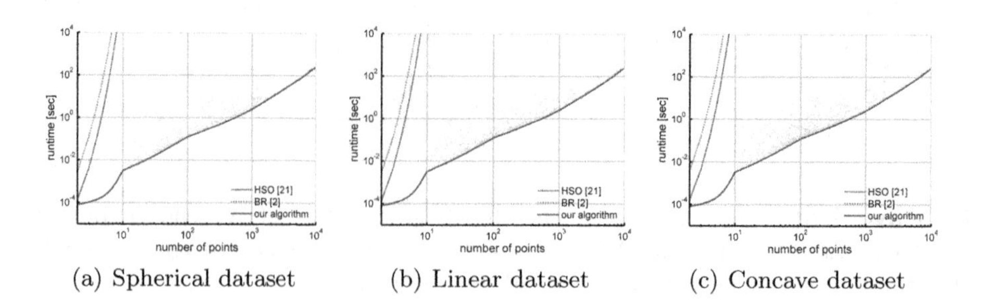

(a) Spherical dataset (b) Linear dataset (c) Concave dataset

Fig. 4. Experimental results for $d = 100$

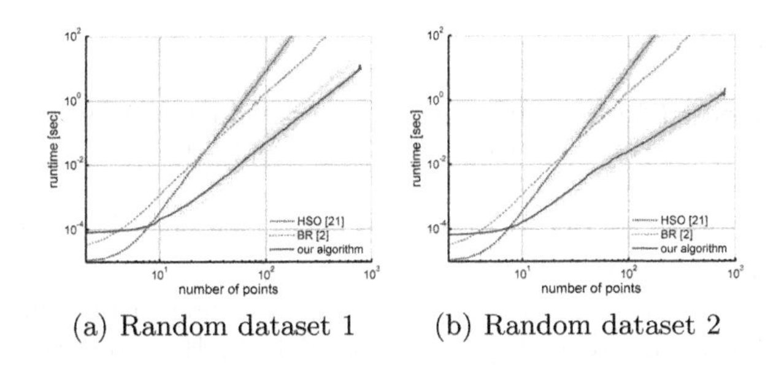

(a) Random dataset 1 (b) Random dataset 2

Fig. 5. Experimental results for random datasets with $d = 5$

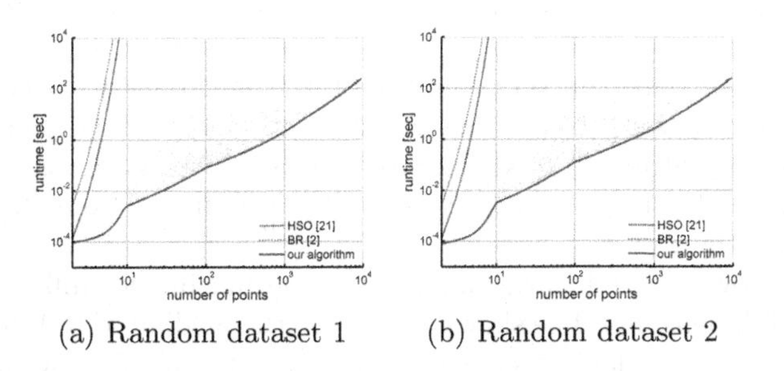

(a) Random dataset 1 (b) Random dataset 2

Fig. 6. Experimental results for random datasets with $d = 100$

For $d = 3$, the surface of the dataset is shown in Figure 1. Additionally to random points lying on a lower-dimensional surface, we have also examined the following two datasets with points sampled from the actual space similar to the random dataset examined by While et al. [17].

Random Dataset 1: We first draw n uniformly distributed points from $[0,1]^d$ and then replace all dominated points by new random points until we have a set of n nondominated points.

Random Dataset 2: Very similar to the previous dataset, we choose random points until there are no dominated points. The only difference is that this time the points are not drawn uniformly, but Gaussian distributed in $\mathbb{R}^d$ with mean 1.

Note that the last two datasets are far from being uniformly distributed. The points of the first set all have at least one coordinate very close to 1 while the points of the second set all have at least one coordinate which is significantly above the mean value. This makes their computation for many points (e.g., $n \geq 100$) in small dimensions (e.g., $d \leq 5$) computationally very expensive as it becomes more and more unlikely to sample a nondominated point.

4.2 Comparison

We have implemented our algorithm in C++ and compared it with the available implementations of HSO by Eckart Zitzler [19] and BR by Nicola Beume [2]. We did not add any further heuristics to both exact algorithms as all published heuristics do not improve the *asymptotic* runtime and even a speedup of a few magnitudes does not change the picture significantly.

It would be better to compare our approximation algorithm with other approximation algorithms instead of exact algorithms. However, the only other published approximation algorithm seems to be [1], which is not publicly available yet. Another reason is that all available optimization algorithms based on the hypervolume indicator use exact calculations and hence our speedup is carried over to them.

All experiments were run on a cluster of 100 machines with two 2.4 GHz AMD Opteron processors, operating in 32-bit mode, running Linux. For our approximation algorithm we used the parameters $\delta = 10^{-6}$ and $\varepsilon = 10^{-2}$. The code used is available upon request and will be distributed from the homepage of the second author.

Figure 2-6 show double-logarithmic plots of the runtime for different datasets and number of dimensions. The shown values are the median of 100 runs each. To illustrate the occurring deviations below and above the median, we also plotted all measured runtimes as ligther single points in the back. As both axes are scaled logarithmically, also the examined problem sizes are distributed logarithmically. That is, we only calculated Pareto sets of size n if $n \in \{\lfloor \exp(k/100) \rfloor \mid k \in \mathbb{N}\}$. We examined dimensions $d = 3, 10, 100$ for the first three datasets and $d = 5, 100$ for the last two datasets.

Independent of the number of solutions and dimension, we always observed that, unless $n \leq 10$, our algorithm outperformed HSO and BR substantially. On the used machines this means that only if the calculation time was insignificant (say, below 10^{-4} seconds), the exact algorithm could compete. On the other hand, the *much* lower median of our algorithm also comes with a much higher empirical standard deviation and interquartile range. In fact, we observed that the upper quartile can be up to five times slower than the median (for the especially degenerated random dataset 1). The highest ratio observed between the maximum runtime and the average runtime is 66 (again for the random dataset 1). This behavior is represented in the plots by the spread of lighter datapoints in the back of the median. However, there are not too many outliers and even their runtime outperforms HSO and BR. The non-monotonicity of our algorithm around $n = 10$ for $d = 10$ is caused by the approximate for the runtimes of the exact algorithms.

For larger dimensions the advantage of our approximation algorithm becomes tremendous. For $d = 100$ we observed that within 100 seconds our algorithm could solve all problems with less than 6000 solutions while HSO an BR could not solve any problem for a population of 6 solutions in the same time. For example for 7 solutions on the 100-dimensional linear front, HSO needed 13 minutes, BR 7 hours while our algorithm terminated within 0.5 milliseconds.

5 Conclusions

We have proven that most natural questions about the hypervolume contribution which are relevant for evolutionary multi-objective optimizers are not only computationally hard to decide, but also hard to approximate. On the other hand, we have presented a new approximation algorithm which works extremely fast for all tested practical instances. It can solve efficiently large high-dimensional instances ($d \geq 10$, $n \geq 100$) which are intractable for all previous exact algorithms and heuristics.

It would be very interesting to compare the algorithms on further datasets. We believe that only when two solutions have contributions of very close value, our algorithm slows down. For practical instances this should not matter as it simply occurs too rarely – but this conjecture should be substantiated by some broader experimental study in the future.

References

[1] Bader, J., Zitzler, E.: HypE: An Algorithm for Fast Hypervolume-Based Many-Objective Optimization. TIK Report 286, Institut für Technische Informatik und Kommunikationsnetze, ETH Zürich (2008)

[2] Beume, N., Rudolph, G.: Faster S-metric calculation by considering dominated hypervolume as Klee's measure problem. In: Proc. Second International Conference on Computational Intelligence (IASTED 2006), pp. 233–238 (2006)

[3] Beume, N., Fonseca, C., López-Ibáñez, M., Paquete, L., Vahrenhold, J.: On the complexity of computing the hypervolume indicator. Technical report CI-235/07, Technical University of Dortmund (2007)

[4] Bringmann, K., Friedrich, T.: Approximating the volume of unions and intersections of high-dimensional geometric objects. In: Hong, S.-H., Nagamochi, H., Fukunaga, T. (eds.) ISAAC 2008. LNCS, vol. 5369, pp. 436–447. Springer, Heidelberg (2008a), `http://arxiv.org/abs/0809.0835`

[5] Bringmann, K., Friedrich, T.: Approximating the least hypervolume contributor: NP-hard in general, but fast in practice (2008a), `http://arxiv.org/abs/0812.2636`

[6] Deb, K., Agrawal, S., Pratap, A., Meyarivan, T.: A fast elitist non-dominated sorting genetic algorithm for multi-objective optimization: NSGA-II. In: Deb, K., Rudolph, G., Lutton, E., Merelo, J.J., Schoenauer, M., Schwefel, H.-P., Yao, X. (eds.) PPSN 2000. LNCS, vol. 1917, pp. 849–858. Springer, Heidelberg (2000)

[7] Deb, K., Thiele, L., Laumanns, M., Zitzler, E.: Scalable multi-objective optimization test problems. In: Proc. IEEE Congress on Evolutionary Computation (CEC 2002), pp. 825–830 (2002)

[8] Klee, V.: Can the measure of $\bigcup[a_i, b_i]$ be computed in less than $O(n \log n)$ steps? American Mathematical Monthly 84, 284–285 (1977)

[9] Knowles, J.D.: Local-Search and Hybrid Evolutionary Algorithms for Pareto Optimization. PhD thesis, Department of Computer Science, University of Reading, UK (2002)

[10] Muller, M.E.: A note on a method for generating points uniformly on n-dimensional spheres. Commun. ACM 2, 19–20 (1959)

[11] Overmars, M.H., Yap, C.-K.: New upper bounds in Klee's measure problem. SIAM J. Comput. 20, 1034–1045 (1991); announced at 29th Annual Symposium on Foundations of Computer Science (FOCS 1988)
[12] Papadimitriou, C.M.: Computational Complexity. Addison-Wesley Publishing Company, Reading (1994)
[13] Press, W.H., Teukolsky, S.A., Vetterling, W.T., Flannery, B.P.: Numerical Recipes: The Art of Scientific Computing, 2nd edn. Cambridge University Press, Cambridge (1992)
[14] Roth, D.: On the hardness of approximate reasoning. Artif. Intell. 82, 273–302 (1996)
[15] Valiant, L.G.: The complexity of computing the permanent. Theor. Comput. Sci. 8, 189–201 (1979)
[16] While, R.L., Bradstreet, L., Barone, L., Hingston, P.: Heuristics for optimizing the calculation of hypervolume for multi-objective optimization problems. In: Proc. IEEE Congress on Evolutionary Computation (CEC 2005), pp. 2225–2232 (2005)
[17] While, R.L., Hingston, P., Barone, L., Huband, S.: A faster algorithm for calculating hypervolume. IEEE Trans. Evolutionary Computation 10, 29–38 (2006)
[18] Zhou, X., Mao, N., Li, W., Sun, C.: A fast algorithm for computing the contribution of a point to the hypervolume. In: Proc. Third International Conference on Natural Computation (ICNC 2007), vol. 4, pp. 415–420 (2007)
[19] Zitzler, E.: Hypervolume metric calculation, 2001. Computer Engineering and Networks Laboratory (TIK), ETH Zurich, Switzerland, `ftp://ftp.tik.ee.ethz.ch/pub/people/zitzler/hypervol.c`
[20] Zitzler, E., Thiele, L.: Multiobjective evolutionary algorithms: a comparative case study and the strength Pareto approach. IEEE Trans. Evolutionary Computation 3, 257–271 (1999); announced at 5th International Conference Parallel Problem Solving from Nature (PPSN V)
[21] Zitzler, E., Laumanns, M., Thiele, L.: SPEA2: Improving the strength pareto evolutionary algorithm for multiobjective optimization. In: Giannakoglou, K., et al. (eds.) Proc. Evolutionary Methods for Design, Optimisation and Control with Application to Industrial Problems (EUROGEN 2001), pp. 95–100. International Center for Numerical Methods in Engineering (CIMNE) (2002)
[22] Zitzler, E., Thiele, L., Laumanns, M., Fonseca, C.M., da Fonseca, V.G.: Performance assessment of multiobjective optimizers: an analysis and review. IEEE Trans. Evolutionary Computation 7, 117–132 (2003)

Effects of 1-Greedy $\mathcal{S}$-Metric-Selection on Innumerably Large Pareto Fronts

Nicola Beume[1], Boris Naujoks[1], Mike Preuss[1], Günter Rudolph[1], and Tobias Wagner[2]

[1] Department of Computer Science (LS11), TU Dortmund University, Germany
{nicola.beume,boris.naujoks,mike.preuss,guenter.rudolph}@tu-dortmund.de
[2] Institute of Machining Technology (ISF), TU Dortmund University, Germany
wagner@isf.de

Abstract. Evolutionary multi-objective algorithms (EMOA) using performance indicators for the selection of individuals have turned out to be a successful technique for multi-objective problems. Especially, the selection based on the $\mathcal{S}$-metric, as implemented in the SMS-EMOA, seems to be effective. A special feature of this EMOA is the greedy $(\mu + 1)$ selection. Based on a pathological example for a population of size two and a discrete Pareto front it has been proven that a $(\mu + 1)$- (or 1-greedy) EMOA may fail in finding a population maximizing the $\mathcal{S}$-metric. This work investigates the performance of $(\mu+1)$-EMOA with small fixed-size populations on Pareto fronts of innumerable size. We prove that an optimal distribution of points can always be achieved on linear Pareto fronts. Empirical studies support the conjecture that this also holds for convex and concave Pareto fronts, but not for continuous shapes in general. Furthermore, the pathological example is generalized to a continuous objective space and it is demonstrated that also $(\mu + k)$-EMOA are not able to robustly detect the globally optimal distribution.

1 Introduction

The main question addressed in this work is concerned with the general suitability of a 1-greedy evolutionary multi-objective algorithm (EMOA) for the approximation of continuous Pareto fronts, which consist of an innumerable number of Pareto optimal solutions. As a 1-greedy EMOA, we denote a steady-state $(\mu+1)$-EMOA that replaces only one individual by greedily selecting the μ best ones according to a preference relation (in the style of the definitions of k-greediness by Zitzler et al. [ZTB08]). The question is of special interest, since—for some time now—we advocate the use of an EMOA that adheres to the 1-greedy scheme using the $\mathcal{S}$-metric or dominated hypervolume as preference relation, namely the SMS-EMOA [BNE07]. In contrast to other EMOA (e.g. NSGA-II [DPAM02]), it accepts only one new individual per generation in order to monotonically improve the quality of the Pareto front approximation. Naturally, one can ask if exchanging only one individual at a time is sufficient to avoid getting stuck in non-optimal configurations. However, past experience with the SMS-EMOA has

M. Ehrgott et al. (Eds.): EMO 2009, LNCS 5467, pp. 21–35, 2009.

nourished the belief that this algorithm is capable of coping with all practically relevant situations, although a general proof in either direction is missing. This kind of general proof, even if restricted to continuous Pareto fronts, is sophisticated, unless impossible. Thus, the aim of this paper is to gradually phase this task using both case-related formal proofs and empirical studies.

Recently, a simple discrete counter-example has been provided, which proved that the 1-greedy scheme based on the dominated hypervolume can fail [ZTB08] (cf. Sec. 2.3). However, the example is extreme in many aspects: It employs a population of only two individuals on a Pareto front of four points. Thus, we would like to know to what extend this phenomenon occurs in more realistic scenarios. We are interested in continuous Pareto fronts and show that the discrete counter example can easily be extended into the (piecewise) continuous domain, with the essential property still holding: For most initializations, a 1-greedy EMOA will fail to obtain the optimal distribution of points on that Pareto front. To further investigate the cause of failure, we optimize the $\mathcal{S}$-metric value of the population directly using a $(1,5)$- and a $(5,10)$-CMA-ES (Covariance Matrix Adaptation Evolution Strategy [HO01]). Our results show that not only 1-greedy EMOA, but also non-elitist EMOA with $\lambda > \mu$ can fail with high probability. This indicates that the problem is indeed very hard since the local optimum is a strong attractor for any kind of optimizer. Studying the structure of the problem, we give generalizing conjectures on the interrelationship of the Pareto front and greediness.

Given the successful applications and assuming that the failure on the mentioned counter examples stem from the extreme constitution of the Pareto front, we investigate the properties on connected simpler shapes. For linear Pareto fronts, it is proved that a 1-greedy hypervolume selection scheme is sufficient to reach the optimal distribution of points with respect to the dominated hypervolume. Regarding convex Pareto fronts, we show that the problem of maximizing the hypervolume with a given number of points is not concave, otherwise the 1-greediness would hold directly (cf. Sec. 4.1). However, the concavity is not a necessary condition for 1-greediness. We perform empirical studies on Pareto fronts of different curvature, which demonstrate that even a simplified SMS-EMOA reaches the global optimum showing that the problem is solvable by a 1-greedy EMOA. Furthermore, these studies give counter-intuitive insights on the optimal distributions of the points and their corresponding hypervolume contributions, i.e., the amount that is disjointly dominated by a point and is lost when the point is removed [BNE07].

The paper is structured as follows. In section 2 the basic definitions, which are used in this paper, are provided and the discrete pathological example for 1-greedy indicator-based EMOA is recapitulated. The continuous variant of the example is derived and, together with another problem with disconnected fronts, empirically studied in section 3. Afterwards, we focus on continuous Pareto fronts by analyzing and empirically studying connected fronts of different curvature in section 4. For simple cases, also formal proofs are provided. Finally, the paper is summarized and the important results are concluded in section 5.

2 Hypervolume Selection and Greediness

2.1 Definitions of Greediness

Zitzler et al. [ZTB08] denote a preference relation as *k-greedy* if

1. for any given set, there exists a finite number of iterations resulting in the optimal set regarding the preference relation, and
2. there is a sequence of improving populations per iteration when exchanging k elements of a population at most.

We denote an EMOA as *greedy* if the selection is performed greedily according to a preference relation, i.e., the best population regarding the preference is selected. Greediness implies elitist selection and k is related to the number of offspring, i.e., the number of possible changes in the population per iteration. Thus, a k-greedy EMOA performs a $(\mu + k)$ selection scheme regarding a pre-defined preference relation. The problem of finding a population for a given optimization problem which is optimally composed regarding the indicator of the preference relation is termed *k-greedy solvable* if from any initial population there exists an improving path to the optimum which can be traversed by changing at most k element of the population per iteration. Note that the selection allows that any problem is μ-greedy solvable for a $(\mu + k)$-EMOA with $k \geq \mu$ assuming that all search points are sampled with positive probability. A problem is *local k-greedy solvable* if the optimum can be obtained by exchanging only with neighboring solutions in the objective space. If a problem is local k-greedy solvable, this implies that it also is k-greedy solvable. We concentrate on 1-greediness regarding the hypervolume indicator thus study if the population which achieves the highest possible hypervolume value given a fixed population size and a reference of the $\mathcal{S}$-metric is reachable by replacing at most one individual per generation by selecting the subset of size μ which obtains the highest $\mathcal{S}$-metric value among all those $\mu + 1$ subsets.

2.2 Considered Test Functions

For the experimental investigation of greedy EMOA, a set of academic minimization problems is considered. This set contains the simple test functions T1-T4, which have a continuous concave or convex Pareto front, where the sign of the second derivative with respect to the first objective does not change. Test function T5 changes its curvature from concave to convex, while still being connected and continuously differentiable. Note that T1 and T4 describe the same Pareto front. The decision variable $x \in [0, 1]$ is bounded to an interval, in which $f_1(x)$ increases and $f_2(x)$ decreases in order to allow only non-dominated individuals.

T1: $f_1(x) = x^2, \quad f_2 = (1 - x)^2$ — Schaffer [Sch85], convex
T2: $f_1(x) = x, \quad f_2 = 1 - x$ — DTLZ1 [DTLZ02], convex
T3: $f_1(x) = \sin((\pi/2)x), \quad f_2 = \cos((\pi/2)x)$ — DTLZ2 [DTLZ02], concave
T4: $f_1(x) = x, \quad f_2(x) = (1 - x^{1/\alpha})^{\alpha}, \quad \alpha = 2$ — convex

T5: $f_1(x) = x, \quad f_2(x) = (1 - x^{1/\alpha})^\alpha, \quad \alpha = 3x/2 + 1/2$ concave-convex

T6: $f_1(x) = x, \quad f_2(x) = \begin{cases} -(1/8)x + 6.125 & x < 5 \\ -x + 8 & x \geq 5 \end{cases}$

T7: $f_1(x) = x, \quad f_2(x) = 1 - \sqrt{x} - x \cdot \sin(10\pi x)$ ZDT3 [ZT98]

For the experiments on T1-T4, the reference point applied in the selection of the EMOA is fixed to $\boldsymbol{R} = (2,2)^T$ (superscript T denotes transposition). The two test functions T6 and T7 are multi-modal with respect to the hypervolume of the population. T6 is the continuous conversion of a pathological example given by Zitzler et al. [ZTB08] (cf. Sec. 2.3). Its decision variable $x \in [1,7]$ is bounded and $\boldsymbol{R} = (10,7)^T$ is used. T7 is defined in the domain of $x \in [0,1]$ being the only function, for which not all x are Pareto optimal leading to a disconnected Pareto front of five convex parts.

2.3 Pathological Example for a Finite Pareto Front

Zitzler et al. [ZTB08] proved that, in general, a 1-greedy EMOA is not able to obtain the set which covers the maximal dominated hypervolume. They showed this by a counter example in a two-dimensional objective space with a Pareto front consisting of four points as reproduced in Fig. 1, where the algorithm shall optimize the distribution of a population of two individuals. When the population is initialized with the two points a and b, the global optimum formed by the points c and d is unreachable for a 1-greedy EMOA. Any combination of either a or b with a different point leads to a worse hypervolume value and is therefore not accepted. Thus, the set $\{a,b\}$ is a local optimum of the hypervolume maximization. The example can easily be extended to a higher number of objectives by choosing all additional coordinates as 1, since multiplying by 1 does not change the hypervolume values.

Note which aspects are necessary to make the problem hard: The reference point is chosen such that the objective values are weighted asymmetrically. Thus, the points on the right have a high hypervolume contribution though being quite close to each other. Furthermore, the second coordinate of the point a is

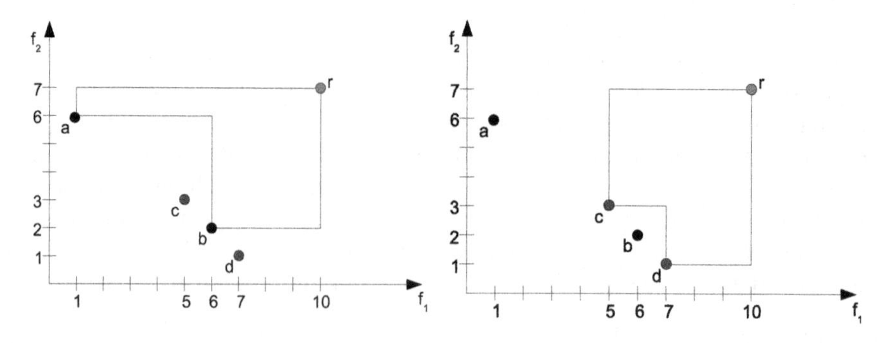

Fig. 1. Pathological example for 1-greedy EMOA with hypervolume-based selection. Points c and d are optimal but the population is initialized with a and b which form a local optimum.

positioned close enough to the reference point to avoid an optimal distribution which includes this point.

3 Pathological Examples with Innumerable Pareto Fronts

The discrete example of section 2.3 may easily be extended to the (piecewise) continuous case by connecting the points of the original configuration by two line segments as shown in Fig. 2 (cf. T6 in Sec. 2.2). The slope of the right segment results in $m_2 = -1$ and for the left one $m_1 = -1/8$ is chosen, to correctly transfer the situation of the discrete case in terms of the optimality properties of the different point distributions. For $m_1 < -0.2$, point a is no longer part of a local optimum and the basin of attraction is shifted to the right. For reasons of simplification, we further on discuss the problem as a two-dimensional parameter optimization problem, whose parameters are the two x-coordinates of the two search points on the Pareto front.

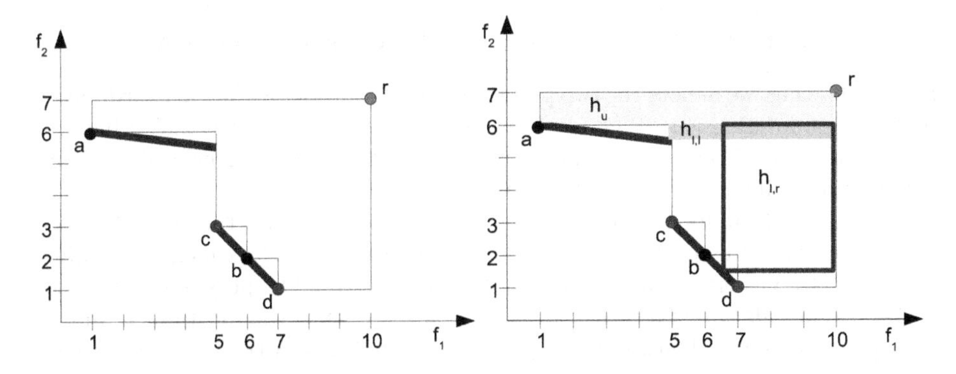

Fig. 2. Left: Conversion of the discrete pathological example for 1-greedy EMOA to the continuous space with a disconnected Pareto front of two linear segments (T6). Still the points c and d are optimal, whereas a and b form a local optimum. Right: Dissection of the hypervolume for one point fixed at a and one moving on either the left ($h_{l,l}$) or right ($h_{l,r}$) line (cf. Sec. 3.1 for details).

In the sense of parameter optimization, one may speak of multi-modality, and the property of 1-greediness translates to the possibility to execute a successful line search parallel to the coordinate axes. Thus, we can have multi-modality while still being able to do a successful step out of a local optimum by moving in parallel to one of the coordinate axes towards a better point. This especially is the case for multi-modal but separable hypervolume landscapes, such as shown for the problem T7 in the right plot of Fig. 3, where it is often necessary to cross large areas of worse values, so that the function is not local 1-greedy solvable.

The hypervolume landscape of problem T6 is depicted in the left plot of Fig. 3. The contour lines indicate that a 1-greedy EMOA is not able to leave the local optimum. A μ-greedy scheme would not encounter this principal difficulty as it allows for steps in any direction. However, it also faces the problem of locating

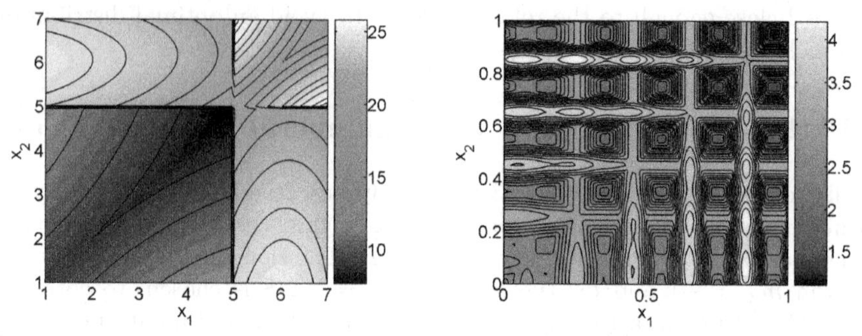

Fig. 3. Hypervolume landscapes for two individuals on the disconnected test functions T6 (left, $\boldsymbol{R} = (10,7)^T$) and T7 (right, $\boldsymbol{R} = (2,2)^T$)

a good area, which may be also difficult. The analysis of both 1-greedy and μ-greedy approaches on T6 is performed in the next subsections.

3.1 Proof: T6 Is Not 1-Greedy Solvable Regarding the $\mathcal{S}$-Metric

In the following, we denote the two points of our population by their x-coordinates so that a population is $x = (x_1, x_2)$. It can be shown that for T6, the global optimum at $x = (5,7)$ can indeed not be reached by doing only 1-greedy steps from the starting point $x = (1,6)$. To accomplish this, we have to look at the different cases resulting from fixing one point and moving the other over the allowed interval. In each case, the resulting hypervolume shall not exceed 25, which is the value for $x = (1,6)$. Otherwise, a 1-greedy successful step would be possible.

Let the population start with $x_1 = 1$ and $x_2 = 6$, thus the locations a and b. If x_1 stays at $x_1 = 1$ then x_2 can either move on the right or the left line. We can always compute the total hypervolume as sum of the volume beyond dominated by $(x_1 = 1, 6)$ ($h_u = 9$, cf. Fig. 2, left), and the contribution of the point at x_2. For x_2 moving on the right line, the contribution $h_{l,r}(x_2)$ is:

$$h_{l,r}(x_2) = (10 - x_2)\,(6 - (-x_2 + 8)) = -x_2^2 + 12x_2 - 20. \tag{1}$$

The maximum of this upside-down parabola obtained by standard calculus is at $x_2 = 6$. That is, for moving on the right line, there is no better point than $x_2 = 6$, which leads to a total hypervolume of 25. If x_2 moves on the left line segment, its contribution $h_{l,l}(x_2)$ is:

$$h_{l,l}(x_2) = (10 - x_2)\left(6 - \left(-\frac{x_2}{8} + \frac{6}{8}\right)\right) \tag{2}$$

This negative parabola has its optimum at $x_2 = \frac{11}{2}$, which is unreachable in our scenario as the largest x_2 value is still below 5. The resulting total hypervolume for this case is, thus, smaller than 13. When fixing $x_2 = 6$, and moving x_1, we can either situate it on the left or the right line segment. For the former, the whole hypervolume evaluates to:

$$h_l(x_1) = (10 - x_1)\left(7 - \left(-\frac{x_1}{8} + \frac{6}{8}\right)\right) + 4\left(\frac{6}{8}\right) - \frac{x_1}{8} - 2 = -\frac{x_1^2}{8} - \frac{x_1}{8} + \frac{25}{4} \quad (3)$$

This parabola has its maximum at $x_1 = -\frac{1}{2}$, the best allowed point is $x_1 = 1$, for which the hypervolume is 25. The latter case considers x_1 moving one the right line segment. Here, we define a helper function $h_{r,r}(y, z)$, which computes the hypervolume of any two points on that line, using the fact that the slope is $m_2 = -1$ so that the desired value is the difference of a large rectangle through both points and the reference point and a small rectangle with both points as diagonal corners:

$$h_{r,r}(y, z) = (10 - y)(7 - (z + 8)) - (z - y)^2 = -y^2 - z^2 + yz + y + 10z - 10. \quad (4)$$

For $x_1 < 6$ ($h_{r,r}(x_1, 6)$) or respectively $x_1 > 6$ ($h_{r,r}(6, x_1)$) this leads to

$$h_{r,r}(x_1, 6) = -x_1^2 + 7\,x_1 + 14 \quad \text{and} \quad h_{r,r}(6, x_1) = -x_1^2 + 16\,x_1 - 40 \quad (5)$$

with the maximum values at

$$\arg\max_{x_1}(h_{r,r}(x_1, 6)) = \frac{3}{2} \qquad \text{and} \qquad \arg\max_{x_1}(h_{r,r}(6, x_1)) = 8$$

with corresponding largest attainable values $h_{r,r}(5, 6) = 24$ and $h_{r,r}(6, 7) = 23$. Consequently, there is no 1-greedy move from $x = (1, 6)^T$ resulting in at least the same hypervolume value of 25. □

3.2 Experiment: How μ-Greedy Solvable Is T6?

Pre-experimental planning: We consider a 1-greedy and a μ-greedy single-objective evolutionary algorithm (EA) moving on the Pareto front only (resembling, e.g., SMS-EMOA and NSGA-II). So, the search space is the Pareto set and the EA directly maximize the $\mathcal{S}$-metric value of the population. A $(1, 5)$- and a $(5, 10)$-CMA-ES are added to the set of algorithms. These do not have existing EMOA counterparts, but shall be tested to see if moving with even more degrees of freedom (non-elitist selection and a surplus of offspring) pays off. In our first runs, we observed that the standard set of termination criteria as well as standard boundary treatment (by quadratic penalties) deteriorate the performance of the CMA-ES. The termination criteria make it stop too early, when there is still a good chance to obtain the optimal solution of $x = (5, 7)$, and the boundary treatment hinders coming near to it. Both have been switched off hereafter.

Task: We expect that the μ-greedy EA performs significantly better than the 1-greedy EA in terms of success rates.

Setup: All four algorithms are run 100 times per mutation step size (0.1 and 0.5) allowing up to 5000 evaluations and a minimum hypervolume value of 25.9 is regarded as success. The start points are scattered uniformly at random over the allowed domain (1 to 7).

Table 1. Success rates (100 repeats) of different algorithm types for detecting the globally optimal distribution of two individuals on the T6 Pareto front. The initial mutation step size as the only free parameter is tested at 0.1 and 0.5. Only the CMA-ES variants adapt it through the run.

Mutation step size	1-greedy EA	μ-greedy EA	$(1, 5)$-CMA-ES	$(5, 10)$-CMA-ES
0.1	12%	15%	55%	100%
0.5	15%	13%	61%	100%

Results/Visualization: The results are given in Table 1 by means of success rates.

Observations: Table 1 documents that the 1-greedy EA indeed fails, but so also does the μ-greedy EA. The CMA-ES solves the problem in more than half of the runs. The effective run length (until stagnation) is very short for the 1-greedy and μ-greedy EA, usually below 1000 evaluations. The CMA-ES often takes much longer. At the same time, it can be observed that it pushes the internally adapted mutation step sizes to very high values.

Discussion: The most surprising fact is surely that also the μ-greedy EA fails. It seems that the small basin of the global optimum is hard to find, even if it is possible to move there. A larger mutation step size could help in jumping out of the vicinity of the local optimum, but it also scatters search steps over a larger area. Furthermore, the attractor at $(1, 6)$ is much stronger than expected. Most runs end here, even if started at far distant points. The CMA-ES uses a very interesting strategy by enlarging the mutation rates. It is finally able to generate offspring over the whole domain of the problem, thereby degenerating (by learning) to a random search. Presumably, this is necessary to hinder premature convergence to the point $x_2 = 6$. Eventually, some points are placed in the vicinity of the global optimum. Therefore, increasing the number of evaluations most likely leads to higher success rates.

From the in-run distribution of the individuals, it is clear that the $(5, 10)$-CMA-ES manages to place some of the 5 individuals in each basin of attraction after some generations. Thus, it approximates the global optimum quite well. However, a population of more than one parent would translate back to a multi-population EMOA.

Summarizing, it shall be stated that although the function is of course μ-greedy solvable, μ-greedy EA without additional features like step-size adaptation have roughly the same chance of getting to the global optimum like 1-greedy EA. Note that the same applies to the original discrete example presented by Zitzler et al. [ZTB08], where, however, a much lower number of points to jump to exists. This means that, where the discrete example does not pose a problem to a μ-greedy scheme, the continuous one does.

3.3 More Points and a Strong Geometrical Argument

The provided example of a non 1-greedy function is fragile: Moving the reference point from $\mathbf{R} = (10, 7)^T$ towards symmetry makes it 1-greedy solvable again. Also note that the whole construction breaks down when going to a three-dimensional problem. Empirical tests with EMOA using a population with size $\mu = 3$ show that the optimal distribution $x = (1, 5, 7)$ will always be obtained, regardless of the chosen method (1-greedy or μ-greedy).

Continuing this line of thought, it is of course possible to build a problem that is also misleading for 1-greedy algorithms with population sizes of three. In fact, reducing the sizes of the basins of attraction in the hypervolume landscape would be a move towards this goal (see Fig. 3). However, such a problem will also become increasingly difficult for a μ-greedy algorithm, as empirically shown in section 3.2.

Conjecture 1. Continuously defined functions, which are not 1-greedy solvable for large population sizes ($\mu \gg 3$) are not generally considerably easier for μ-greedy algorithms.

One may however pose the question if these non 1-greedy solvable functions have to be defined piecewise. From Fig. 3, we may deduct that piecewise definition here is just a matter of construction and not a necessary condition. There is no reason withstanding creation of a continuous and even continuously differentiable non 1-greedy solvable function (so that the boundaries between pieces become flat) except that its analytical formulation may be much more complicated. Remember that, for non 1-greedy solvability, we only have to establish that from *one* point, line searches in all dimensions fail. This leads us to the following conjecture:

Conjecture 2. Non 1-greedy solvable, continuously differentiable functions can be constructed for any finite population size.

4 $\mathcal{S}$-Metric Properties on Continuously Differentiable Pareto Fronts

This section analyzes the convergence of 1-greedy EMOA to the distribution maximizing the $\mathcal{S}$-metric for two special cases in the first part. Afterwards, the properties of 1-greedy EMOA on differently shaped Pareto fronts are empirically studied.

4.1 Theoretical Analysis on Linear and Convex Pareto Fronts

Let $f : \mathbb{R}^2 \to \mathbb{R}^2$ be a bi-objective function to be minimized. We assume that the Pareto front $f(X^*)$ associated with the Pareto set $X^* \subset \mathbb{R}^2$ is a Jordan arc with parametric representation

$$f(X^*) = \left\{ \begin{pmatrix} s \\ \gamma(s) \end{pmatrix} : s \in [0,1] \subset \mathbb{R} \right\}, \tag{6}$$

where $\gamma : [0,1] \to \mathbb{R}$ is twice continuously differentiable. Let $y^{(1)}, \ldots, y^{(\mu)} \in f(X^*)$ be distinct objective vectors on the Pareto front. According to (6), we have $y^{(i)} = (s_i, \gamma(s_i))^T$ for $i = 1, \ldots, \mu$. As a consequence, the $\mathcal{S}$-metric or dominated hypervolume of the points $y^{(1)}, \ldots, y^{(\mu)}$ is given by

$$H(s) = (r_1 - s_1)\,[\,r_2 - \gamma(s_1)\,] + \sum_{i=2}^{\mu} (r_1 - s_i)\,[\,\gamma(s_{i-1}) - \gamma(s_i)\,] \qquad (7)$$

with reference point $\mathbf{R} = (r_1, r_2)^T$, $0 \le s_1 < s_2 < \cdots < s_\mu \le 1 \le r_1$ and $r_2 \ge \gamma(0)$.

Whenever the $\mathcal{S}$-metric is concave, the 1-greedy selection scheme of the SMS-EMOA with fixed reference point and a population of μ individuals on a continuous front will not get stuck prematurely since it is sufficient to move a single variable s_i at each iteration towards ascending values of the $\mathcal{S}$-metric in order to reach its maximum. Furthermore, we are going to use the result that a twice differentiable function is concave if and only if its Hessian matrix is negatively semidefinite. Partial differentiation of (7) leads to

$$\frac{\partial H(s)}{\partial s_1} = \gamma(s_1) - r_2 + (s_1 - s_2)\,\gamma'(s_1) \qquad (8)$$

$$\frac{\partial H(s)}{\partial s_i} = \gamma(s_i) - \gamma(s_{i-1}) + (s_i - s_{i+1})\,\gamma'(s_i) \qquad (i = 2, \ldots, \mu - 1) \qquad (9)$$

$$\frac{\partial H(s)}{\partial s_\mu} = \gamma(s_\mu) - \gamma(s_{\mu-1}) + (r_1 - s_\mu)\,\gamma'(s_\mu) \qquad (10)$$

and finally to

$$\frac{\partial^2 H(s)}{\partial s_i \partial s_{i-1}} = -\gamma'(s_{i-1}) \qquad (i = 1, \ldots, \mu - 1) \qquad (11)$$

$$\frac{\partial^2 H(s)}{\partial s_i \partial s_i} = 2\,\gamma'(s_i) + (s_i - s_{i+1})\,\gamma''(s_i) \qquad (i = 1, \ldots, \mu) \qquad (12)$$

$$\frac{\partial^2 H(s)}{\partial s_i \partial s_{i+1}} = -\gamma'(s_i) \qquad (i = 2, \ldots, \mu) \qquad (13)$$

with $s_{\mu+1} := r_1$. Other second partial derivatives are zero. Thus, the Hessian matrix $\nabla^2 H(s)$ of the $\mathcal{S}$-metric as given in (7) is a tridiagonal matrix.

Linear Pareto Front. Suppose that $\gamma(s) = m\,s + b$ is a *linear function*. Then, $\gamma(\cdot)$ is strongly monotone decreasing with $\gamma'(s) = m < 0$ and $\gamma''(s) = 0$ for all $s \in (0, 1)$. In this case, the Hessian matrix reduces to a tridiagonal matrix with identical diagonal entries $2\,m < 0$ and identical off-diagonal entries $-m > 0$. Recall that a square matrix A is weakly diagonal dominant if $|a_{ii}| \ge \sum_{j \ne i} |a_{ij}|$ for all i and that a weakly diagonal dominant matrix is negatively definite if all diagonal entries are negative. It is easily seen that these conditions are fulfilled. As a result, we have proven:

Theorem 1. *If the Pareto front of a bi-objective minimization problem is linear, then the $\mathcal{S}$-metric or dominated hypervolume of μ distinct points on the Pareto front is a strictly concave function.* □

From this result, we can deduce that it is sufficient to move a single point at a time for reaching the maximal $\mathcal{S}$-metric value in the limit. Next, we try to generalize this result.

Convex Pareto Front. Suppose that $\gamma(\cdot)$ is a *convex function*. Then, $\gamma(\cdot)$ is strongly monotone decreasing with $\gamma'(s) < 0$ and $\gamma''(s) > 0$ for all $s \in (0,1)$. Again, the Hessian matrix is tridiagonal, but the criterion of diagonal dominance of the Hessian does not always hold. Actually, the Hessian is not negatively semidefinite in general. This is easily seen from a counter-example: Let $\gamma(s) = (1 - \sqrt{s})^2$ (T4, $\alpha = 2$) with $\gamma'(s) = 1 - 1/\sqrt{s} < 0$ and $\gamma''(s) = \frac{1}{2}\, s^{-\frac{3}{2}} > 0$ for $s \in (0,1)$ and reference point $\mathbf{R} = (1,1)^T$. Consider three points on the Pareto front with $s_1 = (\frac{1}{10})^2$, $s_2 = (\frac{19}{20})^2$, $s_3 = (\frac{20}{21})^2$ leading to the Hessian matrix

$$\nabla^2 H(s) = \begin{pmatrix} -\frac{1857}{4} & 9 & 0 \\ 9 & -\frac{326392}{3024819} & \frac{1}{19} \\ 0 & \frac{1}{19} & -\frac{2461}{16000} \end{pmatrix},$$

whose leading principal minors are $\Delta_1 < 0$, $\Delta_2 < 0$ and $\Delta_3 > 0$ indicating that the Hessian matrix with this particular choice of points s_1, s_2, s_3 is not negatively semidefinite. On the other hand, if $s = (\frac{1}{100}, \frac{1}{4}, \frac{4}{9})^T$, it is easily verified that $\Delta_1 < 0$, $\Delta_2 > 0$ and $\Delta_3 < 0$ indicating that the Hessian matrix is negatively definite in this particular case. In summary, the Hessian matrix is indefinite and we have proven:

Theorem 2. *If the Pareto front of a bi-objective minimization problem is convex, then the $\mathcal{S}$-metric or dominated hypervolume of μ distinct points on the Pareto front is not a concave function in general.* □

However, this result does imply neither that there are no convex fronts with concave $\mathcal{S}$-metric nor that the 1-greedy selection scheme of the SMS-EMOA gets necessarily stuck on convex fronts.

4.2 Empirical Results of SMS-EMOA on Connected Pareto Fronts

In this section, it is empirically analyzed whether the 1-greedy SMS-EMOA can robustly obtain the $\mathcal{S}$-metric-optimal distribution of points for the approximation of piecewise continuous Pareto fronts with different curvature (convex to concave). It is assumed that the population has already arrived *on* the Pareto front and performs only local refinements. Thereby, the local 1-greediness as defined in section 2.1 of the considered test functions is empirically analyzed. Recall that a local 1-greedy solvable problem is also 1-greedy solvable. To accomplish this, a comprehensive study on the set of simple test functions T1-T5 is conducted (cf. Sec. 2.2). Due to the results in section 3.3, we restrict our analyses on small populations.

Pre-experimental planning: The SMS-EMOA selection operator discards the individual with the lowest hypervolume contribution, i.e., the amount that gets lost when the individual is removed since the hypervolume part is disjointly dominated by that individual. In order to provide a deeper understanding of this selection, the areas of individuals, which enter the population, are visualized in Fig. 5 for a given approximation using exemplarily the one-dimensional test function T1 defined in Sec. 2.2. It can be seen that the areas of success are directly adjacent to the solutions of the current approximation, which would be discarded instead. They therefore indicate the direction in which this solution should be shifted. Fig. 5 has been created based on T1, but the same fact holds for T2-T5. Intuitively, one may assume that the hypervolume contributions of individuals tend to equal values for all points of an optimally distributed set since, otherwise, a solution can move closer to the point with a higher contribution.

In order to investigate this conjecture, we compute these hypervolume contributions for the analytically determined optimal distributions of populations of five individuals, which are shown in Fig. 4 (left). For computing these distributions, test function T4 is considered with $\alpha \in \{1/3, 1/2, 1, 2, 3\}$ resulting in two concave fronts for $\alpha < 1$, convex fronts for $\alpha > 1$, and a linear front for $\alpha = 1$. Furthermore, a reference point $\mathbf{R} = (1.0, 1.0)^T$ positioned exactly at the boundaries of the Pareto front is used. Due to the construction of T4, the distribution is symmetrical to the bisecting line. Thus, the central point of the population lies exactly on this line. Since the hypervolume of the population of SMS-EMOA monotonically increases, the population will tend to these optimal distributions in case of a successful optimization.

The right part of Fig. 4 shows the corresponding hypervolume contributions sorted with respect to the first objective. The contribution values are symmetrical to the point in the middle. It can be observed that the contribution values tend to grow with increasing α, when $\alpha < 3$. On the concave Pareto front, the point in the middle (in the knee region) has the highest contribution and the contribution values are decreasing when going to the boundaries. On the convex Pareto front, the values decrease from the boundaries to the middle, so the point in the knee region has the lowest contribution. On linear fronts, the distribution obtaining the maximal hypervolume value is the set of equally spaced points as proved by Beume et al. [BFLI+07]. Only in this case, the contributions of all points are equal. Therefore, our first intuition was misleading.

To further investigate the effect of single local refinements, a local search SMS-EMOA, which uses only Gaussian mutations of single individuals of the current population with small stepsize $\sigma = 0.01$ to generate new candidates for selection, has been implemented. Fig. 5 plots the run of the decision space variables of this local search (5+1)-SMS-EMOA on T4 with $\alpha = 1/3$, when a fixed reference point $\mathbf{R}' = (2, 2)^T$ is chosen. As starting positions, the optimal distributions for the closer reference point $\mathbf{R} = (1.0, 1.0)^T$ are used. It can be seen that the algorithm is able to guide the solutions from the old to the new optimal positions. A closer look on the resulting population yields that the contribution of the points at the boundaries depends on the choice of the reference point. For the new reference

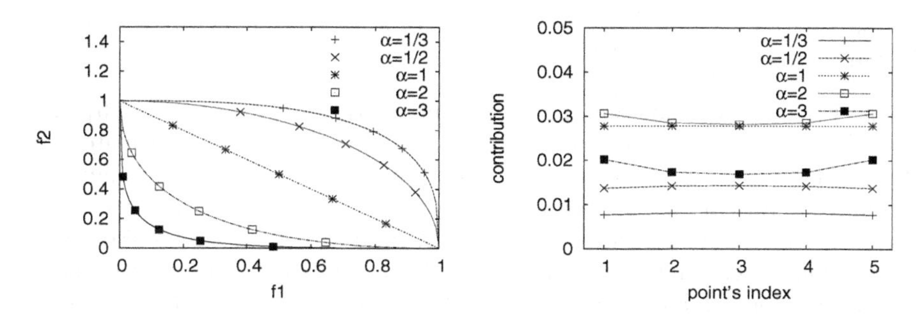

Fig. 4. Optimal positions of points on Pareto fronts of different curvature (left) and the corresponding hypervolume contributions of the points (right). The reference point $\mathbf{R} = (1.0, 1.0)$ is chosen.

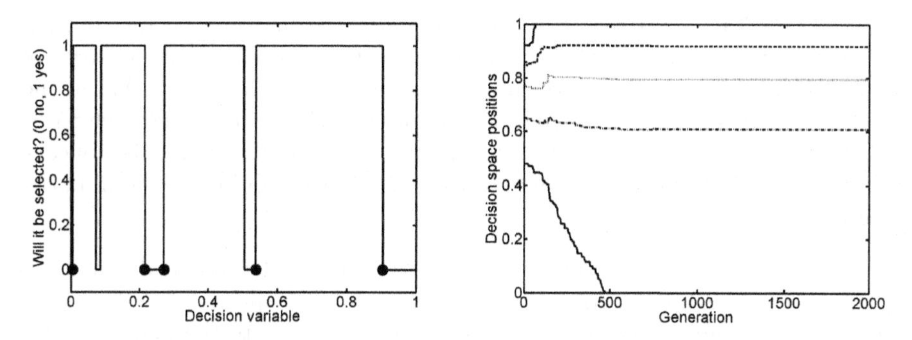

Fig. 5. Left: Acceptance of solutions depending on the decision space variable x for test function T1 and a randomly initialized population. The decision space variables of this population are indicated by black dots. Right: The run of the sorted decision space variables, which result in the population, over the generations of the local search (5 + 1)-SMS-EMOA on the concave test function T4 with $\alpha = 1/3$, when a fixed reference point $\mathbf{R}' = (2, 2)^T$ is chosen.

point $\mathbf{R}'$, which is situated at a greater distance to the Pareto front, the optimal points move closer to the boundaries and the contributions near the boundaries grow[1]. As a consequence, the other points follow these extremal solutions to cover the resulting distance. The optimal distribution does emerge.

The following experiment is conducted to empirically support the arising assumption that even a local search-based (μ+1) SMS-EMOA is able to approximate a set of optimally distributed Pareto-optimal points for continuous problems with convex and concave shaped Pareto fronts.

Task: Check the hypothesis that the local search SMS-EMOA is able to approximate the optimally distributed subset of the Pareto front of the given continuous

[1] For $\mathbf{R}'$ the sorted contributions are $(0.61, 0.015, 0.016, 0.015, 0.61)$.

test problems T1-T5 for fixed population sizes $\mu \in \{1, \ldots, 6\}$ with an accuracy limited by the step size $\sigma = 0.01$.

Setup: For each test function T1-T5 and population sizes $\mu \in \{1, \ldots, 6\}$, approximations for the $\mathcal{S}$-metric-optimal distribution are globally calculated by the MATLAB implementation of the (5,10)-CMA-ES [HO01], where no limit on the function evaluations, but a lower limit on σ_i $(i = 1, \ldots, \mu)$ of 10^{-12} is specified. The successful application of this algorithm for the calculation of optimal distributions, even for multi-modal hypervolume landscapes, has already been shown in section 3.2. For each configuration 10,000 runs of the local search SMS-EMOA are performed using different random initializations and the results after $\mu \cdot 1{,}000$ generations are compared to the approximations found by the global optimization of the CMA-ES. A run is denoted as failed when the hypervolume of the found approximation is below 99% of the approximated optimal one.

Observations: The local search SMS-EMOA detects the optimal distribution in all runs for the convex and concave test function T1-T5 except for 10% of the initializations on the concave-convex Pareto front T5 for $\mu = 1$. When the initial solution is situated close to the left border ($x < 0.1$), the local search SMS-EMOA converges to the left border, which indicates the optimum for the concave part of the Pareto front, instead of detecting the globally optimal position in the inflection point.

Discussion: Based on thorough experimentation, it can be assumed that even a local search SMS-EMOA robustly detects the globally optimal distribution in cases where the sign of the second derivative with respect to the first objective does not change. However, due to emerging effects of the local refinements and their interaction, for higher population sizes, this result seems to hold also for concave-convex Pareto fronts.

5 Conclusions

In this paper, we have investigated how a 1-greedy EMOA performs on different kinds of continuous Pareto fronts using both formal proofs and empirical analyses. So far, only an artificial discrete problem existed to show that a 1-greedy EMOA with hypervolume selection is not able to obtain the set covering the maximal hypervolume. We have shown that this problem can be converted to the continuous space while preserving its important properties. Thereby, it has been demonstrated that the local optimum is not only a singularity, but has an actual attractor, which makes the problem also hard for μ-greedy EMOA.

Furthermore, it has been shown that even a local-search-based 1-greedy EMOA successfully detects the globally optimal distribution for most connected continuous Pareto front types. Failures have only been observed for very small population sizes and we therefore think that the risk of not being 1-greedy decreases with increasing population size. First hints on possible explanations have been provided. Additionally, we have proven that the hypervolume is 1-greedy on linear Pareto fronts and formalize the necessary condition of 1-greediness in general.

In this work, we only consider situations, in which the points are located exactly on the Pareto front, which is not realistic for continuous spaces. We claim that the risk of getting stuck in a local optimum decreases when the population is not close to the Pareto front since there are more improving directions. Future work shall further investigate the influence of the reference point on the properties of the distribution of points and the convergence to the distribution obtaining the optimal $\mathcal{S}$-metric value.

Acknowledgments

This work was supported by the Deutsche Forschungsgemeinschaft (DFG) as part of the Collaborative Research Center 'Computational Intelligence' (SFB 531), the SFB/TR TRR 30, and grant no. RU 1395/3-2. We also acknowledge support by the German *Federal Ministry of Economics and Technology (BMWi)*.

References

[BFLI+07] Beume, N., Fonseca, C.M., López-Ibáñez, M., Paquete, L., Vahrenhold, J.: On the Complexity of Computing the Hypervolume Indicator. Technical Report CI-235/07, Reihe CI, SFB 531 (2007)

[BNE07] Beume, N., Naujoks, B., Emmerich, M.: SMS-EMOA: Multiobjective selection based on dominated hypervolume. European Journal of Operational Research 181(3), 1653–1669 (2007)

[DPAM02] Deb, K., Pratap, A., Agarwal, S., Meyarivan, T.: A Fast and Elitist Multiobjective Genetic Algorithm: NSGA–II. IEEE Transactions on Evolutionary Computation 6(2), 182–197 (2002)

[DTLZ02] Deb, K., Thiele, L., Laumanns, M., Zitzler, E.: Scalable Multi-objective Optimization Test Problems. In: Proc. of the 2002 Congress on Evolutionary Computation (CEC 2002), vol. 1, pp. 825–830. IEEE Press, Piscataway (2002)

[HO01] Hansen, N., Ostermeier, A.: Completely Derandomized Self-Adaptation in Evolution Strategies. IEEE Computational Intelligence Magazine 9(2), 159–195 (2001)

[Sch85] Schaffer, J.D.: Multiple objective optimization with vector evaluated genetic algorithms. In: Grefenstette, J.J. (ed.) Proc. 1st Int'l. Conf. Genetic Algorithms (ICGA), pp. 93–100. Lawrence Erlbaum, Mahwah (1985)

[ZT98] Zitzler, E., Thiele, L.: Multiobjective optimization using evolutionary algorithms - A comparative case study. In: Eiben, A.E., Bäck, T., Schoenauer, M., Schwefel, H.-P. (eds.) PPSN 1998. LNCS, vol. 1498, pp. 292–301. Springer, Heidelberg (1998)

[ZTB08] Zitzler, E., Thiele, L., Bader, J.: On Set-Based Multiobjective Optimization. Technical Report 300, Computer Engineering and Networks Laboratory, ETH Zurich (February 2008)

Noisy Multiobjective Optimization on a Budget of 250 Evaluations

Joshua Knowles[1], David Corne[2], and Alan Reynolds[2]

[1] School of Computer Science, University of Manchester, UK
j.knowles@manchester.ac.uk
http://dbkgroup.org/knowles/TOMO/

[2] School of Mathematics and Computer Science, Heriot-Watt University, UK

Abstract. We consider methods for noisy multiobjective optimization, specifically methods for approximating a true underlying Pareto front when function evaluations are corrupted by Gaussian measurement noise on the objective function values. We focus on the scenario of a limited budget of function evaluations (100 and 250), where previously it was found that an iterative optimization method — ParEGO — based on surrogate modeling of the multiobjective fitness landscape was very effective in the non-noisy case. Our investigation here measures how ParEGO degrades with increasing noise levels. Meanwhile we introduce a new method that we propose for limited-budget and noisy scenarios: TOMO, deriving from the single-objective PB1 algorithm, which iteratively seeks the basins of optima using nonparametric statistical testing over previously visited points. We find ParEGO tends to outperform TOMO, and both (but especially ParEGO), are quite robust to noise. TOMO is comparable and perhaps edges ParEGO in the case of budgets of 100 evaluations with low noise. Both usually beat our suite of five baseline comparisons.

1 Introduction

Real-world optimization problems often involve solutions that are expensive to evaluate, either financially or in time, thus imposing a budget on the number of evaluations that can be done during an optimization procedure. Sometimes, the expense is so acute that only a 'handful' of evaluations is feasible, so that using a latin hypercube design, other experimental designs (DoE approaches), or even a random search may yield better results than iterative approaches, particularly on multiobjective problems. When thousands or more evaluations may be done, however, we would expect iterative sampling methods such as evolutionary algorithms (EAs) to generally outperform random search and/or DoE. But between these two regimes there may lie a third where EAs are ineffective, yet there is the potential to outperform static 'designs' or random search.

There is evidence of this regime in the work done on EAs combined with surrogate modeling, and in the statistics/DoE, direct search and machine learning (ML) literature. Although DoE traditionally concerns itself with static designs, modern techniques also include iterative approaches that augment an initial design, based on the values observed. The EGO method [19] is such a technique: starting from an initial latin hypercube design, it proceeds point by point, always using all previous points to (fit a model and) estimate the point of maximum expected improvement. A similar iterative method

M. Ehrgott et al. (Eds.): EMO 2009, LNCS 5467, pp. 36–50, 2009.

(PB1 [1]) was also proposed for optimization on a very limited evaluation budget, and seems fairly robust to noise — a very common phenomenon in real-world applications.

Here we investigate optimization given the combined difficulties of a very limited evaluations budget and the existence of noise. We aim towards an understanding where MOEAs, random search, DoE and advanced iterative approaches might stand relative to one another, under these conditions. We continue in Section 2 with an account of prior and related research. Then in Section 3 we introduce TOMO in some detail, and briefly outline our set of comparison and baseline methods: ParEGO, DoE, random search, a simple multiple trajectory hillclimber, what we call a 'simple Gaussian model learner', and PESA-II. Section 4 describes our experimental setup, results are presented in Section 5, and Section 6 discusses the results and concludes.

2 Background

Expensive optimization problems are now rather common. Optimizing structural form, guided by the use of accurate simulators, is perhaps the most familiar domain in which such problems arise (e.g. [16,17]), while they also occur in biochemistry and materials science [6,12], robotics (e.g. [14]), and instrument configuration (e.g. [10]).

Typically, expensive fitness functions involve computational fluid dynamics (CFD) or similar simulations or finite element grids. E.g. [16] uses CFD in evaluating candidate shapes for the combustion chamber of a diesel engine, aiming to minimize Nitrous Oxide emissions. Often, real-world testing rather than simulation is involved. E.g. an instrumentation setup problem ([10]) which formed the motivation for ParEGO [22], concerns the efficiency of instruments used to test and monitor biological samples; [10] reports that it took several days to perform 180 evaluations, each of which required manual configuration and testing of an instrument. Similar time may be needed for evolving locomotion controllers for robot gaits, in which physical setup and testing of a configuration is preferable to simulation [14]. In [14], they observed the (very common) complication of noise: some configurations may score well, but be non-robust to slight changes in the evaluation regime.

The effects of noise and ways to deal with it in evolutionary computation have been analysed much (e.g. [2,3]). Unfortunately no comfort is yet to be found in this for those with a very limited evaluations budget. It turns out that effects of noise are highly problem dependent, while the key questions relate to whether increasing exploration (e.g. larger population sizes) or increased resampling (multiple evaluations to better characterize individual solutions) are best. Either way, there is little help here for practitioners on a very limited evaluation budget.

Even ignoring noise, it seems that, particularly in the context of multiobjective optimization, the published work relevant to limited evaluation budgets is sparse. One approach is to attempt to learn a model of the search landscape with neural networks [26,13], enabling predicted fitnesses to replace the need for evaluation in specific phases of an overall control algorithm. The simplest approach to 'guessing' fitness is actually fitness inheritance, first proposed for multiobjective optimization in [4] and later tested by [9]. Meanwhile, the research literature in general is becoming richer in suggestions of modelling techniques to underpin these guesses and consequently provide guidance towards the most promising next point to evaluate, and in multiobjective optimization specifically, this work (some of which is reviewed in [23]) has included the use of Bayesian model-building (e.g. [24,33]), support vector regression [27]) and the use of and the adoption of a Gaussian process model in ParEGO [22], by Emmerich and co-authors [11], and also in EGOMOP [16].

Such models, which attempt to learn to predict fitness via neural networks, Gaussian processes, or otherwise, are generally known as 'meta-models'. Detailed reviews of meta-model based evolutionary algorithms include [30,18]. It seems intuitively right that, when the evaluation budget is limited, algorithms that incorporate sophisticated meta-models should be promising. This is borne out in published work so far. It seems clear that ParEGO's employment of a metamodel leads to significantly better performance in this context than traditional MOEAs, random search, and a few additional alternatives (using less sophisticated models) against which it has so far been tested [21,22].

3 Optimization Methods

3.1 The Tau-Oriented Multiobjective Optimizer (TOMO)

TOMO is based closely on the principles and some of the procedures used in the single-objective optimizer, PB1.

Principles of PB1. This method [1] is based on the assumption that near an optimum of a function there should be a negative correlation between the (single-objective) fitness of points and their distance from the optimum. This leads to a basic procedure for estimating the location of an optimum: given a cloud of previously evaluated points, search for a 'test point' where the correlation reaches its maximum absolute value. PB1 iterates this, as illustrated in Fig. 1a. NB: throughout the paper fitnesses are actually costs, i.e. we assume minimization.

On non-convex problems as well as functions with plateaus or ridges, the correlation between fitness and distance from the optimum may only hold locally. To account for this, PB1 attempts to identify a subset of the points previously evaluated, forming a convex region in decision space, for which the correlation is observed to occur (see below for details). It is this subset of points which is subsequently used to estimate a new test point. Correlations in PB1 are measured using Kendall's Tau (τ), a nonparametric method based on the ranks of the distances and fitnesses.

Empirical tests of PB1 [1] indicate that it can successfully optimize low-dimensional functions in a small number of steps, including multimodal, ridge and plateau functions, and it is relatively robust to noise. On the one hand, PB1's ability to aggressively search a space can be attributed to the fact that it exploits information on the topology of the search landscape gleaned from all previous points. In this regard, PB1 works similarly to EGO (see ParEGO, below). On the other hand, its robustness to noise can be attributed to the fact that the test it uses to 'reason' about the topology is quite weak: a nonparametric correlation value is not disturbed much when noise is added to the points. See Fig. 1 to see how this contrasts with EGO.

Adapting PB1 to the Multiobjective Case. There are two main hurdles to making an effective multiobjective algorithm based on PB1 and its use of Kendall's correlation

Algorithm 1. High-level Algorithm Pseudocode for TOMO and ParEGO

input: a multiobjective optimization problem with k objectives
require: a sequence of scalarizing weight vectors $\langle \boldsymbol{\lambda} \rangle$
distribute initial E points in a latin hypercube design; evaluate each one
while evaluation limit not reached **do**
  draw the next scalarizing weight vector and use it to scalarize all previously evaluated points
  construct a model of the scalarized search landscape based on a subset of (or all) previous points
  search the model iteratively to find a single new candidate point; evaluate this point on the real multiobjective function
end while
output: all visited solutions

measure to orient search. The first concerns how to convert the basic principle of using fitness-distance correlations to work in the multiobjective case. This, we achieve by taking a simple scalarizing approach, very similar to that used in ParEGO, where at each step of the algorithm the next weight vector from a sequence is used to scalarize the fitnesses of all points for that step (see Algorithm 1). The second hurdle derives from the fact that although PB1 can cope with some multimodality it is not designed to find multiple optima. Our initial testing of PB1 showed that it strongly favours one optimum in a multimodal function, and has limited ability to escape local optima. To overcome this, TOMO was equipped with a parameter-space niching method and intermittent generation of explorative (random) search points.

Latin hypercube initialization. The initial solutions are generated in a space-filling design using a latin hypercube routine following a description in [31]. The number of initial solutions is set to $E = 11d - 1$, where d is the parameter space dimension of the function to be optimized, as suggested in [19]. This procedure in TOMO is adopted directly from ParEGO.

The scalarizing weight vectors. TOMO begins by normalizing the k cost functions with respect to the known (or estimated) limits of the cost space, so that each cost function lies in the range $[0, 1]$. Then, at each iteration of the algorithm, a weight vector $\boldsymbol{\lambda}$ is drawn from the set of evenly distributed vectors defined by:

$$\Lambda = \left\{ \boldsymbol{\lambda} = (\lambda_1, \lambda_2, \ldots, \lambda_k) \mid \textstyle\sum_{j=1}^{k} \lambda_j = 1 \wedge \forall j, \lambda_j = l/s, l \in \{0, \ldots, s\} \right\}, \quad (1)$$

with $|\Lambda| = \binom{s+k-1}{k-1}$ (so that the choice of the parameter s determines how many vectors there are in total). The scalar cost of a solution is then computed using the augmented Tchebycheff function (see [25] pp. 100–102) :

$$f_{\boldsymbol{\lambda}}(x) = \max_{j=1}^{k}(\lambda_j.f_j(x)) + \rho \sum_{j=1}^{k} \lambda_j.f_j(x), \quad (2)$$

where ρ is a small positive value which we set to 0.05. The weight vectors are arranged in a sequence using a Gray coding. To select the 'next' vector, an index into this sequence is incremented mod $|\Lambda|$ so that the sequence wraps around.

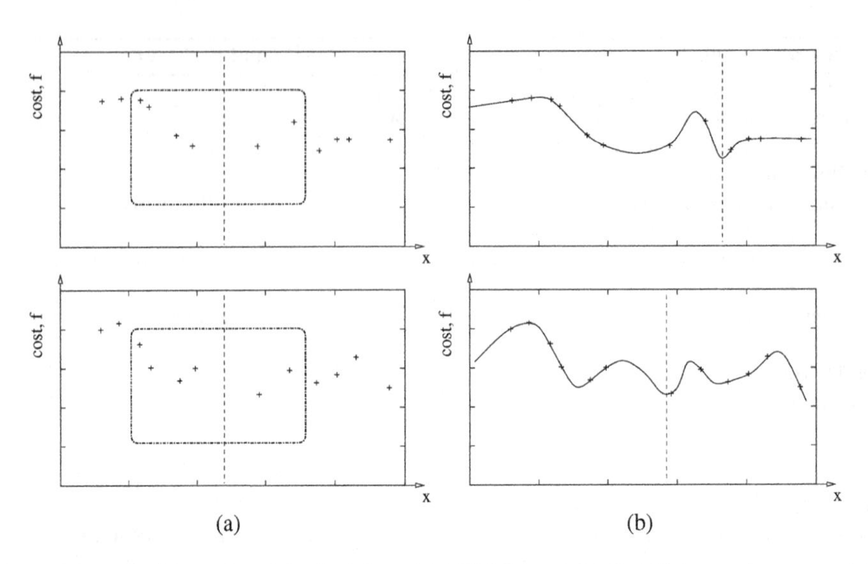

Fig. 1. Schematic illustrating the principles of (a) TOMO and (b) ParEGO on a 1-parameter 1-objective cost function. The upper and lower plots indicate different noisy measurements of the same points. (a) TOMO identifies a region (dashed rectangle) where a statistically significant negative correlation seems to occur between fitness difference and parameter-space distance from a 'test' point (the vertical line). This region and the test point are little changed as a result of noise (compare top and bottom). (b) ParEGO fits a model which interpolates the set of previous points. The model may move significantly under different noisy instances of the same set of measurements (compare top and bottom). The minimum of the (mean) model is shown by the vertical line, however ParEGO does not necessarily move to this minimum, but rather to a point of maximum *expected improvement*, which accounts also for the variance in the model (not shown).

Using niching and random explorative moves. In the original PB1, each main iteration begins by finding a subset of the previously evaluated points possessing a high τ value. This is done by starting with the set of all the points and iteratively removing points on the exterior of the convex hull in decision space, one by one. More precisely, the exterior point furthest from the centroid of the current point cloud is removed in each of these mini-steps. For each subset visited along the way, τ is calculated, using the current centroid as the point to compute τ from, and these values are stored. The subset with the smallest number of points that has a statistically significant τ is then selected for subsequent use. This ensures that τ is calculated over a region from within which no points have been removed, and over which the required fitness-distance correlation holds. To generate a new point for real evaluation, PB1 then searches over a region defined by this subset (specifically, the union of Voronoi regions pertaining to these points) for a point yielding a strong fitness-distance correlation.

We replace PB1's method of obtaining a subset by an approach less likely to converge on a local optimum. Instead of using the centroid of the point cloud during this whittling down process, we use either (i) tournament selection, based on parameter-space niched fitness [7]; or (ii) a randomly generated point. An exploitation (i) step is used with probability ν and a random explore (ii) step with probability $1 - \nu$. We tuned this on a single-objective test function with two local optima (see Fig. 2), which easily

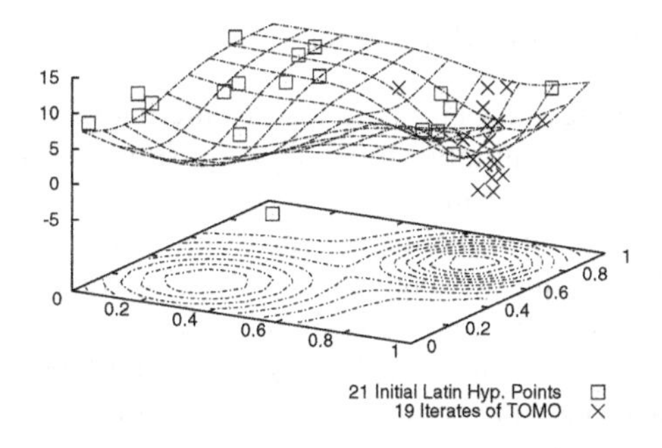

Fig. 2. A single objective two-peak noisy function used to test and tune TOMO. Twenty-one initial points are shown, followed by the nineteen next points TOMO visits. The true objective value of points are indicated by the surface.

misleads the original PB1. The parameters we found to give both rapid convergence and the ability to escape the local optimum were: tournament size 10, Niche radius set according to [7] assuming 10 peaks, and rate of exploitation $\nu = 0.8$. The remaining procedures and parameters of TOMO follow the detailed specification of PB1 in [1].

3.2 ParEGO

The ParEGO (Pareto EGO) algorithm used here is identical to that described in [22], which is essentially a multiobjective translation of EGO [19], making use of scalarizing weight vectors at each step. The high-level pseudocode is given in Algorithm 1. ParEGO works by fitting a Gaussian process stochastic model called DACE [32], to the previously evaluated points, and using this to estimate interesting new points to visit subsequently. The mean of the DACE model interpolates the points, which might make it sensitive to noise, as shown in Fig. 1b. However, because ParEGO uses expected improvement, defined as

$$E[I(\boldsymbol{x})] \equiv E\{\max(y^* - Y_p(\boldsymbol{x}), 0)\},$$

where y^* is the best cost sampled so far, and $Y_p(\boldsymbol{x})$ is a Gaussian distributed random variable representing the model through the point $\boldsymbol{x}$, — a calculation that is based on the *variance* of the model as well as its expected value, this may counteract the problem of interpolating evaluated points to some degree.

3.3 Latin Hypercube and Random Search

As our first two baseline algorithms for comparison, we consider (i) random search (RS) and (ii) the latin hypercube [31] (LH) method used, for initialization, in both TOMO and ParEGO. In the case of random search, a single ‘run’ corresponds to generating and evaluating *numevals* random points in the parameter space, where *numevals* is the maximum number of fitness evaluations allowed in the experiment (i.e. either 100 or 250). Each solution is independently generated in the standard fashion, by choosing a uniform random value from the range of each parameter. In the case of LH, a single run simply applies one iteration of the method of [31] to generate either 100 or 250 solutions.

3.4 Simple Multiple Trajectory Hillclimber

We also considered baselines that allow us to test alternative yet simple strategies. The first of these attempts a multiobjective search by spreading the evaluation budget (almost) equally among $k + 1$ hillclimbers, where k is the number of objectives. Each objective is assigned to one hillclimber, whose fitness is that objective alone; the $k + 1$th hillclimber uses an equally weighted sum of objectives as its fitness function. A hillclimber maintains a 'current' solution c initialized uniformly at random, and then iterates the following: generate mutant m by copying c and then choosing a random parameter and applying a Gaussian perturbation to it with standard deviation s. If m is no worse than c, then c becomes m.

When the overall evaluations budget is n, the k hillclimbers each have a budget of $n/(k + 1)$, maintaining whole numbers by allowing the sum-of-objectives hillclimber to use the surfeit after division by $k + 1$. The result of SMH is then taken to be the non-dominated set of all solutions visited. We test three parameterizations of SMH, namely with $s = 0.1, 0.3, 1.0$.

3.5 Simple Gaussian Model Learner

Our second alternative yet simple baseline is best described as a type of estimation of distribution algorithm for real-valued parameters, although it can equally well be described as a standard type of evolution strategy with multi-parent uniform crossover and a Pareto-oriented truncation selection scheme. Its inspiration comes originally from the considerably growing body of work that finds combinations of learning and exploration to be highly valuable in accelerating progress per evaluation, even when the learning mechanism or model is very simple.

The simple Gaussian model learner (SGM) operates as follows, with a population size P and a standard deviation s. The key part of the algorithm is the way a new point is generated from the current nondominated set of points visited, S. Given S, we generate a new point by doing the following for each parameter j of the new point: choose a member c of S uniformly at random; let parameter j of c be $c[j]$; let parameter j of the new point be drawn from a Gaussian distribution with mean $c[j]$ and standard deviation s. After initializing the population uniformly at random, we evaluate these P points and find their Pareto Set S; then we continue as follows until our evaluation budget is used up. (i) generate and evaluate a new population of P points using the procedure above; (ii) update the Pareto Set S, and return to (i).

So, in SGM, a simple probabilistic model is learned, based on the current approximation to the Pareto Front. Considering only the PF points, each parameter is modelled independently as an equally weighted mixture of Gaussians each with standard deviation s, with one Gaussian per point in the current Pareto Set approximation, centred on that point's value for the parameter in question. In our experiments we set $P = 10$ and use $s = 0.1, 0.3, 1.0$ as for SMH.

3.6 PESA-II

The Pareto Envelope based Selection Algorithm (PESA-II) [5] is one of the several multiobjective evolutionary algorithms that emerged in the resurgence of interest in this field in the late 90s. PESA-II attempts to find good approximations to the Pareto front by maintaining a datastructure that keeps track of the density of solutions across its current Pareto front approximation. The objective space is divided into 'hyperboxes', and selection of points for further exploration is guided by the relative crowding of hyperboxes, preferring to explore (i.e. use as parents for crossover or mutation) areas that currently have low density. We use it here as a convenient example of one of the several MOEAs that, in noise-free unlimited-evaluations budget scenarios at least, is quite proficient. [5] explains PESA-II in detail. We set the key parameters of PESA-II in our experiments as follows: population size $IP = 10$; archive size $EP = 100,\ 250$; number of hyperboxes=10^k, where k is number of objectives; binary representation of parameters (30 bits per parameter); $1/L$ bit-flip mutation rate, where L is bitstring length; uniform crossover applied at a rate of 0.2.

4 Testing Regime and Procedures

Our testing regime and procedures are informed by both real-world problems/ applications of interest to us, and what is available and best practice in the MOEA literature. The numbers of function evaluations available, the number of real-valued variables that we consider (less than 9) and the noise levels are all typical of real problems in mass-spectrometer optimization (e.g. [28]) as well as the optimization of chemical mixtures [6,12] and process optimization problems such as the one used in [1].

Test Functions. Originally from four sources, our test functions are those used and described fully in [22][1]. They range in dimension from 2 to 8 decision variables, and are all 2- or 3-objective problems. Note that the DTLZa functions we use are derived from the popular DTLZ ones, but we have reduced the number of parameters commensurate with the limited number of evaluations we are using, and also further reduced the difficulty of DTLZ1 by lessening the ruggedness of the function. (This lessening of the dimension/difficulty of the functions was done entirely independently and before any optimization was begun.)

Noise Model. We apply additive Gaussian noise to the objective function values before passing the values to the optimization algorithm. Repeated evaluation of the same point would therefore yield different results. We test three noise levels, 0%, 10% and 30%. 10% noise, for example, indicates that the objective value is perturbed by a Gaussian with mean zero and standard deviation of 10% of the cost function's range. When the output of the optimization algorithm is measured and compared, we use the true underlying (noiseless) objective values. This makes sense in the case that the noise represents just (unbiased) measurement error, but that underlying differences are important or, equally, the case that the noise represents natural variation in the measured objective, but we are interested in the expected value of this, e.g. the average yield that a chemical process would give over the long term, given some setting (see [3], section 3.1 Type C uncertainty).

[1] We consider here 8 of the 9, having dropped VLMOP2, due only to space limitations.

KNO1 [22] Features: Two decision variables; two objectives; Fifteen locally optimal Pareto fronts.
OKA1 [29] Features: Two decision variables; two objectives; Pareto optima lie on curve; density of solutions low at PF.
OKA2 [29] Features: Three decision variables; two objectives; Pareto optima lie on spiral-shaped curve; density of solutions very low at PF.
VLMOP3 [34] Features: Two decision variables; three objectives; disconnected Pareto optimal set and PF is a curve 'following a convoluted path through objective space'.
DTLZ1a, adapted from [8] Features: Six decision variables; two objectives; local optima on the way to the PF.
DTLZ2a and DTLZ4a, adapted from [8] Features: Eight decision variables; three objectives; DTLZ4a biases the density distribution of solutions toward the $f_3 - f_1$ and $f_2 - f_1$ planes.
DTLZ7a, adapted from [8] Features: Eight decision variables, three objectives; four disconnected regions in the Pareto front (in objective space).

Fig. 3. Summary of the eight test functions

Performance Assessment. Performance assessment of multiobjective optimizers is well known to be a nontrivial task [20,35] owing largely to the fact that the result of an optimization is a *set* of points, defining an approximation to a Pareto surface, and pairs of such surfaces (e.g. from the results of different algorithms) are commonly incomparable. Following the analysis in [20], we use Jaszkiewicz and Hansen's R metrics, which tend to dominate alternatives in terms of their profile of desirable properties. They tend to avoid being biased in favour of a particular property of a Pareto set approximation, (such as cardinality or uniformity), they do not rely on knowledge of the true Pareto front, and they are relatively scalable to many-objectives. They require using, however, a (relatively arbitrary) reference set of nondominated points for any given problem. Given the reference set and a set of points S output from an optimization run, an R metric provides a single scalar value that estimates the 'utility' of S. We mainly use R_3, but resort to R_2 in two cases where the R_3 measure led to excessive standard deviations, arising from vagaries of the relationship between certain result sets and the chosen reference sets.

5 Results

We compared ParEGO, TOMO, SGM, SMH, PESA-II, LH and RS, with budgets of 100 and 250 evaluations. All parameters of the algorithms have been given in Section 3, but recall SGM and SMH are each tried with three values of their standard deviation parameter: 0.1, 0.3 and 1.0; the other algorithms have no free parameters.

For every (algorithm, test-function, max-evals, noise-level) tuple, 21 independent runs were done. The use of an odd number of runs allows for plots of median attainment surfaces, although space precludes that here. Results tables show the mean and standard deviation of the R_3 metric values for each tuple. Our reference sets[2] and R-metric code are available from the first author's web space, so the tabulated values allow others to directly compare their algorithms with those tested here.

[2] Reference sets were generated using PESA-II runs of 50,000 evaluations.

Table 1. Results for function DTLZ1a, DTLZ2a, DTLZ4a and DTLZ7a - each entry provides mean and standard deviation of Hansen's R_3 metric (R_2 for DTLZ1a) based on 21 runs per algorithm

	100 evals	250 evals	100 evals 10% noise	250 evals 10% noise	100 evals 30% noise	250 evals 30% noise
Function	DTLZ1a					
RS	0.033 (0.01)	0.024 (0.01)	0.033 (0.01)	0.024 (0.01)	0.033 (0.01)	0.024 (0.01)
LH	0.030 (0.01)	0.024 (0.01)	0.030 (0.01)	0.024 (0.01)	0.030 (0.01)	0.024 (0.01)
SMH-0.1	0.013 (0.01)	0.006 (0.00)	0.035 (0.02)	0.021 (0.01)	0.050 (0.02)	0.040 (0.02)
SGM-0.1	0.008 (0.00)	0.003 (0.00)	0.036 (0.02)	0.024 (0.02)	0.071 (0.05)	0.063 (0.04)
PESA2	0.010 (0.01)	**0.000** (0.00)	0.050 (0.02)	0.045 (0.02)	0.112 (0.05)	0.101 (0.06)
ParEGO	**0.001** (0.00)	0.004 (0.00)	**0.002** (0.01)	**0.012** (0.05)	**0.006** (0.00)	**0.006** (0.00)
TOMO	0.011 (0.00)	0.001 (0.00)	0.022 (0.01)	**0.012** (0.01)	0.026 (0.01)	0.017 (0.01)
ParEGO vs TOMO	100/0, 100/30, 250/30 ParEGO wins 99.95; 250/0 TOMO wins 99.95					
SGM-0.1 vs SMH-0.1	100/0, 250/0 SGM wins 97.5; 100/30, 250/30 SMH wins 99.75					
Function	DTLZ2a					
RS	0.367 (0.21)	0.157 (0.33)	0.367 (0.21)	0.157 (0.33)	0.367 (0.21)	0.157 (0.33)
LH	0.232 (0.01)	0.227 (0.11)	0.232 (0.01)	0.227 (0.11)	**0.232** (0.01)	0.227 (0.11)
SMH-0.3	0.305 (0.06)	0.254 (0.04)	0.297 (0.07)	0.294 (0.20)	0.295 (0.07)	0.235 (0.10)
SGM-0.3	0.317 (0.63)	0.057 (0.10)	0.282 (0.09)	0.298 (0.38)	0.443 (0.31)	0.309 (0.05)
PESA2	0.547 (0.53)	0.509 (0.68)	0.535 (0.52)	0.468 (0.67)	0.683 (0.36)	0.646 (0.40)
ParEGO	0.511 (1.14)	-0.032 (0.02)	0.317 (0.63)	**0.057** (0.10)	0.290 (0.26)	0.195 (0.10)
TOMO	**0.182** (0.46)	**-0.137** (1.26)	**0.141** (0.73)	0.289 (0.62)	0.352 (0.32)	**0.136** (0.43)
ParEGO vs TOMO	250/10 ParEGO wins 90					
SGM vs SMH	100/0, 250/0 SGM wins 99.95 ; 100/30, 250/30 SMH wins 99.95					
Function	DTLZ4a					
RS	0.524 (0.08)	0.574 (0.50)	0.524 (0.08)	0.574 (0.50)	0.524 (0.08)	0.574 (0.50)
LH	0.534 (0.11)	0.438 (0.12)	0.534 (0.11)	0.438 (0.12)	0.534 (0.11)	0.438 (0.12)
SMH-0.3	0.264 (0.06)	0.212 (0.04)	0.257 (0.04)	0.221 (0.03)	**0.265** (0.04)	0.245 (0.05)
SGM-0.3	**0.259** (0.07)	0.201 (0.02)	**0.235** (0.04)	**0.211** (0.03)	0.311 (0.09)	0.255 (0.05)
PESA2	0.659 (0.37)	0.549 (0.10)	0.582 (0.10)	0.564 (0.10)	0.598 (0.11)	0.594 (0.04)
ParEGO	0.508 (0.20)	**0.148** (0.13)	0.557 (0.05)	0.223 (0.11)	0.445 (0.13)	**0.201** (0.28)
TOMO	0.616 (0.44)	0.423 (0.25)	0.529 (0.10)	0.529 (0.07)	0.677 (0.43)	0.428 (0.22)
ParEGO vs TOMO	100/30, 250/0, 250/10, 250/30 ParEGO wins 99.5					
SGM-0.3 vs SMH-0.3	100/10 SGM wins 95 ; 100/30 SMH wins 97.5					
Function	DTLZ7a					
RS	2.480 (8.83)	1.940 (9.20)	2.480 (8.83)	1.940 (9.20)	2.480 (8.83)	1.940 (9.20)
LH	0.58 (0.05)	0.489 (0.17)	0.58 (0.05)	0.489 (0.17)	0.58 (0.05)	0.489 (0.17)
SMH-1.0	0.553 (0.09)	0.490 (0.05)	0.613 (0.13)	0.510 (0.19)	0.616 (0.08)	0.508 (0.10)
SGM-1.0	0.538 (0.06)	0.545 (1.12)	0.622 (0.07)	0.580 (0.07)	0.656 (0.08)	0.633 (0.08)
PESA2	0.517 (0.16)	0.212 (0.63)	0.650 (0.32)	0.570 (0.22)	0.681 (0.09)	0.332 (0.19)
ParEGO	0.573 (0.04)	0.232 (0.07)	0.535 (0.13)	0.307 (0.42)	0.561 (0.05)	0.403 (0.15)
TOMO	0.501 (0.11)	**-0.442** (2.71)	**0.497** (0.27)	**0.136** (0.53)	**0.451** (0.87)	**0.323** (1.23)
ParEGO vs TOMO	100/0 TOMO wins 99					
SGM-1.0 vs SMH-1.0	100/30, 250/10, 250/30 SMH wins 90					

We compare ParEGO and TOMO directly, using t-tests (assuming unequal variances) on the R values. Similarly, we compare a selected pair of parameter variants of SGM and SHM per test function (we choose those that performed best with noise). Further statistical comparisons (e.g. ParEGO vs LH) without generating further independent sets of results would amount to multiple testing and be statistically invalid (we could additionally compare LH and RS, but we omit that for reasons of space and salience). Instead, we calculate 'naive' rank orderings of the algorithms for each (test-function,

Table 2. Results for functions VLMOP3, KNO1, OKA1 and OKA2 - each table entry provides mean and standard deviation of Hansen's R_3 metric based on 21 runs per algorithm

	100 evals	250 evals	100 evals 10% noise	250 evals 10% noise	100 evals 30% noise	250 evals 30% noise
Function	VLMOP3					
RS	0.290 (0.15)	0.132 (0.07)	0.290 (0.15)	0.132 (0.07)	0.290 (0.15)	0.132 (0.07)
LH	0.213 (0.137)	0.146 (0.08)	0.213 (0.137)	0.146 (0.08)	0.213 (0.137)	0.146 (0.08)
SMH-0.3	0.269 (0.15)	0.133 (0.08)	0.238 (0.18)	0.173 (0.12)	0.271 (0.17)	0.158 (0.07)
SGM-0.1	0.095 (0.08)	0.042 (0.05)	0.157 (0.11)	0.101 (0.24)	0.277 (0.22)	0.164 (0.16)
PESA2	0.169 (0.22)	0.086 (0.17)	0.260 (0.24)	0.156 (0.22)	0.256 (0.25)	0.181 (0.22)
ParEGO	**0.019** (0.01)	**0.013** (0.00)	0.058 (0.05)	0.029 (0.01)	0.080 (0.08)	**0.033** (0.01)
TOMO	0.024 (0.01)	0.026 (0.03)	**0.031** (0.01)	**0.017** (0.00)	**0.078** (0.09)	0.034 (0.05)
ParEGO vs TOMO	100/0, 250/0 ParEGO wins 90; 100/10, 250/10 TOMO wins 99					
SGM-0.1 vs SMH-0.3	100/0, 100/10 250/0 SGM wins 90					
Function	KNO1					
RS	0.012 (0.08)	-0.126 (0.07)	0.012 (0.08)	-0.126 (0.07)	0.012 (0.08)	-0.126 (0.07)
LH	-0.033 (0.1)	-0.142 (0.08)	-0.033 (0.01)	-0.142 (0.08)	-0.033 (0.1)	-0.142 (0.08)
SMH-0.3	-0.107 (0.10)	**-0.282** (0.08)	**-0.130** (0.09)	-0.258 (0.09)	-0.117 (0.11)	**-0.268** (0.09)
SGM-0.3	-0.106 (0.13)	-0.224 (0.09)	-0.001 (0.14)	-0.096 (0.12)	0.189 (0.18)	0.107 (0.16)
PESA2	0.112 (0.14)	0.085 (0.16)	0.177 (0.13)	0.138 (0.15)	0.299 (0.11)	0.272 (0.14)
ParEGO	-0.049 (0.10)	-0.137 (0.11)	-0.128 (0.11)	**-0.290** (0.07)	**-0.120** (0.11)	-0.265 (0.07)
TOMO	**-0.127** (0.12)	-0.191 (0.14)	-0.129 (0.11)	-0.216 (0.12)	-0.074 (0.10)	-0.196 (0.12)
ParEGO vs TOMO	100/0, 250/0 TOMO wins 90 ; 100/30, 250/10, 250/30 ParEGO wins 90					
SGM-0.3 vs SMH-0.3	100/10, 100/30, 250/0, 250/10, 250/30 SMH wins 97.5					
Function	OKA1					
RS	0.376 (0.04)	0.310 (0.04)	0.376 (0.04)	0.310 (0.04)	0.376 (0.04)	0.310 (0.04)
LH	0.380 (0.03)	0.302 (0.38)	0.380 (0.02)	0.302 (0.38)	0.380 (0.03)	0.302 (0.38)
SMH-0.3	0.339 (0.04)	0.289 (0.03)	0.348 (0.06)	0.302 (0.05)	0.380 (0.06)	0.301 (0.06)
SGM-1.0	0.351 (0.03)	0.321 (0.02)	0.396 (0.05)	0.375 (0.05)	0.440 (0.06)	0.437 (0.05)
PESA2	0.354 (0.07)	0.266 (0.05)	0.447 (0.07)	0.417 (0.08)	0.562 (0.08)	0.540 (0.08)
ParEGO	**0.071** (0.03)	**0.071** (0.03)	**0.211** (0.09)	**0.086** (0.03)	**0.302** (0.04)	**0.195** (0.04)
TOMO	0.273 (0.05)	0.219 (0.06)	0.303 (0.04)	0.198 (0.06)	0.330 (0.12)	0.296 (0.06)
ParEGO vs TOMO	100/0, 100/10, 250/0, 250/10, 250/30 ParEGO wins 99.95					
SGM-1.0 vs SMH-0.3	100/10, 100/30, 250/0, 250/10, 250/30 SMH wins 99.5					
Function	OKA2					
RS	0.458 (0.03)	0.410 (0.03)	0.458 (0.03)	0.410 (0.03)	0.458 (0.03)	0.410 (0.03)
LH	0.440 (0.04)	0.411 (0.03)	0.440 (0.04)	0.411 (0.03)	0.440 (0.04)	0.411 (0.03)
SMH-1.0	0.303 (0.04)	0.241 (0.04)	0.304 (0.06)	0.248 (0.06)	**0.316** (0.05)	0.263 (0.06)
SGM-1.0	0.271 (0.05)	0.237 (0.05)	0.335 (0.08)	0.285 (0.07)	0.410 (0.08)	0.382 (0.05)
PESA2	0.364 (0.09)	0.251 (0.10)	0.488 (0.06)	0.456 (0.07)	0.590 (0.13)	0.555 (0.14)
ParEGO	**0.146** (0.05)	**0.058** (0.04)	**0.245** (0.06)	**0.070** (0.04)	0.330 (0.07)	**0.251** (0.06)
TOMO	0.354 (0.07)	0.307 (0.07)	0.414 (0.04)	0.331 (0.03)	0.444 (0.04)	0.391 (0.04)
ParEGO vs TOMO	100/0, 100/10, 250/0, 250/10, 250/30 ParEGO wins 99.95					
SGM-1.0 vs SMH-1.0	100/0, SGM wins 97.5 ; 100/10, 100/30, 250/10, 250/30 SMH wins 90					

max-evals, noise-level) triple, based only on mean R values, and we build a summary table of mean ranks for each algorithm in different scenarios. This leads to a series of overall indicative observations, and which we feel provide valuable insight and pointers to further work.

Tables 1 and 2 present summary results on each of the test functions. In each case, Random Search (RS) results for the noise cases are shown for convenience, though they are necessarily identical to the non-noise cases. To save space, only the 'best' of the three variants each of SGM and SMH are shown in the tables. To support

Table 3. A broad summary of the effects of number of evaluations and of levels of noise on an algorithm's mean naive rank over the test functions studied

mean rank	RS	LH	SGM	SMH	PESA2	ParEGO	TOMO
overall	5.4	4.6	3.8	3.4	6.0	**2.1**	2.6
100 evals / no noise	6.1	5.3	**2.4**	3.5	5.4	2.6	2.8
100 evals / 10% noise	5.5	4.1	3.4	3.4	6.7	2.8	**2.1**
100 evals / 30% noise	5.4	3.6	4.9	2.6	6.6	**1.8**	3.1
250 evals / no noise	6.3	6.0	3.0	3.5	4.5	**2.3**	2.5
250 evals / 10% noise	4.6	4.5	3.9	3.9	6.6	**1.5**	2.9
250 evals / 30% noise	4.6	4.1	5.3	3.5	6.4	**1.6**	2.4

understanding the tables, we interpret parts of Table 1 as follows: On DTLZ1a, we see that SGM, with standard deviation 0.1, achieved a mean R_2 value of 0.063, (with R values, lower is always better) with a standard deviation of 0.04, in the 250-evaluations limit case with 30% noise. When we compare ParEGO and TOMO on DTLZ1a, we find the following scenarios in which ParEGO outperformed TOMO with statistical confidence at least 90%: 100 evals at no noise and 30% noise, and 250 evaluations at 30% noise - among these cases, the lowest level of confidence was 99.95%. Meanwhile TOMO outperforms ParEGO in the 250-evaluations no-noise case, with confidence 99.95%; in the cases not mentioned (100/10, 250/10) the comparisons were not significant with $\geq$ 90% confidence.

Table 3 provides a broad summary of the observations that we can make on the basis of the naive rank orderings of the algorithms for each (test-function, max-evals, noise-level) scenario. Naive rank orderings are based only on mean R value; in general, they either have no statistical significance, or have significance but at a low confidence level. For a given scenario (e.g. 100 evals, no noise) we rank algorithms from 1 to 7, considering, for any particular problem, only the best of the three SGM variants, and the best of the three SMH variants for that problem. Hence, for example, in the 250-evaluations 30%-noise case on problem OKA1 (table 2), ParEGO is best with rank 1, TOMO has rank 2, SMH has rank 3, and so on, until PESA-II has rank 7. The table indicates the mean ranks for each scenario over the eight test problems.

Running Times. It is worth noting that both TOMO and ParEGO do consume significant resources to compute each solution to evaluate next, and that this time grows with each iteration. On a single-core Pentium III 2.8GHz desktop machine, they require of the order of 10s per evaluation at the end of a 250 evaluation run. Although limited budgets tend to arise when the time to evaluate a solution is considerably larger than this, one can certainly envisage some budget-constrained scenarios where such a lag would be unacceptable. For this reason, the performance of the baselines, which all have negligible runtimes in comparison to TOMO and ParEGO, are of more than incidental interest.

6 Concluding Discussion

As we suggested towards the end of Section 2, it is not surprising that a metamodel-based technique should do well in the limited-evaluations regime. However, the question of performance in the presence of noise is rather less clear *a priori*. Metamodels repeatedly rely on the positions and fitnesses of samples previously visited (each evaluated only once) in order to build a picture that guides choice of the next sample point. Noise can be expected to mislead this model, and with few evaluations available there is little opportunity for recovery from this. ParEGO / EGO is no exception to this, as it does not explicitly account for uncertainty in evaluated points [15], indeed using a model that interpolates between these points and a sampling method which ensures they will *never* be re-evaluated.

As it turns out, however, ParEGO stands up to noise much more successfully than the other techniques tested; from the indications in Table 3, especially so when given at least the luxury of 250 evaluations, and especially at the higher noise levels. In contrast, SGM and PESA-II get very confused by noise. Both may have done a little better using a larger population size (and hence fewer generations) in the noise cases, but it seems that the strategies inherent in both of these techniques place undue trust in the accuracy of points visited so far. TOMO is clearly the second-best technique tested overall, regularly outperformed by ParEGO, although TOMO seems to have the edge over ParEGO at 100 evaluations and 10% noise. This suggests various ways forward for coping with such severely-limited budgets, such as tweaking the EGO model to account for noisy evaluations (as in [15] for single-objective optimization), or basing sampling decisions on evidence obtained from both ParEGO's and TOMO's strategies, weighting them appropriately given the number of evaluations so far. Meanwhile, many further parametric and design variants of TOMO can be explored, perhaps most pressingly an evaluation of the many ways scalarization could be done more adaptively, e.g. using goal programming.

As for SGM and SMH, herein they have fulfilled a need to provide further alternative strategies, continuing to investigate whether the sophistication inherent in ParEGO (and TOMO) can be undermined by a simpler (and possibly faster) alternative. As it turns out, it seems that the SGM approach appears quite useful when noise is absent, at least in the 100-evaluations regime. Parameter dependence limits this observation, however it could be suggested that an adaptive version of SGM may extend its niche of good performance toward 250 evaluations.

Finally, we note that PESA-II's performance seems generally awful; this adds weight to findings elsewhere (e.g. with NSGA-II in [22]) that standard modern MOEAs, designed with perhaps '0,000s or '000,000s of evaluations in mind, are simply inapplicable when much more limited evaluation numbers are available. However, it must be said that PESA-II's parameterization was not optimized here and the use of a binary representation is almost certainly unfair. We mention in passing that a fitness-inheritance based version of PESA-II was tested in preliminary work, and found to work fine when many '000s of evaluations were available, however the beneficial effect of fitness inheritance simply failed to be present below $\sim$ 1,000 evaluations.

References

1. Anderson, B., Moore, A., Cohn, D.: A nonparametric approach to noisy and costly optimization. In: Langley, P. (ed.) Proc. 17th ICML, pp. 17–24. Morgan Kaufmann, San Francisco (2000)
2. Beyer, H.-G.: Evolutionary algorithms in noisy environments: theoretical issues and guidelines for practice. Computer Methods in Applied Mechanics and Engineering 186(2-4), 239–267 (2000)

3. Beyer, H.-G., Sendhoff, B.: Robust optimization: A comprehensive survey. Computer Methods in Applied Mechanics and Engineering 196(33-34), 3190–3218 (2007)
4. Chen, J.-J., Goldberg, D.E., Ho, S.-Y., Sastry, K.: Fitness inheritance in multi-objective optimization. In: Proc. GECCO 2002, pp. 319–326. Morgan Kaufmann, San Francisco (2002)
5. Corne, D., Jerram, N., Knowles, J., Oates, M.: PESA-II: Region-based selection in evolutionary multiobjective optimization. In: GECCO 2001, pp. 283–290. Morgan Kaufmann, San Francisco (2001)
6. Davies, Z.S., Gilbert, R.J., Merry, R.J., Kell, D.B., Theodorou, M.K., Griffith, G.W.: Efficient improvement of silage additives by using genetic algorithms. In: Applied and Environmental Microbiology, pp. 1435–1443 (2000)
7. Deb, K., Goldberg, D.: An Investigation of Niche and Species Formation in Genetic Function Optimization. In: Proc. 3rd International Conference on Genetic Algorithms, pp. 42–50. Morgan Kaufmann, San Francisco (1989)
8. Deb, K., Thiele, L., Laumanns, M., Zitzler, E.: Scalable Test Problems for Evolutionary Multi-Objective Optimization. Technical Report 112, Computer Engineering and Networks Laboratory (TIK), Swiss Federal Institute of Technology (ETH), Zurich, Switzerland (2001)
9. Ducheyne, E.I., De Baets, B., De Wulf, R.: Is fitness inheritance useful for real-world applications? In: Fonseca, C.M., Fleming, P.J., Zitzler, E., Deb, K., Thiele, L. (eds.) EMO 2003. LNCS, vol. 2632, pp. 31–42. Springer, Heidelberg (2003)
10. Dunn, E., Olague, G.: Multi-objective Sensor Planning for Efficient and Accurate Object Reconstruction. In: Raidl, G.R., Cagnoni, S., Branke, J., Corne, D.W., Drechsler, R., Jin, Y., Johnson, C.G., Machado, P., Marchiori, E., Rothlauf, F., Smith, G.D., Squillero, G. (eds.) EvoWorkshops 2004. LNCS, vol. 3005, pp. 312–321. Springer, Heidelberg (2004)
11. Emmerich, M., Naujoks, B.: Metamodel Assisted Multiobjective Optimisation Strategies and their Application in Airfoil Design. In: Parmee, I. (ed.) Adaptive Computing in Design and Manufacture VI, pp. 249–260. Springer, Heidelberg (2004)
12. Evans, J.R.G., Edirisinghe, M.J., Eames, P.V.C.J.: Combinatorial searches of inorganic materials using the inkjet printer: science philosophy and technology. Journal of the European Ceramic Society 21, 2291–2299 (2001)
13. Gaspar-Cunha, A., Vieira, A.S.: A hybrid multi-objective evolutionary algorithm using an inverse neural network. In: Hybrid Metaheuristics (HM 2004) Workshop at ECAI 2004, pp. 25–30 (2004), `http://iridia.ulb.ac.be/~hm2004/proceedings/`
14. Hornby, G.S., Takamura, S., Yamamoto, T., Fujita, M.: Autonomous evolution of dynamic gaits with two quadruped robots. IEEE Transactions on Robots 21(3), 402–410 (2005)
15. Huang, D., Allen, T.T., Notz, W.I., Zeng, N.: Global Optimization of Stochastic Black-Box Systems via Sequential Kriging Meta-Models. Journal of Global Optimization 34(3), 441–466 (2006)
16. Jeong, S., Minemura, Y., Obayashi, S.: Optimisation of combustion chamber for diesel engine using kriging model. Journal of Fluid Science and Technology 1(2), 138–146 (2006)
17. Jeong, S., Suzuki, K., Obayashi, S., Kirita, M.: Improvement of nonlinear lateral characteristics of lifting-body type reentry vehicle using optimization algorithm. In: Proc. of AIAA Infotech-Aerospace Conference 2007, pp. 1–15. AIAA (2007)
18. Jin, Y.: A comprehensive survey of fitness approximation in evolutionary computation. Soft Computing-A Fusion of Foundations, Methodologies and Applications 9(1), 3–12 (2005)
19. Jones, D., Schonlau, M., Welch, W.: Efficient global optimization of expensive black-box functions. Journal of Global Optimization 13, 455–492 (1998)
20. Knowles, J., Corne, D.: On metrics for comparing nondominated sets. In: Congress on Evolutionary Computation (CEC 2002), Piscataway, New Jersey, vol. 1, pp. 711–716. IEEE Service Center, Los Alamitos (2002)
21. Knowles, J., Hughes, E.J.: Multiobjective Optimization on a Budget of 250 Evaluations. In: Coello Coello, C.A., Hernández Aguirre, A., Zitzler, E. (eds.) EMO 2005. LNCS, vol. 3410, pp. 176–190. Springer, Heidelberg (2005)

22. Knowles, J.: ParEGO: A hybrid algorithm with on-line landscape approximation for expensive multiobjective optimization problems. IEEE Trans. Evol. Comp. 10(1), 50–66 (2006)
23. Knowles, J., Nakayama, H.: Meta-Modeling in Multiobjective Optimization. In: Branke, D., Deb, K., Miettinen, S., Słowiński, R. (eds.) Multiobjective Optimization: Interactive and Evolutionary Approaches. LNCS, vol. 5252. Springer, Heidelberg (2008)
24. Laumanns, M., Ocenasek, J.: Bayesian optimization algorithms for multi-objective optimization. In: Guervós, J.J.M., Adamidis, P.A., Beyer, H.-G., Fernández-Villacañas, J.-L., Schwefel, H.-P. (eds.) PPSN 2002. LNCS, vol. 2439, pp. 298–307. Springer, Heidelberg (2002)
25. Miettinen, K.M.: Nonlinear Multiobjective Optimization. Kluwer, Dordrecht (1999)
26. Nain, P.K.S., Deb, K.: A computationally effective multi-objective search and optimization technique using coarse-to-fine grain modeling. Technical Report Kangal Report No. 2002005, IITK, Kanpur, India (2002)
27. Nakayama, H., Yun, Y.: Multi-objective Model Predictive Optimization using Computational Intelligence. In: Artificial Intelligence in Theory and Practice II, pp. 319–328. Springer, Heidelberg (2008)
28. O'Hagan, S., Dunn, W., Knowles, J., Broadhurst, D., Williams, R., Ashworth, J., Cameron, M., Kell, D.: Closed-loop, multiobjective optimization of two-dimensional gas chromatography/mass spectrometry for serum metabolomics. Analytical Chemistry 79(2), 464–476 (2007)
29. Okabe, T., Jin, Y., Olhofer, M., Sendhoff, B.: On Test Functions for Evolutionary Multi-objective Optimization. In: Parallel Problem Solving from Nature - PPSN VIII, pp. 792–802. Springer, Heidelberg (2004)
30. Ong, Y.S., Nair, P.B., Keane, A.J., Zhou, Z.Z.: Surrogate-assisted evolutionary optimization frameworks for high-fidelity engineering design problems. In: Jin, Y. (ed.) Knowledge Incorporation in Evolutionary Computation. Springer, Heidelberg (2004)
31. Press, W., Teukolsky, S., Vetterling, W., Flannery, B.: Numerical Recipes in C: The Art of Scientific Computing. Cambridge University Press, Cambridge (1992)
32. Sacks, J., Welch, W., Mitchell, T., Wynn, H.: Design and analysis of computer experiments (with discussion). Statistical Science 4, 409–435 (1989)
33. Bosman, P.A.N., Thierens, D.: Multi-objective Optimization with the Naive MIDEA. Studies in Fuzziness and Soft Computing 192, 123–157 (2006)
34. van Veldhuizen, D.A., Lamont, G.B.: Multiobjective Evolutionary Algorithm Test Suites. In: Proc. 1999 ACM Symposium on Applied Computing, pp. 351–357. ACM, New York (1999)
35. Zitzler, E., Thiele, L., Laumanns, M., Fonseca, C.M., da Fonseca, V.G.: Performance assessment of multiobjective optimizers: An analysis and review. IEEE Transactions on Evolutionary Computation 7(2), 117–132 (2003)

On Uncertainty and Robustness in Evolutionary Optimization-Based MCDM

Daniel E. Salazar Aponte[1], Claudio M. Rocco S.[2], and Blas Galván[1]

[1] University of Las Palmas de Gran Canaria, Institute SIANI,
Edif. Central Pqe. Científico y Tecnológico, Las Palmas 35017, Spain
danielsalazaraponte@gmail.com, bgalvan@step.es
[2] Central University of Venezuela, Faculty of Engineering
Apartado Postal 47937, Los Chaguaramos. Caracas, Venezuela
crocco@reacciun.ve

Abstract. In this article we present a methodological framework entitled 'Analysis of Uncertainty and Robustness in Evolutionary Optimization' or AUREO for short. This methodology was developed as a diagnosis tool to analyze the characteristics of the decision-making problems to be solved with Multi-Objective Evolutionary Algorithms (MOEA) in order to: 1) determine the mathematical program that represents best the current problem in terms of the available information, and 2) to help the design or adaptation of the MOEA meant to solve the mathematical program. Regarding the first point, the different versions of decision-making problems in the presence of uncertainty are reduced to a few classes, while for the second point possible configurations of MOEA are suggested in terms of the type of uncertainty and the theory used to represent it. Finally, the AUREO has been introduced and tested successfully in different applications in [1].

1 Introduction

In this article we are concerned about the use Multiple Objective Evolutionary Algorithms (MOEA)[1] in Multiple-Criterion Decision-Making (MCDM) under uncertainty.

By MCDM we mean the process of selecting a final alternative from a group of more than one solving actions to a problematic (e.g. choice, sorting, ranking) within a common quality framework made up of various figures of merit called criteria, established by an entity called decision maker (DM). No matter what the problematic is, a rational DM is expected to maximize its level of satisfaction by choosing the alternative that scores best in terms of the criteria. Mathematically, we model it as a program of the form[2]:

$$\begin{array}{ll} & \mathbf{x}^* = \arg\max_{\mathbf{x}} \left(F(\mathbf{x}) = (f_1(\mathbf{x}), f_2(\mathbf{x}), \ldots, f_k(\mathbf{x}))^t \right) \\ \text{s.t.:} & \mathbf{x} \in \Omega \\ & \Omega = \{\mathbf{x} \in \mathbf{X} : G(\mathbf{x}) \leq 0 \ , \ H(\mathbf{x}) = 0\} \end{array} \quad (1)$$

[1] Acronyms for singular and plural forms are spelled the same hereafter.
[2] With no loss of generality, optimality is expressed in terms of maximization hereafter.

M. Ehrgott et al. (Eds.): EMO 2009, LNCS 5467, pp. 51–65, 2009.

where $F : \mathbf{X} \to \mathbf{Y}$ is a vector of criteria $f_i : \mathbb{R}^n \to \mathbb{R}$ that map a vector of n decision variables $\mathbf{x} = (x_1, x_2, \ldots, x_n)^t$ (called also *decision vector* or simply *alternative*) from the *decision space* $\mathbf{X}$, into a k-dimensional *objective vector* $\mathbf{y} = (y_1, y_2, \ldots, y_k)^t$ in the *objective space* $\mathbf{Y} \subseteq \mathbb{R}^k, k \in \mathbb{N}$. Additionally, the *feasible space* Ω is defined by two vectors $G(\mathbf{x})$ and $H(\mathbf{x})$ of inequality and equality constraints respectively.

Such a kind of problems are characterized by some conflict amongst the criteria, so that the set of alternatives cannot be arranged as a total order regarding their quality. Consequently, eq. 1 is not satisfied by a unique alternative but a subset of them called *efficient* or *non-dominated*. Typically the relation used for classifying the alternatives is the Pareto dominance, viz.:

Definition 1 (Pareto Dominance). $\mathbf{x}_1$ *dominates* $\mathbf{x}_2$*, denoted* $\mathbf{x}_1 \succ \mathbf{x}_2$ *, iff* $f_i(\mathbf{x}_1) \geq f_i(\mathbf{x}_2) \wedge F(\mathbf{x}_1) \neq F(\mathbf{x}_2);\ i \in \{1, 2, \ldots, k\}$. *If there is no solution dominating* $\mathbf{x}_1$*, then* $\mathbf{x}_1$ *is called non-dominated.*

In order to solve MCDM problems, MOEA are often used to approximate the set of non-dominated solutions. As a subclass of Evolutionary Algorithms, MOEA are searching methods based upon a population sequential sampling process ruled by heuristics. Such heuristics can be implemented in any fashion but in general they find inspiration in some natural processes (like mating and survival -Genetic Algorithms-, foraging -Ant colonies-, flocking -PSO) as well as mathematic (Differential Evolution) and thermodynamic (Simulated Annealing) principles.

Regardless of the final form given to their instances, all of the MOEA share a common principle of evolving towards a higher level of global fitness as iterations go on. In general MOEA associate fitness with Pareto optimality and approximation sets with spatial even distributions. In practice it is possible by defining a ranking procedure concerned by optimality and density built upon some fitness expression which turns out to be function of the mathematical model of eq. 1. Needless to say that, if $F(x)$ cannot be properly assessed as it happens in the presence of uncertainty, the very foundation of the operation of MOEA could be seriously compromised.

As we shall see later on, several algorithms have been proposed to operate under uncertainty. Regardless the computational efficiency of such existing approaches or any other one to come, the variety of sources, types and targets of uncertainty as well as the current theoretical frameworks to represent it, hinders the ability of MOEA designers to develop approaches valid for a wide range of situations. In response, we propose an analytical methodology called *Analysis of Uncertainty and Robustness in Evolutionary Optimization* or AUREO that allows one to study how to use MOEA in MCDM problems under uncertainty from a broad view. First the effort is oriented towards finding the mathematical formulation that suits best the decision-making problem regarding the characteristics of the uncertainty involved, while later on the analysis focuses on the structure of the MOEA propounded as solving technique, according to the characteristics of the problem formulated beforehand and on its efficiency. The benefit of doing so is double. On the one hand, having uncertainty in MCDM problems does not necessarily imply that the MOEA have to cope with uncertainty. We argue that the definitive element to decide whether the MOEA actually have to is the available information. On the other hand, in the case of dealing with uncertainty, to make a device of the structural requirements of the MOEA in terms of uncertainty handling helps one select amongst the existing instances or design new ones.

The remainder of the article is organized as follows: the next section introduces some problems raised by uncertainty and the possible reasonings available to deal with it. Section 3 brings the methodology proposed with some examples while section 4 gives some concluding remarks.

2 Accounting for Uncertainty in Decision-Making

In this section we give a glance at the notion of uncertainty, its relation with the decision-making and the existing views and reasonings about it.

The term uncertainty is understood in different ways, all of them related to defects of knowledge and information (for further insight see [2]). We adopt the full identification of uncertainty with imperfection of information, data or evidence herein. When the lack of information is originated by the inherent variability of physical systems and thus it cannot be reduced by further empirical efforts we say the uncertainty is *aleatory*. By contrast when the actual state of uncertainty is reducible by additional information of the system or its environment we call it *epistemic* uncertainty. A mixed aleatory-epistemic uncertainty is also possible.

To account for the effect of uncertainty in decision-making, consider the different scenarios depicted in fig. 1. The first target of our uncertainty analysis is the domain. This one can be subject to aleatory (case 1), epistemic (case 3) or mixed uncertainty (not depicted). In all of these cases the uncertainty associated to $\mathbf{x}$ should be propagated through $F(\mathbf{x})$ onto space $\mathbf{Y}$. If $F(\mathbf{x})$ is free of uncertainty, the propagation of epistemic uncertainty will yield epistemic objective vectors $\mathbf{y}$ (trajectory 1-3-5). In this case both the decision and objective vectors will be characterized by bounding sets (usually an interval) enclosing the true but unknown values. Likewise the propagation of aleatory decision vectors through a function free of uncertainty will yield objective vectors that actually are random variables (trajectory 3-6-9). On the other hand, the second target of the uncertainty analysis is the function. Indeed, $F(\mathbf{x})$ can be intrinsically uncertain (e.g. noisy or dynamic functions) or the way we assess it can be subject of aleatory (e.g. Monte Carlo simulation) or epistemic uncertainty (e.g. interval approximation), although the functional expression is deterministic. In such a case the result will be uncertain no matter if the input is (trajectories $\{1,3\}$,5,$\{7,8,9\}$) or not (trajectories 2,5,$\{7,9\}$) uncertain. Notice that we can have objective vectors $\mathbf{y}$ subject to mixed uncertainty, i.e. the result is a set of possible sets of outcomes.

As we just have seen, whether it is epistemic, aleatory or mixed, the presence of uncertainty always entails comparing sets of objective instead of precise vectors. Consequently, one of the challenges risen to decision-making is how to compare and classify alternatives in terms of sets comparisons. We shall consider next the

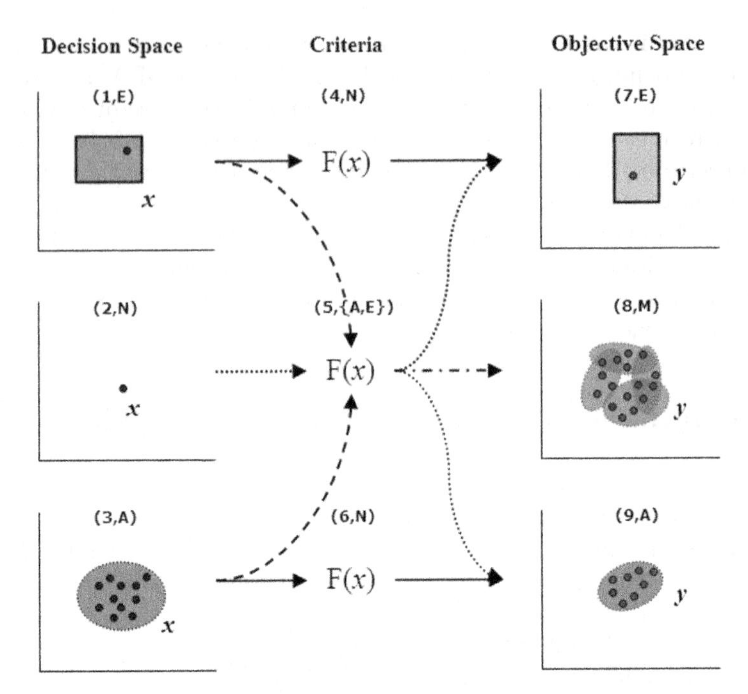

Fig. 1. Effect of Uncertainty in Decision-Making: labels (a,b) indicate the index a and the type of uncertainty b of each element. Uncertainty types are denoted by N (none), A (aleatory), E (epistemic) and M (mixed). Possible scenarios are denoted by different arrow types.

different theoretical frameworks for representing uncertainty and how they can influence decision-making and MOEA design.

2.1 Reasoning about Uncertainty

Theories about uncertainty provide us with logical frameworks to make statements about uncertain quantities. The basic principle that underpins the reasoning about uncertainty is that there is a set called *universe of discourse* denoted by $\mathcal{X}$ herein, that contains all the possible and pertinent states that an uncertain quantity can adopt. For instance, when we define $\mathbf{x}$ with an interval, we intrinsically state that, in principle, the evidence shows that all the values contained by the latter might be adopted by the former. Axioms and logic derivations formulated afterwards about the universe of discourse define the theoretical frameworks.

If $\mathcal{X}$ is continuous we can define it as an interval. Now if our uncertain decision vectors can be treated as intervals, we can use *Interval Arithmetic* to propagate $\mathbf{x}$ though $F(\mathbf{x})$ to assess $\mathbf{y}$. Regardless the nature of the uncertainty, the resulting interval is expected to bound the true value(s) of $\mathbf{x}$. On the other hand, if one has more information about the nature of the uncertainty at her disposal, one should use it.

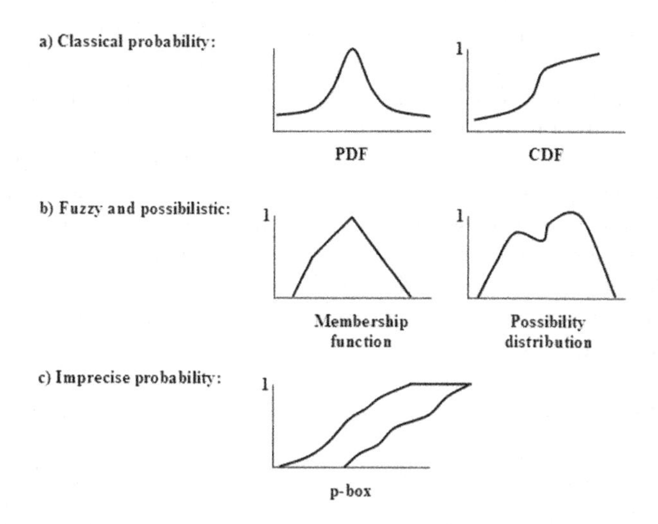

Fig. 2. Elements of the main theories for represetning uncertainty

Fig. 2 sketches some relevant elements of the main theories about uncertainty. In classical probability theory every element of the universe of discourse is assigned a probability. The relation between discrete domains and probabilities are captured by probability mass functions while the probability density function (PDF) are used with continuous domains. In both cases, it is possible characterize tendencies of variation within the universe of discourse by some symmetry axis of such variation (expected value) or its size (variance) amongst other things. It is also possible to assess the probability of the uncertain quantity adopting values within a set (like $P(X \leq x)$): this is expressed as a cumulative probability distribution (CFD).

Sometimes $\mathcal{X}$ is roughly or ill defined, as when one says '*it's cold*' and we know that 'cold' has different meanings according to the person who says it. This kind of uncertainty appears in natural language or when one handles blurred concepts (e.g. when defining the DM's preferences). In this case $\mathcal{X}$ is described by a membership functions that assigns numbers in [0,1] where 0 means the argument is not contained in the set and 1 the opposite. For insight into fuzzy logic see [3].

We can also extend the previous concept to talk about 'the possibility' of an event, using a bivalent logic (it is or not possible) or a graded logic captured by possibility distributions, which are in deep connection with the notion of probability, although saying that something is possible is different than saying it is probable. *Possibility Theory* provides therefore a non-probabilistic framework to represent epistemic uncertainty.

One of the shortcomings of classical probability is that it is not suitable for representing epistemic or mixed uncertainty. For instance, having limited evidence, an agent could make imprecise statements like '*the probability of x is in* $[0.3, 0.5]$' or '*vector* $\mathbf{x}$ *follows a normal PDF with mean in* $[3.26, 4.5]$ *and variance in* $[0.82, 0.97]$. Statements of the such can be captured by p-boxes e.g. saying that every CDF within the p-box is a possible representation for the actual aleatory uncertainty, whereas the epistemic uncertainty is captured by the fact that we don't actually know the true PDF. *Dempster-Shafer Theory* [4], *Walley's Theory* [5] or *p-boxes* [4] support theoretically this kind of approaches.

Table 1 summarizes the main elements of such approaches. From a practical viewpoint the relevant issue is that the theoretical frameworks mentioned previously are best used in certain situations as they cover distinct types and sources of uncertainty. Besides, all of them prescribe propagation methods, which means that in the plausible case of having uncertain domains represented by one of the theories mentioned so far, the outputs and therefore the ranking of alternatives within the MOEA will also be related to such theory.

Table 1. Some relevant elements of theories about uncertainty

Theory	Accounts for	Especially suitable for	Propagation method	Notable elements
Fuzzy logic	Graded membership of elements to sets	Linguistic epistemic uncertainty	Extension principle	Membership functions, core and support sets
Possibility	Binary or graded membership of elements to sets	Epistemic uncertainty	Choquet integrals and extension principle	Possibility distributions, possibility and necessity measures
Classical Probability	Likelihood of events	Aleatory uncertainty	Convolution and Monte Carlo simulation	PDF, percentiles, mean, variance and higher moments
Imprecise Probability	Imprecise probabilities and subjective judgements on sets	Epistemic and aleatory uncertainty	Convolution and Monte Carlo simulation	Belief and plausibility measures. Intervals for distributions and moments

3 Analysis of Uncertain and Robustness in Evolutionary Optimization (AUREO)

In this section we describe a two-stage methodology for the 'Analysis of Uncertain and Robustness in Evolutionary Optimization' (AUREO). The basic premise of this framework is that the analysis of the available information about a problem subject to uncertainty (fig. 3) determines the solving program (stage 1) and the MOEA structure (stage 2).

Consider a refined mathematical program based on eq. 1. Let $F(\mathbf{x}, \mathbf{p})$ represents the DM's criteria in a free-of-uncertainty scenario in terms of the objective vector $\mathbf{x}$ and a vector of environmental parameters $\mathbf{p}$. As discussed in sec. 2 uncertainty may come up as lack of information about the variables $\mathbf{x}$ and $\mathbf{p}$. Besides the environmental parameters in $\mathbf{p}$ are often subject to change in real world. The assessment of $F(\cdot)$ might be a source of uncertainty as well. The first stage of AUREO, summarized in fig. 3, focuses therefore on the form of the MCDM problem considering the existing uncertainties.

If the model is accepted to be adequate the attention centres on the input vectors $\mathbf{x}$ and $\mathbf{p}$. If such vectors are free of uncertainty, no action is required, otherwise the analyst should ask about the type of such uncertainty and further investigate the best theory to represent it. Immediately the attention focuses on the outcomes of $F(\mathbf{x}, \mathbf{p})$ to find out if the $f_i(\mathbf{x}, \mathbf{p})$ are dynamic functions or if they will be assessed through surrogate or approximate models.

1. Analyze the model:

$\mathbf{p} \in \mathbf{P}$

$\mathbf{x} \in \mathbf{X}$, $\mathbf{X} \subseteq \mathbb{R}^n$ → $F(\mathbf{x}, \mathbf{p})$ → $\mathbf{y} \in \mathbf{Y}$, $\mathbf{Y} \subseteq \mathbb{R}^k$

1.1. Check for model adequacy.
1.2. Consider characteristics of the domain and objective functions.

2. Check for uncertain objective functions:

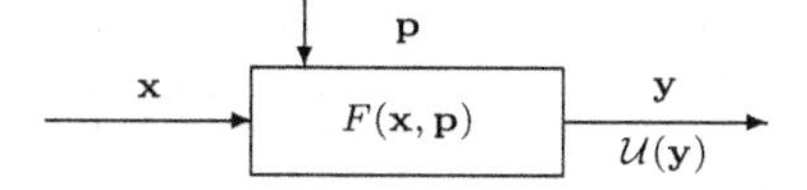

2.1. Do many evaluations of the same argument produce different outcomes?
2.2. Is $F(\mathbf{x}, \mathbf{p})$ a dynamic or stochastic function?
2.3. How is $F(\mathbf{x}, \mathbf{p})$ to be evaluated (surrogate model, approximation, simulation)?
2.4. Is the cardinality of $\mathcal{Y} \subseteq \mathbf{Y}$ reducible to the unit?

3. Check for input uncertainties:

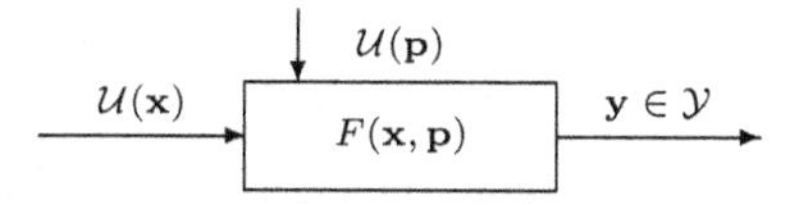

3.1. Is $\mathbf{x}$ subject to uncertainty ($\mathcal{U}(\mathbf{x})$)? If so, what type?
3.2. Are the environmental parameters $\mathbf{p}$ subject to change ($\mathcal{U}(\mathbf{p})$)?
3.3. Is the objective function sensitive to uncertain inputs ($\mathcal{U}(\mathbf{y})$)?

Fig. 3. AUREO Stage 1: Analysis of interactions between model and uncertainties

Once this analysis is ready, the original MCDM problem can be transformed into a new one based on new criteria defined in terms of the -possibly uncertain-input $(\mathbf{x}, \mathbf{p})$, the original criteria $F(\mathbf{x}, \mathbf{p})$, the uncertain outcome $\mathbf{y}$ and the theory employed to represent the uncertainty. For the sake of generality, let $\mathbf{x}$ be a nominal vector denoting a precise alternative and let $\mathcal{U}(\mathbf{x})$ denotes the uncertainty associated to $\mathbf{x}$, i.e. the universe of discourse and other particular elements related to the uncertainty representation (see sec. 2.1). The same notation stands for the uncertainty of $\mathbf{p}$ and $\mathbf{y}$.

Now, in the most general way, the new MCDM problem can be expressed through the following:

Definition 2 (Uncertainty-handling program). *Let $F(\cdot)$ be a measure of performance of a system determined by the decisional vector $\mathbf{x}$ and influenced by a vector of environmental parameters $\mathbf{p}$, each of which is subject to uncertainties $\mathcal{U}(\mathbf{x})$ and $\mathcal{U}(\mathbf{p})$ respectively. Let $C(\cdot)$ be a vector of constraints defined regarding the original constraints for the optimization of $F(\cdot)$. Finally let $I(\cdot)$ be a vector of requirements imposed upon the performance. The resulting uncertainty-handling formulation consists in solving the following program*

$$
\begin{array}{rc}
 & \max\ R\left(F(\mathbf{x},\mathbf{p}),\mathbf{x},\mathbf{p},\mathcal{U}(\mathbf{x}),\mathcal{U}(\mathbf{p})\right) \\
s.t.: & \\
 & \mathbf{x}\in\mathbf{X}\ ,\ \mathbf{p}\in\mathbf{P} \\
 & C(\mathbf{x},\mathbf{p},\mathcal{U}(\mathbf{x}),\mathcal{U}(\mathbf{p}))\leq 0 \\
 & I(F(\mathbf{x},\mathbf{p}),\mathbf{x},\mathbf{p},\mathcal{U}(\mathbf{x}),\mathcal{U}(\mathbf{p}))\leq 0
\end{array} \tag{2}
$$

The new function denoted $R(\cdot)$ is typically an expression of risk, reliability or robustness. For example, in the case of pure uncertain functions, $R(\cdot)$ can be formulated as the original $F(\cdot)$ plus a measure of uncertainty to be minimized. $R(\cdot)$ can also account for reliability or robustness when the input is uncertain, resulting in a *robustness-seeking program.* Vector $I(\cdot)$ on the other hand, accounts for requirements formulated by the DM as additional performance constraints due to uncertainty (e.g. acceptance thresholds for variance or interquartile distances). From the previous program derive two classes of definitions of robustness with well defined solving procedures and a third mixed class that combines the reasoning of the preceding classes in order to solve problems with the least amount of information. Table 2 shows what class is applicable regarding the amount of information available.

Table 2. AUREO Stage 1: Classes of uncertainty-handling formulations according to the available information

Input: $\mathbf{x},\mathcal{U}(\mathbf{x}),\mathbf{p},\mathcal{U}(\mathbf{p})$		**Output:** $\mathbf{y},\mathcal{U}(\mathbf{y})$	
$\mathcal{U}(\mathbf{x})$	$\mathcal{U}(\mathbf{p})$	$I(\cdot)$ definable	$I(\cdot)$ undefinable
None	None	**Pure uncertain functions ≡ Class 1**	
None	Definable	**Class 1**	**Class 1**
None	Undefinable	**Class 2**	**Class 3**
Definable	None	**Class 1**	**Class 1**
Definable	Definable	**Class 1**	**Class 1**
Definable	Undefinable	**Class 2**	**Class 3**
Undefinable	None	**Class 2**	**Class 3**
Undefinable	Definable	**Class 2**	**Class 3**
Undefinable	Undefinable	**Class 2**	**Class 3**

Class 1: Uncertainty Propagating Programs. This class is characterized by a suitable description of $\mathcal{U}(\mathbf{x})$ and $\mathcal{U}(\mathbf{p})$ in such a way that the uncertainty can be propagated through $F(\cdot)$. If the uncertainty is aleatory, $\mathcal{U}(\mathbf{x})$ and $\mathcal{U}(\mathbf{p})$ have associated PDF. For instance, $\mathcal{U}(\mathbf{x})$ could be a normally distributed number $N(\mathbf{x},\sigma)$. On the contrary, if the uncertainty is epistemic, $\mathcal{U}(\mathbf{x})$ might be a crisp or a fuzzy interval, a p-box or something of the like.

Once this program has been identified, the second stage of AUREO consists in defining how the optimality and the density are to be assessed within the MOEA as well as in addressing efficiency issues. As shown in fig. 1 whenever the uncertainty is propagated the outcomes become sets. Thus, the first problem risen is how to decide about optimality using sets. Consider fig. 4: according

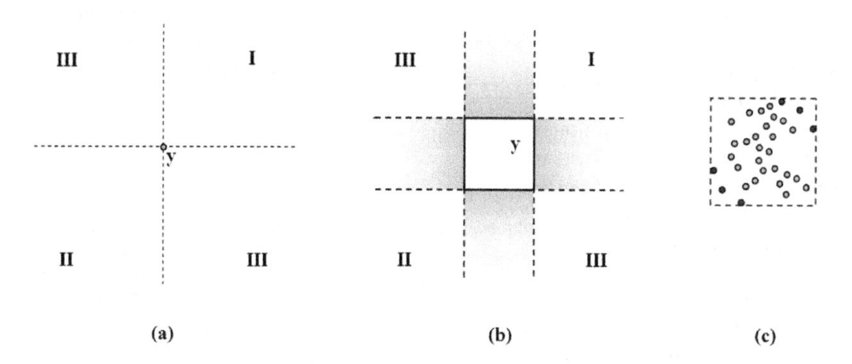

Fig. 4. Spatial dominance relationships regarding $\mathbf{y}$ in (panel a) the absence and (panel b) in the presence of uncertainty. The interval representing $\mathbf{y}$ in panel b can be constructed for discrete sets using the extreme (dark) points (panel c).

to the Pareto optimality (def. 1) in the absence of uncertainty every solution in region I (panel a) dominates $\mathbf{y}$, in region II is dominated by $\mathbf{y}$ and in region III is non-dominated regarding $\mathbf{y}$. In the presence of uncertainty (panel b) there is an additional (colored cross) region such that any set intersecting that area cannot be classified in terms of dominance without an additional criterion. In such a case the comparison of sets is often reduced to a comparison of representative points, but can also be settle using the whole sets. Density can also be assessed using the approaches just mentioned.

Working with representative points: If we can define a few crisp points representing the main features of the uncertain outcomes, we can solve the uncertainty-handling program with a regular application of existing MOEA. Defining such points, however, can be very tricky and computational cumbersome in practice. For representing a set, the extreme points, some symmetry axis and some size measures are commonly used. Let us consider some examples regarding the different theories to represent uncertainty.

Best and worst case are risk criteria that corresponds to extreme points of crisp and fuzzy intervals. Uniform distributions also exhibit finite extreme points, but in general such points are infinite in probability distributions. Nevertheless, extreme quantiles can be used to implement best and worst cases. With imprecise probabilities there are intervals of quantiles so one can use the best of the bests and the worst of the worst of the cases.

The mean value is commonly used as symmetry axis although the median can be used as well. A common approach is to optimize the mean value of the sets, minimizing sometimes its size simultaneously. This is the typical way to implement robustness (see *robust optimization* in tab. 3). In classical probabilistic contexts the size is measured as variance in $R(\cdot)$ although the interquartile range is another possibility. As one only have estimations of means and variances instead of the true statistics most of the time, it is more than desirable to have proper statistical tests supporting MOEA the ranking procedure. On the

Table 3. Some state of the art MOEA for optimization under uncertainty

Realm of study	Authors and works
Probabilistic Dominance:	
Optimization with interval fitness value	Teich [12]
Optimization with noisy fitness function	Hughes [13], Fieldsen et al. [14]
Quality Indicator-Based Procedures:	
Indicator-based optimization	Basseur and Zitzler [15]
Robust Optimization:	
Optimization with uncertainty propagation	Sörensen [16], Ray [17], Deb and Gupta [18], Barrico et al. [19]
Info-gap based robust design	Lim et al. [20]
Reliability-based optimization	Deb et al. [21]
Non-Probabilistic Procedures:	
Optimization with epistemic uncertainty	Limbourg [22] and Salazar A. [23]

other hand, in the case of handling alternative uncertainty theories, the interested reader might also define ranking procedure based on fuzzy and possibilistic means and variances [6,7,8,9] or mean values for imprecise probabilities [10,11].

Working with the whole sets: Treating the whole set instead of a few points is also possible. For probabilistic contexts the concept of probabilistic dominance provides a way of doing this. The basic idea is to assess the probability of one whole outcome dominating another one and to accept dominance if such probability surpass an acceptance threshold. Formulae exist for assessing dominance in classical [24] and imprecise probabilities [25]. The main drawback in practice is that there is no easy way to estimate the probability of dominance if the PDF of the concerned outcomes are not available.

There are also methods for assessing dominance in fuzzy and possibilistic contexts [26]. Such methods solve the problem of optimality but left the density control unattended. One possibility is to resort to representative points like the mean value to assess density, or to maximize the minimal distance between two neighbours.

Table 3 lists some of the existing MOEA that can be used in Class 1 problems. Chronologically speaking, the first attempts implemented probabilistic dominance. These approaches rely upon the assumption that the PDF of all the outcomes are known and share the same shape. In some real problems such an assumption stands but in general it constitutes a limitation. Current robust optimization approaches, on the other hand, do not make a sharp distinction between aleatory and epistemic uncertainty in their proposals. The user may be aware of this to avoid careless uncertainty handling. Other approaches make use of indicator quality measures to handle density assessment [23] or optimality and density as well [15]. Regarding the two referred here, the former were developed to work with intervals and seems suitable for epistemic uncertainty although the optimality is settle by rules that could not be universally accepted. By contrast, the latter approach relies on the assumption of probabilistic outcomes so it seems suitable for aleatory uncertainty albeit the algorithm makes no statistical treatment of the outcomes. A first conclusions is that making better and careful treatments of uncertainty taking into account the theoretical frameworks for reasoning about it is still a challenge for the design of MOEA.

Class 2: Robust Domain-Seeking Programs. Some situations may keep the uncertainty associated with the input variables from being propagated through $F(\cdot)$ and therefore handled as a Class 1 robustness problem. Whether the uncertainty is purely epistemic or mixed in nature, any additional assertion about $\mathcal{U}(\mathbf{x})$ or $\mathcal{U}(\mathbf{p})$ to transform the problem into one of Class 1, entails making some assumptions that could be wrong, leading to identify inadequate alternatives in $\mathbf{Y}$ as optimal solutions.

For example, if the DM knows that the real value of the nominal vector $\mathbf{x}$ is susceptible to vary but ignore the range of such variation, to assume the set of values that $\mathbf{x}$ can take or their likelihood may underestimate the uncertainty. The DM is therefore compelled to maximize the range of 'acceptable' realizations of $\mathbf{x}$ in order to hedge against regrettable consequences. In that sense, robustness is sought by widen the range of possible inputs, or in other words, by maximizing the cardinality of the $\mathcal{U}(\mathbf{x})$.

The Class 2 is therefore characterized by the existence of a constraints vector $I(F(\cdot)) \leq 0$ that constitute desired performance levels of attainment (quality requirements), and a robustness function $R(\cdot)$ that aims at maximizing the range of variation of the input variables that conform with $I(\cdot)$. The robust design problems is a good example of this class. For instance, [27] brings an application where the reliability R_s of a system is a function of the reliability R_i of its components. Since R_i may change, one is interested in knowing the effect of such variations over R_s although the $\mathcal{U}(R_i)$ are unknown. Instead of making assumptions about the $\mathcal{U}(R_i)$, the DM is asked about the desired performance requirements of R_i. This way the DM defines the restriction $I(R_s) = 0.90 \leq R_s \leq 0.99$ and the Class 2 program takes the form of *find* $\{[\underline{R_i}, \overline{R_i}]\} = \arg\max_i \prod_i (\overline{R_i} - \underline{R_i})$ *s.t.* $R_i \in [0.8, 1]$ *and* $I(R_s)$. The second objective is the minimization of the maximal cost that one can incur when selecting components within the range defined by the objective $R(\cdot)$. Notice that such a problem can be handled by a regular MOEA simply using interval arithmetic to check $I(R_s)$. For details about the formulation and the efficiency issues see [27].

Class 3: Mixed Robust-Seeking Procedure. To close this section, let us consider again the general robustness-seeking program formulated in def. 2. If the DM and the analyst are unable to characterize the input uncertainty nor the desired performance levels of attaintment, or alluding def. 2, if they cannot set $\mathcal{U}(\mathbf{x})$, $\mathcal{U}(\mathbf{p})$ nor $I(\cdot)$, the actual definition of the robustness function $R(\cdot)$ is not possible. It is therefore mandatory to generate information to help the DM to make their minds about $\mathcal{U}(\mathbf{x})$, $\mathcal{U}(\mathbf{p})$ or about $I(\cdot)$, in such a way that the problem collapses into a Class 1 or Class 2 program. The first case could be possible by means of further elicitation of uncertainty, while the second requires initial assumptions about $\mathcal{U}(\mathbf{x})$ and $\mathcal{U}(\mathbf{p})$ to roughly approximate the frontier, allowing to set $I(\cdot)$ and to solve the corresponding Class 2 program afterwards.

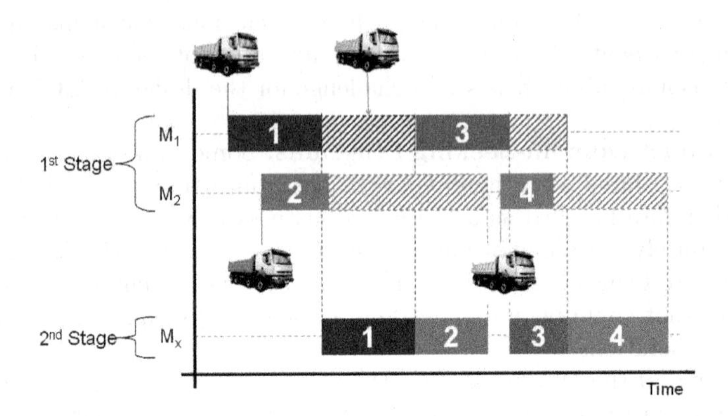

Fig. 5. Scheduling problem in a waste treatment plant: Machines M_m performs the unloading while the mixer M_x performs the waste processing [28]

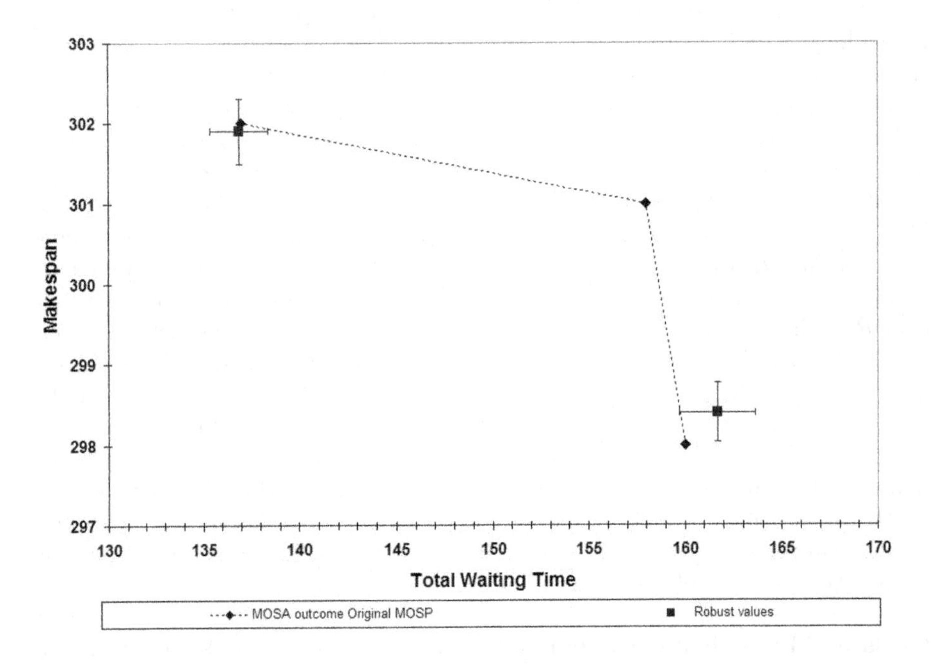

Fig. 6. Robust schedules that satisfy $I(\cdot)$ and maximizes the allowed variations in arrival times (objectives in minutes) [28]

This procedure is explored in [28]. The original problem consists in minimizing the makespan and the total waiting time of a waste treatment plant. The waste is carried by trucks that arrive at scheduled times and the operations are the unloading of the trucks into silos and the transference of the silos' content into a critical machine (see fig. 5). Assuming that the processing times do not change and the sequence of operations cannot be rescheduled on line, the new problem arises when considering uncertain arrival times for the trucks. The goal is to identify the more robust sequence of tasks given that no information about the possible variations of arrival times nor about $I(\cdot)$ were available a priori.

In order to solve this problem as one of the Class 2, a first assumption was made about the arrival time, accepting a uniform variation of $\pm 10'$ around the expected timing. Criteria were assessed using Monte Carlo simulation. The results obtained using MOSA [29] with this assumption showed that it is possible to absorb variations of 10' without deviations from the deterministic results greater than 1'. Constraint $I(\cdot)$ was set to allow at most deviations of 2' from the known deterministic optima. With this in mind the new goal is to find a sequence conforming $I(\cdot)$ that accepts the greater range of variations above 10' in the arrival time. The results are shown in fig. 6.

4 Final Remarks

In this article we offered an analysis of the interaction between MCDM and MOEA, emphasizing the importance of considering the different forms for representing uncertainty. The possible instances of MCDM problems under uncertainty were classified into three classes according to the elements concerned by uncertainty. In the Class 1 it is necessary to deal with uncertain outcomes so the MOEA designed to work with this group of problem have to implement mechanisms to propagate uncertainty and to rank the solutions in terms of optimality and density. Most of the existing approaches lie within this class, but there is still room for more research on the integration of MOEA and uncertainty theories to better treat aleatory and epistemic uncertainty and to cover scenarios not considered yet.

In Class 2 the problem does not require the MOEA to handle uncertain outcomes. Readers interested in this approach are referred to [1,27] for further details. Finally in Class 3 the decision-making problem requires additional efforts to generate information and to reduce it to a Class 1 or Class 2. We briefly exemplified the application of MOEA to solve a real Class 3 problem. For further details see [28].

References

1. Salazar Aponte, D.E.: On Uncertainty and Robustness in Evolutionary Optimization-based Multi-Criterion Decision-Making, PhD Thesis, University of Las Palmas de Gran Canaria, Spain (2008)
2. Jousselme, A.L., Maupin, P., Bossé, E.: Uncertainty in a situation analysis perspective. In: Proceedings of the Sixth International Conference of Information Fusion 2003, pp. 1207–1214 (2003)
3. Klir, G.J., Folger, T.A.: Fuzzy Sets, Uncertainty, and Information. Prentice-Hall International, USA (1988)
4. Ferson, S., Kreinovich, V., Ginzburg, L., Myers, D.S., Sentz, K.: Constructing probability boxes and Dempster-Shafer structures. Technical report SAND2002-4015, Sandia National Laboratories (2002)

5. Walley, P.: Statistical Reasoning with Imprecise Probabilities. Chapman and Hall, London (1991)
6. Kröger, R.: On the variance of fuzzy random variables. Fuzzy Sets and Systems 92(1), 83–93 (1997)
7. Dubois, D., Prade, H.: The mean value of a fuzzy interval. Fuzzy Sets and Systems 24(3), 279–300 (1987)
8. Bronevich, A.G.: The Maximal Variance of Fuzzy Interval. In: Proceedings of the Third International Symposium on Imprecise Probabilities and Their Applications ISIPTA 2003. Electronic proceedings (2003), `http://www.carleton-scientific.com/isipta/2003-toc.html`
9. Carlsson, C., Fullér, R.: On possibilistic mean value and variance of fuzzy numbers. Fuzzy Sets and Systems 122, 315–326 (2001)
10. Kreinovich, V., Xiang, G., Ferson, S.: Computing mean and variance under Dempster-Shafer uncertainty: Towards faster algorithms. International Journal of Approximate Reasoning 42(3), 212–227 (2006)
11. Bruns, M., Paredis, C., Ferson, S.: Computational Methods for Decision Making Based on Imprecise Information. In: NSF Workshop on Reliable Engineering Computing, Savannah, GA (2006)
12. Teich, J.: Pareto-front exploration with uncertain objectives. In: Zitzler, E., Deb, K., Thiele, L., Coello Coello, C.A., Corne, D.W. (eds.) EMO 2001. LNCS, vol. 1993, pp. 314–328. Springer, Heidelberg (2001)
13. Hughes, E.J.: Constraint Handling with Uncertain and Noisy Multi-Objective Evolution. In: Proceedings of the IEEE Congress on Evolutionary Computation 2001 (CEC 2001), pp. 963–970. IEEE Service Center, Piscataway (2001)
14. Fieldsend, J.E., Everson, R.M.: Multi-objective Optimisation in the Presence of Uncertainty. In: Proceedings of the 2005 IEEE Congress on Evolutionary Computation (CEC 2005), pp. 243–250. IEEE Service Center, Edinburgh (2005)
15. Basseur, M., Zitzler, E.: Handling Uncertainty in Indicator-Based Multiobjective Optimization. Int. J. Computational Intelligence Research 2(3), 255–272 (2006)
16. Sörensen, K.: A framework for robust and flexible optimisation using metaheuristics. OR4: A Quarterly Journal of Operations Research 1(4), 341–345 (2003)
17. Ray, T.: Constrained robust optimal design using a multiobjective evolutionary algorithm. In: IEEE Congress on Evolutionary Computation (CEC 2002), pp. 419–424. IEEE Service Center, Honolulu (2002)
18. Deb, K., Gupta, H.: Searching for robust pareto-optimal solutions in multi-objective optimization. In: Coello Coello, C.A., Hernández Aguirre, A., Zitzler, E. (eds.) EMO 2005. LNCS, vol. 3410, pp. 150–164. Springer, Heidelberg (2005)
19. Barrico, C., Henggeler Antunes, C.: Robustness Analysis in Multi-Objective Optimization Using a Degree of Robustness Concept. In: Proceedings of the IEEE Congress on Evolutionary Computation (CEC 2006), pp. 1887–1892. IEEE Service Center, Vancouver (2006)
20. Lim, D., Ong, Y.S., Jin, Y., Sendhoff, B., Lee, B.S.: Inverse Multi-objective Robust Evolutionary Design. Genetic Programming and Evolvable Machines Journal 7(4), 383–404 (2006)
21. Deb, K., Padmanabhan, D., Gupta, S., Mall, A.K.: Reliability-based multi-objective optimization using evolutionary algorithms. In: Obayashi, S., Deb, K., Poloni, C., Hiroyasu, T., Murata, T. (eds.) EMO 2007. LNCS, vol. 4403, pp. 66–80. Springer, Heidelberg (2007)
22. Limbourg, P.: Multi-objective optimization of problems with epistemic uncertainty. In: Coello Coello, C.A., Hernández Aguirre, A., Zitzler, E. (eds.) EMO 2005. LNCS, vol. 3410, pp. 413–427. Springer, Heidelberg (2005)

23. Limbourg, P., Salazar Aponte, D.E.: An Optimization Algorithm for Imprecise Multi-Objective Problem Functions. In: Proceedings of the IEEE Congress on Evolutionary Computation (CEC 2005), pp. 459–466. IEEE Service Center, Edinburgh (2005)
24. Graves, S.: Probabilistic dominance criteria for comparing uncertain alternatives: A tutorial. Omega 37(2), 346–357 (2009)
25. Limbourg, P., Kochs, H.D.: Multi-objective optimization of generalized reliability design problems using feature models–A concept for early design stages. Reliab. Eng. Sys. Saf. 93(6), 815–828 (2008)
26. Hernandes, F., Lamata, M.T., Verdagay, J.L., Yamakami, A.: The shortest path problem on networks with fuzzy parameters. Fuzzy Sets and Systems 158(14), 1561–1570 (2007)
27. Salazar, A., D.E., Rocco, S., C.M.: Solving Advanced Multi-Objective Robust Designs By Means Of Multiple Objective Evolutionary Algorithms (MOEA): A Reliability Application. Reliab. Eng. Sys. Saf. 92(6), 697–706 (2007)
28. Salazar, D., Gandibleux, X., Jorge, J., Sevaux, M.: A Robust-Solution-Based Methodology to Solve Multiple-Objective Problems with Uncertainty. In: Barichard, V., Ehrgott, M., Gandibleux, X., T'Kindt, V. (eds.)) Multiobjective Programming and Goal Programming. LNEMS, vol. 618, pp. 197–207. Springer, Heidelberg (2009)
29. Ulungu, E.L., Teghem, J., Fortemps, P.H., Tuyttens, D.: MOSA Method: A Tool for Solving Multiobjective Combinatorial Optimization Problems. J. Multi-Criteria Decision Analysis 8, 221–236 (1999)

Feedback-Control Operators for Evolutionary Multiobjective Optimization

Ricardo H.C. Takahashi[1], Frederico G. Guimarães[2], Elizabeth F. Wanner[3], and Eduardo G. Carrano[4]

[1] Universidade Federal de Minas Gerais, Department of Mathematics, Brazil
taka@mat.ufmg.br
[2] Universidade Federal de Ouro Preto, Department of Computer Science, Brazil
frederico.guimaraes@yahoo.com.br
[3] Universidade Federal de Ouro Preto, Department of Mathematics, Brazil
efwanner@iceb.ufop.br
[4] Centro Federal de Educação Tecnológica de Minas Gerais, Brazil
egcarrano@deii.cefetmg.br

Abstract. New operators for Multi-Objective Evolutionary Algorithms (MOEA's) are presented here, including one archive-set reduction procedure and two mutation operators, one of them to be applied on the population and the other one on the archive set. Such operators are based on the assignment of "spheres" to the points in the objective space, with the interpretation of a "representative region". The main contribution of this work is the employment of feedback control principles (PI control) within the archive-set reduction procedure and the archive-set mutation operator, in order to achieve a well-distributed Pareto-set solution sample. An example EMOA is presented, in order to illustrate the effect of the proposed operators. The dynamic effect of the feedback control scheme is shown to explain a high performance of this algorithm in the task of Pareto-set covering.

1 Introduction

Two main concerns are involved in the task of designing of Multi-Objective Evolutionary Algorithms (MOEA's): (i) the "quality" of the Pareto-set estimates that are generated, and (ii) the convergence velocity of the algorithm. The first of such concerns is, in itself, multi-dimensional, and there are not, up to now, any definitive standards for measuring such "quality" [1]. A high-quality solution set, anyway, can be defined as a set of samples that [2]:

- Approach the exact Pareto-set (i.e., should be dominated by a subset of the decision variable space that is as small as possible);
- Cover the whole extension of the Pareto-set (i.e., include samples which are spread along the whole range of the Pareto-set, including the regions near the extremes of such Pareto-set);
- Describe in detail the "body" of the Pareto-set (i.e., these samples are "regularly" spread along the Pareto-set).

M. Ehrgott et al. (Eds.): EMO 2009, LNCS 5467, pp. 66–80, 2009.

Notice that a MOEA can be built without a commitment with the search for a Pareto-set estimate in its whole extension, i.e., without referring to such quality measures, in the cases in which some *a priori* or *on line* decision information is available, allowing the concentration of the search in some sub-regions that are identified as being "of interest".

This paper presents new operators that can be used for designing high-performance MOEA's (in the sense of MOEA'S that produce high-quality Pareto-set estimates, as defined above): the *Sphere-Control operators.* Such operators are based on the raw information about the distances between every pair of solution samples in a set – this motivates the denomination of "sphere" operators. The key concept behind the proposed operators is the usage of a *feedback-control* scheme for the purpose of establishing a dynamic equilibrium associated to the high-quality description of the Pareto-set. This means that while such high-quality description is not attained, there will be measured variables that indicate this fact, carrying the information about what control action should be taken in order to enhance such quality [3]. An instance of such feedback-control scheme is represented in Figure 1. In this figure, the measured variable is the error $e = r - a$, which feeds the Proportional-Integral (PI) controller, which in turn determines the value of the control variable ρ. In the equilibrium, $e = 0$ (which means the desired result of $a = r$).

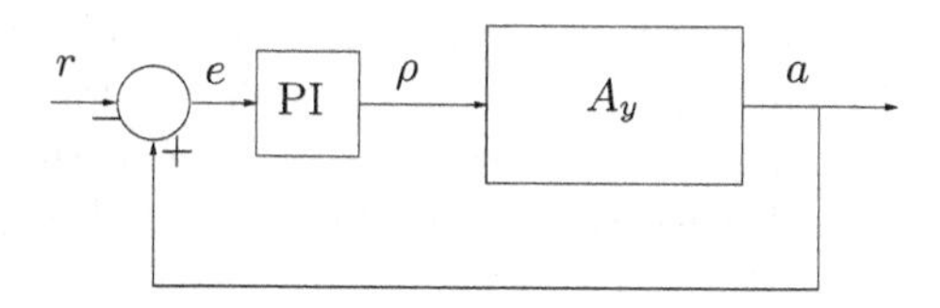

Fig. 1. Diagram of PI feedback control loop for controlling the number of points in the archive set. The variable ρ, the radius of the spheres associated to the points in the archive file, plays the role of the control input variable, while the variable a, the number of points in the archive file, plays the role of the controlled variable. The pre-established reference number of points in the archive, denoted by r, will be attained in the equilibrium, by virtue of the feedback mechanism.

As in other contexts of application of feedback-control techniques, the role of the feedback control scheme is to induce an overall system behavior that presents low sensitivity to variations in the initial conditions and in the algorithm parameter values, delivering rather "stable" results – meaning a high repeatability in the reach of high-quality solution sets [3]. The error variables are defined such that the feedback loop reaches an equilibrium only when a "good description" of the Pareto-set is attained. An unbalanced spread of solutions causes the number of solutions to shrink (by eliminating the more redundant solutions), while the existence of connected areas in the Pareto-set which are not well-covered causes the number of solutions to grow, with the equilibrium being reached only there are no more non-described regions, and the solutions are evenly distributed along the Pareto-set object. The equilibrium itself can be used as a criterion for detecting the end of the task of Pareto-set description, serving as a stop criterion.

Specifically, two sphere-control operators are proposed here: a mutation operator that is applied in the archive set, in order to fill eventual gaps in its Pareto-set description – which is called here the *surface-filling mutation*, and an *archive-set reduction* operator which controls the number of non-dominated solutions that are stored. Feedback control mechanisms, based on a switched controller and on a Proportional-Integral (PI) controller, are employed in the surface-filling mutation and in the archive-set reduction operator, in order to enhance the distribution of solutions along the Pareto surface. This motivates the denomination of "sphere-control" operators. These operators are to be employed together, since their effects are complementary, and their dynamic interaction is necessary in order to achieve the desired behavior. Another mutation operator is also defined here, still employing the "sphere" concept, but not employing the feedback-control scheme. This operator is applied in the current population, and resembles the "hypermutation" operation employed in the Artificial Immune System proposed in [4].

The ideas presented here have connection with the ones presented in [5,2] which employ "sphere" operations which are similar to the archive-set reduction operator presented here, yet without any feedback adaptation scheme. The basic idea, both in that references and here, is that a "sphere" means roughly a domain in which the information gained by a solution point in its center would be representative – with no need of further function evaluations inside such sphere. The references [6,7] also employ the concept of "spheres" for construction of an EMOA, with a dual meaning: in that cases, the "sphere" is the domain in which a local search is conducted, with sub-populations assigned to perform searches inside each sphere.

In the specific formulation that is presented here, the proposed operators are structured for continuous-variable spaces. However, the adaptation for discrete-variable problems can be performed directly, provided that some distance metric becomes defined in the discrete-variable space.

An algorithm that instantiates the application of such operators is constructed: the SCMGA (Sphere-Control Multiobjective Genetic Algorithm). Such algorithm is compared with an NSGA-based algorithm and with an SPEA-based algorithm, in order to illustrate the enhancements of the Pareto-set estimates that can be obtained via the proposed approach. The role of the feedback control operators is analyzed, and the results suggest that such operators interact in order to increasingly enhance the description of the Pareto set. In particular, the gaps in the Pareto surface are systematically filled by the proposed operators – leading to surface descriptions of high definition.

2 Multiobjective Genetic Algorithms

Consider $f(\cdot) : \mathbb{R}^n \mapsto \mathbb{R}^m$ a vector-valued real function. Let $f_i(\cdot)$ denote the *i-th* coordinate of the function in the image space. The multiobjective problems appear from the partial ordering induced by the relation of *dominance*:

$$u \prec v \Leftrightarrow \begin{cases} f_i(u) \le f_i(v) \ \forall\, i = 1, \ldots, m \\ \text{and} \\ \exists\, i \in \{1, \ldots, m\} \text{ such that } f_i(u) < f_i(v) \end{cases} \quad (1)$$

Consider the Pareto-set $\mathcal{P}$ defined by:

$$\mathcal{P} \triangleq \{x^* \in \Omega \mid \nexists\, x \in \Omega \text{ such that } x \prec x^*\} \quad (2)$$

in which $x \in \mathbb{R}^n$ is the decision variable vector, and $\Omega \subseteq \mathbb{R}^n$ is the feasible set. A multiobjective optimization problem is defined as the task of generating samples of the set $\mathcal{P}$.

A Multiobjective Genetic Algorithm (MGA) is a genetic algorithm which is intended to produce a set of samples of $\mathcal{P}$. These algorithms can be stated, in general, as:

Algorithm 1. Pseudocode for generic MGA

```
k ← 0
P_k ← initial_population
E_k ← evaluate_function(P_k)
A_k ← ∅ % the archive set
M_k ← ∅ % the set of points resulting from mutation
C_k ← ∅ % the set of points resulting from crossover
while not stop criterion do
   A_{k+1} ← update_archive(A_k, P_k, E_k)
   F_k ← fitness_assignment(P_k, M_k, C_k, E_k)
   P_{k+1} ← new_population(A_k, P_k, M_k, C_k, F_k)
   M_{k+1} ← mutation(P_{k+1})
   C_{k+1} ← crossover(P_{k+1})
   E_{k+1} ← join( evaluate_function(M_{k+1}, C_{k+1}), E_k)
end while
```

In addition to this "basic" structure, other operators can be added within the main loop. A very common kind of additional operator performs a local search [8]. The algorithm that is to be tested here and the algorithms that are adopted for comparison follow this basic structure. At the end of the execution, the "archive set" A_k contains the algorithm outcome, which constitutes an estimate of the Pareto-set $\mathcal{P}$.

3 The Sphere Operators

Consider the sets A and P, respectively meaning the archive and the current population:

$$A \triangleq \{\tilde{x}_1, \ldots, \tilde{x}_a\}$$
$$P \triangleq \{x_1, \ldots, x_p\} \quad (3)$$

with $|A| = a$ and $|P| = p$. The images of such sets in the objective space are denoted by:

$$A_y = \{\tilde{y}_1, \ldots, \tilde{y}_a\} \triangleq \{f(\tilde{x}_1), \ldots, f(\tilde{x}_a)\}$$
$$P_y = \{y_1, \ldots, y_p\} \triangleq \{f(x_1), \ldots, f(x_p)\} \tag{4}$$

The remainder of this paper considers that both sets A_y and P_y are normalized according to the extreme values of set A_y, such that the minimal value in any dimension receives the value 0, and the maximal value receives the value 1. This means that some values in P_y can fall outside the range $[0, 1]$. A re-normalization is performed in both sets every time the set A_y is updated.

The main idea behind the "sphere operators" is that if the solution samples regularly cover the set $\mathcal{P}$, they should be located in relation to their nearest neighbors such that the distances to them become of the same order of magnitude. Therefore, a parameter ρ, which has the meaning of a *reference domain radius* for each point, must be defined. This parameter is employed in order to guide the algorithm operations, with the intent to generate points which approximately "represent" the region inside the sphere of such radius centered in that point. Any two neighbors should be separated, therefore, by at least 2ρ. The parameter ρ is dynamically adjusted during the algorithm execution, in order to reach a good dispersion of the sample points along the Pareto-set estimate, considering a reference value of the number of sample points that is to be found.

3.1 Archive-Set Reduction

Consider the set $A = \{x_1, x_2, \ldots, x_m, x_{m+1}, \ldots, x_a\}$ in which the individual minima of the m objective functions have been put in the first m positions, and the remainder $a - m$ points have been ordered randomly. The set A_y is ordered correspondingly. The archive set is reduced by the Algorithm 2.

Algorithm 2. Pseudocode for Archive-Set Reduction

1: $i \leftarrow 1$
2: **while** $|A| > i$ **do**
3: $\quad A \leftarrow A - \{x_j \mid \|y_i - y_j\| < 2\rho \ , \ i \neq j\}$
4: $\quad A_y \leftarrow A_y - \{y_j \mid \|y_i - y_j\| < 2\rho \ , \ i \neq j\}$
5: $\quad i \leftarrow i + 1$
6: **end while**

After this operation, there will be no two points in A_y with pairwise distances smaller than 2ρ. This controls the size of the archive set, ensuring that the points will be distributed smoothly. Notice that this operation can be performed directly even in the case of discrete-variable problems, since all information is processed with reference to the objective space.

If this operation was executed with a pre-defined value of ρ, this could cause the size a of set A to become too large or too small, since the exact extension of the set $\mathcal{P}$ is not known *a priori*. This leads to the need of a dynamic adjustment of this parameter.

3.2 Controlling ρ

The adjustment of parameter ρ takes into account a reference number of points that should be stored in the archive set A. Let such reference number be denoted by r, and let the actual number of points in A be denoted by a. The adjustment procedure is described in Algorithm 3. The initial value of ρ is defined such that r spheres of dimension $(m-1)$ occupy a volume equivalent to the unitary simplex of dimension $(m-1)$.

Algorithm 3. Pseudocode for ρ Control

```
1: if a > r then
2:     e ← (a − r)/r
3:     if e > s then
4:         Δ ← s
5:     else
6:         Δ ← K_p × e
7:     end if
8:     ρ ← (1 + Δ) × ρ
9: else
10:    ρ ← K_n × ρ
11: end if
```

Default values can be indicated for the control parameters: $K_p = 0.6$, $s = 0.1$, $K_n = 0.9$. This algorithm resembles the PI (proportional-integral) controllers with control signal saturation, which are employed in industrial control systems. A relative error e is calculated in each step. An incremental control action is calculated on the basis of such error. A diagram of such closed-loop feedback control scheme is presented in Figure 1.

For negative errors (the number of archive points is smaller than the reference one), the size of the *reference domain radius*, ρ, must be reduced, in order to allow that more spheres become defined in the next step. In this case, ρ is multiplied by the parameter K_n which must be chosen such that $0 < K_n < 1$. This tends to be the case when the algorithm is starting: there are few points in the archive set, and the radius is continuously reduced.

In the case of positive errors (the number of archive points becomes greater than the reference one), the size of ρ must be increased, in order to eliminate more points in the operation of archive reduction (which is equivalent to make any point to "represent" a larger sphere around itself). In this case, the increment Δ is proportional to the error, in the case of small errors, and fixed, in the case of large errors, in order to avoid rapid increments of ρ. The need for such saturation is due to stability considerations, in the same sense that appears in the context of industrial control – in this way avoiding the excessive oscillation of the control variable ρ.

It should be noticed that, as the control action over variable ρ is incremental, the net effect has the form of an *integral* control. This integral term in the PI controller is necessary in order to induce an error-less steady behavior in the closed-loop system. Indeed, if a simple P (proportional only) controller were adopted, the steady state value of the controlled variable a would not become equal to the reference value r. This, of course, would lead to the need of *ad-hoc* parameter adjustments in order to obtain a solution set with a pre-established size. In this way, the PI controller structure is the simplest one that fulfills the requirements which are needed here. (Most textbooks on classic control theory, for instance [3], will present a detailed discussion about such effects).

The adoption of such closed-loop control scheme makes the size of ρ to reach an equilibrium point by itself, avoiding the need of an *a priori* knowledge about such parameter, and rendering the multiobjective genetic algorithm robust in relation to this parameter.

Notice that even in the degenerate case of the Pareto-set surface being of dimension less than $m-1$ (one or more objectives being redundant), the algorithm still works as expected, forming an archive set of Pareto estimates which will have still r elements. The only exception would be in the case of a single non-redundant objective, in which the control variable ρ would shrink up to very small values without obtaining the effect of forming an archive set with r elements. Rigorously, an algorithm should have a stop condition related to the detection of such situation.

3.3 Surface-Filling Mutation

Define the function $v(\cdot)$ as

$$v(y_i) = |\{y_j \mid \|y_i - y_j\| < 3\rho\}|$$

which means the cardinality of the set of points from the archive set A_y which are inside a ball of radius 3ρ around the function argument y_j. Without loss of generality, consider that the set A_y is ordered in increasing order of $v(y_i)$:

$$A_y = \{y_i \mid v(y_i) \leq v(y_{i+1})\}$$

The set A receives the corresponding ordering. The *surface-filling mutation* operator is defined by Algorithm 4 (notice that $|A| = a$).

In Algorithm 4, the matrix Γ performs the tasks of adjusting the mutation to different ranges of the decision variables and introducing correlation between the mutation in different variables, if necessary (in ordinary cases, it can be set as the identity matrix). An important notice about this mutation operator is: it is performed over the *archive* set, instead of being performed over the *current population* set. This operator is followed by a controlled adaptation of mutation radius β, as shown in Algorithm 5.

The idea of the *surface-filling mutation* is to generate mutations in the individuals that have less neighbors in the objective space (i.e., less other points at a distance smaller than 3ρ). These individuals are subject to mutations that are intended either to provide local enhancements or to generate new neighbors that fill the gaps in the description of the Pareto surface by the archive set. However, the radius β that should be employed for the Gaussian mutation is not known *a*

Algorithm 4. Pseudocode for Surface-Filling Mutation

1: $k_1 \leftarrow k_2 \leftarrow k_3 \leftarrow k_4 \leftarrow 0$
2: **for** $i = 1$ to $a/3$ **do**
3: Generate $\omega \in \mathbb{R}^n$ with Gaussian distribution (0,1)
4: $\hat{x}_i \leftarrow \beta\Gamma\omega + x_i$
5: $\hat{y}_i \leftarrow f(\hat{x}_i)$
6: **if** $\hat{y}_i \prec y_i$ **then**
7: $x_i \leftarrow \hat{x}_i$
8: $y_i \leftarrow \hat{y}_i$
9: $k_1 \leftarrow k_1 + 1$
10: **else if** $y_i \prec \hat{y}_i$ **then**
11: $k_2 \leftarrow k_2 + 1$
12: **else**
13: $k_3 \leftarrow k_3 + 1$
14: **if** $\|y_i - \hat{y}_i\| > \rho$ **then**
15: $A \leftarrow A + \hat{x}_i$
16: $A_y \leftarrow A_y + \hat{y}_i$
17: $k_4 \leftarrow k_4 + 1$
18: **end if**
19: **end if**
20: **end for**

Algorithm 5. Pseudocode for Mutation-Radius Control

1: **if** $k_3 > 0$ **then**
2: $\alpha = k_4/k_3$
3: **if** $\alpha > 0.8$ **then**
4: $\beta \leftarrow 0.9 \times \beta$
5: **else if** $\alpha < 0.4$ **then**
6: $\beta \leftarrow 1.1 \times \beta$
7: **end if**
8: **end if**

priori, because the distances that define "neighbors" are measured in the objective space, while the mutation must be performed in the decision variable space. Therefore, a dynamic adaptation is necessary.

A too small β would cause the mutated individuals to become near the original ones, leading to distances (in the objective space) smaller than ρ. The dynamic control of β is built in order to produce a reference proportion of new non-dominated individuals outside the sphere of radius ρ around the original ones. In the instance case shown above, if less than 40% of the new non-dominated individuals are outside such sphere, the mutation radius β is increased by 10%. On the other hand, if more than 80% of such individuals become outside such sphere, the radius is reduced to 90% of its value; this is performed in order to guarantee the local search nature of the *surface filling mutation* operator.

Once more, the adaptation of parameter β employs a feedback control scheme. In this case, a switched control action is performed over variable β.

This mechanism gives rise to a local search procedure that performs perturbation steps that are both local and not too small, leading to an enhanced efficiency in the search.

3.4 Inverse-Fitness Mutation

Another mutation operator, to be applied over the *current population* set, is also defined: the *inverse-fitness mutation.* This is a simple Gaussian mutation:

$$\hat{x} \leftarrow x + \eta \Gamma \omega$$

in which $\omega \in \mathbb{R}^n$ is obtained from a Gaussian distribution (0,1). The radius η is variable: each individual has its own radius, that depends on its fitness value. The idea is to rank the fitness of all individuals in the population, and assign an η to each individual as a linear function of its position in the ranking, with the worst individual receiving an $\eta = \sigma$, where σ is the search radius employed in the generation of the initial population. After this, the values of η that are smaller than 2% of σ are changed to that value. Such mutation is similar to the one proposed in [4], in the context of Artificial Immunological Algorithms.

4 An Instance of EMOA: The SCMGA

In order to evaluate the *sphere-control* operators, an instance of multiobjective genetic algorithm is proposed here, using such operators along with some other well-known operators. This algorithm instance is called here the SCMGA, standing for *Sphere-Control Multiobjective Genetic Algorithm.*

The additional operators that are needed are defined as:

- Initial population: generated from a Gaussian distribution, with a given search radius σ, around a given center x_0;
- Crossover: the real-biased scheme presented in [9] is employed here[1];
- Pareto-ranking fitness assignment: the scheme employed by MOGA [10,11] is employed here;
- Selection: a binary tournament is employed, over a set of individuals composed by the old population plus the individuals generated via the mutation and crossover operations;

[1] The real-biased crossover scheme is stated as follows: (i) Consider the parent individuals A and B. Without loss of generality, consider that the fitness of individual A is better than the fitness of individual B (the operator is to be applied after the selection, when the fitness values of all individuals are known). (ii) Take a parametrized line segment that passes over A and B, such that A is parametrized by 0.1 and B is parametrized by 0.9. (iii) Generate two random numbers ϕ_1 and ϕ_2, both from an uniform distribution in the interval $[0, 1]$. (iv) Take $\phi_3 = \phi_1 \times \phi_2$. Clearly, ϕ_3 has a quadratic distribution over the same interval, with greater density values near 0, and smaller density values near 1. (v) One offspring individual is taken as the point parametrized by ϕ_3 (which means that this point has a greater probability of being near A, the best parent, than of being near B). (vi) The other offspring is obtained as an uniform-distribution sample of the same segment.

This algorithm, featuring also all *sphere-control* operators, is employed here with the purpose of performing some preliminary tests on the proposed new operators.

5 Results

5.1 Establishing the SCMGA

A preliminary test is conducted for the purpose of verifying if the proposed algorithm gives Pareto-set estimates that are compatible with the currently employed reference algorithms. For this purpose, algorithms and benchmark functions which are available at the PISA platform[2] are employed here. The functions to be employed are the Kursawe function with 3 and with 8 variables (denoted respectively by Kur-3 and Kur-8), and the Quagliarella & Vicini function with 3 variables (denoted by QV-3), all described in [12]. These particular functions have been chosen because they are the only options in PISA which comply with the requirements of (i) being of continuous-variables, and (ii) not depending strongly on the definition of "bounding boxes" for the decision variables (this last requirement is due to the particularly simple implementation of SCMGA used here, which has not been constructed for dealing with such bounding box constraints).

The reference algorithms are the Nondominated Sorting GA II (NSGA-II) presented in [13], the Strength Pareto Evolutionary Algorithm 2 (SPEA-2) presented in [12], the Indicator-Based Evolutionary Algorithm (IBEA) presented in [14] and the Hypervolume Estimation Algorithm for Multiobjective Optimization (HypE) presented in [15]. All those algorithms have been run as available from PISA.

All algorithms, for all test functions, have been executed with population size of 300 individuals, number of parent individuals and number of offspring individuals per generation both equal to 150, and other specific parameters of each algorithm as defined by default in PISA. The SCMGA has been run with population of 150 individuals and the reference size of the archive set equal to 300. The algorithms have been assigned 30000 function evaluations in all cases.

The comparisons have been performed in the following way. Each algorithm is executed once over each problem. For each problem, the 300 solutions of the five algorithms, are pooled in a single set (the combined Pareto-set estimate), and a non-dominance algorithm is executed over this pool set. After that, the number of solutions coming from each algorithm in the pool are counted, and the results are presented in Table 1.

A visual comparison, which cannot be shown here due to space limitation, also indicates that SCMGA produces solution sets that are well-distributed. As long as such results are "typical" (they are similar in several executions), it seems reasonable to conduct further studies about the SCMGA. We advise the reader that due to the limited number of tests that has been performed so far,

[2] http://www.tik.ee.ethz.ch/pisa/

Table 1. Results of different algorithms on benchmark problems, in a single execution

	pool size	NSGA-II	SPEA-2	IBEA	HypE	SCMGA
Kur-3	376	22	17	27	24	286
Kur-8	288	0	0	0	0	288
QV-3	1049	171	178	184	228	288

the only conclusion that can be drawn is that SCMGA produces results that are compatible with the ones produced by other standard algorithms. Further conclusions with the meaning of a comparison will need much more tests.

5.2 The Structure of Control Action

An experiment has been conducted with the SCMGA, with the same parameters above, in the 3D Kur problem, allowing 160000 function evaluations (which have been performed in 498 generations). Figure 2 presents the evolution of the number of points in the archive set, $|A|$, along with the evolution of the size of the sphere radius ρ which controls such set size. The reference adopted for $|A|$ is 300 points (i.e., the control subroutine will try to reach this reference and stay on it). Interesting observations can be drawn from Figure 2. It should be noticed that the number of points grows in the first phase of the algorithm time evolution. As long as the number of points is smaller than the reference in this

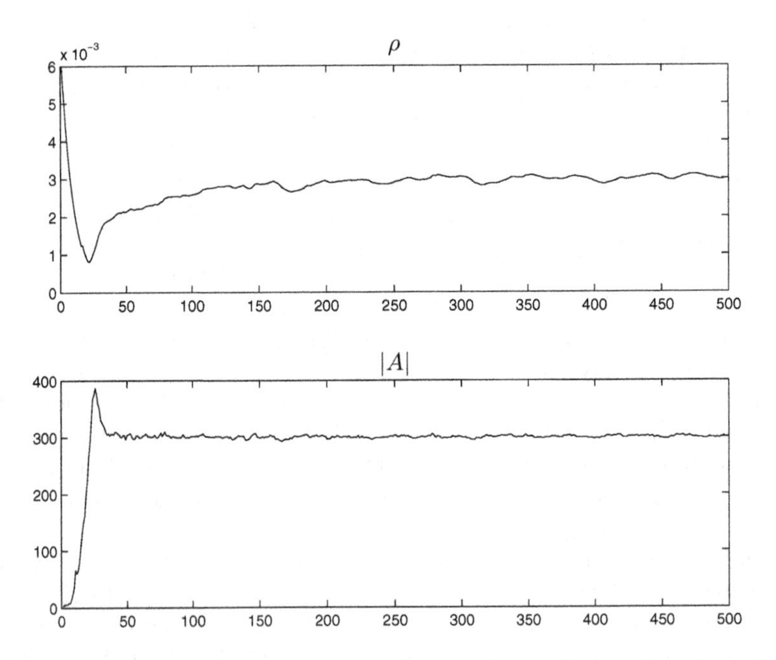

Fig. 2. The evolution of the size of the sphere radius, ρ (upper figure); and of the number of points in the archive set A, denoted by $|A|$ (lower figure), with the number of SCMGA generations

first phase, the control action results in a continuous decreasing of the radius ρ. At the generation 21, the number of points becomes greater than the reference value; the control variable ρ begins to grow. However, as its value has possibly been set to a value smaller than the "equilibrium" one, the controlled variable $|A|$ continues to grow, giving rise to an "overshoot" effect. Further increments in ρ lead the controlled variable $|A|$ to track the reference value nearly at the generation 30. In all generations after that one, $|A|$ presents a small oscillation around the reference. It is important to notice the behavior of the control variable ρ after the moment that $|A|$ reaches the reference value: although $|A|$ becomes nearly constant up to the end of algorithm execution, ρ presents a trend of slow growth from generation 30 up to generation 150. This is related to an adjustment of the samples y_i in A_y, which become increasingly more well-distributed along the Pareto-set of the problem. This leads to a greater "smallest" distance among any two points in A_y. After generation 150, up to the end, the value of control variable ρ becomes approximately constant, with a small oscillation around this "equilibrium".

Figure 3 presents the evolution of the size of the surface-filling mutation radius, β, along with the evolution of the proportion of successful (not too close to the original point) attempts of mutation. The control variable β is switched in order to keep the value of the controlled variable k_4/k_3 between the values 0.4 and 0.8. It can be seen that, at the beginning of algorithm execution, most of successful mutations occur at large distances (the proportion k_4/k_3 is nearly equal to 1). In order to guarantee that the mutation operator performs a local

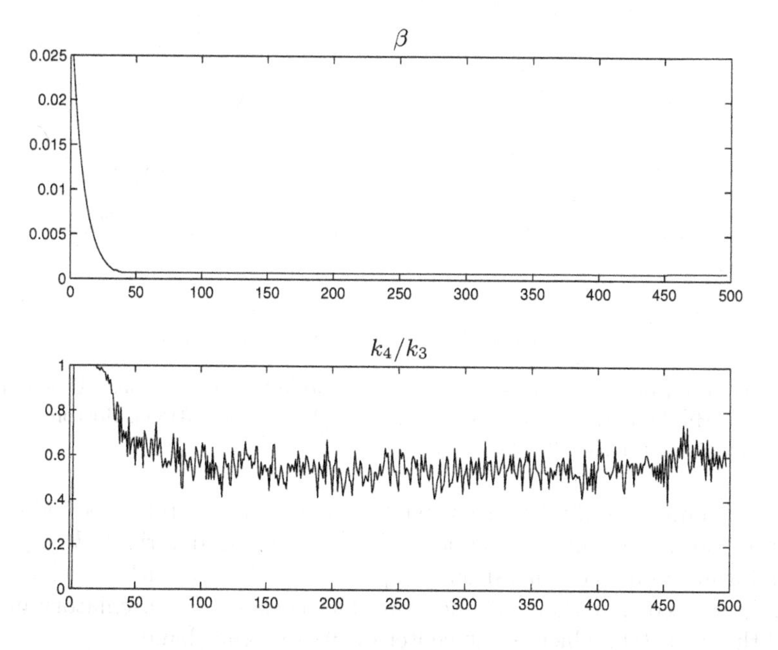

Fig. 3. The size of the mutation radius, β (upper figure); and the proportion k_4/k_3 of successful (not too close) attempts of mutation (lower figure)

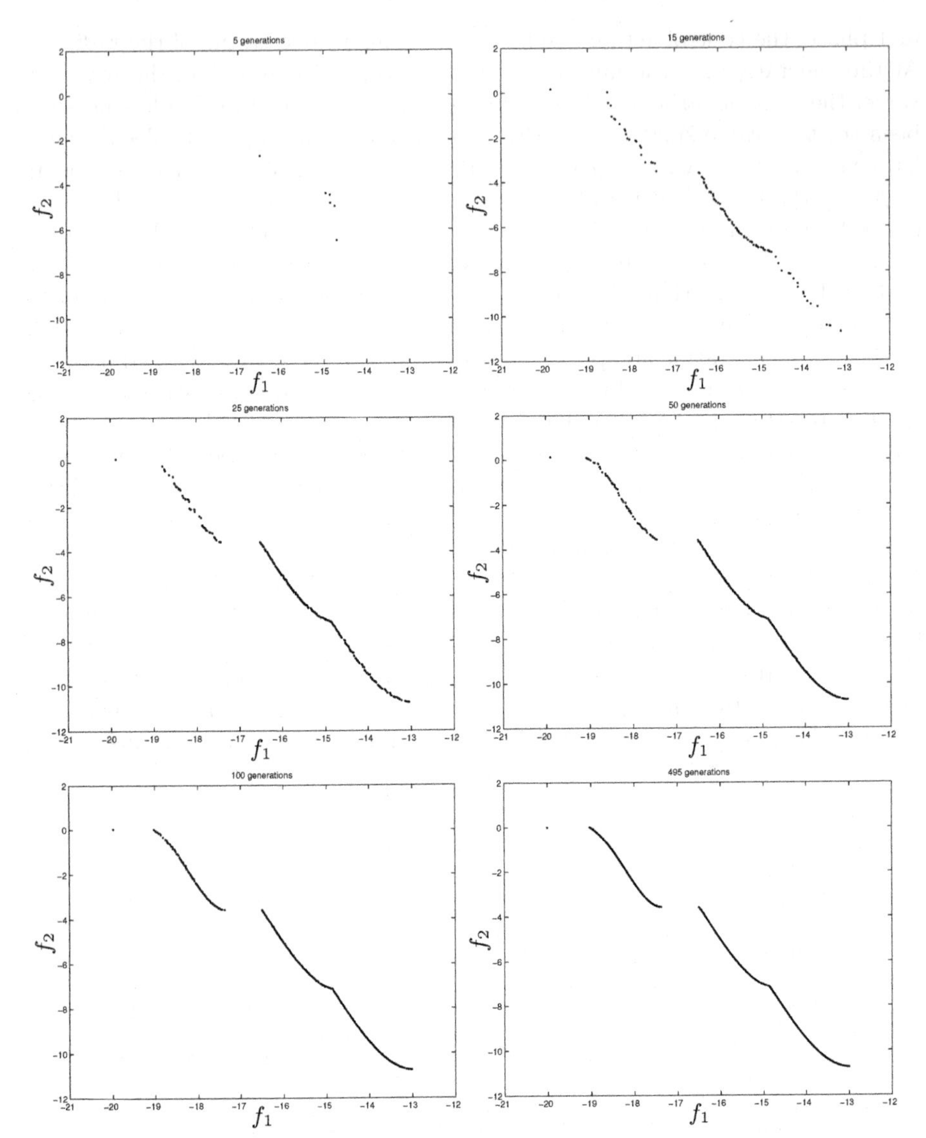

Fig. 4. Evolution of the archive-set image A_y. Top-left: 5 generations. Top-right: 15 generations. Middle-left: 25 generations. Middle-right: 50 generations. Bottom-left: 100 generations. Bottom-right: 495 generations.

search, the mutation radius β is reduced: this causes the controlled variable k_4/k_3 to reduce. Both the control variable β and the controlled variable k_4/k_3 reach an equilibrium from the generation 50 up to the end of algorithm execution. It should be noticed that the controlled variable reaches a fast oscillatory motion around the value 0.6, which is the center of the reference band.

The gradual evolution of the archive set image A_y is depicted in the sequence of frames in Figure 4.

It can be seen that, in the first 5 generations, very few archive points have been generated. A delineation of an estimation of the Pareto-front starts to appear near generation 15. The lower portion of the Pareto-front becomes well-delineated nearly at generation 25, and becomes "completed" nearly at generation 50. The upper portion of the front, however, becomes completed only nearby generation 100. From generation 100 to generation 495, there is no visually perceptible enhancement in the front.

These observations should be compared with the former analysis, which indicated roughly that the variables $|A|$ and β become stable before generation 50, the variable k_4/k_3 reaches equilibrium after generation 50 and before generation 100, and variable ρ becomes in equilibrium nearby generation 100. A very interesting conjecture to be launched is: the equilibrium of such variables seems to be related to the completion of Pareto-set description.

6 Conclusion

The algorithm SCMGA, constructed with the proposed sphere-control operators, has shown effectiveness in finding a well-defined and stable Pareto-set estimate. The algorithm seems to be endowed with a capability of guiding the search toward the "less-defined" regions of the current estimate, until producing a complete estimate. The dynamic interaction between the algorithm internal variables that is induced by a feedback-control scheme seems to play an essential role in constituting such property.

Beyond the specific operators and the algorithm instance that have been presented here, the authors believe that the main contribution of this paper is to introduce the idea of using feedback-control principles for the design of evolutionary computation algorithms. The main concern in using such principles is to define, in a meaningful way, the control variables and the controlled variables of the feedback scheme.

References

1. Zitzler, E., Thiele, L., Laumanns, M., Fonseca, C.M., da Fonseca, V.G.: Performance assessment of multiobjective optimizers: An analysis and review. IEEE Trans. on Evolutionary Computation 7(2), 117–132 (2003)
2. Silva, V.L.S., Wanner, E.F., Cerqueira, S.A.A.G., Takahashi, R.H.C.: A new performance metric for multiobjective optimization: The integrated sphere counting. In: Proc. IEEE Congress on Evolutionary Computation, Singapore (2007)
3. Ogata, K.: Modern Control Engineering, 4th edn. Prentice-Hall, Englewood Cliffs (2001)
4. de Castro, L.N., Timmis, J.: An artificial immune network for multimodal function optimization. In: Proceedings of the 2002 IEEE Congress on Evolutionary Computation, vol. 1, pp. 699–704 (2002)
5. Takahashi, R.H.C., Palhares, R.M., Dutra, D.A., Gonalves, L.P.S.: Estimation of Pareto sets in the mix H_2/H_{inf} control problem. International Journal of Systems Science 35(1), 55–67 (2004)

6. Bui, L.T., Deb, K., Abbass, H.A., Essam, D.: Interleaving guidance in evolutionary multi-objective optimization. Journal of Computer Science and Technology 23(1), 44–63 (2008)
7. Bui, L.T., Abbass, H.A., Essam, D.: Local models – an approach to distributed multi-objective optimization. In: Computational Optimization and Applications (2007) (to appear, published online in 2007), doi:10.1007/s10589-007-9119-8
8. Wanner, E.F., Guimaraes, F.G., Takahashi, R.H.C., Fleming, P.J.: Local search with quadratic approximations into memetic algorithms for optimization with multiple criteria. Evolutionary Computation 16(2), 185–224 (2008)
9. Takahashi, R.H.C., Vasconcellos, J.A., Ramirez, J.A., Krahenbuhl, L.: A multiobjective methodology for evaluation genetic operators. IEEE Trans. on Magnetics 39, 1321–1324 (2003)
10. Fonseca, C.M., Fleming, P.: Genetic algorithms for multiobjective optimization: formulation, discussion and generalization. In: Proceedings of the 5th International Conference: Genetic Algorithms, San Mateo, USA, pp. 416–427 (1993)
11. Fonseca, C.M., Fleming, P.J.: An overview of evolutionary algorithms in multiobjective optimization. Evolutionary Computation 7(3), 205–230 (1995)
12. Zitzler, E., Laumanns, M., Thiele, L.: SPEA2: Improving the Strength Pareto Evolutionary Algorithm, Computer Engineering and Networks Laboratory (TIK), Swiss Federal Institute of Technology (ETH) Zurich, Tech. Rep. 103 (2001)
13. Deb, K., Agrawal, S., Pratap, A., Meyarivan, T.: A fast elitist non-dominated sorting genetic algorithm for multi-objective optimization: NSGA-II. In: Deb, K., Rudolph, G., Lutton, E., Merelo, J.J., Schoenauer, M., Schwefel, H.-P., Yao, X. (eds.) PPSN 2000. LNCS, vol. 1917, pp. 849–858. Springer, Heidelberg (2000)
14. Zitzler, E., Künzli, S.: Indicator-based selection in multiobjective search. In: Yao, X., Burke, E.K., Lozano, J.A., Smith, J., Merelo-Guervós, J.J., Bullinaria, J.A., Rowe, J.E., Tiňo, P., Kabán, A., Schwefel, H.-P. (eds.) PPSN 2004. LNCS, vol. 3242, pp. 832–842. Springer, Heidelberg (2004)
15. Bader, J., Zitzler, E.: HypE: Fast Hypervolume-Based Multiobjective Search Using Monte Carlo Sampling. Institut für Technische Informatik und Kommunikationsnetze, ETH Zürich, TIK Report 286 (October 2006)

A Diversity Management Operator for Evolutionary Many-Objective Optimisation

Salem F. Adra[1] and Peter J. Fleming[2]

[1] Department of Computer Science
[2] Department of Automatic Control and Systems Engineering
University of Sheffield, Mappin Street
Sheffield S1 3JD, UK
{s.adra,p.fleming}@sheffield.ac.uk

Abstract. The proximity of an approximation set to the Pareto-optimal front of a multiobjective optimisation problem and the diversity of the solutions within the approximation set are two essential requirements in evolutionary multiobjective optimisation. These two requirements may be found to be in conflict with each other in *many*-objective optimisation scenarios deploying Pareto-dominance selection alongside active diversity promotion mechanisms. This conflict is hindering the optimisation process of some of the most established MOEAs and introducing problems such as the problem of dominance resistance and speciation. In this study, a diversity management operator (DMO) for controlling and promoting the diversity requirement in *many*-objective optimisation scenarios is introduced and tested on a set of test functions with increasing numbers (6 to 12) of objectives. The results achieved by the proposed strategy outperform results achieved by a reputed and representative MOEA in terms of both criteria: convergence and diversity.

Keywords: Evolutionary Multiobjective optimisation, Diversity requirement.

1 Introduction

Finding a "good" set of solutions to a multiobjective optimisation problem (MOP) consisting of m objectives can be more accurately thought of as an optimisation scenario with $m+2$ objectives. These $m+2$ objectives are divided into m tangible and application-specific objectives and an additional two general objectives. The latter two objectives are the required *convergence* of the solutions to a MOP towards the Pareto front and their *diversity* across the trade-off surface in the objective space. When solving MOPs, the existence of objective preferences and priorities, and their incorporation in the search process is an optional and application-dependent scenario. Nevertheless, one thing is common: the *convergence* criterion is usually prioritised over the *diversity* criterion. As a result, diversity promotion is usually deployed as a secondary consideration to proximity promotion in most MOEAs. This is well justified since, as stated by Bosman and Thierens (2003) [1]:

M. Ehrgott et al. (Eds.): EMO 2009, LNCS 5467, pp. 81–94, 2009.

... the goal is to preserve diversity along an approximation set that is as close as possible to the Pareto optimal front, rather than to preserve diversity in general; the exploitation of diversity should not precede the exploitation of proximity.

The diversity requirement is mainly sought in multiobjective optimisation scenarios with competing objectives where no single optimal solution can be found. In such scenarios, diversity is required to provide the decision maker (DM) with a diversified set of solutions across the objectives. This prioritisation of the convergence requirement over the diversity requirement can be observed in the "selection for variation" and the "selection for survival" procedures of most MOEAs. In NSGA-II [2], for example, when selecting solutions for inclusion in the mating pool, two solutions are first chosen randomly from the population and compared primarily in terms of their non-dominated ranks. The solution with the higher (Pareto-dominance based) rank is selected for inclusion in the mating pool. In the case where the two selected solutions share the same rank, the secondary criterion for the "selection for variation" consists of their crowding measure and, therefore, the diversity requirement. The solution situated in the less dense area of the space would be selected for inclusion in the mating pool, thus promoting diversity.

Similarly, the "selection for survival" process in NSGA-II uses the same hierarchy of selection criteria. Deploying a strategy that maintains a fixed size for the online archive, NSGA-II uses a "selection for survival" procedure that starts by filling the online archive with the highest ranked solutions following a non-dominated sorting strategy. Only in the case where filling the empty (remaining) slots of the archive necessitates the selection of a subset of solutions from a certain non-domination rank does the diversity requirement intervene as a selection criterion.

In this paper the requirement for solution diversity in multiobjective optimisation with *many* (more than 3) conflicting objectives, is explored. In particular, the effect of the diversity extent, rather than uniformity, on many-objective optimisation is investigated. In section 2, the motivation for promoting new methods for diversity promotion and maintenance in evolutionary multiobjective optimization (EMO), especially as the number of conflicting objectives increases, is justified. In section 3, a new strategy for tackling the diversity requirement is introduced. In section 4, the test functions and the performance metrics used to assess the DMO are presented. In section 5, an analysis of the effect of diversity extent on multiobjective optimisation as the number of objectives increase is produced. Experimental results achieved by the introduced strategy for a set of MOPs with an increasing number of conflicting objectives are described. Performance evaluation and conclusions are based on an analytical comparison of the results produced by a well-established and representative MOEA and its coupled version with the suggested diversity management technique.

2 Study Framework and Motivation

The objective of this study is to investigate new strategies for promoting diversity in evolutionary *many*-objective optimisation. This is initially motivated by the outcome of the studies by Purshouse and Fleming [3], [4] which highlight the conflict between the primary MOP requirement for convergence towards the Pareto front, and the secondary requirement for maintaining diversity in the approximation set. This conflict between convergence and diversity requirements in multiobjective optimisation has a detrimental impact on the optimisation process and is particularly aggravated in *many*-objective optimisation. This is due to the fact that the size of the feasible objective space for a certain MOP increases with the dimensionality of the optimisation problem (in terms of the number of objectives). Consequently, the probabilities for the candidate solutions to become non-dominated (in the Pareto sense) increase. In fact, no matter how far from the Pareto front, any solution excelling in one of the objectives will have a high probability of being preferred in terms of Pareto dominance *and* the diversity criterion (as it will be an extreme solution) for "selection for variation" and survival. The increasing proportion of non-dominated solutions explored in *many*-objective hyperspaces leads to diversity promotion mechanisms becoming more emphasized and hence, diversity can gradually become the primary selection criterion. This tends to over-emphasize the diversification process at the expense of the convergence requirement. As a result, when performing "selection for variation", solutions from various distant areas of the hyperspace will have greater chances for recombining and producing lower performance offspring known as *lethals* [5]. The superfluous production of lethals, known as *dominance resistance* [6], will consequently slow down convergence towards the Pareto front.

Having explained the motivations for the study, a new approach for diversity promotion in the *many*-objective optimisation frameworks is suggested, investigated and discussed in the next sections.

3 The Proposed Diversity Management Operator

In this study, the requirement for promoting diversity in MOEAs is envisaged as a local, adaptive and varying requirement rather than a necessity. A diversity management operator (DMO) for controlling and promoting diversity in the *many*-objective optimisation framework is hence introduced and hybridized with NSGA-II. The DMO is an adaptive strategy that promotes the integration of effective diversity indicators, such as the *maximum spread* metric [7] to efficiently guide the search process of an MOEA towards the trade-off surface of a MOP while controlling the diversity requirement. In this study, the DMO has integrated a particular, computationally efficient, diversity metric which is based on the *maximum spread indicator* defined in Equation (1) below.

In Equation (1), D represents the measure of the diagonal of the hypercube formed by the extreme objective values attained in a certain approximation set Z_A, M denotes the number of objectives and z_A is a candidate objective vector solution which belongs to the approximation set Z_A:

$$D = \left[\sum_{m=1}^{M} \left(\max_{z_A \in Z_A} \{ z_{A_m} \} - \min_{z_A \in Z_A} \{ z_{A_m} \} \right)^2 \right]^{1/2} \quad (1)$$

In order to track the diversity quality of the manipulated set of solutions, the value of the spread indicator presented in Equation (1) is normalised with respect to an optimal spread value suggested by the DM. Such optimal spread can correspond to the spread of the set of solutions, Z^*, representing the Pareto or a targeted front of a certain MOP. It is only by knowing the desired conditions, that the undesirable conditions, such as the dispersal of solutions in suboptimal regions of the objective space or alternatively the convergence to Pareto-optimal solutions outside the region of interest, can be defined and avoided. In other words, an application-dependent scale defining the approximate notion of a *low*, *ideal*, and *average* quality of diversity is required to overcome the clash of the requirements (convergence and diversity), which is especially evident in high-dimensional problems (in terms of objective dimensionality).

In the context of the suggested DMO, the DM, usually and preferably an application domain expert, is only required to suggest an approximate estimate of the defining extremities of the *desired* trade-off surface. These extremities will then serve as the vertices of a hypercube containing the ideally sought Pareto front. In an optimisation problem consisting of two objectives, these extremities correspond to the best and the worst approximate values for each of the two objectives. Equation (1) will then be normalised with respect to the length of the diagonal of such a hypercube and the normalised diversity indicator will be defined by Equation (2).

$$I_S = D \Bigg/ \left[\sum_{m=1}^{M} \left(\max_{z_* \in Z_*} \{ z_{*_m} \} - \min_{z_* \in Z_*} \{ z_{*_m} \} \right)^2 \right]^{1/2} \tag{2}$$

The spread indicator, I_S, can take any positive real value. Ideally, an indicator value close to unity ($I_S = 1$) is sought. Indicator values smaller than one ($I_S < 1$) point to a lack of diversity among the derived solutions compared with the desired spread of solutions. This is most likely due to convergence towards an area - potentially a Pareto-optimal sub region - of the solution space outside the region of interest. On the other hand, indicator values larger than one ($I_S > 1$) highlight an excessive dispersal of the solutions in the objective space. This kind of excessive dispersal in the hyperspace most likely causes the divergence of the solutions away from the Pareto-optimal front and frustrates the optimisation process by forcing the MOEA to repeatedly explore previously visited regions of the space [3],[6]. For example, when attempting to downscale the size of a MOEA's active archive to its pre-determined size, *good* locally non-dominated solutions, in terms of proximity towards the Pareto front, might be filtered out at the expense of keeping *good* solutions in terms of diversity, although they may be distant from the Pareto front. However, as the evolutionary search progresses, the MOEA might rediscover this part of the solution space and accept solutions at later iterations that are similar to those previously filtered out.

A pseudocode description of the DMO is presented in Figure 1.

- Calculate the spread indicator I_s for the current approximation set at generation i
- **If** $I_s < 1\text{-}\varepsilon$
  Activate the diversity promotion mechanism in the "selection for variation" and the "selection for survival" process
- **Else If** $I_s \geq 1\text{-}\varepsilon$
  Deactivate the diversity promotion mechanism in the "selection for variation" and the "selection for survival" process

Fig. 1. NSGA-II with the addition of the DMO

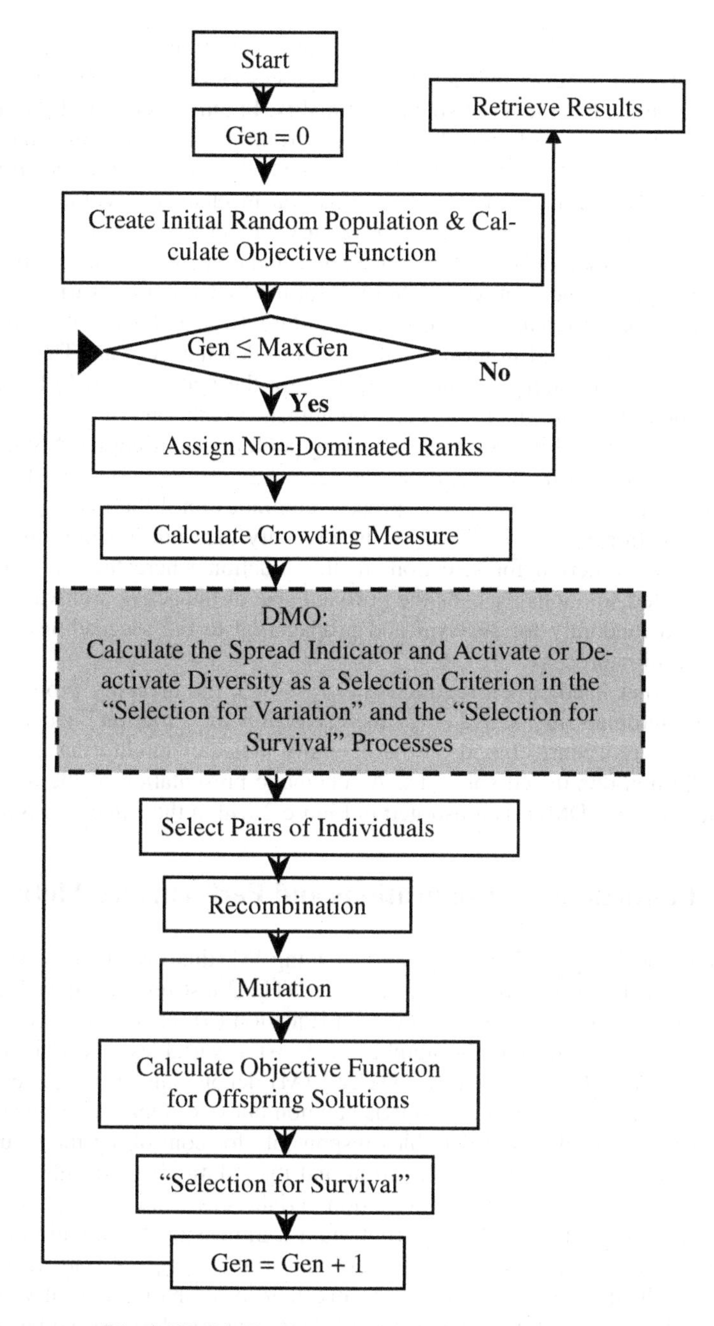

Fig. 2. NSGA-II with the addition of the DMO

The DMO begins by calculating the spread indicator defined in Equation (2) for the local front of non-dominated solutions. The calculation of the spread indicator occurs at every generation of the optimisation process prior to the execution of the genetic operators ("selection for variation" and "selection for survival"). The DMO then adjusts and controls the global search processes of the MOEA in an informed way based on the local level of spread optimality. In other words, if the spread indicator reports an excessive dispersal of the local front of solutions in the objective space (i.e. $I_s \geq 1\text{-}\varepsilon$, ε being an optional, application-dependent tolerance value defined by the DM), the DMO switches off the diversification mechanisms within the subsequent "selection for variation" and "selection for survival" procedures. The goal is to maintain the optimal trade-off between convergence and diversity requirements.

For example, in the context of NSGA-II, at the "selection for variation" stage, deploying the binary tournament selection procedure, two candidate solutions are picked randomly and compared in terms of their Pareto dominance rank. The solution with the highest Pareto dominance rank is inserted in the mating pool. In the case where the two solutions share the same Pareto dominance rank, one of the two solutions is chosen at random and included in the mating pool, disregarding the NSGA-II crowding measure which usually constitutes the secondary criterion for "selection for variation". The "selection for variation" process continues until the mating pool is filled. At the "selection for survival" stage, the diversity measure is again eliminated as a discriminatory criterion for selection. In the situation where the number of locally non-dominated solutions exceeds the prefixed size of the active archive, the solutions are selected randomly for survival and propagation to the succeeding generation of the optimisation process.

On the other hand, when required (i.e. $I_s < 1\text{-}\varepsilon$), the diversity promotion mechanisms are automatically activated in the "selection for variation" and the "selection for survival" procedures based on the diversity indicator monitoring the diversity of the locally non-dominated solutions. A schematic presentation of the diversity management operator (DMO) is illustrated in Figure 2 within the context of NSGA-II.

4 Test Functions, Configurations and Performance Metrics

Similar to the study by Purshouse and Fleming [3], this investigative study experiments with different versions of the scalable DTLZ2 test function introduced by Deb et al [8] and described in Equation (3). In Equation (3), M represents the number of objectives, $n = M + K - 1$ is the number of decision variables, and K is a "difficulty parameter" ($K = 10$ in this study). DTLZ2 (M) denotes an M-objective instance of DTLZ2. DTLZ2 possesses a continuous and non-convex global Pareto front and comprises two types of decision variables responsible for controlling the solution's convergence towards the global Pareto front and the solution's distribution in objective space respectively. The first (m-1) decision variables ($x_1 \ldots x_{m-1}$) control the proximity of the solutions to the true Pareto front via a k-dimensional quadratic bowl, g, with global minimum $x_{m,\ldots n} = 0.5$. The decision variables ($x_m, \ldots x_n$) are responsible for controlling the diversity of the solutions and their location on the positive quadrant of the unit sphere. The different versions of DTLZ2 deployed to test the performance of the DMO vary in terms of the number of competing objectives to be optimised. DTLZ2, with its well-defined Pareto fronts, is a suitable test function for the analysis and the examination of the performance of new optimisation strategies.

$$
\begin{aligned}
&\min && z_1(x) = [1+g(x_M)]\cos(x_1\pi/2)\cos(x_2\pi/2)\ldots\cos(x_{M-2}\pi/2)\cos(x_{M-1}\pi/2),\\
&\min && z_2(x) = [1+g(x_M)]\cos(x_1\pi/2)\cos(x_2\pi/2)\ldots\cos(x_{M-2}\pi/2)\sin(x_{M-1}\pi/2),\\
&\min && z_3(x) = [1+g(x_M)]\cos(x_1\pi/2)\cos(x_2\pi/2)\ldots\sin(x_{M-2}\pi/2),\\
&\vdots && \vdots \quad \vdots\\
&\min && z_{M-1}(x) = [1+g(x_M)]\cos(x_1\pi/2)\sin(x_2\pi/2), \qquad (3)\\
&\min && z_M(x) = [1+g(x_M)]\sin(x_1\pi/2),\\
&w.r.t && x = [x_1,\ldots\ldots,x_n],\\
&where && g(x_M) = \textstyle\sum x_i \in x_M (x_i - 0.5)^2, with \quad x_M = [x_M,\ldots\ldots,x_n],\\
&and && 0 \le x_i \le 1, \quad for\ i = 1,2,\ldots\ldots,n, with\ n = M + \kappa - 1
\end{aligned}
$$

The DMO is incorporated into NSGA-II. The resulting optimisation strategy will be referred to as NSGA-II/DMO. Three different versions of the scalable DTLZ2 test function, featuring 6, 8, and 12 objective optimisation problems, are used to assess the performance of NSGA-II/DMO. These three instances of DTLZ2 will be referred to as DTLZ2 (6), DTLZ2 (8), and DTLZ2 (12) respectively.

The simulated binary crossover (SBX) [9], a two-parent crossover operator that produces two new solutions, is used in this study. Similar to the study by Purshouse and Fleming [3], in this work, each decision variable x is independently considered for undertaking the variation operator. The probability of uniformly applying the variation operator on a certain decision variable is commonly set to a value of 0.5 alongside a distribution parameter value $\eta_c = 15$ [8],[10] and a probability of applying variation to a certain pair of solutions $p_c = 1$.

The polynomial mutation operator [11] was also used and configured with the standard parameters for each of the DTLZ2 functions. The configuration of NSGA-II and NSGA-II/DMO is presented in Table 1 below.

Table 1. NSGA-II and NSGA-II/DMO Configurations

Size of Population	100
Crossover operator	SBX
Mutation Operator	Polynomial Mutation (probability: 1/(number of decision variables)
Number of generations	200
Number of Runs	10

The performance of NSGA-II/DMO is assessed by comparing its results with those obtained by NSGA-II for each of the three versions of DTLZ2 deployed. In order to make a rigorous comparison of the two optimisers, NSGA-II/DMO and NSGA-II were each executed 10 times and their results were compared for each execution. The *dominated distance metric* (DD-Metric) [7] is one of two binary performance metrics used to assess the quality of the approximation sets achieved by NSGA-II/DMO and NSGA-II. The DD-metric calculates the difference of dominated distances between two approximation sets 'A' and 'B' produced by two MOEAs 'A1' and 'A2', for example, in the objective space. The dominated distance between an approximation set 'A' and an approximation set 'B' (ddAB) is the sum of Euclidean distances between each solution A_i in 'A' and the closest solution B_i which belongs to the subset of 'B' that dominates A_i. The dominated distances dd_{AB} and dd_{BA} are calculated respectively, and their difference forms the value of DD-Metric (A, B).

The other binary metric deployed is the coverage metric (*C-metric*) [7], which calculates the percentage of solutions in a certain approximation set that are dominated or equal to any solution in another competing approximation set.

Moreover, the proximity of the solutions achieved by NSGA-II and NSGA-II/DMO towards the Pareto front and their diversity across the region of interest (in this study the region of interest is the whole Pareto front) are assessed. The normalised maximum spread metric (Equation 2) was used to measure the performance of the two optimisers in terms of the diversity quality of their produced results. The ideal diversity measure sought was $I_s = 1$ and represents an intermediate spread measure between the two extreme situations:

1. Dispersal of solutions in sub-optimal regions of the objective space, and
2. An approximation set confined to a specific region of the Pareto front.

The convergence quality of the achieved approximation sets is assessed in terms of their proximity to the well-defined Pareto fronts (k-dimensional quadratic bowl in the positive quadrant of the unit sphere) of the DTLZ2 test function. A specialised proximity metric for DTLZ2 is used to measure the median proximity of the achieved approximation sets (Z_A) to the Pareto fronts of each of the DTLZ2 versions investigated. The proximity metric, presented in Equation (4), is the generational distance (GD) metric [12] for the case of a continuous Pareto optimal reference set Z_*.

$$GD = median_{z_A \in Z_A} \left\{ \left[\sum_{m=1}^{M} (z_{A_m})^2 \right]^{1/2} - 1 \right\} \tag{4}$$

Finally, in order to illustrate the performance of the two optimisers in terms of the desired requirements, the *Non-Dominated Evaluation* metric [13] was used to simultaneously visualise the performance of the optimisers in terms of proximity to the Pareto front and in terms of diversity. The spread metric and the generational distance metric were posed as two objective functions evaluating two competing objectives: *Objective 1: Convergence* and *Objective 2: Diversity*. The problem can then be formulated as a two-objective optimisation scenario optimising (minimising) these two objectives. As a result, the performance of an optimiser **A** would be confidently deemed superior to the performance of another optimiser **B** if its approximation set to the posed bi-objective optimisation problem dominates the approximation set achieved by **B**.

The Non-dominated evaluation metric is illustrated in Figure 3 where it can be inferred that optimiser **A** outperforms optimiser **B** in terms of convergence and diversity but it cannot be concluded that **A** outperforms **C**.

In the following section, the results achieved by NSGA-II/DMO are illustrated and compared with the results achieved by NSGA-II for each of the test functions investigated.

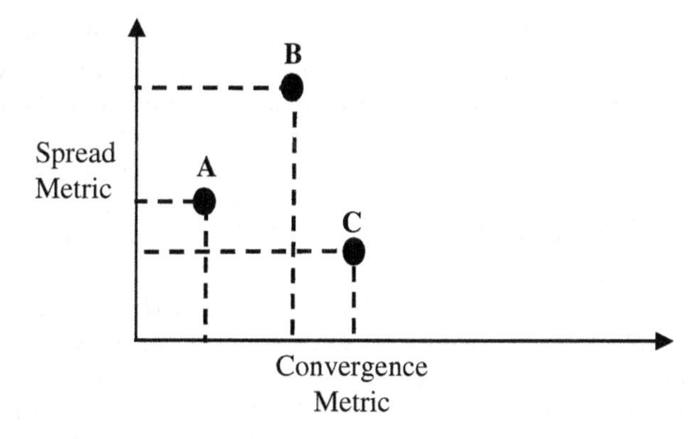

Fig. 3. The Non-Dominated Evaluation Metric

5 Results

The final approximation sets achieved by NSGA-II/DMO and NSGA-II for each of the DTLZ2 test functions after 200 generations are contrasted. Each optimiser was executed 10 times in order to assess the significance of the observed results and to make sure that the observations have not arisen by chance. Table 2 contains the DD-metric results and the C-metric results achieved by NSGA-II/DMO and NSGA-II for DTLZ2 (6). The two optimisers were similarly configured with the standard SBX and polynomial mutation parameters often used in the EMO community. The C-metric is not a symmetric indicator, and therefore in order to have a full appreciation of the relative quality of the two approximation sets, the metric had to be executed twice, switching the order of its input.

Table 2. DD-metric and C-metric results for DTLZ2 (6)

	DTLZ2 (6) A = NSGA-II/DMO AND B = NSGA-II		
Execution Number:	**C-Metric (A, B)**	**C-Metric (B, A)**	**DD-Metric (A, B)**
1	34%	0%	-0.65711
2	35%	0%	-0.58455
3	50%	0%	-0.85758
4	66%	0%	-1.3793
5	54%	0%	-1.017
6	47%	0%	-0.73012
7	44%	0%	-0.70773
8	46%	0%	-0.88246
9	47%	0%	-0.88319
10	51%	0%	-0.81869
Mean Value:	47%	0%	-0.8518

Form Table 2, it can deduced that the approximation sets achieved by NSGA-II/DMO cover an average of 47% of the solutions achieved by NSGA-II over the 10 runs. On the other hand, NSGA-II was repeatedly achieving nil coverage of the solutions achieved by its DMO hybridized counterpart.

The dominated distance metric uniformly produced results that highlight the superior quality of the approximation sets achieved by NSGA-II/DMO. A negative DD-metric value denotes that the first input of the metric (*e.g. Algorithm A in DD-Metric (A, B)*) produced an approximation set which is overall better than and dominates most of the approximation set produced by its second input (*Algorithm B*).

In Figure 4, the black circles, whose (x, y) coordinates are the values of the GD-metric and the maximum spread indicator respectively, represent the values of the non-dominated evaluation metric achieved by NSGA-II/DMO at each of the 10 executions of the optimiser solving DTLZ2 (6). The values of the non-dominated evaluation metric achieved by NSGA-II for the same optimisation scenario are represented by the black squares. The values of the GD metric achieved by NSGA-II/DMO over the 10 runs are consistently lower than 0.1 and are accompanied by spread measures with an approximate ceiling value of 2.

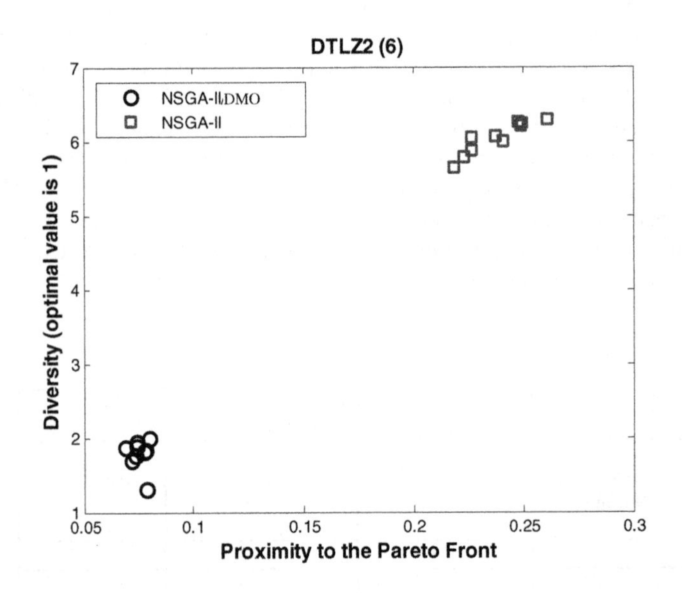

Fig. 4. Comparing NSGA-II/DMO with NSGA-II when optimising DTLZ2 (6)

The observed results highlight good proximity to the Pareto front alongside a simultaneous near-optimal diversity (the optimal value for the diversity is 1 and corresponds to the diversity value of the true Pareto front). On the other hand, the GD values achieved by NSGA-II over the 10 runs consistently exceed the value 0.2 alongside an average diversity measure of 6. From the results illustrated in Figure 4, it is clear that the performance of NSGA-II/DMO is superior to the performance of NSGA-II in terms of both requirements (convergence to the Pareto front and desired diversity). The results achieved by NSGA-II are much more diverse in terms of

Table 3. DD-metric and C-metric results for DTLZ2 (8)

	DTLZ2 (8) A = NSGA-II/DMO AND B = NSGA-II		
Execution Number:	**C-Metric (A, B)**	**C-Metric (B, A)**	**DD-Metric (A, B)**
1	45%	0%	-1.028
2	25%	0%	-0.55142
3	57%	0%	-1.3388
4	59%	0%	-1.3333
5	31%	0%	-0.70815
6	35%	0%	-0.77246
7	40%	0%	-0.84258
8	37%	0%	-0.76733
9	42%	0%	-0.90902
10	19%	0%	-0.42073
Mean Value:	39%	0%	-0.8672

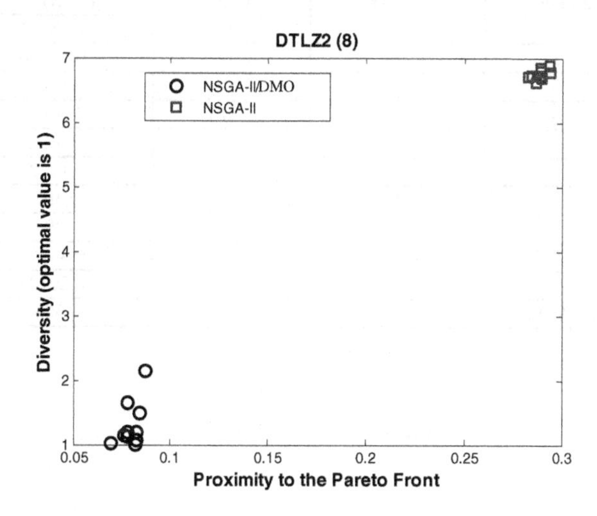

Fig. 5. Comparing NSGA-II/DMO with NSGA-II when optimising DTLZ2 (8)

dispersal in the 6-dimensional objective space. Nevertheless, *absolute* diversity, which can be achieved using a completely random search process, is undesirable in many-objective optimisation scenarios and once again led to the deterioration of the convergence of the search process towards the optimal regions of the space.

Similar to the results achieved for DTLZ2 (6), the results achieved for DTLZ2 (8) highlight a significantly superior performance of NSGA-II/DMO when compared with the performance of NSGA-II. The values of the C-metric and the DD-metric achieved by NSGA-II/DMO and NSGA-II, each optimising DTLZ2 (8), are illustrated in Table 3. The same observations highlight the superior performance of NSGA-II/DMO (higher coverage and lower dominated distances) may be made.

Figure 5 again demonstrates the superiority of NSGA-II/DMO over NSGA-II when a fine-grained analysis of the convergence and the diversity requirements is performed simultaneously. The results achieved by NSGA-II/DMO are significantly better than the results achieved by NSGA-II for the 8-objective optimisation problem, highlighting the beneficial impact and utility of the DMO.

In Table 4, the values of the C-metric and the DD-metric achieved at each execution of A and B are respectively presented. This time the optimisers are solving a 12-objective version of the DTLZ2 test function. The same conclusions drawn from the 6- and 8-objective scenarios are achieved for the 12-objective scenario. The results of the non-dominated evaluation metric for the 12-objective scenarios are presented in Figure 6. The significantly superior performance of the NSGA-II/DMO in terms of convergence and diversity is again apparent.

Table 4. DD-metric and C-metric results for DTLZ2 (12)

	DTLZ2 (12) A = NSGA-II/DMO and B = NSGA-II		
Execution Number:	C-Metric (A, B)	C-Metric (B, A)	DD-Metric (A, B)
1	8%	0%	-0.18421
2	17%	0%	-0.42636
3	7%	0%	-0.17411
4	20%	0%	-0.4973
5	3%	0%	-0.070199
6	6%	0%	-0.15524
7	12%	0%	-0.3052
8	19%	0%	-0.4975
9	17%	0%	-0.44757
10	12%	0%	-0.29128
Mean Value:	12%	0%	-0.3049

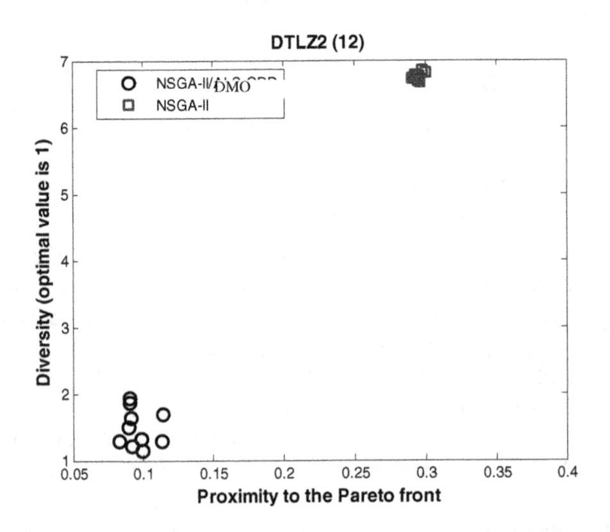

Fig. 6. Comparing NSGA-II/DMO with NSGA-II when optimising DTLZ2 (12)

NSGA-II/DMO significantly out-performs NSGA-II by producing overall better quality approximation sets in terms of convergence and 'good' desired diversity for DTLZ2 (3), (8) and (12).

6 Conclusions

From the set of experiments presented in this study, the success of the diversity management operator was established on a set of *many*-objective optimisation problems. The DMO is a beneficial strategy that addresses the conflict between the EMO requirement for good proximity towards the Pareto front and the requirement for maintaining a diverse set of solutions. Additionally, the DMO was demonstrated to be highly beneficial for controlling the diversity requirement which usually hampers the search process and, therefore, the convergence of the manipulated solutions to the Pareto front of a MOP with many conflicting objectives.

NSGA-II/DMO significantly and repeatedly outperformed NSGA-II by producing solutions closer to the Pareto front and maintaining a near-optimal and desired diversity among the solutions, for a set of *many*-objective optimisation problems (6 to 12 objectives). DMO was exercised on a set of test functions with well-defined Pareto fronts. Nevertheless, the strategy can be used to solve any multiobjective optimisation problem which is tackled by a MOEA. The DMO is a simple and efficient strategy with minimal computational overhead. The decision maker is only required to provide an approximate, targeted or desired, value for the extreme solutions (in terms of each objective) in the objective space. These solutions will serve to define an approximation to the vertices of the hypercube which contains the desired region of interest, and therefore define the notion of a 'good' diversity. Note that these suggested vertices will solely play a role defining the notion of a desired diversity measure in order to efficiently control the diversity promotion mechanisms in a MOEA and guide the search towards the Pareto-optimal front. If necessary, the notion of a 'desired' or 'good' diversity can then be progressively and appropriately modified using a progressive preference articulation technique such as the technique by Fonseca and Fleming [14] or Branke and Deb [15]. Future work will include assessing the performance of the DMO in MOEAs with more elaborate diversity promotion mechanisms such as the Genetic Diversity Evolutionary Algorithm (GDEA) [16] and SPEA2 [17]. Alternatives to the maximum spread metric used in this paper will also be investigated and used within the DMO to control the diversity extent and uniformity.

References

[1] Bosman, P.A.N., Thierens, D.: The balance between proximity and diversity in multiobjective evolutionary algorithms. IEEE Transactions on Evolutionary Computation 7(2), 174–188 (2003)

[2] Deb, K., Pratap, A., Agrawal, S., Meyarivan, T.: A Fast and Elitist Multiobjective Genetic Algorithm: NSGA-II. IEEE Transactions on Evolutionary Computation 6(2), 182–197 (2002)

[3] Purshouse, R.C., Fleming, P.J.: On the evolutionary optimisation of many conflicting objectives. IEEE Trans Evolutionary Computation 11(6), 770–784 (2007)

[4] Purshouse, R.C., Fleming, P.J.: Evolutionary many-objective optimisation: An exploratory analysis. In: IEEE Neural Networks Council (ed.), Proceedings of the 2003 Congress on Evolutionary Computation (CEC 2003). IEEE Service Center, Piscataway (2003)
[5] Deb, K., Goldberg, D.E.: An investigation of niche and species formation in genetic function optimization. In: Schaffer, J.D. (ed.) Proceedings of the Third International Conference on Genetic Algorithms, pp. 42–50. Morgan Kaufmann, San Francisco (1989)
[6] Ikeda, K., Kita, H., Kobayashi, S.: Failure of Pareto-based MOEAs: Does nondominated really mean near to optimal? In: IEEE Neural Networks Council (ed.), Proceedings of the 2001 Congress on Evolutionary Computation (CEC 2001), vol. 2, pp. 957–962. IEEE Service Center, Piscataway (2001)
[7] Zitzler, E.: Evolutionary Algorithms for Multiobjective Optimization: Methods and Applications, PhD thesis, Swiss Federal Institute of Technology (ETH), Zurich, Switzerland (1999)
[8] Deb, K., Thiele, L., Laumanns, M., Zitzler, E.: Scalable Multi-Objective Optimization Test Problems. In: Proceedings of the 2002 Congress on Evolutionary Computation (CEC 2002). IEEE Service Center, Piscataway (2002)
[9] Deb, K., Agrawal, R.B.: Simulated Binary Crossover for Continuous Search Space. Complex Systems 9, 115–148 (1995)
[10] Khare, V., Yao, X., Deb, K.: Performance scaling of multi-objective evolutionary algorithms. In: Fonseca, C.M., Fleming, P.J., Zitzler, E., Deb, K., Thiele, L. (eds.) EMO 2003. LNCS, vol. 2632, pp. 376–390. Springer, Heidelberg (2003)
[11] Deb, K., Goyal, M.: A combined genetic adaptive search (geneAS) for engineering design. Computer Science and Informatics 26(4), 30–45 (1996)
[12] Van Veldhuizen, D.A.: Multiobjective Evolutionary Algorithms: Classifications, Analyses, and New Innovations, PhD thesis, Air Force Institute of Technology (1999)
[13] Deb, K.: Multi-Objective Optimization Using Evolutionary Algorithms. John Wiley & Sons, New York (2001)
[14] Fonseca, C.M., Fleming, P.J.: Multiobjective Optimization and Multiple Constraint Handling with Evolutionary Algorithms - Part I: A Unified Formulation. IEEE Transactions on Systems, Man, and Cybernetics, Part A: Systems and Humans 28(1), 26–37 (1998)
[15] Branke, J., Deb, K.: Integrating user preferences into evolutionary multiobjective optimization. In: Jin, Y. (ed.) Knowledge Incorporation in Evolutionary Computation, pp. 461–477. Springer, Heidelberg (2004)
[16] Toffolo, A., Benini, E.: Genetic Diversity as an Objective in Multi-objective Evolutionary Algorithms. Evolutionary Computation Journal 11(2), 151–167
[17] Zitzler, E., Laumanns, M., Thiele, L.: SPEA2: Improving the Strength Pareto Evolutionary Algorithm for Multiobjective Optimization. In: Giannakoglou, K.C., et al. (eds.) Evolutionary Methods for Design, Optimisation and Control with Application to Industrial Problems (EUROGEN 2001), pp. 95–100 (2001)

Enhancing Decision Space Diversity in Evolutionary Multiobjective Algorithms

Ofer M. Shir[1], Mike Preuss[2], Boris Naujoks[2], and Michael Emmerich[3]

[1] Department of Chemistry, Princeton University
Frick Lab, Princeton NJ 08544, USA
oshir@Princeton.EDU

[2] Technische Universität Dortmund, Lehrstuhl für Algorithm Engineering
44221 Dortmund, Germany
{mike.preuss,boris.naujoks}@uni-dortmund.de

[3] Natural Computing Group, LIACS, Leiden University
Niels Bohrweg 1, 2333 CA Leiden, The Netherlands
emmerich@liacs.nl

Abstract. In multi-criterion optimization, Pareto-optimal solutions that appear very similar in the objective space may have very different pre-images. In many practical applications the decision makers, who select a solution or preferred region on the Pareto-front, may want to know different pre-images of the selected solutions. Especially, this will be the case when they would like to present alternative design candidates in later stages of a multidisciplinary design process.

In this paper we extend an existing CMA-ES niching framework, which has been previously applied successfully to multi-modal optimization, to the multi-criterion domain for boosting decision space diversity. At the same time, we introduce the concept of space aggregation for diversity maintenance in the aggregated spaces, i.e. search/decision and objective space. Empirical results on synthetic multi-modal bi-criteria test problems with known efficient sets and Pareto-fronts demonstrate that the diversity in the decision space can be significantly enhanced without hampering the convergence to a precise and diverse Pareto front approximation in the objective space of the original algorithm.

1 Introduction

Pareto-optimization aims at solving optimization problems with multiple, possibly conflicting, objective functions [1]. The general approach is to find non-dominated solution sets and, especially in continuous spaces, approximate true Pareto-fronts of the problem. It is important in the context of this paper to distinguish between the *Pareto-front* and the *efficient set*. While the former denotes the set of non-dominated points in the objective space, the latter refers to the set of vectors in the search space that are pre-images of the points in the Pareto-front under the mapping of the vector-valued objective function. At the same time, multiple points in the efficient set may be projected onto the same point on the Pareto-front. Moreover, unless certain continuity assumptions on

M. Ehrgott et al. (Eds.): EMO 2009, LNCS 5467, pp. 95–109, 2009.

the objective functions hold, there is no evidence that neighboring points on the Pareto-front stem from the same region of the decision space. This scenario is illustrated in Figure 1. Therefore, attaining a set of solutions that covers the entire Pareto-front does not necessarily guarantee obtaining a set that yields a good coverage of the decision set. Moreover, diversity of an approximation set to the Pareto front in the objective space does not necessarily imply diversity of solutions in its corresponding efficient set approximation, though the latter is desirable.

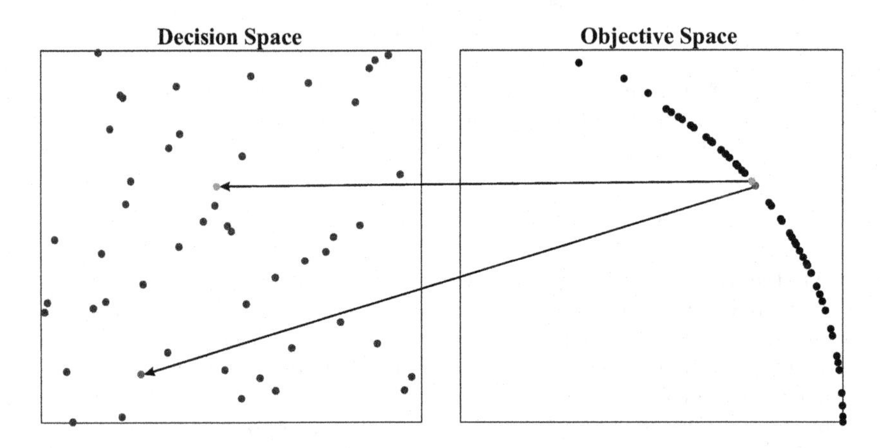

Fig. 1. Diversity for *decision making*: Illustrative example for a scenario where two adjacent points on the Pareto front are mapped onto two points in two completely different regions in the decision space. Units and scales are arbitrary.

1.1 Motivation

Indeed, it has been pointed out recently that not only high diversity of solutions in the objective space but also high diversity of solutions in the efficient set can be of interest for decision makers [2,3]. We choose to furthermore motivate this idea with the following two examples:

- Firstly, let us consider the problem of finding molecules with certain properties that can serve as drug candidates in a de novo drug discovery process [4,5]. Clearly, the approximation of different target properties can be formulated as a multi-objective optimization task. However, once a set of molecules has been found that has a good spread over the Pareto front, there may still be molecules that violate some constraints that had not been considered by the expert. In such cases, alternative solutions with similar properties, would be of interest.
- A second example is multidisciplinary optimization processes in the automotive or aerospace industries [6,7], that follow a restricted design process workflow. Here, different development teams focus on different aspects of a design and come up with a set of solutions that are favorable from the point of view of their discipline to discuss these solutions with experts from other

disciplines until a consensus design is found. Also in this case it would be desirable for the decision maker to identify different possible solutions that map to the preferred region on the Pareto front, as the objectives of the other disciplines cannot be evaluated *a priori*.

1.2 Background

Up to date, there has been very few work that addressed search space diversity in Evolutionary Multiobjective Optimization (see, e.g., [2,6,8,9]). Apart from this, current benchmarks do not consider this issue in the way performance is evaluated. This paper presents a wholehearted attempt to increase decision space diversity in existing state-of-the-art Evolutionary Multiobjective Optimization Algorithms (EMOA).

Based on related studies in multi-modal optimization, the modification of the selection criteria alone is not sufficient to boost diversity in the decision space. This is due to the fact that *Evolutionary Algorithms* (EAs) tend to lose their population diversity for several reasons, such as genetic drift, fast takeover, and disruptive recombination [10]. This problem is typically addressed by *Niching methods*, an extension of EAs to multi-modal optimization [11,12]. These methods allow for parallel convergence into multiple good solutions. Niching has been traditionally investigated within Genetic Algorithms (GAs) [11], but recently there were several studies of niching in Evolution Strategies (ES), especially as combined with the Covariance Matrix Adaptation ES (CMA-ES; See, e.g., [13]). The obtained ES-based niching techniques proved to be robust and efficient strategies for identifying multiple global optima in degenerate landscapes, and were successfully applied to synthetic as well as to real-world high-dimensional problems [14].

1.3 Overview

The new approach reported in this paper introduces two conceptual changes to the selection strategy of EMOA: The *first* is the employment of an aggregated diversity measure that takes into account the local density of solutions in the decision space with the local density in the objective space. However, aggregation alone would not be sufficient to prevent fast takeover and drift effects from occurring. These effects are already known to cause a rapid loss of diversity in ordinary EA/EMOA in early stages of the evolution, where Pareto domination rather than contribution to diversity is still the governing criterion for selection. Therefore, we consider the introduction of *dynamic niching* using *resource sharing*, also referred to as the *dynamic niching framework*, as the second element, due to counteract the aforementioned effects.

As a proof of concept for the new approach, we shall present in this paper empirical results on synthetic multi-modal bi-criteria test problems with known efficient sets and Pareto-fronts. We will demonstrate that diversity in the decision space can be significantly enhanced without hampering the convergence to a diverse Pareto front approximation in the objective space of the original algorithm.

As reference methods, we will report on the performance of the *multi-criterion* version of the CMA-ES (referred to in our notation as CMA-MO) [15], as well as the NSGA-II and its derived variants [16,2] on the same test problems.

The paper is organized as follows: In section 2 we discuss related work. The algorithmic approach is outlined in section 3. Then, in section 4, the proposed scheme is evaluated on test problems. Finally, in section 5 we summarize our findings and suggest directions for future research.

2 Related Work

We review here several related studies to our work. Due to the crossing-branches nature of our work, these treat the topics of *niching* and *multi-objective* optimization.

Niching techniques have been already used in the multi-objective optimization arena, earlier. Horn et al. introduced a niching technique for multi-objective optimization, known as the *niched-Pareto GA (NPGA)* [17]. The algorithm was a variant of the *fitness sharing* niching method, whereas the *niching distance metric was set to consider the objective space only.* Selection was based on so-called *Pareto domination tournaments* or on the minimal niche count, otherwise. The NPGA was a classical example of using an existing single-objective niching technique, in a straightforward manner, for multi-objective optimization – only by redefining the niching distance metric and the selection mechanism. However, its kernel was the simple GA and it lacked any self-adaptation mechanism.

A multi-objective approach aiming for a good diversity in decision as well as in objective space was the GDEA, as introduced by Toffolo and Benini [9]. GDEA invoked two selection criteria, non-dominated sorting as the primary one and a metric for decision space diversity as the secondary one.

Another approach, the so-called *Omni-Optimizer* [2], extended the classical NSGA-II [16] by considering the diversity in the decision space additionally. Its selection is performed with a changing secondary selection criterion, targeting either the decision or the objective space diversity in each generation.

An EMOA approach designed for maintaining diversity in both spaces is the KP1, as proposed by Chan and Ray [8]. Here, two criteria to measure the diversity of solutions in the corresponding spaces are defined and applied in each generation. These are the dominated hypervolume of each individual for the objective space and a neighborhood counting approach for the decision space.

A more structural analysis of the correlation between decision and objective space in multi-objective optimization has been introduced lately [3,18], while focusing on defining different test functions and analyzing the algorithmic behavior on them.

3 The Algorithmic Approach

Before introducing the new framework we would like to review some of its components, and in particular the extension of the CMA-ES into multi-modal domains by means of a specific niching technique.

The CMA-ES (see, e.g., [13]), is a derandomized ES variant that has been successful in treating correlations among object variables by efficiently learning matching mutation distributions. Explicitly, in generation g, λ offspring are generated by means of Gaussian sampling:

$$\boldsymbol{x}_k^{(g)} \sim \mathcal{N}\left(\langle \boldsymbol{x} \rangle_W^{(g-1)}, \sigma^{(g-1)^2} \mathbf{C}^{(g-1)}\right) \quad k = 1, \ldots, \lambda \tag{1}$$

The best μ search points out of these λ offspring undergo weighted recombination and become the parent of the following generation, denoted by $\langle \boldsymbol{x} \rangle_W^{(g)}$. The covariance matrix $\mathbf{C}^{(g)}$ is initialized as the *unity matrix* and is learned during the course of evolution, based on cumulative information of successful past mutations (the so-called *evolution path*). The global step-size, $\sigma^{(g)}$, is updated based on information extracted from *principal component analysis* of $\mathbf{C}^{(g)}$ (the so-called *conjugate evolution path*). For more details we refer the reader to [13].

A *niching framework* for $(1 \overset{+}{,} \lambda)$ derandomized-ES kernels subject to a fixed niche radius has been introduced recently (see, e.g., [14]). This framework considers q search points, which carry their defining strategy parameters (referred to as *CMA-Sets* or *D-Sets*), and correspond to sub-populations operating in different parts of the search space (niches). The niches and their representatives are re-formed in each generation using the dynamic peak identification (DPI) routine [14]. It takes into account both the ranked fitness of the individuals as well as the spatial distance between them; For the spatial selection, a niche radius must be defined *a priori* [14]. Individuals that belong to the same niche are located in a hyper-sphere, defined by that radius, around the central individual, namely the peak individual. Unlike previous CMA-Niching ES, this study will introduce multiple parents in each niche, subject to (μ_W, λ) selection with weighted recombination according to the standard formulas [19]. Sizing the niche population is done with $\lambda = 4 + \lfloor 3 \cdot \ln(n) \rfloor$, $\mu = \lfloor \frac{\lambda}{2} \rfloor$, with n as the search space dimensionality, following the recommendation in [19] (for further argumentation see also [13]). To this end, we choose to define the additional selected offspring as the set of at most $\lfloor \frac{\lambda}{2} \rfloor - 1$ individuals that are within niche radius from the peak individual and share its same parent. This way, it is guaranteed that the ES mutation distribution evolves continuously. Since the value of μ may vary over time, other auxiliary coefficients must be updated accordingly, such as the recombination weights. Algorithm 1 summarizes the Niching-CMA routine.

The proposed multi-objective routine uses the Niching-CMA scheme as it is, with the following modifications:

- ranking of individuals is based upon non-dominated sorting.
- distance between niches is calculated in the aggregated space.
- the estimation of the niche radius is adjusted.

Given the n-dimensional decision vector of individual k, $\boldsymbol{x}_k = (x_{k,1}, ..., x_{k,n})$, with its assigned objective d-dimensional vector, $\boldsymbol{f}_k = (f_{k,1}, ..., f_{k,d})$, and given the equivalent decision and objective vectors of individual l, $(\boldsymbol{x}_l, \boldsymbol{f}_l)$, the distance between individuals k, l is defined as follows:

Algorithm 1. (μ_w, λ)-CMA-ES Niching with Fixed Niche Radius

1: **for** $i = 1, \ldots, q$ search points **do**
2: Generate λ samples based on the CMA-Set of individual i
3: **end for**
4: Evaluate fitness of the population
5: Compute the Dynamic Peak Set (DPS) with the DPI Routine
6: **for** $j = 1..q$ elements of DPS **do**
7: Identify at most $\mu = \lfloor \frac{\lambda}{2} \rfloor$ fittest individuals of niche j with $Parent(peak(j))$
8: Apply weighted recombination on $\boldsymbol{x}_w$ and $\boldsymbol{z}_w$ w.r.t. those individuals
9: Inherit the CMA-Set of $peak(j)$ and update it w.r.t. the variations carried out
10: **end for**
11: **if** N_{dps}=size of $DPS < q$ **then**
12: Generate $q - N_{dps}$ new search points, reset CMA-Sets
13: **end if**

$$d_{k,l} = \sqrt{\frac{1}{n}\sum_{i=1}^{n}(x_{k,i} - x_{l,i})^2 + \frac{1}{d}\sum_{j=1}^{d}(f_{k,j} - f_{l,j})^2} \tag{2}$$

It is implicitly assumed that decision parameters and objective function values are scaled within a common order of magnitude. In order to *select individuals* based on multiple objectives, the selection mechanism was modified. As outlined before, the niches are identified based on their ranked quality, which is implemented here by means of *non-dominated sorting* [16]. Following this, the routine will proceed as usual: Starting with rank 0, a greedy identification of the niches will be carried out, considering the distance with respect to the aggregated objective and decision spaces. If not all q niches are populated, the routine will proceed to rank 1, and so on.

Comparison. The uniqueness of the proposed approach with respect to the mainstream EMOA lies in two main aspects: Firstly, the employment of a *single selection phase*, rather than two, and secondly, the consideration of space aggregation for the sake of diversity measurement. Moreover, this method differs from the CMA-MO algorithm in its ES mechanism: Unlike the elitist single-parent $(1 + \lambda)$-kernel of the CMA-MO, the proposed scheme employs a *comma* multi-parent (μ_W, λ)-kernel, which may be advantageous in certain environments.

Setting a Default Value for the Niche Radius. Since our method aims to approximate the Pareto front by populating it with a uniform distribution of q niches, we can estimate the niche radius, whenever the aim is to distribute the niches evenly across the search space. The following derivations are valid for $2D$ objective spaces, but we believe that they could be generalized to d-dimensional spaces. Consider a connected Pareto front, and assume that we can define its *length*, denoted by l_{FRONT}. Also, let the diameter of the Pareto set be denoted by l_{SET}. Upon demanding a uniform distribution of niches, one may write:

$$2 \cdot \rho \cdot q = \sqrt{l_{FRONT}^2 + l_{SET}^2} \tag{3}$$

Simplified Model. One can consider a simplified model for providing an upper and a lower bounds for ρ, by taking into account only the objective space. For this purpose let us consider the *Nadir* objective vector, denoted here as $\boldsymbol{\zeta}^{(\mathcal{N})} = (f_{1,\mathcal{N}}, f_{2,\mathcal{N}})^T$. In the general d-dimensional objective space, the *Nadir* objective vector is defined as the vector with the *worst objective values of all Pareto optimal solutions* (as opposed to the worst objective values of the entire space):

$$\zeta_i^{(\mathcal{N})} = \max\left\{f_i \,\middle|\, (f_1, \ldots, f_i, \ldots, f_d)^T \in \mathcal{F}_N\right\}. \tag{4}$$

The Nadir objective vector can be computed for $d = 2$ by employing single-objective optimization. For $d > 2$, heuristics are available, but the problem is considered to be computationally hard [20].

Without loss of generality, assume that the objectives $\{f_1, f_2\}$ are assigned with values in the intervals $\{[f_{1,min}, f_{1,\mathcal{N}}], [f_{2,min}, f_{2,\mathcal{N}}]\}$, respectively. The length of the assumably-connected Pareto front has the following lower and upper bounds:

$$\begin{aligned} l_{FRONT,min} &= \sqrt{\left((f_{1,\mathcal{N}} - f_{1,min})^2 + (f_{2,\mathcal{N}} - f_{2,min})^2\right)} \\ l_{FRONT,max} &= |f_{1,\mathcal{N}} - f_{1,min}| + |f_{2,\mathcal{N}} - f_{2,min}| \end{aligned} \tag{5}$$

Hence, upon assuming a uniformly spaced population of the q niches along the front, one can derive

$$\frac{\sqrt{\left((f_{1,\mathcal{N}} - f_{1,min})^2 + (f_{2,\mathcal{N}} - f_{2,min})^2\right)}}{2 \cdot q} \leq \rho \leq \frac{|f_{1,\mathcal{N}} - f_{1,min}| + |f_{2,\mathcal{N}} - f_{2,min}|}{2 \cdot q} \tag{6}$$

The General Case. For the general case, we choose to define the default values as the diameters of the decision or the objective spaces, respectively:

$$r_{SET} = \sqrt{\sum_{i=1}^{n} (x_{i,max} - x_{i,min})^2} \qquad r_{FRONT} = \sqrt{\sum_{j=1}^{d} (f_{j,max} - f_{j,min})^2} \tag{7}$$

and thus

$$\rho = \frac{\sqrt{\sum_{i=1}^{n} (x_{i,max} - x_{i,min})^2 + \sum_{j=1}^{d} (f_{j,max} - f_{j,min})^2}}{2 \cdot q} \tag{8}$$

The niche radius is essentially a crucial parameter of this method, and its estimation or tuning is critical for the algorithmic success.

4 Experimental Analysis

Our aim is to provide a *proof of concept* for the proposed aggregation approach: Concerning the achieved decision space diversity of the generated results, an originally single-objective method enhanced by the aggregation scheme shall generally be competitive to any multi-objective algorithm designed for that purpose, and superior to any standard multi-objective algorithm. We therefore focus our experimental procedure on landscapes with interesting decision space characteristics, that is, functions with several pre-images for certain points in the efficient set (non-injective functions).

4.1 Test Functions: Non-injective Artificial Landscapes

The following set of bi-objective functions is considered in order to test the algorithmic performance. Not many more test problems with these characteristics are known to us, however the chosen four still have very different properties.

1. **Omni-Test by Deb.** Deb et al. constructed a bi-criteria multi-global landscape for testing their Omni-Optimizer [2]. Explicitly, it reads:

$$f_1(\boldsymbol{x}) = \sum_{i=1}^{n} \sin(\pi x_i) \longrightarrow \min, \qquad f_2(\boldsymbol{x}) = \sum_{i=1}^{n} \cos(\pi x_i) \longrightarrow \min \tag{9}$$

   where $\forall i\ x_i \in [0, 6]$. We consider $n = 5$.
2. **EBN.** The EBN family of functions [21] introduced a very basic set of test-problems for multi-objective algorithms. Explicitly, it reads:

$$f_1^{(\gamma)}(\boldsymbol{x}) = \left(\sum_{i=1}^{n} |x_i|\right)^{\gamma} \cdot n^{-\gamma} \rightarrow \min, \quad f_2^{(\gamma)}(\boldsymbol{x}) = \left(\sum_{i=1}^{n} |x_i - 1|\right)^{\gamma} \cdot n^{-\gamma} \rightarrow \min \tag{10}$$

   The EBN problems are attractive in the context of efficient set approximation, as the pre-images of points in the objective space are not single points, but rather line segments on the diagonals of $[0, 1]^n$, excepting the extremal points $(0, 1)^T$ and $(1, 0)^T$ [22]. Each point in $[0, 1]^n$ is efficient. In our study we consider the case of a linear Pareto front, $\gamma = 1$, with $n = 10$.
3. **"Two-on-One".** This test-case was originally introduced in an interesting study of the Pareto-optimal set [18], large parts of which have two pre-images. It is a two-dimensional function, with a 4th-degree polynomial with two minima as f_1 versus the sphere function as f_2:

$$\begin{aligned} f_1(x_1, x_2) &= x_1^4 + x_2^4 - x_1^2 + x_2^2 - g x_1 x_2 + h x_1 + 20 \longrightarrow \min \\ f_2(x_1, x_2) &= (x_1 - k)^2 + (x_2 - l)^2 \longrightarrow \min \end{aligned} \tag{11}$$

   We consider the asymmetric case, with $g = 10$, $h = 0.25$, $k = 0$, and $l = 0$ (case number 3 as reported in [18]).
4. **Lamé Superspheres.** We consider a multi-global instantiation of a family of test problems introduced by Emmerich and Deutz [23], the Pareto fronts of which have a spherical or super-spherical geometry. In contrast to the EBN problem, the set of pre-images of a point on the Pareto front for this instance is finite, and solutions are placed on equidistant parallel line-segments, with

Table 1. Hypervolume of the resulting Pareto fronts of the 5 different algorithms on the 4 test-cases: average and standard-deviation over 30 runs

Hypervolume	Niching-CMA	CMA-MO	NSGA-II	NSGA-II-Agg.	Omni-Opt.
Omni-Test	30.27 ± 0.05	$\mathbf{30.43 \pm 0.002}$	30.17 ± 0.034	29.81 ± 0.2	29.72 ± 0.20
EBN	3.295 ± 0.038	$\mathbf{3.489 \pm 0.001}$	3.30 ± 0.082	2.848 ± 0.173	2.058 ± 0.064
Two-on-One	173.44 ± 0.14	$\mathbf{174.52 \pm 0.005}$	172.59 ± 1.53	171.58 ± 2.1	168.24 ± 7.72
Superspheres	3.172 ± 0.037	$\mathbf{3.205 \pm 0.007}$	3.203 ± 0.001	3.109 ± 0.108	2.481 ± 0.375

Table 2. Decision-space diversity, as defined in Eq. 13, of the 5 different algorithms on the 4 test-cases: average and standard-deviation over 30 runs. See also Figure 2.

Diversity	Niching-CMA	CMA-MO	NSGA-II	NSGA-II-Agg.	Omni-Opt.
Omni-Test	$\mathbf{0.247 \pm 0.061}$	0.042 ± 0.028	0.191 ± 0.085	0.207 ± 0.065	0.0301 ± 0.002
EBN	$\mathbf{0.484 \pm 0.007}$	0.424 ± 0.010	0.412 ± 0.023	0.357 ± 0.027	0.012 ± 0.010
Two-on-One	$\mathbf{0.296 \pm 0.012}$	0.113 ± 0.002	0.183 ± 0.102	0.162 ± 0.088	0.093 ± 0.032
Superspheres	$\mathbf{0.412 \pm 0.022}$	0.115 ± 0.019	0.224 ± 0.046	0.307 ± 0.049	0.0729 ± 0.060

integer distances to each other, each of them being a pre-image of a local Pareto front. Let $\xi = \frac{1}{n-1}\sum_{i=2}^{n} x_i$, and $r = \sin^2(\pi \cdot \xi)$,

$$f_1 = (1+r) \cdot \cos(x_1) \longrightarrow \min \qquad f_2 = (1+r) \cdot \sin(x_1) \longrightarrow \min \tag{12}$$

with $x_1 \in \left[0, \frac{\pi}{2}\right]$, and $x_i \in [1, 5]$ for $i = 2 \ldots n$. We consider here $n = 4$.

4.2 Experiment

For presentation of the experimental results, we adhere to the structured reporting scheme suggested in [24], starting with the scientific question to answer.

Research Question. Does aggregation-niching boost decision space diversity?

Pre-Experimental Planning. Within first test runs, we found that a Pareto front of size 50 provides a meaningful compromise between speed and solution quality, especially for the purpose of visually examining the resulting solution sets. Most of the considered algorithms ran into stagnation after less than 50.000 evaluations, so that we chose this limit for the following experiment.

In order to assess the diversity in decision space, we set up and tested a corresponding quantifier. Given a population of size μ_N, we define the population diversity of the Pareto set as the mean value of the $\frac{\mu_N(\mu_N-1)}{2}$ Euclidean distances between all individuals, normalized by the diameter R of the decision space:

$$D = \frac{2}{R \cdot \mu_N(\mu_N - 1)} \cdot \sum_{A \neq B} \|\boldsymbol{x}_A - \boldsymbol{x}_B\| \tag{13}$$

Table 3. Calculation of the U-Test for the 4 landscapes for the 5 different algorithms. The tables contain calculations for both performance criteria: p-values for the diversity measure are presented in the upper-right part of the table; p-values for the hypervolume measure are presented in the lower-left part. Highlighted values indicate where the null hypothesis cannot be rejected at the 5% significance level (no difference).

Omni-Test

p-values	Niching-CMA	CMA-MO	NSGA-II	NSGA-II-Agg.	Omni-Opt.
Niching-CMA		5.49e-11	0.0117	0.0199	3e-11
CMA-MO	3.02e-11		5.19e-07	5.07e-10	0.0138
NSGA-II	6.01e-08	3.02e-11		**0.684**	7.66e-08
NSGA-II-Agg.	3.02e-11	3.02e-11	3.02e-11		1.94e-10
Omni-Opt.	3e-11	3e-11	3e-11	3e-11	

EBN

p-values	Niching-CMA	CMA-MO	NSGA-II	NSGA-II-Agg.	Omni-Opt.
Niching-CMA		3.02e-11	3.02e-11	3.02e-11	3e-11
CMA-MO	3.02e-11		0.017	3.69e-11	3e-11
NSGA-II	**0.971**	3.02e-11		2.23e-09	3e-11
NSGA-II-Agg.	3.02e-11	3.02e-11	3.02e-11		3e-11
Omni-Opt.	3e-11	3e-11	3e-11	3e-11	

Two-on-One

p-values	Niching-CMA	CMA-MO	NSGA-II	NSGA-II-Agg.	Omni-Opt.
Niching-CMA		3.02e-11	6.36e-05	5.46e-09	3.02e-11
CMA-MO	3.02e-11		0.00868	**0.865**	**0.0701**
NSGA-II	0.000377	9.51e-06		**0.122**	4.94e-05
NSGA-II-Agg.	0.000141	8.48e-09	0.0451		0.00907
Omni-Opt.	3.02e-11	3.02e-11	3.02e-11	3.02e-11	

Super-Spheres

p-values	Niching-CMA	CMA-MO	NSGA-II	NSGA-II-Agg.	Omni-Opt.
Niching-CMA		3.02e-11	3.02e-11	2.37e-10	3.02e-11
CMA-MO	6.72e-10		9.91e-11	3.02e-11	5.86e-06
NSGA-II	1.61e-06	8.48e-09		2.19e-07	2.22e-09
NSGA-II-Agg.	0.00152	8.99e-11	3.02e-11		4.97e-11
Omni-Opt.	3.02e-11	3.02e-11	3.02e-11	3.02e-11	

Task. We demand that the aggregation enhanced algorithms perform better than their non-aggregating counterparts in terms of diversity. Statistically, they should be better in at least 3 of 4 cases (U-test 5% level). Furthermore, they should perform as well as multi-objective algorithms specifically designed for keeping decision space diversity high (not worse at 5%) while keeping the hypervolume metric performance at a competitive level (this task is secondary and therefore not specified in detail).

Setup. We ran the proposed aggregation-enhanced niching method (Niching-CMA) against four reference methods: The CMA-MO [15], the NSGA-II [16], the Omni-Optimizer [2], and a variant of the NSGA-II which considers an aggregated space in the crowding calculations (referred to as *NSGA-II-Agg*). The latter routine is created from the standard NSGA-II in order to assess the importance

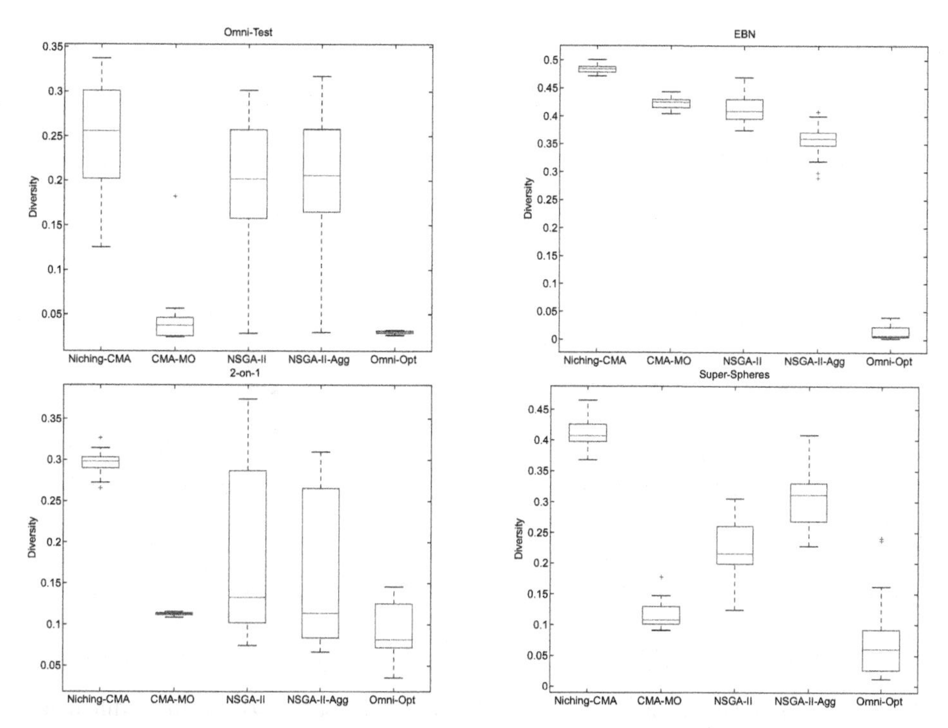

Fig. 2. Measured diversities in 30 runs of each of the 5 algorithms on Omni-test (upper left), EBN (upper right), the Two-on-one (lower left), and Super-Spheres (lower right) test problems

of the aggregation concept for attaining decision space diversity. All 5 methods are run on all 4 test problems of section 4.1 with 30 repeats each. We are aware that the enforced small populations may not be optimal for all algorithms; the Omni-Optimizer, for instance, was reported in [2] to employ a population of 1,000 individuals. However, apart from these settings, we rely on default values.

Experimentation/Visualization. Figures 3 and 4 show typical outcomes of the resulting approximated Pareto-sets and Pareto-fronts. Note that the decision space is represented by plotting x_1 versus x_2, except for the Superspheres test-case where x_1 is plotted versus $\frac{1}{(n-1)} \cdot \sum_{i=2}^{n} x_i$.

Table 1 provides the S-metric results, following $2D$ hypervolume calculations for test-cases 1-4 with reference points $\{(1,1),(2,2),(35,7),(2,2)\}$, respectively; Table 2 presents the calculations of the decision space diversity as defined in Eq. 13. Figure 2 contains the box-plots for the latter table. Furthermore, Table 3 presents the p-values for Mann-Whitney U-Tests for both the hypervolume as well as the diversity criterion, between all 5 algorithms on all 4 test problems.

Observations. In the Omni-Test landscape, Niching-CMA performed very well, while typically obtaining 4 Pareto subsets, in comparison to one or two subsets for each of the other routines. In the EBN landscape, Niching-CMA attained

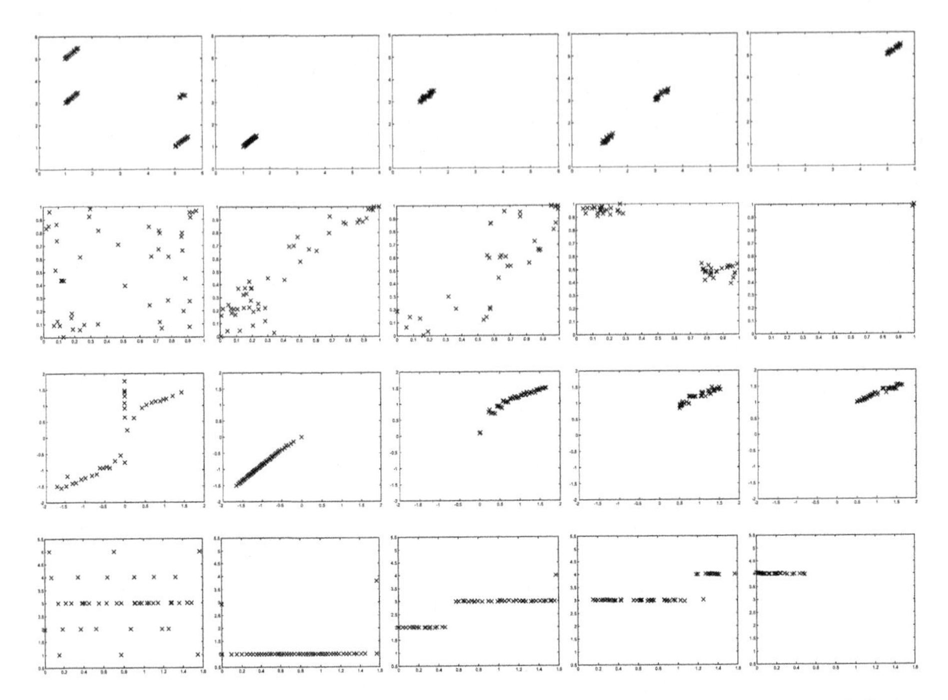

Fig. 3. Final populations of the 5 algorithms in the decision spaces of the 4 different landscapes. Note that the decision space is represented by plotting x_1 versus x_2, except for the Superspheres test-case where x_1 is plotted versus $\frac{1}{(n-1)} \cdot \sum_{i=2}^{n} x_i$. Columns, from left to right, present the algorithms in the following order: Niching-CMA, CMA-MO, NSGA-II, NSGA-II-Agg, Omni-Opt. First row presents the Omni-Test problem, followed by the EBN, 2-on-1, and Superspheres.

a quasi-uniform distribution in the decision space. In the "Two-on-One" landscape, the proposed algorithm managed to explore both branches of the so-called *propeller-shaped Pareto-set* (for more details see [18]), while the other algorithms typically explored either one of the two branches. In the Super-Spheres landscape, Niching-CMA performed extremely well, while obtaining a good distribution of typically 3 Pareto subsets. The other methods, nevertheless, usually obtained a single Pareto subset. This is clearly observed in the fourth row of Figure 3, where the final population of the these algorithms is mostly concentrated along a single line, corresponding to a single Pareto subset.

Discussion. Generally speaking, the proposed algorithm performs in a satisfying manner, obtaining good Pareto-sets with high diversity in the decision space, which are mapped onto well-approximated Pareto-fronts. In terms of the performance criterion in the objective space, the S-metric (hypervolume), CMA-MO did best on all test problems, whereas Niching-CMA and NSGA-II performed slightly worst and equally well, and NSGA-II with aggregation and Omni-Optimizer showed slightly worse performance. Regarding the diversity in the decision space, the proposed algorithm accomplished its goal: It attained

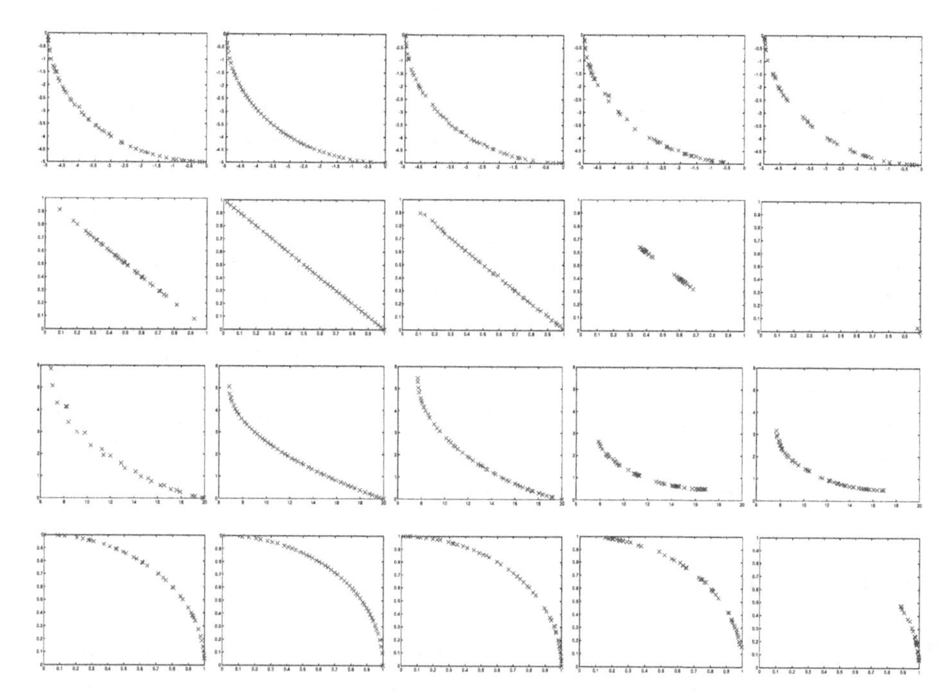

Fig. 4. Final populations of the 5 algorithms in the objective spaces of the 4 different landscapes, f_1 plotted versus f_2. Columns, from left to right, present the algorithms in the following order: Niching-CMA, CMA-MO, NSGA-II, NSGA-II-Agg, Omni-Opt. Rows from top to bottom: Omni-Test problem, EBN, 2-on-1, and Superspheres.

higher decision space diversity in comparison to the other methods on all landscapes. The CMA-MO, the S-metric winner, did not attain high decision space diversity; This is not a surprising result, as it is not meant to target this goal.

It should be noted that introducing the aggregation component into the NSGA-II did improve the attained decision space diversity to some extent on two landscapes, but did not have a considerable contribution. We may conclude that considering the aggregated space by itself does not seem to be sufficient for attaining high diversity in the decision space. We rather consider it as a *bridge* for niching to multi-objective domains. The Omni-Optimizer performed comparably poor in terms of the attained decision space diversity, and it is likely due to being hampered by the small population size.

5 Summary and Outlook

This paper addressed the topic of decision space diversity in the framework of Evolutionary Multi-Objective Algorithms. After providing the reader with the motivation for this study, and reviewing the existing work done on this topic, we outlined a new approach which aims at tackling multi-criterion problems while boosting diversity in the efficient set. The proposed algorithm relied on an existing CMA-based niching technique, which required adjustments in the selection scheme and the diversity measure. Due to the fact that it is a niche-radius based method, we proposed a way to choose a default value for this parameter. The algorithm was applied to a test-bed of non-injective artificial bi-criteria landscapes of various dimensions, and compared to the multi-objective CMA as well as to the classical GA-based EMOA: NSGA-II and its variants. The observed numerical results were satisfying, and provided us with the desired proof of concept for the proposed method. Furthermore, we concluded that employing space aggregation solely does not seem to be sufficient for attaining decision space diversity, and that niching could be the required bridging mechanism for multi-objective optimization. It should be noted that the GA-based methods performed poorly, likely due to the small population sizes that are typically employed by ES-based algorithmic kernels. Future research will be needed to test the approach on higher dimensional objective spaces and to explore various possibilities for parametrization and instantiation of the proposed approach.

Acknowledgments

The authors would like to thank Christian Igel for his dedicated support in using the Shark Library for the CMA-MO calculations. Mike Preuss gratefully acknowledges support from the Deutsche Forschungsgemeinschaft (DFG), grant no. RU 1395/3-2 "Ein Verfahren zur Optimierung von aus Mehrkomponenten bestehenden Schiffsantrieben".

References

1. Coello, C.A.C., Lamont, G.B., Van Veldhuizen, D.A.: Evolutionary Algorithms for Solving Multiobjective Problems. Springer, Berlin (2007)
2. Deb, K., Tiwari, S.: Omni-optimizer: A Procedure for Single and Multi-objective Optimization. In: Coello Coello, C.A., Hernández Aguirre, A., Zitzler, E. (eds.) EMO 2005. LNCS, vol. 3410, pp. 47–61. Springer, Heidelberg (2005)
3. Rudolph, G., Naujoks, B., Preuss, M.: Capabilities of EMOA to Detect and Preserve Equivalent Pareto Subsets. In: Obayashi, S., Deb, K., Poloni, C., Hiroyasu, T., Murata, T. (eds.) EMO 2007. LNCS, vol. 4403, pp. 36–50. Springer, Heidelberg (2007)
4. Nicolaou, C., Brown, N., Pattichis, C.: Molecular optimization using computational multi-objective methods. Current Opinion in Drug Discovery and Development 10, 316–324 (2007)
5. Kruisselbrink, J.W., Bäck, T., IJzerman, A.P., van der Horst, E.: Evolutionary algorithms for automated drug design towards target molecule properties. In: GECCO 2008: Proceedings of the 10th annual conference on Genetic and evolutionary computation, pp. 1555–1562. ACM, New York (2008)
6. Parmee, I.C., Cvetković, D., Watson, A.H., Bonham, C.R.: Multiobjective satisfaction within an interactive evolutionary design environment. ECJ 8(2), 197–222 (2000)

7. Schütze, O., Vasile, M., Coello, C.C.: Approximate solutions in space mission design. In: Rudolph, G., Jansen, T., Lucas, S., Poloni, C., Beume, N. (eds.) PPSN 2008. LNCS, vol. 5199, pp. 805–814. Springer, Heidelberg (2008)
8. Chan, K.P., Ray, T.: An Evolutionary Algorithm to Maintain Diversity in the Parametric and the Objective Space. In: International Conference on Computational Robotics and Autonomous Systems (CIRAS), Centre for Intelligent Control, National University of Singapore (2005) ISSN: 0219-6131
9. Toffolo, A., Benini, E.: Genetic Diversity as an Objective in Multi-Objective Evolutionary Algorithms. Evolutionary Computation 11(2), 151–167 (2003)
10. Preuss, M., Schönemann, L., Emmerich, M.: Counteracting genetic drift and disruptive recombination in $(\mu + /, \lambda)$-EA on multimodal fitness landscapes. In: Genetic and Evolutionary Computation Conference (GECCO), pp. 865–872. ACM Press, New York (2005)
11. Mahfoud, S.W.: Niching Methods for Genetic Algorithms. PhD thesis, University of Illinois at Urbana Champaign (1995)
12. Shir, O.M.: Niching in Derandomized Evolution Strategies and its Applications in Quantum Control. PhD thesis, Leiden University, The Netherlands (2008)
13. Hansen, N., Ostermeier, A.: Completely Derandomized Self-Adaptation in Evolution Strategies. Evolutionary Computation 9(2), 159–195 (2001)
14. Shir, O.M., Bäck, T.: Niching with Derandomized Evolution Strategies in Artificial and Real-World Landscapes. Natural Computing: An International Journal (2008)
15. Igel, C., Hansen, N., Roth, S.: Covariance Matrix Adaptation for Multi-objective Optimization. Evolutionary Computation 15(1), 1–28 (2007)
16. Deb, K.: Multi-Objective Optimization Using Evolutionary Algorithms. Wiley, New York (2001)
17. Horn, J., Nafpliotis, N., Goldberg, D.E.: A Niched Pareto Genetic Algorithm for Multiobjective Optimization. In: Conference on Evolutionary Computation (CEC), pp. 82–87. IEEE Service Center, Piscataway (1994)
18. Preuss, M., Naujoks, B., Rudolph, G.: Pareto Set and EMOA Behavior for Simple Multimodal Multiobjective Functions. In: Runarsson, T.P., Beyer, H.-G., Burke, E.K., Merelo-Guervós, J.J., Whitley, L.D., Yao, X. (eds.) PPSN 2006. LNCS, vol. 4193, pp. 513–522. Springer, Heidelberg (2006)
19. Hansen, N., Kern, S.: Evaluating the CMA Evolution Strategy on Multimodal Test Functions. In: Yao, X., Burke, E.K., Lozano, J.A., Smith, J., Merelo-Guervós, J.J., Bullinaria, J.A., Rowe, J.E., Tiňo, P., Kabán, A., Schwefel, H.-P. (eds.) PPSN 2004. LNCS, vol. 3242, pp. 282–291. Springer, Heidelberg (2004)
20. Ehrgott, M.: Multicriteria Optimization, 2nd edn. Springer, Berlin (2005)
21. Emmerich, M., Beume, N., Naujoks, B.: An EMO algorithm using the hypervolume measure as selection criterion. In: Coello Coello, C.A., Hernández Aguirre, A., Zitzler, E. (eds.) EMO 2005. LNCS, vol. 3410, pp. 62–76. Springer, Heidelberg (2005)
22. Emmerich, M.: Single- and Multi-objective Evolutionary Design Optimization Assisted by Gaussian Random Field Metamodels. PhD thesis, University of Dortmund, Germany (2005)
23. Emmerich, M., Deutz, A.: Test problems based on lamé superspheres. In: Obayashi, S., Deb, K., Poloni, C., Hiroyasu, T., Murata, T. (eds.) EMO 2007. LNCS, vol. 4403, pp. 922–936. Springer, Heidelberg (2007)
24. Preuss, M.: Reporting on Experiments in Evolutionary Computation. Technical Report CI-221/07, University of Dortmund, SFB 531 (2007)

Solving Bilevel Multi-Objective Optimization Problems Using Evolutionary Algorithms

Kalyanmoy Deb* and Ankur Sinha

Department of Business Technology
Helsinki School of Economics
PO Box 1210, FIN-00101, Helsinki, Finland
{Kalyanmoy.Deb,Ankur.Sinha}@hse.fi

Abstract. Bilevel optimization problems require every feasible upper-level solution to satisfy optimality of a lower-level optimization problem. These problems commonly appear in many practical problem solving tasks including optimal control, process optimization, game-playing strategy development, transportation problems, and others. In the context of a bilevel single objective problem, there exists a number of theoretical, numerical, and evolutionary optimization results. However, there does not exist too many studies in the context of having multiple objectives in each level of a bilevel optimization problem. In this paper, we address bilevel multi-objective optimization issues and propose a viable algorithm based on evolutionary multi-objective optimization (EMO) principles. Proof-of-principle simulation results bring out the challenges in solving such problems and demonstrate the viability of the proposed EMO technique for solving such problems. This paper scratches the surface of EMO-based solution methodologies for bilevel multi-objective optimization problems and should motivate other EMO researchers to engage more into this important optimization task of practical importance.

1 Introduction

In evolutionary optimization, a few studies have considered solving bilevel programming problems in which an upper level solution is feasible only if it is one of the optimum of a lower level optimization problem. Such problems are found abundantly in practice, particularly in optimal control, process optimization, transportation problems, game playing strategies, reliability based design optimization, and others. In such problems, the lower level optimization task ensures a certain quality or certain physical properties which make a solution acceptable. Often, such requirements come up as equilibrium conditions, stability conditions, mass/energy balance conditions, which are mandatory for any solution to be feasible. For example. in reliability based design optimization, a feasible design must correspond to a certain specified reliability against failures.

* Also Department of Mechanical Engineering, Indian Institute of Technology Kanpur, PIN 208016, India (deb@iitk.ac.in).

M. Ehrgott et al. (Eds.): EMO 2009, LNCS 5467, pp. 110–124, 2009.

Solutions satisfying such conditions or requirements are not intuitive to obtain, rather they often demand an optimization problem to be solved. These essential tasks are posed as lower level optimization tasks in a bilevel optimization framework. The upper level optimization then must search among such reliable, equilibrium or stable solutions to find an optimal solution corresponding to one or more different (higher level) objectives.

Despite the importance of such problems in practice, the difficulty of searching and defining optimal solutions for bilevel optimization problems [7] exists. Despite the lack of theoretical results, there exists a plethora of studies related to bilevel single-objective optimization problems [1,3,12,15] in which both upper and the lower level optimization tasks involve exactly one objective each. Despite having a single objective in the lower level task, usually in such problems the lower level optimization problem has more than one optimum. The goal of a bilevel optimization technique is then to first find the lower level optimal solutions and then search for the optimal solution for the upper level optimization task. In the context of bilevel multi-objective optimization studies, however, there does not exist too many studies using classical methods [8] and none to our knowledge using evolutionary methods, probably because of the added complexities associated with solving each level. In such problems, every lower level optimization problem has a number of trade-off optimal solutions and the task of the upper level optimization algorithm is to focus its search on multiple trade-off solutions which are members of optimal trade-off solutions of lower level optimization problems.

In this paper, we suggest a viable evolutionary multi-objective optimization (EMO) algorithm for solving bilevel problems. We simulate the proposed algorithm on five different test problems, including a bilevel single-objective optimization problem. This proof-of-principle study shows viability of EMO for solving bilevel optimization problems and should encourage other EMO researchers to pay attention to this important class of practical optimization problems.

2 Description of Bilevel Multi-Objective Optimization Problem

A bilevel multi-objective optimization problem has two levels of multi-objective optimization problems such that the optimal solution of the lower level problem determines the feasible space of the upper level optimization problem. In general, the lower level problem is associated with a variable vector $\mathbf{x}_l$ and a fixed vector $\mathbf{x}_u$. However, the upper level problem usually involves all variables $\mathbf{x} = (\mathbf{x}_u, \mathbf{x}_l)$, but we refer here $\mathbf{x}_u$ exclusively as the upper level variable vector. A general bilevel multi-objective optimization problem can be described as follows:

$$\begin{aligned} \text{minimize}_{(\mathbf{x}_u,\mathbf{x}_l)}\ & \mathbf{F}(\mathbf{x}) = (F_1(\mathbf{x}), \ldots, F_M(\mathbf{x})), \\ \text{subject to}\ & \mathbf{x}_l \in \text{argmin}_{(\mathbf{x}_l)} \left\{ \mathbf{f}(\mathbf{x}) = (f_1(\mathbf{x}), \ldots, f_m(\mathbf{x})) \,\middle|\, \mathbf{g}(\mathbf{x}) \geq \mathbf{0}, \mathbf{h}(\mathbf{x}) = \mathbf{0} \right\}, \\ & \mathbf{G}(\mathbf{x}) \geq \mathbf{0}, \mathbf{H}(\mathbf{x}) = \mathbf{0}, \\ & x_i^{(L)} \leq x_i \leq x_i^{(U)}, \quad i = 1, \ldots, n. \end{aligned} \tag{1}$$

In the above formulation, $F_1(\mathbf{x}), \ldots, F_M(\mathbf{x})$ are the upper level objective functions, and $\mathbf{G}(\mathbf{x})$ and $\mathbf{H}(\mathbf{x})$ are upper level inequality and equality constraints, respectively. The objectives $f_1(\mathbf{x}), \ldots, f_m(\mathbf{x})$ are the lower level objective functions, and functions $\mathbf{g}(\mathbf{x})$ and $\mathbf{h}(\mathbf{x})$ are lower level inequality and equality constraints, respectively. Equality constraints are not considered here in both levels, for simplicity. It should be noted that the lower level optimization problem is optimized only with respect to the variables $\mathbf{x}_l$ and the variable vector $\mathbf{x}_u$ is kept fixed. The Pareto-optimal solutions of a lower level optimization problem become feasible solutions to the upper level problem. The Pareto-optimal solutions of the upper level problem are determined by objectives $\mathbf{F}$ and constraints $\mathbf{G}$, and restricting the search among the lower level Pareto-optimal solutions.

3 Classical Approaches

Several studies exist in determining the optimality conditions for a Pareto-optimal solution to the upper level problem. The difficulty arises due to the existence of the lower level optimization problems. Usually the KKT conditions of the lower level optimization problems are used as constraints in formulating the KKT conditions of the upper level problem. As discussed in [7], although KKT optimality conditions can be written mathematically, the presence of many lower level Lagrange multipliers and an abstract term involving coderivatives makes the procedure difficult to be applied in practice.

Fliege and Vicente [9] suggested a mapping concept in which a bilevel single-objective optimization problem can be converted to an equivalent four-objective optimization problem with a special cone dominance concept. Although the idea can be, in principle, extended for bilevel multi-objective optimization problems, the number of objectives to be considered is large and moreover handling constraints seems to introduce additional difficulties in obtaining resulting objectives. However, such an idea is interesting and can be pursued in the future.

In the context of bilevel single-objective optimization problems, there exists a number of studies, including some useful reviews [3,13], test problem generators [1], and even some evolutionary algorithm (EA) studies [12,11,15,10,14]. However, there does not seem to be too many studies on bilevel multi-objective optimization.

A recent study by Eichfelder [8] suggested a refinement based strategy in which the algorithm starts with a uniformly distributed set of points on $\mathbf{x}_u$. Thereafter, for each $\mathbf{x}_u$ vector, the lower level Pareto-optimal solutions are found using a classical generating based optimization method. The set of such points obtained are said to be an approximation of the induced set. Non-dominated points with respect to the upper level problem are chosen from this set and they provide an approximate idea of the desired upper level Pareto-optimal front. Now, the chosen $\mathbf{x}_u$ vectors are refined in their vicinities and the lower level optimizations are repeated till a good approximation of the Pareto-optimal front is obtained. The difficulty with such a technique is that if the dimension of $\mathbf{x}_u$ is high, generating a uniformly spread $\mathbf{x}_u$ vectors and refining the resulting $\mathbf{x}_u$ vector will be computationally expensive. Definitely, an optimization algorithm for simultaneous selection of $\mathbf{x}_u$ and corresponding optimal $\mathbf{x}_l$ vectors will be more effective.

The greatest challenge in handling bilevel optimization problems seems to lie in the fact that unless a solution is optimal for the lower level problem, it cannot be feasible for the overall problem. This requirement, in some sense disallow any approximate optimization algorithm (including an EA or an EMO) to be used for solving the lower level task. But from all practical point of view near-optimal or near-Pareto-optimal solutions are often acceptable and it is in this spirit for which EA and EMO may have a great potential for solving bilevel optimization problems. EA or EMO has another advantage. Unlike the classical point-by-point approach, EA/EMO uses a population of points in their operation. By keeping two interacting populations, a coevolutionary algorithm can be developed so that instead of a serial and complete optimization of lower level problem for every upper level solution, both upper and lower level optimization tasks can be pursued simultaneously through iterations. In the following, we suggest one such procedure.

4 Proposed Procedure (BLEMO)

The proposed method uses the elitist non-dominated sorting GA or NSGA-II [6], however any other EMO procedures can also be used instead. The upper level population (of size N_u) uses NSGA-II operations for T_u generations with upper level objectives ($\mathbf{F}$) and constraints ($\mathbf{G}$) in determining non-dominated rank and crowding distance values of each population member. However, the evaluation of a population member calls a lower level NSGA-II simulation with a population size of N_l for T_l generations. The upper level population has a special feature. The population has $n_s = N_u/N_l$ subpopulations of size N_l each. Each subpopulation has the same $\mathbf{x}_u$ variable vector. To start the proposed BLEMO, we create all solutions at random, but maintain the above structure. From thereon, the proposed operations ensure that the above-mentioned structure is maintained from one generation to another. In the following, we describe one iteration of the proposed BLEMO procedure. At the start of the upper level NSGA-II generation t, we have a population P_t of size N_u. Every population member has the following quantities computed from the previous iteration: (i) a non-dominated rank ND_u corresponding to $\mathbf{F}$ and $\mathbf{G}$, (ii) a crowding distance value CD_u corresponding to $\mathbf{F}$, (iii) a non-dominated rank ND_l corresponding to $\mathbf{f}$ and $\mathbf{g}$, and (iv) a crowding distance value CD_l using $\mathbf{f}$. For every subpopulation in the upper level population, members having the best non-domination rank (ND_u) are saved as an 'elite set' which will be used in the recombination operator in the lower level optimization task of the same subpopulation.

Step 1: Apply a pair of binary tournament selections on members ($\mathbf{x} = (\mathbf{x}_u, \mathbf{x}_l)$) of P_t using ND_u and CD_u lexicographically. The upper level variable vectors $\mathbf{x}_u$ of two selected parents are then recombined using the SBX operator [5] to obtain two new vectors of which one is chosen at random. The chosen solution is mutated by the polynomial mutation operator [4] to obtain a child vector (say, $\mathbf{x}_u^{(1)}$). We then create N_l new lower level variable vectors

$\mathbf{x}_l^{(i)}$ by applying selection-recombination-mutation operations on entire P_t. Thereafter, N_l child solutions are created by concatenating upper and lower level variable vectors together, as follows: $c_i = (\mathbf{x}_u^{(1)}, \mathbf{x}_l^{(i)})$ for $i = 1, \ldots, N_l$. Thus, for every new upper level variable vector, a subpopulation of N_l lower level variable vectors are created by genetic operations from P_t. The above procedure is repeated for a total of n_s new upper level variable vectors.

Step 2: For each subpopulation of size N_l, we now perform a NSGA-II procedure using lower level objectives ($\mathbf{f}$) and constraints ($\mathbf{g}$) for T_l generations. It is interesting to note that in each lower level NSGA-II, the upper level variable vector $\mathbf{x}_u$ is not changed. For every mating, one solution is chosen as usual using the binary tournament selection using a lexicographic use of ND_l and CD_l, but the second solution is always chosen randomly from the 'elite set'. The mutation is performed as usual. After the lower level NSGA-II simulation is performed for a subpopulation, the resulting solutions are marked with their non-dominated rank (ND_l) and crowding distance value (CD_l). All N_l members from each subpopulation are then combined together in one population (the child population, Q_t). It is interesting to note that in Q_t, there are at least n_s members having $ND_l = 1$ (at least one coming from each subpopulation). Also, in Q_t, there are exactly n_s different $\mathbf{x}_u$ variable vectors.

Step 3: Each member of Q_t is now evaluated with $\mathbf{F}$ and $\mathbf{G}$. Populations P_t and Q_t are combined together to form R_t. The combined population R_t is then ranked according to non-domination and members within an identical non-dominated rank are assigned a crowding distance computed in the $\mathbf{F}$ space. Thus, each member of Q_t gets a upper level non-dominated rank ND_u and a crowding distance value CD_u.

Step 4: From the combined population R_t of size $2N_u$, half of its members are chosen in this step. First, the members of rank $ND_u = 1$ are considered. From them, solutions having $ND_l = 1$ are noted one by one in the order of reducing crowding distance CD_u, for each such solution the entire N_l subpopulation from its source population (either P_t or Q_t) is copied in an intermediate population S_t. If a subpopulation is already copied in S_t and a future solution from the same subpopulation is found to have $ND_u = ND_l = 1$, the subpopulation is not copied again. When all members of $ND_u = 1$ are considered, a similar consideration is continued with $ND_u = 2$ and so on till exactly n_s subpopulations are copied in S_t.

Step 5: Each subpopulation of S_t is now modified using the lower level NSGA-II procedure applied with $\mathbf{f}$ and $\mathbf{g}$ for T_l generations. This step helps progress each lower level populations towards their individual Pareto-optimal frontiers.

Step 6: Finally, all subpopulations obtained after the lower level NSGA-II simulations are combined together to form the next generation population P_{t+1}.

The evaluation of the initial population is similar to the above. First, members of P_0 are created at random with n_s subpopulations, each having an identical $\mathbf{x}_u$ vector for all its subpopulation members. Thereafter, each subpopulation is sent for an update of $\mathbf{x}_l$ vectors to the lower level NSGA-II (with $\mathbf{f}$ and $\mathbf{g}$) for T_l generations. Every member is assigned corresponding ND_l and CD_l values. The resulting subpopulations (from NSGA-II) are combined into one population (renamed as P_0) and evaluated using $\mathbf{F}$ and $\mathbf{G}$. Every member is then assigned a non-dominated rank ND_u and a crowding distance value CD_u.

The good solutions of every generation are saved in an archive (A_t). Initially, the archive A_0 is an empty set. Thereafter, at the end of every upper level generation, solutions having both $ND_u = 1$ and $ND_l = 1$ from P_t is saved in the archive A_t. The non-dominated solutions (with $\mathbf{F}$ and $\mathbf{G}$) of the archive are kept in A_t and rest members are deleted from the archive.

In the above BLEMO, we have used a simple termination rule based on specified number of generations for both lower and upper level tasks. Every lower level task for each subpopulation requires $N_l(T_l + 1)$ solution evaluations and since there are n_s subpopulations in every generation, this requires $n_s N_l(T_l + 1)$ or $N_u(T_l + 1)$ solution evaluations in Step 2. In the initial population evaluation, a final upper level objective and constraint evaluation of N_u is required, but since a solution evaluation refers to both upper and lower level evaluations, this N_u is not extra. For any other generation, Step 5 above requires another $N_u(T_l + 1)$ solution evaluations, thereby totaling $2N_u(T_l + 1)$ solution evaluations. Thus, considering evaluations involved in all generations from $t = 0$ till $t = T_u$, total solution evaluations needed are $N_u(2T_u + 1)(T_l + 1)$.

5 Test Problems and Pareto-Optimal Solutions

In the context of bilevel single-objective optimization, there exists some studies [3,1] which suggest linear, quadratic and transport related problems. However, to our knowledge, there does not exist any systematic study suggesting test problems for bilevel multi-objective optimization. In this study, we borrow a couple of problems used in [8] and suggest a small and a large-dimensional version of a new test problem.

5.1 Problem 1

Problem 1 has a total of three variables with x_1, x_2 belonging to $\mathbf{x}_l$ and y belonging to x_u and is taken from [8]:

$$
\begin{aligned}
&\text{minimize } \mathbf{F}(\mathbf{x}) = \begin{Bmatrix} x_1 - y \\ x_2 \end{Bmatrix}, \\
&\text{subject to } (x_1, x_2) \in \operatorname{argmin}_{(x_1,x_2)} \left\{ \mathbf{f}(\mathbf{x}) = \begin{pmatrix} x_1 \\ x_2 \end{pmatrix} \middle| g_1(\mathbf{x}) = y^2 - x_1^2 - x_2^2 \geq 0 \right\}, \\
&\qquad G_1(\mathbf{x}) = 1 + x_1 + x_2 \geq 0, \\
&\qquad -1 \leq x_1, x_2 \leq 1, \quad 0 \leq y \leq 1.
\end{aligned}
\tag{2}
$$

Both the lower and the upper level optimization tasks have two objectives each. A little consideration will reveal that for a fixed y value, the feasible region of the lower-level problem is the area inside a circle with center at origin ($x_1 = x_2 = 0$) and radius equal to y. The Pareto-optimal set for the lower-level optimization task for a fixed y is the bottom-left quarter of the circle:

$$\{(x_1, x_2) \in \mathbb{R}^2 \quad | \quad x_1^2 + x_2^2 = y^2, x_1 \leq 0, x_2 \leq 0\}.$$

The linear constraint in the upper level optimization task does not allow the entire quarter circle to be feasible for some y. Thus, at most a couple of points from the quarter circle belongs to the Pareto-optimal set of the overall problem. Eichfelder [8] reported the following Pareto-optimal solutions for this problem:

$$\mathbf{x}^* = \left\{(x_1, x_2, y) \in \mathbb{R}^3 \quad \middle| \quad x_1 = -1 - x_2, x_2 = -\frac{1}{2} \pm \frac{1}{4}\sqrt{8y^2 - 4}, y \in \left[\frac{1}{\sqrt{2}}, 1\right]\right\}. \tag{3}$$

The Pareto-optimal front in F_1-F_2 space is given in parametric form, as follows:

$$\left\{(F_1, F_2) \in \mathbb{R}^2 \quad \middle| \quad F_1 = -1 - F_2 - t, F_2 = -\frac{1}{2} \pm \frac{1}{4}\sqrt{8t^2 - 4}, t \in \left[\frac{1}{\sqrt{2}}, 1\right]\right\}. \tag{4}$$

Figure 1 shows the Pareto-optimal front of problem 1. Lower level Pareto-optimal fronts of some representative y values are also shown on the figure, indicating that at most two such Pareto-optimal solutions (such as points B and C for $y = 0.9$) of a lower level optimization problem becomes the candidate Pareto-optimal solutions of the upper level problem. It is interesting to note that in this problem there exists a number of lower level Pareto-optimal solutions (such as solution A marked in the figure) which are infeasible to the upper level task. Thus, if the lower level optimization is unable to find critical Pareto-optimal solutions (such as B or C) which correspond to the upper level Pareto-optimal solutions, but finds solutions like A in most occasions, the lower level task becomes useless. This makes the bilevel optimization task challenging and difficult.

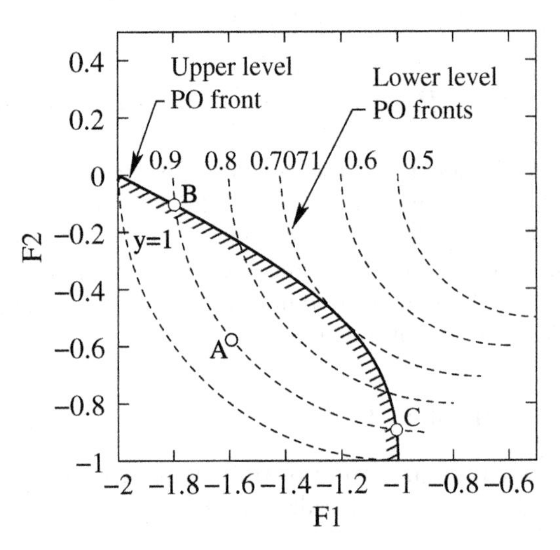

Fig. 1. Pareto-optimal fronts of upper level (complete problem) and some representative lower level optimization tasks are shown for problem 1

5.2 Problem 2

This problem is also taken from [8]:

$$
\begin{array}{ll}
\text{minimize} & \mathbf{F}(\mathbf{x}) = \left\{ \begin{array}{c} x_1 + x_2^2 + y + \sin^2(x_1 + y) \\ \cos(x_2)(0.1 + y)(\exp(-\frac{x_1}{0.1+x_2}) \end{array} \right\}, \\
\text{subject to} & \\
(x_1, x_2) \in & \left\{ \begin{array}{l} \text{argmin}_{(x_1,x_2)} \mathbf{f}(\mathbf{x}) = \left(\begin{array}{c} \frac{(x_1-2)^2+(x_2-1)^2}{4} + \frac{x_2 y+(5-y_1)^2}{16} + \sin(\frac{x_2}{10}) \\ \frac{x_1^2+(x_2-6)^4-2x_1 y_1-(5-y_1)^2}{80} \end{array} \right) \Bigg| \\ g_1(\mathbf{x}) = x_2 - x_1^2 \geq 0, \\ g_2(\mathbf{x}) = 10 - 5x_1^2 - x_2 \geq 0, \\ g_3(\mathbf{x}) = 5 - \frac{y}{6} - x_2 \geq 0, \\ g_4(\mathbf{x}) = x_1 \geq 0. \end{array} \right\}, \\
G_1(\mathbf{x}) \equiv 16 - (x_1 - 0.5)^2 - (x_2 - 5)^2 - (y - 5)^2 \geq 0, & \\
0 \leq x_1, x_2, y \leq 10. &
\end{array}
\tag{5}
$$

For this problem, the exact Pareto-optimal front of the lower or the upper level optimization problem is difficult to derive mathematically. The previous study [8] did not report the true Pareto-optimal front, instead presented a front through their obtained results.

5.3 Problem 3

Next, we create a simplistic bilevel two-objective optimization problem, having $\mathbf{x}_l = (x_1, x_2)$ and $\mathbf{x}_u = (y)$:

$$
\begin{array}{l}
\text{minimize } \mathbf{F}(\mathbf{x}) = \left\{ \begin{array}{c} (x_1 - 1)^2 + x_2^2 + y^2 \\ (x_1 - 1)^2 + x_2^2 + (y - 1)^2 \end{array} \right\}, \\
\text{subject to } (x_1, x_2) \in \text{argmin}_{(x_1,x_2)} \left\{ \mathbf{f}(\mathbf{x}) = \left(\begin{array}{c} x_1^2 + x_2^2 \\ (x_1 - y)^2 + x_2^2 \end{array} \right) \right\}, \\
\quad -1 \leq x_1, x_2, y \leq 2.
\end{array}
\tag{6}
$$

For a fixed value of y, the Pareto-optimal solutions of the lower level optimization problem are given as follows: $\{(x_1, x_2) \in \mathbb{R}^2 | x_1 \in [0, y], x_2 = 0\}$. For example, for $y = 0.75$, Figure 2 shows these solutions (points A through B) in the F_1-F_2 space. The points lie on a straight line and are not conflicting to each other. Thus, only one point (point A with $x_1 = y = 0.75$ and $x_2 = 0$) is a feasible solution to the upper level optimization task for a fixed $y = 0.75$. Interestingly, for a fixed y, the bottom-left boundary of the F_1-F_2 space corresponds to the upper bound of x_1 or $x_1 = 1$. However, solutions having $x_1 = 1$ till $x_1 = y$ are not Pareto-optimal for the overall problem. For $y = 0.75$, solutions on line CA (excluding A) are not Pareto-optimal to both lower and upper level problems. Similarly solutions from B upwards on the '$y = 0.75$' line are also not Pareto-optimal for both levels.

When we plot all solutions for which $x_1 = y$ and $x_2 = 0$, we obtain the dotted line marked with '$x_1 = y$' in the figure. Different lower level Pareto-optimal fronts (for different y values) are shown in the figure with dashed straight lines. It is interesting to note that all solutions on this '$x_1 = y$' curve are not Pareto-optimal to the overall problem. By investigating the figure, we observe that the

Pareto-optimal solutions to the upper-level problem corresponds to following solutions: $\{(x_1, x_2, y) \in \mathbb{R}^3 | x_1 = y, x_2 = 0, y \in [0.5, 1.0]\}$. This problem does not have any constraint in its lower or upper level. If an algorithm fails to find true Pareto-optimal solutions of a lower level problem and ends up finding a solution below the '$x_1 = y$' curve, such as solution C, it can potentially dominate a true Pareto-optimal point (such as point A) thereby making the task of finding true Pareto-optimal solutions difficult.

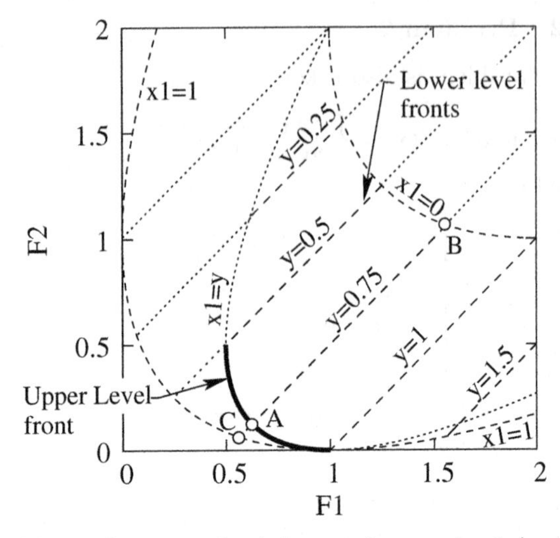

Fig. 2. Pareto-optimal fronts of upper level (complete problem) and some representative lower level optimization tasks are shown for problem 3

5.4 Problem 4

In this problem, we increase the dimension of the variable vector of problem 3 by adding more variables like x_2:

$$\begin{aligned}
&\text{minimize} \quad \mathbf{F}(\mathbf{x}) = \left\{ \begin{array}{c} (x_1-1)^2 + \sum_{i=1}^{K} x_{i+1}^2 + y^2 \\ (x_1-1)^2 + \sum_{i=1}^{K} x_{i+1}^2 + (y-1)^2 \end{array} \right\}, \\
&\text{subject to} \\
&(x_1, x_2, \ldots, x_{K+1}) \in \text{argmin}_{(x_1, x_2, \ldots, x_{K+1})} \left\{ \mathbf{f}(\mathbf{x}) = \left(\begin{array}{c} x_1^2 + \sum_{i=1}^{K} x_{i+1}^2 \\ (x_1 - y)^2 + \sum_{i=1}^{K} x_{i+1}^2 \end{array} \right) \right\}, \\
&-1 \leq x_1, x_2, \ldots, x_{K+1}, y \leq 2.
\end{aligned} \tag{7}$$

This problem has an identical Pareto-optimal front as in problem 3 with $x_i = 0$ for $i = 2, \ldots, (K+1)$, $x_1 = y$ and $y \in [0.5, 1]$. In our simulation here, we use $K = 13$, so that total number of variables is 15.

5.5 Problem 5

To test the proposed BLEMO procedure for bilevel single objective optimization problems, we include one problem from the literature [2] having $\mathbf{x}_l = (x_1, x_2, x_3, x_4)^T$ and $\mathbf{x}_u = (y_1, y_2, y_3, y_4)^T$:

$$
\begin{aligned}
&\text{minimize } F(\mathbf{x}) = -(200 - x_1 - x_2)(x_1 + x_3) - (160 - x_2 - x_4)(x_2 + x_4),\\
&\text{subject to}\\
&\qquad \mathbf{x}_l \in \text{argmin}_{(\mathbf{x}_l)} \{\mathbf{f}(\mathbf{x}) = (x_1 - 4)^2 + (x_2 - 13)^2 + (x_3 - 35)^2 + (x_4 - 2)^2 |\\
&\qquad\quad g_1(\mathbf{x}) = 0.4x_1 + 0.7x_2 \le y_1,\ g_2(\mathbf{x}) = 0.6x_1 + 0.3x_2 \le y_2,\\
&\qquad\quad g_3(\mathbf{x}) = 0.4x_3 + 0.7x_4 \le y_3,\ g_4(\mathbf{x}) = 0.6x_3 + 0.3x_4 \le y_4\},\\
&\qquad G(\mathbf{x}) = y_1 + y_2 + y_3 + y_4 \le 40,\\
&\qquad 0 \le y_1 \le 10,\ 0 \le y_2 \le 5,\ 0 \le y_3 \le 15, 0 \le y_4 \le 20,\\
&\qquad 0 \le x_1 \le 20,\ 0 \le x_2 \le 20,\ 0 \le x_3 \le 40,\ 0 \le x_4 \le 40.
\end{aligned} \tag{8}
$$

The reported solution to this problem [2] is $\mathbf{x}_u^* = (7.36, 3.55, 11.64, 17.45)^T$ and $\mathbf{x}_l^* = (0.91, 10, 29.09, 0)^T$ with $F(\mathbf{x}^*) = -6600.0$ and $f(\mathbf{x}^*) = 57.48$.

6 Proof-of-Principle Results

We use the following parameter settings: $N_u = 400$, $T_u = 200$, $N_l = 40$, and $T_l = 40$. Since lower level search is made interacting with the upper level search, we have run lower level optimization algorithm for a fewer generations and run the upper level simulations longer. The other NSGA-II parameters are set as follows: for SBX crossover, $p_c = 0.9$, $\eta_c = 15$ [5] and for polynomial mutation operator, $p_m = 0.1$, and $\eta_m = 20$ [4].

6.1 Problem 1

Figure 3 shows the obtained solutions using proposed BLEMO. It is clear that the obtained solutions are very close to the theoretical Pareto-optimal solutions of this problem. The lower boundary of the objective space is also shown to indicate that although solutions could have been found lying between the theoretical front and the boundary and dominate the Pareto-optimal points, BLEMO is able to avoid such solutions and find solutions very close to the Pareto-optimal solutions. Also, BLEMO is able to find a good spread of solutions on the entire range of true Pareto-optimal front. Figure 4 shows the variation of $\mathbf{x}$ for these solutions. It is clear that all solutions are close to being on the upper level constraint $G(\mathbf{x})$ boundary ($x_1 + x_2 = -1$) and they follow the relationship depicted in equation 3.

6.2 Balancing Computations between Lower and Upper Levels

For a fixed overall population size N_u, the number of solution evaluations depends on the product $(2T_u + 1)(T_l + 1)$. Thus, a balance between T_u and T_l is needed for the overall BLEMO to work well. A too large T_l will ensure near Pareto-optimality of lower level solutions (thereby satisfying the upper level constraint better), but this will be achieved only at the expense of not having adequate upper level generations. On the other hand, a too small value of T_l means inadequate generations for the lower level task for getting close to Pareto-optimal fronts. To understand the effect of this balance between lower level and

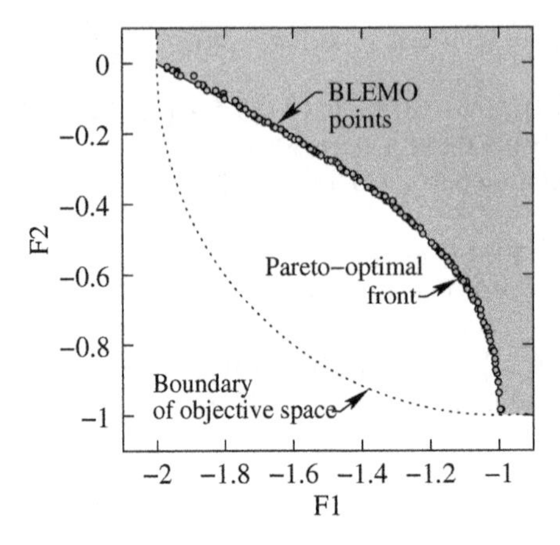

Fig. 3. BLEMO Results for problem 1

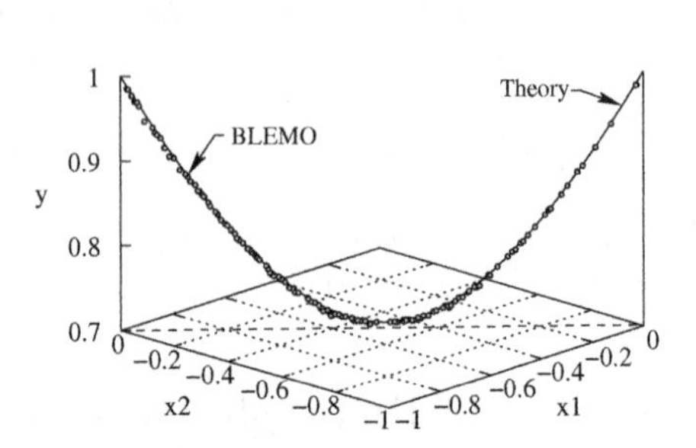

Fig. 4. Variable values of obtained solutions for problem 1. BLEMO solutions are close to theoretical results.

upper level computational efforts, we perform a number of simulations of our algorithm for different T_u-T_l combinations by keeping the overall solution evaluation constant. Table 1 shows the hypervolume values computed for four other T_u-T_l combinations. To not consider the effect of any non-Pareto-optimal solutions, we eliminate all solutions which lie below the theoretical Pareto-optimal curve before we compute the hypervolume. The reference point used in calculating the hypervolume is $(-1,0)^T$. The combination $T_l = 40$ and $T_u = 200$ seems to perform the best. It is clear that hypervolume degrades with an increase in T_l from 40. To keep the solution evaluations the same as before, T_u must be reduced for an increase in T_l. The use of smaller number of upper level generations is detrimental to the overall algorithm. On the other hand, when a smaller T_l (=20) is used, the performance degrades marginally, due to reduced number of lower level generations which did not allow lower level solutions to reach close to their Pareto-optimal sets.

Table 1. Hypervolume values obtained by different T_u-T_l combinations on problem 1

T_l	T_u	Hypervolume
20	391	0.29851
40	**200**	**0.30268**
100	81	0.29716
200	41	0.28358
400	20	0.23796

Table 2. Hypervolume values obtained by different T_u-T_l combinations on problem 2

T_l	T_u	Hypervolume
20	391	0.45034
40	**200**	**0.49256**
100	81	0.47164
200	41	0.46157
400	20	0.43145

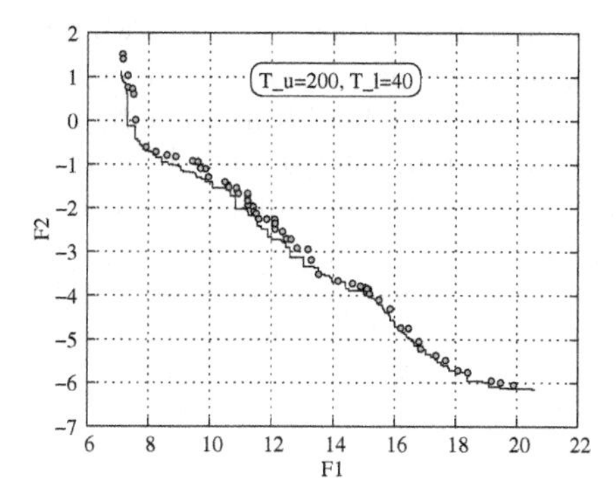

Fig. 5. Results obtained using BLEMO for problem 2

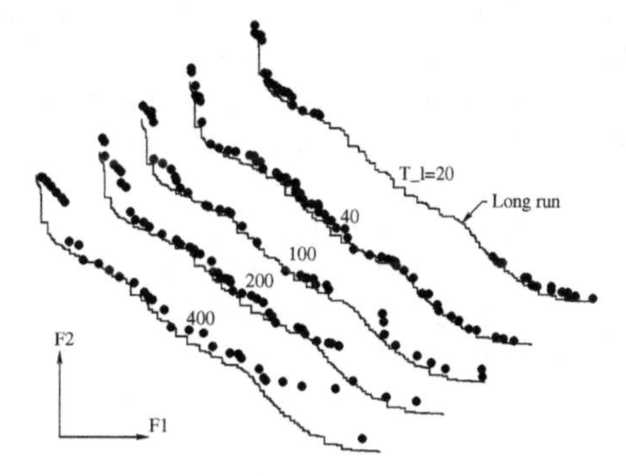

Fig. 6. Results for different T_u-T_l combinations for problem 2

6.3 Problem 2

This problem is more complex than the problem 1 involving non-linearities and periodic functions. We use $N_u = 600$ and $N_l = 60$ for this problem, but use identical termination conditions on generations as before. The number of sub-population is also the same as in problem 1 and is equal to 600/60 or 10. Figure 5 shows the obtained non-dominated points. For this problem, the exact Pareto-optimal front is not known, but our front is similar to that reported in the previous study [8]. We have also plotted the solutions found by a simulation of the proposed algorithm which is run for an exorbitantly long number of generations ($N_u = 2,000$, $T_u = 400$, $N_l = 100$, $T_l = 100$). Although, our limited generation results are not exactly the same as this 'long run', the solutions are close.

Table 2 tabulates the hypervolumes obtained using different T_u-T_l combinations. In this case also, we remove all the points which are below the F_1-F_2 points found by the 'long run'. Again, our setting of $T_u = 200$ and $T_l = 40$ is found to perform the best in terms of the hypervolume measure. Figure 6 shows the obtained solutions of different T_u-T_l combinations with respect to the 'long run' (shown in a solid line). In each case, the lower level non-Pareto-optimal solutions which are below the 'long run' front are deleted from the final front owing to being non-Pareto-optimal in the lower level. The distribution and convergence get worse with an increasing T_l value. For $T_l = 20$, there were too many solutions which were below the 'long run', simply because these solutions were not close to Pareto-optimal front of the corresponding lower level problem. However, 40 generations for the lower level search seems adequate with $T_u = 200$ in this problem as well.

6.4 Problem 3

Figure 7 shows the obtained BLEMO points on problem 3. Although solutions in between this front and the feasible boundary of objective space could have been found for an apparently better non-dominated front, these solutions would be non-Pareto-optimal with respect to the lower level problems and our algorithm has succeeded in eliminating them to appear on the final front. The figure shows that BLEMO is able to find a good distribution of solutions on the entire range of the true Pareto-optimal front. Figure 8 shows that for obtained optimal solutions, the relationship $y = x_1$ in the range $x_1 \in [0.5, 1]$ holds. Additionally, we observed that $x_2 = 0$ for all obtained solutions. These observations match with the theoretical Pareto-optimal solutions on this problem discussed in the previous section.

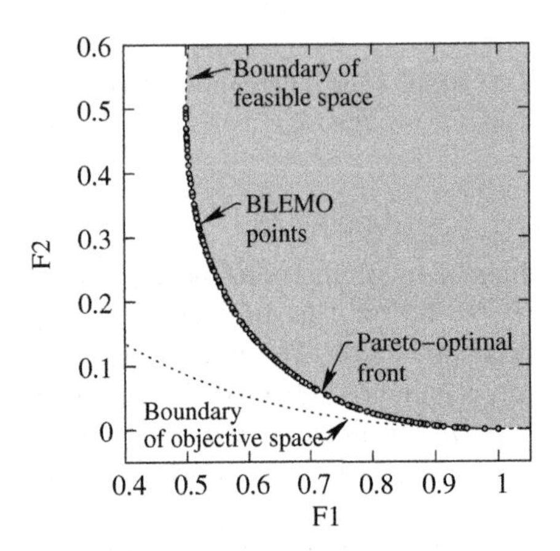

Fig. 7. Results obtained using BLEMO for problem 3

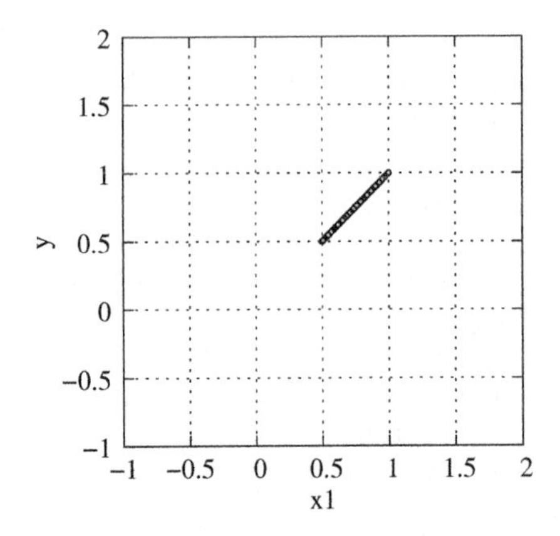

Fig. 8. Variable values of obtained solutions for problem 3

6.5 Problem 4

In this problem, we have 15 variables. Figure 9 shows the obtained BLEMO solutions. An identical Pareto-optimal front to that in problem 3 is obtained here. For all solutions, we observed that $x_i = 0$ for $i = 2, \ldots, 14$. Again, $y = x_1$ in the range $x_1 \in [0.5, 1]$ relationship is obtained for BLEMO solutions.

6.6 Problem 5

For this problem, we have chosen the following parameter setting: $N_u = 400$, $T_u = 40$, $N_l = 40$ and $T_l = 40$. The obtained solution has $F^* = -6599.996$ and $f^* = 57.441$ with variable vectors $\mathbf{x}_l^* = (0.9125, 9.9996, 29.0918, 0.0002)^T$ and $\mathbf{x}_u^* = (7.3601, 3.5516, 11.6400, 17.4520)^T$. With two decimal places of accuracy, this solution is identical to that in [2]. Figure 10 shows the best and average $F(\mathbf{x})$ value with the generation counter. This problem shows that the proposed BLEMO is able to degenerate its multi-objective operations to suit the solution of a bilevel single objective optimization problem.

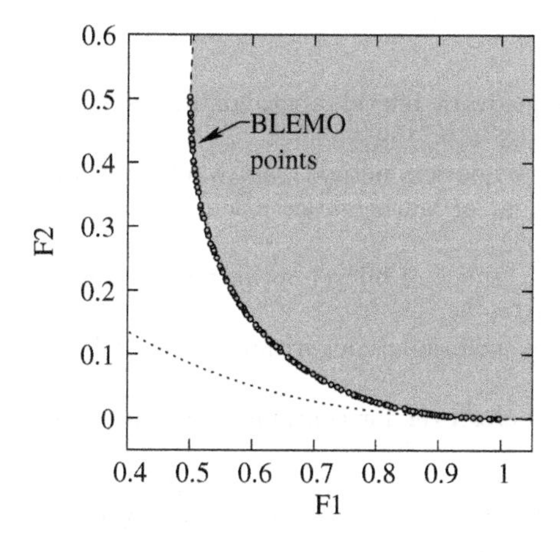

Fig. 9. Results obtained using BLEMO for problem 4

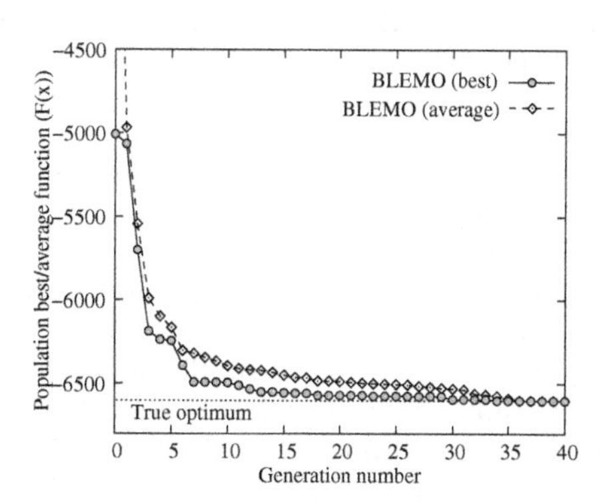

Fig. 10. Average and best function values for $F(\mathbf{x})$ are plotted with generation for problem 5

7 Conclusions

In this paper, we have proposed and simulated a bilevel evolutionary multi-objective optimization (BLEMO) algorithm based on NSGA-II applied to both levels. To coordinate the processing of populations between upper and lower levels we have maintained subpopulations having identical upper level variable values. Although any feasible solution on the upper level must correspond to the Pareto-optimal solutions of the corresponding lower level optimization problem, through simulation studies on a number of problems we have shown that the proposed iterative upper and lower level population processing strategy is able to steer the search close to the correct Pareto-optimal set of the overall problem. In this direction, we have argued and shown through a systematic parametric simulation study that a proper balance between the extent of lower and upper level generations is an important matter for an efficient use of the proposed procedure. Interestingly, we have also shown that the proposed multi-objective algorithm is also able to solve bilevel single-objective optimization problems.

This study has shown one viable way of using an existing EMO methodology for handling bilevel optimization problems. Many other ideas are certainly possible. Hopefully, this study will spur the interest in handling bilevel multi-objective optimization problems more to other interested researchers and practitioners.

Acknowledgments

Authors wish to thank Academy of Finland and Foundation of Helsinki School of Economics (under grant 118319) for their support of this study.

References

1. Calamai, P.H., Vicente, L.N.: Generating quadratic bilevel programming test problems. ACM Trans. Math. Software 20(1), 103–119 (1994)
2. Colson, B.: Bilevel programming with approximation methods: Software guide and test problems. Technical report, Departement of Mathematics, Facultés Universitaires Notre-Dame de la Paix, Brussels (2002)
3. Colson, B., Marcotte, P., Savard, G.: An overview of bilevel optimization. Annals of Operational Research 153, 235–256 (2007)
4. Deb, K.: Multi-objective optimization using evolutionary algorithms. Wiley, Chichester (2001)
5. Deb, K., Agrawal, R.B.: Simulated binary crossover for continuous search space. Complex Systems 9(2), 115–148 (1995)
6. Deb, K., Agrawal, S., Pratap, A., Meyarivan, T.: A fast and elitist multi-objective genetic algorithm: NSGA-II. IEEE Transactions on Evolutionary Computation 6(2), 182–197 (2002)
7. Dempe, S., Dutta, J., Lohse, S.: Optimality conditions for bilevel programming problems. Optimization 55(56), 505–524 (2006)
8. Eichfelder, G.: Soving nonlinear multiobjective bilevel optimization problems with coupled upper level constraints. Technical Report Preprint No. 320, Preprint-Series of the Institute of Applied Mathematics, Univ. Erlangen-Nrnberg, Germany (2007)
9. Fliege, J., Vicente, L.N.: Multicriteria approach to bilevel optimization. Journal of Optimization Theory and Applications 131(2), 209–225 (2006)
10. Koh, A.: Solving transportation bi-level programs with differential evolution. In: 2007 IEEE Congress on Evolutionary Computation (CEC 2007), pp. 2243–2250. IEEE Press, Los Alamitos (2007)
11. Mathieu, R., Pittard, L., Anandalingam, G.: Genetic algorithm based approach to bi-level linear programming. Operations Research 28(1), 1–21 (1994)
12. Oduguwa, V., Roy, R.: Bi-level optimisation using genetic algorithm. In: Proceedings of the 2002 IEEE International Conference on Artificial Intelligence Systems (ICAIS 2002), pp. 322–327 (2002)
13. Vicente, L.N., Calamai, P.H.: Bilevel and multilevel programming: A bibliography review. Journal of Global Optimization 5(3), 291–306 (2004)
14. Wang, Y., Jiao, Y.-C., Li, H.: An evolutionary algorithm for solving nonlinear bilevel programming based on a new constraint-handling scheme. IEEE Transactions on Systems, Man, and Cybernetics, Part C: Applications and Reviews 35(2), 221–232 (2005)
15. Yin, Y.: Genetic algorithm based approach for bilevel programming models. Journal of Transportation Engineering 126(2), 115–120 (2000)

Application of MOGA Search Strategy to SVM Training Data Selection

Tomoyuki Hiroyasu[1], Masashi Nishioka[2], Mitsunori Miki[3], and Hisatake Yokouchi[1]

[1] Faculty of Life and Medical Sciences, Doshisha University, 1-3 Tatara Miyakodani Kyotanabe, Kyoto, Japan
tomo@is.doshisha.ac.jp, hyokouch@mail.doshisha.ac.jp
[2] Graduate School of Engineering, Doshisha University
mnishioka@mikilab.doshisha.ac.jp
[3] Faculty of Science and Engineering, Doshisha University
mmiki@mail.doshisha.ac.jp

Abstract. When training Support Vector Machine (SVM), selection of a training data set becomes an important issue, since the problem of overfitting exists with a large number of training data. A user must decide how much training data to use in the training, and then select the data to be used from a given data set. We considered to handle this SVM training data selection as a multi-objective optimization problem and applied our proposed MOGA search strategy to it. It is essential for a broad set of Pareto solutions to be obtained for the purpose of understanding the characteristics of the problem, and we considered the proposed search strategy to be suitable. The results of the experiment indicated that selection of the training data set by MOGA is effective for SVM training.

1 Introduction

Support Vector Machine (SVM) is a pattern classification technique introduced by V. Vapnik et al. [1]. The basic idea of SVM is to map an input vector x into a high dimensional feature space H by Φ and construct an optimal separating hyperplane in this space [2]. SVM has been applied to various pattern recognition cases, such as digit recognition [1], text categorization [3], and face detection [4].

The goal of SVM is to achieve the best generalization performance by learning on a given training data set. For this purpose, it is important for the problem of overfitting [5] to be considered in the training of SVM. Generally, all examples in the training data set are treated equally and used in the training process of SVM, however there are examples with more information or those that can be misleading. Therefore, there are existing researches on data categorization to group examples based on their usefulness [6].

With this background, we considered to handle the selection of training data set for SVM as a multi-objective optimization problem, and solve it by applying multi-objective genetic algorithms (MOGAs). There are researches on application of multi-objective optimization to SVM in forms of evolutionary SVM [5].

M. Ehrgott et al. (Eds.): EMO 2009, LNCS 5467, pp. 125–139, 2009.

As these researches focus on the learning mechanism of SVM, our approach is to optimize the training data set using MOGAs.

There are two objectives to this training data selection problem, which are training error and the confidence margin [6]. Training error is to be minimized, and confidence margin is to be maximized in this case. Minimization of the training error is likely to result in overfitting, whereas maximizing the confidence margin prevents overfitting, and a trade-off relationship is expected between these two objectives. One characteristic of this training data selection problem is the importance of the optimal solution of each objective. In order to provide a decision maker with good understanding of the trade-off relationship between two objectives, extreme solutions must be included in the final Pareto solutions.

There are many multi-objective genetic algorithms (MOGAs) developed to date [7, 8, 9, 10, 11] with the purpose to find Pareto optimal solutions. In multi-objective optimization, it is desirable for the obtained solutions to be high quality regarding accuracy, uniform distribution, and broadness. Accuracy is how close the obtained solutions are to the true Pareto front, and uniform distribution is how evenly located the solutions are without concentrating in certain areas. Broadness is how widespread the solutions are and is decided by the optimal solutions of each objective located at the edge of the Pareto front.

Many MOGAs have mechanisms to improve accuracy and uniform distribution of the solutions. However, not many mechanisms are available to improve broadness of the solutions. NSGA-II [10] and SPEA2 [11] are two well-known MOGAs today, but both algorithms only have mechanisms included to preserve the obtained broadness of the solutions. Same can be said about other algorithms as well, and few are capable of improving broadness of the solutions.

As formerly mentioned, it is important for broad solutions to be obtained when understanding characteristics of the optimization problem. For this reason, Okuda et al. proposed the Distributed Cooperation Scheme [12], which utilizes single-objective GA (SOGA) along with MOGA. SOGA is utilized to search for the optimal solutions of each objective, which leads to improvement of broadness. It was confirmed that the Distributed Cooperation Scheme is capable of deriving broader solutions than conventional MOGAs. However, preliminary experiments have also indicated that the convergence speed is reduced because the solutions are broadened from the beginning of the search.

Because it is difficult to simultaneously improve convergence and broadness of the solutions, we consider dividing the search into two phases in our proposed search strategy. The first phase in the proposed search strategy is to improve convergence of the solutions, and the second is to improve the broadness of the solutions. This search strategy is capable of deriving broader solutions compared to conventional MOGAs without deterioration of accuracy. Therefore, we consider applying this search strategy to the selection of SVM training data.

In this paper, we first introduce our proposed search strategy consisting of two search phases. The search strategy is tested on test problems to verify its performance compared to conventional MOGAs. Then we adapted this search strategy to SVM training data selection problem.

2 Search Strategy for Multi-objective Genetic Algorithm with Consideration of Accuracy and Broadness

2.1 Importance of Broadness

The search strategy we propose considers accuracy and broadness of the solutions. Although, conventional MOGAs attempt to derive Pareto optimal solutions, there are not many mechanisms to improve broadness of the solutions. Lack of broadness becomes a problem especially in real-world optimization problems where a decision maker selects a solution based upon the given solutions. It is important to understand the possible range of solutions for the problem, and deriving solutions in a limited portion of the Pareto front is not enough. Obtained solutions cannot be considered to be as broad as possible without a mechanism to actively improve broadness. Therefore, it is essential for a broadness improving mechanism to be included in the search strategy.

The proposed search strategy consists of two search phases as shown in Fig. 1. The first phase is a search to improve convergence, and the second is for broadness. The search phases are in this order, as the final solutions obtained must be comparable to conventional MOGAs regarding accuracy and also be broad. Especially, in cases where the search time is limited, it becomes important to ensure the accuracy of the solutions first.

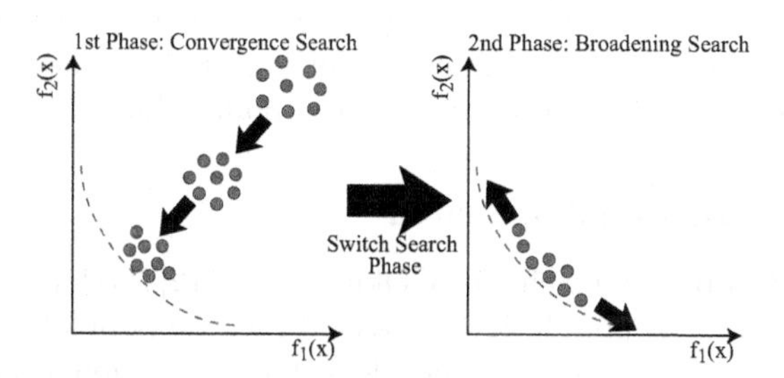

Fig. 1. Concept of the Proposed Search Strategy

2.2 1st Phase: Convergence Search

In the convergence search, preference of a decision maker is adopted in form of a reference point [13]. This reference point can be located in both feasible and infeasible regions. Conventional MOGAs base their search on the dominance relationship of the solutions, but the proposed search method bases its search on the distance information. That is, solutions closer to the reference point are prioritized in the search, which leads to convergence of the solutions around the reference point. The concept of this search is illustrated in Fig. 2.

The proposed search strategy is based on conventional MOGAs, and the distance information is utilized in the selection criterion of the mating selection. The mating selection method is described below, and the archive size here is N.

Step 1: Sort archive solutions in ascending order of the Euclidean distance from the reference point.
Step 2: Add top $\frac{N}{2}$ solutions to the search population.
Step 3: Select remaining solutions by tournament selection based on their rank. If multiple solutions with same rank exist, select the solution with the smallest Euclidean distance.

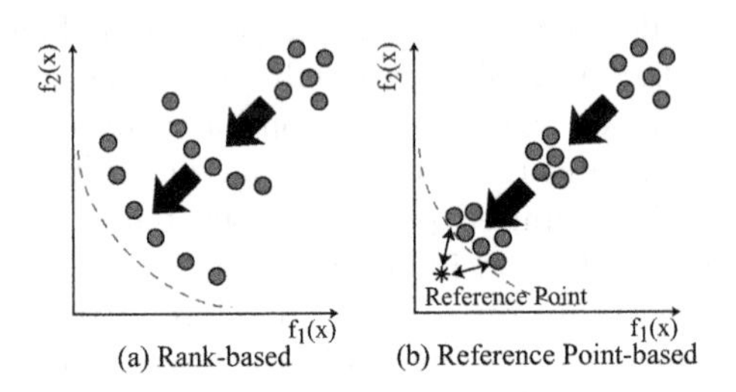

Fig. 2. Concept of the Reference Point-based Search

$\frac{N}{2}$ closest solutions to the reference point are copied to the search population in Step 2, because these solutions are not guaranteed to be selected using methods such as tournament selection. Copying these solutions to the search population should result in improvement of convergence. In addition, both rank and Euclidean distance are considered in the tournament selection at Step 3, which allows selection of non-dominated solutions close to the reference point, and the search is directed towards the reference point while preserving diversity.

2.3 2nd Phase: Broadening Search

The Distributed Cooperation Scheme of Okuda et al. [12] is adopted in the broadening search. The search population is divided into subpopulations that search using MOGA and SOGA in this scheme. Henceforth, subpopulations that search with MOGA and SOGA are called MOGA population and SOGA population, respectively. When there are k objectives, the search population is divided into $k + 1$ subpopulations: one MOGA population and k SOGA populations. The concept of this scheme is illustrated in Fig. 3. As this is a scheme, any MOGA or SOGA methodology can be adopted. In this study, NSGA-II [10] and SPEA2 [11] are each adopted as the MOGA population, and DGA [14] as the SOGA population.

MOGA and SOGA populations search in a parallel manner in the Distributed Cooperation Scheme, and the best solutions from each population are exchanged every interval generations, which was set to 25 generations in this study. The best solution of the f_i SOGA population is the solution with the best f_i objective value. On the other hand, best solutions of the MOGA population are non-dominated solutions with best f_i objective value, and k solutions exist in a k-objective problem. Migration of solutions in a two-objective problem is shown in Fig. 4.

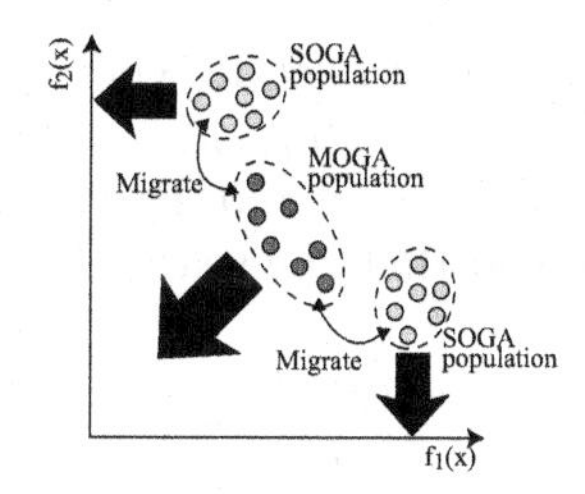

Fig. 3. Concept of the Distributed Cooperation Scheme

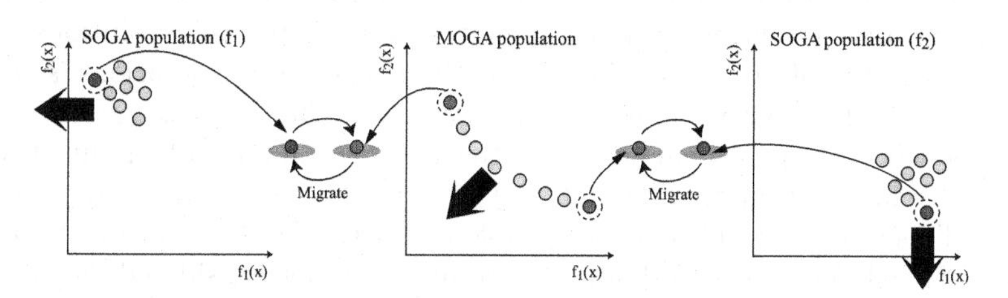

Fig. 4. Concept of Migration in the Distributed Cooperation Scheme

The algorithm of the Distributed Cooperation Scheme with population size of N in the k-objective problem is shown below.

Step 1: Randomly generate N individuals.
Step 2: Divide the individuals into MOGA and k SOGA populations with $\frac{2N}{k+2}$ individuals in MOGA population and $\frac{N}{k+2}$ individuals in SOGA population.
Step 3: Search for non-dominated solutions in the MOGA population and optimal solutions of each objective in the SOGA populations.
Step 4: Collect solutions from all populations and update archive.
Step 5: Exchange best solutions between MOGA and SOGA populations every interval generations.
Step 6: End if terminate criterion is met; else go back to Step 3.

2.4 Search Strategy

In the proposed search strategy, the convergence search described in section 2.2 is conducted, followed by the broadening search described in section 2.3. When to switch the search phase becomes important in this case. It is preferable that the search be switched when the solutions have converged. We consider the following two cases of convergence:

- Advancement of the search towards the Pareto optimal front is little.
- Newly derived non-dominated solutions contribute little to accuracy.

Therefore, we adopt two convergence indicators in switching the search phase. The first indicator is the one utilized in MRMOGA [15]. It is an average ratio of non-dominated solutions in the archive that are dominated by the derived solutions over several generations. This ratio will be high when the search is advancing and low when converged. In detail, when non-dominated solution set of the archive at the ith generation is $PF_{known}(i)$, the ratio of $PF_{known}(i)$ that is dominated ($dominated_i$) can be calculated. Based on the average ratio over g generations, it can be determined that the search has converged if criterion (1) is met.

$$\sum_{i=1}^{g} \frac{dominated_i}{g} \leq \epsilon \tag{1}$$

With MRMOGA, the value of $\epsilon = 0.05$ is used and we used this in our research for two-objective problems as well. $\epsilon = 0.025$ is used for three-objective problem, because it becomes more difficult to dominate other solutions with increasing number of objectives. Moreover, the period of g generations is set to be the same as the migration interval in section 2.3, which was 25 generations.

The second convergence indicator is the average number of archived non-dominated solutions that are dominated by each newly generated non-dominated solution. This indicator will cover the problem of MRMOGA's indicator that the number of newly derived non-dominated solutions is not considered. For example, average value of 1 means that each new non-dominated solution dominates 1 archived non-dominated solution. This indicator shows the effectiveness of the newly generated solutions for advancing the search. Lower average value means that the search is shifting to improvement of diversity instead of accuracy. Therefore, the search can be determined as converged if this indicator value is low.

We take average value of this indicator over g generations and determine that the search has converged if criterion (2) is met. Here, μ_i is the average number of archived non-dominated solutions that are dominated by each new non-dominated solution at ith generation, and we used $g = 25$.

$$\sum_{i=1}^{g} \frac{\mu_i}{g} \leq \alpha \tag{2}$$

We used $\alpha = 0.5$ for two-objective and $\alpha = 0.25$ for three-objective problems as the criterion in our research, since it showed good results in the preliminary experiments. Using the two indicators mentioned above, we switch the search phase when either criterion is met. The process of the search strategy for a k-objective problem is shown below.

Step 1: Initialize the archive.
Step 2: Conduct convergence search as described in section 2.2.
Step 3: Check criterion (1) and (2) every g generations. Go to Step 4 if either criterion is met, else go back to Step 2.
Step 4: Divide solutions stored in archive into $k + 1$ populations.
Step 5: Conduct broadening search as described in section 2.3.
Step 6: End if terminate criterion is met; else go to Step 5.

3 Verification of Search Strategy's Performance

A numerical experiment was performed to verify the effectiveness of the proposed search strategy by comparison with NSGA-II and SPEA2. The MOGA methodology of the proposed search strategy is NSGA-II and SPEA2, and DGA was adopted as the SOGA population. The test problems used in this experiment were KUR [16] and multi-objective knapsack problems. KUR is a two-objective continuous problem with 100 design variables [16]. KP500-2 (i.e., 2 objectives, 500 items), KP750-2, and KP750-3 [9] were selected as multi-objective knapsack problems.

We adopted inverted generational distance (IGD) [17], hypervolume (HV) [18], and spread [19] as the metrics in this experiment: IGD is the average distance from each solution of the Pareto optimal front to the closest obtained solution, and is a metric of accuracy and broadness; HV is a metric of overall performance; and spread, calculated as the sum of differences between maximum and minimum values of each objective within the obtained Pareto front, is a metric of broadness. The Pareto optimal front must be known to calculate IGD, but is unknown for KUR, KP750-2, and KP750-3 problems. Therefore, we obtained near Pareto optimal solutions beforehand using a much greater population size and generations, and used them for IGD calculation.

For both the proposed search strategy and conventional MOGAs, population size is set to 120 for problems other than KP750-3, and 150 for KP750-3. The maximum generation is 1000, and the number of evaluations is the same for all methods. In addition, 2-point crossover is utilized with crossover rate of 1.0, and the mutation rate is 1/Chromosome Length. The parameters specific to the DGA used in the proposed search strategy are as follows: sub population size is 10, tournament selection with tournament size of 4, migration rate is 0.5, and migration interval is 5. The topology of migration is random ring. In the proposed search strategy, a reference point must be set for each problem. Several locations of reference points are tested for each problem.

3.1 Results

50% attainment surfaces of KUR, KP500-2,and KP750-2 by the proposed search strategy, NSGA-II, and SPEA2 in 30 trials are shown in Figs. 5 to 7. In these figures, the reference points of the search strategy are set as (-1000, -400), (21000, 21000), and (30000, 30000) for KUR, KP500-2, and KP750-2, respectively.

The search results in Figs. 5 to 7 indicate that the search strategy obtained broader solutions than NSGA-II or SPEA2. Broader solutions provide more information of the Pareto front, which is important especially in problems such as KUR and KP750-2 where the optimal front is unknown.

In addition, transitions of the 50% attainment surfaces of KP750-2 by the search strategy with three different reference points are shown in Fig. 8. NSGA-II is utilized in the search strategy, and the search was switched from the first phase to the second phase at the average of 600th or 575th generation depending on the reference point used.

As shown in Fig. 8, solutions converge to different regions depending on the location of the reference point. The resulting 50% attainment surfaces are biased toward the edge of the Pareto front in Fig. 8 (b) and (c), but still broader solutions are obtained compared to NSGA-II shown in Fig. 8 (d). Moreover, it

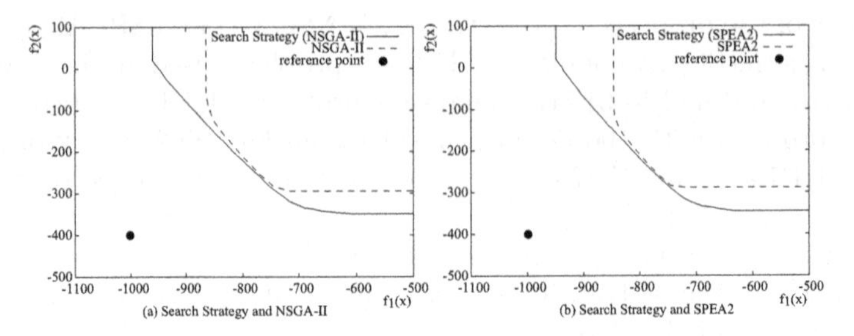

Fig. 5. 50% attainment surfaces of KUR (30 Trials)

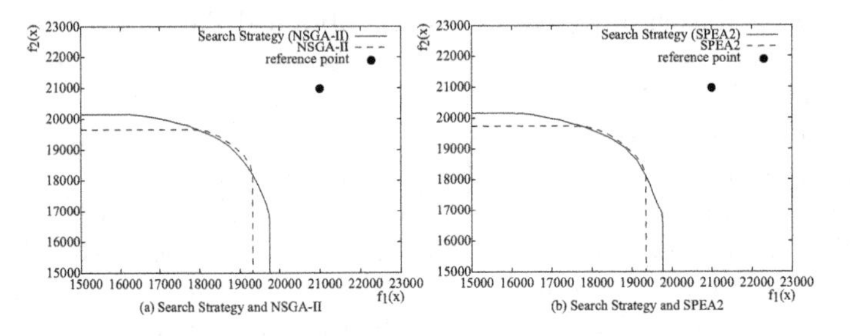

Fig. 6. 50% attainment surfaces of KP500-2 (30 Trials)

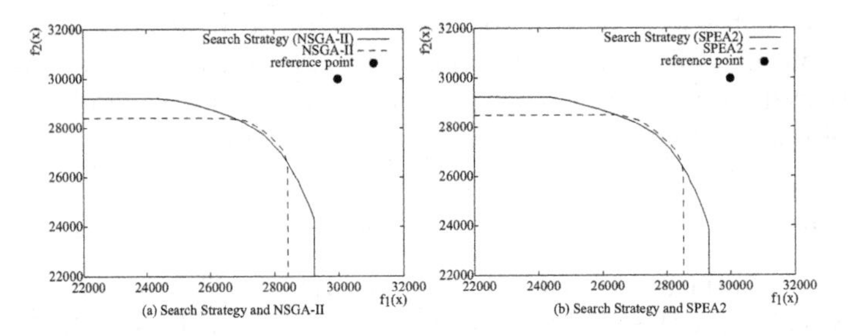

Fig. 7. 50% attainment surfaces of KP750-2 (30 Trials)

can be seen that the broadness of the solutions improve greatly after the search phase is switched. Similar results were seen in other test problems as well, and these results indicate that the proposed search strategy is successful in first converging and then broadening solutions.

Next, the mean values and the standard deviation of IGD, spread, and HV are shown in Tables 1 to 3. For IGD in Table 1, the obtained solutions are closer to the Pareto optimal front when the value is close to 0. On the other hand, solutions with greater values of spread and HV are better. For the search strategy, reference points are set at (-1000, -750), (21000, 21000), (30000, 30000), and (30000, 30000, 30000) for KUR, KP500-2, KP750-2, and KP750-3, respectively.

The mean values of IGD in Table 1 indicate that both implementation of the search strategy is performing equivalent to or better than NSGA-II or SPEA2. Therefore, the search strategy is comparable to both NSGA-II and SPEA2 with

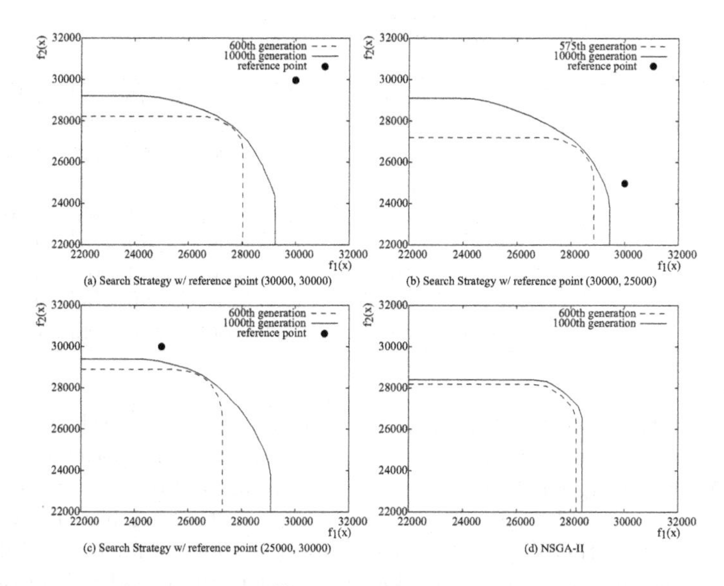

Fig. 8. Transition of the 50% attainment surfaces of KP750-2 (30 Trials)

Table 1. Inverted Generational Distance

		KUR	KP500-2	KP750-2	KP750-3
Search Strategy (NSGA-II):	mean	0.04088	0.01311	0.01548	0.06773
	SD	0.00868	0.00112	0.00130	0.00270
NSGA-II:	mean	0.08846	0.02862	0.02853	0.07602
	SD	0.01061	0.00220	0.00167	0.00314
Search Strategy (SPEA2):	mean	0.04271	0.01322	0.01726	0.06567
	SD	0.00731	0.00099	0.00189	0.00233
SPEA2:	mean	0.10841	0.02478	0.02574	0.06655
	SD	0.01358	0.00176	0.00158	0.00210

Table 2. Spread

		KUR	KP500-2	KP750-2	KP750-3
Search Strategy (NSGA-II):	mean	682.04	6401.10	9134.77	14579.27
	SD	15.30	356.46	489.59	1520.48
NSGA-II:	mean	321.02	2497.23	3130.30	7845.73
	SD	23.06	226.85	236.35	448.87
Search Strategy (SPEA2):	mean	677.28	6568.13	9650.20	13789.67
	SD	15.90	338.40	577.90	2263.41
SPEA2:	mean	263.42	3000.57	3771.07	5176.10
	SD	25.39	200.25	316.30	412.81

regard to accuracy. IGD also describes how close the obtained solutions are to the optimal front regarding broadness. Consequently, the solutions obtained by NSGA-II and SPEA2 are not sufficiently broad.

Table 3. Hypervolume

		KUR	KP500-2	KP750-2	KP750-3
Search Strategy (NSGA-II):	mean	2.86E+05	3.95E+08	8.47E+08	2.41E+13
	SD	7.81E+03	1.23E+06	3.22E+06	2.89E+11
NSGA-II:	mean	2.41E+05	3.79E+08	8.06E+08	2.19E+13
	SD	6.53E+03	1.36E+06	2.79E+06	1.60E+11
Search Strategy (SPEA2):	mean	2.84E+05	3.95E+08	8.49E+08	2.40E+13
	SD	7.42E+03	1.54E+06	3.84E+06	5.45E+11
SPEA2:	mean	2.35E+05	3.81E+08	8.10E+08	2.16E+13
	SD	7.57E+03	1.47E+06	2.99E+06	1.22E+11

The spread values shown in Table 2 also indicate that the search strategy obtained broader solutions. From this, it can be said that the approach to broaden solutions after converging them is successful in obtaining broad solutions. Mean HV values shown in Table 3 also show better results for the search strategy. These results indicate that the proposed search strategy is effective for maintaining accuracy comparable to conventional MOGAs and deriving broader solutions.

4 Application to SVM Training Data Set Selection Problem

When training SVM, it is recommended to select the examples to include in the training data set, because some examples are more useful than others. Data categorization is used for this purpose [6]. In this section, the proposed search strategy is applied to the optimization of training data set selection for SVM. There are two objectives to be optimized in this problem, and they are:

- Minimize training error (f_1)
- Maximize minimum confidence margin (f_2)

Training error is measured with the trained SVM on the entire data set, and is to be minimized. The confidence margin for an example $(\mathbf{x}_i, y_i)$ is measured by $y_i\tilde{g}(\mathbf{x}_i)$, where $\tilde{g}(\mathbf{x})$ is the SVM decision function [6]. We calculate the confidence margin for all examples of the data set, and considered to maximize its minimum value. Examples with negative confidence margins are excluded here, because they are mislabeled. It is expected for a trade-off to exist between these two objectives, because improvement of training error can result in overfitting, which leads to smaller confidence margin.

We applied this SVM training data selection on three data sets from the UCI machine learning repository [20]. The data sets used in this experiment are shown in Table 4. All features of the data sets are scaled to the range of [-1, 1]. For the purpose of examining the generalization performance of the trained SVM, we randomly sampled 20% of the given data set as the hold-out test set. This test set is not used in the training of SVM, therefore the number of examples available for training is 80% of the entire data set. C-SVM with RBF kernel is

Table 4. The data sets used in the experiment. n is the number of data and m is the number of features.

Data set	n	m	classes	C	γ
diabetes	768	8	2	32.0	0.03125
heart	270	13	2	2048.0	0.00049
liver-disorders	345	6	2	512.0	0.03125

used, and the parameters C and γ are decided in advance using cross validation by parallel grid search [21].

This problem is designed in a similar manner to multi-objective knapsack problems, and each example from a data set is represented by 0/1 bit. The example is included in the training data set if the bit value is 1, and not if the bit value is 0. Therefore the length of a chromosome in MOGA is the same as the number of examples in the data set. By this implementation, the number of examples in the training data set and the examples included are decided at the same time.

Population size of 120 and maximum generation of 250 are used in this experiment, and the other basic parameters of the proposed search strategy are the same as the previous experiment. Individuals are initialized randomly, and the number of examples in a training data set range from 0 to the entire data set. Moreover, the reference point is set at $(0, 1)$ in this experiment, because an example with a confidence margin of less than 1 is considered to be a support vector [6].

4.1 Results

10 trials were conducted on each data set, and Figs. 9 to 11 show the 50% attainment surface of diabetes, heart, and liver-disorders data sets, respectively. Trade-off relationship is confirmed between the two objectives in all cases.

From the attainment surfaces shown in Figs. 9 to 11, it was confirmed that the improvement of training error results in smaller confidence margin, and vice versa. The Pareto fronts obtained for the diabetes and liver-disorders data sets were sparse in this experiment, which resulted in the nonsmooth front.

Next, we examined the generalization performance of the trained SVMs using the hold-out test set. Figs. 12 and 13 show the solutions for the diabetes and liver-disorders data set obtained in a single run. In both figures, the left figure shows the obtained Pareto solutions, and the right figure shows the generalization performance of the solutions from the left figure. The x-axis in both left and right plots show the training error, so the solutions in both plots correspond to each other on the x-axis. In addition, the solutions shown here are grouped according to the number of examples included in their training data set. The result of the SVM trained with the entire data set is plotted for comparison as well.

Figs. 12(a) and 13(a) show that selecting the training data set, rather than using the entire data set can obtain SVMs with better training error and the

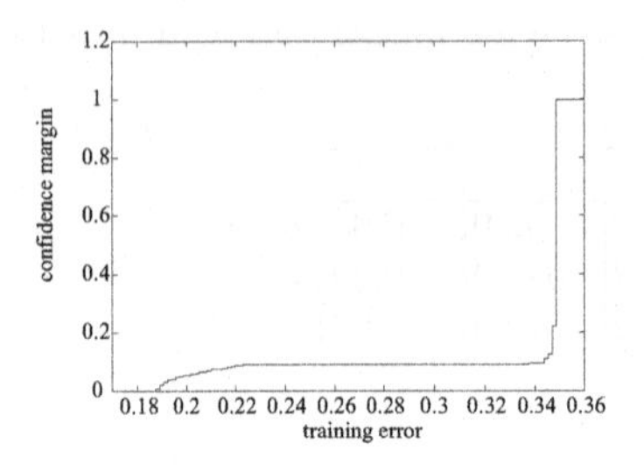

Fig. 9. 50% attainment surface of diabetes data set (10 Trials)

Fig. 10. 50% attainment surface of heart data set (10 Trials)

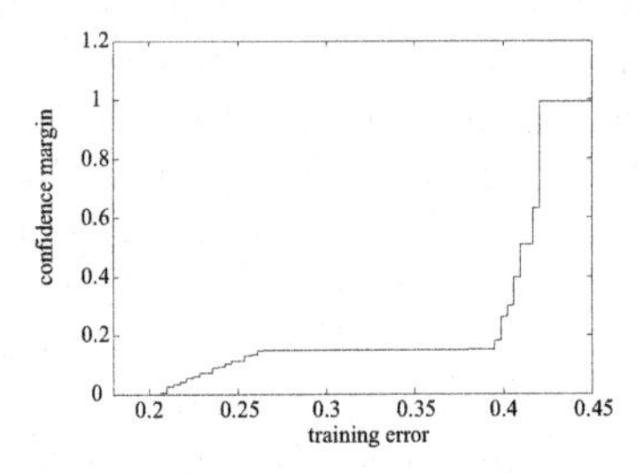

Fig. 11. 50% attainment surface of liver-disorders data set (10 Trials)

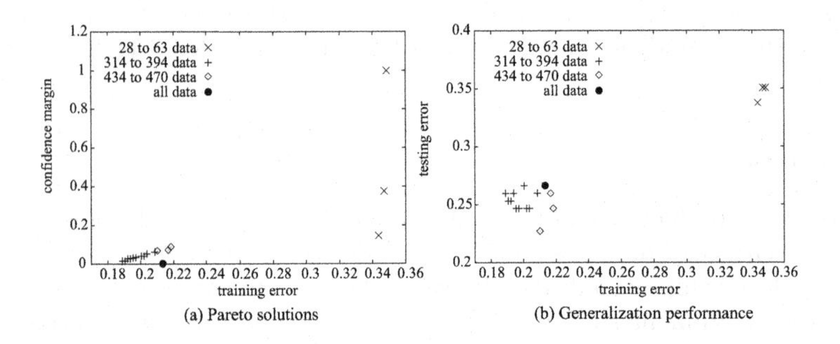

Fig. 12. Example of solutions obtained for diabetes data set

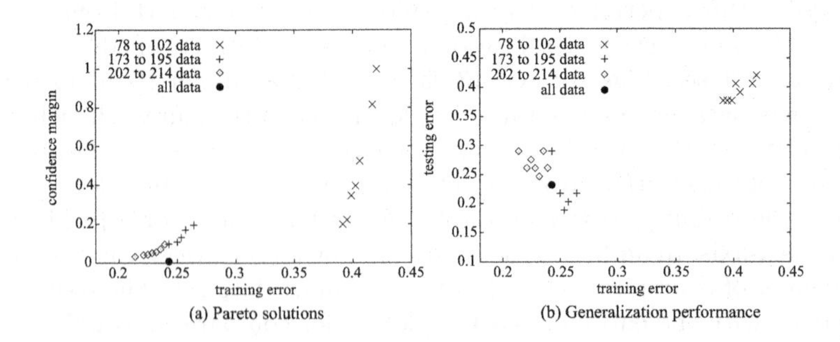

Fig. 13. Example of solutions obtained for liver-disorders data set

confidence margin. The same can be said about the generalization performance shown in Figs. 12(b) and 13(b) as well, because there are SVM models with better test errors. The confidence margin value of the SVM model trained with the entire data set is small compared to other SVM models, and it is likely to be overfit. Therefore, we can understand that the selection of the training data set is beneficial.

If we compare the distribution of the solutions for diabetes data set in Fig. 12, we can see that the best SVM model regarding the training error does not perform best with the test data, which may be caused by overfitting. On the other hand, SVM models with large confidence margins showed very poor performance regarding both the training error and test error. Similar results were observed with the results of liver-disorder data set shown in Figs 13 as well. It is important for the extreme solutions to be obtained in such a case, because they provide the information on the possible range for the SVM's performance. For this reason, we consider the multi-objective approach combined with the hold-out test data to be effective for reducing the possibility of overfitting when selecting the training data set.

Another point we focused on is the number of training data used in each SVM model. Comparing the distribution of the solutions grouped according to the number of training data in Figs. 12 and 13, we observed that the training error is generally low when many examples are used in the SVM training, and high when less examples are used. Although these results show that the increasing number of training data leads to improvement of training error in general, further research is still needed. We assume that cases exist where the SVM parameters of C and γ used in this experiment are not proper for that particular training data set. Therefore, we will consider including C and γ as design variables of MOGA and optimize them for each training data set in the future research.

5 Conclusions

In this paper, we handled the selection of training data set for SVM as a multi-objective optimization problem, and applied MOGA search strategy to it. The proposed search strategy consists of two search phases to separately improve convergence and broadness of the solutions. The first phase improves the convergence of the solutions, and a reference point specified by a decision maker is adopted for this purpose. In the second phase, the solutions are broadened using the Distributed Cooperation Scheme. Through a numerical experiment, we confirmed that the proposed search strategy is capable of deriving broader solutions compared to conventional MOGAs without deterioration of accuracy.

The search strategy was applied to the SVM training data selection problem, and the results showed that there exists trade-off relationship between training error and the confidence margin. SVM models trained with selected training data showed better performance compared to the model trained with the entire data set. From this result, we confirmed the importance of selecting the training data set when training SVM. In future research, we will consider to optimize the SVM parameters along with the selection of the training data set.

References

[1] Cortes, C., Vapnik, V.: Support-Vector Networks. Machine Learning 20(3), 273–297 (1995)

[2] Chapelle, O., Vapnik, V., Bousquet, O., Mukherjee, S.: Choosing Multiple Parameters for Support Vector Machines. Machine Learning 46(1-3), 131–159 (2002)

[3] Joachims, T.: Text Categorization with Support Vector Machines: Learning with Many Relevant Features. In: Proceedings of the European Conference on Machine Learning, pp. 137–142 (1997)

[4] Osuna, E., Freund, R., Girosi, F.: Training Support Vector Machines: An Application to Face Detection. In: Proceedings of 1997 IEEE Computer Society Conference on Computer Vision and Pattern Recognition, pp. 130–136 (1997)

[5] Mierswa, I.: Controlling Overfitting with Multi-Objective Support Vector Machines. In: GECCO 2007: Proceedings of the 9th annual conference on Genetic and Evolutionary Computation, pp. 1830–1837 (2007)

[6] Li, L., Pratap, A., Lin, H.-t., Abu-Mostafa, Y.S.: Improving generalization by data categorization. In: Jorge, A.M., Torgo, L., Brazdil, P.B., Camacho, R., Gama, J. (eds.) PKDD 2005. LNCS (LNAI), vol. 3721, pp. 157–168. Springer, Heidelberg (2005)

[7] Goldberg, D.E.: Genetic Algorithms in search, optimization and machine learning. Addison-Wesly, Reading (1989)

[8] Fonseca, C.M., Fleming, P.J.: Genetic algorithms for multiobjective optimization: Formulation, discussion and generalization. In: Proceedings of the 5th international conference on genetic algorithms, pp. 416–423 (1993)

[9] Zitzler, E., Thiele, L.: Multiobjective Evolutionary Algorithms: A Comparative Case Study and the Strength Pareto Approach. IEEE Transactions on Evolutionary Computation 3(4), 257–271 (1999)

[10] Deb, K., Agarwal, S., Pratap, A., Meyarivan, T.: A Fast Elitist Non-Dominated Sorting Genetic Algorithm for Multi-Objective Optimization: NSGA-II. KanGAL report 200001, Indian Institute of Technology, Kanpur, India (2000)

[11] Zitzler, E., Laumanns, M., Thiele, L.: SPEA2: Improving the Performance of the Strength Pareto Evolutionary Algorithm. Technical Report 103, Computer Engineering and Communication Networks Lab (TIK), Swiss Federal Institute of Technology (ETH) Zurich (2001)

[12] Okuda, T., Hiroyasu, T., Miki, M., Watanabe, S.: DCMOGA: Distributed Cooperation model of Multi-Objective Genetic Algorithm. In: Advances in Nature-Inspired Computation: The PPSN VII Workshops, pp. 25–26 (2002)

[13] Deb, K., Sundar, J.: Reference Point Based Multi-Objective Optimization Using Evolutionary Algorithms. In: GECCO 2006: Proceedings of the 8th annual conference on Genetic and evolutionary computation, pp. 635–642 (2006)

[14] Tanese, R.: Distributed Genetic Algorithms. In: Proc. 3rd ICGA, pp. 434–439 (1989)

[15] Jaimes, A.L., Coello, C.A.C.: MRMOGA: Parallel Evolutionary Multiobjective Optimization using Multiple Resolutions. In: 2005 IEEE Congress on Evolutionary Computation (CEC 2005), pp. 2294–2301 (2005)

[16] Kursawe, F.: A Variant of Evolution Strategies for Vector Optimization. In: Parallel Problem Solving from Nature. 1st Workshop, PPSN I, pp. 193–197 (1991)

[17] Sato, H., Aguirre, H., Tanaka, K.: Local Dominance Using Polar Coordinates to Enhance Multi-objective Evolutionary Algorithms. In: Proc. 2004 IEEE Congress on Evolutionary Computation, pp. 188–195 (2004)

[18] Knowles, J., Thiele, L., Zitzler, E.: A Tutorial on the Performance Assessment of Stochastic Multiobjective Optimizers. TIK Report 214, Computer Engineering and Networks Laboratory (TIK), ETH Zurich (2006)
[19] Ishibuchi, H., Shibata, Y.: Mating Scheme for Controlling the Diversity-Convergence Balance for Multiobjective Optimization. In: Proc. of 2004 Genetic and Evolutionary Computation Conference, pp. 1259–1271 (2004)
[20] Asuncion, A., Newman, D.J.: UCI Machine Learning Repository. Irvine, CA: University of California, School of Information and Computer Science (2007), http://www.ics.uci.edu/~mlearn/MLRepository.html
[21] Chang, C.-C., Lin, C.-J.: LIBSVM: a library for support vector machines (2001), http://www.csie.ntu.edu.tw/~cjlin/libsvm

On Using Populations of Sets in Multiobjective Optimization

Johannes Bader, Dimo Brockhoff, Samuel Welten, and Eckart Zitzler

Computer Engineering and Networks Lab, ETH Zurich, 8092 Zurich, Switzerland
firstname.lastname@tik.ee.ethz.ch
http://www.tik.ee.ethz.ch/sop/

Abstract. Most existing evolutionary approaches to multiobjective optimization aim at finding an appropriate set of compromise solutions, ideally a subset of the Pareto-optimal set. That means they are solving a set problem where the search space consists of all possible solution sets. Taking this perspective, multiobjective evolutionary algorithms can be regarded as hill-climbers on solution sets: the population is one element of the set search space and selection as well as variation implement a specific type of set mutation operator. Therefore, one may ask whether a 'real' evolutionary algorithm on solution sets can have advantages over the classical single-population approach. This paper investigates this issue; it presents a multi-population multiobjective optimization framework and demonstrates its usefulness on several test problems and a sensor network application.

1 Motivation

Most multiobjective evolutionary algorithms (MOEAs) proposed in the literature are designed towards approximating the set of Pareto-optimal solutions [7]. In contrast to single-objective optimizers that look for a single optimal solution, these algorithms aim at identifying a *set* of optimal compromise solutions, i.e., they actually operate on a set problem. With such a set problem, the search space consists of all solution sets and often a set quality measure like the hypervolume indicator [20] is used as a corresponding objective function on sets. From this perspective, current MOEAs can be regarded as hill climbers or $(1,1)$-strategies on solution sets, cf. [21]. The population represents a solution set and as such one element of the set search space. The usual sequence of operations, i.e., mating selection, variation including mutation and recombination, and environmental selection, serves the purpose of generating a new set; therefore, it can be considered as a (complex) set mutation operator. Since the newly generated population usually replaces the old population without a direct comparison and check, one can speak of a $(1,1)$-strategy in this context.

The above observation leads to the question of whether the use of an evolutionary algorithm on sets may be beneficial in this multiobjective setting. In other words: can maintaining a population of solution sets in combination with appropriate set selection and set variation operators have advantages over using

M. Ehrgott et al. (Eds.): EMO 2009, LNCS 5467, pp. 140–154, 2009.

a single solution set only? If we consider classical MOEAs as $(1,1)$-strategies on the corresponding set problem, then we are here interested in extending them to (μ,λ)- or $(\mu+\lambda)$-strategies. To our best knowledge, this issue has not been addressed so far, although there is an interesting close link to parallel MOEAs. Some types of parallel MOEAs, in particular those based on island models, make use of multiple populations that evolve simultaneously and from time to time exchange individuals [15,13,1,16,4,14,5,18]. However, these approaches can in general not be regarded as full evolutionary algorithms on solution sets, as they usually implement only some aspects of set-based fitness, set-based variation, and set-based selection. The considerations presented in the following are independent of the type of implementation, be it sequential or parallel.

This paper investigates the issue of whether a multiobjective optimizer can benefit from utilizing a population of *solution sets* instead of relying on a single population of *solutions*—mainly in terms of the quality of the generated Pareto set approximation, but also with respect to the computing time. As a basis, we consider the optimization scenario where a set with N solutions is sought that maximizes the hypervolume of the dominated objective space. The specific contributions are:

- A general framework for a population-based evolutionary algorithm operating on solution sets; the framework resembles the island model known for parallel evolutionary algorithms.
- The design of a new recombination operator on solution sets which is tailored to the hypervolume indicator, but the principle of which can be generalized to other unary indicator functions.
- A systematic comparison of the classical MOEA scheme and the multi-population scheme on several test problems with up to four objectives.

The following section provides a brief survey of set-based multiobjective optimization, in particular of hypervolume-based multiobjective search, and a background of related work in the area of parallel evolutionary algorithms. Section 3 introduces our general framework of a multi-population MOEA including the new recombination operator. Section 4 presents and discusses the experimental results, and Sec. 5 contains conclusions and future research directions.

2 Background

2.1 Set-Based Multiobjective Optimization

Given an optimization scenario where: X is the decision space; $x \in X$ denotes a solution or decision vector; k objective functions $f = (f_1, \ldots, f_k)$ are to be minimized; $x \preceq y$ denotes weak dominance of y by x and $x \prec y$ denotes strict dominance[1], the goal is usually to find a set A which represents a good approximation of the Pareto-optimal set. Many ways to assess the quality of a Pareto

[1] A solution x is said to weakly dominate a solution y ($x \preceq y$) if it is at least as good as y in all objectives, i.e., $\forall 1 \leq i \leq k : f_i(x) \leq f_i(y)$. If additionally there exists an objective function f_j with $f_j(x) < f_j(y)$ then x is strictly dominating y ($x \prec y$).

set approximation A exist. One is to consider the space that is weakly dominated by the objective vectors $f(A)$ and bounded by a user defined reference set R:

$$H(A, R) := \{h \mid \exists a \in A \, \exists r \in R : f(a) \leq h \leq r\} \quad (1)$$

Let the corresponding *hypervolume indicator* be the hypervolume of this space $I_H(A, R) := \lambda(H(A, R))$, where λ denotes the Lebesgue measure. Of all the numerous measures, the hypervolume indicator or $\mathcal{S}$-metric [3] is one of the most popular; mainly because it is the only known indicator that reflects Pareto dominance, i.e., if a solution set dominates another, the hypervolume indicator of the former is greater than the one of the latter. The goal of hypervolume indicator-based MOEAs can be formalized as finding the solution set A^* that maximizes the indicator value, usually imposing a maximum cardinality of $|A^*| \leq \mu$. By the nature of I_H, the set maximizing I_H is a subset of the Pareto set.

The classical view of MOEAs is illustrated in the upper left corner of Fig. 1. Mating selection, mutation, crossover, and environmental selection operate on single solutions and thereby generate a new—hopefully better—set of solutions. Summarized, one can state that classical MOEAs operate on elements of X and deliver an element of $\mathcal{P}(X)$, where $\mathcal{P}(X)$ denotes the power set of X.

Definition 1. *We refer to an optimizer that operates on elements of the decision space U and returns an element of V as a U/V-optimizer.*

Hence, MOEAs are, from a classical EA perspective, $X/\mathcal{P}(X)$ optimizer. On the other hand, multiobjective algorithms using aggregation are considered as X/X-optimizers. However, the individual steps (fitness assignment, mating selection, mutation/crossover, and environmental selection) of the MOEA, that lead to a modified set, can be abstracted as a set mutation, see the upper right corner of Fig. 1—they are in fact $\mathcal{P}(X)/\mathcal{P}(X)$-hillclimbers [21].

In the following, we propose a general $\mathcal{P}(X)/\mathcal{P}(X)$ evolutionary algorithm as depicted in the lower half of Fig. 1. The question arises, how the corresponding operators (set mutation, set crossover, set mating and set environmental selection) can be created and if they are beneficial for search. To this end, we propose set operators based on the hypervolume indicator. There are already many algorithms using this indicator (e.g., [3,12]), but they are all $X/\mathcal{P}(X)$-optimizers.

To our knowledge, no study has used the set perspective on evolutionary algorithms explicitly, but parallel evolutionary algorithms can be considered as optimizers operating on sets, as demonstrated in the following subsection.

2.2 Parallel Evolutionary Algorithms

The increasing complexity of large scale problems and the availability of large computer clusters and multiprocessor systems were the first incitement to paralle

The *master-slave* approach uses a master processor that performs all operations on one global population except for fitness evaluations which are delegated to different slave processors [17]. Since this parallelization does not change the algorithm itself, master-slave MOEAs can be either seen as $X/\mathcal{P}(X)$-optimizers or interpreted as $\mathcal{P}(X)/\mathcal{P}(X)$-hillclimbers, see Fig. 1.

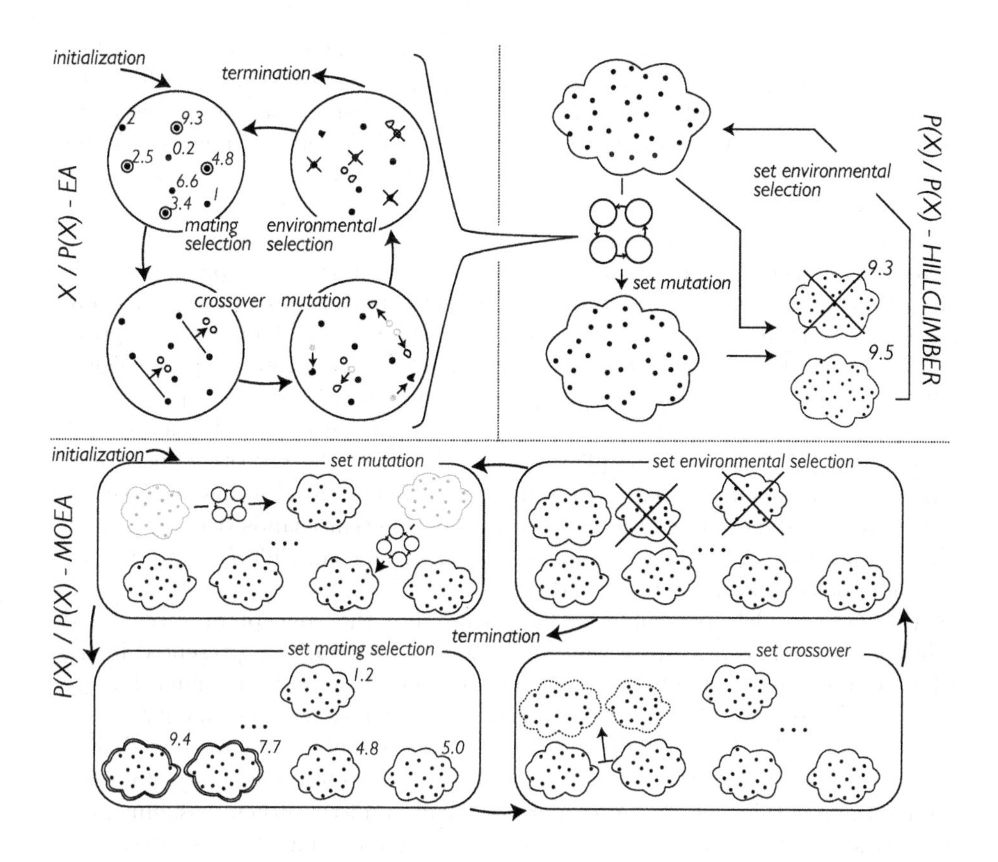

Fig. 1. Illustration of different types of MOEAs: (top left) usual view of a MOEA where the operators work on solutions; (top right) a set-based view of the same algorithm; (bottom) an evolutionary algorithm working on sets, i.e., a $\mathcal{P}(X)/\mathcal{P}(X)$-optimizer

The second major category of parallel MOEAs—the *island model*—however, can be seen as $\mathcal{P}(X)/\mathcal{P}(X)$-optimizer that use more than one set. An island model MOEA divides the overall population into different islands or independent solution sets. Hence, when abstracting away from parallelization, the island model can be interpreted as an algorithm operating on a population of sets. Each of these sets represents one island which is optimized by a separate EA. This enables running different islands on several computers at the same time.

An island model without any exchange of individuals between islands corresponds to a multi-start approach, where each island represents one run, using different seeds or even different optimization strategies [14]. Such an algorithm mainly benefits from increased robustness of obtained solutions and corresponds to a $\mathcal{P}(X)/\mathcal{P}(X)$-optimizer (see Fig. 1) where each set is mutated and no recombination and environmental selection takes place.

Most island models, however, use a cooperative approach. Although the subpopulations evolve independently most of the time, solutions are exchanged once in a while between islands by *migration*. A well designed migration lets information of good individuals pass among islands and at the same time helps to preserve diversity by isolation of the islands. In contrast to the approaches mentioned above, this paradigm also uses recombination of sets (by migration) and can therefore be advantageous not only in terms of runtime and robustness, but also in terms of quality of the obtained Pareto-optimal solutions [5].

There exist many aspects of migration strategy. (a) The way islands are selected for migration (the set mating selection from a set based perspective) is often done deterministically according to the way islands are arranged [16], where the topology is an important parameter which has to be adapted to the problem structure [15]. (b) The way the population is divided into subpopulations. Often each island corresponds to a different region of the objective space determined manually, e.g., by using cones [4] or by assigning each island to a different subproblem [13]. Instead of explicitly, the division of the objective space into different regions can also happen implicitly, e.g., when distributing the best individuals according to one objective function to different islands [10]. (c) Islands are optimized either by the very same optimizer or by using different parameters. For more details, we refer to [5] and [18].

All island models presented so far do not use the concept of a set-based fitness measure and operators. One exception is the algorithm presented in [1], where islands are randomly selected and both mutation and recombination are applied to subpopulations rather than to single solutions. The quality of the newly generated subpopulations as well as their parents is then assessed by a fitness value and the better sets are kept (set environmental selection). However, the environmental selection only operates locally and the fitness assignment is not a true set fitness since it corresponds to the sum of single fitness values that are determined on basis of a global population.

In this paper, we give first insights on how to use the set-based view on island models to propose a general $\mathcal{P}(X)/\mathcal{P}(X)$ MOEA. In the next section, we systematically investigate which extensions are needed and propose a novel recombination scheme on sets using the hypervolume indicator.

3 A General Framework for a Set-Based Evolutionary Algorithm

In this section, we propose a general framework of a $\mathcal{P}(X)/\mathcal{P}(X)$-optimizer for multiobjective optimization the basis of which is a population-based evolutionary algorithm. In contrast to most island-based MOEAs, this new optimizer uses all known operators—mating selection, recombination, mutation, and environmental selection—of a usual evolutionary algorithm, working on sets of solutions. How these operators on sets of solutions can look like is the main focus of this section. In the following, we first describe the general framework and later on present different operators on solution sets.

Algorithm 1. A $\mathcal{P}(X)/\mathcal{P}(X)$-optimizer with $(\mu \dagger \lambda)$-selection

Require: number of solution sets in population μ, number of solutions in each solution set N, number of offspring λ, maximum number of generations $g_{\max}$

Init: choose population $\mathcal{S}$ uniformly at random as μ sets of N solutions from X each

$i \leftarrow 1$ {set generation counter}
while $i \leq g_{\max}$ **do**
  $\mathcal{M} \leftarrow \emptyset$
  for all $A \in \mathcal{S}$ **do**
    $\mathcal{M} \leftarrow \mathcal{M} \cup \{\text{setMutate}(A)\}$
  end for
  $\mathcal{M}' \leftarrow \text{setMatingSelection}(\mathcal{M}, \lambda)$
  $\mathcal{M}'' \leftarrow \emptyset$
  for all $(A_p, A_q) \in \mathcal{M}'$ **do**
    $\mathcal{M}'' \leftarrow \mathcal{M}'' \cup \{\text{setRecombine}(A_p, A_q)\}$
  end for
  $\mathcal{S} \leftarrow \text{setEnvironmentalSelection}(\mathcal{S}, \mathcal{M}'')$
  $i \leftarrow i + 1$
end while

3.1 A $(\mu \dagger \lambda)$-EA as a $\mathcal{P}(X)/\mathcal{P}(X)$-Optimizer

Algorithm 1 shows a general $\mathcal{P}(X)/\mathcal{P}(X)$-optimizer that mainly follows the scheme of Fig. 1. The algorithm resembles an island-based MOEA as discussed in Sec. 2.2 with additional mating and environmental selection. Mutation, recombination, and selection on single solutions are considered as mutations on solution sets and the migration operator is regarded as recombination operator on sets.

The algorithm starts by choosing the first population $\mathcal{S}$ of μ sets (of N solutions each) uniformly at random. Then, the optimization loop produces new sets until a certain number $g_{\max}$ of generations are performed. To this end, every set A in the population $\mathcal{S}$ is mutated to a new set by the operator setMutate(A) and λ pairs of sets are selected in the set mating selection step to form the parents of λ recombination operations. Note that the operator "$\cup$" is the union between two multisets; since the population of evolutionary algorithms usually contains duplicate solutions, we also do not restrict the population of Algorithm 1 to sets. In the environmental selection step, the new population is formed by selecting μ sets from the union of the previous population and the varied solution sets. Figure 1 illustrates the steps performed in one generation graphically.

3.2 Mutation of Solution Sets

As mutation operator on solution sets, we propose to use a simple $X/\mathcal{P}(X)$-optimizer with $(N + N)$-selection on single solutions that aims at optimizing the hypervolume indicator of the final solution set directly. This corresponds to a run of a normal hypervolume-based MOEA, as for example [3] or [12], for

G generations. The used $X/\mathcal{P}(X)$-optimizer starts with a set of N solutions that is obtained from the overall $\mathcal{P}(X)/\mathcal{P}(X)$-optimizer's population. For G generations, N solutions of the current set are selected in a mating selection step, these solutions undergo SBX crossover and polynomial mutation as described in [8] and in the environmental selection step, the best solutions from the previous population and the new solutions are selected to form the new population.

The fitness of a solution in the selection steps is a generalization of the hypervolume loss as proposed in [2]. Instead of using the hypervolume that is solely dominated by a single solution as fitness value as it is done in many hypervolume-based MOEAs, e.g., [3,12], the fitness $I_h^l(a, A, R)$ of a solution $a \in A$ is computed as the expected hypervolume loss if the solution itself and $l-1$ randomly selected other solutions in the population are removed.

Definition 2. *Let A be a solution set, $R \subset Z$ the reference set of the hypervolume indicator, and $l \in \{0, 1, \ldots, |A|\}$ the number of solutions that are to be removed from A. Let $H_i(a, A, R)$, in addition, be the portion of the objective space that is dominated by $a \in A$ and exactly $i-1$ other solutions in A and that itself dominates the reference set R. Then the function*

$$I_h^l(a, A, R) := \sum_{i=1}^{l} \frac{\alpha_i}{i} \lambda(H_i(a, A, R)) \quad \textit{where} \quad \alpha_i := \prod_{j=1}^{i-1} \frac{l-j}{|A|-j} \tag{2}$$

gives the fitness of a solution $a \in A$.

In the mating selection step, we choose $I_h^{|A|}(a, A, R)$ as the fitness in a binary tournament selection whereas in the environmental selection step, we choose $l = N$ since N solutions in the $(N+N)$-selection have to be removed to build the new population. For environmental selection, non-dominated sorting on the set of $N+N$ solutions is performed. Then, non-dominated fronts are added completely to the new population according to their rank until the population size is reached. If the number of selected solutions exceeds N, the solutions with worst fitness are iteratively removed until the population size N is reached again. For a motivation of this new fitness assignment scheme and an evaluation of its usefulness, we refer to [2].

3.3 Recombination of Solution Sets

Since we aim at maximizing the hypervolume indicator in multiobjective search, a recombination operator on sets should also aim at producing offspring with large hypervolume. Therefore, we propose a new recombination operator on solution sets A and B that is targeted at maximizing the hypervolume of the offspring C, see Fig. 2 for an illustrative example. The idea behind the operator is to iteratively delete the worst solution in the first parent and add the best individual from the second parent until no hypervolume improvement is possible. In more detail, the process runs as follows.

In a first step, all solutions in the first set $A = \{a_1, \ldots, a_{|A|}\}$ are ranked according to their fitness $I_h^l(a, A, R)$ in Eq. 2 with $l = 1$ (upper left figure in

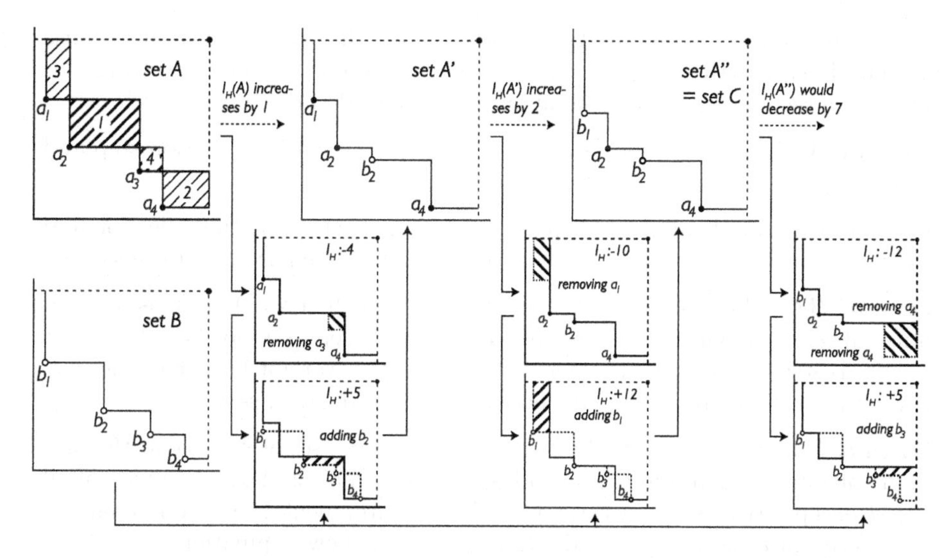

Fig. 2. Illustration of the hypervolume-based recombination operator on solution sets: two exemplary sets A and B with four solutions each are recombined to a set C. First, the solutions in A are ranked according to their hypervolume losses. Then, iteratively, the solution in A with smallest loss is deleted (middle row) and the solution in B that maximizes the hypervolume indicator is added to A (last row) until no hypervolume improvement is possible. For each step, the changes in hypervolume are annotated in the top right corner of the corresponding figure.

Fig. 2). In this case, the fitness of a solution corresponds to the hypervolume that is solely dominated by this solution, in other words, its hypervolume loss. Then, the new set C results from A by iteratively removing the solution a_i with smallest fitness that is not yet removed (ties are resolved randomly, see middle row in Fig. 2) and adding the solution $b \in B$ that maximizes the hypervolume indicator of the new set (last row in Fig. 2). The replacement of solutions stops before the next exchange would decrease the hypervolume of the new set.

Since the fitness values of the solutions in A are only calculated once in the beginning, at most $|A| \cdot |B| + 1$ hypervolume indicator values of $|A|$ points have to be computed in each recombination. Note that the recombination operator can also be seen as a hypervolume-based migration strategy for island model based MOEAs where each island obtains solutions from a neighboring island as long as its hypervolume increases. Another important aspect, we would like to mention is the asymmetry of the recombination operator, i.e., $\text{setRecombine}(A_p, A_q) \neq \text{setRecombine}(A_q, A_p)$. This asymmetry is the reason for selecting ordered pairs in the set mating selection step of Alg. 1.

3.4 Mating and Environmental Selection

In the following, we present four different variants of mating and environmental selection combinations. Two variants choose sets for recombination directly from the mutated sets (we call them A-variants) whereas the other two variants choose one mutated set as the first parent and the set containing all solutions of all other sets as the second parent for recombination (called B-variants):

Variant A1. randomly selects μ pairs of sets in the mating selection step and uses (μ, μ)-selection in its environmental selection step.

Variant A2. selects all possible $\mu \cdot (\mu - 1)$ pairs of sets in mating selection and selects the best μ out of the $\mu \cdot (\mu - 1)$ new sets in the environmental selection.

Variant B1. selects one pair of sets only, where the first set $A_1 \in M$ is selected uniformly at random and the second set A_2 is chosen as union of all $A \in M$ except A_1 itself. In the environmental selection step, variant B1 copies the only new set μ times to create the new population of μ identical sets.

Variant B2. selects μ pairs of sets by choosing every set of M once as the first set A_1 of a parent pair and the second set A_2 of the pair is chosen as union of all $a \in M$ except A_1 itself as in variant B1. The environmental selection of variant B2 chooses all μ newly generated sets to create the new population.

Note that all variants perform the mating selection independent of the hypervolume indicator the consideration of which may improve the optimizer further. Note also that parallel MOEAs, when interpreted as $\mathcal{P}(X)/\mathcal{P}(X)$-optimizers, usually do not perform environmental selection and select the individuals for mating according to a fixed scheme given by the neighborhood of the islands.

4 Experiments

The experiments described in this section serve three main goals. First, we extensively compare four $\mathcal{P}(X)/\mathcal{P}(X)$-optimizer variants with a standard MOEA on various test problems. Then, we study the set mutation operator, in particular, the length G of a set mutation step. Third, we apply the $\mathcal{P}(X)/\mathcal{P}(X)$-optimizer to a sensor network application and compare it with the standard MOEA.

4.1 Experimental Setup

As the baseline standard MOEA, we use the algorithm described in Sec. 3.2 and [2]. Single solutions are mutated by polynomial mutation and recombined via SBX crossover [7]. In addition, we consider four $\mathcal{P}(X)/\mathcal{P}(X)$-optimizer variants A1, A2, B1, and B2 named after the used selection scheme as described in Sec. 3.4. The set mutation and set recombination operators are the same in all variants and implemented as described in Sec. 3. For a fair comparison, the standard MOEA is also used as set mutation operator in all four $\mathcal{P}(X)/\mathcal{P}(X)$-optimizer variants. Note that the implementation of the set mutation step is parallelized, i.e., the μ set mutation operations can be performed in parallel as μ independent runs of the standard MOEA if the algorithm is run on a machine with more than one core. Unless otherwise stated, we always use the same parameters for all algorithms. The hypervolume indicator is computed exactly for all bi-objective problems; otherwise, 10,000 samples are used to approximate it; the reference point is chosen as 40^k such that all solutions of the considered problems have a positive hypervolume contribution. For comparing the algorithms, the standard MOEA runs for 500 generations with a population size of 200—the $\mathcal{P}(X)/\mathcal{P}(X)$-optimizer variants use the same number of function evaluations within $g_{\max} = 25$ generations where the $\mu = 10$ sets of $N = 20$ solutions each are mutated for $G = 20$ generations of the standard MOEA.

4.2 Comparison between Four $\mathcal{P}(X)/\mathcal{P}(X)$-Optimizer Variants and a Standard MOEA

To compare the four $\mathcal{P}(X)/\mathcal{P}(X)$-optimizer variants of Sec. 3 and the standard MOEA with the parameters described above, 30 runs are performed for each of the test problems DTLZ2, DTLZ5, DTLZ7 [9], WFG3, WFG6, and WFG9 [11] with 2, 3, and 4 objectives. Figure 3 shows the boxplots of the normalized hypervolume in the last generation, i.e., the hypervolume indicator of the set containing all single solutions in the last population. In addition, Fig. 5 shows the running times of the different algorithms on a 64bit AMD linux machine with 4 cores (2.6GHz) averaged over all 6 test problems.

There are two main observations: On the one hand, the $\mathcal{P}(X)/\mathcal{P}(X)$-optimizer variants are faster than the standard MOEA. On the other hand, the quality of the solution sets obtained by the $\mathcal{P}(X)/\mathcal{P}(X)$-optimizer variants are, in part, better than the standard MOEA in terms of hypervolume indicator values.

As to the running time, a speed-up is not surprising due to the parallel implementation of the $\mathcal{P}(X)/\mathcal{P}(X)$-optimizer variants. However, the speed-ups are higher than the number of cores except for the A2 variant which indicates that there will be a speed-up even on a single processor. The reason is mainly the faster hypervolume computation. To substantiate the statement that the speed-up is not only caused by the parallelization, we compare the hypervolume indicator improvements of the A1 variant and the standard MOEA over the performed function evaluations. Figure 4 shows the overall hypervolume of all solutions and the hypervolume of a randomly selected set of variant A1 together with the hypervolume of the standard MOEA averaged over 30 runs. After a certain number of function evaluations, the A1 variant outperforms the standard MOEA even for the same number of function evaluations which indicates that also a non-parallelized $\mathcal{P}(X)/\mathcal{P}(X)$-optimizer variant A1 outperforms the standard MOEA. This result is even more surprising since the standard MOEA operates on a set of 200 solutions whereas the $\mathcal{P}(X)/\mathcal{P}(X)$-optimizer operates on much smaller sets of 20 solutions only. In the latter case, the diversity between the sets of solutions is not guaranteed; in the former case, the hypervolume indicator ensures a good spread of all 200 solutions which explains the higher hypervolume in the beginning of the optimization.

As to the solution quality, we can make two observations, that are both supported by statistical tests[2]. The B1 and B2 variants obtain, statistically signifi-

[2] We used the non-parametric Kruskal-Wallis test followed by the Conover-Inman procedure with a p-value of 0.01 as described in [6] on p. 288ff.

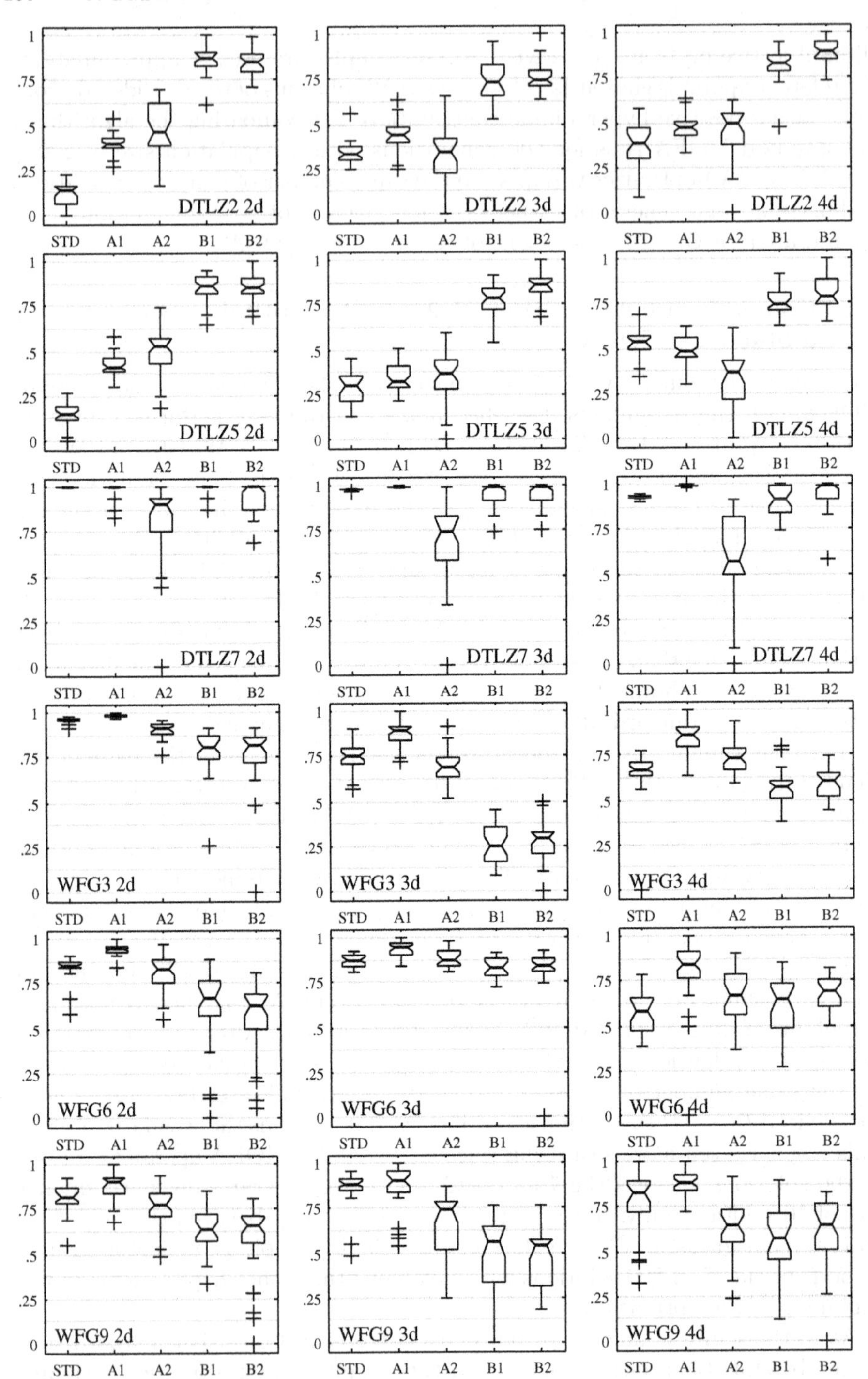

Fig. 3. Box plots of the normalized hypervolume indicator values for the four $\mathcal{P}(X)/\mathcal{P}(X)$-optimizer variants and the standard MOEA (STD) for six test problems with 2 (left), 3 (middle), and 4 (right) objectives. Higher values are better.

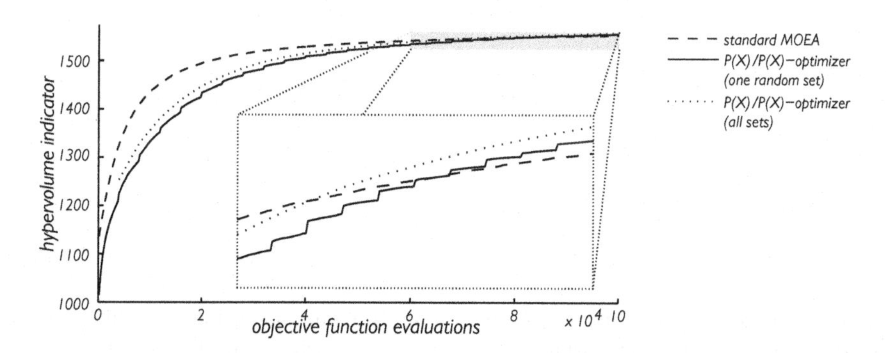

Fig. 4. Hypervolume indicator values of the $\mathcal{P}(X)/\mathcal{P}(X)$-optimizer variant A1 in comparison to the standard MOEA averaged over 30 runs over time. For the A1 variant, both the overall hypervolume of all solutions and the indicator value of one randomly picked set are shown. The insert shows a detailed view of the last time period.

cantly, better hypervolume values than the standard MOEA on all DTLZ2 and DTLZ5 instances. No general conclusion over all problems can be made for the A2, B1, and B2 variants. The A1 variant, however, yields for 16 of the 18 problems better results than the standard MOEA (except for 4-objective DTLZ5 and 2-objective DTLZ7). Hence, the A1 variant is used in all further investigations.

The huge differences between the DTLZ and the WFG problems for the different $\mathcal{P}(X)/\mathcal{P}(X)$-optimizer variants may be caused by the different characteristics of elitism: a good solution is more likely to be contained in *all* solution sets after recombination within the variants A2, B1, and B2 in comparison to the A1 variant, i.e., the diversity is lower. In addition, the diversity of solutions is also higher in the A1 variant because of its random mating selection. This low diversity between single solutions might be the reason why the three variants A2, B1, and B2 are not performing as good as the A1 variant on the WFG problems. For the DTLZ problems, however, the small diversity seems to cause no problems for the search due to the structure of the problems.

4.3 Comparing Different Mutations on Sets

In order to study the influence of the parameter G, i.e., the length of a mutation step, we run the $\mathcal{P}(X)/\mathcal{P}(X)$-optimizer variant A1 with different values for G—all other parameters are kept the same except the number of generations which is changed to keep the overall number of objective function evaluations the same. Figure 6 shows the normalized hypervolume values averaged over 30 runs together with the standard deviation. Although the influence of G is small compared to the hypervolume of the standard MOEA, we observe a tendency towards better results if the mutation length is smaller. This gives evidence that the used set recombination is a powerful operator. However, using the set recombination more frequently results in a higher running time. Although further investigations on the choice of G are needed, our choice makes a first compromise between running time and solution quality.

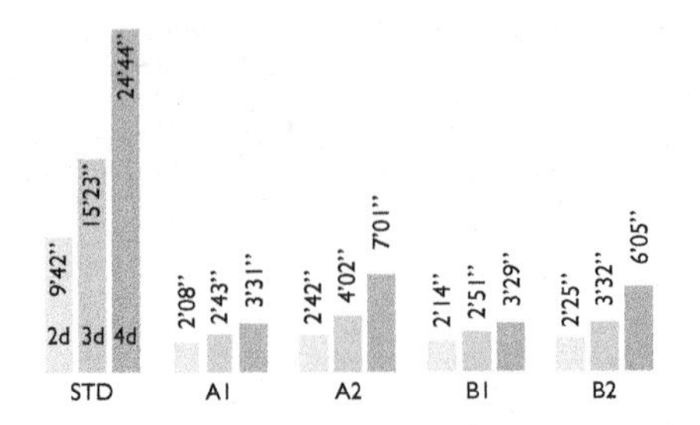

Fig. 5. Averaged running times of the four $\mathcal{P}(X)/\mathcal{P}(X)$-optimizer variants and the standard MOEA

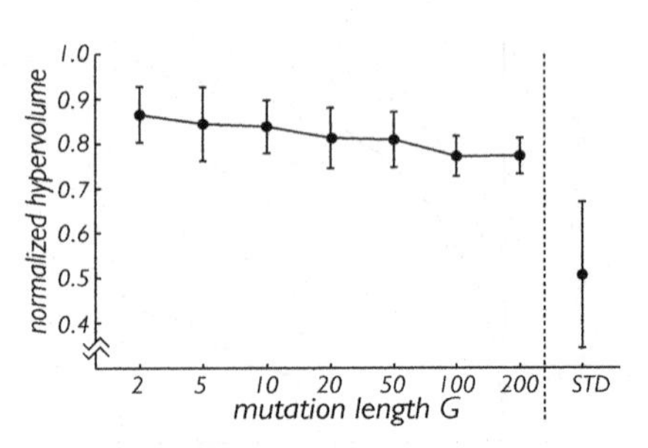

Fig. 6. Comparison between different set mutations with respect to the number G of generations (mutation length)

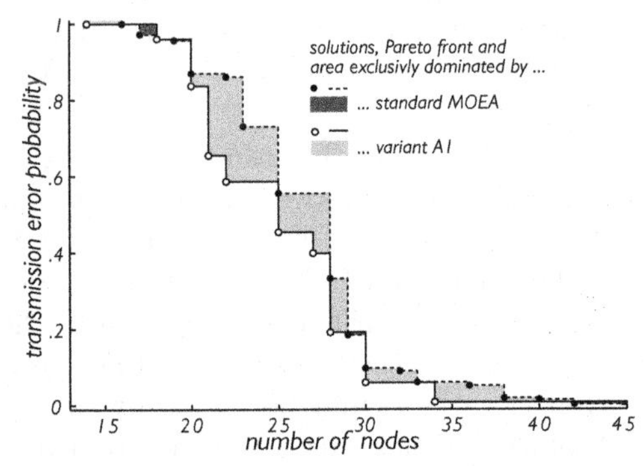

Fig. 7. Illustration of non-dominated points of the wireless sensor network application after 2,000 function evaluations

4.4 Application to Wireless Sensor Networks Deployment

Woehrle et al. [19] tackled the problem of placing wireless sensor nodes to monitor certain regions of interest by using a MOEA based on a new mutation operator. This problem of finding a sensor network deployment with a minimal number of nodes that minimizes the transmission error probability while the constraint of covering a certain region of interest is fulfilled, here, serves as an example application. Compared to the test problems investigated above, the evaluation time of a single solution is long: one evaluation takes up to several seconds per solution in comparison to milliseconds for the test problems.

When comparing the $\mathcal{P}(X)/\mathcal{P}(X)$-optimizer variant A1[3] and the standard MOEA with the same number of 2,000 function evaluations exemplary for one run[4], it turns out that A1 outperforms the standard MOEA in terms of

[3] As parameter values, $G = 2$, $\mu = 4$, $N = 10$, and $g_{\max} = 50$ are used. Furthermore, the mutation and recombination operators on single solutions within the set mutation as well as the objective functions are implemented as described in [19].

[4] We are aware of the small significance of one run, but know that changing random seeds or problem instances does not change the results qualitatively—presenting results for more runs would, however, lengthen the paper beyond the page limit.

hypervolume. The indicator value of the union of all solutions sets in the last population equals 28.376 for the A1 variant compared to 26.595 for the standard MOEA (reference point at $(50, 1.1)$). Also the plot of all achieved non-dominated solutions in objective space (Fig. 7) indicates that the solutions found are of higher quality for a decision maker. With respect to the running time, the A1 variant reaches a speed-up of about 3 on a 2-processor machine with 4 cores.

5 Conclusions

This paper has demonstrated that maintaining a population of solution sets in combination with appropriate set variation and selection operators can have advantages over classical MOEAs—in a setting where the hypervolume is the set measure to be optimized. The experimental results not only show that the quality of the generated Pareto set approximations can be largely improved, but that also the overall computation time can be reduced. As to the former, set recombination seems to play a major role, while the latter is mainly because the hypervolume indicator is faster to compute for small solution sets.

The present study represents just a first step towards evolutionary algorithms for sets and there are different promising directions for future research. In particular, the choice of the parameters (solution set size, population size, etc.) and the effects of different set operators need to be investigated. Moreover, it would be worthwhile to see whether similar results can be observed for other types of set optimization criteria.

Acknowledgments

Dimo Brockhoff has been supported by the Swiss National Science Foundation (SNF) under grant 112079. Johannes Bader has been supported by the Indo-Swiss Joint Research Program IT14.

References

1. Aherne, F.J., Thacker, N.A., Rockett, P.I.: Optimising Object Recognition Parameters using a Parallel Multiobjective Genetic Algorithm. In: Conference on Genetic Algorithms in Engineering Systems: Innovations and Applications (GALESIA 1997), pp. 1–6. IEEE Press, Los Alamitos (1997)
2. Bader, J., Zitzler, E.: HypE: An Algorithm for Fast Hypervolume-Based Many-Objective Optimization. TIK Report 286, Computer Engineering and Networks Laboratory (TIK), ETH Zurich (November 2008)
3. Beume, N., Naujoks, B., Emmerich, M.: SMS-EMOA: Multiobjective selection based on dominated hypervolume. European Journal on Operational Research 181, 1653–1669 (2007)
4. Branke, J., Schmeck, H., Deb, K., Reddy, M.: Parallelizing Multi-Objective Evolutionary Algorithms: Cone Separation. In: Congress on Evolutionary Computation (CEC 2004), pp. 1952–1957. IEEE Press, Los Alamitos (2004)

5. Coello Coello, C.A., Lamont, G.B., Van Veldhuizen, D.A.: Evolutionary Algorithms for Solving Multi-Objective Problems. Springer, Heidelberg (2007)
6. Conover, W.J.: Practical Nonparametric Statistics, 3rd edn. John Wiley, Chichester (1999)
7. Deb, K.: Multi-objective optimization using evolutionary algorithms. Wiley, Chichester (2001)
8. Deb, K., Agrawal, S., Pratap, A., Meyarivan, T.: A Fast Elitist Non-Dominated Sorting Genetic Algorithm for Multi-Objective Optimization: NSGA-II. In: Deb, K., Rudolph, G., Lutton, E., Merelo, J.J., Schoenauer, M., Schwefel, H.-P., Yao, X. (eds.) PPSN 2000. LNCS, vol. 1917, pp. 849–858. Springer, Heidelberg (2000)
9. Deb, K., Thiele, L., Laumanns, M., Zitzler, E.: Scalable Test Problems for Evolutionary Multi-Objective Optimization. In: Evolutionary Multiobjective Optimization: Theoretical Advances and Applications, pp. 105–145. Springer, Heidelberg (2005)
10. Hiroyasu, T., Miki, M., Watanabe, S.: The new model of parallel genetic algorithm in multi-objective optimization problems—divided range multi-objective genetic algorithm. In: Congress on Evolutionary Computation (CEC 2000), pp. 333–340. IEEE Press, Los Alamitos (2000)
11. Huband, S., Hingston, P., Barone, L., While, L.: A Review of Multiobjective Test Problems and a Scalable Test Problem Toolkit. IEEE Transactions on Evolutionary Computation 10(5), 477–506 (2006)
12. Igel, C., Hansen, N., Roth, S.: Covariance Matrix Adaptation for Multi-objective Optimization. Evolutionary Computation 15(1), 1–28 (2007)
13. Lee, J., Hajela, P.: Parallel Genetic Algorithm Implementation in Multidisciplinary Rotor Blade Design. Journal of Aircraft 33(5), 962–969 (1996)
14. Mezmaz, M., Melab, N., Talbi, E.-G.: Using the Multi-Start and Island Models for Parallel Multi-Objective Optimization on the Computational Grid. In: eScience, p. 112. IEEE Press, Los Alamitos (2006)
15. Poloni, C.: Hybrid GA for Multi-Objective Aerodynamic Shape Optimization. In: Genetic Algorithms in Engineering and Computer Science, pp. 397–416. John Wiley & Sons, Chichester (1995)
16. Sawai, H., Adachi, S.: Effects of Hierarchical Migration in a Parallel Distributed Parameter-free GA. In: Congress on Evolutionary Computation (CEC 2000), Piscataway, NJ, pp. 1117–1124. IEEE Press, Los Alamitos (2000)
17. Stanley, T.J., Mudge, T.: A Parallel Genetic Algorithm for Multiobjective Microprocessor Design. In: International Conference on Genetic Algorithms, pp. 597–604. Morgan Kaufmann Publishers, San Francisco (1995)
18. Talbi, E.-G., Mostaghim, S., Okabe, T., Ishibuchi, H., Rudolph, G., Coello Coello, C.A.: Parallel Approaches for Multiobjective Optimization. In: Branke, J., others (eds.) Multiobjective Optimization: Interactive and Evolutionary Approaches, pp. 349–372. Springer, Heidelberg (2008)
19. Woehrle, M., Brockhoff, D., Hohm, T., Bleuler, S.: Investigating Coverage and Connectivity Trade-offs in Wireless Sensor Networks: The Benefits of MOEAs. TIK Report 294, Computer Engineering and Networks Laboratory (TIK), ETH Zurich (October 2008); accepted for publication at MCDM 2008 conference
20. Zitzler, E., Thiele, L.: Multiobjective Evolutionary Algorithms: A Comparative Case Study and the Strength Pareto Approach. IEEE Transactions on Evolutionary Computation 3(4), 257–271 (1999)
21. Zitzler, E., Thiele, L., Bader, J.: SPAM: Set Preference Algorithm for Multiobjective Optimization. In: Rudolph, G., Jansen, T., Lucas, S., Poloni, C., Beume, N. (eds.) PPSN 2008. LNCS, vol. 5199, pp. 847–858. Springer, Heidelberg (2008)

Recombination for Learning Strategy Parameters in the MO-CMA-ES

Thomas Voß[1], Nikolaus Hansen[2], and Christian Igel[1]

[1] Institut für Neuroinformatik
Ruhr-Universität Bochum
44780 Bochum, Germany
{thomas.voss,christian.igel}@neuroinformatik.rub.de
[2] Université de Paris-Sud
Centre de recherche INRIA Saclay – Île-de-France
F-91405 Orsay Cedex, France
hansen@lri.fr

Abstract. The multi-objective covariance matrix adaptation evolution strategy (MO-CMA-ES) is a variable-metric algorithm for real-valued vector optimization. It maintains a parent population of candidate solutions, which are varied by additive, zero-mean Gaussian mutations. Each individual learns its own covariance matrix for the mutation distribution considering only its parent and offspring. However, the optimal mutation distribution of individuals that are close in decision space are likely to be similar if we presume some notion of continuity of the optimization problem. Therefore, we propose a lateral (inter-individual) transfer of information in the MO-CMA-ES considering also successful mutations of *neighboring individuals* for the covariance matrix adaptation. We evaluate this idea on common bi-criteria objective functions. The preliminary results show that the new adaptation rule significantly improves the performance of the MO-CMA-ES.

1 Introduction

The multi-objective covariance matrix adaptation evolution strategy (MO-CMA-ES) recently presented in [1,2,3] extends the single-objective CMA-ES [4,5,6] to real-valued vector optimization. The algorithm in [2] considers a population of individuals subject to multi-objective, indicator-based selection. Each of the individuals adapts its own variable-metric for generating offspring. Up until now, the update of the strategy parameters, that is, the covariance matrix and a global step-size parameter, considers only information within the genealogical tree of each individual. This work presents a new covariance matrix update procedure that incorporates information from multiple genealogies. The performance of the modified MO-CMA-ES with the enhanced adaptation scheme is empirically evaluated and compared to the performance of the original MO-CMA-ES.

The remainder of this work is organized as follows. Section two briefly describes the original MO-CMA-ES. In Section three, the new covariance matrix

M. Ehrgott et al. (Eds.): EMO 2009, LNCS 5467, pp. 155–168, 2009.

adaptation procdure is presented. The empirical evaluation is summarized in section four. We close with the final conclusions and suggestions for future research directions.

2 The MO-CMA-ES

In the following, we briefly outline the MO-CMA-ES according to [1,2]. For a detailed description and a performance evaluation on bi-objective benchmark functions we refer to [1] (see also [2,3]).

The MO-CMA-ES relies on the non-dominated sorting selection scheme [7]. As in the SMS-EMOA [8], the hypervolume-indicator serves as second-level sorting criterion to rank individuals at the same *level of non-dominance*. In the following, we first describe the general ranking procedure and then summarize the other parts of the MO-CMA-ES.

Let A be a population, and let a, a' be two individuals in A. Let the non-dominated solutions in A be denoted by $\text{ndom}(A) = \{a \in A \,|\, \nexists a' \in A : a' \prec a\}$, where $\prec$ denotes the Pareto-dominance relation. The elements in $\text{ndom}(A)$ are assigned a level of non-dominance of 1. The other ranks of non-dominance are defined recursively by considering the set A without the solutions with lower ranks [7]. Formally, let $\text{dom}_0(A) = A, \text{dom}_l(A) = \text{dom}_{l-1}(A) \setminus \text{ndom}_l(A)$, and $\text{ndom}_l(A) = \text{ndom}(\text{dom}_{l-1}(A))$ for $l \geq 1$. For $a \in A$ we define the level of non-dominance $\text{rank}(a, A)$ to be i iff $a \in \text{ndom}_i(A)$.

The hypervolume measure or $\mathcal{S}$-metric was introduced in [9] in the domain of evolutionary MOO. It can be defined as the Lebesgue measure Λ (i.e., the volume) of the union of hypercuboids in the objective space [10]:

$$\mathcal{S}_{a_{\text{ref}}}(A') = \Lambda\left(\bigcup_{a \in \text{ndom}(A')} \{(f_1(a'), \ldots, f_M(a')) \,|\, a \prec a' \prec a_{\text{ref}}\} \right) , \tag{1}$$

where a_{ref} is an appropriately chosen reference point. The contributing hypervolume of a point $a \in A' = \text{ndom}(A)$ is given by

$$\Delta_{\mathcal{S}}(a, A') = \mathcal{S}_{a_{\text{ref}}}(A') - \mathcal{S}_{a_{\text{ref}}}(A' \setminus \{a\}) . \tag{2}$$

Now we define the *contribution rank* $\text{cont}(a, A')$ of a. This is again done recursively. The element, say a, with the largest contributing hypervolume is assigned contribution rank 1. The next rank is assigned by considering $A' \setminus \{a\}$ etc. More precisely, let $c_1(A') = \text{argmax}_{a \in A'} \Delta_{\mathcal{S}}(a, A')$ and

$$c_i(A') = c_1 \left(A' \setminus \bigcup_{j=1}^{i-1} \{c_j(A')\} \right) \tag{3}$$

for $i > 1$. For $a \in A'$ we define the contribution rank $\text{cont}(a, A')$ to be i iff $a = c_i(A')$. In the ranking procedure ties are broken at random.

Finally, the following relation between individuals $a, a' \in A$ is defined:

$$a \prec_A a' \Leftrightarrow \text{rank}(a, A) < \text{rank}(a', A) \vee \\ \left[\text{rank}(a, A) = \text{rank}(a', A) \wedge \text{cont}(a, \text{ndom}(A)) < \text{cont}(a', \text{ndom}(A))\right] \quad (4)$$

In the $\mu_{\text{MO}} \times (1+1)$-MO-CMA-ES, a candidate solution $a_i^{(g)}$, $i \in \{1, \ldots, \mu_{\text{MO}}\}$, in generation g is a tuple $\left[\mathbf{x}_i^{(g)}, \bar{p}_{\text{succ},i}^{(g)}, \sigma_i^{(g)}, \mathbf{p}_{i,c}^{(g)}, \mathbf{C}_i^{(g)}\right]$, where

$\mathbf{x}_i^{(g)}$ is the current search point,
$\bar{p}_{\text{succ},i}^{(g)}$ is the smoothed success probability,
$\sigma_i^{(g)}$ is the global step-size,
$\mathbf{p}_{i,c}^{(g)}$ is the cumulative evolution path,
$\mathbf{C}_i^{(g)}$ is the covariance matrix of the search distribution.

The standard version of the $\mu_{\text{MO}} \times$ (1+1)-MO-CMA-ES is given in Algorithm 1. The indicator function $\mathbb{I}(\cdot)$ evaluates to one if its argument is true and to zero otherwise.

Algorithm 1. $\mu_{\text{MO}} \times (1+1)$-MO-CMA-ES

1 $g \leftarrow 0$, initialize $a_k^{(g)}$ for $k \in \{1, \ldots, \mu_{\text{MO}}\}$
2 **repeat**
3   **for** $k = 1, \ldots, \mu_{\text{MO}}$ **do**
4     $a'^{(g+1)}_k \leftarrow a_k^{(g)}$
5     $\mathbf{x}'^{(g+1)}_k \sim \mathbf{x}_k^{(g)} + \sigma_k^{(g)} \mathcal{N}\left(\mathbf{0}, \mathbf{C}_k^{(g)}\right)$
6   $Q^{(g)} \leftarrow \left\{a'^{(g+1)}_k, a_k^{(g)}\right\}$
7   **for** $k = 1, \ldots, \mu_{\text{MO}}$ **do**
8   `σ-update` $\left(a'^{(g+1)}_k, \mathbb{I}\left(\mathbf{x}'^{(g+1)}_k \prec_{Q^{(g)}} \mathbf{x}_k^{(g)}\right)\right)$
9   `rank-one-update` $\left(a'^{(g+1)}_k, \frac{\mathbf{x}'^{(g+1)}_k - \mathbf{x}_k^{(g)}}{\sigma_k^{(g)}}\right)$
10   `σ-update` $\left(a_k^{(g)}, \mathbb{I}\left(\mathbf{x}'^{(g+1)}_k \prec_{Q^{(g)}} \mathbf{x}_k^{(g)}\right)\right)$
11   **for** $i \in \{1, \ldots, \mu_{\text{MO}}\}$ **do**
12     $a_i^{(g+1)} \leftarrow Q^{(g)}_{\prec:i}$
13 **until** stopping criterion is met

For each of the μ_{MO} individuals, one offspring is sampled (lines 3–5). The decision whether a new candidate solution is better than its parent is made in the context of the population $Q^{(g)}$ of parent and offspring individuals due to the indicator-based selection strategy implemented in the algorithm. The covariance matrix of each offspring is adapted (line 9, see the procedure `rank-one-update`). Subsequently, the step-sizes $\sigma_k^{(g)}$ and $\sigma'^{(g+1)}_k$ of parent and offspring individuals

$a_k^{(g)}$ and $a_k^{(g+1)}$ are updated (line 8 and 10, see the procedure σ-`update`). Finally, the new parent population is selected from the set of parent and offspring individuals according to the indicator-based selection scheme (lines 11–12). Here, $Q^{(g)}_{\prec:i}$ denotes the ith best individual in $Q^{(g)}$ ranked by non-dominated sorting and the contributing hypervolume according to (4) (see also [1]).

Procedure σ-`update`($a = [\mathbf{x}, \bar{p}_{\text{succ}}, \sigma, \mathbf{p}_c, \mathbf{C}]$, p_{succ})

1 $\bar{p}_{\text{succ}} \leftarrow (1 - c_p)\bar{p}_{\text{succ}} + c_p \bar{p}_{\text{succ}}$

2 $\sigma \leftarrow \sigma \exp\left(\frac{1}{d}\frac{\bar{p}_{\text{succ}} - p^{\text{target}}_{\text{succ}}}{1 - p^{\text{target}}_{\text{succ}}}\right)$

The (external) strategy parameters are the population size, initial global step size, target success probability $p^{\text{target}}_{\text{succ}}$, step-size damping d, success rate averaging parameter c_p, cumulation time horizon parameter c_c, and covariance matrix learning rate c_{cov}. Default values as given in [1] and used in this paper are: $d = 1 + n/2$, $p^{\text{target}}_{\text{succ}} = (5 + \sqrt{1/2})^{-1}$, $c_p = p^{\text{target}}_{\text{succ}}/(2 + p^{\text{target}}_{\text{succ}})$, $c_c = 2/(n + 2)$, $c_{\text{cov}} = 2/(n^2 + 6)$ and $p_{\text{thresh}} = 0.44$. The initial global step sizes $\sigma_i^{(0)}$ are set dependent on the problem (e.g., in the case of box constraints, see below, with $x_i^u - x_i^l = x_j^u - x_j^l$ for $1 \le i, j \le n$ to $0.6 \cdot (x_1^u - x_1^l)$).

Procedure `rank-one-update`($a = [\mathbf{x}, \bar{p}_{\text{succ}}, \sigma, \mathbf{p}_c, \mathbf{C}]$, $\mathbf{z} \in \mathbb{R}^n$)

1 **if** $\bar{p}_{\text{succ}} < p_{\text{thresh}}$ **then**

2 $\quad\mathbf{p}_c \leftarrow (1 - c_c)\,\mathbf{p}_c + \sqrt{c_c(2 - c_c)}\mathbf{z}$

3 $\quad\mathbf{C} \leftarrow (1 - c_{\text{cov}})\,\mathbf{C} + c_{\text{cov}}\mathbf{p}_c\mathbf{p}_c^{\mathrm{T}}$

4 **else**

5 $\quad\mathbf{p}_c \leftarrow (1 - c_c)\,\mathbf{p}_c$

6 $\quad\mathbf{C} \leftarrow (1 - c_{\text{cov}})\,\mathbf{C} + c_{\text{cov}}\left(\mathbf{p}_c\mathbf{p}_c^{\mathrm{T}} + c_c\,(2 - c_c)\,\mathbf{C}\right)$

When in this study the MO-CMA-ES is applied to a benchmark problem $\mathbf{f}$ with box constraints, we consider a penalized fitness function

$$\mathbf{f}^{\text{penalty}}(\mathbf{x}) = \mathbf{f}(\text{feasible}(\mathbf{x})) + \alpha\|\mathbf{x} - \text{feasible}(\mathbf{x}))\|_2^2 \tag{5}$$

in the search process, where

$$\text{feasible}(\mathbf{x}) = (\min(\max(x_1, x_1^l), x_1^u), \ldots, \min(\max(x_n, x_n^l), x_n^u))^{\mathrm{T}} \tag{6}$$

and x_i^l and x_i^u are the lower and upper bound of the ith component of the search space. We set ad-hoc $\alpha = 10^{-6}$.

3 A New Covariance Matrix Update

The recombination of information provided by candidate solutions is a powerful variation operator that is present in most current single- and multi-objective evolutionary algorithms. Currently, the MO-CMA-ES as proposed in [1] lacks this feature. This section introduces a method for recombining neighbouring individuals to further speed up the strategy parameter adaptation in the MO-CMA-ES.

3.1 Incorporation of Information from Successful Offspring

The basic idea is that appropriate covariance matrices (i.e., appropriate coordinate systems) are similar for individuals that are in the same region of the decision space if we presume some notion of continuity of the objective function (more precisely, we presume that the principle of *strong causality* [11] is not too often violated). Thus, combining information about the topology of the search space gathered by neighbouring individuals is expected to speed up the learning of the covariance matrix. In the following, we realize this idea in the MO-CMA-ES.

Consider the set of parent individuals $P^{(g)}$ and the set of newly generated candidate solutions $Q^{(g)}$ in generation g. Let $Q'^{(g)} \subseteq Q^{(g)}$ be the set of successful offspring (i.e., $Q'^{(g)} \subseteq P^{(g+1)}$). The covariance matrix of each individual in $Q'^{(g)}$ is updated. The standard $\mu_{\text{MO}} \times (1+1)$-MO-CMA-ES relies on the "isolated" rank-one-update with cumulative evolution path that solely exploits the step from the parent to its offspring.

Let $a_i^{(g)}$ and $a'^{(g+1)}_i$ be an individual and its offspring, respectively. Let us assume that $a'^{(g+1)}_i$ is successful and therefore $a'^{(g+1)}_i \in Q'^{(g)}$. The rank-one-update of the covariance matrix of $a'^{(g+1)}_i$ is given by

$$\mathbf{C}'^{(g+1)}_i = (1 - c_{\text{cov}})\mathbf{C}^{(g)}_i + c_{\text{cov}}\mathbf{p}'^{(g+1)}_i \left(\mathbf{p}'^{(g+1)}_i\right)^{\mathrm{T}} , \tag{7}$$

where $\mathbf{p}'^{(g+1)}_i$ is the updated cumulative evolution path of $a'^{(g+1)}_i$. Our modification of the adaptation method reads:

$$\mathbf{C}'^{(g+1)}_i = (1 - c_{\text{cov}})\left[\left(1 - \sum_{j=1}^{\mu_{\text{MO}}} w_{ij}^{(g+1)}\right)\mathbf{C}^{(g)}_i \right.$$
$$\left. + \underbrace{\sum_{j=1}^{\mu_{\text{MO}}} w_{ij}^{(g+1)} \frac{\mathbf{x}'^{(g+1)}_j - \mathbf{x}^{(g)}_j}{\sigma^{(g)}_j}\left(\frac{\mathbf{x}'^{(g+1)}_j - \mathbf{x}^{(g)}_j}{\sigma^{(g)}_j}\right)^{\mathrm{T}}}_{=\mathbf{Z}^{(g+1)}}\right] + \underbrace{c_{\text{cov}}\mathbf{p}'^{(g+1)}_i \mathbf{p}'^{(g+1)\mathrm{T}}_i}_{\text{rank-one-update}} , \tag{8}$$

Here $w_{ij}^{(g+1)}$ is a weighting coefficient assigned to the j-th offspring individual $a'^{(g+1)}_j$. The weight is calculated anew in each generation. It is different for each individual in the offspring population. If the individual is not selected for the next parent generation, it is assigned a weight $w_{ij}^{(g+1)} = 0$.

The matrix $\mathbf{Z}^{(g+1)}$ aggregates information from the selected new candidate solutions and is of rank $\min\{\mu_{\text{MO,succ}}, n\}$ with probability one, where $\mu_{\text{MO,succ}}$

denotes the number of successful offspring. Thus, the new adaptation method is referred to as rank-$\mu_{\mathrm{MO,succ}}$-update.

In contrast to the combination of the rank-μ- and rank-one-update in the non-elitist CMA-ES (see [4]), no constant μ_{cov} is used to balance the impact of the two different update rules. The blending of old information $\mathbf{C}^{(g)}$ and new information $\mathbf{Z}^{(g+1)}$ is controlled by considering the sum of weighting coefficients.

3.2 Weighting of Neighbouring Individuals

The following assumptions underlie the calculation of the weighting coefficients:

- Non-successful offspring individuals do not represent promising sampling directions and are assigned a weight of zero.
- Information contributed by individuals that are closer to the individual to be updated is more important, as the chance for a similar topology of the search space is higher for individuals nearby.
- Distances between individuals should be measured in terms of the metric learnt by the individual to be updated.

Now we take the point of view of an offspring individual $a'^{(g+1)}_i$, $i \in \{1, \dots, \mu_{\mathrm{MO}}\}$, whose covariance matrix $\mathbf{C}'^{(g+1)}_i$ needs to be updated. A weighting coefficient $w^{(g+1)}_{ij}$ for each offspring $a'^{(g+1)}_j$, $j \in \{1, \dots, \mu_{\mathrm{MO}}\}$ is determined. The weight reflects the relevance of $a'^{(g+1)}_j$ for the covariance matrix update of $a'^{(g+1)}_i$. The importance of $a'^{(g+1)}_j$ depends on its distance to $a'^{(g+1)}_i$. The closer $a'^{(g+1)}_j$ is situated to $a'^{(g+1)}_i$, the higher the weight it is assigned. An individual $a'^{(g+1)}_j$ is considered near in the search space if $a'^{(g+1)}_i$ can reach $a'^{(g+1)}_j$ by a small number of mutative steps. That is, the individual $a'^{(g+1)}_j$ is close to $a'^{(g+1)}_i$ if the probability to sample a point close to $a'^{(g+1)}_j$ is high according to the search distribution

$$\mathcal{N}\left(\mathbf{x}'^{(g+1)}_i, \sigma'^{(g+1)^2}_i \mathbf{C}'^{(g+1)}_i\right) , \tag{9}$$

of the individual $a'^{(g+1)}_i$.

Accordingly, the distance calculation is carried out w.r.t. the shape of the search distribution and the step-size of $a'^{(g+1)}_i$ using the Mahalanobis distance based on the covariance matrix of the individual to be updated (9):

$$d_{\mathrm{M}}\left(a'^{(g+1)}_i, a'^{(g+1)}_j\right) = \frac{\sqrt{\left(\mathbf{x}'^{(g+1)}_i - \mathbf{x}'^{(g+1)}_j\right)^T \mathbf{C}'^{(g+1)^{-1}}_i \left(\mathbf{x}'^{(g+1)}_i - \mathbf{x}'^{(g+1)}_j\right)}}{\sigma'^{(g+1)}_i} \tag{10}$$

Note that $d_{\mathrm{M}}\left(a'^{(g+1)}_i, a'^{(g+1)}_j\right)$ is not symmetric as the difference vector $\mathbf{x}'^{(g+1)}_i - \mathbf{x}'^{(g+1)}_j$ is transformed into the coordinate system of $a'^{(g+1)}_i$ by multiplying with the inverse of $\sigma'^{(g+1)^2}_i \mathbf{C}'^{(g+1)}_i$.

The (MO-)CMA-ES explores the search space by means of mutative steps. Therefore, we normalize the distance w.r.t. this "unit of measurement". The scaling of the Euclidean norm of an $\mathcal{N}(\mathbf{0}, \mathbf{I})$-distributed random vector with the dimension of the search space needs to be addressed to render the distance calculation independent of the search space dimension n. To this end, the expected length of an $\mathcal{N}(\mathbf{0}, \mathbf{I})$-distributed is approximated by

$$\mathrm{E}\left(||\mathcal{N}(\mathbf{0}, \mathbf{I})||_2\right) = \sqrt{n} + \mathcal{O}(1/n) \approx \sqrt{n} \ . \tag{11}$$

Thus, the comparison of the distance $d_{\mathrm{M}}\left(a'^{(g+1)}_i, a'^{(g+1)}_j\right)$ with the expected length $\mathrm{E}\left(||\mathcal{N}(\mathbf{0}, \mathbf{I})||_2\right)$ corresponds approximately to a division by $\sqrt{n}$. This can be viewed as a normalization by the unit of measurement "reachable in *one* mutative step". Now this normalization is extended to d_{steps} mutative steps. Basically, the distance $d_{\mathrm{M}}\left(a'^{(g+1)}_i, a'^{(g+1)}_j\right)$ is compared to the expected length of a random vector $\mathbf{x} = \sum_{k=1}^{d_{\mathrm{steps}}} \mathbf{x}_k$, where $\mathbf{x}_1, \ldots, \mathbf{x}_{d_{\mathrm{steps}}}$ are independently $\mathcal{N}(\mathbf{0}, \mathbf{I})$-distributed. The variance of $\mathbf{x}$ is equal to the sum of variances of the independent steps $\mathbf{x}_k$, $k \in \{1, \ldots, d_{\mathrm{steps}}\}$,

$$\mathbf{x} = \sum_{k=1}^{d_{\mathrm{steps}}} \mathbf{x}_k \sim \mathcal{N}(\mathbf{0}, d_{\mathrm{steps}}\, \mathbf{I}) \sim \sqrt{d_{\mathrm{steps}}}\, \mathcal{N}(\mathbf{0}, \mathbf{I}) \tag{12}$$

and therefore the expected length of $\mathbf{x}$ distributed according to $\mathcal{N}(\mathbf{0}, d_{\mathrm{steps}}\, \mathbf{I})$ is

$$\mathrm{E}\left(||\mathcal{N}(\mathbf{0}, d_{\mathrm{steps}}\, \mathbf{I})||_2\right) = \sqrt{d_{\mathrm{steps}}}\ \mathrm{E}\left(||\mathcal{N}(\mathbf{0}, \mathbf{I})||_2\right) \approx \sqrt{d_{\mathrm{steps}}\, n} \ . \tag{13}$$

We determine the weights based on this distance measure. A weight $w_{ij}^{(g+1)}$ is computed by

$$w_{ij}^{(g+1)} = w'^{(g+1)}_{ij} \min\left\{1, \frac{2\mu_{\mathrm{MO,eff}} - 1}{(n+2)^2 + \mu_{\mathrm{MO,eff}}}\right\} \tag{14}$$

using the intermediate weights

$$w''^{(g+1)}_{ij} = \begin{cases} h\left(d_{\mathrm{M}}\left(a'^{(g+1)}_i, a'^{(g+1)}_j\right) / \sqrt{d_{\mathrm{steps}} n}\right) & \text{if } a'^{(g+1)}_j \prec_{Q^{(g)}} a_j^{(g+1)} \\ 0 & \text{otherwise} \end{cases} \tag{15}$$

$$w'^{(g+1)}_{ij} = \frac{w''^{(g+1)}_{ij}}{\mu_{\mathrm{MO}} - \mu_{\mathrm{MO,succ}} + \sum_{k=1}^{\mu_{\mathrm{MO}}} w''^{(g+1)}_{ik}} \ . \tag{16}$$

Only successful offspring individuals shall be considered in the update, and therefore an intermediate weight of zero is assigned to non-successful offspring. For the others, the intermediate weight $w''^{(g+1)}_{ij}$ is calculated by applying a monotonically decreasing *distance weighting function* $h : \mathbb{R}^{\geq 0} \to \mathbb{R}$ to the distance $d_{\mathrm{M}}\left(a'^{(g+1)}_i, a'^{(g+1)}_j\right) / \sqrt{d_{\mathrm{steps}} n}$. Here, $h(\cdot)$ has been chosen as

$$h : \mathbb{R}^{\geq 0} \to \mathbb{R}, \quad x \mapsto e^{-x} \ . \tag{17}$$

Thus, the neighbourhood of $a'^{(g+1)}_i$ that is considered important for the covariance matrix update depends "smoothly" on the distance measured by d_{M}. Our goal is to fuse the information encoded in the covariance matrix $\mathbf{C}'^{(g+1)}_i$ and in the matrix $\mathbf{Z}^{(g+1)}$, which contains information about successful steps:

$$\mathbf{Z}^{(g+1)} = \sum_{j=1}^{\mu_{\mathrm{MO}}} w'^{(g+1)}_{ij} \frac{\mathbf{x}'^{(g+1)}_j - \mathbf{x}^{(g)}_j}{\sigma^{(g)}_j} \left(\frac{\mathbf{x}'^{(g+1)}_j - \mathbf{x}^{(g)}_j}{\sigma^{(g)}_j} \right)^{\mathrm{T}} \tag{18}$$

The sum of all final weights determines how much emphasis we put on $\mathbf{Z}^{(g+1)}$ in the covariance matrix update (8). The "more information" is contained in $\mathbf{Z}^{(g+1)}$ the larger the sum of the final weights can be. To account for that, we first normalize the intermediate weights $w''^{(g+1)}_{ij}$ by the number of successful offspring individuals $\mu_{\mathrm{MO,succ}}$ according to Eq. (16).

Consider the case of all offspring individuals being selected for the next generation. Then the sum $\sum_{j=1}^{\mu_{\mathrm{MO}}} w'^{(g+1)}_{ij}$ evaluates to one. If these weights were used in Eq. (8), the covariance matrix $\mathbf{C}'^{(g+1)}_i$ would be replaced by the newly estimated covariance matrix $\mathbf{Z}^{(g+1)}$. This shows that we have to be careful not to put too much emphasis on $\mathbf{Z}^{(g+1)}$. The matrix $\mathbf{Z}^{(g+1)}$ has a rank of at most $\mu_{\mathrm{MO,succ}}$, which is likely to be less than n, and therefore just considering $\mathbf{Z}^{(g+1)}$ would lead to a degenerate covariance matrix.

The amount of "information" contained in $\mathbf{Z}^{(g+1)}$ clearly depends on the number of successful offspring $\mu_{\mathrm{MO,succ}}$. But due to the weighting, one can observe a "loss of variance" in $\mathbf{Z}^{(g+1)}$ we want to account for. To this end, we rely on the *variance effective selection mass*

$$\mu_{\mathrm{MO,eff}} = \frac{\left(\sum_{j=1}^{\mu_{\mathrm{MO}}} w'^{(g+1)}_{ij} \right)^2}{\sum_{j=1}^{\mu_{\mathrm{MO}}} \left(w'^{(g+1)}_{ij} \right)^2} \tag{19}$$

as a measure for the "amount of information" contained in $\mathbf{Z}^{(g+1)}$ [6]. The dependence of $\mu_{\mathrm{MO,eff}}$ on i is not indicated to keep the notation uncluttered.

To get an idea of this measure, let us assume that successful steps are distributed independently according to $\mathcal{N}(\mathbf{0}, \mathbf{I})$. Then the weighted sum of successful steps is distributed according to

$$\sum_{j=1}^{\mu_{\mathrm{MO}}} w'^{(g+1)}_{ij} \mathcal{N}(\mathbf{0}, \mathbf{I}) \quad , \tag{20}$$

with variance $\sum_{j=1}^{\mu_{\mathrm{MO}}} \left(w'^{(g+1)}_{ij} \right)^2$. As $\sum_{j=1}^{\mu_{\mathrm{MO}}} \left(w'^{(g+1)}_{ij} \right)^2 \leq \left(\sum_{j=1}^{\mu_{\mathrm{MO}}} w'^{(g+1)}_{ij} \right)^2$ we in general loose variance due to the weighting, and this is captured by $\mu_{\mathrm{MO,eff}}$. The value of $\mu_{\mathrm{MO,eff}}$ is always greater than one and less than or equal to $\mu_{\mathrm{MO,succ}}$. It is equal to μ_{MO} for $w'_{i1} = \cdots = w'_{i\mu_{\mathrm{MO}}} = 1/\mu_{\mathrm{MO}}$ and goes to one if all but one weights go to zero.

Finally, the relation between the information contributed by all selected steps and the information required to prevent from a degenerated covariance matrix is evaluated. A covariance matrix is a symmetric matrix with $n(n+1)/2$ degrees of freedom. With $\mu_{\text{MO,eff}}$ providing a measure of information within the offspring population, the term

$$\frac{\mu_{\text{MO,eff}}}{n(n+1)/2} = \frac{2\mu_{\text{MO,eff}}}{n(n+1)} , \tag{21}$$

gives an estimate of the relation between "contributed" and "required" knowledge. For constant n and $\mu_{\text{MO,eff}}$ large enough, the term evaluates to a value greater than one, thus indicating that a re-estimate of the covariance matrix based on the offspring population is possible without degenerating. If the value is less than one, the newly generated offspring do not exhibit enough information. In [4] a slightly different expression based on the same idea has been found and validated empirically. It reads

$$\frac{2\mu_{\text{MO,eff}} - 1}{(n+2)^2 + \mu_{\text{MO,eff}}} . \tag{22}$$

The calculation of the weights is now completed by incorporating this "estimate of information". It is given by:

$$w_{ij}^{(g+1)} = w'^{(g+1)}_{ij} \min\left\{1, \frac{2\mu_{\text{MO,eff}} - 1}{(n+2)^2 + \mu_{\text{MO,eff}}}\right\} . \tag{23}$$

The weight $w'^{(g+1)}_{ij}$ is rescaled if the information contained within the successful offspring is not sufficient to prevent the covariance matrix from degenerating.[1]

If none of the offspring individuals is successful, all weights are equal to zero and no rank-$\mu_{\text{MO,succ}}$-update of the covariance matrix is carried out. If all individuals of the offspring population are successful and enough information is provided by the selected steps, the old covariance matrix is discarded and re-estimated from scratch. The standard rank-one-update is always applied, see Eq. (8).

Now we have all ingredients for the modified MO-CMA-ES with "recombination" (in the sense that information from several offspring are combined) for learning strategy parameters. It is referred to as $(\mu_{\text{MO}}+\mu_{\text{MO}})$-MO-CMA-ES. Only a small modification of the original $\mu_{\text{MO}}\times(1+1)$-MO-CMA-ES (see Algorithm 1) is necessary. Before the `rank-one-update` is carried out (Algorithm 1, line 9), the covariance matrices of the individuals are updated according to the Procedure `rank-`$\mu_{\text{MO,succ}}$`-update`.

Choosing the right neighbourhood size by setting the parameter d_{steps} is crucial for the performance of the $(\mu_{\text{MO}}+\mu_{\text{MO}})$-MO-CMA-ES. A value that works reliably across different fitness functions is desired, but there is no obvious heuristic for the selection of d_{steps}. For this reason, an empirical investigation of the performance of the $(\mu_{\text{MO}}+\mu_{\text{MO}})$-MO-CMA-ES with different values for d_{steps} has been conducted in the context of this study. The bi-criteria benchmark function

[1] The expression $\min\left\{1, \frac{2\mu_{\text{MO,eff}}-1}{(n+2)^2+\mu_{\text{MO,eff}}}\right\}$ evaluates to one only if $\mu_{\text{MO,eff}}$, which is bounded from above by μ_{MO}, is larger than n^2+4n+5.

Procedure rank-$\mu_{\text{MO,succ}}$-update($a = [\mathbf{x}, \bar{p}_{\text{succ}}, \sigma, \mathbf{p}_c, \mathbf{C}], Q^{(g)}$)

1 $\mu_{\text{MO,succ}} \leftarrow 0$

2 **for** $k \leftarrow 1, \ldots, \mu_{\text{MO}}$ **do**

3 $\quad d_k \leftarrow \frac{\left\| \left[\mathbf{x}'^{(g+1)}_k - \mathbf{x}\right]^{\mathrm{T}} \mathbf{C}^{-1} \left[\mathbf{x}'^{(g+1)}_k - \mathbf{x}\right] \right\|_2}{\sigma}$

4 $\quad w_i \leftarrow \mathbb{I}\left(a'^{(g+1)}_k \prec_{Q^{(g)}} a^{(g)}_k\right) h\left(\frac{d_k}{\sqrt{n\, d_{\text{steps}}}}\right)$

5 $\quad \mu_{\text{MO,succ}} \leftarrow \mu_{\text{MO,succ}} + \mathbb{I}\left(a'^{(g+1)}_k \prec_{Q^{(g)}} a^{(g)}_k\right)$

6 $\mu_{\text{MO,eff}} \leftarrow \frac{\left(\sum_{i=1}^{\mu_{\text{MO}}} w_i\right)^2}{\sum_i^{\mu_{\text{MO}}} w_i^2}$

7 **for** $k \leftarrow 1, \ldots, \mu_{\text{MO}}$ **do**

8 $\quad w_k \leftarrow \frac{w_k}{\mu_{\text{MO}} - \mu_{\text{MO,succ}} + \sum_{i=1}^{\mu_{\text{MO}}} w_i} \min\left\{1, \frac{2\mu_{\text{MO,eff}} - 1}{(n+2)^2 + \mu_{\text{MO,eff}}}\right\}$

9 $\mathbf{C} \leftarrow \left(1 - \sum_{i=1}^{\mu_{\text{MO}}} w_i\right) \mathbf{C} + \sum_{k=1}^{\mu_{\text{MO}}} w_k \frac{\mathbf{x}'^{(g+1)}_k - \mathbf{x}^{(g)}_k}{\sigma} \left(\frac{\mathbf{x}'^{(g+1)}_k - \mathbf{x}^{(g)}_k}{\sigma}\right)^{\mathrm{T}}$

ELLI1, CIGTAB1, ELLI2, CIGTAB2 [1] and different search space dimensions n have been considered. Based on the results we derived the preliminary rule $d_{\text{steps}} = n + 3$.

4 Empirical Evaluation

This section presents a performance evaluation of the new (μ_{MO}+μ_{MO})-MO-CMA-ES on a set of common multi-objective benchmark functions. Our goal is to answer the question whether the "recombination" of strategy parameters improves the performance of the MO-CMA-ES on a broad range of bi-objective fitness functions. Therefore we compare the (μ_{MO}+μ_{MO})-MO-CMA-ES to the results of the original $\mu_{\text{MO}}\times$(1+1)-MO-CMA-ES. For comparisons of this baseline algorithm with alternative multi-objective optimization methods we refer to previous studies [1,2,3]. The experiments have been conducted using the Shark machine learning library [12].

4.1 Experimental Setup

We compare the (μ_{MO}+μ_{MO})- and the original $\mu_{\text{MO}}\times$(1+1)-MO-CMA-ES presented in [1] on three classes of two-objective benchmark functions. Both algorithms rely on the hypervolume-indicator as second-level sorting criterion. The constrained benchmark functions ZDT1, ZDT2, ZDT3 and ZDT6 (see [13]) and their rotated variants IHR1, IHR2, IHR3 and IHR6 (see [1]) have been chosen for the performance evaluation. Moreover, the set of test problems has been augmented by the unconstrained and rotated functions ELLI1, ELLI2, CIGTAB1 and CIGTAB2 (see [1]), with the distance of the optima of the single objectives

Table 1. Results of the performance comparison of the $(\mu_{\text{MO}}+\mu_{\text{MO}})$-MO-CMA-ES and the original $\mu_{\text{MO}}\times(1+1)$-MO-CMA-ES, respectively. The table shows the median of 50 trials after 250 and 500 generations, respectively, of the hypervolume-indicator (the lower the better). The better value in each row is printed in bold. The superscripts indicate whether the $(\mu_{\text{MO}}+\mu_{\text{MO}})$-MO-CMA-ES$_{n+3}$ performs significantly better than the $\mu_{\text{MO}}\times(1+1)$-MO-CMA-ES, respectively (two-sided Wilcoxon rank sum test, ** indicates a significance level of 0.001 and * a significance level of 0.01). It is important to note that different reference sets were used for computing the values after 250 and 500 generations, respectively, and that therefore the absolute values after 250 and 500 generations can not be compared directly.

	$(\mu_{\text{MO}}+\mu_{\text{MO}})$-MO-CMA-ES$_{n+3}$	$\mu_{\text{MO}}\times(1+1)$-MO-CMA-ES
	250 Generations	
ZDT1	**0.000564****	*0.000592*
ZDT2	**0.000304****	*0.000462*
ZDT3	**0.000279****	*0.000621*
ZDT6	**0.000006***	*0.000017*
IHR1	**0.000242***	*0.000443*
IHR2	**0.000783****	*0.000922*
IHR3	**0.000047****	*0.000066*
IHR6	**0.000009****	*0.000045*
ELLI1	**0.017792***	*0.018896*
ELLI2	**0.007773***	*0.008844*
CIGTAB1	**0.006947****	*0.007956*
CIGTAB2	**0.000549****	*0.000599*
	500 Generations	
ZDT1	**0.000134****	*0.000201*
ZDT2	**0.000314***	*0.000429*
ZDT3	**0.000025****	*0.000172*
ZDT6	**0.000379***	*0.000421*
IHR1	**0.000006****	*0.000023*
IHR2	**0.005776***	*0.005111*
IHR3	**0.000067****	*0.000193*
IHR6	**0.000572****	*0.000977*
ELLI1	**0.000791***	*0.000844*
ELLI2	**0.004592****	*0.006392*
CIGTAB1	**0.000253****	*0.000542*
CIGTAB2	**0.003194***	*0.003978*

set to the default value two. The default search space dimension for constrained and non-rotated benchmark functions has been chosen to be 30. In case of rotated benchmark functions, the search space dimensions has been chosen to be 10.

The value of the parameter d_{steps} of the $(\mu_{\text{MO}}+\mu_{\text{MO}})$-MO-CMA-ES has been set to the empirically validated choice of $n+3$. The number of parent and offspring individuals has been set $\mu_{\text{MO}} = 100$. We conducted 50 independent trials and evaluated the algorithms after 250 and 500 generations.

Table 2. Results of the performance comparison of the $(\mu_{MO}+\mu_{MO})$-MO-CMA-ES and the original $\mu_{MO}\times(1+1)$-MO-CMA-ES. The table shows the median of 50 trials after 250 and 500 generations of the ε_+-indicator (the lower the better). The superscripts indicate whether the $(\mu_{MO}+\mu_{MO})$-MO-CMA-ES$_{n+3}$ performs significantly better than the $\mu_{MO}\times(1+1)$-MO-CMA-ES, respectively (two-sided Wilcoxon rank sum test, ** indicates a significance level of 0.001 and * a significance level of 0.01). Different reference sets were used for computing the values after 250 and 500 generations, respectively, and therefore the absolute values after 250 and 500 generations can not be compared directly.

	$(\mu_{MO}+\mu_{MO})$-MO-CMA-ES$_{n+3}$	$\mu_{MO}\times(1+1)$-MO-CMA-ES
	250 Generations	
ZDT1	**0.013756****	*0.015349*
ZDT2	**0.222876***	*0.400001*
ZDT3	**0.076653***	*0.199655*
ZDT6	**0.000112****	*0.000231*
IHR1	**0.004432****	*0.005654*
IHR2	**0.002005****	*0.010001*
IHR3	**0.000003***	*0.000339*
IHR6	**0.002134****	*0.002667*
ELLI1	**0.027492***	*0.039816*
ELLI2	**0.000573****	*0.000742*
CIGTAB1	**0.001947****	*0.004753*
CIGTAB2	**0.002239****	*0.003333*
	500 Generations	
ZDT1	**0.013756****	*0.015349*
ZDT2	**0.222876***	*0.400001*
ZDT3	**0.076653***	*0.199655*
ZDT6	**0.000112****	*0.000231*
IHR1	**0.004432****	*0.005654*
IHR2	**0.002005****	*0.010001*
IHR3	**0.000003***	*0.000339*
IHR6	**0.002134****	*0.002667*
ELLI1	**0.027492***	*0.039816*
ELLI2	**0.000573****	*0.000742*
CIGTAB1	**0.001947***	*0.004753*
CIGTAB2	**0.002239***	*0.003333*

The evaluation procedure adheres to the suggestions given in [14]. We briefly outline the process and refer to [9,15] for a detailed description of the methods. We consider the unary hypervolume-indicator and the unary additive ϵ_+-indicator as performance measures. Before indicator values are computed, the data are normalized. We want to compare $k = 2$ algorithms on a particular optimization problem $\mathbf{f}$ after g fitness evaluations and we assume that we have conducted t trials. We consider the non-dominated individuals of the union of all $k \cdot t$ populations after g evaluations. These individuals make up the reference set

$\mathcal{R}$. Their objective vectors are normalized by an affine linear transformation such that for every objective the smallest and largest objective function values are mapped to 1 and 2, respectively. The value for the unary hypervolume-indicator is compared to the hypervolume of the reference set $\mathcal{R}$, which is also used to compute the ϵ_+-indicator value. Therefore lower indicator values indicate better performance.

We used different reference sets for the evaluation after 250 and 500 generations. Therefore, the absolute values of the results of these two lines of experiments can not be compared.

4.2 Results

The results of the performance evaluation after 25,000 and 50,000 evaluations are presented in Tables 1 and 2. Although only small differences between the two algorithms can be observed, the $(\mu_{\text{MO}}+\mu_{\text{MO}})$-MO-CMA-ES performed statistically significantly better than the $\mu_{\text{MO}}\times(1+1)$-MO-CMA-ES in all our experiments. This shows that the strategy parameter adaptation of the MO-CMA-ES is clearly improved by considering the information contributed by selected offspring individuals. The results suggest that the choice of the parameter d_{steps} linearly dependent on the search space dimension n indeed results in a robust behaviour of the $(\mu_{\text{MO}}+\mu_{\text{MO}})$-MO-CMA-ES across different classes and types of benchmark problems.

5 Conclusions and Future Work

We presented a new, more elaborate covariance matrix update scheme for the multi-objective covariance matrix adaptation evolution strategy (MO-CMA-ES). The difference from the original update method is that each individual additionally considers successful mutations of neighboring individuals. Such a lateral information transfer was not considered in the MO-CMA-ES so far, and it allows for faster adaptation of the covariance matrix. Our preliminary empirical evaluation on common bi-criteria benchmark functions shows that the new update scheme significantly improves the performance in all cases.

There is the need for further investigation and room for improvements of the proposed algorithm. For example, the choice of the parameter d_{steps} should be studied in more detail and different choices for the distance weighting function $h(\cdot)$, see Eq. (17), could be considered. The empirical evaluation should include additional benchmark functions, for instance, with a larger number of objectives. In particular, we are searching for functions where the covariance matrix update scheme presented here is outperformed by the original rank-one-update procedure.

The evaluation of the algorithms after a fixed number of evaluations – although standard in the empirical analysis of evolutionary multi-objective algorithms – may be misleading. In future work we will study the evolution of the absolute hypervolume over the whole optimization process.

References

1. Igel, C., Hansen, N., Roth, S.: Covariance matrix adaptation for multi-objective optimization. Evolutionary Computation 15(1), 1–28 (2006)
2. Igel, C., Suttorp, T., Hansen, N.: Steady-state selection and efficient covariance matrix update in the multi-objective CMA-ES. In: Obayashi, S., Deb, K., Poloni, C., Hiroyasu, T., Murata, T. (eds.) EMO 2007. LNCS, vol. 4403, pp. 171–185. Springer, Heidelberg (2007)
3. Voß, T., Beume, N., Rudolph, G., Igel, C.: Scalarization versus indicator-based selection in multi-objective CMA evolution strategies. In: IEEE Congress on Evolutionary Computation 2008 (CEC 2008), pp. 3041–3048. IEEE Press, Los Alamitos (2008)
4. Hansen, N., Ostermeier, A.: Completely derandomized self-adaptation in evolution strategies. Evolutionary Computation 9(2), 159–195 (2001)
5. Hansen, N., Müller, S.D., Koumoutsakos, P.: Reducing the time complexity of the derandomized evolution strategy with covariance matrix adaptation (CMA-ES). Evolutionary Computation 11(1), 1–18 (2003)
6. Hansen, N.: The CMA Evolution Strategy: A Tutorial, `http://www.bionik.tu-berlin.de/user/niko/cmatutorial.pdf`
7. Deb, K., Pratap, A., Agarwal, S., Meyarivan, T.: A fast and elitist multiobjective genetic algorithm: NSGA-II. IEEE Transactions on Evolutionary Computation 6, 182–197 (2002)
8. Beume, N., Naujoks, B., Emmerich, M.: SMS-EMOA: Multiobjective selection based on dominated hypervolume. European Journal of Operational Research 181(3), 1653–1669 (2007)
9. Zitzler, E., Thiele, L.: Multiobjective optimization using evolutionary algorithms - A comparative case study. In: Eiben, A.E., Bäck, T., Schoenauer, M., Schwefel, H.-P. (eds.) PPSN 1998. LNCS, vol. 1498, pp. 292–301. Springer, Heidelberg (1998)
10. Coello Coello, C.A., Van Veldhuizen, D.A., Lamont, G.B.: Evolutionary Algorithms for Solving Multi-Objective Problems. Kluwer Academic Publishers, Dordrecht (2002)
11. Rechenberg, I.: Evolutionsstrategie 1994. Werkstatt Bionik und Evolutionstechnik. Frommann-Holzboog, Stuttgart (1994)
12. Igel, C., Glasmachers, T., Heidrich-Meisner, V.: Shark. Journal of Machine Learning Research 9, 993–996 (2008)
13. Zitzler, E., Deb, K., Thiele, L.: Comparison of multiobjective evolutionary algorithms: Empirical results. Evolutionary Computation 8(2), 173–195 (2000)
14. Bleuler, S., Laumanns, M., Thiele, L., Zitzler, E.: PISA – A platform and programming language independent interface for search algorithms. In: Fonseca, C.M., Fleming, P.J., Zitzler, E., Deb, K., Thiele, L. (eds.) EMO 2003. LNCS, vol. 2632, pp. 494–508. Springer, Heidelberg (2003)
15. Zitzler, E., Thiele, L., Laumanns, M., Fonseca, C.M., Grunert da Fonseca, V.: Performance assesment of multiobjective optimizers: An analysis and review. IEEE Transactions on Evolutionary Computation 7(2), 117–132 (2003)

Multi-Objective Optimisation Problems: A Symbolic Algorithm for Performance Measurement of Evolutionary Computing Techniques

Sameh Askar* and Ashutosh Tiwari

Manufacturing Department, Decision Engineering Centre, School of applied Science, Cranfield University, Cranfield MK 43 0AL, UK
Tel.: +44(0)1234750111 Ext.: 5656
{s.e.a.askar,a.tiwari}@cranfield.ac.uk

Abstract. In this paper, a symbolic algorithm for solving constrained multi-objective optimisation problems is proposed. It is used to get the Pareto optimal solutions as functions of KKT multipliers $\vec{\lambda}$ for multi-objective problems with continuous, differentiable, and convex/pseudo-convex functions. The algorithm is able to detect the relationship between the decision variables that form the exact curve/hyper-surface of the Pareto front. This algorithm enables to formulate an analytical form for the true Pareto front which is necessary in absolute performance measurement of evolutionary computing techniques. Here the proposed technique is tested on some test problems which have been chosen from a number of significant past studies. The results show that the proposed symbolic algorithm is robust to find the analytical formula of the exact Pareto front.

Keywords: Multi-objective optimisation, Evolutionary algorithms, Symbolic algorithm, Pareto front.

1 Introduction

Due to the importance of multi-objective optimisation problems (MOOP) for scientists and engineering designers several mathematical approaches and evolutionary algorithms (EA) have been proposed. In mathematics, the Karush-Kuhn-Tucker conditions (also known as the Kuhn-Tucker or the KKT conditions) are necessary for a solution in MOOP to be optimal. Many valuable theoretical results have been gained and have drawn much attention over the past several years since Kuhn and Tucker published their paper [1] in 1950.

Optimisation algorithms such as evolutionary or particle swarm algorithms are heuristic techniques that have been recently used to deal with multi-objective optimisation problems [2]. They have adequately demonstrated their usefulness in finding a well-converged and a well-distributed set of near Pareto-optimal solutions [3] and [4]. Because of the extensive studies and the available source codes both commercially and freely of these algorithms, they have been popularly applied in various

* Corresponding author.

M. Ehrgott et al. (Eds.): EMO 2009, LNCS 5467, pp. 169–182, 2009.

problem-solving tasks and have received great attention [5]. However recent studies [6] have shown that multi-objective optimisation with fitness assignment based on Pareto-domination leads to long processing times for large population sizes. This has motivated a considerable amount of research and a wide variety of approaches have been suggested in the last few years [6]. Deb et al. have suggested a verification procedure based on KKT conditions to build confidence about the near-optimality of solutions obtained using an evolutionary optimisation procedure [11].

The aim of this paper is to present a proposed symbolic algorithm which is able to solve constrained multi-objective optimisation analytically. This new algorithm can be used to get an analytical form of the curve/hyper-surface of the Pareto front for a certain class of multi-objective problems. This class involves the set of continuous, differentiable, and convex/pseudo-convex objective functions. Moreover, for this class of functions the algorithm guides to the relationship between the decision variables which describes the Pareto front surface exactly. It is not clear from the mathematical description of the multi-objective optimisation problem (1) what would be the analytical relationship between the decision variables for the solutions to be on the true Pareto front. There is no doubt that such relationship between the decision variables need careful analysis so that one can guarantee that the solutions provided by them are Pareto optimal solutions. The observations emanated from such relationship would be very important for a designer. With such observations, the designer may be able to switch from one optimal solution to another by simple changes in the design, achieving different trade-off requirements of the objectives. This information is not only important for operational purposes; it could also provide vital insight into the problem at hand and may guide evolutionary computing techniques to finding stopping criteria and reduce the time consumption for converging to the true Pareto front. Both the analytical formula of the exact Pareto front and the relationship between the decision variables responsible for constructing this analytical formula are not provided by the state-of-the-art evolutionary algorithm, NSGA-II. The central part of the algorithm is the Karush-Kuhn-Tucker (KKT) theorem [1] which can handle high dimensionality. With this symbolic algorithm one can apply several metrics which need an analytical formula for the exact Pareto front to be known so that one can measure the performance of Evolutionary algorithms.

The layout of the paper is as follows: in section 2 some basic concepts required throughout the paper are presented. A description of the proposed symbolic algorithm is given in section 3. In section 4 some test problems to be solved using the proposed algorithm are described and the results which obtained in the experiments that performed using the algorithm are presented and discussed as well. Section 5 shows NSGA-II performance using the generational distance metric and the analytical formula of the exact Pareto front provided by the symbolic algorithm. Finally, in section 6 some conclusions are presented.

2 Preliminaries

This section highlights some definitions and notations that will be used throughout the paper. The n-dimensional Euclidean space is denoted by R^n . The constrained multi-objective optimisation problem to be handled here in this paper takes the following form [12]:

$$\begin{aligned} \min \quad & f(x) = (f_1(x), f_2(x), \dots, f_m(x))^T \\ \text{s.t.} \quad & g(x) = (g_1(x), g_2(x), \dots, g_p(x))^T \leq 0 \\ & x \in X \subseteq R^n, x^{(L)} \leq x_k \leq x^{(U)}, k = \{1,2,\dots,n\} \end{aligned} \tag{1}$$

where, $f_M : X \to R, M = \{1,2,\dots,m\}$ and $g_j : X \to R, j = \{1,2,\dots,p\}$. In this formulation, $f_i(\vec{x})$ denotes the ith objective function, $g_j(\vec{x})$ denotes inequality type of constraints. The ultimate goal is simultaneous minimisation or maximisation of all given objective functions. When the objective functions conflict each other there may be a set of many alternative solutions. This family of possible solutions cannot improve all the objective functions concurrently. This is called Pareto optimality [12] and the definition is given below. Note that any maximisation objective function can be converted into a minimisation objective by changing its sign.

Definition 2.1. A point $\hat{x} \in X$ is said to be a Pareto optimal solution (or non-inferior or efficient) to the problem (1) if and only if there is no $x \in X$ such that $f(x) \leq f(\hat{x})$.

Definition 2.2. A point $\hat{x} \in X$ is said to be a weak Pareto optimal solution to the problem (1) if and only if there is no $x \in X$ such that $f(x) < f(\hat{x})$.

Roughly speaking, a point $\hat{x} \in X$ is Pareto optimal to problem (1) if and only if one can improve (in the sense of minimisation) the value of one of the objective functions only at the cost of making at least one of the remaining objective function(s) worse; it is weak Pareto optimal if and only if one can not improve all of the objective functions simultaneously.

Definition 2.3. For any $x = (x_1, x_2, \dots, x_n)^T$ and $y = (y_1, y_2, \dots, y_n)^T \in R^n$, one can define the following [13]:

(i) $x = y$ if and only if $x_i = y_i$ for all $i = 1,2,\dots,n$;
(ii) $x \leq y$ if and only if $x_i \leq y_i$ for all $i = 1,2,\dots,n$;
(iii) $x < y$ if and only if $x_i < y_i$ for all $i = 1,2,\dots,n$

Definition 2.4. A subset $X \subseteq R^n$ is said to be a convex set if for any two points $x, y \in X$ the segment $\alpha\, x + (1-\alpha)\, y \in X$ and $\alpha \in [0,1]$.

Definition 2.5. A function $f : X \subseteq R^n \to R$ is convex if for all $x, y \in X$ is valid that $f(\alpha\, x + (1-\alpha)\, y) \leq \alpha\, f(x) + (1-\alpha)\, f(y)$ for all $\alpha \in [0,1]$.

Definition 2.6. A function $f : X \subseteq R^n \to R$ is differentiable at $\hat{x} \in X$ if $f(\hat{x} + d) - f(\hat{x}) = \nabla f(\hat{x})^T d + \|d\| \varepsilon(\hat{x}, d)$, where $\nabla f(\hat{x})$ is the gradient of f at $\hat{x}$ and $\varepsilon(\hat{x}, d) \to 0$ as $\|d\| \to 0$.

Definition 2.7. Let the function $f : X \subseteq R^n \to R$ be differentiable at every $x \in X$. Then it is pseudo-convex function if for all $x, y \in X$ such that $\nabla f(x)^T (y-x) \geq 0$, we have $f(y) \geq f(x)$.

Theorem 2.1. Suppose $f(\vec{x})$ has continuous second-order partial derivatives at $\vec{x} \in C$ on some open convex set C in R^n. If the Hessian H of $f(\vec{x})$ is positive semi-definite ($H \geq 0$) on C, then $f(\vec{x})$ is convex on C (for a proof one can see [7]).

Theorem 2.2. Let the objective and the constraint functions of problem (1) be convex and continuously differentiable at a decision vector $\hat{x} \in X$. A sufficient condition for $\hat{x}$ to be Pareto optimal is that there exist multipliers $0 < \lambda \in R^m$ and $0 \leq \mu \in R^p$ such that

$$\left.\begin{array}{l} \sum_{i=1}^{m} \lambda_i \nabla f_i(\hat{x}) + \sum_{j=1}^{p} \mu_j \nabla g_j(\hat{x}) = 0 \\ \mu_j \, g_j(\hat{x}) = 0, \; j = 1,2,\ldots, p \end{array}\right\} \quad (2)$$

Proof. See [8].

3 The Algorithm

The following steps of the algorithm are directly motivated by the KKT theorem. They have been automated and coded in Mathematica® Symbolic Toolbox and run step by step [10]. Later the algorithm is applied to some test problems to illustrate its performance. After that the same problems are solved using the state-of-the-art stochastic algorithm NSGA-II to validate the symbolic algorithm. Below are the steps of the symbolic algorithm:

Step 1. Define the objective functions f_i, $i=1,2,\ldots,M$ to be minimised.

Step 2. Calculate the Hessian Matrix $H(f_i)$ for each function separately.

If $H \geq 0$ then go to step 4, otherwise go to step 3.

Step 3. Check if the condition $f(\alpha x + (1-\alpha) y) \leq \max[f(x), f(y)]$ or $\nabla f(x)^T (y-x) \leq 0$ for an arbitrary y in the feasible space is satisfied. If yes go to step 4, otherwise terminate.

Step 4. Solve the system $\sum_{i=1}^{m} \lambda_i \nabla f_i(\hat{x}) + \sum_{j=1}^{p} \mu_j \nabla g_j(\hat{x}) = 0$ to get $x = x(\lambda, \mu)$.

Step 5. Use the system $\mu_j \, g_j(\hat{x}) = 0, \; j = 1,2,\ldots, p$ to get $\mu_j, j = 1,2,\ldots, p$ as a function of $\lambda_i, i = 1,2,\ldots, m$ and substitute in step 4.

Step 6. Substitute the result from step 5 in step 4 to obtain $x = x(\lambda)$.

Step 7. Construct the analytical formula between $f_i, i = 1,2,\ldots, m$ using $x = x(\lambda)$.

Step 8. End.

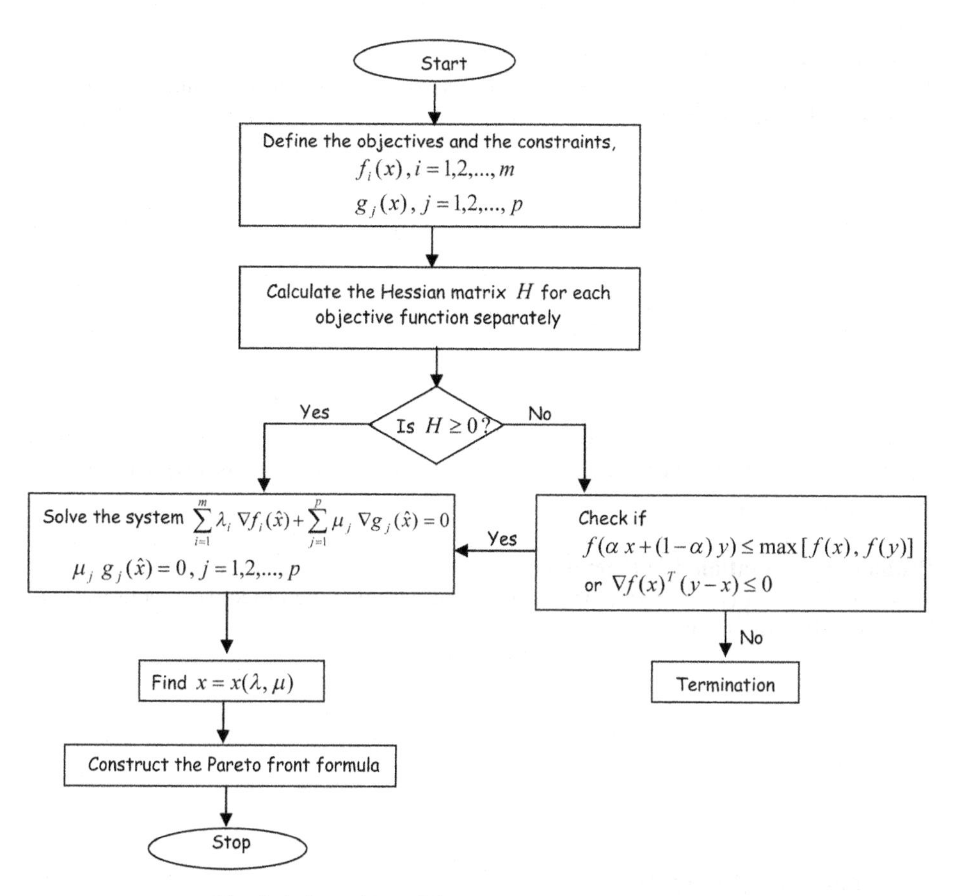

Fig. 1. A flow chart of the proposed symbolic algorithm

4 Test Problems and Results

For the validity of the new symbolic algorithm, some test problems that have been solved using the stat-of-the-art stochastic algorithms were selected from the literature to be solved by the symbolic algorithm. A complete description of these problems is shown below first and after that come a complete discussion for the results provided for each test problem separately:

Problem Formulation 4.1: (Fonseca and Fleming [3])
This problem is a typical multi-objective evolutionary algorithm (MOEA) benchmark problem. It consists of two objective functions and n decision variables as follows:

$$\min \begin{cases} f_1 = 1 - \exp\left(- \sum_{\ell=1}^{n} (x_\ell - \frac{1}{\sqrt{n}})^2 \right) \\ f_2 = 1 - \exp\left(- \sum_{\ell=1}^{n} (x_\ell + \frac{1}{\sqrt{n}})^2 \right) \end{cases} . \tag{3}$$

Restrictions: $-4 \leq x_\ell \leq 4, \quad \ell = 1,2,\ldots,n$

Problem Formulation 4.2: (Deb [4])
This problem is a two-variable problem. It consists of two objective functions which have the following form:

$$
\begin{aligned}
&\min \quad F = (f_1(x), f_2(x)), \textit{ where} \\
&f_1(x) = f(x) \\
&f_2(x) = g(y) \cdot h(f, y) \\
&\textit{and} \\
&f(x) = x \\
&g(y) = 1 + 10\, y \\
&h(f, y) = 1 - \left(\frac{f}{g}\right)^{\alpha} - \left(\frac{f}{g}\right) \cdot \sin\,(2\pi \cdot q \cdot f)
\end{aligned}
\tag{4}
$$

where q defines the number of lags in the interval $[0,1]$ and $\alpha = 2$ is a typical choice.
Restrictions: $0 \leq x, y \leq 1$

Problem Formulation 4.3: (Viennet [3])
This problem is a two-variable problem. It consists of three objective functions that have the following form:

$$
\min \begin{cases}
f_1(x, y) = x^2 + (y-1)^2, \\
f_2(x, y) = x^2 + (y+1)^2 + 1, \\
f_3(x, y) = (x-1)^2 + y^2 + 2
\end{cases}
\tag{5}
$$

Restrictions: $-2 \leq x, y \leq 2$

Problem Formulation 4.4: (Constrained problem [4])
This problem is a two-variable problem. It consists of two objective functions and two inequality constraints. It has the following form:

$$
\begin{cases}
f_1(x) = x_1, \\
f_2(x) = \dfrac{1 + x_2}{x_1}, \\
g_1(x) \equiv x_2 + 9x_1 \geq 6, \\
g_2(x) \equiv -x_2 + 9x_1 \geq 1
\end{cases}
\tag{6}
$$

Restrictions: $0.1 \leq x_1 \leq 1$ and $0 \leq x_2 \leq 5$

After executing the symbolic algorithm, the following results and observations are obtained:

Problem Analysis 4.1: For this problem at $n = 2$ the analytical formula of the exact Pareto front is:

$$f_2 = 1 - \exp\left(-\left[2 - \sqrt{-\ell n\,(1 - f_1)}\right]^2\right) \text{ and } 0 \le f_1 < 1 - \exp\left(-2\left[4 + \frac{1}{\sqrt{2}}\right]^2\right) \tag{7}$$

Although both the objective functions are convex functions, the exact Pareto front is non-convex curve as can be seen in Figure 2 (black bold curve). This curve is constructed by the linearity relationship between the decision variables, $x_1 = x_2$ in the interval $[-1.4389, 0.707]$. As can be shown from Figure 3, not all the linearity relationships between the decision variables are used to construct the exact Pareto front. This interesting observation will help the designer to switch from one optimal solution to another. In addition, this linear relationship can be written as functions of KKT multipliers as follows:

$$\left\{ x_1 = \frac{\sqrt{2}\,\lambda_1 - \sqrt{2}\,\lambda_2}{2\,(\lambda_1 + \lambda_2)},\quad x_2 = \frac{\sqrt{2}\,\lambda_1 - \sqrt{2}\,\lambda_2}{2\,(\lambda_1 + \lambda_2)} \right\}$$

This problem has been solved using – a state of the art evolutionary technique – NSGA-II [5] with population size 100 and 100 generations using standard parameters. The result is plotted in Figure 2 (Red squares). It is shown that the robust of NSGA-II in finding uniform solutions on the exact curve of the Pareto front. Here raises the robust of the symbolic algorithm in providing a connected curve of the Pareto front. Furthermore, the symbolic algorithm is guided to the relationship between the decision variables responsible for constructing that curve. In addition, this problem has been solved using the symbolic algorithm at $n = 3$ and it yielded the following:

$$f_2 = 1 - \exp\left(-\left[2 - \sqrt{-\ell n\,(1 - f_1)}\right]^2\right) \text{ and } 0 \le f_1 < 1 - \exp\left(-3\left[4 + \frac{1}{\sqrt{3}}\right]^2\right) \tag{8}$$

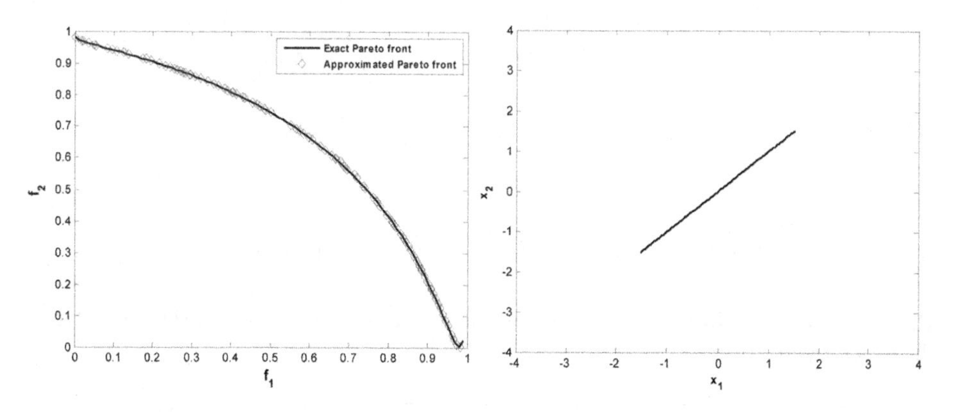

Fig. 2. Objective space of problem 4.1

Fig. 3. Decision space of problem 4.1

This means that the dimension has no impact on the shape of the Pareto front; only the constraint imposed on f_1 changed. The algorithm shows also that the Pareto optimal solutions for this problem satisfy at $x_1 = x_2 = x_3$. Again the linearity between the decision variables is the responsible for constructing the analytical formula of the exact Pareto front.

Problem Analysis 4.2: For this problem at $q = 12$ the symbolic algorithm yielded the following results:

Case 1. $x = y = 1 \Rightarrow \vec{f} = (1, 10.9091)$, This is local Pareto front point.

Case 2. $x = y = 0 \Rightarrow \vec{f} = (1, 1)$, This is local Pareto front point.

Case 3. $y = 0$ and x satisfies the equation, $2x + \sin(24\pi x) + 24\pi x \cos(24\pi x) = \frac{\lambda_1}{\lambda_2}$.

In this case the analytical formula that involves the exact Pareto front takes the form:

$$f_2 = 1 - f_1^2 - f_1 \sin(24\pi f_1) \text{ and } 0 < f_1 < 1 \tag{9}$$

This formula is plotted in Figure 4 (Black bold curve). As can be seen from Figure 4 it is a disconnected Pareto front. Cases 1 and 2 are dominated by points on this curve. This problem has been solved using NSGA-II with population size 100 and 100 generations using standard parameters. The result is plotted in Figure 4 (Red circles) and same observations found like the previous problem.

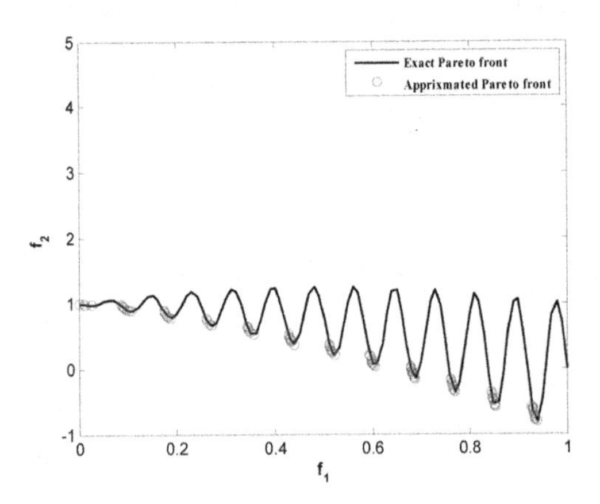

Fig. 4. Exact and approximated Pareto front to problem 4.2

Problem Analysis 4.3: For this problem the equation of the hyper-surface involved the exact Pareto front is:

$$f_3 = \left(\frac{1}{4}\sqrt{10f_2 + 2f_1(3 + f_2) - f_1^2 - f_2^2 - 25} - 1\right)^2 + \frac{1}{16}(f_1 - f_2 + 1)^2 + 2 \tag{10}$$

$$\text{and } 0 \leq f_1 \leq 13,\ 0 \leq f_2 \leq 14$$

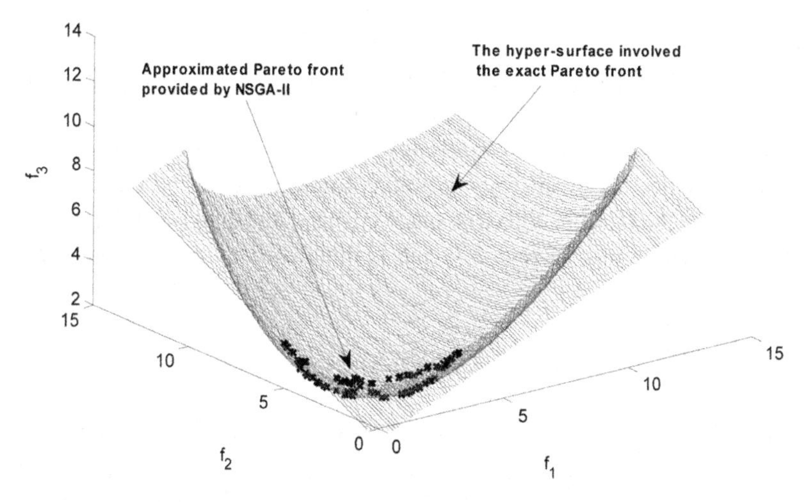

Fig. 5. Exact and approximated Pareto front

The eq. 10 is plotted in Figure 5 (the coloured hyper-surface). This equation is constructed by the following relationship between the decision variable:

$$y = \frac{\lambda_1 - \lambda_2}{\lambda_3} x$$

This problem has been solved using NSGA-II with population size 100 and 100 generations using standard parameters. The result is plotted in Figure 5 (Black crosses) and same observations found like the two previous problems.

Problem Analysis 4.4: For this problem step 4 of the symbolic algorithm yielded:

$$x_1 = \frac{\lambda_2}{\mu_1 - \mu_2 + \mu_5 - \mu_6} \quad , \quad x_2 = \frac{\lambda_2[\lambda_1 - 9(\mu_1 + \mu_2) - \mu_3 + \mu_4]}{(\mu_1 - \mu_2 + \mu_5 - \mu_6)^2} - 1 \qquad (11)$$

with 19 cases for $\mu_1, \mu_2, \mu_3, \mu_4, \mu_5$ and μ_6 have been obtained by step 5. Eight cases are accepted as they make Eq. 11 within its range. The other cases are rejected as they make the decision variables out of their ranges and give complex values for x_1 and x_2. In addition, these rejected cases make the constraints unsatisfied. The accepted cases are:

Case 1. $\mu_1 = \frac{\lambda_1}{18}$, $\mu_2 = \frac{\lambda_1}{18} - \frac{18\lambda_2}{7}$, $\mu_3 = \mu_4 = \mu_5 = \mu_6 = 0.$

This case yields the point $(0.3889, 2.5)$ on the border of the feasible decision space (Bold line, Figure 6). It satisfies the inequality constraints imposed on the problem. The corresponding point in the feasible objective space is $(0.3889, 8.9997)$ on the region A (Bold curve, Figure 7).

Case 2

$$\mu_1 = 0.111111\lambda_1 - 0.25\lambda_2,\ \mu_2 = \mu_3 = \mu_4 = \mu_6 = 0,\ \mu_5 = 0.111111\lambda_1 + 1.75\lambda_2.$$

This case yields the point $(0.8, 0.44)$ in the feasible decision space (Bold point, Figure 6). It satisfies the inequality constraints imposed on the problem. The corresponding point in the feasible objective space is $(0.8, 1.8)$ (Bold point, Figure 7). This point is better than points on region C and is dominated by points from regions A and B.

Case 3. $\mu_1 = \sqrt{\frac{\lambda_1 \lambda_2}{7}},\ \mu_2 = \mu_3 = \mu_4 = \mu_5 = \mu_6 = 0.$

This case yields: $x_1 = \sqrt{\frac{7\lambda_2}{\lambda_1}},\ x_2 = 6 - 9x_1$ and $0.3889 \le x_1 \le 0.6667$. This relationship between x_1 and x_2 represents the bold line A in the decision space (Figure 6). All the points satisfying this line A are used to construct the formula, $f_2 = \frac{7}{f_1} - 9,\ 0.3889 \le f_1 \le 0.6667$. This formula is the first part of the exact Pareto front (Bold curve A, Figure 7)

Case 4. $\mu_1 = \mu_2 = \mu_3 = \mu_4 = \mu_6 = 0, \mu_5 = \sqrt{\lambda_1 \lambda_2}.$

This case yields: $x_1 = \sqrt{\frac{\lambda_2}{\lambda_1}},\ x_2 = 0$ and $0.6667 \le x_1 \le 1$. This relationships for x_1 and x_2 represent the bold line B in the decision space (Figure 6). All the points satisfying this line B are used to construct the formula, $f_2 = \frac{1}{f_1},\ 0.6667 \le f_1 \le 1$. This formula is the second part of the exact Pareto front (Bold curve B, Figure 7)

Case 5. $\mu_1 = 0,\ \mu_2 = 0.111111\lambda_1 - 1.5\lambda_2,\ \mu_3 = \mu_4 = \mu_5 = 0,\ \mu_6 = -0.111111\lambda_1.$
This case yields the point $(0.6667, 5)$ in the feasible decision space (Bold point on the bold line C, Figure 6). It satisfies the inequality constraints imposed on the problem. The corresponding point in the feasible objective space is $(0.6667, 8.9996)$ (Bold point on the curve C, Figure 7). This point is dominated by points on the curve A.

Case 6. $\mu_1 = \mu_2 = \mu_3 = \mu_4 = \mu_5 = 0,\ \mu_6 = -\sqrt{\frac{\lambda_1 \lambda_2}{6}}.$

This case yields: $x_1 = \sqrt{\frac{6\lambda_2}{\lambda_1}},\ x_2 = 5$ and $0.6667 \le x_1 \le 1$. This relationships for x_1 and x_2 represent the bold line C in the decision space (Figure 6). All the points satisfying this line C are used to construct the formula, $f_2 = \frac{6}{f_1},\ 0.6667 \le f_1 \le 1$. This formula creates the part C (Red curve in Figure 7). It is a local Pareto front and is dominated by both curves A and B.

Case 7. $\mu_1 = \mu_2 = \mu_3 = 0$, $\mu_4 = -\lambda_1 + \lambda_2$, $\mu_5 = \lambda_2$, $\mu_6 = 0$.
This case yields the point $(1,0)$ on the border of the feasible decision space (Bold line B, Figure 6). It satisfies the inequality constraints imposed on the problem. The corresponding point in the feasible objective space is $(1,1)$ on the border of the region B (Bold curve, Figure 7).

Case 8. $\mu_1 = \mu_2 = \mu_3 = 0$, $\mu_4 = -\lambda_1 + 6\lambda_2$, $\mu_5 = 0$, $\mu_6 = -\lambda_2$.
This case yields the point $(1,5)$ on the border of the feasible decision space (Bold line C, Figure 6). It satisfies the inequality constraints imposed on the problem. The corresponding point in the feasible objective space is $(1,6)$ on the border of the region C (Red bold curve, Figure 7). This point is dominated by the point $(1,1)$.

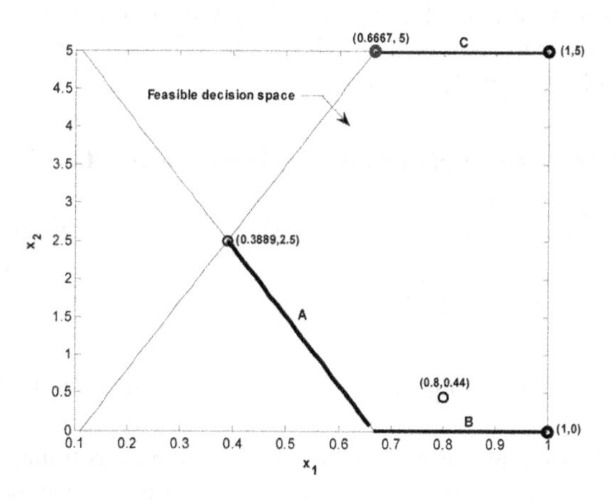

Fig. 6. Feasible search region for problem 1 in the decision variable space

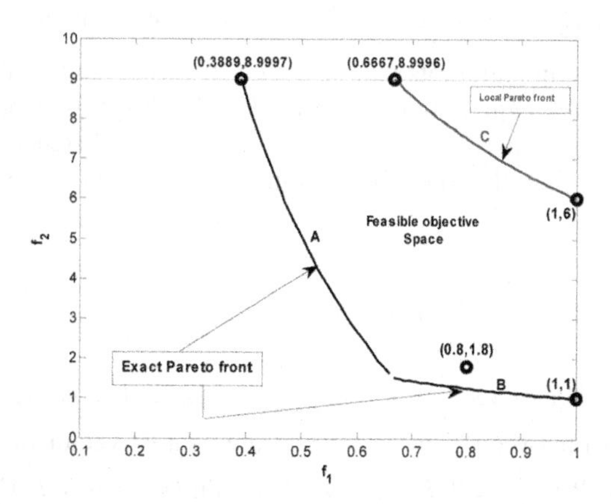

Fig. 7. The exact and local Pareto front for problem 1 using the proposed symbolic algorithm

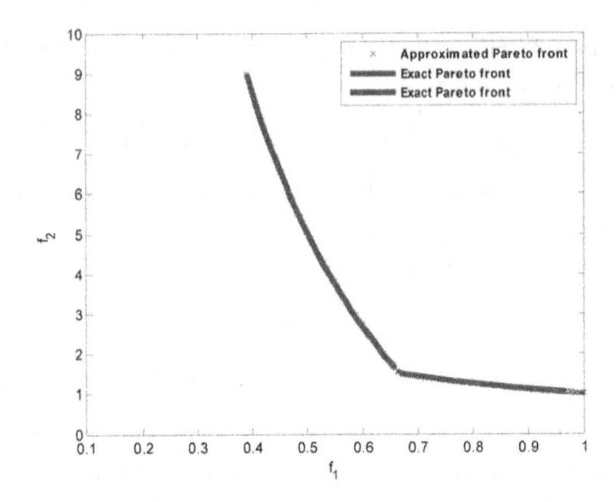

Fig. 8. The approximated Pareto front for problem 1 using NSGA-II algorithm

This problem has been solved using NSGA-II with population size 500 and 500 generations using standard parameters. The result is plotted in Figure 8 and same observations found like the three previous problems.

5 NSGA-II Absolute Performance Measurement

The term performance is always involved when comparing different optimisation techniques experimentally. In the case of multi-objective optimisation, the definition of quality is substantially complex because the optimisation goal itself consists of multiple objectives [9]:

- The distance of the resulting non-dominated set to the Pareto front should be minimised.
- A good (in most cases) uniform distribution of the solutions found is desirable. The assessment of this criterion might be based on a certain distance metric.
- The extent of the obtained non-dominated front should be maximized, i.e., for each objective, a wide range of values should be covered by the non-dominated solutions.

In the literature, some attempts can be found to formalize the above definition (or parts of it) by means of quantitative metrics [2]. Within this paper the generational distance (GD) metric is used. This metric is the average distance from the obtained Pareto front (PF_{known}) to the true Pareto front (PF_{true}) and is defined as follows [2]:

$$GD = \frac{\left(\sum_{i=1}^{n} d_i^p\right)^{\frac{1}{p}}}{n} \tag{12}$$

where n is the number of vectors in PF_{known}, $p=2$ and d_i is the Euclidean distance (in objective space) between each vector and the nearest vector of PF_{true}. The result $GD=0$ indicates $PF_{known} = PF_{true}$; any other value indicates PF_{known} deviates from PF_{true}.

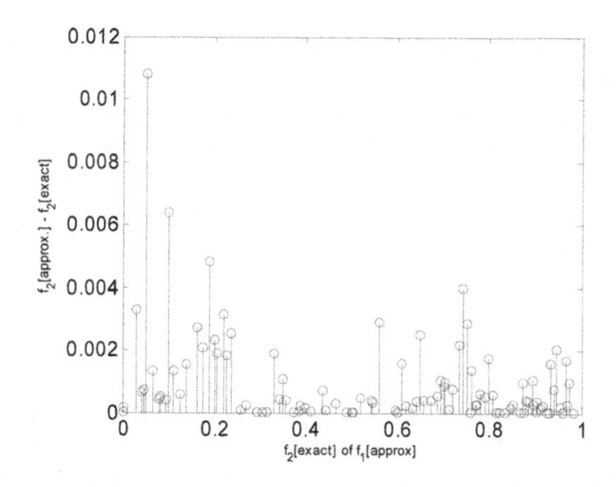

Fig. 9. The residual plot between the approximated Pareto front and the exact Pareto front for problem 4.1

The performance of NSGA-II is shown in Figure 9. This figure illustrates the residuals between the approximated Pareto front and the true Pareto front. As expected for a stochastic technique NSGA-II is able to find some points but not all the points on the true Pareto front in the final generation. There are also some minor deviations from the true Pareto front as shown in the left part of Figure 9.

The NSGA-II with 100 generations and 100 individuals in each generation has been executed 10 times on Fonseca and Fleming problem. By choosing 100 values on the true Pareto front provided by the paper's algorithm near to the 100 individuals obtained by NSGA-II, the generational distance metric has been calculated in each experiment separately using eq. 12.

The experiments show that with a minimum $GD = 0.000829839$ the NSGA-II can approximate the Pareto front in some runs quite well. However, in other runs the approximated Pareto front obtained by NSGA-II is not perfect since maximum $GD = 0.001065864$. The mean and standard deviation of all the 10 experiments are $\mu = 0.982912E\text{-}3$ and $\sigma = 7.23078E\text{-}5$, respectively. The small standard deviation shows that the GD values after 100 generations of the NSGA-II are already quite close to the mean. However, a GD different from zero indicated an ongoing approximation process.

The proposed KKT-based algorithm providing a closed formula for the Pareto front curve allows for a very precise statistical analysis of the performance of stochastic multi-objective optimisation techniques such as NSGA-II using an absolute performance measure such as the GD.

6 Conclusions

A symbolic algorithm for multi-objective optimisation problems was proposed. It has been applied on some test problems. Exact solutions for these problems have been found by this algorithm. The analytical form of the exact Pareto front has been formulated using the algorithm for these problems as well. Furthermore, a linear relationship between the decision variables has been formulated as a function of KKT multipliers.

This relationship itself is responsible for constructing the true Pareto front. As has been mentioned within this paper, this relationship has a significant contribution in innovation. It guides the designer to switch from one optimal solution to other. Furthermore, it helps to measure the performance of evolutionary algorithms. In addition, it might be used to form the stopping criteria for evolutionary algorithms. The generational distance metric was used to evaluate the performance of the NSGA-II algorithm using the analytical formula of the exact Pareto front found by the proposed algorithm.

References

1. Kuhn, H.W., Tucker, A.W.: Nonlinear Programming. In: Proceedings of the Second Berkeley Symposium on Mathematical Statistics and Probability, pp. 481–491 (1950)
2. Collette, Y., Siarry, P.: Multiobjective Optimisation: Principles and Case Studies. Springer, Berlin (2003)
3. Coello, C.A.C., Van Veldhuizen, D.A., Lamont, G.B.: Evolutionary Algorithms for Solving Multi-Objective Problems. Kluwer Academic Publishers, New York (2002)
4. Deb, K.: Multi-Objective Optimisation using Evolutionary Algorithms. Wiley, Chichester (2001)
5. Deb, K., Sundar, J.: Reference Point Based Multi-Objective Optimisation Using Evolutionary Algorithms. In: Proceedings of the 8th annual conference on Genetic and evolutionary computation, Seattle, Washington, USA, pp. 635–642 (2006)
6. Jensen, M.K.: Reducing the Run-Time Complexity of Multi-objective EAs: The NSGA-II and Other Algorithms. IEEE Transactions on Evolutionary Computation 7(5), 503–515 (2003)
7. Arora, J.S.: Introduction to Optimum Design. Elsevier, Academic Press, UK (2004)
8. Miettinen, K.: Nonlinear Multiobjective Optimisation. Kluwer Academic Publishers, Boston (1999)
9. Zitzler, E., Deb, K., Thiele, L.: Comparison of Multi-objective Evolutionary Algorithms: Empirical Results. Evolutionary Computation 8(2), 173–195 (2002)
10. Askar, S.S., Tiwari, A., Mehnen, J., Ramsden, J.: Solving Real-Life Multi-objective Optimisation Problems: A Mathematical Approach. In: Cranfield Multi-Strand Conference: Creating Wealth Through Research and Innovation (CMC 2008), UK (accepted, 2008)
11. Deb, K., Tewari, R., Dixit, M., Dutta, J.: Finding Trade-off Solutions Close to KKT Points Using Evolutionary Multi-objective Optimisation. In: IEEE Congress on Evolutionary Computation (CEC 2007), pp. 2109–2116 (2007)
12. Sawaragi, Y., Nakayama, H., Tanino, T.: Theory of Multi-Objective Optimisation. Academic Press Inc., London (1985)
13. Ehrgott, M.: Multicriteria Optimisation: Lecture notes in Economics and Mathematical Systems. Springer, Germany (2000)

On the Effect of the Steady-State Selection Scheme in Multi-Objective Genetic Algorithms

Juan J. Durillo, Antonio J. Nebro, Francisco Luna, and Enrique Alba

Department of Computer Science, University of Málaga, Spain
{durillo,antonio,flv,eat}@lcc.uma.es

Abstract. Genetic Algorithms (GAs) have been widely used in single-objective as well as in multi-objective optimization for solving complex optimization problems. Two different models of GAs can be considered according to their selection scheme: generational and steady-state. Although most of the state-of-the-art multi-objective GAs (MOGAs) use a generational scheme, in the last few years many proposals using a steady-state scheme have been developed. However, the influence of using those selection strategies in MOGAs has not been studied in detail. In this paper we deal with this gap. We have implemented steady-state versions of the NSGA-II and SPEA2 algorithms, and we have compared them to the generational ones according to three criteria: the quality of the resulting approximation sets to the Pareto front, the convergence speed of the algorithm, and the computing time. The results show that multi-objective GAs can profit from the steady-state model in many scenarios.

Keywords: Genetic Algorithms, Comparative Study, Generational and Steady-State Selection Scheme.

1 Introduction

Genetic Algorithms (GAs) have been widely applied for solving optimization problems in many areas. Since the appearance of the first multi-objective genetic algorithm (MOGA), the *Multiple Objective Optimization with Vector Evaluated Genetic Algorithms* (VEGA) [16], until today, there has been a growing interest in these kinds of algorithms for problems with two or more objectives. GAs are very popular in multi-objective optimization in part because they can obtain a front of solutions in one single run. Thus, the most well-known algorithms in this area are GAs: NSGA-II [5] and SPEA2 [20].

Based on the selection scheme, there exist two main models of GAs: generational and steady-state. In the generational model, the algorithm creates a population of individuals from an old population using the typical genetic operators (selection, crossover, and mutation); this new population becomes the population of the next generation. On the other hand, a steady-state GA creates typically only one new member which is tested to be inserted in the population at each step of the algorithm.

In this paper we study the aforementioned selection schema and their effect in MOGAs. In order to investigate this issue, we have used steady-state versions of

M. Ehrgott et al. (Eds.): EMO 2009, LNCS 5467, pp. 183–197, 2009.

NSGA-II and SPEA2 and we have compared them to the original generational ones. For making a broad comparison of these algorithms, we have evaluated them by using test functions belonging to three different benchmarks (ZDT [19], DTLZ [6], and WFG [11]) and we have considered three different criteria. First, we have assessed the quality of the Pareto fronts obtained by those algorithms applying the additive unary epsilon ($I_{\epsilon}^{1}+$) [13], spread (Δ) [5], and hypervolume (HV) [21] quality indicators. Second, we have studied the convergence speed of them, i.e., the number of functions evaluations required by the algorithms to converge towards the optimal Pareto front. Finally, we have measured the computing time they require for solving the considered problems.

The remainder of this paper is structured as follows. The next section presents some previous implementations of steady-state multi-objective algorithms in the literature. Section 3 describes the steady-state versions of NSGA-II and SPEA2. In Section 4, we detail the experimentation we have carried out, while Section 5 is devoted to showing and discussing the obtained results. Finally, Section 6 draws the main conclusions and future lines of work.

2 Related Work

In this section we analyze previous works which have made use of a steady-state scheme in MOGAs. Many algorithms based on such a scheme have been proposed in the last few years; we focus here on some representative proposals.

One of the first steady-state MOGAs described in the literature was the *Pareto Converging Genetic Algorithm* (PCGA) [14]. PCGA used a $(\mu + 2)$ scheme and a novel mechanism based on histograms of ranks for assessing convergence to the Pareto front. It was found to produce diverse sampling of the Pareto front without niching and with significantly less computing effort than NSGA, the previous version of NSGA-II. Nevertheless, the algorithms were evaluated using only three test problems and no comparisons with PCGA using a generational scheme were reported.

The *Simple Evolutionary Algorithm for Multi-Objective Optimization* (SEAMO) was proposed in [18]. It was a simple steady-state approach following a $(\mu + 1)$ scheme that used only one population and depended entirely on the replacement policy used: no rankings, subpopulations, niches or auxiliary approach were required. Due to that a generational version of SEAMO makes no sense, it was only compared with NSGA-II and SPEA2 using as benchmark the multiple knapsack problem.

Deb et *al.* proposed in [4] an ϵ-Domination Based Steady State MOEA, which was also evaluated in [3]. This algorithm used a $(\mu + 1)$ scheme and it was composed of a population and an archive, which used an ϵ-Domination mechanism. In each generation, one parent from the population and one from the archive were selected to create new offsprings, which were tested to be inserted in both the population and the archive using different strategies. It was compared with several state-of-the-art multi-objective algorithms using both bi-objective and three-objective optimization problems. No comparisons with the same algorithm using a generational scheme were reported.

Two steady-state MOGAs were presented in [1]: the *Objective Exchanging Genetic Algorithm for Design Optimization* (OEGADO), and the *Objective Switching Genetic Algorithm for Design Optimization* (OSGADO). The former proposal consisted of several steady-state single-objective optimization GAs which periodically exchanged information about the objectives; the second algorithm was also composed of multiple single objective optimization algorithms, but in this case these algorithms periodically switched the objective they optimized. Both algorithms were compared to NSGA-II using a benchmark composed of four academic problems, and two engineering problems. In this work, neither OSGADO nor OEGADO were evaluated using generational single-objective GAs.

Emmerich et *al.* presented in [10] the so-called *S metric selection EMOA* (SMS-EMOA), which was a hypervolume based steady-state GA. It had also a (μ+1) scheme. The paper included a theoretical analysis in which the advantages of using a steady-state scheme in terms of the complexity of this kind of algorithms were proved. The algorithm was evaluated using the ZDT benchmark, and it was compared to NSGA-II, SPEA2, and the above described ϵ-MOEA. As with the above described proposals, no comparisons using a generational scheme were reported.

Srinivasan et *al.* proposed in [17] a new version of the NSGA-II algorithm. This algorithm used a ($\mu + \lambda$) scheme, like the original NSGA-II. The main difference was that once all the individuals have been generated, they were considered to update the population in a steady-state model. The new proposal was evaluated using a benchmark composed of 9 problems and compared to the original NSGA-II algorithm.

Igel et *al.* studied in [12] the effect of two different steady-state schemes, (μ+1) and ($\mu_<$+1), for the *Multi-objective covariance matrix adaption evolution strategy* (MO-CMA-ES). The latter steady-state scheme did not consider all the population for selecting the parents. These different approaches were compared to a generational scheme, NSGA-II and SPEA2, using a benchmark composed of constrained and unconstrained test functions.

Durillo et *al.* proposed in [8] a steady-state version of the NSGA-II algorithm with a ($\mu + 1$) scheme and they compared it to the original one. The reported results showed that, by using such scheme, the algorithm was able to outperform the original one in terms of convergence towards the optimal Pareto front and spread of the resulting fronts of solutions.

Summarizing this section, many of the works in the literature present new steady-state algorithms and compare them against the state-of-the-art MOGAs; comparisons with the same algorithm using a generational scheme are scarce. Furthermore, many of these proposals are only evaluated using a benchmark composed of a small number of problems, and they take into account only the quality of the final front obtained, paying no attention to other issues such as the convergence speed of the algorithms.

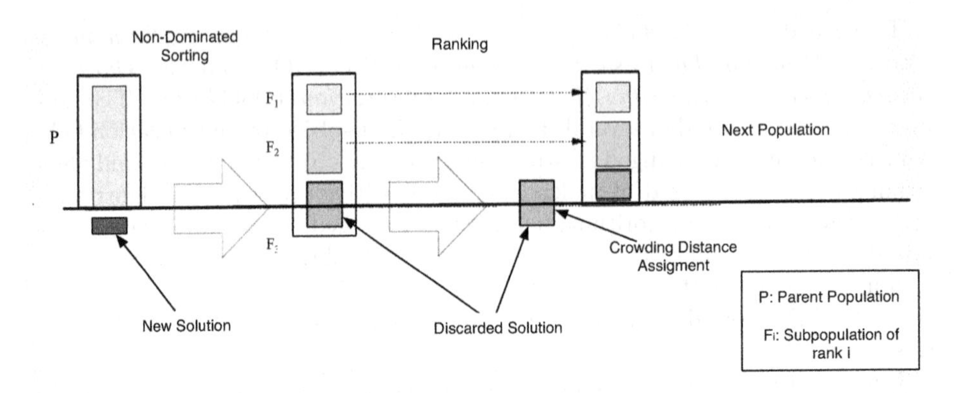

Fig. 1. The NSGA-II$_{ss}$ procedure. Only one offspring is generated and tested to be inserted at each step.

3 Steady State Versions

In this section we present the steady-state versions of NSGA-II and SPEA2. We describe briefly the original (generational) algorithms, and we will only go deeply into details related to the steady-state proposals.

The NSGA-II algorithm was proposed by Deb *et al.* [5]. It is based on a current population that is used to create an auxiliary one (the offpring population); after that, both populations are combined to obtain the new current population. The procedure is as follows: the two populations are sorted according to their rank, and the best solutions are chosen to create a new population. In the case of having to select some individuals with the same rank, a density estimation based on measuring the crowding distance to the surrounding individuals belonging to the same rank is used to get the most promising solutions. Typically, both the current and the auxiliary populations have the same size.

A steady-state version of NSGA-II can be easily implemented by using an offspring population of size 1. In this way, the new individual is immediately incorporated into the evolutionary cycle. However, this also means that the ranking and crowding procedures have to be applied each time a new individual is created, so the time required by the algorithm will increase notably. The procedure of this version is shown in Fig. 1. In the rest of this work we will refer to the steady-state version as NSGA-II$_{ss}$, and to the original one as NSGA-II$_{gen}$.

SPEA2 was proposed by Zitzler *et al.* in [20]. This algorithm uses a population and an archive. It assigns to each individual a fitness value that is the sum of its strength raw fitness plus a density estimation. In each generation the non-dominated individuals of both the original population and the archive are used to update the archive; if the number of non-dominated individuals is greater than the population size, a truncation operator based on calculating the distances to the k-th nearest neighbor is used. This way, the individuals having the minimum distance to any other individual are chosen. All this procedure is known as *Environmental Selection.* Then, the algorithm applies the selection, crossover,

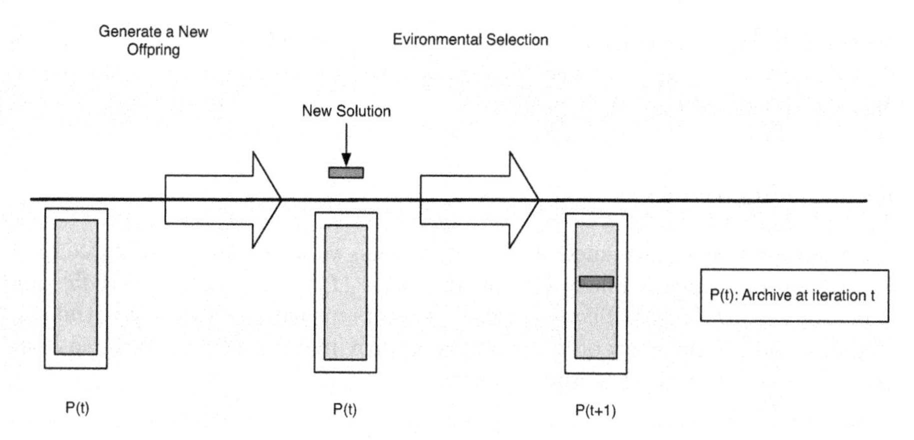

Fig. 2. The SPEA2$_{ss}$ procedure. Only one individual is tested to be inserted in the archive at each step.

and mutation operators to members of the archive in order to create a new population of offsprings which becomes the population of the next generation.

As in the previous case, the steady-state version of SPEA2 can be defined by using a population of size 1; only the initial population must be of the same size as the archive, because in the first generation the archive is filled by all the members belonging to the population. Thus, each time a new individual is created, the environmental selection takes place, and the individual is tested to update the archive for the next generation. Figure 2 shows the steady-state SPEA2 procedure. Since the environmental selection is time consuming, this procedure also increases notably the computing time of the algorithm. The steady-state and generational versions of SPEA2 will be referred to as SPEA2$_{ss}$ and SPEA2$_{gen}$, respectively, in the rest of the paper.

4 Experimentation

In this section we describe the experiments we have carried out. First, we start by presenting the problems families that we have considered. Then, we describe the quality indicators we have chosen for assessing the performance of the algorithms. After that, we introduce the convergence criterion we have used for measuring the convergence speed of the algorithms. Later, we present the parameterization and the stopping criteria used in the experiments. Finally, we describe the statistical tests we have performed to provide the results with statistical confidence.

4.1 Test Problems

The problems we have used are those belonging to three well-known benchmarks: the Zitzler-Deb-Thiele (ZDT) problem family [19], the Deb-Thiele-Laumanns-Zitzler (DTLZ) group of problems [6], and the Walking-Fish-Group (WFG) benchmark. We have considered the bi-objective formulation in the case of the DTLZ and WFG families. All the problems have been configured as in the papers they are described in.

4.2 Quality Indicators

For assessing the performance of the algorithms, we have considered three quality indicators: additive unary epsilon indicator ($I^1_{\epsilon+}$) [13], spread (Δ) [5], and hypervolume (HV) [21]. The first two indicators measure, respectively, the convergence and the diversity of the resulting approximation sets, while the last one measures both convergence and diversity.

4.3 Convergence Speed Criterion

Since one of our main interests is to analyze the convergence speed of the studied MOGAs, it is important to define first what we mean by convergence in this case, and to ensure that such a definition allows us to measure it in a quantitative and meaningful way. We have studied and defined in [15] a stopping condition based on the *high quality* of the approximation of the Pareto front found. We have used the HV quality indicator for that purpose. In Fig. 3 we show different approximations to the Pareto front for the problem ZDT1 with different percentages of HV. We can observe that a front with a hypervolume of 98.26% represents a reasonable approximation to the optimal Pareto front in terms of convergence and diversity of solutions. So, we have taken 98% of the hypervolume of the optimal Pareto front as a criterion to consider that a problem has been successfully solved. In this way, we mean by convergence speed the number of functions evaluations to achieve this termination condition. Those algorithms requiring fewer function evaluations to achieve this termination condition can be considered to be more efficient or faster.

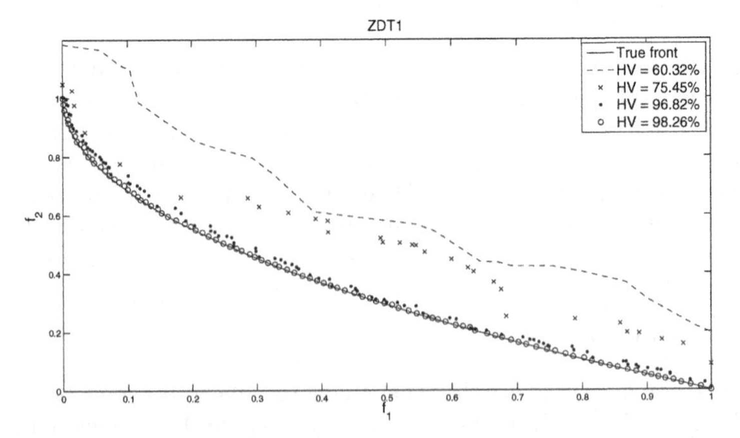

Fig. 3. Fronts with different HV values obtained for problem ZDT1

4.4 Implementation and Parameterization

All the algorithms have been implemented using jMetal [9], a Java-based framework aimed at the development, experimentation, and study of metaheuristics for solving multi-objective optimization problems.

The parameter settings used are the following: the population size is 100 individuals in all the algorithms except in SPEA2$_{ss}$, where the population size is 1. The archives of SPEA2$_{gen}$ and SPEA2$_{ss}$ also have a maximum size of 100 individuals. We have used SBX and polynomial mutation [2] as operators for crossover and mutation operations, respectively, and the distribution indices for both operators are $\eta_c = 20$ and $\eta_m = 20$, respectively. The crossover probability is $p_c = 0.9$ and the mutation probability is $p_m = 1/L$, where L is the number of decision variables.

The stopping criterion is to reach $25,000$ function evaluations in the experiments performed for assessing the quality of the obtained solution sets. The quality indicators are calculated after the algorithms have finished their executions.

In the experiments carried out to study the convergence speed, the stopping criterion is to reach $10,000,000$ of function evaluations or a front with 98% of the HV of the optimal Pareto front. If an algorithm stops according to the first condition, we consider that it has failed when solving the problem. In these experiments, the HV is measured at every 100 evaluations. Therefore, we have considered the nondominated solutions at each generation in the original algorithms, and each 100 generations in the steady-state versions.

For measuring the computing time of the algorithms we have performed $25,000$ evaluations of each algorithm for solving each problem. The experiments have been executed on a PC equipped with an Intel Core 2 Duo 3GHz processor, running Debian 4.0, and using Java JDK 1.6 update 5.

4.5 Statistical Tests

Since we are dealing with stochastic algorithms and we want to provide the results with confidence, we have made 100 independent runs of each experiment, and we show the median, $\tilde{x}$, and interquartile range, IQR, as measures of location (or central tendency) and statistical dispersion, respectively. The following statistical analysis has been performed throughout this work [7]. Firstly, a Kolmogorov-Smirnov test was performed in order to check whether the values of the results follow a normal (gaussian) distribution or not. If the distribution is normal, the Levene test checks for the homogeneity of the variances. If samples have equal variance (positive Levene test), an ANOVA test is done; otherwise a Welch test is performed. For non-gaussian distributions, the non-parametric Kruskal-Wallis test is used to compare the medians of the algorithms. Figure 4 summarizes the statistical analysis.

We always consider in this work a confidence level of 95% (i.e., significance level of 5% or p-value under 0.05) in the statistical tests, which means that the differences are unlikely to have occurred by chance with a probability of 95%. Successful tests are marked with '+' symbols in the last column in all the tables

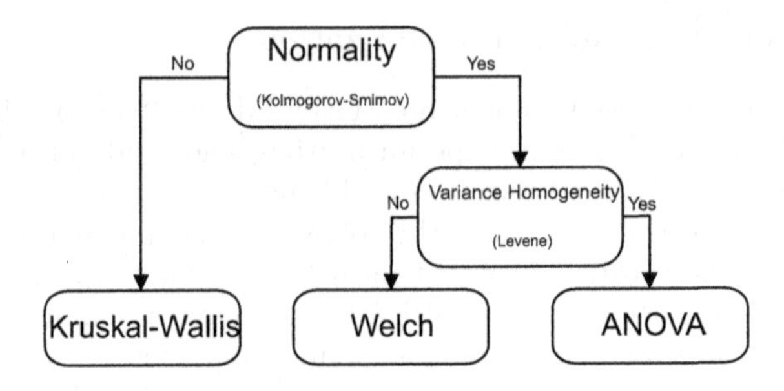

Fig. 4. Statistical analysis performed in this work

containing the results; conversely, '-' means that no statistical confidence was found (p-value > 0.05). For the sake of a better understanding, the best result for each problem has a gray colored background.

5 Results

This section is devoted to evaluating and analyzing the result of the experiments. We first consider the quality of the obtained results through the value of the $I^1_{\epsilon+}$, Δ, and HV quality indicators. After that, we pay attention to the convergence speed, and finally, to the computing time of the algorithms.

5.1 Quality Assessment

We start by analyzing the results of the $I^1_{\epsilon+}$ indicator obtained by the two versions of NSGA-II (Table 1). We can see that NSGA-II$_{ss}$ has obtained better (lower) values than its generational counterpart in 18 out of the 21 problems evaluated according to this indicator. Furthermore, the statistical tests have indicated that the differences are significant in all the problems but in four cases (DTLZ3, WFG1, WFG2, WFG8). It is worth mentioning that these four cases include the three problems in which NSGA-II$_{gen}$ obtained lower $I^1_{\epsilon+}$ values than NSGA-II$_{ss}$. Attending to SPEA2 (Table 2), the steady-state version outperformed the generational one in 17 out of the 21 evaluated problems. The same comments about the statistical test on the results obtained by NSGA-II for this indicator hold for SPEA2: the four problems for which SPEA2$_{gen}$ has produce better values than SPEA2$_{ss}$ are included in those cases in which no statistical differences can be assured by the tests at 95% of confidence level.

The values of the Δ indicator are included in Tables 3 and 4. Proceeding as before, we start by analyzing NSGA-II (Table 3). According to this indicator, NSGA-II$_{ss}$ has outperformed NSGA-II$_{gen}$ in all the solved problems. The statistical tests have confirmed that in all the problems the differences are significant, but in DTLZ3. Something similar has happened with SPEA2 (Table 4). The steady-state version (SPEA2$_{ss}$) has got the best (lowest) values of this indicator in all the problems but WFG1. In this case, statistical confidence has been found in 18 out of the 21 evaluated problems.

Table 1. Median and interquartile range of the $I^1_{\epsilon+}$ quality indicator for NSGA-II_{gen} and NSGA-II_{ss}

Problem	NSGA-II_{gen} $\tilde{x}_{IQR}$	NSGA-II_{ss} $\tilde{x}_{IQR}$	
ZDT1	$1.37e-2_{3.0e-3}$	$5.81e-3_{6.3e-4}$	+
ZDT2	$1.28e-2_{2.3e-3}$	$5.79e-3_{5.5e-4}$	+
ZDT3	$8.13e-3_{1.9e-3}$	$5.24e-3_{5.4e-4}$	+
ZDT4	$1.49e-2_{3.0e-3}$	$9.78e-3_{2.6e-3}$	+
ZDT6	$1.47e-2_{2.8e-3}$	$7.02e-3_{7.6e-4}$	+
DTLZ1	$7.13e-3_{1.6e-3}$	$4.62e-3_{1.9e-3}$	+
DTLZ2	$1.11e-2_{2.7e-3}$	$5.13e-3_{3.6e-4}$	+
DTLZ3	$1.04e+0_{1.2e+0}$	$9.63e-1_{1.4e+0}$	-
DTLZ4	$1.13e-2_{9.9e-1}$	$5.24e-3_{9.9e-1}$	+
DTLZ5	$1.05e-2_{2.5e-3}$	$5.14e-3_{3.4e-4}$	+
DTLZ6	$4.39e-2_{3.4e-2}$	$3.07e-2_{2.5e-2}$	+
DTLZ7	$1.04e-2_{2.8e-3}$	$5.13e-3_{4.1e-4}$	+
WFG1	$3.52e-1_{4.6e-1}$	$4.98e-1_{5.3e-1}$	-
WFG2	$7.10e-1_{7.0e-1}$	$7.10e-1_{7.0e-1}$	-
WFG3	$2.00e+0_{5.8e-4}$	$2.00e+0_{4.3e-4}$	+
WFG4	$3.26e-2_{6.7e-3}$	$1.52e-2_{1.5e-3}$	+
WFG5	$8.41e-2_{8.3e-3}$	$6.41e-2_{1.5e-3}$	+
WFG6	$4.14e-2_{1.6e-2}$	$2.50e-2_{2.8e-2}$	+
WFG7	$3.47e-2_{8.1e-3}$	$1.51e-2_{1.5e-3}$	+
WFG8	$3.38e-1_{2.3e-1}$	$5.08e-1_{2.2e-1}$	-
WFG9	$3.73e-2_{7.5e-3}$	$1.80e-2_{3.7e-3}$	+

Table 2. Median and interquartile range of the $I^1_{\epsilon+}$ quality indicator for SPEA2_{gen} and SPEA2_{ss}

Problem	SPEA2_{gen} $\tilde{x}_{IQR}$	SPEA2_{ss} $\tilde{x}_{IQR}$	
ZDT1	$8.69e-3_{1.1e-3}$	$6.30e-3_{4.5e-4}$	+
ZDT2	$8.73e-3_{1.4e-3}$	$6.45e-3_{7.5e-4}$	+
ZDT3	$9.72e-3_{1.9e-3}$	$8.79e-3_{1.5e-3}$	+
ZDT4	$3.42e-2_{7.9e-2}$	$4.31e-2_{9.4e-2}$	-
ZDT6	$2.42e-2_{5.2e-3}$	$1.01e-2_{1.9e-3}$	+
DTLZ1	$5.89e-3_{2.8e-3}$	$4.88e-3_{1.7e-3}$	+
DTLZ2	$7.34e-3_{1.1e-3}$	$5.33e-3_{4.8e-4}$	+
DTLZ3	$2.28e+0_{1.9e+0}$	$1.63e+0_{1.7e+0}$	+
DTLZ4	$7.66e-3_{9.9e-1}$	$5.48e-3_{9.9e-1}$	+
DTLZ5	$7.47e-3_{1.2e-3}$	$5.32e-3_{6.6e-4}$	+
DTLZ6	$3.03e-1_{5.3e-2}$	$2.79e-2_{2.1e-2}$	+
DTLZ7	$9.09e-3_{1.4e-3}$	$7.85e-3_{1.1e-3}$	+
WFG1	$9.92e-1_{2.1e-1}$	$1.04e+0_{2.4e-1}$	-
WFG2	$7.10e-1_{6.9e-1}$	$7.10e-1_{6.9e-1}$	-
WFG3	$2.00e+0_{1.1e-3}$	$2.00e+0_{8.7e-4}$	-
WFG4	$2.52e-2_{4.0e-3}$	$1.75e-2_{1.3e-3}$	+
WFG5	$7.27e-2_{2.9e-3}$	$6.59e-2_{9.5e-4}$	+
WFG6	$3.11e-2_{1.4e-2}$	$3.07e-2_{2.6e-2}$	-
WFG7	$2.54e-2_{3.0e-3}$	$1.76e-2_{1.7e-3}$	+
WFG8	$5.11e-1_{1.9e-1}$	$5.17e-1_{1.5e-1}$	-
WFG9	$2.92e-2_{5.9e-3}$	$2.20e-2_{6.4e-3}$	+

Table 3. Median and interquartile range of the Δ quality indicator for NSGA-II_{gen} and NSGA-II_{ss}

Problem	NSGA-II_{gen} $\tilde{x}_{IQR}$	NSGA-II_{ss} $\tilde{x}_{IQR}$	
ZDT1	$3.70e-1_{4.2e-2}$	$7.52e-2_{1.5e-2}$	+
ZDT2	$3.81e-1_{4.7e-2}$	$7.80e-2_{1.3e-2}$	+
ZDT3	$7.47e-1_{1.8e-2}$	$7.03e-1_{3.5e-3}$	+
ZDT4	$4.02e-1_{5.8e-2}$	$1.27e-1_{2.9e-2}$	+
ZDT6	$3.56e-1_{3.6e-2}$	$1.05e-1_{1.5e-2}$	+
DTLZ1	$4.03e-1_{6.1e-2}$	$1.18e-1_{4.0e-2}$	+
DTLZ2	$3.84e-1_{3.8e-2}$	$1.10e-1_{1.6e-2}$	+
DTLZ3	$9.53e-1_{1.6e-1}$	$9.52e-1_{3.4e-1}$	-
DTLZ4	$3.95e-1_{6.4e-1}$	$1.13e-1_{9.0e-1}$	+
DTLZ5	$3.79e-1_{4.0e-2}$	$1.11e-1_{1.6e-2}$	+
DTLZ6	$8.64e-1_{3.0e-1}$	$1.81e-1_{5.3e-2}$	+
DTLZ7	$6.23e-1_{2.5e-2}$	$5.19e-1_{1.9e-3}$	+
WFG1	$7.18e-1_{5.4e-2}$	$5.81e-1_{5.8e-2}$	+
WFG2	$7.93e-1_{1.7e-2}$	$7.47e-1_{1.0e-2}$	+
WFG3	$6.12e-1_{3.6e-2}$	$3.71e-1_{7.2e-3}$	+
WFG4	$3.79e-1_{3.9e-2}$	$1.40e-1_{2.0e-2}$	+
WFG5	$4.13e-1_{5.1e-2}$	$1.38e-1_{1.5e-2}$	+
WFG6	$3.90e-1_{4.2e-2}$	$1.23e-1_{3.2e-2}$	+
WFG7	$3.79e-1_{4.6e-2}$	$1.11e-1_{1.9e-2}$	+
WFG8	$6.45e-1_{5.5e-2}$	$5.63e-1_{5.7e-2}$	+
WFG9	$3.96e-1_{4.1e-2}$	$1.52e-1_{2.1e-2}$	+

Table 4. Median and interquartile range of the Δ quality indicator for SPEA2_{gen} and SPEA2_{ss}

Problem	SPEA2_{gen} $\tilde{x}_{IQR}$	SPEA2_{ss} $\tilde{x}_{IQR}$	
ZDT1	$1.52e-1_{2.2e-2}$	$7.26e-2_{1.7e-2}$	+
ZDT2	$1.55e-1_{2.7e-2}$	$9.51e-2_{1.9e-2}$	+
ZDT3	$7.10e-1_{7.5e-3}$	$7.07e-1_{5.1e-3}$	+
ZDT4	$2.72e-1_{1.6e-1}$	$2.48e-1_{1.7e-1}$	-
ZDT6	$2.28e-1_{2.5e-2}$	$1.50e-1_{2.0e-2}$	+
DTLZ1	$1.81e-1_{9.8e-2}$	$1.45e-1_{6.5e-2}$	+
DTLZ2	$1.48e-1_{1.6e-2}$	$8.57e-2_{1.8e-2}$	+
DTLZ3	$1.07e+0_{1.6e-1}$	$1.05e+0_{1.5e-1}$	-
DTLZ4	$1.48e-1_{8.6e-1}$	$8.96e-2_{9.2e-1}$	+
DTLZ5	$1.50e-1_{1.9e-2}$	$8.65e-2_{2.1e-2}$	+
DTLZ6	$8.25e-1_{9.3e-2}$	$3.93e-1_{4.0e-1}$	+
DTLZ7	$5.44e-1_{1.3e-2}$	$5.34e-1_{7.7e-3}$	+
WFG1	$6.51e-1_{4.8e-2}$	$6.68e-1_{7.7e-2}$	+
WFG2	$7.53e-1_{1.3e-2}$	$7.51e-1_{1.2e-2}$	-
WFG3	$4.39e-1_{1.2e-2}$	$3.72e-1_{1.2e-2}$	+
WFG4	$2.72e-1_{2.5e-2}$	$2.16e-1_{2.7e-2}$	+
WFG5	$2.79e-1_{2.3e-2}$	$2.13e-1_{2.3e-2}$	+
WFG6	$2.49e-1_{3.1e-2}$	$2.01e-1_{2.9e-2}$	+
WFG7	$2.47e-1_{1.8e-2}$	$1.89e-1_{2.0e-2}$	+
WFG8	$6.17e-1_{8.1e-2}$	$5.98e-1_{6.3e-2}$	+
WFG9	$2.92e-1_{2.0e-2}$	$2.41e-1_{2.1e-2}$	+

The results of the *HV* indicator (Tables 5 and 6) have confirmed those obtained by the previous indicators. In these tables, a "-" symbol means that the *HV* in this problem has a value of 0, meaning that the solution sets obtained by the algorithms are outside the limits of the Pareto front. The original versions of

Table 5. Median and interquartile range of the HV quality indicator for NSGA-II$_{gen}$ and NSGA-II$_{ss}$

Problem	NSGA-II$_{gen}$ $\tilde{x}_{IQR}$	NSGA-II$_{ss}$ $\tilde{x}_{IQR}$	
ZDT1	$6.59e-1_{4.4e-4}$	$6.62e-1_{1.4e-4}$	+
ZDT2	$3.26e-1_{4.3e-4}$	$3.28e-1_{1.6e-4}$	+
ZDT3	$5.15e-1_{2.3e-4}$	$5.16e-1_{8.0e-5}$	+
ZDT4	$6.56e-1_{4.5e-3}$	$6.57e-1_{4.0e-3}$	+
ZDT6	$3.88e-1_{2.3e-3}$	$3.96e-1_{1.1e-3}$	+
DTLZ1	$4.88e-1_{5.5e-3}$	$4.89e-1_{6.5e-3}$	-
DTLZ2	$2.11e-1_{3.1e-4}$	$2.12e-1_{4.3e-5}$	+
DTLZ3	-	-	+
DTLZ4	$2.09e-1_{2.1e-1}$	$2.11e-1_{2.1e-1}$	+
DTLZ5	$2.11e-1_{3.5e-4}$	$2.12e-1_{3.7e-5}$	+
DTLZ6	$1.75e-1_{3.6e-2}$	$1.73e-1_{2.8e-2}$	-
DTLZ7	$3.33e-1_{2.1e-4}$	$3.34e-1_{3.9e-5}$	+
WFG1	$5.23e-1_{1.3e-1}$	$4.90e-1_{1.9e-1}$	+
WFG2	$5.61e-1_{2.8e-3}$	$5.62e-1_{2.6e-3}$	+
WFG3	$4.41e-1_{3.2e-4}$	$4.42e-1_{1.8e-4}$	+
WFG4	$2.17e-1_{4.9e-4}$	$2.19e-1_{2.4e-4}$	+
WFG5	$1.95e-1_{3.6e-4}$	$1.96e-1_{6.7e-5}$	+
WFG6	$2.03e-1_{9.0e-3}$	$2.03e-1_{1.9e-2}$	-
WFG7	$2.09e-1_{3.3e-4}$	$2.11e-1_{1.4e-4}$	+
WFG8	$1.47e-1_{2.1e-3}$	$1.48e-1_{1.6e-3}$	+
WFG9	$2.37e-1_{1.7e-3}$	$2.40e-1_{1.9e-3}$	+

Table 6. Median and interquartile range of the HV quality indicator for SPEA2$_{gen}$ and SPEA2$_{ss}$

Problem	SPEA2$_{gen}$ $\tilde{x}_{IQR}$	SPEA2$_{ss}$ $\tilde{x}_{IQR}$	
ZDT1	$6.60e-1_{3.9e-4}$	$6.61e-1_{1.8e-4}$	+
ZDT2	$3.26e-1_{8.1e-4}$	$3.28e-1_{3.8e-4}$	+
ZDT3	$5.14e-1_{3.6e-4}$	$5.15e-1_{2.3e-4}$	+
ZDT4	$6.51e-1_{1.2e-2}$	$6.52e-1_{1.1e-2}$	-
ZDT6	$3.79e-1_{3.6e-3}$	$3.92e-1_{1.5e-3}$	+
DTLZ1	$4.89e-1_{6.2e-3}$	$4.89e-1_{5.7e-3}$	-
DTLZ2	$2.12e-1_{1.7e-4}$	$2.12e-1_{5.6e-5}$	+
DTLZ3	-	-	-
DTLZ4	$2.10e-1_{2.1e-1}$	$2.11e-1_{2.1e-1}$	+
DTLZ5	$2.12e-1_{1.7e-4}$	$2.12e-1_{6.1e-5}$	+
DTLZ6	$9.02e-3_{1.4e-2}$	$1.83e-1_{2.8e-2}$	+
DTLZ7	$3.34e-1_{2.2e-4}$	$3.34e-1_{7.9e-5}$	+
WFG1	$3.85e-1_{1.1e-1}$	$3.58e-1_{1.2e-1}$	+
WFG2	$5.62e-1_{2.8e-3}$	$5.62e-1_{2.8e-3}$	-
WFG3	$4.42e-1_{2.0e-4}$	$4.42e-1_{2.1e-4}$	+
WFG4	$2.18e-1_{3.0e-4}$	$2.19e-1_{1.8e-4}$	+
WFG5	$1.96e-1_{1.8e-4}$	$1.96e-1_{7.3e-5}$	+
WFG6	$2.04e-1_{8.6e-3}$	$2.01e-1_{1.6e-2}$	+
WFG7	$2.10e-1_{2.4e-4}$	$2.10e-1_{1.4e-4}$	+
WFG8	$1.47e-1_{2.2e-3}$	$1.47e-1_{2.3e-3}$	-
WFG9	$2.39e-1_{2.3e-3}$	$2.39e-1_{2.1e-3}$	-

Table 7. Median and interquartile of the number of evaluations computed by NSGA-II$_{gen}$ and NSGA-II$_{ss}$

Problem	NSGA-II$_{gen}$ $\tilde{x}_{IQR}$	NSGA-II$_{ss}$ $\tilde{x}_{IQR}$	
ZDT1	$1.43e+4_{8.0e+2}$	$1.16e+4_{9.0e+2}$	+
ZDT2	$2.46e+4_{1.6e+3}$	$1.77e+4_{1.3e+3}$	+
ZDT3	$1.28e+4_{8.5e+2}$	$1.09e+4_{1.0e+3}$	+
ZDT4	$2.24e+4_{5.9e+3}$	$1.98e+4_{5.4e+3}$	+
ZDT6	$2.93e+4_{1.4e+3}$	$2.28e+4_{1.2e+3}$	+
DTLZ1	$2.49e+4_{8.4e+3}$	$2.22e+4_{8.6e+3}$	+
DTLZ2	$8.15e+3_{1.2e+3}$	$5.30e+3_{7.0e+2}$	+
DTLZ3	$1.12e+5_{5.3e+4}$	$8.27e+4_{3.5e+4}$	+
DTLZ4	$8.65e+3_{1.3e+3}$	$5.50e+3_{7.0e+2}$	+
DTLZ5	$8.30e+3_{1.4e+3}$	$5.15e+3_{6.0e+2}$	+
DTLZ6	-	-	-
DTLZ7	$1.36e+4_{9.0e+2}$	$1.06e+4_{9.0e+2}$	+
WFG1	$4.31e+4_{5.4e+4}$	$3.71e+4_{1.5e+4}$	+
WFG2	$1.70e+3_{4.0e+2}$	$1.40e+3_{5.0e+2}$	+
WFG3	-	-	-
WFG4	$2.05e+4_{8.8e+3}$	$8.20e+3_{2.9e+3}$	+
WFG5	-	-	-
WFG6	$7.72e+6_{8.7e+6}$	$1.46e+6_{1.0e+7}$	+
WFG7	$1.68e+5_{2.5e+5}$	$1.03e+4_{2.6e+3}$	+
WFG8	-	$1.00e+7_{9.0e+6}$	+
WFG9	-	$1.00e+7_{9.9e+6}$	+

Table 8. Median and interquartile range of the number of evaluations computed by SPEA2$_{gen}$ and SPEA2$_{ss}$

Problem	SPEA2$_{gen}$ $\tilde{x}_{IQR}$	SPEA2$_{ss}$ $\tilde{x}_{IQR}$	
ZDT1	$1.60e+4_{1.1e+3}$	$1.26e+4_{9.3e+2}$	+
ZDT2	$2.48e+4_{1.9e+3}$	$1.91e+4_{2.0e+3}$	+
ZDT3	$1.52e+4_{1.0e+3}$	$1.21e+4_{1.1e+3}$	+
ZDT4	$2.52e+4_{6.0e+3}$	$2.48e+4_{6.5e+3}$	-
ZDT6	$3.33e+4_{1.0e+3}$	$2.65e+4_{1.2e+3}$	+
DTLZ1	$2.40e+4_{7.5e+3}$	$2.26e+4_{6.4e+3}$	+
DTLZ2	$7.40e+3_{8.0e+2}$	$5.72e+3_{6.3e+2}$	+
DTLZ3	$1.00e+5_{3.0e+4}$	$9.22e+4_{3.1e+4}$	+
DTLZ4	$7.80e+3_{5.0e+6}$	$6.15e+3_{1.0e+7}$	+
DTLZ5	$7.50e+3_{7.0e+2}$	$5.66e+3_{5.7e+2}$	+
DTLZ6	-	-	-
DTLZ7	$1.58e+4_{1.1e+3}$	$1.23e+4_{8.7e+2}$	+
WFG1	$1.09e+5_{7.7e+5}$	$2.51e+5_{4.8e+6}$	+
WFG2	$2.00e+3_{7.0e+2}$	$1.54e+3_{5.8e+2}$	+
WFG3	-	-	-
WFG4	$1.28e+4_{4.6e+3}$	$9.08e+3_{3.1e+3}$	+
WFG5	-	-	-
WFG6	$8.10e+6_{9.0e+6}$	$9.23e+6_{9.1e+6}$	-
WFG7	$1.77e+4_{5.4e+3}$	$1.17e+4_{4.0e+3}$	+
WFG8	-	-	-
WFG9	$1.00e+7_{3.9e+6}$	$1.00e+7_{9.8e+6}$	-

the algorithms, NSGA-II$_{gen}$ and SPEA2$_{ss}$, have only yielded the best (highest) values in two and three out of the 21 problems, respectively. The statistical tests have also shown confidence in most of the cases for this indicator.

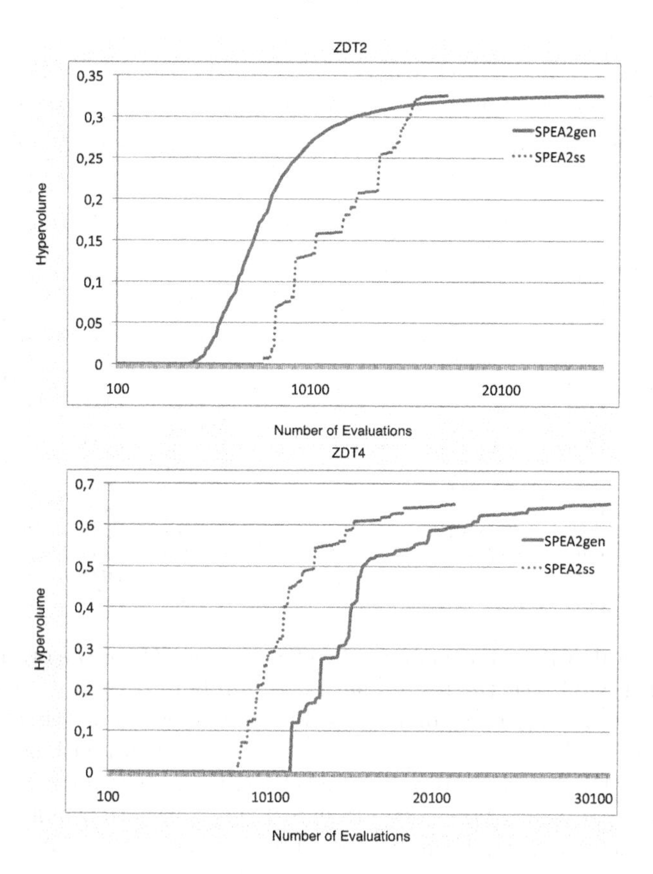

Fig. 5. Evolution of the HV indicator in SPEA2$_{gen}$ and SPEA2$_{ss}$ in problems ZDT2 (top) and ZDT4 (bottom)

5.2 Convergence Speed

The previous section has shown that the steady-state versions of both NSGA-II and SPEA2 have performed better than the original algorithms on most of the problems considered. In this section we discuss the results obtained when measuring the convergence speed.

Tables 7 and 8 contain the evaluations required by the two versions of NSGA-II and SPEA2, respectively, to reach a Pareto front with 98% of the HV of the optimal Pareto front. There are cases in which a "-" is reported in the tables when the algorithm have required more than 10, 000, 000 function evaluations.

In the case of NSGA-II, its steady-state version has been faster (lower number of evaluations) than the original one in all the problems in which the two approaches have reached the convergence criterion of the stopping condition. These results have been supported by statistical confidence.

Considering SPEA2, we can see that SPEA2$_{ss}$ has outperformed the original one in 14 out of the 21 problems evaluated, while SPEA2$_{gen}$ has been faster only in three out of those 21 problems. There are also four problems in which the two

Table 9. Computing time of NSGA-II$_{gen}$ and NSGA-II$_{ss}$. Time is given in ms.

Problem	NSGA-II$_{gen}$	NSGA-II$_{ss}$
ZDT1	1227	13155
ZDT2	1227	12915
ZDT3	1297	12945
ZDT4	1219	13226
ZDT6	1151	12521
WFG1	1326	13259
WFG2	1203	12554
WFG3	1311	12355
WFG4	1201	12783
WFG5	1271	12076
WFG6	1259	13497
WFG7	1170	13220
WFG8	1369	12547
WFG9	1316	12752
DTLZ1	1191	12737
DTLZ2	1158	13274
DTLZ3	1210	16070
DTLZ4	1135	12963
DTLZ5	1206	12626
DTLZ6	1192	12621
DTLZ7	1212	12706

Table 10. Computing time of SPEA2$_{gen}$ and SPEA2$_{ss}$. Time is given in ms.

Problem	SPEA2$_{gen}$	SPEA2$_{ss}$
ZDT1	5204	73077
ZDT2	4537	68485
ZDT3	5065	71291
ZDT4	3907	62163
ZDT6	3926	65040
WFG1	4488	77184
WFG2	6108	82147
WFG3	7269	95962
WFG4	5621	82328
WFG5	6338	95223
WFG6	5476	79464
WFG7	6111	87409
WFG8	4478	63991
WFG9	5943	83399
DTLZ1	3982	61438
DTLZ2	5360	78446
DTLZ3	2899	59185
DTLZ4	5201	77541
DTLZ5	5160	78277
DTLZ6	3025	60424
DTLZ7	4957	71982

versions of the algorithm have not reached 98% of the HV of the optimal Pareto front. As with NSGA-II, the results have been supported by statistical tests.

We analyze now two particular cases. Figure 5 shows how the HV value in SPEA2$_{gen}$ and SPEA2$_{ss}$ has evolved over the different evaluations carried out for problems ZDT2 (Fig. 5 (top)) and ZDT4 (Fig. 5 (bottom)). SPEA2$_{ss}$ has reached the target value of HV faster than SPEA2$_{gen}$ in both problems; however, two different behaviors can been observed. On the one hand, we observe that the HV value in ZDT4 has been better (higher) in the steady-state version since the beginning of the search. On the other hand, in the case of problem ZDT2, we observe that the original (SPEA2$_{gen}$) has been better than SPEA2$_{ss}$ until $13,000$ evaluations have been computed. This way, if we define a less strong termination condition (e.g., we could consider that a front with 95% of the HV of the optimal Pareto front can represent an accurate approximation), the generational SPEA2 algorithm would be faster than the steady-state version for this problem. This indicates that if we only report the number of evaluations to reach the stopping condition we can lose valuable information about the behavior of the algorithms, which suggests the need of further studies in this issue.

5.3 Computing Time

The complexity of the steady-state versions of the algorithms is higher than that of the generational version. In this section we quantify the differences between them by solving all the problems using 25,000 function evaluations as the stopping condition.

The execution times (in milliseconds) are included in Tables 9 and 10. We can observe that NSGA-II$_{gen}$ has been between 9.4 times (the worst case) and 11.4 times (the best case) faster than NSGA-II$_{ss}$. Regarding the results of SPEA2, the original algorithm ($SPEA2_{gen}$) has been between 13.2 times (the worst case) and 20.4 times (the best case) faster than $SPEA2_{ss}$.

We have used the profiling tool provided by the Java IDE Netbeans 6.1 to analyze the execution of the algorithms when solving the ZDT1 problem. The profiling report has shown that the computational time required to evaluate the problem is less than the 1% of the total time. Then, the values which appear in Tables 9 and 10 for this problem represent an accurate approximation of the times required only to run the algorithm, because the function evaluation time is negligible. To solve the benchmark problems used in this study, the steady-state versions are not a sensible choice: it takes about 1.2 secs to solve ZDT1 with $NSGA_{gen}$, while $SPEA2_{ss}$ requires more than 73 secs. However, in a real scenario, where the computing of the objective functions can take minutes and even hours, it is up to those responsible for optimizing the problem to choose the more appropriate algorithm.

6 Conclusions and Future Work

We have presented a first study of the effect of applying a steady-state selection scheme to two *state-of-the-art* multi-objective optimization algorithms, NSGA-II and SPEA2. Both the original and the steady-state versions have been evaluated using a benchmark composed of 21 bi-objective problems for comparing the performance of the algorithms in terms of the quality of the obtained solutions sets, their speed converging towards the optimal Pareto front, as well as the running time they require.

The obtained results have shown that, in the context of the problems, the quality indicators, and the parameter settings used, the use of a steady-state scheme has improved on the results obtained by the generational scheme in most of the problems. Most of these results have been also supported by statistical confidence. The main disadvantage of the evaluated steady-state algorithms is that they have been between 10 and 20 times slower than their generational counterpart.

A line of future work is to deepen into the study and application of the steady-state scheme to other algorithms, and also for solving benchmarks composed of rotated problems and with more than two objectives. Another interesting line of research lies in the study and development of theoretical models which support the experimental results in order to help in the understanding of the effect of applying the steady-state scheme in multi-objective optimization genetic algorithms.

Acknowledgments

The authors acknowledge funding from the "Consejería de Innovación, Ciencia y Empresa", Junta de Andalucía under contract P07-TIC-03044 DIRICOM project, `http://diricom.lcc.uma.es`. Juan J. Durillo is supported by grant AP-2006-003349 from the Spanish Ministry of Education and Science. Francisco Luna acknowledges support from the Spanish Ministry of Education and Science and FEDER under contract TIN2005-08818-C04-01 (the OPLINK project).

References

1. Chafekar, D., Xuan, J., Rasheed, K.: Constrained multi-objective optimization using steady state genetic algorithms. In: Proceedings of Genetic and Evolutionary Computation Conference, pp. 813–824. Springer, Heidelberg (2003)
2. Deb, K.: Multi-Objective Optimization Using Evolutionary Algorithms. John Wiley & Sons, Chichester (2001)
3. Deb, K., Mohan, M., Mishar, S.: Evaluating the ϵ-Domination Based Multi-Objective Evolutionary Algorithm for a Quick Computation of Pareto-Optimal Solutions. Evolutionary Computation 13(4), 501–525 (2005)
4. Deb, K., Mohan, M., Mishra, S.: Towards a Quick Computation of Well-Spread Pareto-Optimal Solutions. In: Fonseca, C.M., Fleming, P.J., Zitzler, E., Deb, K., Thiele, L. (eds.) EMO 2003. LNCS, vol. 2632, pp. 222–236. Springer, Heidelberg (2003)
5. Deb, K., Pratap, A., Agarwal, S., Meyarivan, T.: A Fast and Elitist Multiobjective Genetic Algorithm: NSGA-II. IEEE Transactions on Evolutionary Computation 6(2), 182–197 (2002)
6. Deb, K., Thiele, L., Laumanns, M., Zitzler, E.: Scalable Test Problems for Evolutionary Multiobjective Optimization. In: Abraham, A., Jain, L., Goldberg, R. (eds.) Evolutionary Multiobjective Optimization. Theoretical Advances and Applications, pp. 105–145. Springer, Heidelberg (2005)
7. Demšar, J.: Statistical Comparisons of Classifiers over Multiple Data Sets. J. Mach. Learn. Res. 7, 1–30 (2006)
8. Durillo, J.J., Nebro, A.J., Luna, F., Alba, E.: A Study of Master-Slave Approaches to Parallelize NSGA-II. In: IEEE Iinternational Symposium on Parallel and Distributed Processing, 2008 - IPDPS 2008, pp. 1–8 (2008)
9. Durillo, J.J., Nebro, A.J., Luna, F., Dorronsoro, B., Alba, E.: jMetal: A Java Framework for Developing Multi-Objective Optimization Metaheuristics. Technical Report ITI-2006-10, Dept. de Lenguajes y Ciencias de la Computación, University of Málaga (December 2006)
10. Emmerich, M., Beume, N., Naujoks, B.: An EMO Algorithm Using the Hypervolume Measure as Selection Criterion. In: Coello Coello, C.A., Hernández Aguirre, A., Zitzler, E. (eds.) EMO 2005. LNCS, vol. 3410, pp. 62–76. Springer, Heidelberg (2005)
11. Huband, S., Hingston, P., Barone, L., While, R.L.: A Review of Multiobjective Test Problems and a Scalable Test Problem Toolkit. IEEE Trans. Evolutionary Computation 10(5), 477–506 (2006)
12. Igel, C., Suttorp, T., Hansen, N.: Steady-State Selection and Efficient Covariance Matrix Update in the Multi-objective CMA-ES. In: Obayashi, S., Deb, K., Poloni, C., Hiroyasu, T., Murata, T. (eds.) EMO 2007. LNCS, vol. 4403, pp. 171–185. Springer, Heidelberg (2007)
13. Knowles, J., Thiele, L., Zitzler, E.: A Tutorial on the Performance Assessment of Stochastic Multiobjective Optimizers. Technical Report 214, Computer Engineering and Networks Laboratory (TIK), ETH Zurich (2006)
14. Kumar, R., Rockett, P.: Improved Sampling of the Pareto-Front in Multiobjective Genetic Optimizations by Steady-State evolution: A Pareto Converging Genetic Algorithm. Evolutionary Computation 10(3), 283–314 (2002)
15. Nebro, A.J., Durillo, J.J., Coello, C.A., Luna, F., Alba, E.: A Study of Convergence Speed in Multi-objective Metaheuristics. In: Rudolph, G., Jansen, T., Lucas, S., Poloni, C., Beume, N. (eds.) PPSN 2008. LNCS, vol. 5199, pp. 171–185. Springer, Heidelberg (2008)

16. Schaffer, J.D.: Multiple Objective Optimization with Vector Evaluated Genetic Algorithms. In: Grefenstette, J.J. (ed.) Proceedings of the 1st International Conference on Genetic Algorithms, pp. 93–100 (1985)
17. Srinivasan, D., Rachmawati, L.: An Efficient Multi-objective Evolutionary Algorithm with Steady-State Replacement Model. In: GECCO 2006, pp. 715–722 (2006)
18. Valenzuela, C.L.: A Simple Evolutionary Algorithm for Multi-Objective Optimization (SEAMO). In: IEEE Congress on Evolutionary Computation, 2002 - CEC 2002, pp. 717–722 (2002)
19. Zitzler, E., Deb, K., Thiele, L.: Comparison of Multiobjective Evolutionary Algorithms: Empirical Results. Evolutionary Computation 8(2), 173–195 (2000)
20. Zitzler, E., Laumanns, M., Thiele, L.: SPEA2: Improving the Strength Pareto Evolutionary Algorithm. Technical Report 103, Computer Engineering and Networks Laboratory (TIK), Swiss Federal Institute of Technology (ETH), Zurich, Switzerland (2001)
21. Zitzler, E., Thiele, L.: Multiobjective Evolutionary Algorithms: A Comparative Case Study and the Strength Pareto Approach. IEEE Transactions on Evolutionary Computation 3(4), 257–271 (1999)

OCD: Online Convergence Detection for Evolutionary Multi-Objective Algorithms Based on Statistical Testing

Tobias Wagner[1], Heike Trautmann[2], and Boris Naujoks[3]

[1] Institute of Machining Technology (ISF), TU Dortmund University, Germany
wagner@isf.de
[2] Faculty of Statistics, TU Dortmund University, Germany
trautmann@statistik.tu-dortmund.de
[3] Chair of Algorithm Engineering, TU Dortmund University, Germany
boris.naujoks@tu-dortmund.de

Abstract. Over the last decades, evolutionary algorithms (EA) have proven their applicability to hard and complex industrial optimization problems in many cases. However, especially in cases with high computational demands for fitness evaluations (FE), the number of required FE is often seen as a drawback of these techniques. This is partly due to lacking robust and reliable methods to determine convergence, which would stop the algorithm before useless evaluations are carried out. To overcome this drawback, we define a method for online convergence detection (OCD) based on statistical tests, which invokes a number of performance indicators and which can be applied on a stand-alone basis (no predefined Pareto fronts, ideal and reference points). Our experiments show the general applicability of OCD by analyzing its performance for different algorithmic setups and on different classes of test functions. Furthermore, we show that the number of FE can be reduced considerably – compared to common suggestions from literature – without significantly deteriorating approximation accuracy.

1 Introduction

In real-world industrial problems and engineering applications, improvements, e.g., in simulation techniques, machines, tools, and materials, constantly offer increasing productivity. However, in order to completely exploit these potentials, an appropriate setup of the inherent parameters is necessary. Due to the numerous requirements of modern processes, these problems are mainly multi-objective, which supports the application of evolutionary multi-objective algorithms (EMOA). Nevertheless, their applicability is still put into question, even though EMOA have already been successfully applied to these kinds of problems.

A possible reason, for instance when compared to mathematical programming methods, may be the lack of convergence criteria for EMOA. More specific, the performance of an a-posteriori multi-objective optimization algorithm can be expressed in simple terms by two objectives:

M. Ehrgott et al. (Eds.): EMO 2009, LNCS 5467, pp. 198–215, 2009.

1. maximize the quality of the Pareto-front approximation and
2. minimize the number of function evaluations or computation time, respectively.

In the last decade, many EMOAs have been introduced to achieve one or both of the above objectives. For instance, the use of performance indicators [1,2,3], which evaluate the quality of the current Pareto-front approximation, has turned out to be successful in achieving the first objective [4]. The second objective has recently been approached by integrating modeling methods into the EMOA framework [5,6,7]. However, in the evaluation of all these methods, the number of allowed function evaluations (FE) is fixed at a predefined level, which is high (30k-500k FE [8,9]) when the main objective is a good approximation and low for model-assisted approaches (130-250 FE [6,7]). In order to perform the optimization in an efficient manner, the EMOA should be stopped when

1. no improvement can be gained by further iterations or
2. the approximation quality has reached the desired level.

Right now, these stopping criteria are only applied for single-objective approaches. Nevertheless, the detection of convergence is an equally important issue for EMOA since further evaluations are a waste of computational resources and may lead to a loss of diversity by means of genetic drift [10]. Multi-objective performance indicators allow the reduction of a multi-objective optimization (MOO) problem to a single-objective problem on sets [3]. Thereby, the above criteria can be transferred to MOO. Furthermore, multiple indicators can be used to reliably detect different kinds of improvement in the set.

In this paper, an approach for online convergence detection (OCD) is introduced. Due to the stochastic nature of evolutionary algorithms, OCD is based on systematic statistical testing. The number of parameters is low, it can be combined with any set-based EMOA, and the selection of the considered preference indicators is up to the user. Thus, OCD is an intuitive, yet flexible tool to guarantee an effective use of EMOA, which may promote the industrial application of these methods.

In section 2, the state of the art in multi-objective convergence detection is summarized. Afterwards, OCD is detailed, and the algorithmic steps are presented (section 3). The applicability of OCD is demonstrated by comprehensive experiments, which are described and analyzed in section 4. Finally, conclusions are drawn and the results are summarized in section 5.

2 State of the Art

For the application of EMOA on new industrial problems, where no sufficient a-priori knowledge exists, it is generally hard to find a suitable termination criterion. Therefore, the most frequently used limit is the maximum number of generations or FE. Hybrid EMOA using quadratic programming methods have been developed to guarantee (local) optimality of solutions [11,12]. These approaches are formally converged as soon as Karush-Kuhn-Tucker (KKT) points for a given set of aggregation or reference-point-based distance functions have been identified, but can not guarantee the quality of the set of solutions, e.g., in terms of diversity and spread. This is accomplished by recent approaches, which compute the gradient of the hypervolume for a set of solutions [13]. Note that all these approaches require sufficient accuracy in the approximation of the Hessian.

Deb and Jain [14] investigate so-called running performance metrics for convergence and diversity of solutions to be monitored in the course of the algorithm. Thereby, the algorithm may be stopped when convergence is observed. However, therein the authors focus on performance evaluation and algorithm comparison. An automated procedure for detecting convergence has not been proposed. For this purpose, Rudenko and Schoenauer [15] survey possible online termination criteria for elitist EMOA, such as the disappearance of all dominated individuals or the deterioration of the number of newly produced non-dominated individuals. Finally, they suggest a technique for determining stagnation based on stability of the maximum crowding distance, which requires the determination of a threshold, which depends on the scale of the objectives as well as the population size. Furthermore, its application is only tested with NSGA2, which uses the crowding distance as selection criterion [16]. It is an open question whether a stability of the maximum crowding distance can be observed in EMOA, which do not directly use this measure in the selection process.

The basic idea of using dominance-related metrics to compare sets [17] has recently been used to reduce the multi-objective to a single-objective problem on sets [3]. This allows to use convergence criteria from single-objective theory. Furthermore, a method for offline detection of the expected generation, in which the EMOA converges, has been introduced [18]. This method is based on statistical testing of the similarity in the distribution of performance measures for consecutive generations relying on multiple parallel runs of the EMOA. In this paper, the main ideas of both contributions are transferred to online convergence detection.

3 Online Convergence Detection

In the progression of OCD, two different analyses are carried out. It is sequentially tested whether the variance of the performance indicator values decreases below a predefined limit ($VarLimit$) or whether no significant trend of the performance indicators can be detected over the last generations. The EMOA terminates if at least one of these conditions is met.

All algorithmic steps of the proposed OCD approach and the required subroutines are given in Algorithms 1, 2, and 3. These steps are described in depth to ensure a straightforward implementation of OCD. The required input parameters for Algorithm 1 can be set easily, even by inexperienced users. The variance limit $VarLimit$ corresponds to the desired approximation accuracy in single-objective optimization, but does not require knowledge about the actual minima of the objectives. The algorithm stops when the standard deviation of the indicator values over the given time window of $nPreGen$ generations is significantly below $\sqrt{VarLimit}$. Thus, the user can exactly determine how many

Algorithm 1. OCD: Algorithm for Online Convergence Detection

```
Require: VarLimit                                           /* maximum variance limit */
         nPreGen          /* number of preceding generations for comparisons */
         α                                         /* significance level of the tests */
         MaxGen                                   /* maximum generation number */
         (PI_1,...,PI_n)  /* vector of performance indicators, e.g., (HV, ε, R2) */
 1. i = 0                                         /* initialize generation number */
 2. for all i ∈ {1,...,nPreGen}, j ∈ {1,...,n} do
 3.   pChi2(j,i) = 1                    /* initialize p-values of the χ²-variance Test */
 4.   pReg(i) = 0        /* initialize p-values of the t-Test on regression coefficient */
 5. end for
 6. lb = []                                       /* initialize lower bound vector */
 7. ub = []                                       /* initialize upper bound vector */
 8. repeat
 9.   i = i + 1
10.   Compute d-objective Pareto front PF_i of i-th EMOA generation
11.   lb = min(lb ∪ PF_i)                            /* update lower bound vector */
12.   ub = max(ub ∪ PF_i)                            /* update upper bound vector */
13.   if (i > nPreGen) then
14.     PF_i = 1 + (PF_i − lb)/(ub − lb)               /* normalize PF_i to [1,2]^d */
15.     for all k ∈ {i − nPreGen,...,i − 1} do
16.       Compute Pareto front PF_k of k-th EMOA generation
17.       PF_k = 1 + (PF_k − lb)/(ub − lb)             /* normalize PF_k to [1,2]^d */
18.     end for
19.     for all j ∈ {1,...,n} do
20.       PI_{j,i} = (PI_j(PF_{i−nPreGen}, PF_i, 1, 2.1),...,(PI_j(PF_{i−1}, PF_i, 1, 2.1)))
            /* compute PI_j for PF_{i−nPreGen},...,PF_{i−1} using PF_i as reference set,
                                            1 as ideal, and 2.1 as reference point */
21.       pChi2(j,i) = call Chi2(PI_{j,i}, VarLimit)            /* p-value of χ² test */
22.     end for
23.     pReg(i) = call Reg(PI_{1,i},...,PI_{n,i})
                 /* p-value of the t-Test on the generation's effect on the PI_{j,i} */
24.   end if
25. until ∀j ∈ {1,...,n} : (pChi2(j,i) ≤ α/n) ∧ (pChi2(j,i − 1) ≤ α/n)
          ∨    (pReg(i) > α) ∧ (pReg(i − 1) > α)
          ∨    i = MaxGen
26. Terminate EMOA
27. return {MaxGen, Chi2, Reg}          /* criterion which terminates the EMOA */
             i                          /* generation in which the criterion holds */
```

generations the EMOA is maximally allowed to compute with average changes in the indicator values significantly below the specified limit. The user also has to specify a significance level α for each statistical test procedure. Established levels for α, such as 0.05 (standard) and 0.01 (conservative), exist. The maximum generation number $MaxGen$ ensures that the resources required by the algorithm cope with the restrictions of the individual application, especially in the case where no convergence of the EMOA can be detected. However, the maximum number of function evaluations has to be specified for most known EMOA as well. The number and types of desired performance indicators (PI) have to be selected in order to evaluate the solution quality at each generation with respect to the requirements of the user, which allows him to express his own preferences on the final Pareto-front approximation [3]. Users, who are not familiar with multi-objective performance assessment, can resort to the standard set of PI as defined by Knowles et al. [19], which comprises the hypervolume, the additive ε-, and the R2 indicator. Only these indicators meet the requirement of strict compliance with the Pareto dominance relation.

After the first *nPreGen* generations, convergence is checked after each generation i. The n indicator values of the vector $\boldsymbol{PI}_{j,i}$ $(j = 1, \ldots, n)$ are computed for the objective sets of generations $i - nPreGen, \ldots, i - 1$ using the Pareto-front approximation of generation i as reference set. Thus, no a-priori knowledge about the true Pareto front is required, making the method applicable to practical problems. If a specific indicator PI_j does not use a reference set and evaluates each set separately (e.g., the hypervolume indicator), the difference between the indicator value of the preceding and the current set is calculated and stored in $\boldsymbol{PI}_{j,i}$.

The sets are normalized to the interval $[1, 2]^d = [1, 2] \times \ldots \times [1, 2] \subset \mathbb{R}^d$ as it is also implemented in PISA [20], where d is the number of objective dimensions. This is done in order to avoid problems within the indicator calculation based on objectives which are negative, equal to zero, or extremely large [19]. Since the actual bounds of the non-normalized objectives are not a-priori known, they are updated at each generation. The Pareto-front approximations of the *nPreGen* preceding generations are also normalized based on the current objective-bound approximations. Due to the normalization, $\mathbf{1} = (1, \ldots, 1) \in \mathbb{R}^d$ and $\mathbf{2.1} = (2.1, \ldots, 2.1) \in \mathbb{R}^d$ can be used as ideal point and (anti-optimal) reference point for the PI calculation, respectively.

The resulting *nPreGen* vectors of n indicator values at each generation are then – separately for each indicator – checked against the alternative hypothesis that the variance of these values is lower than the predefined threshold $VarLimit$ using the χ^2-variance test [21] (cf. Algorithm 2). This parametric test is used being aware of its sensitivity to the normality assumption of the underlying sample as no nonparametric test for this problem exists. Due to the multiple testing, a Bonferroni correction on α is performed [22] resulting in an individual significance level of α/n for each test. The α-correction ensures that at each generation the global desired significance level is met. However, a correction with respect to the sequential testing over all generations is impossible concerning a reasonable applicability of OCD.

Additionally, a regression analysis is performed in order to check the significance of the descending linear trend (cf. Algorithm 3). Unfortunately, a test for $H_0 : \beta \neq 0$ vs. $H_1 : \beta = 0$ cannot be constructed. Thus, the test has to be performed with interchanged hypotheses, and the generation, in which the null hypothesis cannot be rejected anymore, has to be determined. Additionally, the decreasing linear trend has been checked via the negative sign of the estimator $\hat{\beta}$.

Algorithm 2. *Chi2*: One-Sided χ^2-variance test for

$$H_0: \quad \text{var}(\boldsymbol{PI}) \geq VarLimit \quad \text{vs.} \quad H_1: \quad \text{var}(\boldsymbol{PI}) < VarLimit$$

Require: $\boldsymbol{PI}$ /* *vector of performance indicator values* */
$VarLimit$ /* *variance limit* */
1. $N = length(\boldsymbol{PI}) - 1$ /* *determine degrees of freedom* */
2. $Chi = [\text{var}(\boldsymbol{PI}) * N]/VarLimit$ /* *compute test statistic* */
3. $p = \chi^2(Chi, N)$ /* *look up* χ^2 *distribution function with* N *degrees of freedom* */
4. **return** p

Strictly speaking, the α-error for the desired decision cannot be controlled by α, but equals $1 - power$(t-test), where the *power* of a statistical test is the probability that the test will reject a false null hypothesis. As a result, an overall significance level at generation i cannot be maintained since the χ^2-variance test initiates the EMOA termination in the case of H_0 being rejected whereas the t-test initiates it in the opposite case. Thus, no combination of the α-levels can be performed relating to multiple test theory [22] although both tests are simultaneously performed on the same data. However, the main focus when setting up α is not on correctly controlling the α-error, but on finding reasonable critical values for the test statistics in order to make OCD applicable and successful within industrial applications.

Algorithm 3. *Reg*: Two-sided t-test on the significance of the linear trend

$$H_0: \quad \beta = 0 \quad \text{vs.} \quad H_1: \quad \beta \neq 0$$

Require: $\boldsymbol{PI}_j, \quad j = (1, \ldots, n)$ /* *vectors of performance indicator values* */
1. $N = n \cdot length(\boldsymbol{PI}^*) - 1$ /* *determine degrees of freedom* */
2. **for all** $j \in \{1, \ldots, n\}$ **do**
3. $\quad \boldsymbol{PI^*}_j = (\boldsymbol{PI}_j - \overline{\boldsymbol{PI}_j})/\sigma_{\boldsymbol{PI}_j}$ /* *standardize* */
4. **end for**
5. $\boldsymbol{PI}^* := \text{concatenate}(\boldsymbol{PI^*}_1, \ldots, \boldsymbol{PI^*}_n)$ /* *row vector of all* $\boldsymbol{PI}_j$ */
6. $X = \underbrace{(1, \ldots, length(\boldsymbol{PI}^*), \ldots, 1, \ldots, length(\boldsymbol{PI}^*))}_{n \text{ times}}$
/* *row vector of generations corresponding to* $\boldsymbol{PI}^*$ */
7. $\hat{\beta} = (X * X^T)^{-1} * X * (\boldsymbol{PI}^*)^T$ /* *linear regression without intercept* */
8. $\varepsilon = \boldsymbol{PI}^* - X * \hat{\beta}$ /* *compute residuals* */
9. $s^2 = (\varepsilon * \varepsilon^T)/N$ /* *mean squared error of regression* */
10. $t = \frac{\hat{\beta}}{\sqrt{s^2 (X * X^T)^{-1}}}$ /* *compute test statistic* */
11. $p = 2 \cdot min(\text{t}_N(t), 1 - \text{t}_N(t))$
/* *look up p-value from t distribution with* N *degrees of freedom* */
12. **return** p

For performing the t-test, all indicator values $\boldsymbol{PI}_j$ are standardized, i.e., linearly transformed to mean zero and standard deviation one. The standardization of $\boldsymbol{PI}_j$ provides two benefits: the regression can be performed for all indicators at once and no intercept (constant term) is required. The least squares estimator $\hat{\beta}$ of the actual slope β is determined in line 7 [23]. Afterwards, the fit is calculated via the mean squared error of the linear model, and a standard error of the estimator is computed [23]. Based on these measures, the t-statistic, i.e., the standardized regression coefficient, and the p-value can be computed using a standard statistical library.

The algorithm stops if either the variance test or the regression analysis indicates the convergence of the EMOA for generations i and $(i-1)$. OCD returns the stopping generation i and the method that initiated the EMOA termination. Thereby, the user is informed about the final state of the algorithm. In the case of termination based on the maximum number of generations, the user knows that the EMOA has not yet converged and further generations may further improve the Pareto-front approximation.

Additional Runtime for OCD

The update, normalization, and standardization of the objective sets within each iteration can be performed in $O(N)$, where N denotes the population size. The calculation of the Pareto front requires $O(N \log^{d-1} N)$ [24], but is already part of most known EMOA. Thus, the calculation of the indicator values is the crucial part of OCD. Especially when the hypervolume is used, the runtime is in $O(N^{d/2+1})$ for $d > 3$ [25]. For hypervolume-based algorithms, such as SMSEMOA [2], this is not critical since the selection procedure is in the same complexity as OCD. Also for expensive real-world problems, the time, which can be saved by an appropriate termination, is considerably higher than the additional runtime. Nevertheless, the approach can be efficiently used for time-critical optimization as well by using performance measures in $O(Nd)$, such as the R2 indicator.

4 Experiments

The experiments are conducted to analyze the proposed OCD applied to modern EMOA. At present, online convergence detection can only be performed by a human decision maker, who inspects the running metrics, i.e., the PI, and terminates the algorithm when convergence is observed. For a successfully automatized application, the time when OCD indicates convergence has to be in agreement with the intuitive understanding of the decision maker. Thus, the first experiments focus on the correspondence of OCD and a human decision maker. In order to analyze the applicability of the statistical tests separated from the whole OCD framework, OCD is additionally computed using pre-calculated Pareto front discretizations as well as the known ideal and anti-ideal points. Apart from the OCD version in Algorithm 1, we will refer to the latter as OCD with full information. Finally, the results received by standard OCD are compared to the common termination criterion from EMOA literature, i.e., a fixed number of FE. Here, we focus on the reduction of the number of evaluations as well as the loss of quality by stopping the evolution earlier.

Research Question. The main question of the analysis is whether or not the proposed OCD algorithm helps to reduce FE without resulting in an uncontrollable loss of quality. Therefore, we evaluate the results received regarding both approximation quality and the required number of FE and compare them to the ones we receive after applying the number of FE, which are originally proposed in standard EMO literature. Moreover, we are interested in the criterion which first indicates convergence and how this is motivated by the $\boldsymbol{PI}_{i,j}$ characteristics over time. In order to inspect the behavior of OCD more closely, it is also analyzed whether OCD, with the reference set and the ideal and anti-ideal point approximated on the fly, performs similar to the case of full information. Last but not least, we want to demonstrate that the time, when OCD indicates convergence, matches with an intuitive observation of the running metrics.

Pre-experimental planning. NSGA2 [16] and SMSEMOA [2] are considered since NSGA2 is the industrially most popular EMOA and recent studies motivate the use of the hypervolume contribution during selection [4]. The test functions are chosen to represent different kinds of problem characteristics, such as dimension in decision and in objective space, the number of local optima, and the shape of the Pareto front. The population sizes used on the problems vary in order to allow for different problem characteristics and evaluate OCD for a wider variety of algorithmic setups.

Initial preparative analyses of OCD indicate that the time window *nPreGen* should span at least seven, but better ten, generations to permit an adequate calculation of the p-values in the tests. In this context, it has to be considered that the tests will not indicate convergence until the $\boldsymbol{PI}_{j,i}$ stagnate over a large span of this time window. Thus, when it is reviewed whether OCD's indication matches with the generation determined by a human decision maker, the delay of *nPreGen* generations has to be accepted within the assessment.

Task. Check if OCD provides a robust and reliable termination of EMOA on several test cases. Compare the results of OCD with an intuitive understanding of termination and with the results provided in standard EMO literature. Furthermore, systematical deviations between the proposed approach and the one with full information are to be identified, which may occur due to a inaccurate approximation of the true Pareto front.

Setup. NSGA2 and SMSEMOA are analyzed on the four bi-objective test functions Fonseca [27], ZDT1, ZDT2, and ZDT4 [28] as well as on the three-objective DTLZ2 [29] test function. Different population sizes $\mu \in \{60$ (Fonseca), 100 (ZDT1, ZDT2, DTLZ2), 200 (ZDT4)$\}$ and selection strategies – $(\mu + \mu)$ in the NSGA2 and $(\mu + 1)$ in the SMSEMOA – are incorporated, where, for the sake of comparability, a generation of SMSEMOA equals a sequence of μ FE. For each combination of EMOA and test function, ten independent runs are performed.

The variance bound for the χ^2-variance test is set to $VarLimit = 0.001^2$, the significance level for both tests is set to $\alpha = 0.05$, and the time window is of size

Table 1. Parameter settings within the experiments

test problem	$MaxGen$	$MaxGen2$
Fonseca	66	66
ZDT1, ZDT2	120	200
ZDT4	200	100
DTLZ2	300	300

algorithm	implem.	p_c	p_m	η_c	η_m	p_{swap}
NSGA2	R [26][a]	0.7	0.2	20	20	0
SMSEMOA	PISA [20]	0.9	1/length($\boldsymbol{x}$)	15	20	0.5

[a] NSGA2 is taken from the package *mco* (http://cran.r-project.org/web/packages/mco/index.html).

$nPreGen = 10$. The different numbers of FE allowed within our experiments ($MaxGen$) and within the standard literature (MaxGen2) [8] as well as the parameters used in the simulated binary crossover and polynomial mutation [30] are displayed in Tab. 1. For measuring the performance of the algorithms, the following PI have been invoked: hypervolume (HV) [31], additive ε (Eps) [17], and R2 [32]. Recall that OCD as well as OCD with full information terminate if and only if one of the tests (χ^2-variance or t-test) simultaneously indicates convergence with respect to all three metrics. The reference fronts used within OCD with full information have been calculated via equidistant sampling of the known Pareto fronts.

Experimentation/Visualization. Several ways of visualization are used to demonstrate our findings. In the first plots, the PI behavior is inspected over the generations of the EMOA on the ZDT4 (cf. Fig. 1) and the DTLZ2 test function (cf. Fig. 2), where the median run with respect to the difference between the full information-based performance metrics and OCD is plotted semi-logarithmically. The black and light-gray solid lines indicate the generation, in which either the χ^2-variance or the regression criterion detect convergence in case of the reference set and objective bounds being approximated online. The black and light-gray dashed lines indicate the generation, in which convergence is detected for the given combination of EMOA and test problem within the full information approach.

The differences in performance are visualized using boxplots. The subsequent figures present the differences between the $\boldsymbol{PI}_{j,i}$ after the number of FE recommended in literature ($i = MaxGen2$) and after OCD indicated convergence. One box is shown for each PI_j and each considered test case, in Fig. 3 for the NSGA2 and in Fig. 4 for the SMSEMOA. Due to different scales, the displayed area had to be changed for some of the test cases, i.e., DTLZ2 for NSGA2 and ZDT1 as well as ZDT2 for the SMSEMOA. For the combinations of EMOA and test function, in which the variance criterion initiated termination for most of the runs, the interval $[-\sqrt{VarLimit}, \sqrt{VarLimit}]$ is highlighted in order to assist inspecting the effect of $VarLimit$ on the final approximation quality. Fig. 5 splits the runs for all test problems into two categories: runs being terminated by the regression criterion and by the χ^2-variance test. This analysis is done separately for NSGA2 and SMSEMOA in order to show the two different types of EMOA behavior and how OCD copes with these challenges.

Statistic details of the boxplots can be found in Tab. 2. Here, the median differences are listed with respect to the corresponding algorithm/test case combination. Note that all median differences are given multiplied by 10^{-3}. Besides

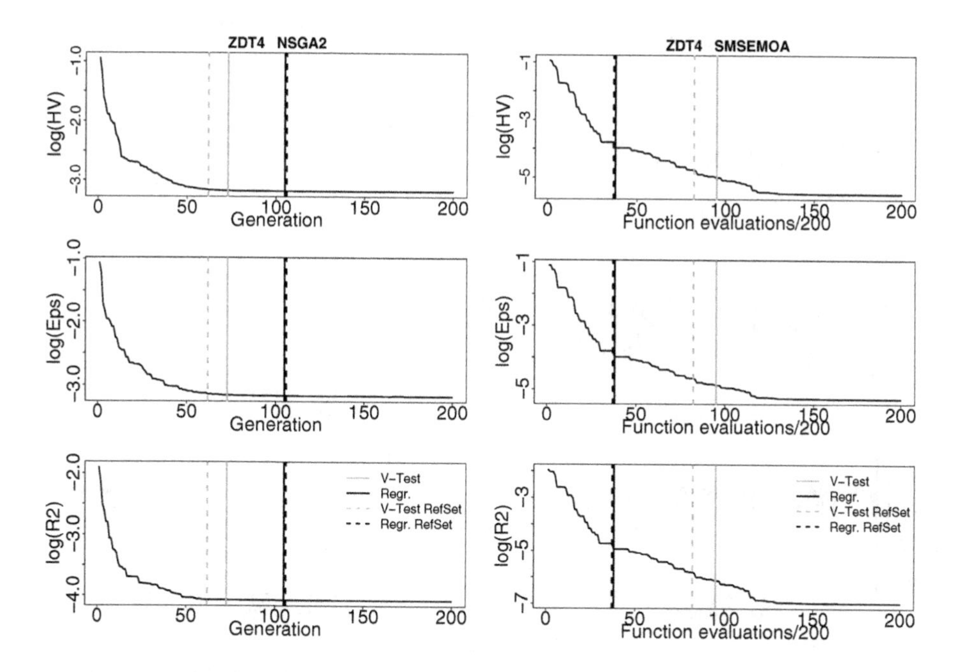

Fig. 1. The run of the metrics with respect to the reference set for NSGA2 and SMSEMOA on ZDT4. Exemplary, the run is chosen which obtains the median difference between the approaches with and without full information. The vertical lines indicate the generations, in which the different tests and variants of OCD would stop the algorithm.

the results regarding the received quality, the additional rows within Tab. 2 indicate the number of generations OCD terminated the algorithm earlier in contrast to the generation number suggested in the literature ($MaxGen2$) [8]. Furthermore, the number of saved function evaluations and their percentage of MaxGen2 are calculated to emphasize what is saved by using OCD with only the given median loss in quality.

In the line plots of Fig. 6 and Fig. 7, the values of each run with and without full information are compared. By these means, systematic deviations can easily be observed. Since OCD terminates the EMOA when the first of the tests indicates convergence, it is also labeled which of the tests initiates the termination of each run using different symbols. Fig. 6 shows the results for NSGA2 on each test case whereas Fig. 7 provides these for SMSEMOA.

Observations. OCD efficiently copes with two different types of convergence. In case the variance test terminates the EMOA (cf. Fig. 5, subfigures 1 and 3), the standard deviation of all $\boldsymbol{PI}_{i,j}$ is significantly below $\sqrt{VarLimit} = 0.001$. Fig. 5 shows that the $\boldsymbol{PI}_{j,i}$ differences between OCD Stop and $MaxGen2$ are approximately in the range of $[-0.001, 0.001]$ for the EMOA runs, which have been terminated by the χ^2-variance-test. Furthermore, big differences to the runs, which are terminated by the regression criterion (cf. Fig. 5, subfigures 2

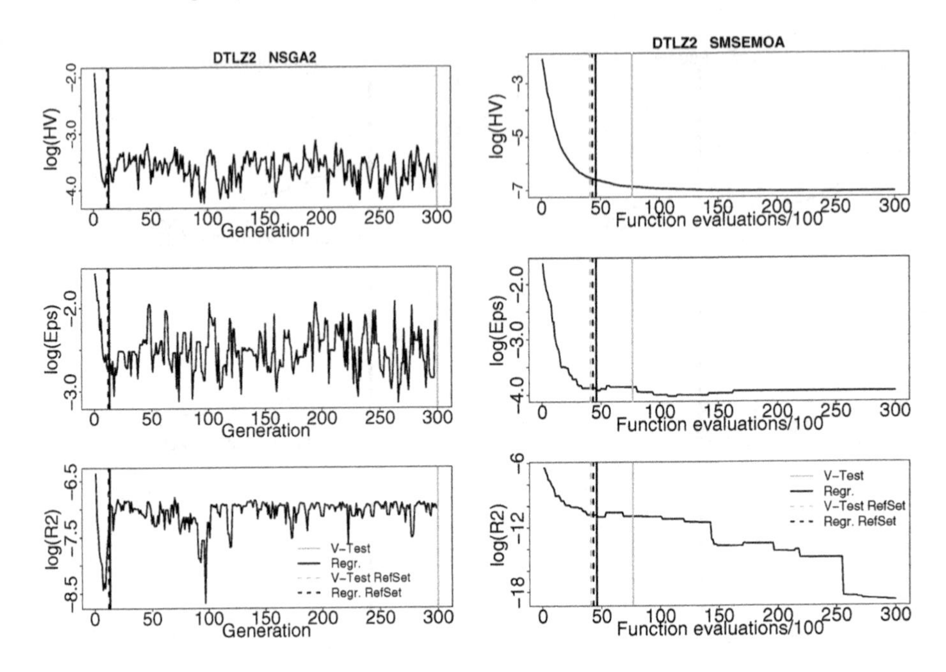

Fig. 2. The run of the metric with respect to the reference set for NSGA2 and SMSEMOA on DTLZ2. Exemplary, the run is chosen which obtains the median difference between the approaches with and without full information. The vertical lines indicate the generations, in which the different tests and variants of OCD would stop the algorithm.

and 4), can be observed. In these cases the differences between the approximation quality of OCD Stop and *MaxGen*2 are much higher, strictly positive for the SMSEMOA and balanced between positive and negative values for NSGA2.

The basic results from above can also be recognized in the boxplots for NSGA2 (cf. Fig. 3) and SMSEMOA (cf. Fig. 4). However, systematic differences between the NSGA2 and the SMSEMOA results can be detected on ZDT4 and DTLZ2. For NSGA2 on ZDT4, the variance criterion indicates convergence much earlier than the regression criterion. This is different from the findings for SMSEMOA, where the regression criterion terminates the algorithm earlier. The progressions of $\boldsymbol{PI}_{j,i}$ on DTLZ2 are strongly distorted for NSGA2 with alternating phases of convergence and divergence. The ones of SMSEMOA are much smoother. In both cases, the regression criterion is able to identify convergence very early in the run, but due to the rough structure, the variance test is not able to do so for NSGA2, while for SMSEMOA the variance criterion terminates the optimization about 25 to 30 generations later than the regression criterion.

The differences in generations between the ones proposed by OCD and *MaxGen*2 range from rather small (18 for NSGA2 on ZDT4) to very large (287 for NSGA2 on DTLZ2). In the latter case, only less than 5% of the evaluations are needed to find better solutions compared to the ones found after the complete optimization run with the termination criterion proposed in the literature.

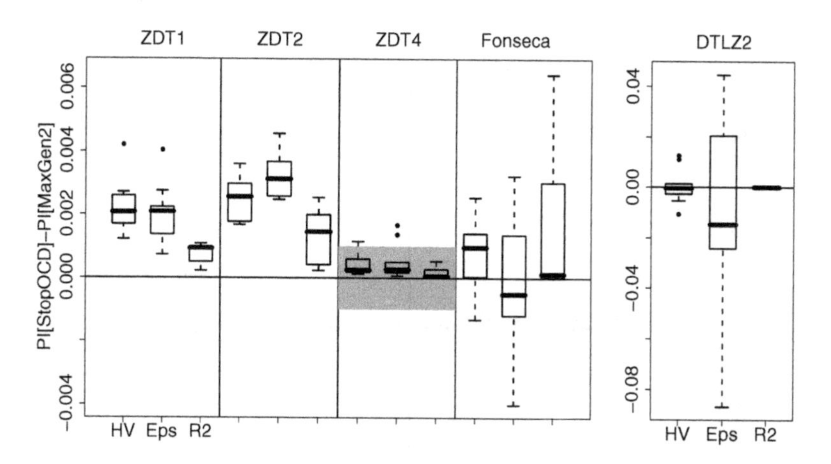

Fig. 3. Boxplots of PI differences at OCD StopGen and $MaxGen2$ for NSGA2. The interval $[-\sqrt{VarLimit}], \sqrt{VarLimit}]$ is highlighted in gray where appropriate.

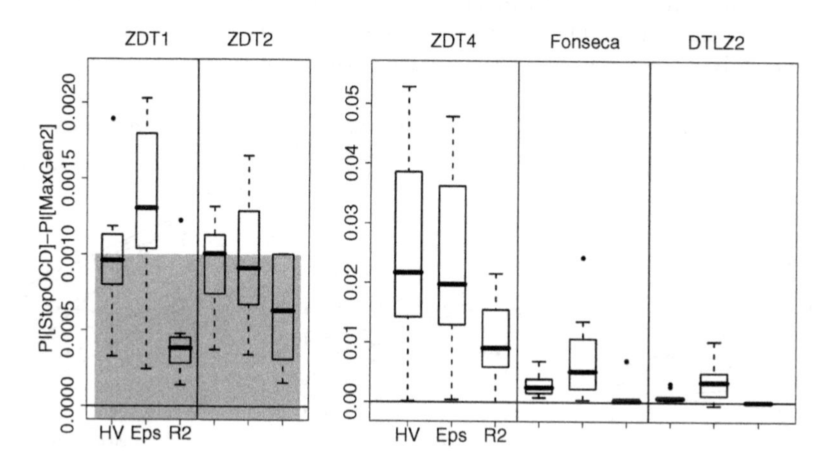

Fig. 4. Boxplots of PI differences at OCD StopGen and $MaxGen2$ for SMSEMOA. The interval $[-\sqrt{VarLimit}], \sqrt{VarLimit}]$ is highlighted in gray where appropriate.

In most cases, slightly more than 50% of the generations can be saved. This results in over 10,000 unnecessary evaluations for the high-dimensional problems. Even in the worst case, more than 2,900 evaluations can be saved.

The coincidence of both tested OCD variants are indicated in the line plots in Fig. 6 for NSGA2 and Fig. 7 for SMSEMOA. The differences between OCD and its full-information variant are strongly depending on the EMOA in use. For SMSEMOA the results with full information and approximated reference sets are well-correlated and no general trend can be observed. The median differences between the indications of convergence in both situations are within one to five generations (cf. Fig. 7). This is different to NSGA2, which shows a trend to overestimate the stop generation for the high-dimensional problems. Furthermore, some outliers

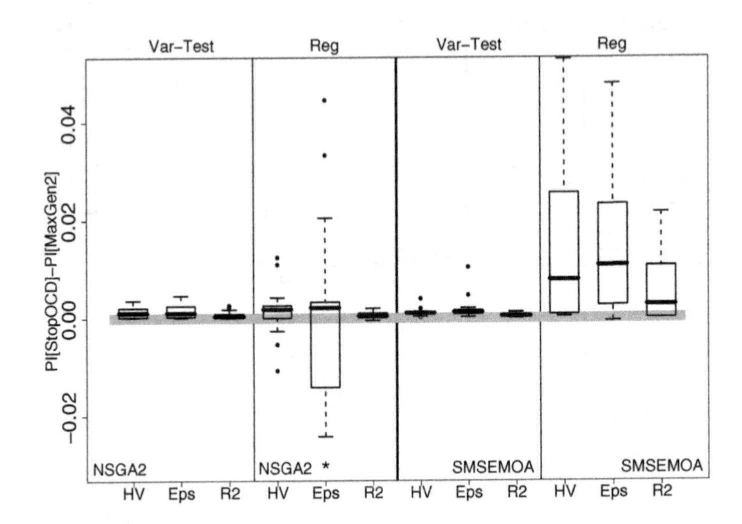

Fig. 5. Separated boxplots of PI differences between OCD StopGen and *MaxGen*2 for the runs of the SMSEMOA which are terminated by the χ^2-variance test and the regression analysis, respectively. *Two extreme outliers not shown.

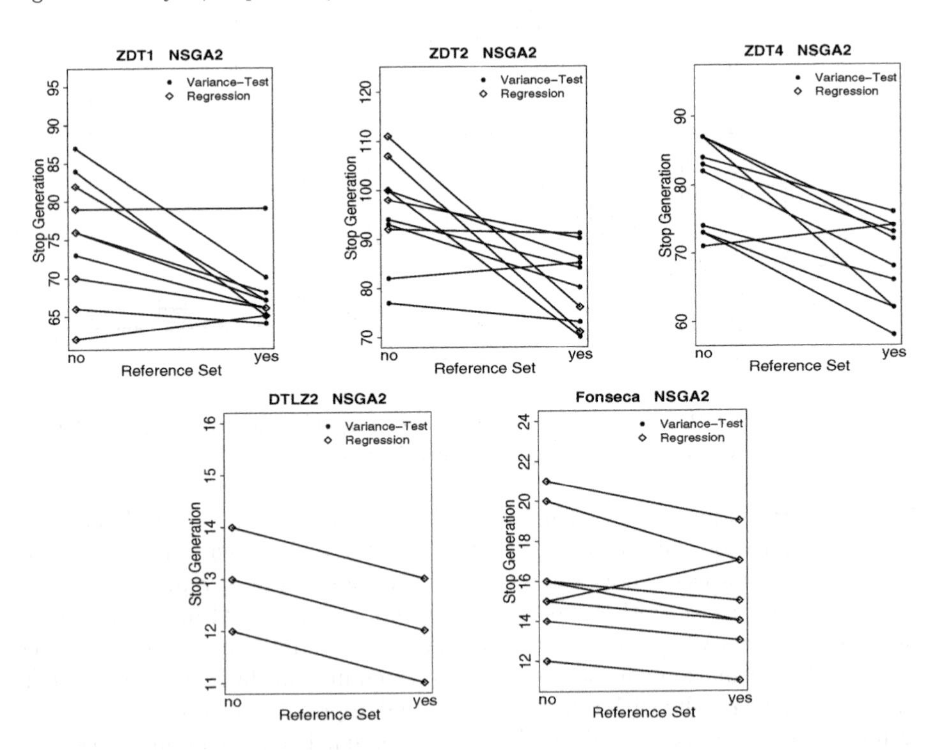

Fig. 6. In the line plots, the generations of NSGA2, in which the OCD stopping criterion is first met (left), are connected to the corresponding generations of OCD with full information (right). Furthermore, the test, which initiates the termination, is indicated by a specific symbol.

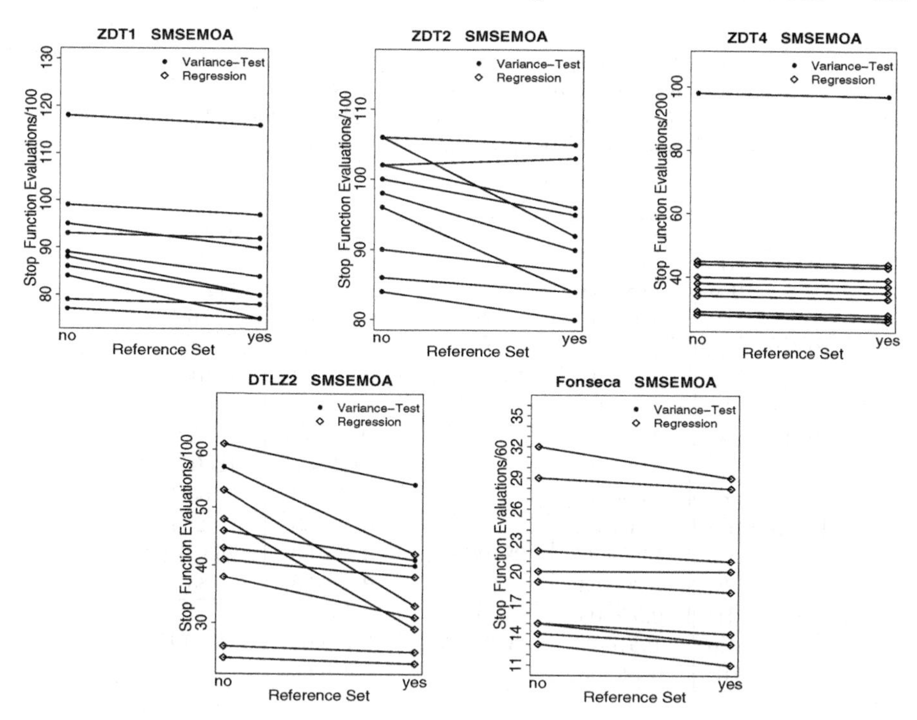

Fig. 7. In the line plots, the generations of SMSEMOA, in which the OCD stopping criterion is first met (left), are connected to the corresponding generations of OCD with full information (right). Furthermore, the test, which initiates the termination, is indicated by a specific symbol.

with extreme differences can be detected (cf. Fig. 6). Nevertheless, the generations proposed by OCD are matching the subjective localization of the termination generation with an accuracy of approximately $nPreGen = 10$ generations.

Discussion. The χ^2-variance test as well as the test on the regression coefficient are necessary to successfully detect convergence of EMOA. While the former indicates a low level of improvement in cases of successful optimization, e.g., on ZDT1 and ZDT2, the latter is extremely important when the high variance in the indicator values does not provide further improvements due to cyclic deterioration effects. These effects can be observed for NSGA2 on DTLZ2 and Fonseca. In contrast, on ZDT4 phases of temporary stagnation lead to the termination of the SMSEMOA based on the regression criterion. Due to a lower selection pressure, NSGA2 can avoid these phases and is therefore stopped by the variance criterion after global convergence.

Another important observation is that, in cases, in which OCD terminates the EMOA based on the χ^2-variance test, the value of $\sqrt{VarLimit} = 0.001$ is close to the differences in approximation quality compared to the one after the commonly proposed $MaxGen2$ FE. Thus, the user can approximately adjust the desired level of approximation accuracy ϵ by choosing $VarLimit = \epsilon^2$. However, the figures show that the value $VarLimit = 0.001^2$ is suitable for the considered test cases.

Table 2. Summary of $\boldsymbol{PI}_{i,j}$ and generation differences at the stop generation of OCD denoted as OCDStop and $MaxGen2$, where PIDiff = $\boldsymbol{PI}_{j,OCDStop} - \boldsymbol{PI}_{j,MaxGen2}$ and GenDiff = $MaxGen2$–OCDStop (j = {HV, EPS, R2}). Additionally, the number of saved function evaluations and their percentage of MaxGen2 are calculated.

		NSGA2		SMSEMOA	
problem	PI	med(PIDiff)	med(GenDiff)	med(PIDiff)	med(GenDiff)
ZDT1	HV	2.07e-03	124	0.96e-03	112
	Eps	2.08e-03	12400 FE	1.31e-03	11200 FE
	R2	0.93e-03	62%	0.38e-03	56%
ZDT2	HV	2.56e-03	104	1.01e-03	101
	Eps	3.13e-03	10400 FE	0.91e-03	10100 FE
	R2	1.46e-03	52%	0.63e-03	51%
ZDT4	HV	0.26e-03	18	21.72e-03	63
	Eps	0.28e-03	3600 FE	19.75e-03	12600 FE
	R2	0.06e-03	18%	9.07e-03	63%
DTLZ2	HV	-0.39e-03	287	0.72e-03	256
	Eps	-14.76e-03	28700 FE	3.37e-03	25600 FE
	R2	0.06e-03	96%	0.02e-03	85%
Fonseca	HV	0.97e-03	50	2.49e-03	49
	Eps	-0.51e-03	3000 FE	5.14e-03	2940 FE
	R2	0.14e-03	76%	0.21e-03	74%

The experiments document the general ability of the statistical tests within OCD to detect convergence based on performance indicator values. The delayed detection of convergence on the Fonseca problem is due to the time window of preceding generations and the very fast convergence of the EMOA. For a faster detection of stagnation, $nPreGen$ has to be decreased. However, the time of convergence as indicated by OCD can be accounted as premature for SMSEMOA on ZDT4 and DTLZ2 regarding the run of the metrics in further generations. In such situations, a larger time window allows longer phases of stagnation and provides the EMOA with the possibility to escape from local optima. In summary, a conflict between a fast detection of convergence and robustness with respect to short phases of stagnation exists. Therefore, the specification of the length of the time windows $nPreGen$ allows the user of OCD to express his own preferences based on the expected kind of problem.

The problem of the overestimation of the generation, in which stagnation occurs, when OCD is applied within NSGA2 can be explained by the selection that is implemented within this EMOA. Due to the high number of non-dominated solutions in the already converged population, the individuals are mainly evaluated by means of the crowding distance [16]. Thus, in combination with the $(\mu + \mu)$ selection, the population is still in motion. Since the reference set itself is part of this motion, a high variance in the indicator values is likely to appear. In contrast, SMSEMOA does only accept solutions, which increase the hypervolume of the current population. Thereby, a monotonic improvement can be expected, which also guarantees appropriate reference sets for OCD.

5 Conclusion

In this paper, a robust and reliable method for convergence detection within evolutionary multi-objective optimization algorithms has been introduced. This method is based on two statistical tests, namely the t-test on the regression coefficient and the χ^2-variance test, which guarantee an accurate convergence detection in all the considered examples. The proposed method is able to invoke different performance indicators, and it was investigated using the three recommended metrics from the EMO field. This way, we have been able to save half of the function evaluations for common test cases without having to accept a considerable loss of quality. However, the application of OCD to optimization scenarios, which include temporary phases of stagnation, such as in discrete optimization, could result in a premature indication of convergence.

In addition, we tried OCD on an already solved practical example [33], which is not shown due to a lack of space. This test indicated that the former analysis wasted many computational resources. Processing this hint by means of comprehensive evaluations of OCD on real-world problems is a task for the near future.

Furthermore, the technique of OCD offers a way for algorithm comparison. For this purpose, all EMOA parameters and operators have to be set to comparable values, and a high number of parallel runs of each benchmarked EMOA has to be performed. This way, a proper statistical analysis on the distributions of the stop generations proposed by OCD combined with the internally used performance indicators becomes possible. In this context, a comparison to an approach for offline convergence detection, which has been recently proposed by one of the authors [18], seems revealing.

Acknowledgements. This paper is based on investigations of the collaborative research center SFB/TR TRR 30, which is kindly supported by the *Deutsche Forschungsgemeinschaft (DFG)*. Moreover, we acknowledge financial support from the German *Federal Ministry of Economics and Technology (BMWi)*.

References

1. Zitzler, E., Künzli, S.: Indicator-based selection in multiobjective search. In: Yao, X., Burke, E.K., Lozano, J.A., Smith, J., Merelo-Guervós, J.J., Bullinaria, J.A., Rowe, J.E., Tiňo, P., Kabán, A., Schwefel, H.-P. (eds.) PPSN 2004. LNCS, vol. 3242, pp. 832–842. Springer, Heidelberg (2004)
2. Beume, N., Naujoks, B., Emmerich, M.: SMS-EMOA: Multiobjective selection based on dominated hypervolume. European Journal of Operational Research 181(3), 1653–1669 (2007)
3. Zitzler, E., Thiele, L., Bader, J.: SPAM: Set preference algorithm for multiobjective optimization. In: Rudolph, G., Jansen, T., Lucas, S., Poloni, C., Beume, N. (eds.) PPSN 2008. LNCS, vol. 5199, pp. 847–858. Springer, Heidelberg (2008)
4. Wagner, T., Beume, N., Naujoks, B.: Pareto-, aggregation-, and indicator-based methods in many-objective optimization. In: Obayashi, S., Deb, K., Poloni, C., Hiroyasu, T., Murata, T. (eds.) EMO 2007. LNCS, vol. 4403, pp. 742–756. Springer, Heidelberg (2007)

5. Emmerich, M., Giannakoglou, K., Naujoks, B.: Single- and multi-objective evolutionary optimization assisted by gaussian random field metamodels. IEEE Trans. on Evolutionary Computation 10(4), 421–439 (2006)
6. Knowles, J.: ParEGO: A hybrid algorithm with on-line landscape approximation for expensive multiobjective optimization problems. IEEE Trans. on Evolutionary Computation 10(1), 50–66 (2006)
7. Ponweiser, W., Wagner, T., Biermann, D., Vincze, M.: Multiobjective optimization on a limited amount of evaluations using model-assisted $\mathcal{S}$-metric selection. In: Rudolph, G., Jansen, T., Lucas, S., Poloni, C., Beume, N. (eds.) PPSN 2008. LNCS, vol. 5199, pp. 784–794. Springer, Heidelberg (2008)
8. Deb, K., Mohan, M., Mishra, S.: A fast multi-objective evolutionary algorithm for finding well-spread pareto-optimal solutions. KanGAL report 2003002, Indian Institute of Technology, Kanpur, India (2003)
9. Huang, V.L., Qin, A.K., Deb, K., Zitzler, E., Suganthan, P.N., Liang, J.J., Preuss, M., Huband, S.: Problem definitions for performance assessment of multi-objective optimization algorithms. Technical report, Nanyang Technological University (2007)
10. Rudolph, G., Naujoks, B., Preuss, M.: Capabilities of EMOA to detect and preserve equivalent pareto subsets. In: Obayashi, S., Deb, K., Poloni, C., Hiroyasu, T., Murata, T. (eds.) EMO 2007. LNCS, vol. 4403, pp. 36–50. Springer, Heidelberg (2007)
11. Kumar, A., Sharma, D., Deb, K.: A hybrid multi-objective optimization procedure using PCX based NSGA-II and sequential quadratic programming. In: Michalewicz, Z., Reynolds, R.G. (eds.) Congress on Evolutionary Computation (CEC). IEEE Press, Piscataway (2007)
12. Deb, K., Lele, S., Datta, R.: A hybrid evolutionary multi-objective and SQP based procedure for constrained optimization. In: Kang, L., Liu, Y., Zeng, S. (eds.) ISICA 2007. LNCS, vol. 4683, pp. 36–45. Springer, Heidelberg (2007)
13. Emmerich, M., Deutz, A., Beume, N.: Gradient-based/Evolutionary relay hybrid for computing pareto front approximations maximizing the S-metric. In: Bartz-Beielstein, T., Blesa Aguilera, M.J., Blum, C., Naujoks, B., Roli, A., Rudolph, G., Sampels, M. (eds.) HCI/ICCV 2007. LNCS, vol. 4771, pp. 140–156. Springer, Heidelberg (2007)
14. Deb, K., Jain, S.: Running performance metrics for evolutionary multi-objective optimization. In: Simulated Evolution and Learning (SEAL), pp. 13–20 (2002)
15. Rudenko, O., Schoenauer, M.: A steady performance stopping criterion for pareto-based evolutionary algorithms. In: Multi-Objective Programming and Goal Programming (2004)
16. Deb, K., Pratap, A., Agarwal, S.: A fast and elitist multi-objective genetic algorithm: NSGA-II. IEEE Trans. on Evolutionary Computation 6(8) (2002)
17. Zitzler, E., Thiele, L., Laumanns, M., Fonseca, C., Fonseca, V.: Performance assessment of multiobjective optimizers: An analysis and review. IEEE Trans. on Evolutionary Computation 8(2), 117–132 (2003)
18. Trautmann, H., Ligges, U., Mehnen, J., Preuss, M.: A convergence criterion for multiobjective evolutionary algorithms based on systematic statistical testing. In: Rudolph, G., Jansen, T., Lucas, S., Poloni, C., Beume, N. (eds.) PPSN 2008. LNCS, vol. 5199, pp. 825–836. Springer, Heidelberg (2008)
19. Knowles, J., Thiele, L., Zitzler, E.: A tutorial on the performance assessment of stochastic multiobjective optimizers. 214, Computer Engineering and Networks Laboratory (TIK), Swiss Federal Institute of Technology (ETH) Zurich (2005)

20. Bleuler, S., Laumanns, M., Thiele, L., Zitzler, E.: PISA – A platform and programming language independent interface for search algorithms. In: Fonseca, C.M., Fleming, P.J., Zitzler, E., Deb, K., Thiele, L. (eds.) EMO 2003. LNCS, vol. 2632, pp. 494–508. Springer, Heidelberg (2003)
21. Sheskin, D.J.: Handbook of Parametric and Nonparametric Statistical Procedures, 2nd edn. Chapman and Hall, Boca Raton (2000)
22. Dudoit, S., van der Laan, M.: Multiple Testing Procedures with Applications to Genomics. Springer, Berlin (2008)
23. Stapleton, J.H.: Linear Statistical Models. Wiley Series in Probability and Statistics. Wiley, New York (1995)
24. Jensen, M.T.: Reducing the run-time complexity of multiobjective EAs: The NSGA-II and other algorithms. IEEE Trans. on Evolutionary Computation 7(5), 503–515 (2003)
25. Beume, N., Rudolph, G.: Faster S-metric calculation by considering dominated hypervolume as Klee's measure problem. In: International Conference on Computational Intelligence (CI 2006) (2006)
26. Ihaka, R., Gentleman, R.: R: A language for data analysis and graphics. Journal of Computational and Graphical Statistics 5, 299–314 (1996)
27. Fonseca, C.M., Fleming, P.J.: Multiobjective genetic algorithms made easy: Selection, sharing, and mating restriction. In: Genetic Algorithms in Engineering Systems: Innovations and Applications, pp. 42–52 (1995)
28. Zitzler, E., Deb, K., Thiele, L.: Comparison of multiobjective evolutionary algorithms: Empirical results. Evolutionary Computation 8(2), 173–195 (2000)
29. Deb, K., Thiele, L., Laumanns, M., Zitzler, E.: Scalable multi-objective optimization test problems. In: Congress on Evolutionary Computation (CEC), vol. 1, pp. 825–830. IEEE Press, Piscataway (2002)
30. Deb, K.: Multi-objective Optimization using Evolutionary Algorithms. Wiley, Chichester (2001)
31. Zitzler, E., Thiele, L.: Multiobjective optimization using evolutionary algorithms - A comparative case study. In: Eiben, A.E., Bäck, T., Schoenauer, M., Schwefel, H.-P. (eds.) PPSN 1998. LNCS, vol. 1498, pp. 292–301. Springer, Heidelberg (1998)
32. Hansen, M.P., Jaszkiewicz, A.: Evaluating the quality of approximations to the non-dominated set. Technical Report IMM-REP-1998-7 (1998)
33. Wagner, T., Michelitsch, T., Sacharow, A.: On the design of optimisers for surface reconstruction. In: Thierens, D., et al. (eds.) 9th Annual Genetic and Evolutionary Computation Conference (GECCO 2007), Proc., London, UK, pp. 2195–2202. ACM, New York (2007)

Spread Assessment for Evolutionary Multi-Objective Optimization

Miqing Li and Jinhua Zheng

Institute of Information Engineering, Xiangtan University,
411105, Hunan, China
limit1008@126.com, jhzheng@xtu.edu.cn

Abstract. Convergence, uniformity and spread are three basic issues in comparing the performance of multi-objective evolutionary algorithms. However, most of metrics pay more attention on former two performance indices. In this paper, we introduce a metric for evaluating the spread of non-dominated solutions. Unlike existed metrics only calculating the extreme solutions in objective space, this metric defines boundary concept of non-dominated set. And it evaluates the extent of boundary solutions by projecting them on low-dimensional spaces. Moreover, the centroid of solutions set is introduced to avoid the impact of different convergence result of algorithms. From a comparative study on several test problems, the metric is examined to assess spread of non-dominated solutions set in objective space.

Keywords: Multi-objective optimization, Performance assessment, Hypervolume, Spread, Boundary solution.

1 Introduction

There has been a growing interest in applying heuristic search techniques to multi-objective optimization problem (MOP) and various efficient algorithms have been proposed during the past couple years[1,2]. These algorithms usually generate a set of solutions to approximate the Pareto front of a multi-objective optimization problem. Naturally, performance assessment of these generated sets gained much attention. One obvious way to compare algorithms is to simply visualize the final sets of solutions and rely on intuitive judgments to estimate superiority of algorithms. However, this way is only possible for 2 and 3 objectives, and also as discussed by Van Veldhuizen and Lamont[3], intuitive and visual assessment is not a reliable tool for comparison of different multi-objective optimization algorithms.

Recently, several quality metrics have been emerging in the literature to evaluate the quality of observed solution sets[4,5,6]. They mainly assess three criterions that the algorithms draw their strength to optimize: 1) The distance of the obtained non-dominated set to the Pareto optimal front; 2) The uniformity of the obtained non-dominated set; 3) The distribution extent of the obtained non-dominated set. In fact, researchers paid significant attention on

M. Ehrgott et al. (Eds.): EMO 2009, LNCS 5467, pp. 216–230, 2009.

the first and second criterions. For example, there are several metrics in the literature that are claimed to assess "uniformity of solutions" in one way or another, including "Spacing Metric[7]", "Niche-based[8]", "Grid[9]" and "Uniformity Assessment[10]". As the same time, "Generational Distance[11]", "Seven Points Average Distance[7]" and "Coverage of Two Sets[12]" are claimed to assess "convergence of solutions". However, relative to the former two criterions, the third one obtained less attention. One of the important reasons may be the range of the objective function value is hard to measure, and different functions have different shapes. More recently, some metrics has been proposed to assess the comprehensive performance of the obtained non-dominated set. Hypervolume[12] and Inverted Generational Distance[13] were proposed to assess the convergence, uniformity and spread.Furthermore, Entropy Metric[14] and another Spacing Metric proposed by Deb[1] appeared to evaluate both uniformity and spread. These methods will have a detailed discussion in Section 3.

In this paper, we present a metric to assess the spread of solutions set. This metric defines boundary concept of obtained non-dominated set, and projects this set on (M-1) dimensional spaces, then estimates the range of this set by Hypervolume measure. The organization of the rest of paper is as follows. In Section 2, relevant notations and definitions are reviewed. Section 3 presents the existing metrics. Section 4 is dedicated to the spread measurement. Several experiments and results are explained in Section 5 and Section 6 concludes the paper with a summary.

2 Definition and Terminology

The multi-objective optimization problem may be stated as a minimization problem (without loss of generality) as follows:

The formulation of a typical multi-objective design optimization problem with m objective functions is shown below in Eq. (1).

$$\begin{aligned}&\text{Minimize} \quad f(x)=\{f_1(x), f_2(x), \cdots, f_m(x)\}\\&\text{subject to}: \quad x \in D\\&D=\{x \in \Re^n : g_k(x) \leq 0, k=1, \cdots, K; h_l(x)=0, l=1, \cdots, L\}\end{aligned} \tag{1}$$

where x is a design vector containing n components of design variables, $f_i(x)$ is the ith objective function, $g_k(x)$ is the kth in equality constraint and $h_l(x)$ is the lth equality constraint. The set of all design vectors which satisfies all constrains is denoted by D. The n-dimensional space wherein its coordinate axes are design variables is referred to as the "variable space". The m-dimensional space wherein its coordinate axes are design objective functions is referred to as the "objective space".

Definition 1 (***Pareto Dominance and Pareto Optimality set***). *The objective vector $f(x_a)$ is said to dominate the objective vector $f(x_b)$, denoted $f(x_a) \prec f(x_b)$, if $f_i(x_a) \leq f_i(x_b)$ for all $i \in \{1, 2, \cdots, m\}$ and $f_j(x_a) < f_j(x_b)$ for some $j \in \{1, 2, \cdots, m\}$. A point x^* is said to be Pareto optimal solution for the*

MOP if and only if there does not exist $x \in D$ *such that* $f(x) \prec f(x^*)$. *In other words, a point* x^* *is Pareto solution if there does not exist a point* x *in D that would achieve a better value for one of the objectives without worsening at least another. The solution set is termed as Pareto optimality set that is collective of all Pareto solutions.*

Definition 2 (Pareto Optimality Front). *The Pareto optimal front (POF) is the image of the Pareto optimal set. POF contains all those objective vectors that are not dominated by any vector in the objective space.*

Definition 3 (Solution Set and Non-dominated set). *The set of solutions found by an optimizer is known as Solution Set S. The solutions in S that are not dominated by others in the set define the Non-dominated set (NDS). Since in most cases, only non-dominated solutions will be generated, we will not distinguish S and NDS hereafter unless explicitly indicated.*

Definition 4 (Non-dominated Extreme Solution). *Non-dominated Extreme solution (NDES) is the solution that has maximum value for one or more objectives among NDS.*

Note that the solution which has maximum value for one or more objectives among *NDS* is called as boundary solution in some other papers. As *boundary solution* has other definition in next section, we call it as Non-dominated Extreme solution in this paper.

3 Some Existing Metrics

In this section, several quality metrics are introduced to assess the spread of solutions. Now we consider some comprehensive metrics first.

3.1 Metrics for Convergence, Uniformity and Spread

Inverted Generational Distance (*IGD*)[13] and Hypervolume[12] are used to assess comprehensive performance of convergence, uniformity and spread. Let P^* be a set of uniformly distributed points in the objective space along the *POF*, the Inverted Generational Distance from P^* to *NDS* is defined as

$$IGD(P^*, NDS) = \frac{\sum_{v \in P^*} d(v, NDS)}{|p^*|} \quad (2)$$

Where d(v,*NDS*) is the minimum Euclidean distance between v and points in *NDS*. Since P^* is enough to represent the *POF* well, *IGD* could measure the convergence, uniformity and spread of *NDS* in a sense. To have a low value of *IGD*, *NDS* must be close to the *POF* and have good range and uniformity. A drawback of *IGD* is *POF* must be given, which may be trivial for some real world applications.

The Hypervolume measure was proposed by Zitzler and Thiele[12]. It calculates the hypervolume of the multi-dimensional region enclosed by *NDS* and a "reference point". The major shortcoming is that an accurate calculation of the hypervolume requires a normalized and a careful choice of the reference point. Knowles and Corne had a detailed analysis in[15]. Another problem is Hypervolume and *IGD* can not assess the spread of solutions alone, Figure 1 gives an example of distribution.

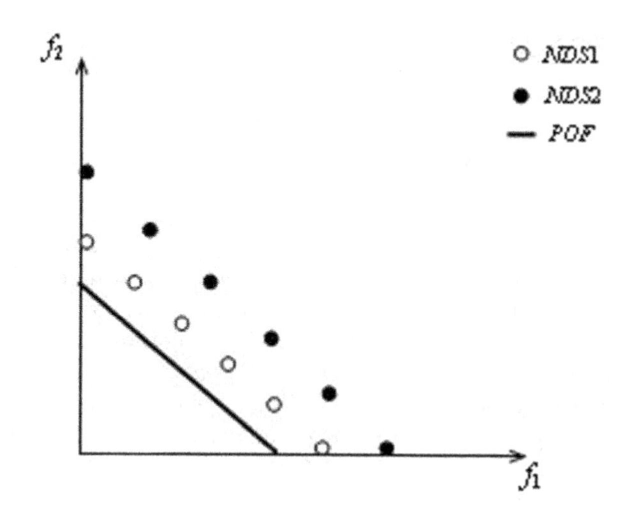

Fig. 1. An example of distribution. Assume *NDS*1 and *NDS*2 are the obtained solution by two algorithms. Obviously, they have same uniformity and spread, while only difference is *NDS*1 is closer to the *POF*. The results of Hypervolume and *IGD* are that *NDS*1 has better value than *NDS*2. Naturally, the methods can not give accurate assessment of spread for solutions.

3.2 Metrics for Uniformity and Spread

The metrics in this subsection include information about both uniformity and spread. Deb et al.[1] proposed a metric to assess the diversity of solutions.

$$\Delta = \frac{d_f + d_l + \sum_{i=1}^{|NDS|-1} |d_i - \overline{d}|}{d_f + d_l + (|NDS| - 1)\overline{d}} \quad (3)$$

where d_f and d_l are the Euclidian distances between the extreme solutions in *POF* and *NDS*. $\overline{d}$ is the average of all distance d_i , $i \in [1, |NDS|-1]$. The drawback of this metric is Δ works only for 2 objective problems.

Another diversity metric based on entropy has been proposed by Farhang et al.[14] The basic idea is that each solution provides some information about its neighborhood modeled by Gaussian distribution. A Density Function has also been calculated by the sum of all Gaussian distribution form all solutions. The peaks and valleys of density function correspond to the dense areas and the sparse areas respectively. Deb et al. gave a particular analysis and pointed out four deficiencies in[9]. In addition, since distance between two solutions is influenced by convergence more or less, these methods can not assess the situation in Figure 1 accurately.

3.3 Metrics Only for Spread

In this subsection, we introduce the metrics only for spread of solutions. *Maximum Spread* proposed in[16] shows the distance between *NDESs*, a bigger value indicates better spread of solutions.

$$D = \sqrt{\sum_{m=1}^{M} (\max_{i=1}^{|NDS|} f_m^i - \min_{i=1}^{|NDS|} f_m^i)^2} \quad (4)$$

A similar metric called *Overall Pareto Spread* (*OS*) is adopted in[17]. Instead of the diagonal line, *OS* calculated the size of the hyper-rectangle. Thus, only the representative values are different. Indeed, these methods aren't fit for the solutions in Figure 1. The assessment result using these methods would mislead that *NDS2* has the better spread value than *NDS1*. Furthermore, another drawback of the metrics is that they, only considering *NDES*, do not give enough information to address the range of solutions, Figure 2 is an example of distribution range.

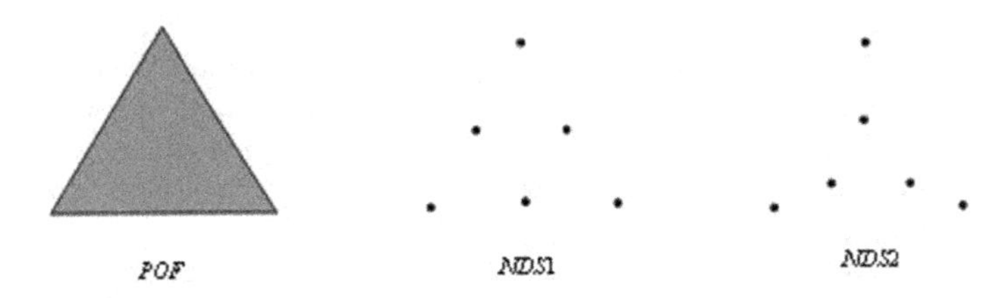

Fig. 2. An example of distribution. Shadow triangle is an ichnography of *POF* for a 3-objective function, and *NDS1* and *NDS2* are the obtained solution by two algorithms for this function. Assume that they have same convergence. Considering the identity of *NDES* in two sets, Maximum Spread or *OS* would have same assessment value. However, the ranges of two sets are different. *NDS1* spreads over a wider extent.

4 Proposed Metric for Spread

An observed solution set that spreads over a wider range of the objective function values provides the designer with broader optimized design choices. A metric criterion provided enough information of extent of solution set, yet general, would be desirable. Bearing this idea in mind, we give some definitions of extent.

Definition 5 (Beyond). *The vector* $f(x_a)$ *is said to beyond the vector* $f(x_b)$ *in* $f_1(x), f_2(x), \cdots, f_m(x)$ *space, denoted* $f(x_a) \rhd f(x_b)$*, if* $f_i(x_a) \geq f_i(x_b)$ *for all* $i \in \{1, 2, \cdots, m\}$ *and* $f_i(x_a) > f_i(x_b)$ *for some* $j \in \{1, 2, \cdots, m\}$.

Note that the definition of *beyond* is equal to that of dominance in maximization MOP. Here, *beyond* is introduced to address the concept of extent of *NDS*.

Definition 6 (Boundary Solution and Boundary Solution in $f_i(x) = 0$ Space). *A vector $f(\overline{x}) \in NDS$ is considered as* Boundary Solution in $f_i(x) = 0$ Space *(called as BSi), if $f(\overline{x})$ is not Beyond by any member of NDS for subsets $f_1(x), \cdots, f_{i-1}(x), f_{i+1}(x), \cdots, f_m(x)$ of the objectives. In other words, a vector $f(\overline{x})$ is Boundary Solution in $f_i(x) = 0$ Space if and only if there does not exist a vector $f(x^\bullet) \in NDS$ that $f(x^\bullet) \triangleright f(\overline{x})$ in $f_1(x), \cdots, f_{i-1}(x), f_{i+1}(x), \cdots, f_m(x)$ space. A vector f(x) is said to be Boundary Solution (BS) for NDS if it is one of vector in $BS_1 \bigcup BS_2 \bigcup \cdots \bigcup BS_{m-1} \bigcup BS_m$.*

Figure 3 and 4 illustrate the boundary solution for a set of points in a 3- and 4-objective space. It is clear that the extent of solutions set only depends on boundary solutions. In addition, extreme solutions are included in boundary solutions, e.g. extreme solution in f_1 direction is boundary solution in $BS_2, BS_3, \cdots,$ BS_{m-1}, BS_m. Especially, for bi-objective problem, the extreme solutions are equal to boundary solutions. Next, a main idea of our metric for evaluating extent of boundary solution is given.

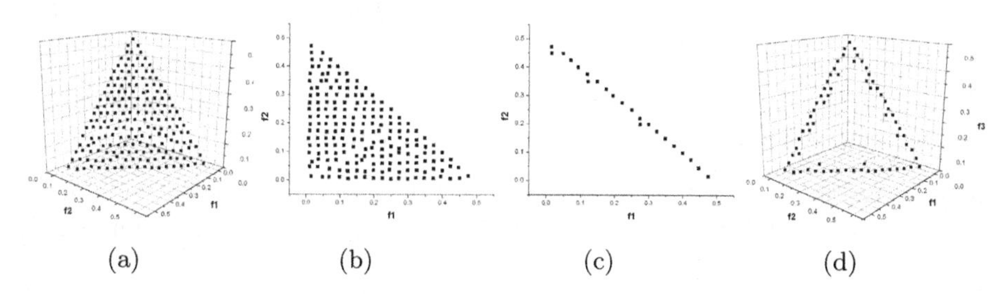

Fig. 3. A 3-objective example of *Boundary Solution* for a set obtained by $\varepsilon - MOEA$[22] on DTLZ1: (a) an obtained solutions set by $\varepsilon - MOEA$ on DTLZ1; (b) a projection of obtained solutions set on (f_1, f_2) objective space; (c) a projection of BS_3 on (f_1, f_2) objective space; (d) *BS* of obtained solutions set

The proposed metric quantifies how widely *NDS* spreads over the objective space. The essential idea is that projecting *NDS* on $(m-1)$ objective spaces and calculating Lebesgue measure of *BS* on projected hyper-plane. Moreover, the centroid is considered to avoid the impact of different convergence results of algorithms. The algorithm procedure of metric is shown as following:

Step 1: Project the obtained *NDS* on m $(m-1)$-objective spaces, respectively (called as NDS_j for f_j=0 objective space).

Step 2: For each $(m-1)$-objective space (here described as f_j=0 objective space), find projected BS_j on f_j=0 objective space.

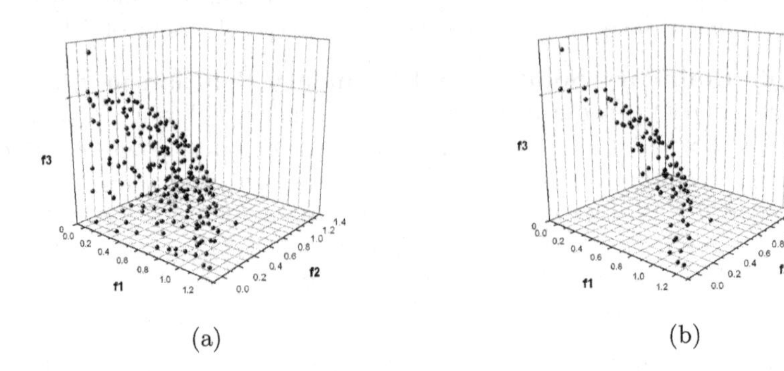

Fig. 4. A 4-objective example of *Boundary Solution* for a set obtained by SPEA2 on DTLZ2: (a) a projection of obtained solutions set on (f_1, f_2, f_3) objective space; (b) a projection of BS_4 on (f_1, f_2, f_3) objective space

Step 3: For each $(m-1)$-objective space, calculate the size of the space covered of projected BS_j by Lebesgue measure.

$$S(BS_j) = \Lambda(\{\bigcup_i h_{ji} | b_{ji} \in BS_j\}) \quad (5)$$

where h_{ji} is the union of hypercubes defined by projected boundary solution b_{ji} and the reference point in Lebesgue measure is coordinate origin.

Step 4: For each $(m-1)$ -objective space ,calculate centroid $c(w_1, \cdots, w_{j-1}, w_{j+1}, \cdots, w_m,)$ of NDS_j, where $W_i = \sum_{k \in NDS} f_i(x_k)/N, i \neq j,$,N is number of *NDS*.

Step 5: For each $(m-1)$-objective space, the assessment value V_j on f_j=0 objective space is calculated as follows:

$$V_j = \frac{S(BS_j)}{\prod_{i=1, i \neq j}^{m} |w_i|} \quad (6)$$

Step 6: The total assessment result V for *NDS* can be expressed as:

$$V = \sqrt[m]{\prod_{j=1}^{m} V_j} \quad (7)$$

The value of V shows the extent of obtained solutions set. When comparing V of two sets, the designer prefers larger V with a wider spread. In Step 3 of this algorithm, The Hypervolume measure is introduced to calculate the size of space covered. Just as presented in 3.1 Section, it is an accurate metric to assess the volume enclosed by solutions. Recently, some researchers give deeper analysis in[18,19,20,26]. The only difference is that coordinate origin instead of reference point be chosen in objective space. In the algorithm, centroid is used to offset the larger value of hypervolume when the solutions have worse convergence result. Figure 5 illustrates intuitive explanation of the algorithm.

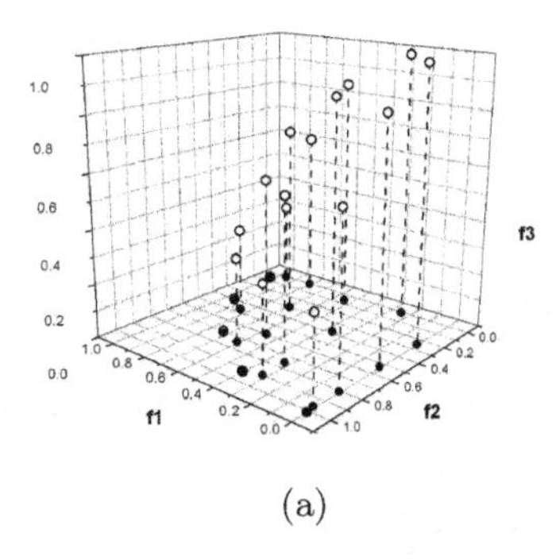

(a)

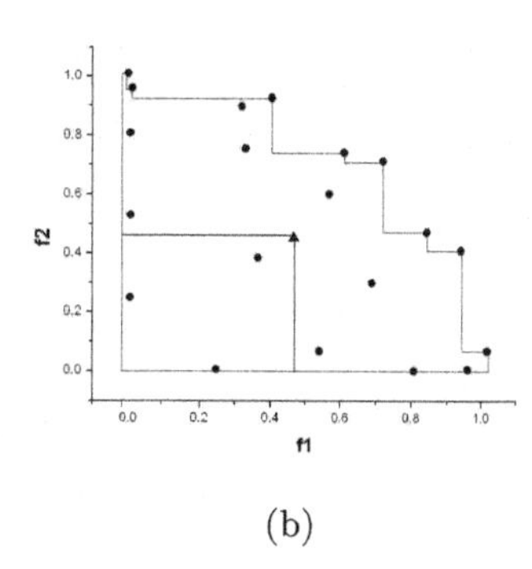

(b)

Fig. 5. An example for calculating V of a set obtained by SPEA2 on DTLZ2: (a) an illustration that 3-dimension *NDS* is projected on f_3=0 2-dimension plane. Box points are original *NDS* obtained in solid space. Solid points are the projected points in this plane; (b) an illustration of calculating V_3 on $(f_1, f_2,)$ objective space. Black points are corresponding to the points in (a). Triangle is the centroid of these black points. The convex polytope is enclosed by BS_3 and coordinate origin. The rectangle is enclosed by centroid and coordinate origin. Therefore, V_3 is the ratio of the convex polytope to the rectangle.

5 Simulation Results

In this section, we apply the above metric on four different methods (NSGA-II[21],ε-MOEA[22], SPEA2[23], PESA-II[24]), in view of the spread of these methods is representational. All MOEAs are given real-valued decision variables. A crossover probability of P_c=0.9 and a mutation probability P_m=1/n (where n is the number of decision variables) are used. The operators for crossover and mutation are simulated binary crossover (SBX)[1] and polynomial mutation, with distribution indexes of η_c=11,and η_m=17, respectively.

Our metric is applied to results of different test problems with 2, 3 and 4 objectives. ZDT1 and ZDT2[16] are 2-objective test problems, and DTLZ1, DTLZ2, DTLZ3, DTLZ5 and DTLZ7[25] with variable number of objectives. In our experiment, the objective number is 3, 4 and 6. For PESA-II, we have set 32×32 hyper-boxed for 2-objective problems, 8×8×8 for 3-objective problems, and 6×6×6×6 for 4-objective problems. Table 1 is some parameters of experiments and Table 2 is the ε value of ε-MOEA . The setting of ε can make the number of solutions accord with population size of other algorithms. Next, the results of different test problems are analyzed separately.

Table 1. The parameters of the experiments

Object number	2	3	4
Population size	100	200	300
Evaluation number	20000	100000	300000

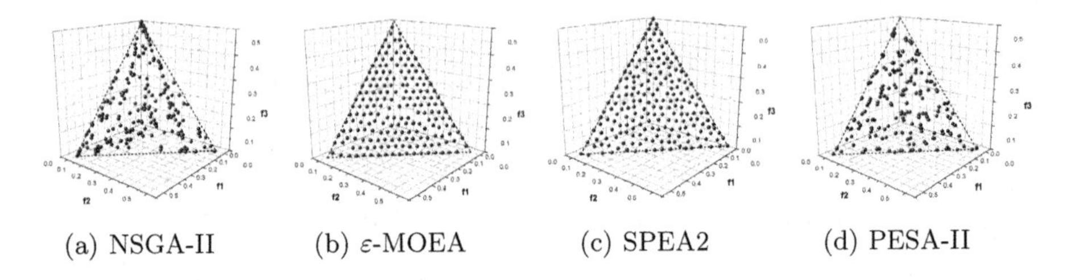

(a) NSGA-II (b) ε-MOEA (c) SPEA2 (d) PESA-II

Fig. 6. The *NDS* obtained by four MOEAs on DTLZ1. The dashed line is the boundary of *POF*.

Table 2. The ε setting of ε-MOEA

Object number	2	3	4
ZDT1	0.0075	\	\
ZDT2	0.0075	\	\
DTLZ1	\	0.025	\
DTLZ2	\	0.042	0.115
DTLZ3	\	0.042	\
DTLZ5	\	0.0025	0.027
DTLZ7	\	0.033	0.076

Table 3. Spread measure of DTLZ1 and DTLZ3

Test problem	NSGA-II	ε-MOEA	SPEA2	PESA-II
DTLZ1	3.97383	4.12704	4.24122	3.95855
	0.07275	0.03121	0.05965	0.08205
DTLZ3	3.06745	2.82745	3.17421	2.96639
	0.05035	0.01164	0.04724	0.04586
DTLZ3	3.06120	\	\	\
(P_m=0.005)	0.04877	\	\	\

5.1 DTLZ1 Test Problem

First, we consider the three-objective DTLZ1 test problem with n=7 variables. The Pareto optimal solutions lie on a three-dimensional plane satisfying $f_1 + f_2 + f_3 = 0.5$. Figure 6 shows the results of methods. It can be observed that SPEA2 has most wide distribution, and the *BS* usually is located in boundary of *POF*. ε-MOEA takes the second place, the next is NSGA-II, and PESA-II has the worst result that quite a part of *BS* is inside red line. The result of the

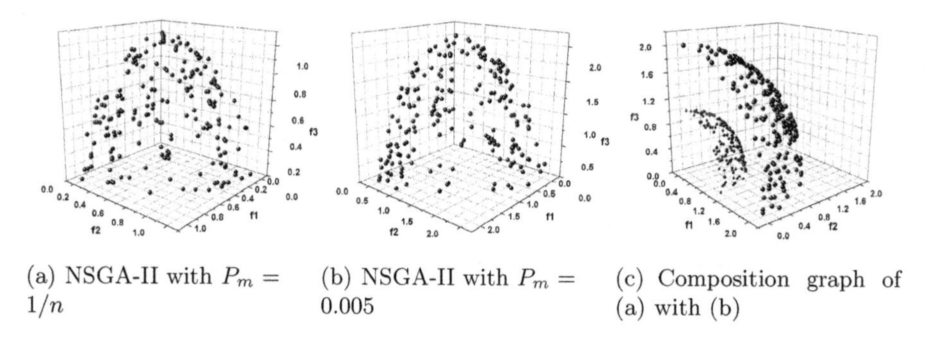

(a) NSGA-II with $P_m = 1/n$

(b) NSGA-II with $P_m = 0.005$

(c) Composition graph of (a) with (b)

Fig. 7. The *NDS* obtained by different mutation probability for NSGA-II on DTLZ3

proposed metric for DTLZ1 is recorded in Table 3. For each problem in result, we have carried out 20 independent runs, these tables include the average and standard deviation, and the upper values in each row of table are the average values. The larger average values indicate a better distribution extent. Indeed in table 3, it is observed that SPEA2 has the best spread metric value, and the following are ε-MOEA, NSGA-II and PESA-II, which confirm the observations from Figure 6.

5.2 DTLZ3 Test Problem

Here, we concentrate on the 12-variable 3-objective DTLZ3 test problem, which has $(3^{12}-1)$ local Pareto optimal fronts[25]. The *POF* satisfies the equation $f_1^2+f_2^2+f_3^2=1$. This problem is chosen to test the measure value impacted by different convergence result of algorithms. Figure 7 gives a result of *NDS* obtained by different mutation probability for NSGA-II on DTLZ3. The NDS obtained with mutation probability $P_m=1/12$ in Figure 7(a) has converged to *POF*. The *NDS* obtained with mutation probability $Pm=0.005$ in Figure 7(b) can only converge to local Pareto optimal front $f_1^2+f_2^2+f_3^2=2^2$. But they have quite similar shapes, uniformity and spread. The only difference is the convergence of algorithms. The result of the proposed metric is recorded in Table 3. It is observed that they have much similar results, which confirm the observations from Figure 7. Figure 8 also shows the results of ε-MOEA and SPEA2. We observe that the solutions obtained by SPEA2 have wider extent than those obtained by ε-MOEA in Figure 8(c). Indeed, observing *BS* in $f_1=0$ plane, blue points have more extensive distribution than red points, which accord with the results in Table 3.

5.3 Other Test Problems

Figure 9 and 10 show the *NDS* obtained by two 2-objective problems (ZDT1, ZDT2), and assessment results can be observed in Table 4, NSGA-II and SPEA2

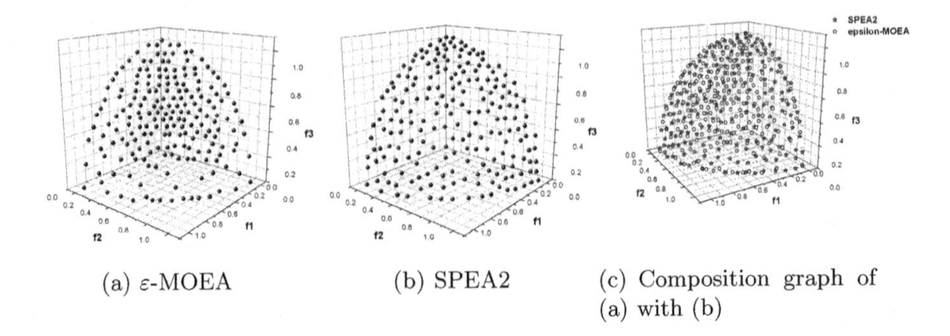

(a) ε-MOEA (b) SPEA2 (c) Composition graph of (a) with (b)

Fig. 8. The *NDS* obtained by ε-MOEA and SPEA2 on DTLZ3

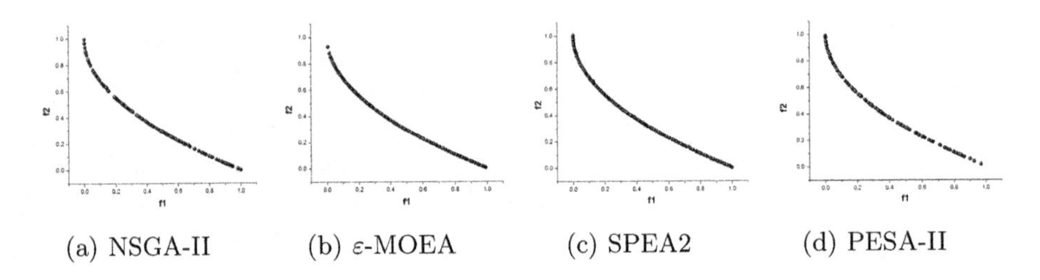

(a) NSGA-II (b) ε-MOEA (c) SPEA2 (d) PESA-II

Fig. 9. The *NDS* obtained by four MOEAs on ZDT1

Table 4. Spread measure of other test problems

Test problem	NSGA-II	ε-MOEA	SPEA2	PESA-II
ZDT1	2.40015	2.33097	2.38568	2.36806
	0.00927	0.00581	0.01181	0.02771
ZDT2	1.71481	1.69658	1.71548	1.69788
	0.01121	0.00291	0.00445	0.00761
DTLZ2(3-obj)	3.06368	2.82859	3.17661	2.96993
	0.03241	0.00650	0.01535	0.04397
DTLZ5(3-obj)	2.10775	1.95871	2.08195	2.02985
	0.03018	0.00662	0.00906	0.04895
DTLZ7 (3-obj)	2.62506	2.72128	2.59497	2.61399
	0.05869	0.016137	0.08814	0.10903
DTLZ2 (4-obj)	6.05588	5.78005	7.06089	5.50716
	0.10151	0.02156	0.11098	0.13846
DTLZ5 (4-obj)	5.06555	4.30551	4.59924	4.25687
	0.16895	0.08473	0.12963	0.23641
DTLZ7 (4-obj)	4.79397	4.31676	5.55387	4.71153
	0.17115	0.06234	0.23839	0.17370

provide better spread compared to PESA-II and ε-MOEA.Furthermore, NSGA-II has best value on ZDT1, and SPEA2 is predominated on ZDT2. In addition, PESA-II has slightly better than ε-MOEA. The reason may be that ε-MOEA

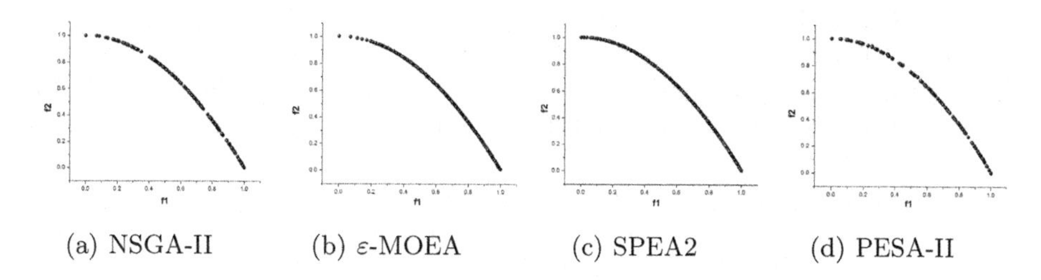

(a) NSGA-II (b) ε-MOEA (c) SPEA2 (d) PESA-II

Fig. 10. The *NDS* obtained by four MOEAs on ZDT2

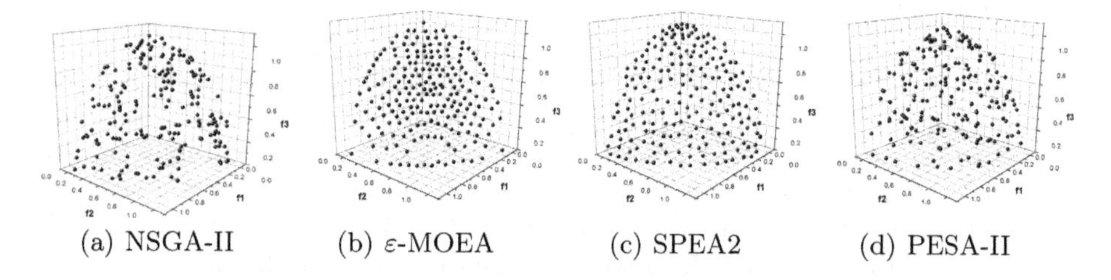

(a) NSGA-II (b) ε-MOEA (c) SPEA2 (d) PESA-II

Fig. 11. The *NDS* obtained by four MOEAs on DTLZ2

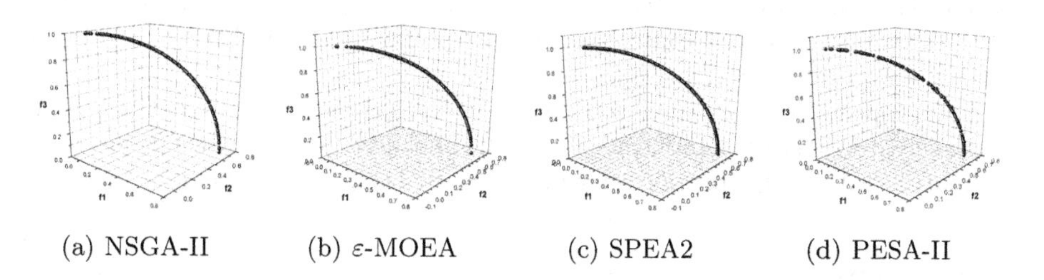

(a) NSGA-II (b) ε-MOEA (c) SPEA2 (d) PESA-II

Fig. 12. The *NDS* obtained by four MOEAs on DTLZ5

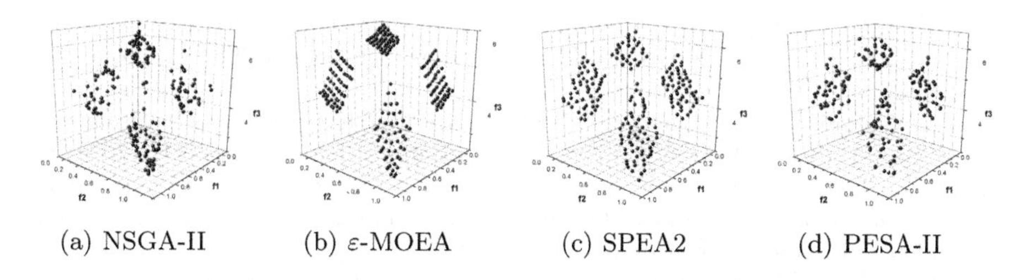

(a) NSGA-II (b) ε-MOEA (c) SPEA2 (d) PESA-II

Fig. 13. The *NDS* obtained by four MOEAs on DTLZ7

which preserves the closest solution to reference point can affect spread in a certain extent. Figure 11-13 show the *NDS* obtained by three 3-objective problems (DLTZ2, DTLZ5, DTLZ7), and assessment results can be observed in

Table 4. SPEA2 has the best result on DTLZ2 and NSGA-II provides the best extent on DTLZ5. Unexpectedly,ε-MOEA which has worst values on DTLZ2 and DTLZ5 do provides better result than other three algorithms on DTLZ7. Table 4 also shows the results for DTLZ2, DTLZ5, and DTLZ7 on 4-objective problems. Likely 3-objective problems, SPEA2 and NSGA-II have best results on DTLZ2 and DTLZ5 respectively. Interestingly, for DTLZ7, SPEA2 which has worst result in 3-objective space holds best value in 4-objective space. In conclusion, by this study of different objective problems, SPEA2 and NSGA-II are demonstrated that can exploit wider range of solutions in most problems.

6 Conclusion

Obtaining an extensively-distributed set of solutions is one of the goals of multi-objective optimization. In this paper, we have introduced and studied a metric for assessing the spread of a non-dominated solutions set. By defining boundary concept of obtained non-dominated set, this metric can provide accurate information of extent for it. Moreover, centroid is introduced into this method to balance the impact that different convergence results of algorithms bring to spread assessment. A test on several problems was done with this method, and the results of four MOEAs (NSGA-II,ε-MOEA, SPEA2, and PESA-II) are compared. The evaluations on DTLZ1 and DTLZ3 confirm the observations from the illustration, and the evaluations on other 2-, 3- and 4-objective problems show the capability of algorithms to win extensive distribution.

The proposed metric using Hypervolume to assess extent of solutions is demanding in terms of computational load. The advance of time efficiency for estimating extent of solutions set will be a subject of future research.

References

1. Deb, K.: Multi-Objective Optimization using Evolutionary Algorithms. John Wiley and Sons, Chichester (2001)
2. Coello Coello, C.A., Van Veldhuizen, D.A., Lamont, G.B.: Evolutionary Algorithms for Solving Multi-Objective Problems. Kluwer Academic Publishers, New York (2002)
3. Van Veldhuizen, D.A., Lamont, G.B.: Multiobjective Evolutionary Algorithms: Analyzing the State-of-the-Art. Evolutionary Computation 8(2), 125–147 (2000)
4. Zitzler, E., Thiele, L., Laumanns, M., Fonseca, C.M., Fonseca, V.G.: Performance Assessment of Multiobjective Optimizers: An Analysis and Review. IEEE Transactions on Evolutionary Computation 7(2), 117–132 (2003)
5. Okabe, T., Jin, Y., Sendhoff, B.: A Critical Survey of Performance Indices for Multi-Objective Optimization. In: Proceedings of the 2003 Congress on Evolutionary Computation (CEC 2003), pp. 878–885 (2003)
6. Knowles, J., Thiele, L., Zitzler, E.: A Tutorial on the Performance Assessment of Stochastic Multiobjective Optimizers, Technical Report No. 214, Computer Engineering and Networks Laboratory (TIK), ETH Zurich, Switzerland (February 2006)

7. Schott, J.R.: Fault Tolerant Design Using Single and Multicriteria Genetic Algorithm Optimization. Master's thesis, Department of Aeronautics and Astronautics, Massachusetts Institute of Technology (1995)
8. Srinivas, N., Deb, K.: Multiobjective Optimization Using Nondominated Sorting in Genetic Algorithms. Evolutionary Computation 2(3), 221–248 (1994)
9. Deb, K., Jain, S.: Running Performance Metrics for Evolutionary Multi-Objective Optimization. In: Proceedings of the 4th Asia-Pacific Conference on Simulated Evolution and Learning (SEAL 2002), pp. 13–20 (2002)
10. Li, M., Zheng, J., Xiao, G.: Uniformity Assessment for Evolutionary Multi-Objective Optimization. In: Proceedings of IEEE Congress on Evolutionary Computation (CEC 2008), pp. 625–632 (2008)
11. Van Veldhuizen, D.A., Lamont, G.B.: Evolutionary Computation and Convergence to a Pareto Front. In: Koza, J.R. (ed.) Late Breaking Papers at the Genetic Programming Conference, pp. 221–228 (1998)
12. Zitzler, E., Thiele, L.: Multiobjective Evolutionary Algorithms: A Comparative Case Study and the Strength Pareto Approach. IEEE Transactions on Evolutionary Computation 3(4), 257–271 (1999)
13. Zhang, Q., Zhou, A., Jin, Y.: RM-MEDA: A Regularity Model-Based Multiobjective Estimation of Distribution Algorithm. IEEE Transactions on Evolutionary Computation 12(1), 41–63 (2008)
14. Farhang-Mehr, A., Azarm, S.: An Information-Theoretic Entropy Metric for Assessing Multiobjective Optimization Solution Set Quality. Transactions of the ASME, Journal of Mechanical Design 125(4), 655–663 (2003)
15. Knowles, J., Corne, D.: Properties of an Adaptive Archiving Algorithm for Storing Nondominated Vectors. IEEE Transactions on Evolutionary Computation 7(2), 100–116 (2003)
16. Zitzler, E., Deb, K., Thiele, L.: Comparison of Multiobjective Evolutionary Algorithms: Empirical Results. Evolutionary Computation 8(2), 173–195 (2000)
17. Wu, J., Azarm, S.: Metrics for Quality Assessment of a Multiobjective Design Optimization Solution Set. Transactions of the ASME, Journal of Mechanical Design 123, 18–25 (2001)
18. Fleischer, M.: The measure of pareto optima. In: Fonseca, C.M., Fleming, P.J., Zitzler, E., Deb, K., Thiele, L. (eds.) EMO 2003. LNCS, vol. 2632, pp. 519–533. Springer, Heidelberg (2003)
19. While, L., Hingston, P., Barone, L., Huband, S.: A Faster Algorithm for Calculating Hypervolume. IEEE Transactions on Evolutionary Computation 10(1), 29–38 (2006)
20. Zitzler, E., Brockhoff, D., Thiele, L.: The hypervolume indicator revisited: On the design of pareto-compliant indicators via weighted integration. In: Obayashi, S., Deb, K., Poloni, C., Hiroyasu, T., Murata, T. (eds.) EMO 2007. LNCS, vol. 4403, pp. 862–876. Springer, Heidelberg (2007)
21. Deb, K., Pratap, A., Agarwal, S., Meyarivan, T.: A Fast and Elitist Multiobjective Genetic Algorithm: NSGA-II. IEEE Transactions on Evolutionary Computation 6(2), 182–197 (2002)
22. Deb, K., Mohan, M., Mishra, S.: Towards a quick computation of well-spread pareto-optimal solutions. In: Fonseca, C.M., Fleming, P.J., Zitzler, E., Deb, K., Thiele, L. (eds.) EMO 2003. LNCS, vol. 2632, pp. 222–236. Springer, Heidelberg (2003)

23. Zitzler, E., Laumanns, M., Thiele, L.: SPEA2: Improving the strength Pareto evolutionary algorithm. TIK-Report 103 (2001)
24. Corne, D., Jerram, N., Knowles, J., Oates, M.: PESA-II: Region-based Selection in Evolutionary Multiobjective Optimization. In: Proceedings of the Genetic and Evolutionary Computation Conference (GECCO 2001), pp. 283–290 (2001)
25. Deb, K., Thiele, T., Laumanns, M., Zitzler, E.: Scalable Test Problems for Evolutionary Multiobjective Optimization. In: Abraham, A., Jain, L., Goldberg, R. (eds.) Evolutionary Multiobjective Optimization. Theoretical Advances and Applications, pp. 105–145 (2005)
26. Beume, N., Naujoks, B., Emmerich, M.: SMS-EMOA: Multiobjective selection based on dominated hypervolume. European Journal of Operational Research 181(3), 1653–1669 (2007)

An Improved Version of Volume Dominance for Multi-Objective Optimisation

Khoi Le, Dario Landa-Silva, and Hui Li

Automated Scheduling, Optimisation and Planning Research Group
School of Computer Science, The University of Nottingham
{kxl,jds,hzl}@cs.nott.ac.uk

Abstract. This paper proposes an improved version of volume dominance to assign fitness to solutions in Pareto-based multi-objective optimisation. The impact of this revised volume dominance on the performance of multi-objective evolutionary algorithms is investigated by incorporating it into three approaches, namely SEAMO2, SPEA2 and NSGA2 to solve instances of the 2-, 3- and 4- objective knapsack problem. The improved volume dominance is compared to its previous version and also to the conventional Pareto dominance. It is shown that the proposed improved volume dominance helps the three algorithms to obtain better non-dominated fronts than those obtained when the two other forms of dominance are used.

1 Introduction

The application of heuristic and evolutionary techniques to solve difficult real-world multi-objective optimisation problems is a very active research area. In Pareto-based multi-objective optimisation, a set of non-dominated solutions, also known as Pareto front, is sought so that the decision-maker can select the most appropriate one. Evolutionary algorithms and other population-based heuristics seem well suited to deal with Pareto based multi-objective optimisation problems because they can evolve a population of solutions towards the Pareto-optimal front in a single run. A good multi-objective evolutionary algorithm (MOEA) should be able to obtain Pareto fronts that are both well-distributed and well-converged. When designing a MOEA an important issue is how to establish superiority between solutions within the population. That is, how to compare solution fitness in a multi-objective sense. Most modern MOEAs adopt the conventional Pareto dominance relationship. There are few papers that propose different types of dominance relationship such as α-dominance, ϵ-dominance, fuzzy dominance and volume dominance (these approaches are reviewed in Section 2). These alternative forms of dominance aim to help finding solutions in difficult areas (like the extremes of the tradeoff front) or attempt to combine convergence and diversity in order to achieve better Pareto fronts in difficult problems. It has been shown that these alternative forms of dominance can help to obtain better quality Pareto fronts (e.g. [1,2,3]). In this paper, we present an improved

M. Ehrgott et al. (Eds.): EMO 2009, LNCS 5467, pp. 231–245, 2009.

version of the *volume dominance* proposed by Le and Landa-Silva [4]. This volume dominance compares two solutions with respect to the objective space volume that each of them dominates. The revised version takes into consideration the current non-dominated front as the search progresses and incorporates a crowding technique. We compare the performance of some well-known MOEAs when using the revised volume dominance, the previous volume dominance and the conventional Pareto dominance. Our experiments are conducted using the multi-objective knapsack problem because benchmark results are available for this problem. We also show that the proposed improved volume dominance could be used within other MOEAs.

Section 2 presents a short literature review of Pareto dominance and alternative forms of dominance. Section 3 describes the new volume dominance which is a modification of the one proposed earlier in [4]. Section 4 describes our experiments to assess the impact on the performance of three MOEAs when incorporating the new volume dominance proposed here. We discuss our results in Section 5 while Section 6 gives conclusions and proposes future work.

2 Dominance Relationship

In general, the multi-objective optimisation problem with m-objectives to be maximised can be written as

$$maximise \quad \{f_1(\boldsymbol{x}), f_2(\boldsymbol{x}), \ldots, f_m(\boldsymbol{x})\} \tag{1}$$

subject to the decision vector $\boldsymbol{x} = (x_1, x_2, \ldots, x_n)^T$ belonging to the feasible region S. Then, the objective vector of $\boldsymbol{x}$ is

$$\boldsymbol{f}(\boldsymbol{x}) = (f_1(\boldsymbol{x}), f_2(\boldsymbol{x}), \ldots, f_m(\boldsymbol{x})) \tag{2}$$

2.1 Pareto Dominance

Vilfredo Pareto proposed the concept of Pareto dominance (Pareto optimum) in 1896 [5]. Since then, this concept has been extensively used to establish the superiority between solutions in multi-objective optimisation. In Pareto dominance, a solution $\boldsymbol{x}$ is considered to be better than a solution $\boldsymbol{x}^*$ if and only if the objective vector of $\boldsymbol{x}$ dominates the objective vector of $\boldsymbol{x}^*$. More formally:

Pareto Dominance. A solution $\boldsymbol{x} \in S$ dominates a solution $\boldsymbol{x}^* \in S$ ($\boldsymbol{x} \succeq \boldsymbol{x}^*$) if and only if $\boldsymbol{x}$ is not worse than $\boldsymbol{x}^*$ in all objectives ($f_i(\boldsymbol{x}) \geq f_i(\boldsymbol{x}^*)\ \forall i = 1, \ldots, m$) and $\boldsymbol{x}$ is strictly better than $\boldsymbol{x}^*$ in at least one objective ($f_i(\boldsymbol{x}) > f_i(\boldsymbol{x}^*)$ for at least one $i = 1, \ldots, m$).

We can also distinguish between *weak dominance* and *strong dominance* [3] or *loose dominance* and *strict dominance* [6] respectively.

Weak dominance. This is often simply referred to as Pareto dominance. A solution $\boldsymbol{x}$ weakly dominates a solution $\boldsymbol{x}^*$ ($\boldsymbol{x} \succeq \boldsymbol{x}^*$) if $\boldsymbol{x}$ is better than $\boldsymbol{x}^*$ in at least one objective and is as good as $\boldsymbol{x}^*$ in all other objectives.

Strong dominance. A solution $\boldsymbol{x}$ strongly dominates a solution $\boldsymbol{x}^*$ ($\boldsymbol{x} \succ \boldsymbol{x}^*$) if $\boldsymbol{x}$ is strictly better than $\boldsymbol{x}^*$ in all objectives.

Non-dominance. If neither $\boldsymbol{x}$ dominates $\boldsymbol{x}^*$ nor $\boldsymbol{x}^*$ dominates $\boldsymbol{x}$ (weakly or strongly), then both solutions are said to be incomparable or mutually non-dominated. In this case, no solution is clearly preferred over the other.

Pareto-optimal front is the set F consisting of all non-dominated solutions $\boldsymbol{x}$ in the whole search space. Hence, a solution $\boldsymbol{x} \in F$ if there is no solution $\boldsymbol{x}^* \in S$ that dominates $\boldsymbol{x}$, i.e. if x is non-dominated with respect to S. A set of non-dominated solutions that approximates the Pareto optimal front is usually called current Pareto front or known Pareto front.

2.2 Alternative Forms of Dominance

Pareto dominance is widely adopted in multi-objective optimisation algorithms. Several alternative forms of dominance have been proposed recently. It has been shown that *relaxing* the conventional Pareto dominance can improve the performance of multi-objective optimisation algorithms. Some of these relaxed forms of Pareto dominance are more effective in finding solutions in the extremes of the feasible region S and in tackling optimisation problems with irregular Pareto-optimal fronts or problems for which it is difficult to generate feasible solutions.

In general, relaxed forms of Pareto dominance allow a solution $\boldsymbol{x}$ to *dominate* another solution $\boldsymbol{x}^*$ for which $\boldsymbol{x}$ does not Pareto-dominate $\boldsymbol{x}^*$. Relaxed forms of Pareto dominance include: structure domination [7], α-dominance [8], ϵ-dominance [1], extended Pareto dominance [9], the fuzzification of Pareto dominance [2,10] and contracting/expanding Pareto dominance [11]. Le and Landa-Silva proposed a relaxed form of Pareto dominance, named *volume dominance* [4]. This form of dominance is based on the volume of the objective space that a solution dominates. This property makes volume dominance distinguisable from conventional Pareto dominance and other relaxed forms of dominance which directly compare the objective vector of solutions in one way or another.

The volume dominance relationship between $\boldsymbol{x}$ and $\boldsymbol{x}^*$ is based on comparing their corresponding dominated volumes, $V(\boldsymbol{x})$ and $V(\boldsymbol{x}^*)$ respectively, to a reference volume called *shared dominated volume* [4]. The dominated volume of $\boldsymbol{x}$, $V(\boldsymbol{x})$, and the shared dominated volume of $\boldsymbol{x}$ and $\boldsymbol{x}^*$, $SV(\boldsymbol{x}, \boldsymbol{x}^*)$, are calculated with respect to the reference point $\boldsymbol{r} = (r_1, r_2, \ldots, r_m)$ as follows:

$$V(\boldsymbol{x}) = \prod_{i=1}^{m}(f_i(\boldsymbol{x}) - r_i) \tag{3}$$

$$SV(\boldsymbol{x}, \boldsymbol{x}^*) = \prod_{i=1}^{m}(min(f_i(\boldsymbol{x}), f_i(\boldsymbol{x}^*)) - r_i) \tag{4}$$

It is said that for a ratio rSV, $\boldsymbol{x}$ volume-dominates $\boldsymbol{x}^*$ ($\boldsymbol{x} \succeq_V \boldsymbol{x}^*$) if either:

- $V(\boldsymbol{x}^*) = SV(\boldsymbol{x}, \boldsymbol{x}^*)$ and $V(\boldsymbol{x}) > SV(\boldsymbol{x}, \boldsymbol{x}^*)$ or
- $V(\boldsymbol{x}) > V(\boldsymbol{x}^*) > SV(\boldsymbol{x}, \boldsymbol{x}^*)$ and $\frac{V(\boldsymbol{x}) - V(\boldsymbol{x}^*)}{SV(\boldsymbol{x}, \boldsymbol{x}^*)} > rSV$

Le and Landa-Silva clearly identified the difference between the fundamental principles of volume dominance and the $\mathcal{S}$ metric proposed by Zitzler and Thiele [12] which look very similar at first sight. They also proved that volume dominance covers Pareto dominance. This could not be the case with some other relaxed forms of Pareto dominance such as ϵ-dominance and extended Pareto dominance. Another interesting property of volume dominance is that it is normalised (see proof by Le and Landa-Silva [4]). That is, volume dominance is able to prevent bias towards some directions in cases with non-commensurable objective functions. Another crucial difference between volume dominance and other alternative forms of dominance in the literature is that volume dominance combines all objectives into a single unit vector to establish the superiority between solutions instead of directly comparing each objective in turn. This allows volume dominance to evaluate the whole objective vector to compensate improvement and detriment between objectives [4].

3 Volume Dominance

Volume dominance shows promising results when compared to the conventional Pareto dominance. Volume dominance is able to obtain results driven by different criteria such as better coverage, better size of space covered or better distribution of the objective values. This can be done by adjusting the rSV ratio [4]. However, the volume dominance presented earlier also has some drawbacks. It requires a preset and fixed reference point $\boldsymbol{r}$ in order to calculate the dominated volume of a solution. As the search progresses, the population moves away from the reference point which could lead to a significant increase in the dominated volume of a solution. Hence, volume dominance could be highly effective at the start of the search but less and less influent as the search progresses. Therefore, we propose to update the reference point to reflect the evolution of the population. In other words, the reference point is defined based on some characteristics of the current population.

Another issue of volume dominance is that it does not take into account the current Pareto front. This issue is illustrated in Figure 1. For both cases 1(a) and 1(b) in Figure 1, $\boldsymbol{x} \succeq_V \boldsymbol{x}^*$ for some ratio rSV by using the volume dominance in [4]. However, one can easily point out that $\boldsymbol{x} \succeq_V \boldsymbol{x}^*$ should not be true in 1(b) because both $\boldsymbol{x}$ and $\boldsymbol{x}^*$ seem equally good (close to the Pareto front) and should be regarded as non-volume-dominated solutions. In order to overcome this issue, we modify volume dominance to consider the current Pareto front when establishing superiority between two solutions. We propose a clustering technique as an additional feature incorporated into volume dominance.

3.1 Dynamic Reference Point

Volume dominance requires a reference point $\boldsymbol{r}$ to calculate the dominated volume of a solution. Le and Landa-Silva [4] proposed a simple strategy to define the reference point $\boldsymbol{r}$ as a fixed point, the origin of coordinates in the objective

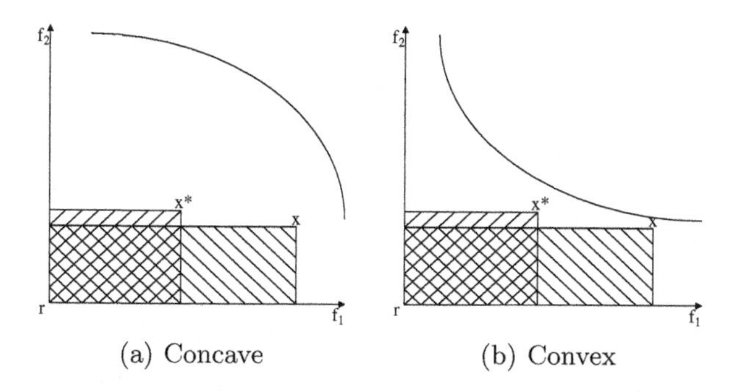

(a) Concave (b) Convex

Fig. 1. Previous Version of Volume Dominance [4]

space. As mentioned above, this simple strategy has the drawback of degrading the effectiveness of volume dominance as the search progresses because the dominated volume of solutions becomes larger and larger. We propose a more elaborate, yet more effective, strategy to estimate the reference point $\boldsymbol{r}$. The strategy designates a reference point for each solution in the current population P and it also defines two common reference points r^{inf} and r^{sup} for all individuals in P. This strategy also considers the current state of P when determining the reference point. The reference point $\boldsymbol{r}^{\boldsymbol{x}} = (r_1^x, r_2^x, \ldots, r_m^x)$ for solution $\boldsymbol{x} \in P$ is as follows:

$$r_i^x = f_i(\boldsymbol{x}) - (r_i^{\text{sup}} - r_i^{\text{inf}}) \tag{5}$$

where $\forall i = 1, 2, \ldots, m$

$$r_i^{\text{inf}} = \inf\{f_i(\boldsymbol{x}^*) \mid \boldsymbol{x}^* \in P\} \tag{6}$$

$$r_i^{\text{sup}} = \sup\{f_i(\boldsymbol{x}^*) \mid \boldsymbol{x}^* \in P\} \tag{7}$$

The estimation of $\boldsymbol{r}^{\boldsymbol{x}} = (r_1^x, r_2^x, \ldots, r_m^x)$ is illustrated in Figure 2.

3.2 Considering the Current Pareto Front

Without considering the current Pareto front during the search, Figure 1 illustrates the drawback of the previous version of volume dominance. While establishing superiority between solutions $\boldsymbol{x}$ and $\boldsymbol{x}^*$, the strength of solution $\boldsymbol{x}$ is defined as the ratio of the dominated volume of $\boldsymbol{x}$ ($V(\boldsymbol{x})$) to the *shared dominated volume* of $\boldsymbol{x}$ and $\boldsymbol{x}^*$ ($SV(\boldsymbol{x}, \boldsymbol{x}^*)$) with respect to the reference point $\boldsymbol{r}$. Then, the dominance of $\boldsymbol{x}$ over $\boldsymbol{x}^*$ (or vice versa) is determined based on comparing the difference between their strengths to a ratio rSV.

We propose a different approach in defining the strength of solution $\boldsymbol{x}$. The strength of $\boldsymbol{x}$ is the ratio between the dominated volume of $\boldsymbol{x}$ ($V(\boldsymbol{x})$) and the volume that *fairly* represents the status of the current Pareto set. With respect to $\boldsymbol{x}$, this *fair* representation of the current Pareto set is the subset consisting

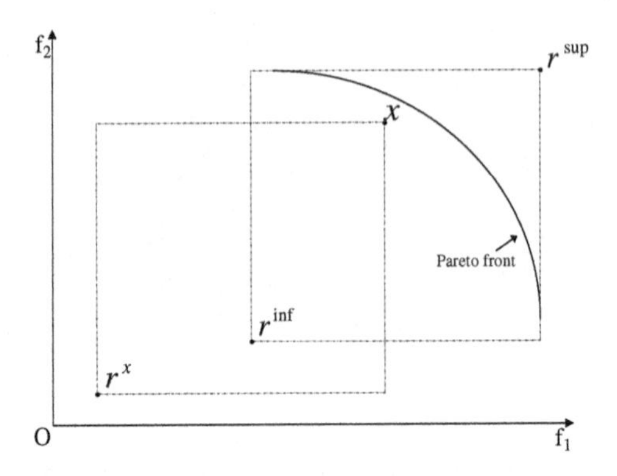

Fig. 2. Volume Dominance - Reference Point

of non-dominated solutions that Pareto-dominate $\boldsymbol{x}$. However, determining the dominated volume of a solution set could be computationally expensive. Therefore, the dominated volume of this solution set is estimated as the dominated volume of the solution that least Pareto-dominates that solution set. Let name this estimated dominated volume (w.r.t $\boldsymbol{x}$) as the reference volume of $\boldsymbol{x}$, $V^{ref}(\boldsymbol{x})$.

$$V^{ref}(\boldsymbol{x}) = \prod_{i=1}^{m}(x_i^{ref} - r_i) \tag{8}$$

$$x_i^{ref} = \inf\{\{f_i(\boldsymbol{x})\} \bigcap \{f_i(\boldsymbol{x}^*) \mid \boldsymbol{x}^* \succeq \boldsymbol{x} \ \wedge \ \boldsymbol{x}^* \in ParetoFront\}\} \tag{9}$$

The strength of $\boldsymbol{x}$ is then defined as follows:

$$Str(\boldsymbol{x}) = \frac{V(\boldsymbol{x})}{V^{ref}(\boldsymbol{x})} \tag{10}$$

Therefore, $\boldsymbol{x}$ volume-dominates $\boldsymbol{x}^*$ ($\boldsymbol{x} \succeq_V \boldsymbol{x}^*$) if and only if the following condition holds for a positive ratio r_{Str}:

$$Str(\boldsymbol{x}) - Str(\boldsymbol{x}^*) \geq r_{Str} \tag{11}$$

Additionally, to ensure the improvement of the Pareto set, condition (11) is relaxed whenever $Str(\boldsymbol{x}) = 1$ and $Str(\boldsymbol{x}^*) < 1$. In other words, if none of non-dominated solutions dominates $\boldsymbol{x}$, $x_i^{ref} = f_i(\boldsymbol{x})$ implying $V^{ref}(\boldsymbol{x}) = V(\boldsymbol{x})$, and it is not the case for $\boldsymbol{x}^*$ then $\boldsymbol{x} \succeq_V \boldsymbol{x}^*$.

It is noted that the current Pareto front is required in order to apply volume dominance. Therefore, one must use conventional Pareto dominance to obtain the Pareto front as well as $\boldsymbol{x}^{ref}$ to estimate the reference volume of $\boldsymbol{x}$, $V^{ref}(\boldsymbol{x})$.

3.3 Clustering Strategy

We also propose a clustering strategy as part of volume dominance to improve the distribution of non-dominated solutions w.r.t the objective space. This strategy is only considered when two solutions $\boldsymbol{x}$ and $\boldsymbol{x}^*$ are regarded as non-volume-dominated. It means that both condition (11) and its relaxation do not hold for solutions $\boldsymbol{x}$ and $\boldsymbol{x}^*$. Solution $\boldsymbol{x}$ is said to dominate solution $\boldsymbol{x}^*$ if $\boldsymbol{x}$ is in a less crowded area when comparing to $\boldsymbol{x}^*$. The degree of crowding of $\boldsymbol{x}$ is measured as the number of neighbours of $\boldsymbol{x}$ which is the number of non-dominated solutions in the current Pareto set that ϵ-dominate $\boldsymbol{x}$. The ϵ-dominance deployed here is slightly different from the one proposed by Laumanns et al. [1] and other variants of ϵ-dominance in the literature. To the best of our knowledge, variants of ϵ-dominance either use a dynamic adaptation of the ϵ value or use a different ϵ_i value for each objective i. The variant of ϵ-dominance employed in this paper takes the advantage of both approaches, a different dynamic adaptive ϵ_i value for each objective. The ϵ_i value is estimated based on the current Pareto front as follows:

$$\epsilon_i = (r_i^{\text{sup}} - r_i^{\text{inf}}) \times \mu \tag{12}$$

where r_i^{inf}, r_i^{sup} were given in (6), (7) respectively and μ is a positive constant. Within the context of volume dominance, it is said that $\boldsymbol{x}$ ϵ-dominates $\boldsymbol{x}^*$ ($\boldsymbol{x} \succeq_{\epsilon_v} \boldsymbol{x}^*$) if and only if $f_i(\boldsymbol{x}) \geq f_i(\boldsymbol{x}^*) - \epsilon_i \; \forall i = 1, \ldots, m$ and $f_i(\boldsymbol{x}) > f_i(\boldsymbol{x}^*) - \epsilon_i$ for at least one i. Then, the number of neighbours of $\boldsymbol{x}$ is defined as follows:

$$N(\boldsymbol{x}) = |\{\boldsymbol{x}^* \mid \boldsymbol{x}^* \succeq_{\epsilon_v} \boldsymbol{x} \, \wedge \, \boldsymbol{x}^* \in CurrentParetoFront\}| \tag{13}$$

It is then said that if condition (11) and its relaxation do not hold for either $(\boldsymbol{x}, \boldsymbol{x}^*)$ or $(\boldsymbol{x}^*, \boldsymbol{x})$, then $\boldsymbol{x}$ volume-dominates $\boldsymbol{x}^*$ ($\boldsymbol{x} \succeq_V \boldsymbol{x}^*$) for a positive constant τ if and only if

$$N(\boldsymbol{x}^*) - N(\boldsymbol{x}) \geq \tau \tag{14}$$

4 Experimental Design

Most alternative forms of dominance proposed in the literature attempt to search and maintain extreme points in the objective space and/or points that are difficult to obtain and maintain with Pareto dominance. Few other forms of relaxed dominance attempt to combine convergence and diversity into a single criterion when distinguishing between solutions. These relaxed forms of dominance have been proposed as an integral part of specific multi-objective optimisation algorithms (see [1,2,3,7,8,10]) with the exception of the contracting/expanding Pareto dominance of Sato et al. [11] which was tested on one existing approach, namely NSGA2. To the best of our knowledge, none of these forms of relaxed dominance has been tested on different multi-objective algorithms.

4.1 Brief Description of the MOEAs Considered

This paper presents experimental results showing that the proposed improved volume dominance performs well on different multi-objective evolutionary algorithms such as SEAMO2 [13], SPEA2 [14] and NSGA2 [15] when solving the multi-objective knapsack problem instances proposed by Zitzler and Thiele [12].

SEAMO2 uses a steady-state population and a simple elitist replacement strategy. Each solution of the population, in turn, acts as the first parent once and a second parent is chosen at random. Offspring is produced by applying cycle crossover on the two parents followed by a single mutation. If the offspring's objective vector improves on any *best-so-far* objective function, it replaces one of the parents and the objective's *best-so-far* is updated. Otherwise, if the offspring dominates one of the parents, it replaces that parent (unless it is a duplicate, then the offspring is deleted). If neither the offspring dominates the parents nor the parents dominate the offspring, the offspring replaces a random solution in the population that the offspring dominates. SPEA2 employs a fixed size archive to store non-dominated solutions in addition to a population. SPEA2 deploys a fine-grained fitness assignment strategy which for each individual p takes into account the number of other individuals that dominate p and that are dominated by p. A nearest neighbour density estimation for environmental selection is used to deal with two situations: when either the archive is too small or too large. The best dominated individuals in the previous archive and the population are copied to the new archive in the first case. In the latter situation, non-dominated individuals in the archive are iteratively removed until the archive's size is not exceeded. The removal of non-dominated individuals from the archive is carefully managed by using an archive truncation method that guarantees the preservation of boundary solutions. NSGA2 uses a fast non-dominated sorting algorithm to classify a population into different non-domination levels. NSGA2 also uses a crowding technique based on the density of solutions surrounding a particular solution to preserve the diversity of the population.

4.2 Enhancing MOEAs with Volume Dominance

All of these 3 algorithms, SEAMO2 [13], SPEA2 [14] and NSGA2 [15], were implemented according to their original description. Parameter settings for tackling the multi-objective knapsack problem with SPEA2 and NSGA2 were kindly provided by Marco Laumanns by means of email-based discussions. Then, in our experiments we replace the conventional Pareto dominance with the revised volume dominance and analyse the impact on the performance of these three algorithms. We aim to investigate the performance of the improved volume dominance within the three evolutionary approaches with minimum alteration to the original algorithms. The replacement of the conventional Pareto dominance with the volume dominance in each algorithm is described below.

In SEAMO2, we replace Pareto dominance with improved volume dominance to decide on the replacement of offspring by one of its parents or a random solution. This is the only stage where solutions are compared for dominance relationship in SEAMO2. However, this is not the case for SPEA2 and NSGA2. In both SPEA2 and NSGA2, there are three possible stages in which the improved volume dominance could be applied. These are: the fitness assignment to individuals, the environmental selection and the mating selection stages. In fact, Pareto dominance is only applied during the fitness assignment stage. The environmental selection and the mating selection stages use computed individual fitnesses to compare solutions for superiority or dominance relationship. However, the individual fitness computed during the fitness assignment stage heavily relies on Pareto dominance. Therefore, to some extent, the environmental selection and the mating selection stage in SPEA2 and NSGA2 also involve Pareto dominance. We think that the fitness assignment strategy is an integral part in these two algorithms, SPEA2 and NSGA2. It makes more sense that we do not alter their fitness assignment strategy to preserve their main characteristics. We also suggest that the mating selection stage is less significant than the environmental selection stage. This is because the environmental selection strategy, which uses an archive truncation operator in SPEA2 and front extraction in NSGA2, decides the survival of individuals into the next generation whereas the mating selection strategy choose random parents based on a binary tournament selection operator to produce offspring. Therefore, for the preliminary investigation in this paper, we apply the improved volume dominance to the mating selection stage in SPEA2 and NSGA2. In other words, we replace the comparison of individual fitnesses with the improved volume dominance in order to decide on the superiority between individuals during the mating selection stage.

4.3 Benchmark Problems, Parameters Setting and Metrics

We use the instances with 750 items and 2, 3, and 4 objectives of the knapsack problem proposed in [12]. We executed short and long runs using different values of rSV to investigate the improved volume dominance. The population size used for the 2-, 3-, and 4-objective instances are 250, 300 and 350 individuals respectively. We use the same number of generations for a short run (500 generations) and a long run (1920 generations) as used by Zitzler et al. [14], Deb et al. [15] and Mumford [13]. For the improved volume dominance, we use 5 different values of $rSV = \{0.025, 0.05, 0.075, 0.10, 0.15\}$, $\mu = 0.01$ in equation (12) and $\tau = 5$ in inequality (14). We also replicated the results obtained by applying the previous volume dominance proposed by Le and Landa-Silva [4] with 4 different values of $rSV = \{0.10, 0.15, 0.20, 0.25\}$. We summarise and discuss the results from 30 independent runs. The results in Section 5 are based on $rSV = 0.075$ for the new improved volume dominance proposed in this paper and $rSV = 0.15$ for the previous approach proposed in [4].

We use four metrics to evaluate the non-dominated fronts produced. The first metric is the $\mathcal{S}$ hypervolume proposed by Zitzler and Thiele [12] which measures the overall size of the objective space covered by all the non-dominated solutions. Here, $\mathcal{S}$ is scaled as the percentage of the volume created by the origin and the reference point (39822, 41166), (41968, 41298, 41402), (41841, 40790, 39651, 41630) which is the sum profits of all items in each objective for 2-, 3- and 4-objective instance respectively. The boxplots in Figure 3 represent the distribution of the complement of the $\mathcal{S}$ hypervolume metric $(1 - \mathcal{S})$. The vertical axes of the boxplots measure the percentage of the non-dominated objective space. The horizontal axes present Pareto Dominance (pd), the previous volume dominance (vd1) and the improved volume dominance proposed here (vd2) when applied to three different evolutionary algorithms SEAMO2 (se), SPEA2 (sp) and NSGA2 (ns). The second metric used is the *cluster* CL_μ proposed by Wu and Azarm [16]. The CL_μ cluster metric measures the average number of *indistinct* solutions in each small grid which size is specified by $1/\mu$. The ideal case is when $CL_\mu = 1$ which means that every obtained Pareto solution is distinct. In all other cases, CL_μ is greater than 1. The higher the value of CL_μ, the more clustered the solution set is, and therefore, the less preferred the solution set. We use $\mu = 0.01$ or in other words, $1/\mu = 100$ units in the objective space. The third metric used is the average distance from the obtained non-dominated front to the approximation of the true Pareto front. The lower the value of this metric, the closer the obtained non-dominated front is to the true Pareto front. Finally, the size of the obtained non-dominated fronts is also computed. The higher the value, the better as more non-dominated solutions have been found.

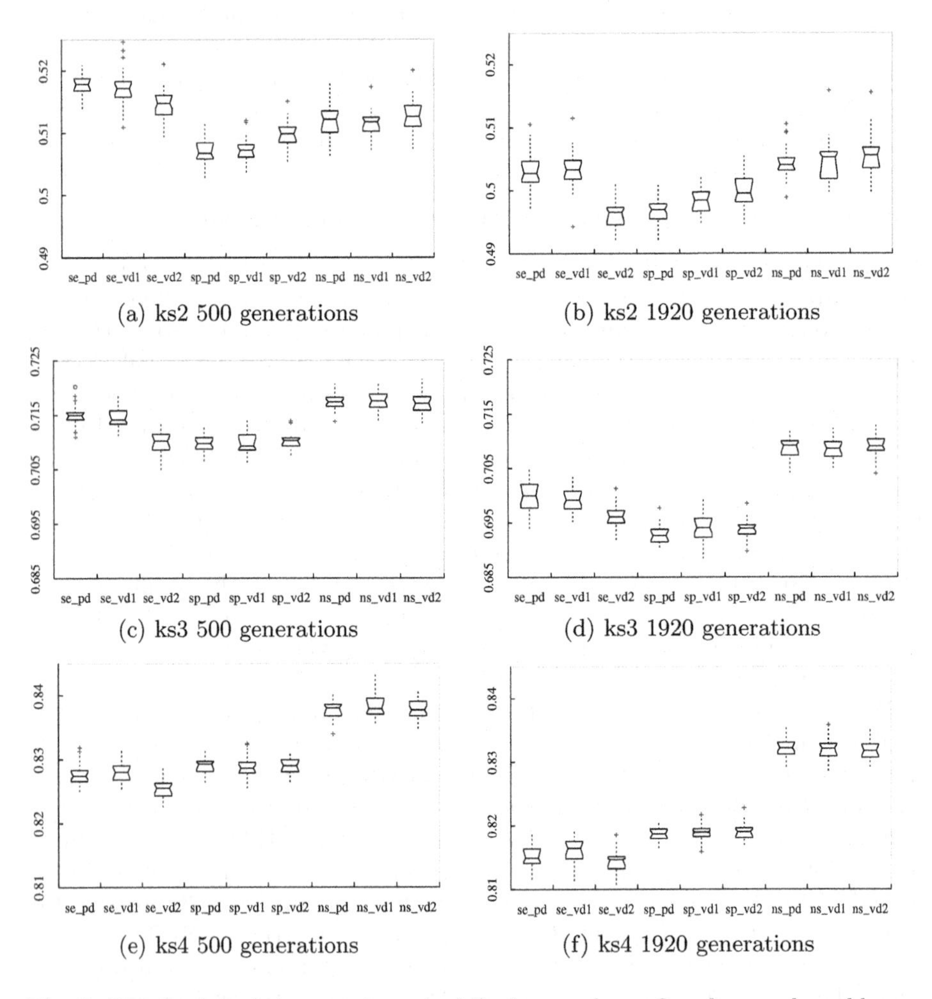

Fig. 3. Distribution of the complement of the hypervolume S on knapsack problems

Table 1. Average values (standard deviation) of the size of the non-dominated set

(a) SEAMO2

knapsack	no. of generations	se_pd	se_vd1	se_vd2
2	500	59.37 (8.1)	59.53 (8.44)	82.9 (9)
2	1920	101.67 (11.37)	104.5 (8.87)	198.9 (16.51)
3	500	199 (19.04)	202.73 (16.29)	299.87 (0.57)
3	1920	244.7 (10.54)	244.87 (11.11)	300 (0)
4	500	284.5 (15.87)	286.77 (15.79)	350 (0)
4	1920	321.27 (8.45)	316.2 (10.24)	350 (0)

(b) SPEA2

knapsack	no. of generations	sp_pd	sp_vd1	sp_vd2
2	500	76.6 (8.81)	77.27 (8.28)	70.9 (8.94)
2	1920	134.63 (10.42)	131.9 (14.22)	126.1 (14.35)
3	500	300 (0)	300 (0)	300 (0)
3	1920	300 (0)	300 (0)	300 (0)
4	500	350 (0)	350 (0)	350 (0)
4	1920	350 (0)	350 (0)	350 (0)

(c) NSGA2

knapsack	no. of generations	ns_pd	ns_vd1	ns_vd2
2	500	66 (8.29)	66.1 (8.19)	62.27 (7.73)
2	1920	85.83 (12.5)	73.2 (21.44)	73.63 (16.96)
3	500	257.37 (9.92)	255.7 (12.55)	256.47 (8.17)
3	1920	272.07 (7.52)	269 (5.38)	266.87 (6.43)
4	500	335.7 (5.64)	335.8 (6.72)	334.37 (5.88)
4	1920	338.7 (3.19)	338.9 (2.45)	338.07 (4.81)

5 Results and Discussion

For the hypervolume $\mathcal{S}$ (Figure 3), the improved volume dominance incorporated into SEAMO2 (se_vd2) outperforms not only the Pareto dominance (se_pd) but also the previous volume dominance (se_vd1) for all knapsack instances both in the short and long runs. The improved volume dominance when incorporated into NSGA2 (ns_vd2) is slightly worst than ns_pd and ns_vd1 in 2-knapsack instance but ns_vd2 is able to compete against ns_pd and ns_vd1 in higher dimension knapsack instances (3 and 4 objectives). We observe a similar result in SPEA2 as in NSAG2 when comparing se_vd2 to se_pd and se_vd1.

We also traced the progress of $\mathcal{S}$ during the evolutionary search. Regarding SPEA2 and NSGA2, vd2, vd1 and pd perform quite similar. However, vd2 incorporated into SEAMO2 is better than vd1 and pd for all 3 knapsack instances from the early stage of the evolutionary search. We omit the graphs related to this results due to space limitations in this paper. The complete results can be obtained on request.

The performance of vd2 when incorporated in SPEA2 and NSGA2 is quite similar to vd1 and pd with respect to the size of the non-dominated set

(Table 1(b), 1(c)) and with respect to the cluster CL_μ of the non-dominated set (Table 2(b), 2(c)). However, for SEAMO2, vd2 is noticably better than vd1 and pd regarding both the size of the non-dominated set (Table 1(a)) and the cluster CL_μ of the non-dominated set (Table 2(a)) in almost all knapsack instances for both short and long runs, except for the 2-knapsack instance when vd2 is worse than vd1 and pd in term of the cluster CL_μ of the non-dominated set.

Table 2. Average values (standard deviation) of the cluster metrics CL_μ

(a) SEAMO2

knapsack	no. of generations	se_pd	se_vd1	se_vd2
2	500	5.78 (0.79)	5.69 (0.7)	6.1 (0.8)
2	1920	6.16 (1.05)	6.49 (0.94)	8.74 (0.74)
3	500	7.23 (1.08)	7.25 (0.82)	3.95 (0.38)
3	1920	6.6 (0.71)	6.29 (0.6)	3.31 (0.23)
4	500	5.8 (0.66)	6.1 (0.63)	3.06 (0.31)
4	1920	5.2 (0.35)	5.29 (0.43)	2.77 (0.16)

(b) SPEA2

knapsack	no. of generations	sp_pd	sp_vd1	sp_vd2
2	500	4.15 (0.45)	4.34 (0.52)	4.32 (0.54)
2	1920	6.17 (0.56)	6.16 (0.61)	6.34 (0.72)
3	500	2.09 (0.16)	2.14 (0.25)	2.16 (0.21)
3	1920	1.66 (0.09)	1.69 (0.08)	1.72 (0.09)
4	500	1.45 (0.06)	1.43 (0.06)	1.46 (0.06)
4	1920	1.36 (0.05)	1.38 (0.06)	1.35 (0.05)

(c) NSGA2

knapsack	no. of generations	ns_pd	ns_vd1	ns_vd2
2	500	4.52 (0.57)	4.52 (0.55)	4.48 (0.46)
2	1920	5.04 (0.69)	4.26 (1.02)	4.42 (0.89)
3	500	2.99 (0.25)	3 (0.3)	3.22 (0.36)
3	1920	1.6 (0.07)	1.58 (0.07)	1.6 (0.07)
4	500	1.97 (0.16)	1.9 (0.13)	1.99 (0.17)
4	1920	1.78 (0.14)	1.77 (0.13)	1.82 (0.12)

We should point out here that se_vd2, comparing to se_vd1 and se_pd, is able not only to find more non-dominated solutions but also to reduce the clustering in the non-dominated set. In other words, se_vd2 is able to obtain more diverse solution sets and better extreme solutions. We believe that this promising result is due to the clustering strategy deployed in the improved volume dominance but further experimentation is required.

The average distance of the non-dominated set found by vd2 when incorporated into SEAMO2 is higher than when using vd1 and pd. We argue that this is because se_vd2 is able to find more extreme solutions than se_vd1 and se_pd.

Finding more extreme solutions could be intepreted as obtaining more solutions that are slightly away from the available approximated Pareto fronts. This is the reason for se_vd2 not being competitive with se_vd1 and se_pd with respect to the average distance from the non-dominated set to the approximated Pareto front (Table 3(a)). However, Table 3(b), 3(c) show that vd2 clearly outperforms vd1 and pd, when incorporated into SPEA2 and NSGA2 in all knapsack instances for both short and long runs.

Table 3. Average values (standard deviation) of the distance from the non-dominated set to the approximation of the true Pareto Front

(a) SEAMO2

knapsack	no. of generations	se_pd	se_vd1	se_vd2
2	500	498.88 (51.62)	491.97 (40.5)	602.42 (70.93)
2	1920	389.18 (35.34)	387.93 (34.69)	432.57 (43.93)
3	500	1381.9 (58.15)	1392.34 (68.96)	1506.43 (72.99)
3	1920	1250.8 (41.71)	1269.06 (35.85)	1318.78 (50.35)
4	500	744.97 (51.94)	733.9 (50)	808.94 (48.64)
4	1920	677.02 (55.06)	668.83 (65.52)	695.78 (33.22)

(b) SPEA2

knapsack	no. of generations	sp_pd	sp_vd1	sp_vd2
2	500	701.4 (53.14)	676.4 (54.15)	638.26 (50.76)
2	1920	466.5 (43.39)	468.89 (53.54)	450.24 (43.8)
3	500	1985.72 (86.65)	1984.3 (68.79)	1960.08 (80.6)
3	1920	1682.92 (47.05)	1687.89 (34.42)	1692.77 (42.44)
4	500	1825.8 (100.38)	1761.69 (98.71)	1769.54 (115)
4	1920	1605.61 (55.04)	1567.62 (67.82)	1571.16 (70.83)

(c) NSGA2

knapsack	no. of generations	ns_pd	ns_vd1	ns_vd2
2	500	613.85 (39.39)	623.58 (43.67)	592.18 (54.07)
2	1920	429 (42.49)	454.39 (39.32)	429.03 (50.83)
3	500	2029.53 (80.73)	2026.49 (90.53)	1933.72 (99.73)
3	1920	1739.44 (74.74)	1729.7 (77.4)	1687.68 (69.35)
4	500	1681.5 (135.55)	1711.23 (143.16)	1640.75 (110.29)
4	1920	1316.23 (95.44)	1290.11 (89.45)	1285.49 (91.13)

Figure 4 shows the offline results for the 2-knapsack instance. They are the combined non-dominated solutions from 30 runs. For better visualisation, we show the non-dominated fronts in a lower density. See that vd2, vd1 and pd are quite similar when incorporated into SPEA2 and NSGA2 (Figure 4(b), 4(c)) but Figure 4(a) shows a better performance of vd2 over vd1 and pd.

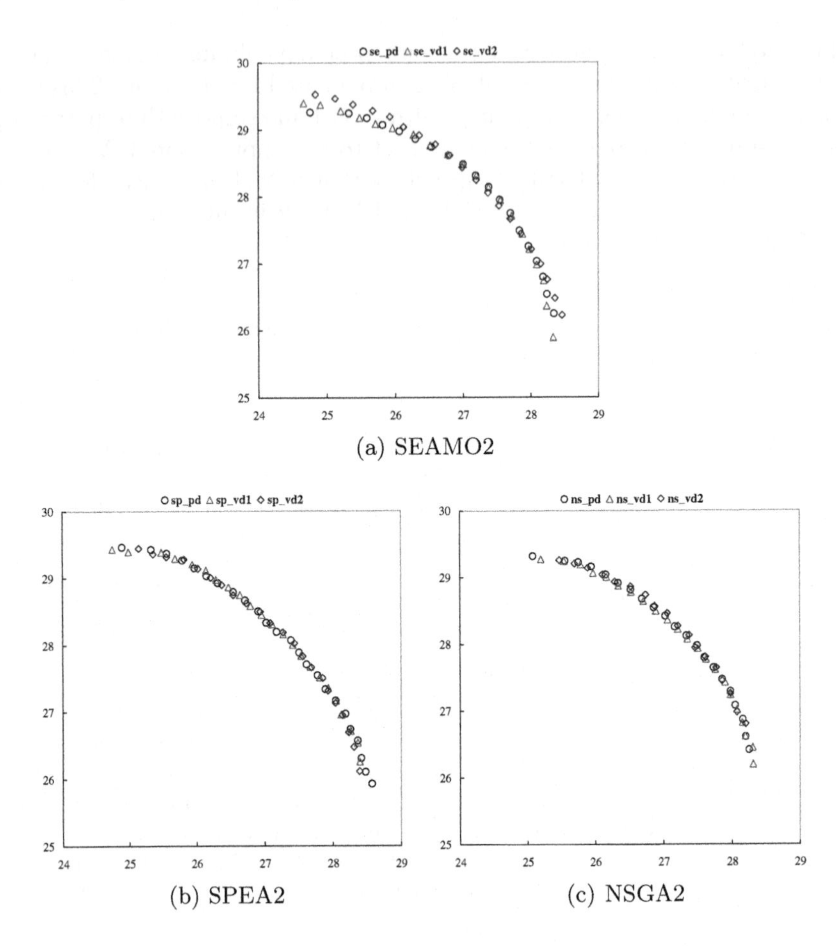

Fig. 4. Offline Result

6 Final Remarks

This paper proposed an improved volume dominance to the one originally proposed by Le and Landa-Silva [4]. We presented extensive experiments to compare the preformances of this improved volume dominance approach, the previous one and the conventional Pareto dominance using three MOEAs: SEAMO2, SPEA2 and NSGA2. The results show that in most of the cases, using 3 different knapsack instances using short and long runs, the new volume dominance approach performs better than the Pareto dominance and the previous proposed volume dominance. The improved volume dominance is more effective when incorporated into SEAMO2 than when incorporated into SPEA2 and NSGA2. This could be due to the fact that SEAMO2 is a very simple strategy whereas SPEA2 and NSGA2 already deploy more elaborate mechanisms. We believe that this revised volume dominance could be used as a new strategy to assign fitness to solutions in multi-objective evolutionary algorithms.

References

1. Laumanns, M., Thiele, L., Deb, K., Zitzler, E.: Combining Convergence and Diversity in Evolutionary Multi-Objective Optimisation. Evolution Computation 10(3), 263–282 (2002)
2. Koppen, M., Vicente-Garcia, R., Nickolay, B.: Fuzzy-Pareto-Dominance and its Application in Evolutionary Multi-objective Optimization. In: Coello Coello, C.A., Hernández Aguirre, A., Zitzler, E. (eds.) EMO 2005. LNCS, vol. 3410, pp. 399–412. Springer, Heidelberg (2005)
3. Burke, E.K., Landa-Silva, D.: The Influence of the Fitness Evaluation Method on the Performance of Multiobjective Search Algorithms. European Journal of Operational Research 169(3), 875–897 (2006)
4. Le, K., Landa-Silva, D.: Obtaining Better Non-Dominated Sets Using Volume Dominance. In: Proceedings of the 2007 Congress on Evolutionary Computation (CEC 2007), pp. 3119–3126. IEEE Press, Los Alamitos (2007)
5. Pareto, V.: Cours D'Economie Polotique. F. Rouge, Lausanne (1896)
6. Dasgusta, P., Chakrabarti, P.P., DeSarkar, S.C.: Multiobjective Heuristic Search: An Introduction to Intelligent Search Methods for Multicriteria Optimization. Computational Intelligence (1999)
7. Yu, P.L.: Cone Convexity, Cone Extreme Points, and Nondominated Solutions in Decision Problems with Multiobjectives. Journal of Optimization Theory and Applications 14(3), 319–377 (1974)
8. Kokolo, I., Hajime, K., Shigenobu, K.: Failure of Pareto-based MOEAS: Does Nondominated Really Mean Near to Optimal? In: Proceedings of the 2001 Congress on Evolutionary Computation (CEC 2001), pp. 957–962. IEEE Press, Los Alamitos (2001)
9. Jin, H., Wong, M.L.: Adaptive Diversity Maintenance and Convergence Guarantee in Multiobjective Evolutionary Algorithms. In: Proceedings of the 2003 Congress on Evolutionary Computation (CEC 2003), pp. 2498–2505. IEEE Press, Los Alamitos (2003)
10. Peng, J., Mok, H.M., Tse, W.: Fuzzy Dominance Based on Credibility Distributions. In: Wang, L., Jin, Y. (eds.) FSKD 2005. LNCS, vol. 3613, pp. 295–303. Springer, Heidelberg (2005)
11. Sato, H., Aguirre, H., Tanaka, K.: Controlling Dominance Area of Solutions and Its Impact on the Performance of MOEAs. In: Obayashi, S., Deb, K., Poloni, C., Hiroyasu, T., Murata, T. (eds.) EMO 2007. LNCS, vol. 4403, pp. 5–20. Springer, Heidelberg (2007)
12. Zitzler, E., Thiele, L.: Multiobjective Evolutionary Algorithms: A Comparative Case Study and the Strength Pareto Approach. IEEE Transactions on Evolutionary Computation 3(4), 257–271 (1999)
13. Mumford, C.: Simple Population Replacement Strategies for a Steady-State Multiobjective Evolutionary Algorithm. In: Deb, K., et al. (eds.) GECCO 2004. LNCS, vol. 3102, pp. 1389–1400. Springer, Heidelberg (2004)
14. Zitzler, E., Laumanns, M., Thiele, L.: SPEA2: Improving the Strength Pareto Evolutionary Algorithm for Multiobjective Optimization. In: Evolutionary Methods for Design, Optimisation and Control with Application to Industrial Problems (EUROGEN 2001), pp. 95–100 (2002)
15. Deb, K., Pratap, A., Agarwal, S., Meyarivan, T.: A Fast and Elitist Multiobjective Genetic Algorithm: NSGA-II. IEEE Transactions on Evolutionary Computation 6(2), 182–197 (2002)
16. Wu, J., Azarm, S.: Metrics for Quality Assessment of a Multiobjective Design Optimization Solution Set. Journal of Mechanical Design 123(1), 18–25 (2001)

Solving Bi-objective Many-Constraint Bin Packing Problems in Automobile Sheet Metal Forming Processes

Madan Sathe, Olaf Schenk, and Helmar Burkhart

Department of Computer Science, High Performance and Web Computing Group, University of Basel, Switzerland
{madan.sathe,olaf.schenk,helmar.burkhart}@unibas.ch

Abstract. The solution of bi-objective bin packing problems with many constraints is of fundamental importance for a wide range of engineering applications such as wireless communication, logistics, or automobile sheet metal forming processes. When the bi-objective bin packing problem is single-constrained, state-of-the-art multi-objective genetic algorithms such as NSGA-II combined with standard constraint handling techniques can be used. In the case of many-constraint bin packing problems, problems with thousand of additional constraints, it is not easy to solve this kind of problem accurately and fast with classical methods. Our approach relies on two key ingredients, NSGA-II and a clustering algorithm in order to generate always feasible solutions independent of the number of constraints. The method allows to tackle bi-objective many-constraint bin packing problems. We will present results for challenging artificial bin packing problems which model typical bi-objective bin packing problems with many constraints arising in the automobile industry.

1 Introduction

Many-constraint bin packing problems occur in many real world applications as for instance in automobile sheet metal forming processes. The task in the single-objective single-constraint one-dimensional bin packing problem (SO1DBPP) is to pack items with different weights in a minimum number of bins such that for each bin the total capacity of items does not exceed its capacity. We deal with bi-objective many-constraint one-dimensional bin packing problems (MO1DBPP) where, in contrast to the SO1DBPP, conflicting objectives must be optimized such that many constraint types have to be fulfilled. The challenging task is to solve bi-objective problems with a huge number of constraints. We tackle the many-constraint problem with a clustering procedure and establish the convergence against the Pareto front with a state-of-the-art multi-objective genetic algorithm.

Common constraint handling techniques based on evolutionary algorithms [3, 14] can be classified into five categories:

1. Penalty functions
2. Special representations and operators

M. Ehrgott et al. (Eds.): EMO 2009, LNCS 5467, pp. 246–260, 2009.

3. Repair algorithms
4. Separation of objectives and constraints
5. Hybrid Methods

In general, the most common constraint handling technique is based on penalty functions. Penalty functions, which add a penalty term to the fitness function value, are the simplest way to deal with constraints. The aim is to punish infeasible solutions as to favor feasible solutions for the selection and replacement process. The main drawback of penalty functions is the determining of the penalty factors due to the problem dependency of the approach.

A different method is to use multi-objective optimization concepts and to transform a constrained single-objective problem into an unconstrained multi-objective problem. The transformation can be mainly realized by two ways. The first one summarizes the constraint violation as a second objective, the second one adds an objective function per constraint. One challenge of existing constraint handling techniques is to deal with many constraints in multi-objective combinatorial problems.

We present a novel algorithm which combines an agglomerative hierarchical clustering strategy [8] with the well-known NSGA-II [5] to treat many-constraint grouping problems. An agglomerative hierarchical clustering algorithm can be characterized by the following procedure: start with n items in n singletons (bins), calculate pairwise distances between items, merge items with the shortest distance until all items are clustered into a group. In order to solve bi-objective problems the NSGA-II establishes the convergence towards the Pareto front. The main advantage of the clustering NSGA-II is the capability to deal with a huge number of constraints, and to generate always feasible solutions.

We embed a representation form and clustering strategy in the NSGA-II which is dedicated to many-constraint grouping problems. The basic idea of the method is to provide a new constraint handling technique which is easy to understand, aims at solving many-constraint grouping problems, and performs well with standard genetic algorithm operators.

The genetic algorithm incorporates a pairwise list of all feasible item combinations as the representation. The list with pairs of items will be processed from left to right, starting at the first position until each possible pair has been checked for the clustering process. At the end we obtain a feasible bin packing.

We validate our clustering NSGA-II with an artificial test problem which models a grouping process in automobile sheet metal forming processes. The main assignment in the real world application is to build a specific part from a blank sheet on a press line. As a preprocessing step several items such as holes or plungings can be grouped into a single unit so that the new group can be processed in one press step. In order to generate the group of items, many constraint types have to be fulfilled. This grouping problem can be modeled as a MO1DBPP. The challenging task is to solve the many-constraint NP-hard problem with two conflicting objectives: minimum costs and minimum number of groups.

2 Related Work

The literature in the field of multi-objective bin packing problems is scarce [13, 10]. The idea to reformulate a clustering problem in a binary mathematical programming is treated by e.g. [12]. Useful modeling tips are given by [20].

In general, solving packing and cutting problems is very difficult. Nevertheless, there are common exact approaches to explore the decision space and to try to obtain an optimal solution for the bin packing problem [19].

An additional approach is to use approximation algorithms [7] such as first fit decreasing [4] or metaheuristics [2] such as evolutionary algorithms [9, 16, 18] in the solution process of the bin packing problem.

For information about theoretical work in multi-objective evolutionary combinatorial optimization see [15] and the references therein.

Common constraint handling techniques used in evolutionary algorithms are discussed in [3, 14].

Solving bin packing problems with a general cost structure will be presented in [1]. However, in our case, items are cost dependent on other items in the same bin and the total costs can not be easily calculated.

3 Real World Many-Constraint One-Dimensional Bin Packing Problem

We are motivated by the fact that automobile sheet metal forming processes can be formalized as a many-constrained bin packing problem. The task in the SO1DBPP is to pack items $I = \{1, \ldots, n\}$ with different weights w_i in a minimum number of bins $J = \{1, \ldots, m\}$ such that for each bin the total capacity of items does not exceed its capacity W.

3.1 Single-Objective Single-Constraint One-Dimensional Bin Packing Problem

In general, the classical SO1DBPP can be formulated as

$$\min \quad \sum_{j=1}^{m} y_j, \tag{1}$$

$$\text{s.t.} \quad \sum_{j=1}^{m} x_{ij} y_j = 1, \qquad i \in I \quad \text{(Singularity)} \tag{2}$$

$$\sum_{i=1}^{n} w_i x_{ij} \leq W y_j, \qquad j \in J \quad \text{(Weight Capacity)} \tag{3}$$

$$x_{ij}, y_j \in \{0, 1\}, \qquad i \in I,\ j \in J \tag{4}$$

where

$$x_{ij} = \begin{cases} 1 & \text{if item } i \text{ is packed into bin } j, \\ 0 & \text{otherwise.} \end{cases}$$

and

$$y_j = \begin{cases} 1 & \text{if bin } j \text{ is used,} \\ 0 & \text{otherwise.} \end{cases}$$

The objective function (see Eq. 1) minimizes the number of used bins j. The singularity constraints (see Eq. 2) ensure that each item i is assigned to exactly one bin j. Each bin j has a weight capacity restriction W. The sum over the weights w_i of items i in a bin should be smaller than the capacity W of each bin (see Eq. 3). The decision variables are restricted to be binary-valued.

3.2 Bi-objective Many-Constraint One-Dimensional Bin Packing Problem in Automobile Sheet Metal Forming Processes

In the mathematical model of automobile sheet metal forming processes we change equation 3, add many constraint types to the original SO1DBPP, and introduce a second objective function.

The main assignment in the real world application is to build a specific part from a blank sheet on a press line with minimized costs. A blank sheet will be processed over a fixed number of stages of a press line until a formed and manufactured part is obtained. Each press has a special die type installed, which processes the items of a sheet. In a simulation tool, a geometry analyzer examines the well-formed part to detect several items such as holes, openings and plungings (see Fig. 1). Each item has to be processed by different processing sequences, such as punching with a cam and subsequent reforming. Each processing sequence consists of a fixed number and type of processing units. For instance, the item plunging with the processing sequence mentioned before, includes the following procedure: first, punch a hole with a cam and then reform the area around the hole. Each group of similar items can be considered as a cluster.

Each processing unit of the item consists of component types to manufacture the item. For each component type, different specific components are available, depending on the installation space and the item size, respectively. Manufacturing each component results in labor and machine costs as well as material costs. Some components can be shared between items in certain circumstances, so that costs for a shared component will only occur once.

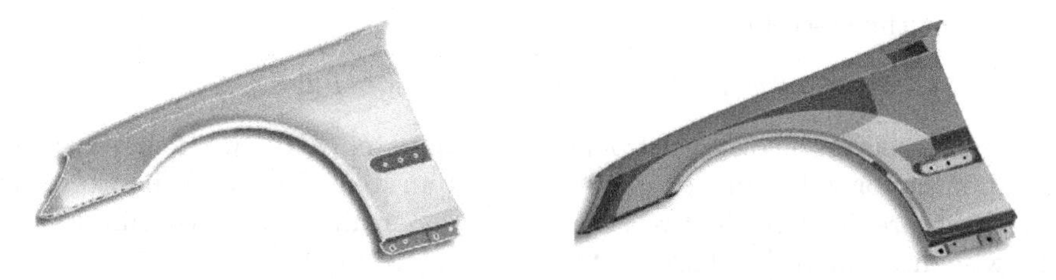

Fig. 1. Automobile sheet metal forming part with several items. Left: Manufactured and processed part at the end of a press line. Right: A geometry analyzer detects several items of a designed part.

Each item within the blank sheet is specified by a 3D position, i.e. $x, y, z \in \mathbb{R}$. The sheet can change the position at a press stage, so that the processing direction of an item will also change. In general, the sheet is non-planar, thus, items can be at different levels of the sheet. For instance, an item can be found at the bottom of a sheet whereas another item will be found at the top of a sheet.

The goal is to find a clustering of items, such that components at a processing unit can be shared, and thus save material costs and time. When building a group of holes, called a cluster, components can be shared between them, if and only if certain constraints are fulfilled. In the following, we will describe several constraint types in a mathematical form and explain the relevance to the real world application. In the following the term distance d_{ik} is defined as

$$d_{ik} = d(i,k) = \| \left((x_i, y_i, z_i)^T - (x_k, y_k, z_k)^T\right) \|$$

where $\| \cdot \|$ is an arbitrary norm.

Note 1. In order to avoid calculating distances between items i and k twice, we consider the case $i < k$. Hence, the distance between two items is symmetric and reflexive.

On the one hand, the distance between two items should not be too small as the installed dies cover some space to manufacture one item. Any overlap of non sharing dies is prohibited to avoid crashing dies (see Eq. 5).

$$|d_{ik}(x_{ij} + x_{kj} - 1)| + |D_{j_{\min}}(x_{ij} + x_{kj} - 2)| \geq D_{j_{\min}} y_j \qquad i,k \in I \text{ and } j \in J. \quad (5)$$

Note 2. The term $x_{ij} + x_{kj} - 1$ is the linearization of $x_{ij} \cdot x_{kj}$.

On the other hand, the distance between two items should be smaller than a certain maximum value because dies have a limited length (see Eq. 6).

$$d_{ik}(x_{ij} + x_{kj} - 1) \leq D_{j_{\max}} y_j \qquad i,k \in I \text{ and } j \in J. \quad (6)$$

An item should also not be located too far away from the next item according to the sharing properties of dies (see Eq. 7).

$$\exists k : d_{ik}(x_{ij} + x_{kj} - 1) \leq D_{j_{\text{next}}} y_j \qquad i \in I \text{ and } j \in J. \quad (7)$$

If the difference in the height is too much, dies are not able to process both items at the same time (see Eq. 8).

$$d_z(x_{ij} + x_{kj} - 1) \leq D_{j_{\text{height}}} y_j \qquad i,k \in I \text{ and } j \in J \quad (8)$$

where d_z measures the distance from the z coordinates of the two items. The last limitation factor is the processing direction of an item. The angle between the processing directions of the two items should be smaller than a certain degree, otherwise dies cannot be shared between items (see Eq. 9).

$$\sphericalangle_{ij}(x_{ij} + x_{kj} - 1) \leq P_j y_j \qquad i,k \in I \text{ and } j \in J \quad (9)$$

where $\sphericalangle_{ij}$ calculates the angle between items i and j.

The sharing option is modeled by the dependency of items in the same bin (see Eqs. 10 - 15). The user can save money by making use of the ability to share some components among items. However, the sharing property can result in increasing costs if the item is not properly located in respect to others. This fact is modeled by the positive or negative dependency matrix entry.

Our modeling approach calculates the final dependency value of an item by a two step procedure. First, the values of the dependency matrix of items in the associated bin has to be summed up. Then, the obtained value must be multiplied with a factor which scales the dependency matrix value of items. The scaling factor will be 1 if all items are in the same bin. If an item is not clustered with any other item in a bin, then the full costs of the item will be integrated in the total costs. The cost dependency of item i in bin j is represented by

$$\delta(i,j) = \left(\sum_{k=1}^{n} x_{kj}\phi(i,k)\right) x_{ij} \qquad i \in I \text{ and } j \in J \tag{10}$$

where $\phi : I \times I \longrightarrow [-1, 1]$ and $\delta : I \times J \longrightarrow \mathbb{R}$.

$\phi(\cdot,\cdot)$ represents the symmetric dependency matrix of every two items and $\delta(\cdot,\cdot)$ the dependency value of an item in a bin. The case $\phi(\cdot,\cdot) = 0$ indicates the fact that items are independent of each other. If the entry has a negative value, then items have a negative influence on each other, or otherwise a positive influence (see Eq. 10). The negative value indicates an increase of the costs for the bin, a positive value indicates a decrease of the costs. The calculation has to be done for each item i in the current bin j and will result in a final cost dependency for each item i to the associated bin j.

All dependencies are summarized between an item and other items in the same bin. The number of dependent items ($\phi(i,k) \neq 0$) is calculated by

$$z_i = \#\{k : \phi(i,k)(x_{ij} + x_{kj} - 1) > 0\} \qquad i \in I \text{ and } j \in J. \tag{11}$$

To obtain the final cost for item i in the current bin j a dependency factor $\gamma(\cdot)$ (see Fig. 2) has to be assessed where $\gamma : I \longrightarrow [0, 1]$ with

$$\gamma(i) = 1 - \alpha^{-z_i} \qquad i \in I \tag{12}$$

where $\alpha = 100^{1/|I|}$. α converges towards 0 if the number of items increases and ensures that $\gamma(\cdot)$ will be bounded by 0 and 1. Now, the cost factor for item i in bin j can be calculated by

$$\epsilon_i = \delta(i,j) x_{ij} \gamma(i) \qquad i \in I \text{ and } j \in J. \tag{13}$$

Finally, the costs per bin j can be assessed by

$$c_j = \sum_{i=1}^{n} (\eta_i(1 - \epsilon_i)) x_{ij} \qquad j \in J \tag{14}$$

where c_j are the total costs of bin j, and η_i the original costs of item i.

Table 1. Distinction of the cases for $\epsilon_i, \delta(i,j)$ and $\gamma(i)$ and the effect on the costs c_j (see Eqs. 10 - 14)

ϵ_i	$\delta(i,j) : \gamma(i)$	c_j
$\epsilon_i > 1 \Leftrightarrow$	$\delta(i,j)\gamma(i) > 1 \Rightarrow$	$c_j \downarrow$
$\epsilon_i = 1 \Leftrightarrow$	$\delta(i,j)\gamma(i) = 1 \Rightarrow$	$c_j = 0$
$\epsilon_i = 0 \Leftrightarrow$	$\delta(i,j) = 0 \vee \gamma(i) = 0 \Rightarrow$	$c_j^{=}$
$\epsilon_i \in]0,1[\Leftrightarrow$	$0 < \delta(i,j)\gamma(i) < 1 \Rightarrow$	$c_j \in]0, \eta_i[$
$\epsilon_i < 0 \Leftrightarrow$	$\delta(i,j) < 0 \Rightarrow$	$c_j \uparrow$

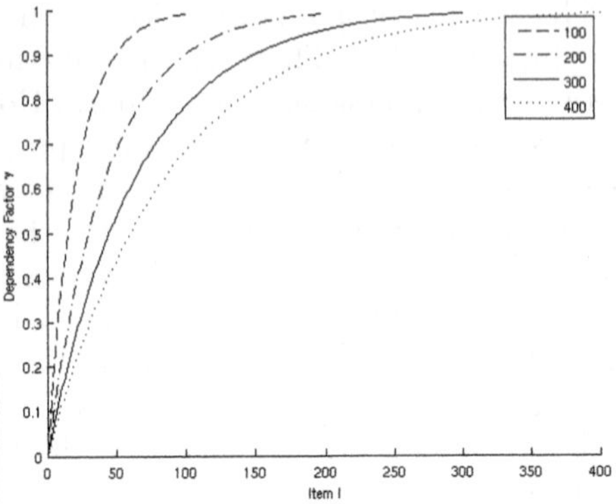

Fig. 2. Visualization of the dependency factor for 100 to 400 items. If the number of dependent items increases, the cost value will decrease accordingly.

The coherency between equations 10 - 14 can be described by a distinction of the cases (see Table 1).

The objectives in the real world problem are to minimize the added costs over all bins and the number of used bins, simultaneously (see Eq. 15).

$$\min \sum_{j=1}^{m} c_j y_j \quad \text{and} \quad \min \sum_{j=1}^{m} y_j \tag{15}$$

In summary, the real world problem can be briefly described by the following: Each item i has its own representation form such as an oval or circle. Also, items have 3D coordinates, i.e. $x, y, z \in \mathbb{R}$, which locate the center of gravity of the item. The second attribute is the normal which is an orthographic vector at the center of the items, i.e. $\tilde{x}, \tilde{y}, \tilde{z} \in \mathbb{R}$ (see Fig. 3). Each item i produces costs η_i. The summarized costs of two items i, k in the same bin can be positively or negatively influenced by a dependency factor of the two items. Thus, the costs of several items in a bin j can not be easily summarized to obtain the final cost of the bin. The items are called cost dependent which will be briefly formalized by $\eta_i \sim \eta_k$. The properties of the bins are the maximum diameter and the cost evaluation function f_{EVAL}. In a bin, the maximum distance between two items has to be smaller than the diameter. The cost evaluation function checks the dependencies between items i and k and calculates the dependent costs (see Fig. 4).

The constraints will be active if at least two items are assigned to the same bin. Otherwise, there is only one item per bin and there is no decision to make. On the one hand, the challenging task is to determine the cost optimal clustering of items into bins, although items are cost dependent on each other. On the other hand, the minimum number of used bins should be decreased as much as possible.

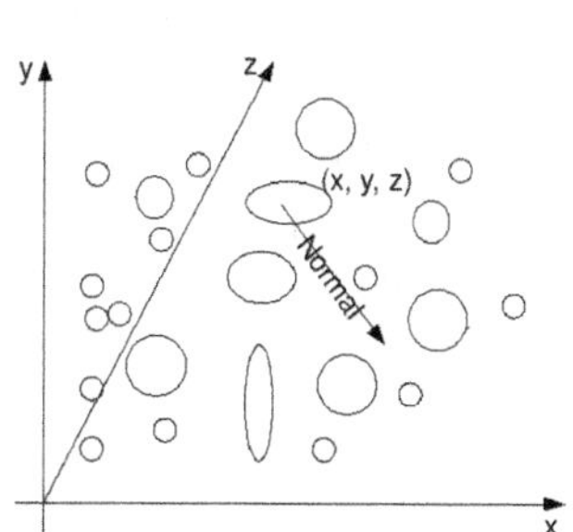

Fig. 3. Set of different sizes of items to be clustered. Each item has properties such as 3D coordinates (x, y, z) and a normal which is a orthographic vector of each item.

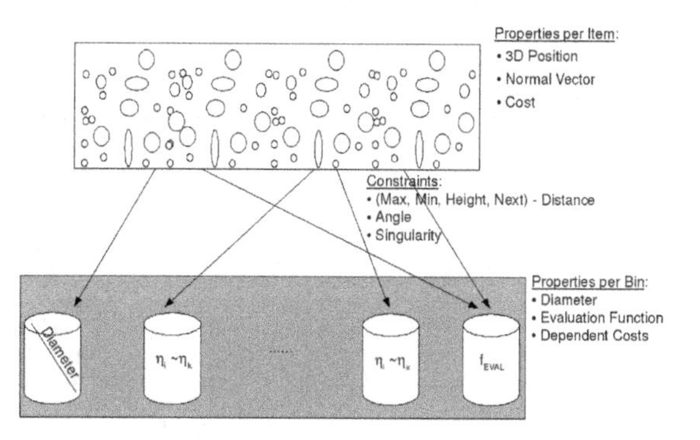

Fig. 4. Overview of the bin packing problem with the properties of items and bins

3.3 Artificial Test Problem Class MOSCS-b

In order to reproduce the results of the clustering NSGA-II in the real world application we introduce an artificial test problem, called MOSCS-b. MOSCS-b is a slim and simplified version of the real world application but it will demonstrate a basic scenario of the optimization problem. The test problem is based on an undirected graph with n items and k edges without self-loops (see Fig. 5). The cluster size b indicates the maximum size of the bin. The problem can be scaled by two parameters: the number of items and the bin size. The goal is to cluster all items in a minimum number of bins with the maximum bin capacity $b \in \{1, \ldots, n\}$ and with minimum costs. The main advantage of the graph representation is that you can easily determine the optimal number of bins and test your algorithm regarding effectiveness and performance. The optimal number of bins can be determined by $\lceil |I|/b \rceil$.

Algorithm 1 initializes the general distance matrix to generate a variable test problem with the two input parameters: the maximal number of items in a bin (nb) and the number of items (ni). For the MOSCS-b we perform the first fit increasing algorithm combined with the clustering NSGA-II. The cost function (see Eqs. 12 - 15) and the minimum number of used bins are our fitness functions which must be optimized, simultaneously. Therefore, we initialize the dependency matrix as in algorithm 1, but the distance matrix entry value 1 is substituted by 0.9 and entry value 2 by -0.9 in the dependency matrix, respectively. The cost array entries for each item are initialized with 1 unit. Thus, we would obtain that the global optimum cost value is far away from the minimum number of bins. The active three constraint types are $D_{\max} = 2$, $D_{\min} = 1$ and $D_{\text{next}} = 1$. Thus, the distance values with an initial value of zero are violating the two constraint types $D_{\min}$ and D_{next}, so that the related items cannot be merged with each other.

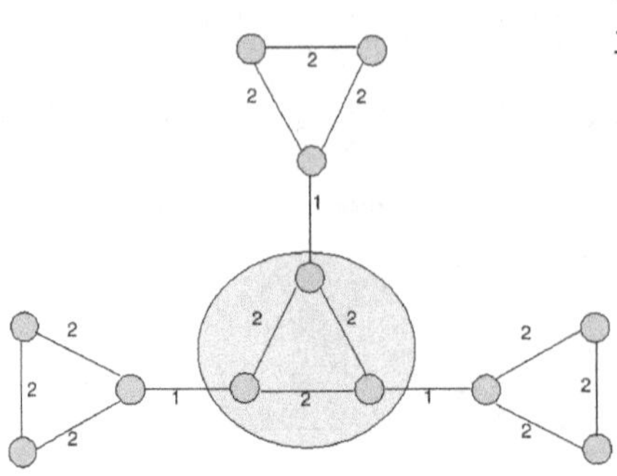

Fig. 5. MOSCS-3 test instance with 12 items and 15 weighted edges. The goal is to find a clustering with a minimum number of bins, here with $b = 3$.

Algorithm 1. initDistanceMatrix(ni, nb)

```
double [][] distanceMatrix;
k ← 1, shift ← 0;
for (i = 0 to (ni/nb) + shift) do
    if (nb · k < ni) then
        if ( i % nb ≠ 0 or i = 0) then
            distanceMatrix[i][nb · k] ← 1;
            k ← k + 1;
        else
            shift ← shift + 1;
        end if
    end if
end for
for (i = 0 to i < ni - 1) do
    if (i % nb ≠ nb - 1) then
        for (j = i + 1 to
            (j < (i/nb+1)nb and j < ni)) do
            distanceMatrix[i][j] ← 2;
        end for
    end if
end for
```

4 Solution Techniques

4.1 Agglomerative Hierarchical Clustering

In the modeling, the initial number of bins is not known before the optimization process starts but it is limited by the number of items n. In realistic scenarios, the number m of bins is less than the number n of items. A specific clustering strategy can be used to determine a good approximated starting solution which can be described by the paradigm first fit increasing (see Alg. 2). The algorithm is very similar to the first fit decreasing algorithm. The first fit decreasing approach sort weights of items in a decreasing order and pack items from the heaviest to the lightest subject to the bin capacity. We are sorting item pairs in an increasing order, cluster them from the shortest to the longest distance subject to various constraint types. Hence, in our problem the weights of the items are interpreted as distances between items, and short distance values should be preferred to longer ones. In the first fit increasing approach the distance values indicate which pair of items should be the next attempt to cluster them. The distances between each item to all other items are calculated and stored once, i.e. the distance relation is reflexive and symmetric. All distances are sorted in an increasing order so that the first entry has the smallest distance value in the list. The distance list can be reduced by all pair combinations violating the constraints (5) - (9). Additionally, the related items are stored in the list. The idea of first fit increasing is to serve the pair of clustering candidates with the smallest distance value at first which is locally the best choice. At the beginning, each item represents one cluster. The items with the smallest distance value are merged if the constraints are fulfilled and they have not been previously merged. If any one constraint is violated, the current distance pair can be skipped. If the item is moved into a group during a previous merge step, all items in the group have to fulfill the restrictions. If the restrictions are not met, the two candidates are skipped as long as there is a successor element in the list. Once all items

Algorithm 2. First Fit Increasing Clustering Algorithm

Initialization: Data, Cluster C;
Assign each item f_i to cluster C_i: $C_1 \leftarrow \{f_1\}, \ldots, C_n \leftarrow \{f_n\}$;
Calculate upper distance matrix for all feasible item combinations:
  $d(f_i, f_k) \qquad \forall i, k$ with $i < k$ and $f_i \in C_i$ and $f_k \in C_k$;
Sort item pair combination in increasing order:
  list $\leftarrow$ sort.asc ($\min_{st}(d(f_s, f_t))$);
while list.next **do**
  Determine indices p, q and relating cluster:
    $f_p \in C_i$ and $f_q \in C_k$;
  if $(i \neq k)$ and (Constraints ((2), (5) - (9)) are true) **then**
    Merge the cluster combination and delete cluster:
    $C_i \leftarrow \{C_i \cup C_k\}$ and delete C_k;
  end if
end while
return Clustering;

have been considered, the algorithm can stop and present the final clustering. The final solution of the clustering algorithm may only be a local optimum of the problem but nevertheless a good starting solution for the following genetic algorithm.

4.2 Genetic Algorithms

Genetic algorithms [17] are random search heuristics to obtain a (near) optimal solution for (combinatorial) optimization problems. In every evolutionary algorithm, several operators (variation, selection), the fitness function, the representation of the individuals, an appropriate population size, and the maximum evaluation number have to be determined. Additionally, the constraint handling by e.g. a penalty function or a repair operator [3, 14] is an issue. We will consider each of these points and present our adjustments.

Representation and Clustering: An intuitive representation for the chromosome of an individual is to use a row of integers. The position of the integer represents the item, and the integer indicates the number of the bin where the item is clustered, e.g. 1 2 1 1 2 means that items 1, 3, and 4 are in bin one and others in bin two. However, there are several difficulties with the representation. For instance, it can happen that possible bin assignments are ambiguous, e.g. 2 1 2 2 1 would generate the same bin assignment. One idea to avoid this problem is to add also the bin number to the representation and then create new solutions [9].

We propose a representation based on an order of pairs of items where the length of the representation is increased from $\mathcal{O}(n)$ to $\mathcal{O}(n^2)$.

We represent our genotype as a list of all feasible item combinations (i, j) generated from the upper distance matrix. The list with pairs of items will be processed from left to right, starting at the first position until each possible pair has been checked.

The first pair (i_1, j_1) of the list is always clustered. The next pair (i_2, j_2) can be clustered if each item of the pair has not been clustered before, i.e. $i_1 \neq i_2, j_2$ and $j_1 \neq i_2, j_2$. Then we generate the two clusters (i_1, j_1) and (i_2, j_2). Otherwise, in the case of $i_1 = i_2, j_2$ or $j_1 = i_2, j_2$, we have also to check the constraints between the pairs (i_1, j_1) and (i_2, j_2).

If no constraint is violated, a new cluster with the items (i_1, j_1, i_2, j_2) is generated. Otherwise, we obtain the two clusters (i_1, j_1), and i_2 or j_2. Again, the next pair has the two possibilities to cluster them if they are independent from the already clustered items, or to check further constraints with other clusters. This process terminates if all pairs in the list have been considered once. At the end the clustering always results in a feasible bin packing.

For instance, in the case of three items we will get a list such as (1,2), (2,3), (1,3) meaning that item 1 and item 2 could be clustered, followed by item 2 and 3 and last but not least items 1 and 3. If items 1 and 2 were clustered at the beginning, item 3 from the next pair has also to check the constraints with item 1, because items 1 and 2 are located in the same bin. The list has a size of $\frac{1}{2}n(n-1)$ and stores all combinations of items to other items once.

For the sorting procedure (see Alg. 2), the corresponding distance value between the two items will be stored. The size of the list can be reduced by item combinations violating a constraint. Thus, the algorithm can be applied to problems with a huge number of constraints. The goal is to find an order of the genotype representation which optimizes the fitness functions.

Fitness Function: In many bin packing applications it is difficult to define an appropriate fitness function. For the real world scenario we will use the minimum number of bins and the cost function (see Eq. 15) as our conflicting fitness functions.

Crossover, Mutation, and Selection: The crossover operator mixes two individuals to generate two new ones. In the literature [11] there are several order-based crossover available which can also perform on this kind of representation. We received the best results with the uniform order preserving crossover (UOPX) operator ($p_{\mathrm{UOPX}} = 0.9$) applied to the individuals. This operator maintains the relative order of genes by the two parents. It works with a randomly generated binary mask, that determines, which parent is used to fill a given position in the chromosome of the offspring with an item pair.

Mutation operates just on one individual and in general, changes the genotype of an individual with a small probability. We will consider a combination of swap mutation with insertion mutation where swap mutation will be applied with a probability p_{swap} and insertion mutation with a probability $(1 - p_{\mathrm{swap}})$. Swap mutation randomly exchanges two positions of a gene with each other. Insertion mutation randomly draws two positions of a parent and inserts the item pair of the first position into the second position. We received the best test results with a value $p_{\mathrm{swap}} = 0.9$. Mutation and crossover change only the order of the pairs of items in the representation of the offspring. Thus, the processing of the pair list will again generate a feasible clustering. The selection operator is the well-known binary tournament selection.

Population Size and Maximal Evaluations: The population size and the maximal evaluation number depend on the number of items of the bin packing problem. For our problem type the population size is fixed at 100 individuals. The evaluation number is determined by the user of the real world application, because the user would like to obtain results very quickly and will only wait a certain period of time. However, we will test our problem on higher evaluation numbers to show the performance of the algorithm.

Constraint Handling: In evolutionary algorithms, the constraint handling is a challenging task. Typically, combinatorial optimization problems have a lot of constraints such as in our case $\mathcal{O}(n^3)$ as a worst case.

Even a bin packing problem with 100 items results in an overall constraint number of one million constraints, if every constraint is evaluated. However, most of the constraints are not active, e.g. if the bin is not used or the items are clustered into a different bin, so the number of constraints can be decreased. Obviously, checking active constraints will be sufficient for the feasibility check of the current clustering and additionally, if one constraint is violated, the current pair can be skipped as in the algorithm 2. Thus, the complexity can be reduced to the average case of $\mathcal{O}(n^2)$.

4.3 Clustering NSGA-II

For the multi-objective optimization test problem (MOSCS-b) we combine the clustering approach with the NSGA-II. The genetic algorithm is based on an order-based vector with pairs of all item combinations. The clustering method is applied to the genotype of the individuals and it guarantees to generate feasible solutions. The non-dominated sorting procedure ensures the convergence towards the approximated Pareto front. With the support of the first fit increasing strategy, an individual will be generated with a good solution for at least one fitness function – minimize the number of used bins. The rest of the population will be created randomly. Each individual obtains a permutation of the pairwise vector of item combinations and the items are clustered according to the order of the vector from the left hand side to the right hand side as in the while loop of algorithm 2. After the non-dominance check, new individuals are again generated and clustered. The algorithm stops if the maximum number of evaluations has been reached. At the end we present few solutions to the user who chooses the final solution according to his/her preference.

5 Results

We provide results to the problem class MOSCS-b, $b = 3$, to give an insight into the new test problem class and the performance of the new clustering NSGA-II. The results of our tests were obtained using an SMP machine powered by Intel Xeon two quad-core CPUs running at 2.33 GHz. The implementation uses the Java programming language and extends the current version 1.8 of the jMetal

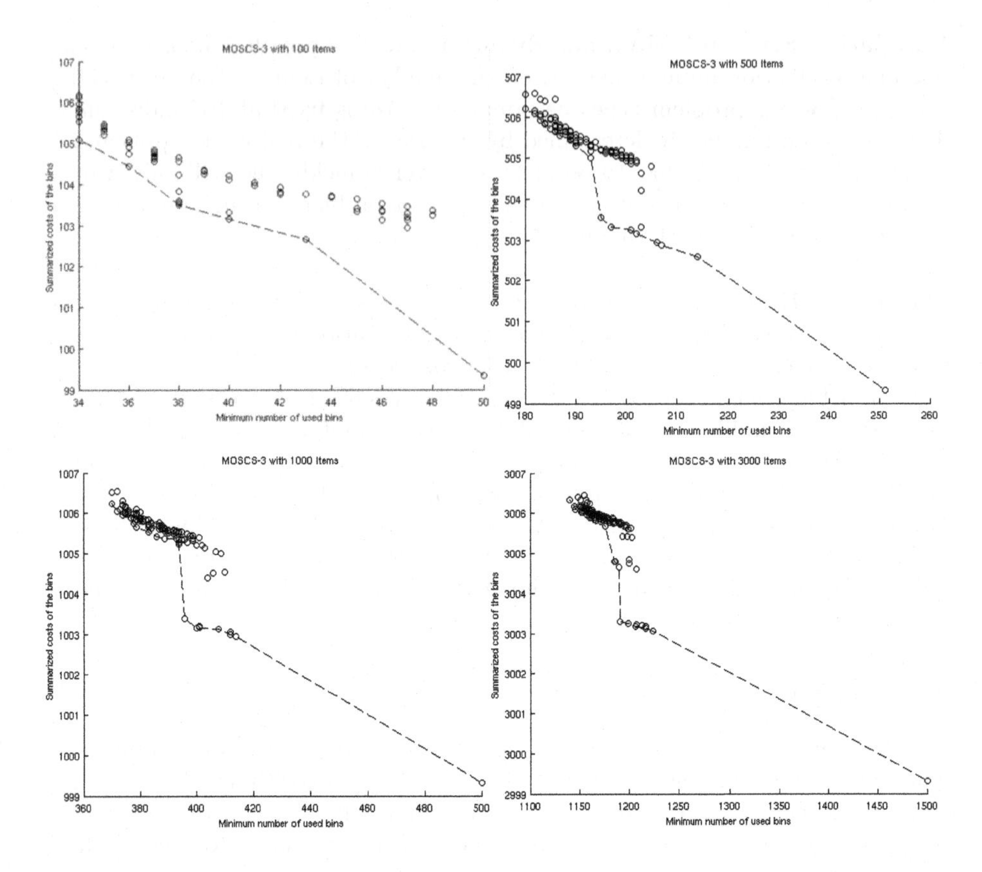

Fig. 6. Approximated Pareto fronts of the MOSCS-3 test problem with 100, 500, 1000, and 3000 items

package [6]. We tested the algorithm with a maximum evaluation number of 20000 for the clustering NSGA-II and different number of items starting from 100 to 3000. We performed 10 iterations per item class. We present the approximated Pareto fronts of the MOSCS-3 test problem for all iterations with 100, 500, 1000, and 3000 items (see Fig. 6). The required average time per evaluation is of quadratic complexity.

At the original application the user obtained the approximated Pareto front and can choose the most compromising solution. On average, the user can save 10 percent of the original costs with a maximum problem size of 150 items. In the case of 100 items, the clustering NSGA-II can calculate the approximated Pareto front very fast and obtains a good result. If the number of items increases, the population of the clustering NSGA-II is concentrated on the upper part of the approximated Pareto front. The spread of the population is an issue for this kind of problem. Thus, the approximated Pareto front can not be easily assessed with the given number of evaluations.

6 Conclusion

Bi-objective bin packing problems with many constraints represent significant challenging problems for standard genetic algorithms. Typically, only convergence to non-feasible solutions is obtained due the high number of additional constraints. In this paper we proposed to use the NSGA-II with standard genetic algorithms operators to solve the bi-objective many-constraint bin packing problem. In addition, a clustering algorithm is used to generate always feasible NSGA-II solutions independent of the number of constraints. The resulting method is able to solve challenging bi-objective bin packing problems with thousands of additional constraints. The new algorithm has been tested with an artificial many-constrained bin packing problem MOSCS-*b*. The problem MOSCS-*b* is designed in such a way that it models typical many-constrained bin packing problems arising in automobile sheet metal forming processes. The clustering NSGA-II is able to recover the Pareto front for MOSCS-*b* accurately and with a reasonable number of objective function evaluations.

Acknowledgments

We thank AutoForm Engineering GmbH for providing the images of the automobile sheet metal forming parts. This work was supported by the Swiss Commission for Technology and Innovation (CTI) under grant CTI No. 8582.1 ESPP-ES.

References

[1] Anily, S., Bramel, J., Simchi-Levi, D.: Worst-case analysis of heuristics for the bin packing problem with general cost structures. Operations Research 42(2), 287–298 (1994)
[2] Blum, C., Roli, A.: Metaheuristics in combinatorial optimization: Overview and conceptual comparison. ACM Computing Surveys 35(3), 268–308 (2003)
[3] Coello Coello, C.: Theoretical and numerical constraint-handling techniques used with evolutionary algorithms: A survey of the state of the art. Computer Methods in Applied Mechanics and Engineering 191(11–12), 1245–1287 (2002)
[4] Coffman Jr., Garey, E.G., Johnson, D.S.: Approximation algorithms for bin packing: A survey. In: Approximation Algorithms for NP-hard Problems, pp. 46–93. PWS (1997)
[5] Deb, K., Pratap, A., Agarwal, S., Meyarivan, T.: A Fast and Elitist Multiobjective Genetic Algorithm: NSGA–II. IEEE Transactions on Evolutionary Computation 6(2), 182–197 (2002)
[6] Durillo, J.J., Nebro, A.J., Luna, F., Dorronsoro, B., Alba, E.: jMetal: A Java Framework for Developing Multi-Objective Optimization Metaheuristics. Technical Report ITI-2006-10, Departamento de Lenguajes y Ciencias de la Computación, University of Málaga, E.T.S.I. Informática, Campus de Teatinos (2006)
[7] Ehrgott, M., Gandibleux, X.: Approximative solution methods for multiobjective combinatorial optimization. TOP: An Official Journal of the Spanish Society of Statistics and Operations Research 12(1), 1–63 (2004)

[8] Everitt, B.S., Landau, S., Leese, M.: Cluster Analysis. Arnold, London (2001)
[9] Falkenauer, E.: A new representation and operators for genetic algorithms applied to grouping problems. Evolutionary Computation 2(2), 123–144 (1994)
[10] Geiger, M.J.: Bin packing under multiple objectives - a heuristic approximation approach. In: The Fourth International Conference on Evolutionary Multi-Criterion Optimization: Late Breaking Papers, Matsushima, Japan, pp. 53–56 (2007)
[11] Goldberg, D.E., Lingle Jr., R.: Alleles loci and the traveling salesman problem. In: Grefenstette, J.J. (ed.) Proceedings of the International Conference on Genetic Algorithms, Pittsburgh, pp. 154–159. Lawrence Erlbaum Associates, Mahwah (1985)
[12] Hansen, P., Jaumard, B.: Cluster analysis and mathematical programming. Mathematical programming 79, 191–215 (1997)
[13] Liu, D.S., Tan, K.C., Goh, C.K., Ho, W.K.: On solving multiobjective bin packing problems using particle swarm optimization. In: Yen, G.G., et al. (eds.) Proceedings of the 2006 IEEE Congress on Evolutionary Computation, Vancouver, BC, Canada, pp. 2095–2102. IEEE Press, Los Alamitos (2006)
[14] Mezura-Montes, E., Coello Coello, C.: A survey of constraint-handling techniques based on evolutionary multiobjective optimization. In: Workshop paper at PPSN 2006, Iceland (2006)
[15] Neumann, F., Witt, C.: Computational complexity of evolutionary computation in combinatorial optimization. In: Rudolph, G., Jansen, T., Lucas, S., Poloni, C., Beume, N. (eds.) PPSN 2008. LNCS, vol. 5199. Springer, Heidelberg (2008)
[16] Reeves, C.: Hybrid genetic algorithms for bin-packing and related problems. Annals of Operations Research 63, 371–396 (1996)
[17] Sathe, M.: Interactive Evolutionary Algorithms for Multi-Objective Optimization: Design and Validation of a Hybrid Interactive Reference Point Method. VDM-Verlag Dr. Müller (2008)
[18] Stawowy, A.: Evolutionary based heuristic for bin packing problem. Computers & Industrial Engineering 55(2), 465–474 (2008)
[19] Wäscher, G., Hauner, H., Schumann, H.: An improved typology of cutting and packing problems. European Journal of Operational Research 183(3), 1109–1130 (2007)
[20] Williams, H.P.: Model Building in Mathematical Programming. Wiley, Chichester (1999)

Robust Design of Noise Attenuation Barriers with Evolutionary Multiobjective Algorithms and the Boundary Element Method

David Greiner, Blas Galván, Juan J. Aznárez, Orlando Maeso, and Gabriel Winter

Institute of Intelligent Systems and Numerical Applications in Engineering (SIANI),
35017, University of Las Palmas de Gran Canaria, Spain
{dgreiner,jaznarez,omaeso}@iusiani.ulpgc.es,
{bgalvan,gabw}@step.es

Abstract. Multiobjective shape design of acoustic attenuation barriers is handled using a boundary element method modeling and evolutionary algorithms. Noise barriers are widely used for environmental protection near population nucleus in order to reduce the noise impact. The minimization of the acoustic pressure and the minimization of the cost of the barrier -considering its total length- are taken into account. First, a single receiver point is considered; then the case of multiple receiver locations is introduced, searching for a single robust shape design where the acoustic attenuation is minimized simultaneously in different locations using probabilistic dominance relation. The case of Y-shaped barriers with upper absorbing surface is presented here. Results include a comparative between the strategy of introducing a single objective optimum in the initial multiobjective population (seeded approach) and the standard approach. The methodology is capable to provide improved robust noise barrier designs successfully.

Keywords: Engineering Design, Evolutionary Multiobjective Optimization, Noise Barriers, Acoustic Attenuation, Uncertainty, Computational Acoustics.

1 Introduction

Shape optimization has been performed in recent years applied to various fields of computational mechanics, such as aeronautics or solid mechanics using evolutionary algorithms [4,5]. Automatically generated optimum designs are possible by using coupled evolutionary computation with accurate numerical modeling.

Noise barriers are widely used for environmental protection in the boundaries of high traffic roads, airports, etc, in the vicinity of population nucleus in order to reduce the noise impact. Here we perform shape optimum design of Y-shape noise barriers using the Boundary Element Method (BEM) [9] to model the sound propagation and NSGA-II [7] for optimization. The aim is to improve the design shape of noise barriers achieving simultaneously higher noise attenuation and also minimizing the cost. The barrier length is considered as representative of the raw material cost and its minimization also leads to limiting its environmental impact.

The paper describes in the second section the acoustic attenuation modeling using BEM, following with the Y-shaped noise barrier optimum design methodology and

M. Ehrgott et al. (Eds.): EMO 2009, LNCS 5467, pp. 261–274, 2009.

problem description, test case, results and discussion. Finally it ends with the conclusions and references.

2 Noise Barriers Acoustic Attenuation Modelling

Sound propagation calculation can be performed efficiently and successfully with the Boundary Element Method (BEM). The main advantages of BEM [9] over other methods based on a geometrical theory of diffraction approach are its flexibility (arbitrary shapes and surface acoustic properties can be accurately represented) and accuracy (a correct solution of the governing equations of acoustics to any required accuracy can be produced providing a boundary element size with small enough fraction of a wavelength). Nowadays, both BEM and the Finite Element Method are the most extended state of the art discretization methods in the computational acoustics field [26]. Concretely, to estimate the efficiency of noise barriers with complex shapes, the BEM has been used from the 80s [6,17,23] and it is still a field of research interest. In recent years, design of noise barriers has been taken into account using BEM, see e.g. [24].

The integral equation for a boundary point i, to be solved numerically by BEM, can be written as:

$$c_i p_i = p_o^* - \int_{\Gamma_b} \left(\frac{\partial p^*}{\partial n} + \mathbf{i} k \beta_b p^* \right) p \, d\Gamma \tag{1}$$

where:

p: acoustic pressure field on the barrier surface (Γ_b) of generic admittance β_b.

p^*: half-space fundamental solution. Acoustic pressure field due to a source at collocation point i over a plane with admittance β_g (ground surface). This fundamental solution only requires the discretization of the barrier boundary (Γ_b). For perfectly reflecting surfaces (barrier or ground), $\beta=0$. If the surface is absorbent, the evaluation of β is obtained from the complex admittance of Delany and Bazley [8] knowing the covering material thickness and its air flow resistivity.

c_i: the local free term at collocation point i: $c_i = \theta / 2\pi$, where θ is the angle subtended by the tangents to the boundary at this point (rads). c_i =0.5 for smooth boundaries.

p_o^*: half-space fundamental solution at problem source due to collocation at point i.

$k = \omega/c$ is the wave number (c: sound wave velocity, ω: angular frequency) and $\mathbf{i}$ the imaginary unit.

The numerical solution of Eq. 1 is possible after a discretization process. A linear system of equations is obtained from this process and lead to values of acoustic pressure over the barrier boundary. The BEM code in this paper uses quadratic elements with three nodal points. For more details about the used model, see [21][22].

3 Y-Noise Barriers Shape Design Optimization

In recent years, noise barrier optimum design has been solved using evolutionary computation. Some works related with single objective optimization are [1,3,10,14].

The simultaneous minimization of two conflicting objectives corresponding to a noise barrier design is performed in this paper. First, a fitness function related with the increase of the acoustic attenuation of the barrier. Concretely, the first fitness function which has to be minimized is:

$$F1 = \sum_{j}^{NFreq} \left(IL_j - IL_j^R\right)^2 \quad (2)$$

where:

IL_i : insertion loss in the third octave band centre frequency for the Y-barrier profile evaluated. Being the *insertion loss* (*IL*), defined as stated in Equation 3 (being dBA the units of IL):

$$IL = -20\log\left(\frac{P_S}{P_B}\right) dBA \quad (3)$$

and calculated at one-third octave band spectra, where P_B and P_S are the acoustic pressure at the receiver with and without the presence of the barrier respectively. This parameter is an accepted estimation of the acoustic efficiency of the analyzed profile.

IL_i^R: insertion loss reference curve in the third octave band centre frequency. When choosing a reference with high IL values, a high efficient attenuation barrier fitting is searched.

The optimum monocriteria design using this first fitness function was previously described in Greiner et al. [14]. It solves an inverse problem, where the objective IL curve at certain frequencies is known (IL^R) and it allows to search for the corresponding barrier design whose IL curve fits IL^R. In [14] was shown the capability to increase a certain percentage the acoustic efficiency of a certain Y-shape barrier taken as original design.

The second fitness function (F2) to be minimized is the noise barrier length, representative of the raw material cost. The higher its value, the easier the noise attenuation capacity of the barrier, and therefore, the easier to fit the searched reference curve. On the contrary, the lower its value, the lower the cost and better environmental impact produced by the barrier.

Here, a multiobjective optimization noise barrier design with evolutionary algorithms is introduced. Concretely, the procedure searches for the barrier shape design which most fits IL^R for each barrier length value.

The modelling approach included in the paper follows the test case implementation of the previous related referenced works and is intentionally chosen because of the simultaneous capability to cover the design space and also to reduce the number of variables of the search optimization (could be interpreted as helping the search including engineering knowledge). The Y-barrier shape is modeled using the two extreme points of the arms and their join point. The x coordinate of the extreme points is supposed fixed in the extremes of the barrier, where only y-coordinate varies. The join point has variable x- and y-coordinates. The evolutionary algorithm variables are set in a transformed space with perpendicular axis and square shape in contrast to the geometric trapezoidal shape limited by b and the sloped line (see Fig. 1). So, four design variables are required to define each shape (the x coordinate vary from -0.5 to +0.5 and the three y coordinates vary from 0 to 1 in the transformed space). For more details, see [14].

With this geometry and for a given source position, the boundary element program calculates the acoustic pressure at the receiver position (r). A maximum element length not bigger than λ / 4 (being λ the wavelength) is necessary to obtain an appropriate accurate solution. With the acoustic pressure, the IL corresponding to each frequency is obtained.

In case we want to consider not a single receiver location, but a certain zone where to minimize the acoustical impact, then various receiver locations are needed and a robust design is pretended, considering the minimization of function F1 at each receiver. Therefore we deal not with a single value, but with a set of F1 values (a distribution estimation). Uncertainty handling in evolutionary optimization has been covered in recent years as a growing field of interest, see e.g. [2, 11, 19]. We follow here the proposal of Teich [25], including the probabilistic dominance relation in the NSGA-II as shown in [20]. So, the F1 objective is not a number, but a random variable with values bounded by an interval evaluated as the average of the F1 values at the receiver points plus and minus their typical deviation.

4 Test Case

The parameters considered in the test case used in the following experiments are according to Fig. 1: b=1m. and d=10m. (noise source distance to the barrier base) We will compare the single-point and multi-point receiver cases. In case of a single receiver, r= 50m. In case of multiple receivers, three receiver positions are taken into account (r=25, 50 and 100m., respectively). The ILref curve is obtained from a straight barrier of 4.5 m height with reflecting surfaces, versus the maximum effective height allowed of our Y-shaped designs of 3.0 m. We will consider only reflecting surfaces, with the exception of the upper boundary of the design (inner surfaces of the arms), which are

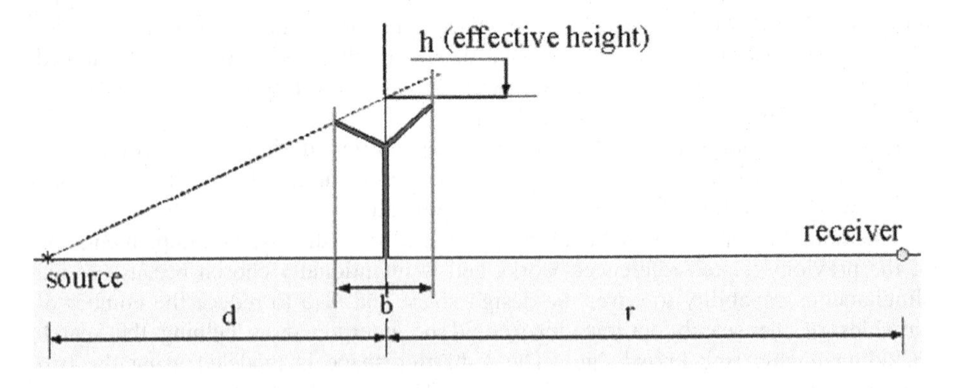

Fig. 1. Problem topology representation

absorbing surfaces. A thickness of 10 cm and an air flow resistivity of 20000 are considered for the calculations described in section 2. A total of 13 frequencies at one-third octave centre band spectra frequency are evaluated: 100, 125, 160, 200, 250, 315, 400, 500, 630, 800, 1000, 1250 and 1600 Hz. The CPU time cost of one F1 fitness function evaluation is 12 seconds in one Pentium IV-3GHz processor.

5 Results and Discussion

Twelve independent runs of the evolutionary optimization design were executed in each case. A population size of 100 individuals and 3% mutation rate were used in a Gray coded [27] NSGA-II algorithm with uniform crossover and probabilistic dominance relation (α=0.5).

Two cases are analyzed: 1. The single point receiver case. 2. The multi-point receiver case. Each one has been solved comparing two different initial population strategies: a) A seeded approach, where a solution of high quality is inserted into the initial population; e.g., see [15]. b) The standard no-seeded initial random population approach.

5.1 About the Initial Population Strategy

The inserted high-quality design is obtained performing a single-objective steady-state evolutionary algorithm optimization on F1. Each of the twelve independent runs obtained the same final value, which will be considered as the optimum in terms of F1. The number of evaluations required to reach the optimum for each run is shown in Table 1. The average values in obtaining the optimum for the single-point and multi-point receiver cases are 4346 and 4526 function evaluations. Since the average values computed are principally influenced by the greatest values of Table 1, if we delete the best and worst values, avoiding the excessive influence of extremes, then the average values are 4005 and 3053, respectively; showing in average that the multi-point receiver case needs less function evaluations.

Table 1. Number of evaluations required to reach the optimum value in the single objective optimization (F1) and average (in italic type)

Single-Point Receiver	3786	6050	3806	4164	2020	3212	*Average*
	4706	2904	10080	4240	3758	3428	*4346*
Multi-Point Receiver	2826	4606	3044	21844	2970	2344	*Average*
	2044	3790	2392	1934	3760	2758	*4526*

In contrast, the best values in terms of F1 obtained after 45000 fitness function evaluations with the multiobjective no-seeded search are shown in Table 3: Only one out of 24 runs were capable to achieve this F1 best solution design.

To compare the outcome of the whole front, we will evaluate the S-metric (hypervolume, originally proposed by Zitzler [28]) of various attainment surfaces. Concretely, we use the S-metric proposal of Fonseca et al. [12][1]. The attainment

[1] Source code available at: http://sbe.napier.ac.uk/~manuel/hypervolume

Table 2. S-Metric (Hypervolume) Results, with Reference Point (2000, 9), including the attainment surfaces 1, 3, 5 and 7 over 12. The constrained space results consider only solutions with F2 values greater than 3.6 m.

	S Metric (Unconstrained Space)		S Metric (Constrained Space)	
Initial Population Strategy – Number of Evaluations	Single Point Receiver	Multi Point Receiver	Single Point Receiver	Multi Point Receiver
	Attainment Surface 1	Attainment Surface 1	Attainment Surface 1	Attainment Surface 1
Noseed - 15000	14420.7833	14412.6969	10790.7287	10789.0082
Noseed - 30000	14425.8523	14421.1657	10790.9215	10789.1444
Noseed - 45000	**14428.0118**	**14423.7072**	10790.3783	**10790.4215**
Seed - 10000	14407.9250	14406.2951	10791.8617	10788.3595
Seed - 25000	14417.2164	14420.7298	10792.5499	10788.7074
Seed - 40000	14420.7381	14423.1665	**10792.6384**	10788.7788
	Attainment Surface 3	Attainment Surface 3	Attainment Surface 3	Attainment Surface 3
Noseed - 15000	14394.8957	14392.7734	10790.3061	10788.6518
Noseed - 30000	14402.1018	14397.0153	10790.6493	10788.9019
Noseed - 45000	**14407.4146**	14400.8746	10790.2065	**10790.2490**
Seed - 10000	14390.6412	14385.8869	10791.4474	10788.0735
Seed - 25000	14398.1735	14396.3221	10792.2399	10788.4565
Seed - 40000	14400.7514	**14401.7005**	**10792.4660**	10788.6195
	Attainment Surface 5	Attainment Surface 5	Attainment Surface 5	Attainment Surface 5
Noseed - 15000	14382.4308	14381.0581	10789.5308	10788.2375
Noseed - 30000	14387.7540	14385.4327	10790.4415	10788.7214
Noseed - 45000	**14391.4587**	**14387.9532**	10790.0582	**10790.0954**
Seed - 10000	14379.4652	14373.8816	10790.7003	10787.8379
Seed - 25000	14386.4227	14382.8896	10792.0495	10788.2732
Seed - 40000	14388.7318	14386.0760	**10792.2480**	10788.4440
	Attainment Surface 7	Attainment Surface 7	Attainment Surface 7	Attainment Surface 7
Noseed - 15000	14371.1840	14370.0224	10788.7894	10787.9192
Noseed - 30000	14376.9554	14375.0176	10790.2124	10788.5177
Noseed - 45000	**14379.5126**	**14376.3961**	10789.8960	**10789.8911**
Seed - 10000	14368.6691	14362.9460	10790.4835	10787.4946
Seed - 25000	14374.9322	14372.3565	10791.9008	10788.0779
Seed - 40000	14378.1628	14374.9925	**10792.0855**	10788.2863

surface concept in multiobjective optimization was introduced in [13,16] and we use here the approach suggested in Knowles [18][2].

We will consider four attainment surfaces, 1 (100%), 3 (83%), 5 (67%) and 7 (50%) out of 12 (total number of independent runs per case), and evaluate its hypervolume after 15000, 30000 and 45000 fitness evaluations in case of no-seeded strategy and 10000, 25000 and 40000 fitness evaluations in case of seeded strategy

[2] Source code available at: http://dbkgroup.org/knowles/plot_attainments

(a fair comparison to take into account the cost of the included solution). As reference point in S-metric calculation, a point sufficiently high has been selected, whose coordinate values of F1 and F2 are respectively, 2000 and 9. In the multi-point receiver case, the average of F1 has been considered for hypervolume calculation. Results are shown in table 2. In this problem the decision maker region of interest is located in the left part of the search space (low F1 values and high barrier length, being the higher F1 values not useful). So, we have also evaluated the S-metric in a constrained design space over a barrier length greater than 3.6 meters. The important information of Table 2 has been put in bold style.

Table 3. Values of the best F1 solutions achieved each run in the standard no-seeded population strategy

	Single Point Receiver		Multi Point Receiver	
	Best F1 value	Corresponding F2 value	Best F1 value	Corresponding F2 value
Run Number 1	0.793816	5.11963	1.01041	5.16681
Run Number 2	0.789535	5.12731	0.99224	5.14777
Run Number 3	0.792179	5.13680	0.993157	5.15853
Run Number 4	0.792731	5.13644	0.99390	5.15735
Run Number 5	0.815023	5.16626	0.991989	5.14819
Run Number 6	0.830437	5.13460	1.00177	5.15647
Run Number 7	0.963581	5.18668	1.00676	5.17805
Run Number 8	0.796091	5.15574	1.00847	5.17743
Run Number 9	0.787557	5.11747	1.01499	5.15646
Run Number 10	0.796999	5.16681	0.994803	5.13744
Run Number 11	0.792993	5.09903	0.994803	5.13744
Run Number 12	0.794859	5.14713	1.00739	5.11768
Best Value	0.787557	5.11747	0.991989	5.14819
Seeded Value	0.787300	5.11796	0.991989	5.14819

Considering the unconstrained space S-metric results, in all cases minus one (3rd attainment surface of multi-point receiver case at 40000 evaluations: 14400.8746 < 14401.7005), the no-seeded strategy achieves a better (higher) hypervolume. The introduced bias towards the optimum may be detrimental to the evolution. In the constrained space, there are manifested two opposite behaviors: in case of the single-point receiver runs, the seeded approach is better in all circumstances over the no-seeded strategy; but in the multi-point receiver case, the no-seeded approach is better in all circumstances over the seeded strategy. That is an indicator of how this multi-point receiver problem has a different landscape than the single-point receiver one.

5.2 Single-Point versus Multi-point Receiver Cases

The accumulated optimum non-dominated solutions are represented in Figures 2 and 3 in search space, showing independently the single-point (crosses) and multi-point (circles) receiver cases. We have focused on the left functional search space part, because it is the region of interest for the designer. In this multi-point receiver problem, only the average of F1 is plotted for clarity.

Table 4. Fitness functions and transformed coordinates values corresponding to the seven selected optimum designs of the single-point receiver case

SinglePoint Receiver Design	F1	F2	y-Coord1	x-Coord2	y-Coord2	y-Coord3
Design 1	0.7873	5.11796	0.972549	-0.04902	0.262745	1.0000
Design 2	1.37468	4.93885	0.976471	0.013725	0.333333	1.0000
Design 3	1.8395	4.71724	0.976471	0.045098	0.419608	1.0000
Design 4	1.89081	4.09768	0.988235	-0.272549	0.737255	1.0000
Design 5	2.551	3.95284	0.952941	-0.296078	0.733333	0.964706
Design 6	5.24347	3.85126	0.94902	-0.272549	0.745098	0.937255
Design 7	7.25369	3.73438	0.917647	-0.194118	0.72549	0.917647

Table 5. Fitness functions and transformed coordinates values corresponding to the seven selected optimum designs of the multi-point receiver case

MultiPoint Receiver Design	F1 Average	F2	y-Coord1	x-Coord2	y-Coord2	y-Coord3
Design 1'	0.991989	5.14819	0.968627	-0.041176	0.247059	1.0000
Design 2'	1.88108	4.87903	0.976471	0.288235	0.380392	1.0000
Design 3'	2.54528	4.63693	0.968627	0.02549	0.439216	0.996078
Design 4'	2.60501	4.01638	0.952941	-0.272549	0.701961	0.972549
Design 5'	3.15836	3.94612	0.941176	-0.268627	0.717647	0.968627
Design 6'	5.99898	3.85323	0.94902	-0.3.0000	0.74902	0.933333
Design 7'	7.57026	3.7397	0.909804	-0.217647	0.721569	0.921569

Table 6. Fitness function F1 value at each receiver of the seven selected designs, average and variance corresponding to both the single and multi-point receiver case (It is hignlighted in italic type the value used as search criterion in the optimization process)

i: SinglePoint Rcptor Design i': MultiPoint Rcptor Design	F1 at Receiver 1	F1 at Receiver 2	F1 at Receiver 3	F1 Average	F1 Variance
Design 1	1.061216	*0.787300*	1.260585	1.036367	0.037642
Design 2	1.909043	*1.374685*	1.919018	1.734248	0.064660
Design 3	2.163635	*1.839499*	2.304096	2.102410	0.037849
Design 4	3.725551	*1.890806*	2.313488	2.643282	0.615430
Design 5	5.594426	*2.550998*	3.511254	3.885559	1.613795
Design 6	7.447395	*5.243466*	5.302573	5.997811	1.051228
Design 7	7.952338	*7.253690*	8.081641	7.762557	0.132259
Design 1'	0.748688	0.847302	1.379976	*0.991989*	*0.076888*
Design 2'	2.074511	1.624732	1.943987	*1.881077*	*0.035696*
Design 3'	2.148006	2.402341	3.085492	*2.545280*	*0.156696*
Design 4'	3.007588	1.899501	2.907952	*2.605014*	*0.250528*
Design 5'	2.731014	2.577981	4.166087	*3.158361*	*0.511659*
Design 6'	7.334568	5.286570	5.375808	*5.998982*	*0.893222*
Design 7'	6.832536	7.135412	8.742823	*7.570257*	*0.702745*

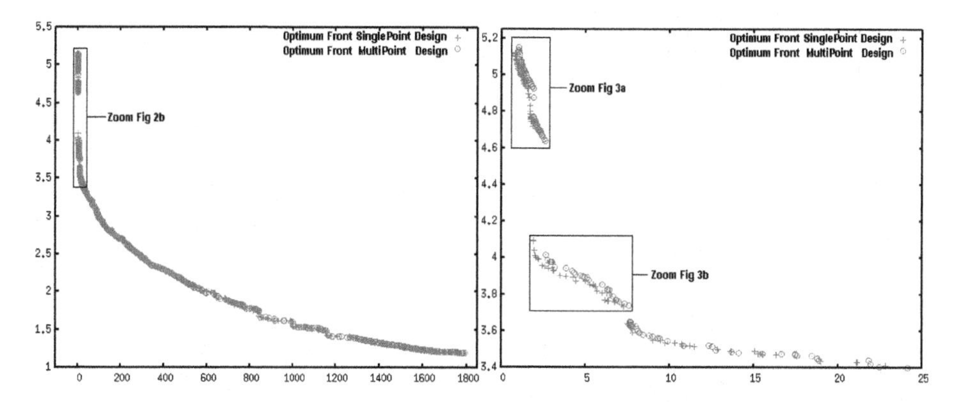

Fig. 2. Non-Dominated final accumulated optimum front function evaluations, including both single-point (crosses) and multi-point (circles) receiver cases. The total front (2a, left) and zoomed left portion (2b, right) are shown. F1 in x-axis and F2 in y-axis.

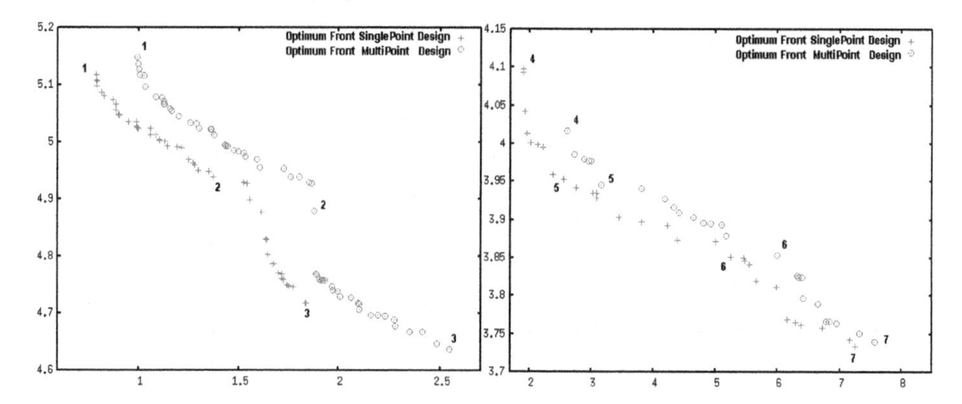

Fig. 3. Zoomed portions (3a, left) and (3b, right) of the non-dominated final accumulated optimum front function evaluations, including the numbering of seven selected designs for both single-point (crosses) and multi-point (circles) receiver cases. F1 in x-axis and F2 in y-axis.

Seven designs (1 to 7 in the single-receiver and 1' to 7' in the case of the multi-receiver) have been chosen along the decision-maker region of interest. They have been marked in the non-dominated front in Figure 3 and their shape designs are represented in Figure 4 (single-receiver) and Figure 5 (multi-receiver). The numerical values of their fitness functions and design variables are shown in Table 4 (single-receiver) and Table 5 (multi-receiver). The single-point receiver front dominates the multi-point receiver front, as can be seen in Figure 3. The need of a robust behaviour when considering various receiver locations implies higher average values of the fitting of the ILref curve. In Table 6 the values of F1 corresponding to the three receiver points are represented for the fourteen designs. In Table 6, we observe in detail the best F1 solutions of both approaches: Design 1 (D1) and Design 1' (D1').

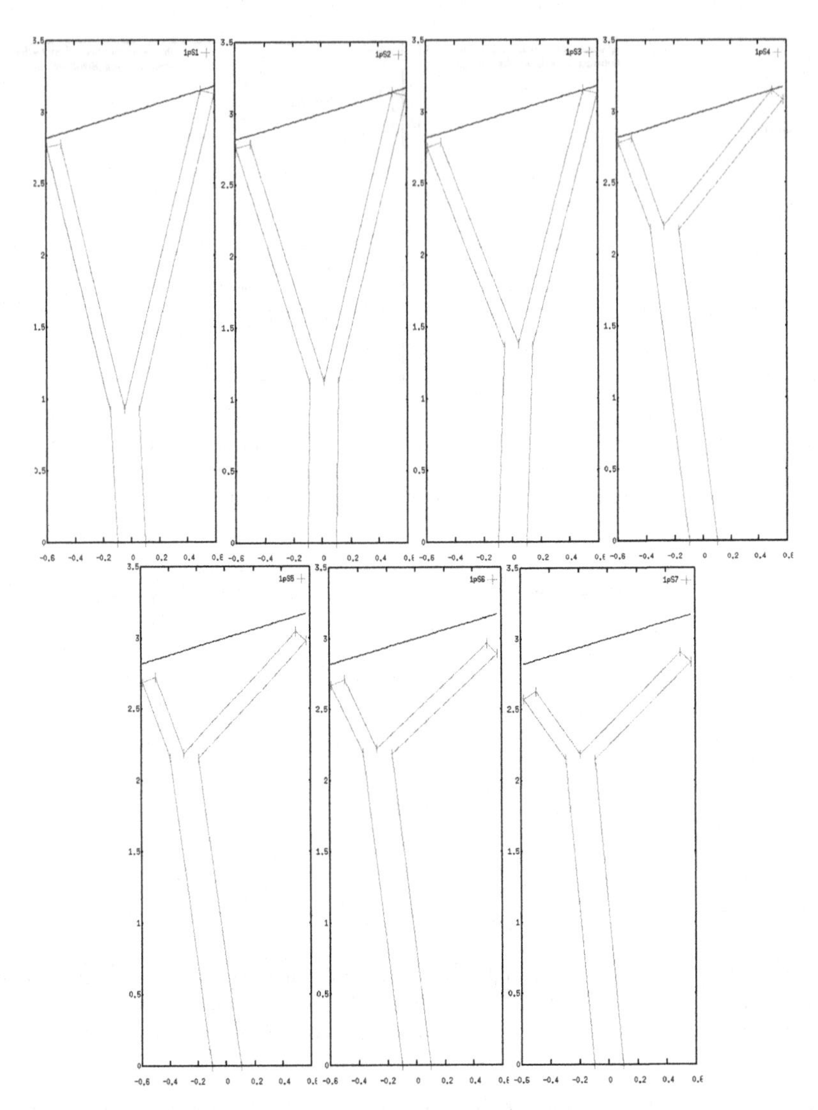

Fig. 4. Shapes of the 7 selected designs, from 1 (left) to 7 (right), single-point receiver case

D1 has the best F1 value in receiver point 2 (distance to the barrier base = 50m.), but an F1 average of 1.036, which is worse than the best value of D1' (0.991989). By the other hand, the value of F1 at receiver 2 of D1' is worse (0.847302 > 0.78730) than the value of D1.

In Figure 6 both the Reference IL curve (corresponding to a 4.5 straight barrier with reflecting surfaces) and the best fitted solutions D1 and D1' are represented for each receiver point. In the x axis the third octave centre spectra frequency is

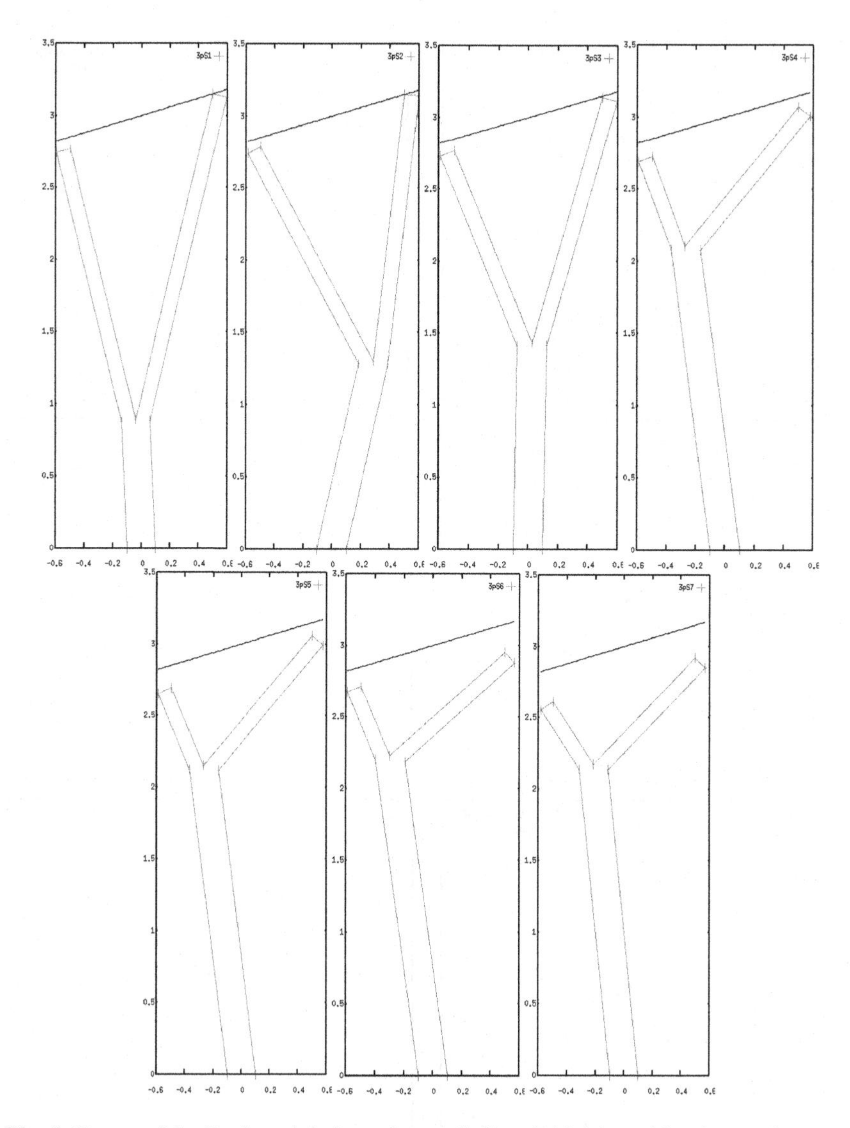

Fig. 5. Shapes of the 7 selected designs, from 1 (left) to 7 (right), multi-point receiver case

represented in Hertz in logarithmic scale. In the y axis the IL is represented in dbA units. As can be seen in the figures, the obtained designs fit accurately the searched IL reference curve, and their differences are really low. Therefore, that means that the same acoustic attenuation efficiency of a 4.5 meters effective height straight barrier can be achieved with a 3.0 meters effective height Y-shaped barrier with absorbent treatment in the inner surface of its arms. The multiobjective approach allows also to locate for each barrier length the barrier that fits most precisely the former noise attenuation capability (the lower the length, the worse the IL curve fit).

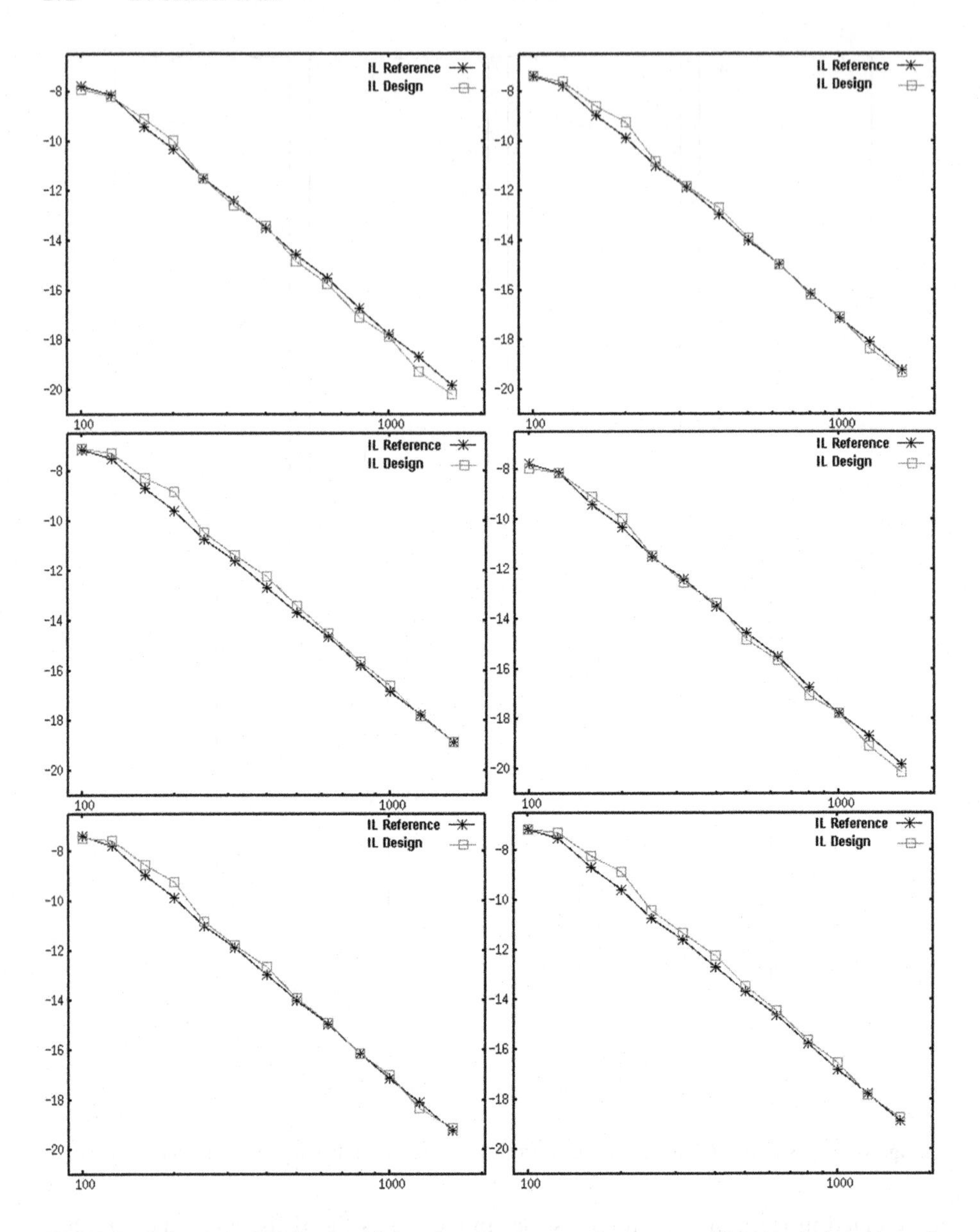

Fig. 6. Insertion loss (IL) in the third octave band centre frequency of barrier design (square) and reference (crossed lines). From left to right and up to down, the first three graphics include the best single-point design and the last three graphics the best multi-point design in terms of F1, being the reference curves those corresponding to receiver points 1, 2 and 3 respectively in each case of the 4.5-height straight barrier. (Frequencies (Hz) in log-scale in x-axis and IL values (dbA) in y-axis)

6 Conclusions

Concerning the problem of multiobjective optimum design of noise barriers, a methodology for considering various receiver points has been introduced in this paper successfully, allowing to obtain robust optimum shape designs that fit various IL reference curves (each receiver represent a IL reference) simultaneously.

Related to the initial population strategy, it has been shown that in certain cases (here the single-point receiver case) the seeded approach introducing one high quality solution design into the initial population, can be useful to obtain improved final fronts. Nevertheless, the reasons that justify when this strategy is profitable or not, should be further investigated.

Taking into account the obtained results in terms of qualitative design information, it is remarkable that introducing the robust design methodology does not lead to new shape designs, being only slight variations of coordinates along the non-dominated front respect the single-point receiver optimization designs.

References

1. Aznárez, J.J., Greiner, D., Maeso, O., Winter, G.: A methodology for optimum design of Y-shape noise barriers. In: 19th International Congress on Acoustics (September 2007)
2. Basseur, M., Zitzler, E.: A preliminary study on handling uncertainty in indicator-based multiobjective optimization. In: Rothlauf, F., Branke, J., Cagnoni, S., Costa, E., Cotta, C., Drechsler, R., Lutton, E., Machado, P., Moore, J.H., Romero, J., Smith, G.D., Squillero, G., Takagi, H. (eds.) EvoWorkshops 2006. LNCS, vol. 3907, pp. 727–739. Springer, Heidelberg (2006)
3. Baulac, M., Defrance, J., Jean, P.: Optimization of multiple edge barriers with genetic algorithms coupled with a Nelder-Mead local search. Journal of Sound and Vibration 300(1-2), 71–87 (2007)
4. Coello, C., Van Veldhuizen, D., Lamont, G.: Evolutionary Algorithms for solving multi-objective problems. GENA Series. Kluwer Academic Publishers, Dordrecht (2002)
5. Coello, C.: Evolutionary Multiobjective Optimization: A Historical View of the Field. IEEE Computational Intelligence Magazine 1(1), 28–36 (2006)
6. Crombie, D.H., Hothersall, D.C.: The performance of multiple noise barriers. Journal of Sound and Vibration 176(9), 447–459 (1994)
7. Deb, K., Pratap, A., Agrawal, S., Meyarivan, T.: A fast and elitist multiobjective genetic algorithm NSGA-II. IEEE Transactions on Evolutionary Computation 6(2), 182–197 (2002)
8. Delany, M.E., Bazley, E.N.: Acoustical properties of fibrous absorbent materials. Applied Acoustics 3, 105–116 (1970)
9. Domínguez, J.: Boundary Elements in Dynamics, Computational. Mechanics Publications: Southampton and Elsevier Applied Science, New York (1993)
10. Duhamel, D.: Shape optimization of noise barriers using genetic algorithms. Journal of Sound and Vibration 297, 432–443 (2006)
11. Everson, R., Fieldsend, J.: Multiobjective Optimization of Safety Related Systems: An application to short-term conflict alert. IEEE Transactions on Evolutionary Computation 10(2), 187–198 (2006)
12. Fonseca, C., Paquete, L., López-Ibáñez, M.: An improved dimension-sweep algorithm for the hypervolume indicator. IEEE Congress on Evolutionary Computation, 1157–1163 (2006)

13. Fonseca, C., Fleming, P.: On the Performance Assessment and Comparison of Stochastic Multiobjective Optimizers. In: Ebeling, W., Rechenberg, I., Voigt, H.-M., Schwefel, H.-P. (eds.) PPSN 1996. LNCS, vol. 1141, pp. 584–593. Springer, Heidelberg (1996)
14. Greiner, D., Aznárez, J.J., Maeso, O., Winter, G.: Shape Design of Noise Barriers using Evolutionary Optimization and Boundary Elements. In: Topping, B., Montero, G., Montenegro, R. (eds.) Proceedings of the Fifth International Conference on Engineering Computational Technology, Civil-Comp Press (September 2006)
15. Greiner, D., Emperador, J.M., Winter, G.: Single and Multiobjective Frame Optimization by Evolutionary Algorithms and the Auto-adaptive Rebirth Operator. Computer Methods in Applied Mechanics and Engineering 193, 3711–3743 (2004)
16. Grunert da Fonseca, V., Fonseca, C., Hall, A.: Inferential performance assessment of stochastic optimisers and the attainment function. In: Zitzler, E., Deb, K., Thiele, L., Coello Coello, C.A., Corne, D.W. (eds.) EMO 2001. LNCS, vol. 1993, pp. 213–225. Springer, Heidelberg (2001)
17. Hothersall, D.C., Chandler-Wilde, S.N., Hajmirzae, M.N.: Efficiency of single noise barriers. Journal of Sound and Vibration 146(2), 303–322 (1991)
18. Knowles, J.: A summary-attainment-surface plotting method for visualizing the performance of stochastic multiobjective optimizers. IEEE Intelligent Systems Design and Applications –ISDAV (2005)
19. Jin, Y.: Evolutionary Optimization in uncertain environments – A survey. IEEE Transactions on Evolutionary Computation 9(3), 303–317 (2005)
20. Limbourg, P., Kochs, H.D.: Multi-objective optimization of generalized reliability design problems using feature models – A concept for early design states. Reliability Engineering and System Safety 93, 815–828 (2008)
21. Maeso, O., Greiner, D., Aznárez, J.J., Winter, G.: Design of noise barriers with boundary elements and genetic algorithms. In: 9th International Conference on Boundary Element Techniques (July 2008)
22. Maeso, O., Aznárez, J.: Strategies for reduction of acoustic impact near highways. An application of BEM. University of Las Palmas of GC (2005), `http://contentdm.ulpgc.es/cdm4/item_viewer.php?CISOROOT=/DOCULPGC&CISOPTR=2389&CISOBOX=1&REC=2`
23. Seznec, R.: Diffraction of sound around barriers: use of Boundary Elements Technique. Journal of Sound and Vibration 73, 195–209 (1980)
24. Suh, S., Mongeau, L., Bolton, J.S.: Application of the Boundary Element Method to prediction of highway noise barrier performance. Sustainability and Environmental Concerns in Transportation, Transportation Research Record, 65–74 (2002)
25. Teich, J.: Pareto-front exploration with uncertain objectives. In: Zitzler, E., Deb, K., Thiele, L., Coello Coello, C.A., Corne, D.W. (eds.) EMO 2001. LNCS, vol. 1993, pp. 314–328. Springer, Heidelberg (2001)
26. Von Estorff, O.: Numerical methods in acoustics: facts, fears, future. Revista de Acústica 38(3-4), 83–101 (2007)
27. Whitley, D., Rana, S., Heckendorn, R.: Representation Issues in Neighborhood Search and Evolutionary Algorithms. In: Quagliarella, D., Périaux, J., Poloni, C., Winter, G. (eds.) Genetic Algorithms and Evolution Strategies in Engineering and Computer Science, pp. 39–57. John Wiley & Sons, Chichester (1997)
28. Zitzler, E., Thiele, L.: Multiobjective optimization using evolutionary algorithms - A comparative case study. In: Eiben, A.E., Bäck, T., Schoenauer, M., Schwefel, H.-P. (eds.) PPSN 1998. LNCS, vol. 1498, pp. 292–301. Springer, Heidelberg (1998)

Bi-objective Optimization for the Vehicle Routing Problem with Time Windows: Using Route Similarity to Enhance Performance

Abel Garcia-Najera and John A. Bullinaria

School of Computer Science, University of Birmingham,
Birmingham B15 2TT, United Kingdom
{A.G.Najera,J.A.Bullinaria}@cs.bham.ac.uk

Abstract. The Vehicle Routing Problem with Time Windows is a complex combinatorial optimization problem which can be seen as a fusion of two well known sub-problems: the Travelling Salesman Problem and the Bin Packing Problem. Its main objective is to find the lowest-cost set of routes to deliver demand, using identical vehicles with limited capacity, to customers with fixed service time windows. In this paper, we consider the minimization of the number of routes and the total cost simultaneously. Although previous evolutionary studies have considered this problem, none of them has focused on the similarity of solutions in the population. We propose a method to measure route similarity and incorporate it into an evolutionary algorithm to solve the bi-objective VRPTW. We have applied this algorithm to a publicly available set of benchmark instances, resulting in solutions that are competitive or better than others previously published.

Keywords: Vehicle routing problem, multi-objective optimization, evolutionary algorithm, similarity of solutions.

1 Introduction

There are many theoretical combinatorial problems that can be directly applied to real-life, one of them being the Vehicle Routing Problem (VRP) [19], which is relevant to transportation logistics such as post, parcel and distribution services.

The VRP's main objective is to obtain the lowest-cost set of routes to deliver demand to customers, but we can also think about reducing the cardinality of the set of routes. In addition, we can contemplate other objectives like the makespan, workload balance, etc. [8]. This means that it is often useful to consider the VRP as a multi-objective problem. Moreover, VRP has several important variants of increased difficulty, in particular, the one with time windows (VRPTW) which has time as well as capacity constraints, and is the main problem to be studied in this paper.

Optimal solutions for small instances of VRPTW can be obtained using exact methods, but the computation time required increases considerably for larger instances [5]. This is why many published studies have made use of heuristic

M. Ehrgott et al. (Eds.): EMO 2009, LNCS 5467, pp. 275–289, 2009.

methods. The recent surveys by Bräysy and Gendreau [2,3] provides a complete list of studies of VRPTW and a comparison of the results obtained.

Over the years, there have been several publications employing evolutionary algorithms to solve VRPTW as a single-objective optimization problem [13,18,20,1]. Recently, Le Bouthillier and Crainic [9] presented a parallel cooperative multi-search method for VRPTW, based on the solution warehouse strategy, in which several search threads cooperate by asynchronously exchanging information on the best solutions identified. Each of these search methods implements a different meta-heuristic, an evolutionary algorithm or a tabu search procedure. Homberger and Gehring [7] proposed a two-phase hybrid meta-heuristic to solve VRPTW, where the first phase was aimed at the minimization of the number of routes by means of a (μ,λ)-evolution strategy, whereas in the second phase the total distance is minimized using a tabu search algorithm.

In the past few years, a couple of studies have been published that are of special relevance to us because they considered VRPTW as a bi-objective optimization problem, minimizing the number of vehicles and the total travel distance, and used a genetic algorithm for solving it. The first is due to Tan et al. [17], who used the dominance rank scheme to assign fitness to individuals. They designed a crossover operator for the specific problem called *route-exchange crossover* and used a multi-mode mutation which considered swapping, splitting and merging of routes. They also used three local search heuristics which were applied every 50 generations. The second is the study of Ombuki et al. [12]. They also proposed the problem-specific genetic operators *best cost route crossover* and *constrained route reversal mutation*, which is an adaptation of the widely used inversion method. However, unlike the work we present later in this paper, neither of these two studies considered using a method to measure similarity of solutions and hence preserve diversity.

It is also worth noting that many existing well known and successful multi-objective evolutionary approaches, such as SPEA2 [21] and NSGA-II [4], are not suitable here because they require the definition of niche spaces, which would be problematic since most good solutions of the VRPTW reside in a very small portion of the vehicle number dimension [12].

In our preliminary work [6], it became clear that the lack of population diversity was leading our algorithm to become stuck in suboptimal solutions, and so we proposed a method to restrict the number of clones in the population. This algorithm eventually forced the population to have no clones at all, but the solutions were still not good enough. Consequently, in this paper, we look for a mechanism to improve further both the quality and diversity of solutions.

The approach we shall follow will focus on the similarity of solutions, based in the genotype space. Several methods for calculating the similarity between two solutions using a permutation representation exist in the literature [14,10,15], but, as we are not using a permutation encoding, we cannot apply any of them. So, the need to find a suitable similarity measure arose.

The work presented in this paper is concerned with the solution of VRPTW as a bi-objective problem using an evolutionary algorithm (BiEA), which incorporates a similarity measure applied in the genotype space, based on Jaccard's similarity coefficient, to select parents for the recombination process, leading to the finding of good solutions to the problem. We have tested this algorithm on publicly available benchmark instances, and when our results are compared with those from recent publications, our algorithm appears very competitive.

The remainder of this paper is organized as follows. First, in Section 2, we introduce VRPTW in more detail. In Section 3 we present our proposed similarity measure for solutions to VRPTW. Our proposed BiEA for solving VRPTW as a bi-objective problem is described in Section 4. In Section 5 we present the results achieved by our algorithm, as well as the comparison with some others already published. Finally, we give our conclusions in Section 6.

2 The Vehicle Routing Problem with Time Windows

The Vehicle Routing Problem (VRP) is a complex combinatorial optimization problem, which can be seen as a fusion of two well-known problems: the Travelling Salesman Problem (TSP) and the Bin Packing Problem (BPP). So, it is at least as difficult as each of them. VRP has several variants of increased difficulty, in particular, the one with time windows (VRPTW) which has both capacity and time constraints.

Before defining VRPTW, we need to specify the information involved in an instance of this problem. First of all, we have a set $\mathcal{V} = \{v_1, \ldots, v_n\}$ of vertices, called customers. We know that customer v_i, $\forall\, i \in \{1, \ldots, n\}$, is geographically located at position (x_i, y_i), has a demand of goods $d_i > 0$, has a time window $[b_i, e_i]$ during which it has to be supplied, and requires a service time s_i to unload goods. There exists a special vertex v_0, called the *depot*, located at (x_0, y_0), with $d_0 = 0$, and time window $[0, e_0 \geq \max\{e_i : i \in \{1, \ldots, n\}\}]$, from which customers are serviced utilising a fleet of identical vehicles with capacity $Q \geq \max\{d_i : i \in \{1, \ldots, n\}\}$.

The travel between vertices v_i and v_j has an associated symmetric cost $c_{ij} = c_{ji}, \forall\, i, j \in \{0, \ldots, n\}$, which is usually considered to be the Euclidean distance. In addition to distance, time also plays an important role, as it is not possible to supply a customer before or after its time window. A vehicle could arrive early at the customer location, but then it has to wait until the beginning of the time window. Arriving late is not allowed. It is common to take the time t_{ij} to travel between vertexes v_i and v_j to simply be $t_{ij} = c_{ij}$.

The problem consists of designing a minimum-cost set of routes, so that each route begins and ends at the depot, and each customer is serviced by exactly one vehicle. Thus, each vehicle is assigned a set of customers that it has to supply, but the sum of their demands can not exceed the vehicle capacity Q.

Let us denote as $r_k = \langle u_1^k, \ldots, u_{n_k}^k \rangle$ the k-th designed route that supplies n_k customers, with u_i^k the i-th vertex to visit in the route. Note that in this notation we are omitting the depot, but we have to consider it before the first customer and after the last customer. Then, the customers demand D_k associated with route r_k is given by

$$D_k = \sum_{i=1}^{n_k} d_{u_i^k} \leq Q \,. \tag{1}$$

Likewise, we can define the cost C_k associated with route r_k as

$$C_k = c_{0u_1^k} + \sum_{i=1}^{n_k-1} c_{u_i^k u_{i+1}^k} + c_{u_{n_k}^k 0} \,. \tag{2}$$

Once we have defined the problem, we can identify at least two objective functions that could be minimized. If $\mathcal{R} = \{r_1, \ldots, r_m\}$ is the set of designed routes, we can consider minimizing the number of routes

$$f_1(\mathcal{R}) = |\mathcal{R}| \tag{3}$$

and the total cost

$$f_2(\mathcal{R}) = \sum_{k=1}^{|\mathcal{R}|} C_k \,. \tag{4}$$

It is these two objectives that we concentrate on in this paper

3 Measuring Similarity of Solutions to the VRPTW

Different solution representations require different distance measures. For example, for binary representations, the *Hamming distance* is the most common measure, for representations using a vector of real numbers, a variation of the *Minkowski-r-distance* can be employed [15], and we can find in the literature many methods for solutions represented as a permutation, like the *exact match distance* and *deviation distance* [14], the *R-type distance* [10] and the *edit distance* [15]. Since we are not using any of these representations, and considering distance in the phenotype space is likely to be misleading, we need to identify a suitable new similarity measure.

Taking the above into consideration, we have designed a new similarity measure, based on *Jaccard's similarity coefficient.* This is applied in the genotype space, and consequently provides a more reliable diversity measure than a phenotype-oriented method. Moreover, probably most interestingly and importantly, this new similarity measure is independent of the solution encoding.

The Jaccard's similarity coefficient is a statistic used for comparing the similarity of two sets. It is defined as the cardinality of the intersection of the sets divided by the cardinality of the union of them, i.e.

$$J(A, B) = \frac{|A \cap B|}{|A \cup B|} \,. \tag{5}$$

It is easy to see that if sets A and B contain the same elements, $A = B = A \cap B = A \cup B$, so Jaccard's similarity coefficient $J(A, B) = 1$. On the other hand, if sets A and B do not share any element at all, $|A \cap B| = 0$, so $J(A, B) = 0$.

Now we can define the similarity between two solutions to VRPTW, according to the Jaccard's similarity coefficient, simply as the ratio of the number of shared arcs to the number of total arcs used in both solutions.

Let $y_{ijk} = 1$ if arc (i, j) from vertex i to vertex j is used by any vehicle in solution r_k, and $y_{ijk} = 0$ otherwise. Then the similarity ς_{pq} between solutions p and q is

$$\varsigma_{pq} = \frac{\sum_{i=0}^{n} \sum_{j=0}^{n} y_{ijp} \cdot y_{ijq}}{\sum_{i=0}^{n} \sum_{j=0}^{n} \operatorname{sign}\left(y_{ijp} + y_{ijq}\right)} , \tag{6}$$

where $y_{ijp} \cdot y_{ijq} = 1$ iff arc (i, j) is used by both solutions, and $\operatorname{sign}(y_{ijp} + y_{ijq}) = 1$ if any of the solutions use it. If solutions p and q are the same, the sum in the numerator will equal the sum in the denominator, and therefore $\varsigma_{pq} = 1$. On the other hand, if they are two completely different solutions with no arc in common, the numerator will equal 0, and then $\varsigma_{pq} = 0$.

In the same manner, if we want to compute the similarity $\mathcal{S}_p$ of individual p with the rest of the population $\mathcal{P}$ of size $N - 1$, we have to calculate the average similarity of p with every other individual $q \in \mathcal{P}$, that is

$$\mathcal{S}_p = \frac{1}{N-1} \sum_{q \in \mathcal{P}} \varsigma_{pq} . \tag{7}$$

4 Evolutionary Approach

We present in this section our proposed EA for solving VRPTW as a bi-objective problem. We detail the encoding of the solutions, and the stages of processing involved. We also describe our main contribution, which is the incorporation of the similarity measure method presented above.

4.1 Solution Encoding

We are using a tree representation, in which each node has at most two children. The left child represents the following customer to visit in a route. The right child points to the next route in the solution. A solution to an example instance and its representation are shown in Figure 1. The allocation of customers to routes, and the sequence they will be serviced within each route, proceeds as follows: customers 1, 2 and 3 to the first route, customers 4 and 5 to the second, 6, 7 and 8 to the third, and 9 and 10 to the fourth.

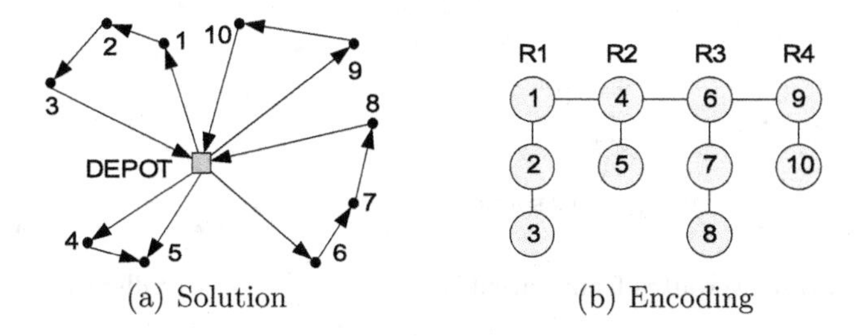

(a) Solution (b) Encoding

Fig. 1. Solution to an example instance of the VRPTW and its encoding

4.2 Fitness Assignment

When solving a single-objective problem using an evolutionary algorithm, fitness is assigned to an individual according to its objective function evaluation. In the multi-objective case, this assignment cannot be done straightforwardly, due to there being not only one objective function, but at least two of them. We have used in this work the non-dominance sort criteria [4] to assign fitness to solutions, where the population is divided into several non-dominated fronts and the depth specifies the fitness of the individuals belonging to them. In this case, the lower the front, the fitter the solution.

4.3 Evolutionary Process

Our algorithm starts with a set of feasible random solutions, each containing a set of randomly generated routes. These routes are constructed using the following process: First, a customer is selected and placed as the first location to visit on that route. Then, a second customer is chosen and, if capacity and time constraints are met, it is placed after the previous one. If any of the constraints are not met, a new route is created and this customer will be the first location to visit in the new route. This process is repeated until all customers have been assigned to a route.

Then, the objective functions are evaluated for every solution in the population and these are assigned a fitness value. Finally, the similarity of each solution with respect to the rest of the collection is computed.

The evolution proceeds with the recombination of two parents that are selected using a standard tournament method, but under different criteria: fitness is used to select the first parent and similarity the second. The recombination of two example parents is shown in Figure 2(a). Here, the algorithm aims at preserving routes from both parents. First, a random number of routes are chosen from the first parent and copied into the offspring. Next, all those routes from the second parent which are not in conflict with the customers already copied from the first, are replicated into the offspring. In this case, both routes on the left from the first parent were selected to be copied into the offspring, and we can only copy from the second parent the route on the right, as the other two

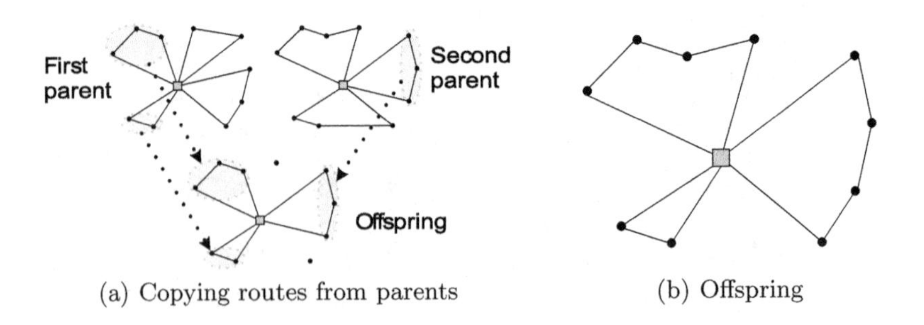

(a) Copying routes from parents (b) Offspring

Fig. 2. The recombination process

contain customers already present in the offspring. If there remain unassigned customers, these are allocated, in the order they appear in the second parent, to the route where the lowest travel distance is achieved, like in the example given in Figure 2(b). If a solution would become infeasible after inserting such a customer, a new route is created. This means that there is no need for a repair process to correct invalid individuals.

Once an offspring has been generated, it is submitted to the mutation process. In our algorithm, we have five possible mutation operators, which can be categorised as inter- and intra-route. In the former, the algorithm will perform changes between two routes, thus modifying the assignment of customers to routes, and in the latter, the changes will be done within a route, hence affecting the travel sequence.

In the first category we can identify two viable processes which are: (*i*) removing a sequence of customers from a route and *inserting* it into another, and (*ii*) *swapping* two sequences of customers from different routes. In the case of intra-route, we use three operations: (*i*) the *inversion* of the sequence of a sub-route, (*ii*) the *shift* of one customer, and (*iii*) *splitting* a route. Examples of these operations are shown graphically in Figure 3. The dotted lines in each figure represent the changes in the sequences. In Figure 3(b), customer 10 was removed from the route on the left and has been inserted in the route on the right. Figure 3(c) shows the swap of customer 4 with customers 2 and 3. In Figure 3(d) we have the inversion of the sequence of customers 7, 8 and 9. Figure 3(e) shows how customer 9 has been shifted between customers 6 and 7. Finally, in Figure 3(f), the route on the right has been split between customers 7 and 8.

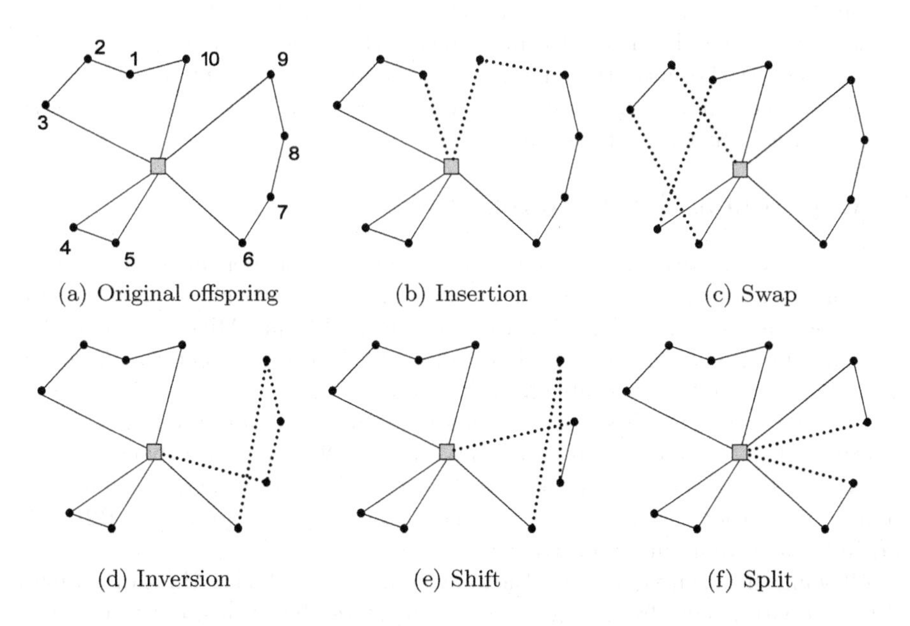

(a) Original offspring (b) Insertion (c) Swap

(d) Inversion (e) Shift (f) Split

Fig. 3. The mutation operators used in BiEA

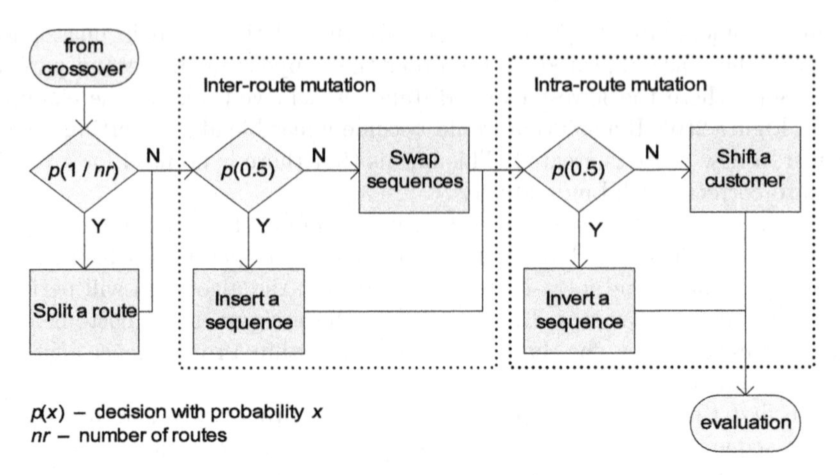

Fig. 4. The full mutation process

Not all of the mutation operators are applied each time an offspring is mutated. First, the split operator is performed with a probability equal to the inverse of the number of routes in the solution. Then the solution is submitted to one of the inter-route operators. The decision of which to apply is random. Finally, one of the intra-route operators is applied to the solution. The complete mutation process is shown in Figure 4.

After this process, the algorithm evaluates the objective functions for each solution in the offspring population and combines both parent and offspring populations to assign fitness. Those solutions having the highest fitness are taken to the next generation. If one front is in conflict with the population size, similarity is computed for those solutions in that front, and the less common are considered for the next iteration. Similarity is computed again and the whole process is repeated for *maxGen* generations.

5 Experimental Set-Up and Results

To test our new algorithm, we used the publicly available benchmark set due to Solomon [16], which includes 56 instances of size $n = 100$. These instances are categorised as Clustered (C1, C2), Random (R1, R2), and Mixed (RC1, RC2). Solomon [16] provides a complete description of the test data, and the data-sets themselves are publicly available from his web site[1].

These data-sets have been previously studied in detail, and a recent analysis by Tan et al. [17] suggests that categories C1 and C2 have positively correlating objectives, which means that the travel cost of a solution increases with the number of vehicles. However, many of the instances in categories R1, R2, RC1 and RC2 have conflicting objectives.

Following the discussion above, the analysis of our results has three objectives: (*i*) to compare results from single- and bi-objective algorithms, (*ii*) to examine

[1] `http://w.cba.neu.edu/~msolomon/home.htm`

the difference between results from our algorithm with and without considering the similarity measure, and (*iii*) to study the performance of our algorithm when compared with others previously published.

We ran our algorithm 30 times for every instance and recorded the solutions in the Pareto approximation each time. The parameters of our algorithm were set to convenient round numbers that worked well as follows:

population size = 100 crossover rate = 0.9
number of generations = 500 mutation rate = 0.1
tournament size = 10

5.1 Single- and Bi-objective Optimization

In this section we compare the results from our algorithm with those from a single-objective genetic algorithm (GA), namely the version of our BiEA which only minimizes one of the two objective functions. For simplicity, the one that minimizes the number of routes will be called GAR, and the one that minimizes the total cost will be called GAC. Note that we are performing this comparison to first determine whether VRPTW really does behave as a multi-objective problem, and then, if it does, to determine whether we can achieve better results by considering it as a multi-objective problem.

In Table 1 is presented the average best results grouped by category set. We have averaged the best results found over all iterations, and averaged these over the set category. We show for each algorithm and instance set the average number of routes (upper) and the average total cost (lower). The last column presents the total accumulated sum, indicating the total number of routes and the total cost for all 56 instances. In this table we are also showing the results from the version of our algorithm that does not include the similarity measure (BiEA_{NS}), but we will not analyse these until the following section.

We can see that the number of routes from GAR is lower than that from GAC in four of the six categories, and only in two compared with BiEA. Our BiEA algorithm achieved fewer routes in all cases compared with GAC. Comparing

Table 1. Comparison of the average best results, averaged by set category, from the single- and bi-objective algorithms

Alg.	C1	C2	R1	R2	RC1	RC2	Accum.
GAR	10.43	3.07	13.08	3.14	12.91	3.60	441.95
	1685.22	898.33	1550.45	1448.03	1742.10	1769.78	84982.41
GAC	10.05	3.01	13.52	4.00	13.51	4.81	467.33
	908.21	601.42	1273.83	954.24	1464.37	1111.69	59376.27
BiEA	10.00	3.00	12.96	3.79	12.65	4.53	448.70
	842.29	597.52	1222.13	968.89	1378.40	1136.16	57800.63
BiEA_{NS}	10.03	3.00	13.26	3.63	13.26	4.28	453.52
	918.86	604.23	1279.74	977.27	1471.58	1143.58	60131.52

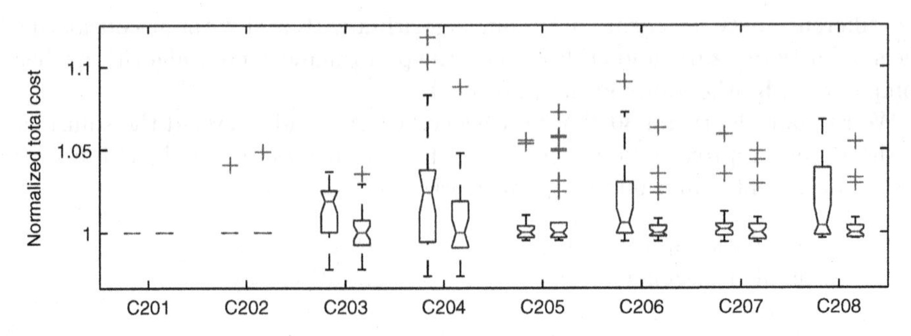

Fig. 5. Box-and-whisker plot for instances in category C2

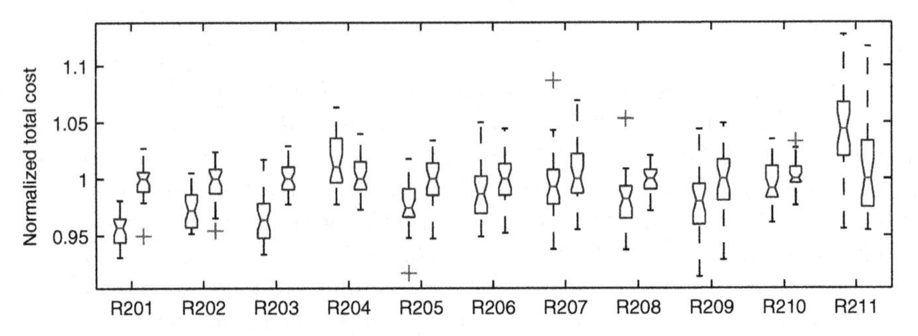

Fig. 6. Box-and-whisker plot for instances in category R2

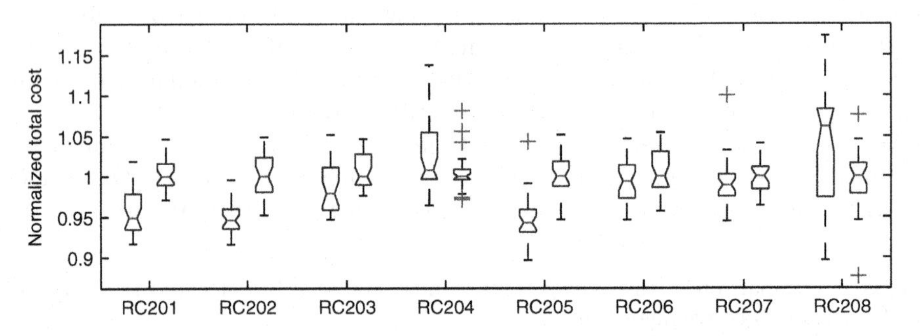

Fig. 7. Box-and-whisker plot for instances in category RC2

the total cost, GAR obtained the highest costs, while our algorithm surpassed GAC in four of the six groups. The last column in the table shows the overall accumulated results for both objectives. In this matter, GAR has the lowest accumulated number of routes, and BiEA the lowest accumulated total cost, reducing by 47% and 3% that achieved by GAR and GAC, respectively.

Given the narrow difference in the total cost from GAC and BiEA for sets C2, R2 and RC2, we decided to analyze in more detail the performance of these algorithms. In Figures 5 to 7 we present box-and-whisker plots for all instances in these sets, with the total cost normalized to the median of the results from BiEA.

For each instance there are two boxes, with the one on the left corresponding to the results from GAC, and the one on the right to BiEA. The boxes have lines at the lower quartile, median, and upper quartile values. Notches display the variability of the median. The width of the notches are computed so that box plots which notches do not overlap have different medians at the 5% significance level [11].

In the case of the instances in category C2, depicted in Figure 5, there is not a clear difference for instances C201, C202, C205 and C207 where the median and its variability appears to be the same for both algorithms. For instances C206 and C208, despite notches that are overlapping, the variability of the median of the results from GAC is larger than that of the results from BiEA. In the case of instances C203 and C204, boxes from BiEA are lower and shorter that those from GAC.

For instances in category R2, shown in Figure 6, we can observe that notches from five out of 11 instances are overlapping (R204, R306, R207, R209 and R210), where the variability of the median is pretty much similar. The median of the results from GAC is lower in five of the six remaining instances and only in one (R212) the median from BiEA is below that from GAC.

Finally, for instances in category RC2, in Figure 7, we can see overlapping notches for half of the instances (RC203, RC204, RC206 and RC207), lower boxes from GAC, consequently lower medians and their variability, for three of the remaining instances, and only in one instance (RC208) results from BiEA surpassed that from GAC.

5.2 Effect of the Similarity Measure

The same series of experiments was carried out using our algorithm but without the similarity measure (BiEA_{NS}). In this case, the selection of parents for the recombination process took only the fitness into account. The purpose of this analysis was two-fold: to determine the performance of BiEA with and without the similarity measure, and to determine whether the similarity measure is accomplishing the goal of diversifying the population.

In Table 1, we can see that BiEA_{NS}, compared with BiEA, could find, on average, solutions with a lower number of routes only for instances in category R2 and RC2, and higher in all others. These solutions have a higher total cost for all categories. The difference in these results is the effect of including the similarity measure, as we are selecting one of the parents to be not so similar to the rest of the population, and hence looking for solutions in other areas of the search space. This selection could result, after recombination, in an offspring with good quality and different from current individuals.

In Figure 8 we consider the solutions in the Pareto approximations, from one of the repetitions. Because of the space limitations, only four instances are shown, two in category R and two in category RC. In each case, the horizontal axis shows the number of routes, and the vertical axis shows the total cost. Values for the total cost are normalized to the minimum value across both sets. It can be seen that solutions from BiEA are dispersed over at least two values

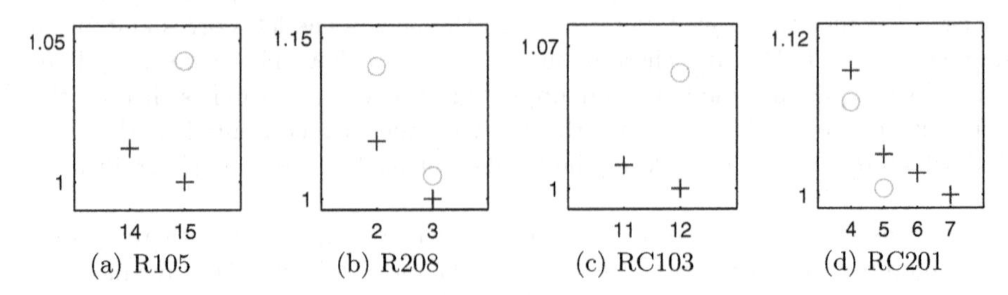

(a) R105 (b) R208 (c) RC103 (d) RC201

Fig. 8. Solutions in the Pareto approximation for four instances. Solutions from BiEA are represented with '+', and solutions from BiEA_{NS} with 'o'.

in the number of routes, in contrast with the solutions found by BiEA_{NS}, which are in two at most. We can also observe that, in three of the four instances, the approximation set from BiEA completely cover that from BiEA_{NS}. Although we are displaying the results for only four instances, nearly all of the 39 instances in categories R and RC exhibit similar results. This characteristic is due to the fact that BiEA searches over a wide range of values, no matter if the number of vehicles is higher, as long as the similarity of solutions remains low.

5.3 Comparison with Recently Published Results

Unfortunately, other recent publications dealing with the same problem have not presented Pareto approximations, even when their authors considered their approach to be multi-objective. Consequently, we cannot compare our results with theirs from a proper multi-objective point of view. Instead, we have averaged the best result in all iterations over the instances in each category set, as this appears to be the most common way in the literature to present and compare results. Our results are shown in Table 2, which has the same format as Table 1. This table also includes the results from four recent studies that minimize both objectives, the number of routes and the total cost, either one after another [7,9] or simultaneously [17,12]. Additionally, in the last two rows, we show the percentage difference between our results and the lowest total cost (ΔL) and the highest total cost (ΔH) for each instance set.

Analysing the results in Table 2, we can see that, for instance set C1, our algorithm obtained, on average, the highest cost, but the gap between this and the lowest result is very narrow, only 0.27%. On the other hand, for set RC1, our algorithm managed to find the lowest costs, and the difference between ours and the highest is 2.71%. For the other categories, although the results from our algorithm are not the overall best, they do show considerable improvement over some of the other algorithms. In fact, they occupy the second or third place among the five. Moreover, in the case of the accumulated total cost, the difference between our results and the lowest is 0.69%, second best, and the difference with the highest is 2.29%, despite our algorithm using a larger number of routes. Another interesting observation is that our results for sets R, RC, and Accumulated confirm that VRPTW really is a multi-objective problem, in the

Table 2. Comparison of the best results, averaged by set category, with others previously published

Author	C1	C2	R1	R2	RC1	RC2	Accum.
[9]	10.00	3.00	12.08	2.73	11.50	3.25	407.00
	828.38	589.86	1209.19	960.95	1386.38	1133.30	57412.37
[7]	10.00	3.00	11.91	2.73	11.50	3.25	405.00
	828.38	589.38	1212.73	955.03	1386.44	1108.52	57192.00
[17]	10.00	3.00	12.92	3.55	12.38	4.25	441.00
	828.74	590.69	1187.35	951.74	1355.37	1068.26	56290.48
[12]	10.00	3.00	13.17	4.55	13.00	5.63	471.00
	828.48	590.60	1204.48	893.03	1384.95	1025.31	55740.33
BiEA	10.00	3.00	12.50	3.18	12.38	4.00	430.00
	830.64	589.86	1191.22	926.97	1349.81	1080.11	56125.35
ΔL	0.27	0.08	0.33	3.80	0.00	5.35	0.69
ΔH	0.00	0.14	1.81	3.67	2.71	4.92	2.29

sense that we obtained lower costs, compared with other the authors, despite using more routes.

Finally, although we do not have room to show the details of our results in this paper, we can note that our algorithm has found 11 new best solutions and another 36 similar to the best published.

6 Conclusions

We have proposed in this paper an evolutionary algorithm for solving VRPTW as a bi-objective problem, simultaneously minimizing the number of routes and the total cost. Importantly, this EA includes a similarity measure, which is used to select one of the two parents for the recombination process. As the other parent is selected according to the quality of the solution, the resulting offspring inherits the quality from this parent, while searching for solutions in a non-common area of the search space. As a consequence, solutions in the resulting population are diverse, covering more than one value in the number of routes dimension.

We have compared the results from our algorithm with those from a single-objective genetic algorithm, with those from our algorithm without the similarity measure, and with algorithms from four recent publications by other authors.

In the first case, we showed that, because the single-objective GA focuses on only one of the two objectives, they do not take into account the likely improvement that could be achieved if the other objective would also be minimized, like the bi-objective evolutionary algorithm does. For this reason, it is highly possible that the single-objective algorithm gets stuck in a suboptimal solution.

In the second case, we have demonstrated the high importance that the similarity measure has, in the sense that the exploration of the search space is wider.

Moreover, the solutions from our algorithm are more dispersed, for example, over two values in the number of routes, in contrast with those obtained without considering similarity, which are concentrated in only one.

In the last case, we have shown that, when compared with other algorithms from recent publications, although our results are not the overall best, they are better than some, and, on average, competitive. Our algorithm also managed to find solutions such that the accumulated travel distance is better than others, despite the number of routes being larger, indicating the multi-objective nature of VRPTW.

Given the promising performance of our algorithm, we are now looking at further ways to exploit the similarity measure, further similarity measures, and more rigorous comparisons of our results with other evolutionary multi-criterion optimization methods using multi-objective performance metrics such as coverage and convergence. We are also exploring the extension of our approach to the minimization of at least one more objective, which could be the makespan or the waiting time. Finally, we are also planning to apply our BiEA to other variants of the VRP, such as one with pick-ups and deliveries.

Acknowledgements

The first author acknowledges support from the Mexican National Council of Science and Technology (CONACYT) through scholarship 179372 to pursue his graduate studies. Both authors are thankful for the valuable comments from the anonymous reviewers.

References

1. Berger, J., Barkaoui, M., Bräysi, O.: A Route-directed Hybrid Genetic Approach for the Vehicle Routing Problem with Time Windows. INFOR 41, 179–194 (2003)
2. Bräysy, O., Gendreau, M.: Vehicle Routing Problem with Time Windows, Part I: Route Construction and Local Search Algorithms. Transport. Sci. 39(1), 104–118 (2005)
3. Bräysy, O., Gendreau, M.: Vehicle Routing Problem with Time Windows, Part II: Metaheuristics. Transport. Sci. 39(1), 119–139 (2005)
4. Deb, K., Pratap, A., Agarwal, S., Meyarivan, T.: A Fast and Elitist Multiobjective Genetic Algorithm: NSGA-II. IEEE T. Evolut. Comput. 6(2), 182–197 (2002)
5. Desrochers, M., Desrosiers, J., Solomon, M.: A New Optimization Algorithm for the Vehicle Routing Problem with Time Windows. Oper. Res. 40(2), 342–354 (1992)
6. Garcia-Najera, A., Bullinaria, J.A.: A Multi-Objective Density Restricted Genetic Algorithm for the Vehicle Routing Problem with Time Windows. In: 2008 UK Workshop on Computational Intelligence. Leicester, UK (2008)
7. Homberger, J., Gehring, H.: A Two-phase Hybrid Metaheuristic for the Vehicle Routing Problem with Time Windows. Eur. J. Oper. Res. 162, 220–238 (2005)
8. Jozefowicz, N., Semet, F., Talbi, E.: Multi-objective Vehicle Routing Problems. Eur. J. Oper. Res. 189, 293–309 (2008)
9. Le Bouthillier, A., Crainic, T.G.: A Cooperative Parallel Metaheuristic for Vehicle Routing with Time Windows. Comput. Oper. Res. 32, 1685–1708 (2005)

10. Martí, R., Laguna, M., Campos, V.: Scatter Search vs. Genetic Algorithms: An Experimental Evaluation with Permutation Problems. In: Rego, C., Alidaee, B. (eds.) Metaheuristic Optimization Via Adaptive Memory and Evolution: Tabu Search and Scatter Search, pp. 263–282. Kluwer Academic, Dordrecht (2005)
11. McGill, R., Tukey, J.W., Larsen, W.A.: Variations of Box Plots. Am. Stat. 32(1), 12–16 (1978)
12. Ombuki, B., Ross, B.J., Hanshar, F.: Multi-Objective Genetic Algorithms for Vehicle Routing Problem with Time Windows. Appl. Intell. 24(1), 17–30 (2006)
13. Potvin, J.Y., Bengio, S.: The Vehicle Routing Problem with Time Windows — Part II: Genetic Search. INFORMS J. Comp. 8, 165–172 (1996)
14. Ronald, S.: More distance functions for order-based encodings. In: 1998 IEEE International Conference on Evolutionary Computation, pp. 558–563. IEEE Press, Piscataway (1998)
15. Sörensen, K.: Distance Measures Based on the Edit Distance for Permutation-type Representations. J. Heuristics 13(1), 35–47 (2007)
16. Solomon, M.M.: Algorithms for the Vehicle Routing and Scheduling Problems with Time Window Constraints. Oper. Res. 35(2), 254–265 (1987)
17. Tan, K.C., Chew, Y.H., Lee, L.H.: A Hybrid Multiobjective Evolutionary Algorithm for Solving Vehicle Routing Problem with Time Windows. Comput. Optim. Appl. 34(1), 115–151 (2006)
18. Tavares, J., Machado, P., Pereira, F.B., Costa, E.: On the Influence of GVR in Vehicle Routing. In: 2003 ACM Symposium on Applied Computing, pp. 753–758. ACM Press, New York (2003)
19. Toth, P., Vigo, D.: The Vehicle Routing Problem. SIAM, Philadelphia (2001)
20. Zhu, K.Q.: A Diversity-Controlling Adaptive Genetic Algorithm for the Vehicle Routing Problem with Time Windows. In: 15th IEEE International Conference on Tools with Artificial Intelligence, pp. 176–183. IEEE Computer Society Press, Washington (2003)
21. Zitzler, E., Laumanns, M., Thiele, L.: SPEA2: Improving the Strength Pareto Evolutionary Algorithm for Multiobjective Optimization. In: Giannakoglou, K., Tsahalis, D., Periaux, J., Papailiou, K., Fogarty, T. (eds.) Evolutionary Methods for Design, Optimisation and Control, pp. 19–26. CIMNE (2002)

Multi-criteria Curriculum-Based Course Timetabling—A Comparison of a Weighted Sum and a Reference Point Based Approach

Martin Josef Geiger

University of Southern Denmark
Department of Business & Economics
Campusvey 55, DK-5230 Odense M, Denmark
mjg@sam.sdu.dk

Abstract. The article presents a solution approach for a curriculum-based timetabling problem, a complex planning problem found in many universities.

With regard to the true nature of the problem, we treat it as multi-objective optimization problem, trying to balance several aspects that simultaneous have to be taken into consideration. A solution framework based on local search heuristics is presented, allowing the planner to identify compromise solutions. Two different aggregation techniques are used and studied. First, a weighted sum aggregation, and second, a reference point based approach.

Experimental investigations are carried out for benchmark instances taken from track 3 of the *International Timetabling Competition ITC 2007*. After having been invited to the finals of the competition, held in August 2008 in Montréal, and thus ranking among the best five approaches world-wide, we here extend our work towards the use of reference points.

Keywords: Multi-criteria timetabling, iterated local search, threshold accepting, reference point approach.

1 Introduction

Timetabling describes a variety of notoriously difficult optimization problems with a considerable practical impact. Important areas within this context include employee timetabling, sports timetabling, flight scheduling, and timetabling in universities and other institutions of (often higher) education [1].

Generally, timetabling is concerned with the assignment of activities to resources. In more detail, these resources provide timeslots (time intervals) to which the activities have to be assigned. In contrast to classical scheduling problems [2], the available time is therefore already discretized into these slots. The result of timetabling is the construction of a *timetable* which defines for each activity when it has to be executed using which resource.

M. Ehrgott et al. (Eds.): EMO 2009, LNCS 5467, pp. 290–304, 2009.

In the case of educational timetabling problems, events are either lectures or examinations, and resources are lecturers and rooms in which the classes are held.

Obviously, the construction of such a timetable has to be done with respect to problem-specific side constraints. Prominent examples comprise:

- All activities must be assigned to timeslots.
- A timeslot may be assigned to at most a single activity.
- Some resources may be unavailable during certain timeslots.
- Individual requirements of the certain activities: Not all resources may be equally suitable for all events.

On the other hand, timetables must be designed such that they optimize several, often conflicting criteria and address the requirements of different groups of stakeholder. This means in case of educational timetabling problems, that both the needs of the students as well as those of the members of staff should be respected.

It is interesting to see that most scientific investigations of educational timetabling implicitly consider the problems to be of multi-objective nature [3, 4]. In contrast to the established terminology from multi-criteria decision making however, desired properties of solutions usually are not expressed using *criteria* nor measured involving a set of *objective functions*. In timetabling, the common approach is to introduce so-called 'soft constraints', each of which computes a score with respect to a particular aspect of the problem. While a violation of these soft constraint is possible, it is penalized introducing a penalty (cost) function. The overall objective may then be derived by minimizing an aggregated overall cost function.

Examples of criteria/ soft constraints include:

- As several lectures share students, they should not be assigned to timeslots of the same period.
- Precedence relations between certain activities: Some activities should be scheduled before or after others.
- Preference of lecturers for certain timeslots or rooms.
- As students have to commute from one room to the other, lectures sharing some students should be placed in rooms which are close to each other.
- Consideration of (meal) breaks.
- Other patterns: In order to ensure a certain compactness of the timetables, any lecture should be for any student adjacent to another lecture.

Most variants of timetabling problems unfortunately are $\mathcal{NP}$-hard [5], and moreover, most problem instances comprise a rather large number of decision variables. Consequently, the application of exact optimization approaches is problematic, given the fact that the solution has to be done within limited time. While there are some exact optimization approaches on the basis of integer linear programming [6], the vast amount of problem solution techniques are based on heuristics, and more recently, metaheuristics.

Heuristic approaches can be categorized with respect to how the two classes of side constraints, hard constraints and soft constraints, are treated. Three classes are commonly considered [7]:

1. *One-stage approaches.*
   One-stage approaches combine the penalty functions of hard and soft constraint violations into a single evaluation function. As generally the minimization of hard constraints is considered to be more important in comparison to the minimization of the violation of soft constraints, a considerable higher weight is given to this aspect.
2. *Two-stage approaches.*
   Contrary to one-stage approaches, two-stage algorithms divide the search for an optimal solution into two disjunct phases. In a first step, a feasible assignment of all events to timeslots is computed, and feasibility is here understood with respect to the set of hard constraints only. The succeeding phase of two-stage approaches is devoted to the minimization of the soft constraint penalties while maintaining feasibility.
3. *Relaxation-based approaches.*
   Feasibility of timetables in relaxation-based solution approaches is assured by either relaxing certain side constraints such that all events may be assigned to timeslots, or by leaving some events unassigned. In this sense, the obtained timetables are feasible not for the initial but with respect to some modified side constraints. When optimizing the timetables for the soft constrain penalties, it is then tried to accommodate all initially defined hard constraints into the solution with the ultimate goal of reaching feasibility. Left aside events are put into the timetable, etc.

Independent from the implemented strategy, most modern approaches are based on the principles of local search. The particular characteristics of timetabling problems imply that neighborhoods are generally composed of moves that unassign and reassign some events, and thus relocate particular activities in the current solution.

An extensive comparison of algorithms has been done in the first International Timetabling Competition ITC 2002 (`http://www.idsia.ch/Files/ttcomp2002/`). Metaheuristics that turned out to be especially effective for timetabling problems are Simulated Annealing, deterministic variants of Simulated Annealing, and Tabu Search. As for most operations research problems, other techniques have been used, too, including Evolutionary Algorithms, Ant Colony Optimization, and Greedy Randomized Adaptive Search. For an extensive listing of references we may refer to [7].

While previous research has primarily been considering an additive aggregation of the penalty functions, our aim is to extend this work towards other ideas from multi-criteria decision making. In detail, we study the influence of the aggregation methodology on the performance and the outcomes of the solution approach. A solution framework (also) allowing the integration of reference points is therefore presented in Section 3 and tested on benchmark data from the ITC 2007.

2 The Curriculum-Based Course Timetabling Problem

The curriculum-based course timetabling problem [8] is a particular variant of an educational timetabling problem. It consists of constructing a weekly timetable by assigning lectures for several university courses to a given number of rooms and time periods. The sketched situation can be found in many universities, where so-called *curricula* are used to describe sets of courses/ lectures that share common students. The underlying logic is based on the assumption, that students who are enrolled in the same program progress together through their studies. Consequently, these students are supposed to attend a previously well-defined set of lectures. This can be seen contrary to post-enrollment based course timetabling problems, where students explicitly register for courses they wish to attend. While in this setting, detailed information about any particular student can be obtained, in curriculum-based course timetabling, registrations of students for courses are not required. The available constraints are solely based on the definition of the curricula.

The data for the curriculum-based course timetabling problem is comparably easy to obtain. Once the curricula are defined, they usually do not change very often, and timetabling can consider to some extend last years plans. On the other hand, students not following the definitions of their curricula may end up having a problem, as several lectures will be scheduled in parallel. The application of the curriculum-based course timetabling problem therefore requires the students to follow the structure of their program.

Besides its practical relevance for many universities, the curriculum-based course timetabling problem has been chosen as one of the competition tracks of the ITC 2007 (`http://www.cs.qub.ac.uk/itc2007/`). The competition invited researchers to propose and submit novel approaches for the solution of timetabling problems. Comparison of results is possible by means of a set of newly released benchmark instances.

A technical description of the problem of the ITC 2007 is given in [8]. Important hard constraints require that no lectures belonging to a common curriculum, as well as lectures being taught by the same professor, are scheduled in parallel. Also, a set of given unavailability constraints has to be respected. These constraints define times when teachers are unavailable. Another common constraint requires that at most one lecture can be assigned to a single room at a time.

Besides these hard constraints, four soft constraints/ objectives sc_1, sc_2, sc_3, sc_4 are relevant that measure desirable properties of the solutions.

1. Objective 1 (sc_1): A room capacity soft constraint tries to ensure that the number of students attending a lecture does not exceed the room capacity. Assignments of lectures to rooms of smaller capacity are penalized with the number of students above the room capacity, multiplied with a penalty weight w_1.
2. Objective 2 (sc_2): The lectures of the courses must be spread into a minimum number of days, penalizing timetables in which lectures appear on too few distinct days. Each day below the minimum is penalized with w_2 points of penalty.
3. Objective 3 (sc_3): The curricula should be compact, meaning that isolated lectures, that is lectures without another adjacent lecture, should be avoided.

For any given curriculum, the number of isolated lectures is computed. Each isolated lecture in a curriculum is penalized with w_3 points of penalty.
4. Objective 4 (sc_4): All lectures of a course should be held in exactly one room. Each distinct room used for the lectures of a course, but the first, counts as w_4 points of penalty.

The overall evaluation of a timetable is then based on some aggregate function $SC = f(sc_1, sc_2, sc_3, sc_4)$.

3 Proposition of a Solution Approach

Figure 1 illustrates the elements of the solution framework in which the following entities play a role:

- A *human decision maker* communicates via a *graphical user interface (GUI)* with the system. Communication includes the definition of decision variables, constraints, objectives and preferences. Also, the penalization (weighting) of particular timetabling patterns and the aggregation of the objective functions are obtained in this process.
- The formal *model* of the current situation is formulated and stored in a database.
- After a *preprocessing stage*, in which some problem-specific properties are pre-computed and structured, two method bases are employed to construct and improve timetables for the quantitative model:
  1. A *constructive approach* is used to obtain a first feasible assignment of all lectures. In this phase, the chosen objective functions are not yet considered, but the method focuses on the hard constraints of the problem.

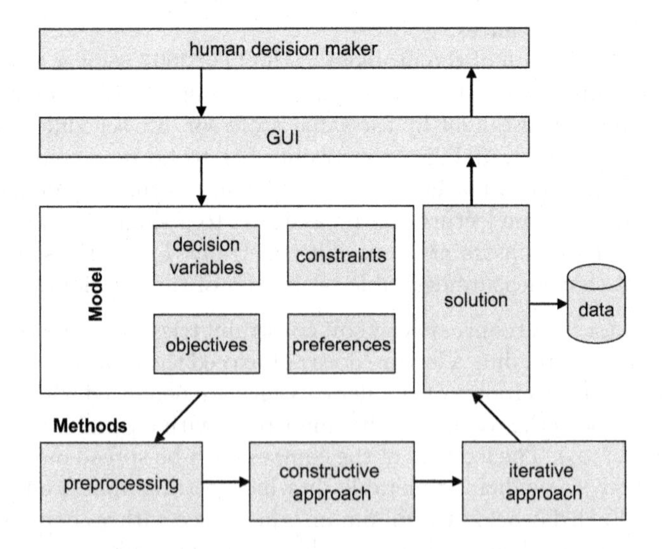

Fig. 1. Elements of the solution concept and their interaction

2. An *iterative approach* is then subsequently executed to improve the timetable obtained from the constructive approach. During this phase, reallocations of lectures are carried out with the final goal of identifying an optimal timetable.

Constructive and iterative phases follow subsequently. In this sense, the solution framework implements a two-phase-concept in which feasibility of the timetable is first considered, and optimality following later.

3.1 Preprocessing

Prior to the computation of a first solution, some preprocessing is carried out. In brief, some problem-specific characteristics are employed, adding some additional structure to the problem.

For each given lecture L_i, events $E_{i1}, \ldots, E_{ie}$ are created which are later assigned to timeslots. The number of events e is given in the problem instances. Creating events for each lecture leads to a more general problem description, and the solution approach only needs to concentrate on the assignment of all events, one to a single timeslot, as opposed to keeping track of assigning a lecture to e timeslots.

Second, we categorize for each lecture L_i (and thus for each event belonging to lecture L_i) the available rooms in three disjunct classes $\mathcal{R}_{i1}, \mathcal{R}_{i2}, \mathcal{R}_{i3}$.

$\mathcal{R}_{i1}$ refers to the rooms in which the lecture fits best, that is the rooms R_k with the minimum positive or zero value of $c_k - s_i$, c_k being the room capacity, s_i the number of students of lecture L_i. The class $\mathcal{R}_{i2}$ stores the rooms in which lecture L_i fits, that is $s_i < c_k$, but not best, and $\mathcal{R}_{i3}$ contains the rooms in which lecture L_i does not fit. With respect to the given problem statement, events of lectures may be assigned to timeslots of rooms in $\mathcal{R}_{i3}$, this however results in a penalty.

The underlying assumption of the classification of the rooms is that events are preferably assigned to timeslots belonging to a room of class $\mathcal{R}_{i1}$, followed by $\mathcal{R}_{i2}$ and $\mathcal{R}_{i3}$. It has to be mentioned however, that this cannot be understood as a binding, general rule but rather should be seen as a recommendation. A randomized procedure is therefore going to be implemented when assigning events to timeslots (see Section 3.3), allowing a certain deviation from the computed room order.

3.2 Constructive Approach

Based on the results for the initial constructive approach, we propose a reactive procedure that self-adapts to the set of unassigned events from previous runs. The logic behind this approach is that the constructive procedure 'discovers' events that are difficult to assign, giving them priority in successive runs. To some extend, we borrow ideas from the *squeaky wheel optimization* approach [9]. In this concept, alternatives are repetitively constructed and analyzed. Unfavorable aspects of the current solution are discovered in each analysis, which are then resolved in the successive iteration. In this sense, the 'squeaky wheel gets the grease'. Events that have not been assigned in previous construction runs are here considered to be the unfavorable aspects of the current solution.

In the following, let $\mathcal{E}^p$ be the set of prioritized events, $\mathcal{E}^{\neg p}$ the set of non-prioritized events, and $\mathcal{E}^u$ the set of events that have not been assigned during the construction phase. It is required that $\mathcal{E}^p \subseteq \mathcal{E}$, $\mathcal{E}^{\neg p} \subseteq \mathcal{E}$, $\mathcal{E}^p \cap \mathcal{E}^{\neg p} = \emptyset$, and $\mathcal{E}^p \cup \mathcal{E}^{\neg p} = \mathcal{E}$.

Algorithm 1 describes the reactive construction procedure.

Algorithm 1. Reactive construction

Require: $Maxloops$

1: Set $\mathcal{E}^p = \emptyset$, $\mathcal{E}^u = \emptyset$, $loops = 0$
2: **repeat**
3: $\quad \mathcal{E}^p \leftarrow \mathcal{E}^u$
4: $\quad \mathcal{E}^u \leftarrow \emptyset$
5: $\quad \mathcal{E}^{\neg p} \leftarrow \mathcal{E} \backslash \mathcal{E}^p$
6: $\quad$ **while** $\mathcal{E}^p \neq \emptyset$ **do**
7: $\quad\quad$ Select the most critical event E from $\mathcal{E}^p$, that is the event with the smallest number of available timeslots
8: $\quad\quad$ **if** E can be assigned to at least one timeslot **then**
9: $\quad\quad\quad$ Select some available timeslot T for E
10: $\quad\quad\quad$ Assign E to the timeslot T
11: $\quad\quad$ **else**
12: $\quad\quad\quad \mathcal{E}^u \leftarrow \mathcal{E}^u \cup E$
13: $\quad\quad$ **end if**
14: $\quad\quad \mathcal{E}^p \leftarrow \mathcal{E}^p \backslash E$
15: $\quad$ **end while**
16: $\quad$ **while** $\mathcal{E}^{\neg p} \neq \emptyset$ **do**
17: $\quad\quad$ Select the most critical event E from $\mathcal{E}^{\neg p}$, that is the event with the smallest number of available timeslots
18: $\quad\quad$ **if** E can be assigned to at least one timeslot **then**
19: $\quad\quad\quad$ Select some available timeslot T for E
20: $\quad\quad\quad$ Assign E to the timeslot T
21: $\quad\quad$ **else**
22: $\quad\quad\quad \mathcal{E}^u \leftarrow \mathcal{E}^u \cup E$
23: $\quad\quad$ **end if**
24: $\quad\quad \mathcal{E}^{\neg p} \leftarrow \mathcal{E}^{\neg p} \backslash E$
25: $\quad$ **end while**
26: $\quad loops \leftarrow loops + 1$
27: **until** $\mathcal{E}^u = \emptyset$ **or** $loops = Maxloops$

As given in the pseudo-code, the construction of solutions is carried out in a loop until either a feasible solution is identified or a maximum number of iterations $Maxloops$ is reached. When constructing a solution, a set of events $\mathcal{E}^u$ is kept for which no timeslot has been found. When reconstructing a solution, these events are prioritized over the others. In that sense, the constructive approach is biased by its previous runs, identifying events that turn out to be difficult to assign.

After a maximum number of at most *Maxloops* iterations, the construction procedure returns a solution that is either feasible ($\mathcal{E}^u = \emptyset$) or not ($\mathcal{E}^u \neq \emptyset$).

It becomes clear that the reactive procedure is principally based on the previous simple greedy heuristic. The choice of the most critical event, as well as the choice of the timeslot is left unchanged. Only the prioritization of the events by dividing them into two disjunctive subsets is an additional feature of the revised method.

3.3 Improvement Procedures

An iterative process continues with the alternative found in the constructive approach, searching for an optimal solution with respect to the soft constraint penalties.

In each step of the algorithm, a number of randomly chosen events is unassigned from the timetable and reinserted in the set $\mathcal{E}^u$. A reassignment phase follows. Contrary to the constructive approach, where events are selected based on whether they are critical with respect to the available timeslots, events are now randomly chosen from $\mathcal{E}^u$, each event with equal probability. The choice of the timeslot for the event is based on the logic described in the preprocessing phase, prioritizing timeslots of particular room classes. We use two possible preference structures of rooms, $\mathcal{R}_{i1}$ over $\mathcal{R}_{i2}$ over $\mathcal{R}_{i3}$, and $\mathcal{R}_{i2}$ over $\mathcal{R}_{i1}$ over $\mathcal{R}_{i3}$. Each of them is randomly chosen with probability 0.5.

The following different variants of local search have been implemented and tested:

- Hillclimbing (HC).
  In this local search variant, only improving reassignments of events are accepted. It can be expected that this strategy does not lead to the best results. However, for comparison reasons, an application will be interesting, simply because the effectiveness of alternative local search strategies can be studied in contrast to this relatively simple algorithm.
- Iterated Local Search (ILS).
  Iterated Local Search [10] is based on a hillclimbing algorithm, which is first used to compute a locally optimal solution. Then, after converging to this alternative, an escape mechanism is triggered, consisting of a worsening perturbation move by means of some neighborhood. Search continues from this perturbed alternative, again executing a hillclimbing run.
- Threshold Accepting (TA).
  The idea of Threshold Accepting has been introduced in [11]. It describes a deterministic variant of Simulated Annealing [12]. Worsening moves are accepted up to a certain threshold, thus allowing an escape of the search procedure from local optima. Throughout the execution of the local search approach, the threshold is subsequently decreased, similar to what is referred to as a 'cooling schedule' in Simulated Annealing.

  As previous research has shown that simplifications of Simulated Annealing may be very effective for timetabling problems [13, 14], we suspect that this approach turns out successful for the problem at hand.

4 Experimental Investigation

4.1 Weighted Sum Aggregation

For the first part of our experimental investigation, we consider a weighted sum aggregation of the objective functions sc_i as given in Expression (1). The values of the weights w_i have been chosen as proposed for the ITC 2007, thus $w_1 = 1$, $w_2 = 5$, $w_3 = 2$, $w_4 = 1$.

$$SC = \sum_{i=1}^{4} w_i sc_i \tag{1}$$

Different configurations of the algorithms have been tested on the benchmark data from the ITC 2007. The running time of each test run has been chosen in accordance with the regulations of the competition, allowing 375 seconds of computing time on an Intel Q6600 processor. While experiments with significantly longer running times are reported in [15], we here favor an experimental setting in which only comparably little running time is permitted. This reflects the practical circumstances of many real-world timetabling problems, in which a planner expects results in comparably short time.

The number of reassigned events in each iteration has been set to five for all variants of the improvement procedure. It should be noticed that other numbers from 1 to 10 have been tested, too. Based on some preliminary tests, in which a reassignment set of five events gave reasonable results for all benchmark instances, this number has been chosen and fixed in the following experiments.

Three configurations of the Iterated Local Search approach have been implemented. The first variant, ILS-3k, starts perturbing after 3,000 non-improving moves, the second, ILS-10k, after 10,000 moves, and the last one, ILS-25k, after 25,000 non-improving moves. Perturbations are done by a random reassignment of five events. Contrary to the usual acceptance rule with respect to the cost function SC, the perturbed alternative is accepted in any case, and search continues from this new solution.

Two different configurations of the Threshold Accepting algorithm have been tested. First, an algorithm with a threshold of 1% of the overall evaluation of the alternatives SC. Second, an algorithm using a threshold of 2% of SC. The choice of a percentage of SC as a threshold has the advantage that the algorithm performs an automated cooling when approaching small values of SC, yet maintaining high thresholds for large values of SC. This idea stands somewhat in contrast to other approaches, in which explicit cooling schedules are needed. Nevertheless, an appropriate choice of the percentage has to be made.

Average results of the metaheuristics
The following Table 1 gives the average values of the soft constraint penalties SC for the different local search strategies over ten independent test runs.

The hillclimbing algorithm never leads to best results, which already has been suspected prior to the investigation. For all benchmark instances, either an Iterated Local Search approach and/ or a Threshold Accepting algorithm leads to better average results.

Table 1. Average values for the weighted sum formulation

Instance	HC	ILS-3k	ILS-10k	ILS-25k	TA 1%	TA 2%
comp01	6.3	8.0	7.1	6.8	6.6	6.0
comp02	158.2	157.2	155.5	147.1	135.0	157.2
comp03	151.4	150.7	144.7	147.3	140.6	150.7
comp04	90.0	95.9	92.3	87.1	84.0	79.0
comp05	581.5	494.8	539.1	516.4	505.5	495.9
comp06	137.5	137.7	135.8	130.2	108.4	127.6
comp07	99.1	122.2	113.5	103.8	73.9	121.1
comp08	102.4	98.5	98.4	92.3	83.3	90.6
comp09	169.2	172.5	161.8	164.7	165.7	176.5
comp10	108.3	113.4	106.7	107.5	84.2	82.8
comp11	0.8	2.6	1.7	0.7	0.3	0.1
comp12	534.0	504.8	497.6	499.6	480.3	472.5
comp13	127.9	132.3	122.3	122.1	112.3	122.6
comp14	130.6	113.4	114.6	111.8	101.3	97.6

When analyzing and comparing the different Iterated Local Search approaches, the variant with a rather high number of evaluations before perturbing the solutions, ILS-25k, turns out to be superior to the other configurations. In nine of the 14 data sets, best results are obtained by this approach. Despite the fact that some counterexamples have been found, such as `comp05`, it is possible to conclude that a sufficient number of evaluations is needed before applying perturbations. In this context, 25,000 describes this 'sufficiently large number' better than 3,000 or 10,000.

Comparing the two Threshold Accepting algorithms, both TA 1% and TA 2% lead to best average result in seven of the 14 data sets. Also it can be seen, that some differences are rather small, such as in case of `comp01` and `comp11`, while others are considerable larger (`comp06`, `comp07`, `comp08`, `comp13`). On the basis of this observation, we conclude that the comparably smaller threshold of 1% leads to better results than the larger one of 2%.

For twelve out of 14 instances, Threshold Accepting proves to be superior to Iterated Local Search. Some counterexamples exist, but the overall conclusions are rather strong in favor of Threshold Accepting.

Best obtained results

Table 2 shows the best results of the Threshold Accepting algorithm with a threshold of 1%. The results are based on 30 trials with different random seeds.

It can be seen that the TA 1% approach leads to competitive results. For some instances, `comp01` and `comp11`, particularly good solutions are identified. Others such as `comp05` and `comp12` have best found alternatives with soft constraint penalties that are still quite large. We suspect that some properties of the data sets induce these results. Benchmark instance `comp05` possesses a rather small average teacher availability, and so does instance `comp12` [8]. It is therefore fair to assume that this property leads to the considerable differences in terms of

Table 2. Best results (out of 30 trials)

Instance	SC	w_1sc_1	w_2sc_2	w_3sc_3	w_4sc_4	Instance	SC	w_1sc_1	w_2sc_2	w_3sc_3	w_4sc_4
comp01	5	4	0	0	1	comp08	75	0	5	56	14
comp02	108	0	5	92	11	comp09	153	0	35	92	26
comp03	115	0	35	68	12	comp10	66	0	0	40	26
comp04	67	0	5	48	14	comp11	0	0	0	0	0
comp05	408	0	175	218	15	comp12	430	2	205	196	27
comp06	94	0	10	58	26	comp13	101	0	25	62	14
comp07	56	0	5	18	33	comp14	88	0	15	60	13

the best found values of SC to the other data sets. Obviously, relatively difficult side constraints complicate the identification of timetables with a small overall evaluation.

Another aspect of the obtained results is the detailed component-wise analysis of the individual objectives $sc_i, i = 1, \ldots, 4$. Recalling that the assignment of events to timeslots made use of a certain pre-computed order, we suspect that this may influence the characteristics of the results. In particular sc_1, which measures the assignment of lectures to rooms of smaller capacity, should be addressed rather well by this assignment logic.

A detailed overview of the best obtained results SC and the individual objective function values $sc_i, i = 1, \ldots, 4$ is given in Table 2. With the only counterexamples of instances comp01 and comp11, sc_1 turns out to be better addressed in comparison to the other objectives. In particular for sc_2 and sc_3, the numbers are rather relatively high. It also should be noticed, that the scoring for these two objectives employed higher penalties w_i as for the others. Nevertheless, and although a weighted sum approach has been used, a considerable bias in terms of a preference of sc_1 over the other objectives becomes obvious.

4.2 Experiments Based on a Reference Point

As pointed out when analyzing the result in Section 4.1, the weighted sum aggregation approach of the objective functions $sc_i, i = 1, \ldots, 4$ leads to a situation in which objective sc_1 is significantly better addressed in comparison to the others. It can be suspected that these results not always represent the individually desired solutions. Often, the human decision maker requires a different methodology for expressing his/ her preferences, for example by stating a reference point that represents the desired outcomes for each objective function. The solution of the problem then lies in the minimization of the distance of the solution to this point.

Expression (2) reformulates the aggregated evaluation of the timetables by introducing a reference point $R = (r_1, r_2, r_3, r_4)$. By minimizing SC^{ref}, we minimize the distance of the outcomes to the given reference values r_i. Using this approach, the decision maker may primarily guide the search towards a preferred solution by stating desired values of r_i.

$$SC^{ref} = \max_i w_i \left(sc_i - r_i \right) \tag{2}$$

The experiments as described in Section 4.1 have been repeated, now using the revised formulation of the aggregated evaluation function. Again, a hillclimbing algorithm, the three variants of the Iterated Local Search algorithm, and the two Threshold Accepting algorithms have been tested. Experiments have been conducted on the identical computer hardware, once more permitting a maximum running time of 375 seconds for each trial. The reference point has been assumed with $r_i = 0, i = 1, \ldots, 4$ as we know that 0 is the smallest possible outcome for each objective $sc_i, i = 1, \ldots, 4$.

Average results of the metaheuristics
Average results for ten independent test runs are given in the following Table 3.

Table 3. Average results of SC^{ref} in the reference point based formulation

Instance	HC	ILS-3k	ILS-10k	ILS-25k	TA 1%	TA 2%
comp01	4.0	7.0	6.2	5.2	4.0	4.0
comp02	72.1	75.0	66.4	72.1	69.0	66.9
comp03	64.7	68.9	66.4	64.3	64.8	66.3
comp04	41.5	51.3	47.9	47.5	38.2	46.6
comp05	248.2	232.1	236.6	226.5	238.2	226.0
comp06	58.5	71.0	65.7	63.9	60.1	80.2
comp07	50.0	73.5	64.8	61.9	71.0	135.8
comp08	41.5	57.0	50.5	50.2	48.6	59.9
comp09	71.3	79.3	76.5	74.1	73.1	79.5
comp10	44.0	59.6	57.0	53.5	54.8	88.0
comp11	0.1	7.2	5.2	4.8	0.0	0.0
comp12	248.5	240.8	240.1	221.0	228.4	211.0
comp13	55.0	65.0	61.0	60.7	59.2	62.6
comp14	50.5	57.7	54.1	52.0	50.8	52.1

For the Iterated Local Search algorithm, we observe again that the variant with the largest number of evaluations before perturbing the alternative, ILS-25k, obtains better results than the two configurations with a faster perturbation. Also, TA 1% leads in more instances to better results than the one with the larger threshold. In addition to that, Threshold Accepting outperforms Iterated Local Search in most instances. So far, the analysis is in line with the one of the weighted sum approach.

What really is remarkable is the behavior of the hillclimbing approach. In comparison to both the Iterated Local Search as well as the Threshold Accepting algorithm, better results are obtained in many cases. This is counterintuitive, especially as the algorithm does not possess any strategy of escaping local optima while the others do. It appears as if the presence of local optima is less problematic for the chosen evaluation function. In brief, and in contrast to the experiments of Section 4.1 we are able to observe that not a single approach is suitable for all possible data sets and variants of evaluation functions, as it has been already pointed out in other contexts [16].

Best obtained results
Another analysis is concerned with the best found solutions of the reference point based evaluation SC^{ref} and the values of the objectives $sc_i, i = 1, \ldots, 4$. The best solutions are reported in the following Table 4.

Table 4. Best results of SC^{ref} and the values for each objective sc_i (out of 10 trials)

Instance	SC^{ref}	w_1sc_1	w_2sc_2	w_3sc_3	w_4sc_4	Instance	SC^{ref}	w_1sc_1	w_2sc_2	w_3sc_3	w_4sc_4
comp01	4	4	0	2	4	comp08	40	0	40	40	40
comp02	45	0	45	44	45	comp09	66	0	65	66	66
comp03	52	0	50	52	52	comp10	35	0	35	34	35
comp04	30	0	30	30	30	comp11	0	0	0	0	0
comp05	190	190	190	190	41	comp12	200	182	200	198	80
comp06	55	0	55	54	55	comp13	45	0	45	44	44
comp07	44	27	40	44	44	comp14	40	0	40	40	40

Obviously, the values are better balanced in comparison to the ones of the weighted sum aggregation. While sc_1 still often has the smallest value, it is possible to see that for example for instance comp05, rather high values of sc_1 are achieved in order to minimize the other objective functions further, and so to minimize the aggregation. A similar observation can be made for the instances comp07 and comp12.

The results indicate that the prioritization of the timeslots when assigning events generally does favor objective sc_1 to some extent. However, there are cases in which the quality of timetables with respect to sc_1 is sacrificed for an improvement in the other objectives, and we may conclude that the solution approach does not generally overfit the investigated problem.

5 Conclusions

The article presented an optimization approach for multi-criteria timetabling problem with an application for curriculum-based course timetabling. On the basis of a general solution framework, a two-stage optimization approach has been proposed that first constructs feasible alternatives and subsequently searches for optimal solutions. For the latter purpose, different local search strategies have been implemented, and experiments have been conducted for benchmark instances taken from the literature.

By comparing a weighted sum and a reference point based approach, considerable differences in the obtained results became obvious. For the weighted sum aggregation, balancing the four objectives appears to be problematic. Even when assigning higher weights on particular objectives, they do not turn out to be better addressed than others. On the other hand, the reference point aggregation led to significantly better balanced outcomes with respect to an equal treatment of all four objectives.

In brief the experiments indicate that an interaction with the presented system employing reference points is more direct and overall preferable, a conclusion that similarly has been derived in [17]. Also, the relative performance of the different metaheuristics is affected by the chosen aggregation procedure. Although the results are overall competitive, as we have been able to demonstrate in the finals of the International Timetabling Competition ITC 2007, future research should therefore be dedicated towards the proposition of more robust solvers.

Besides, a comparison to multi-objective optimization approaches that approximate the whole Pareto-front should be carried out as part of future research. While the approach presented in this article aims to identify an alternative that optimizes the aggregated function only, the entire Pareto-front could be obtained for comparison reasons. Knowing that the weighted sum approach can be problematic, as it only allows the generation of supported efficient solutions, the effect and relevance of this potential drawback could further be explored.

Also, alternative approaches of integrating the reference point should be explored, such as described in e. g. [18]. Such approaches would allow to ensure the efficiency of the final solution, which is not necessarily the case in the implemented min-max achievement function.

References

1. Carter, M.W.: Timetabling. In: Gass, S., Harris, C. (eds.) Encyclopedia of Operations Research and Management Science, 2nd edn., pp. 833–836. Kluwer Academic Publishers, Dordrecht (2001)
2. Pinedo, M.: Scheduling: Theory, Algorithms, and Systems, 2nd edn. Prentice-Hall, Upper Saddle River (2002)
3. Landa-Silva, J.D., Burke, E.K., Petrovic, S.: An introduction to multiobjective metaheuristics for scheduling and timetabling. In: Gandibleux, X., Sevaux, M., Sörensen, K., T'kindt, V. (eds.) Metaheuristics for Multiobjective Optimisation. Lecture Notes in Economics and Mathematical Systems, vol. 535, pp. 91–129. Springer, Heidelberg (2004)
4. Burke, E.K., Petrovic, S.: Recent research directions in automated timetabling. European Journal of Operational Research 140(2), 266–280 (2002)
5. van den Broek, J., Hurkens, C., Woeginger, G.: Timetabling problems at the TU Eindhoven. European Journal of Operational Research (2008) (article in press)
6. Schimmelpfeng, K., Helber, S.: Application of a real-world university-course timetabling model solved by integer programming. OR Spectrum 29(4), 783–803 (2007)
7. Lewis, R.: A survey of metaheuristic-based techniques for university timetabling problems. OR Spectrum 30, 167–190 (2008)
8. Di Gaspero, L., McCollum, B., Schaerf, A.: The second international timetabling competition (ITC 2007): Curriculum-based course timetabling (track 3). Technical Report QUB/IEEE/Tech/ITC2007/CurriculumCTT/v1.0/1 (August 2007)
9. Joslin, D.E., Clements, D.P.: "Squeaky wheel" optimization. Journal of Artificial Intelligence Research 10, 353–373 (1999)
10. Lourenço, H.R., Martin, O., Stützle, T.: Iterated local search. In: Glover, F., Kochenberger, G.A. (eds.) Handbook of Metaheuristics. International Series in Operations Research & Management Science, vol. 57, pp. 321–353. Kluwer Academic Publishers, Dordrecht (2003)

11. Dueck, G., Scheuer, T.: Threshold accepting: A general purpose optimization algorithm appearing superior to simulated annealing. Journal of Computational Physics 90, 161–175 (1990)
12. Kirkpatrick, S., Gellat, C.D., Vecchi, M.P.: Optimization by simulated annealing. Science 220(4598), 671–680 (1983)
13. Burke, E.K., Bykov, Y., Newall, J.P., Petrovic, S.: A time-predefined approach to course timetabling. Yugoslav Journal of Operations Research (YUJOR) 13(2), 139–151 (2003)
14. Landa-Silva, D., Obit, J.H.: Great deluge with non-linear decay rate for solving course timetabling problems. In: Proceedings of the 2008 IEEE Conference on Intelligent Systems (IS 2008), pp. 8.11–8.18. IEEE Press, Los Alamitos (2008)
15. Geiger, M.J.: An application of the threshold accepting metaheuristic for curriculum based course timetabling. In: Proceedings of the 7th International Conference on the Practice and Theory of Automated Timetabling, Montréal, Canada (August 2008)
16. Wolpert, D.H., Macready, W.G.: No free lunch theorems for optimization. IEEE Transactions on Evolutionary Computation 1, 67–82 (1997)
17. Petrovic, S., Bykov, Y.: A multiobjective optimisation technique for exam timetabling based on trajectories. In: Burke, E.K., De Causmaecker, P. (eds.) PATAT 2002. LNCS, vol. 2740, pp. 149–166. Springer, Heidelberg (2003)
18. Wierzbicki, A.P.: The Use of Reference Objectives in Multiobjective Optimization. In: Fandel, G., Gal, T. (eds.) Multiple Criteria Decision Making Theory and Application, pp. 469–486. Springer, Heidelberg (1980)

Optimizing the DFCN Broadcast Protocol with a Parallel Cooperative Strategy of Multi-Objective Evolutionary Algorithms

Carlos Segura[1], Alejandro Cervantes[2], Antonio J. Nebro[3], María Dolores Jaraíz-Simón[4], Eduardo Segredo[1], Sandra García[2], Francisco Luna[3], Juan Antonio Gómez-Pulido[4], Gara Miranda[1], Cristóbal Luque[2], Enrique Alba[3], Miguel Ángel Vega-Rodríguez[4], Coromoto León[1], and Inés M. Galván[2,⋆]

[1] Department of Statistics, O.R. and Computation, University of La Laguna
csegura@ull.es
[2] Computer Science Department, University Carlos III of Madrid
[3] Computer Science Department, University of Málaga
[4] Department of Technologies of Computers and Communications, University of Extremadura

Abstract. This work presents the application of a parallel cooperative optimization approach to the broadcast operation in mobile ad-hoc networks (MANETs). The optimization of the broadcast operation implies satisfying several objectives simultaneously, so a multi-objective approach has been designed. The optimization lies on searching the best configurations of the DFCN broadcast protocol for a given MANET scenario. The cooperation of a team of multi-objective evolutionary algorithms has been performed with a novel optimization model. Such model is a hybrid parallel algorithm that combines a parallel island-based scheme with a hyperheuristic approach. Results achieved by the algorithms in different stages of the search process are analyzed in order to grant more computational resources to the most suitable algorithms. The obtained results for a MANETs scenario, representing a mall, demonstrate the validity of the new proposed approach.

1 Introduction

Mobile ad-hoc networks (MANETs) [1] are fluctuating, self-configuring networks of mobile hosts, called *nodes* or *devices*, connected by wireless links. This kind of network has numerous applications because of its capacity of auto-configuration and its possibilities of working autonomously or connected to a larger network. No static network infrastructure is needed to support the communications between nodes, which are free to move arbitrarily. Devices in MANETs are usually

⋆ This work has been supported by the EC (FEDER) and the Spanish Ministry of Education and Science inside the 'Plan Nacional de I+D+i' (TIN2005-08818-C04) and (TIN2008-06491-C04-02). The work of Gara Miranda has been developed under grant FPU-AP2004-2290.

M. Ehrgott et al. (Eds.): EMO 2009, LNCS 5467, pp. 305–319, 2009.

laptops, PDAs, or mobile phones, equipped with network cards featuring wireless technologies. This implies that devices communicate within a limited range and also that they can move while communicating.

Broadcasting is a common operation at the application level and also widely used for solving many network layer problems. It is expected to be performed very frequently, serving also as a last resort to provide multicast services. Hence, having a well-tuned broadcast strategy results in a major impact in network performance. The optimization implies satisfying several objectives simultaneously: the number of reached devices (*coverage*) must be maximized, a minimum usage of the network (*bandwidth*) is desirable, and the process must take a time as short as possible (*duration*). These objectives are conflicting among them, so we are dealing with a multi-objective optimization problem (MOP).

Since exact approaches are practically unaffordable for real world MOPs, a wide variety of approximated algorithms have been designed. Among them, metaheuristics are a family of techniques which have become popular to solve both single and multi-objective problems. They can be considered as high-level strategies that guide a set of simpler heuristic techniques in the search of an optimum [2]. Among these techniques, evolutionary algorithms for solving MOPs are very popular [3] giving raise to a wide variety of algorithms, such as NSGA-II [4] and SPEA2 [5]. Other family of metaheuristics widely applied in multi-objective optimization is particle swarm optimization or PSO [6].

This work presents an optimization of the broadcast operation for a real MANET instance. In order to provide an efficient, and robust approach, applicable to a wide range of problem instances, a new parallel evolutionary model has been applied[1]. The model is based on the hybridization of parallel island-based evolutionary algorithms and hyperheuristics. In particular, eight different multi-objective algorithms comprising genetic algorithms, differential evolution, evolutionary strategies, and PSO have been combined in the island scheme.

The remaining content is structured in the following way: Section 2 presents the broadcast optimization problem in MANETs. The sequential approaches applied in this work are presented in section 3. The proposed parallel model for multi-objective optimization is described in detail in section 4. The computational study is presented in section 5. Finally, the conclusions and some lines of future work are given in section 6.

2 Broadcast Operation in MANETS

This work focuses on the study of the broadcast operation in a particular kind of MANETs, the *metropolitan* MANETs. These MANETs have some specific features that hinders the testing in real environments: the network density is heterogeneous and it is continuously changing because devices in a metropolitan area move and/or appear/disappear from the environment. For this reason, many simulation tools have been developed [7]. In this work the *Madhoc* simulator [8]

[1] We will use the most familiar term *evolutionary algorithm* instead of *metaheuristic* throughout the paper, although in the algorithms we study there is a PSO.

was the choice. This tool provides a simulation environment for several levels of services based on different types of MANETs technologies and for a wide range of MANET real environments. It also provides implementations of several broadcast algorithms [9]. From the existing broadcast protocols, the *Delayed Flooding with Cumulative Neighbourhood* (DFCN) [10] has been selected because it was specifically designed to deal with metropolitan MANETs.

DFCN is a deterministic and totally localized algorithm. It uses heuristics based on the information from one hop. Thus, it achieves a high scalability. The behaviour of each device when using DFCN is driven by three events: the reception of a message (*reactive behaviour*), the expiration of the *random delay for rebroadcasting* (RAD) of a message, and the arrival of a new neighbour to its covered area (*proactive behaviour*). Although DFCN has shown good behaviour with metropolitan MANETs, the task of configuring such parameters is not trivial, and the proper operation of the protocol is sensitive to such configuration. The set of parameters that must be configured is:

- *minG*: minimum gain for forwarding a message.
- [*lowerRAD*, *upperRAD*]: range values for the RAD.
- *proD*: maximum density for which it is still necessary to use proactive behaviour for complementing the reactive behaviour.
- *safeDensity*: maximum density below which DFCN always rebroadcasts.

Given the values for the five DFCN configuration parameters and a MANET scenario, the *Madhoc* tool does the corresponding simulation and provides an estimate for the three objectives: duration, coverage, and bandwidth. One possibility to find the most suitable configuration is to systematically vary each of the five DFCN parameters. However, the possible parameter combinations are too large and evaluations in the simulator are computationally expensive. So, such technique is unable to obtain good quality solutions in a reasonable time. Other alternative relies on deeply analysing the problem to extract information to define a heuristic strategy, but the complexity and stochastic behaviour of the given problem hinders it. For these reasons, one usual way of affording this problem is through evolutionary techniques [11].

3 Applied Sequential Approaches

The aim of this section is to present the sequential algorithms used in this work for solving the proposed broadcast optimization problem. Algorithms previously used in [12], as well as other new alternatives has been applied. In this work, all these algorithms were also used in parallel, following some standard island-based models and applying the method explained in section 4.

3.1 Non-dominated Sorting Genetic Algorithm II (NSGA-II)

NSGA-II [4] is a non-dominated sorting based multi-objective evolutionary algorithm. Two of the most important characteristics which differentiates NSGA-II of

NSGA and other non-dominated sorting based approaches are the following. First, a fast non-dominated sorting approach with reduced computational complexity ($O(mN^2)$). Second, a selection operator which combines previous populations with new generated child populations, ensuring elitism in the approach.

The procedure is as follows: the two populations are sorted according to their rank, and the best solutions are chosen to create a new population. In the case of having to select some individuals with the same rank, a density estimation based on measuring the crowding distance to the surrounding individuals belonging to the same rank is used to get the most promising solutions.

Algorithm 1. NSGA-II Pseudocode

1: Initialization: Generate an initial population P_0 with N individuals. Assign $t = 0$.
2: **while** (not stopping criterion) **do**
3: Fitness assignment: Calculate fitness values of individuals in P_t. Use the non-domination rank in the first generation, and the crowded comparison operator in other generations.
4: Mating selection: Perform binary tournament selection on P_t in order to fill the mating pool.
5: Variation: Apply genetic operators to the mating pool to create a child population CP.
6: Combine P_t and CP selecting the best individuals using the crowding operator to constitute P_{t+1}.
7: $t = t + 1$
8: **end while**

3.2 Evolution Strategy with NSGA-II (ESN)

The ESN algorithm is based on the hybridization of Evolution Strategies and NSGA-II. The algorithm uses the standard Evolution Strategies' steps [13], replacing the selection process by the NSGA-II [4] selection process. The main difference between Evolution Strategies and Genetic Algorithms is that crossover operators are not used in ES, and each parent produces one offspring only by mutation. The mutation process implemented was the standard $(\mu + \lambda)$ process explained in [14], although in our case, $\lambda = \mu$.

Algorithm 2. ESN Pseudocode

1: Initialize population P of μ individuals
2: Initialize variance σ for each individual $I \in P$
3: **while** (not stopping criterion) **do**
4: $P' = \emptyset$
5: **for** each $I = (x_1, ..., x_n, \sigma) \in P$ **do**
6: $\sigma' = \sigma e^{N(0,\Delta)}$
7: Create $I' = (N(x_1, \sigma'), N(x_2, \sigma'), ..., N(x_n, \sigma'), \sigma')$
8: $P' = P' \cup \{I'\}$
9: **end for**
10: $P = P \cup P'$
11: Calculate front F_1 as Non-dominated individuals of P
12: **for** $i = 2$ to n **do**
13: Generate fronts F_i as Non-dominated individuals of $P \setminus (F_1 \cup ... \cup F_{i-1})$
14: **end for**
15: Sort solutions in each F_i $(i = 1, \ldots, n)$ using the crowding distance
16: Delete the worst μ individuals in population P
17: **end while**

3.3 Strength Pareto Evolutionary Algorithm 2 (SPEA2)

SPEA2 was proposed by Zitzler *et al.* [5]. This algorithm uses a population and an archive. It assigns to each individual a fitness value that is the sum of its strength raw fitness plus a density estimation. In each generation the non-dominated individuals of both the original population and the archive are used to update the archive; if the number of non-dominated individuals is greater than the population size, a truncation operator based on calculating the distances to the k-th nearest neighbor is used. All this procedure is known as *Environmental Selection.* Then, the algorithm applies the selection, crossover, and mutation operators to members of the archive in order to create a new population of offsprings which becomes the population of the next generation.

Algorithm 3. SPEA2 Pseudocode

1: Initialization: Generate an initial population P_0 and create the empty archive $\overline{P}_0$.
2: **while** (not stopping criterion) **do**
3: Fitness assignment: Calculate fitness values of individuals in P_t and $\overline{P}_t$.
4: Environmental selection: Copy nondominated individuals in P_t and $\overline{P}_t$ to $\overline{P}_{t+1}$. if $|\overline{P}_{t+1}| > \overline{N}$ reduce $\overline{P}_{t+1}$; otherwise, fill $\overline{P}_{t+1}$ with dominated individuals in P_t and P_{t+1}.
5: Mating selection: Perform binary tournament selection on $\overline{P}_{t+1}$.
6: Variation: Apply crossover and mutation operators to the mating pool and set $\overline{P}_{t+1}$ to the resulting population.
7: **end while**

3.4 Indicator-Based Evolutionary Algorithm (IBEA)

The IBEA algorithm [15] allows to define the optimization goal in terms of a performance measure or *quality indicator.* This measure is used directly for fitness calculation. The IBEA algorithm allows the use of different binary quality indicators. In this work the binary multiplicative ϵ-indicator [16] was used. There exists two versions of IBEA, the basic one and a more robust version known as adaptive. In the adaptive version, objectives values are normalized, and the indicator values are adaptively scaled. Both versions of the algorithm have been implemented.

Algorithm 4. IBEA Pseudocode (Adaptive Version)

1: Initialization: Generate an initial population P with N individuals.
2: **while** (not stopping criterion) **do**
3: Fitness assignment: calculate the fitness values using the quality indicator.
   1. Calculate indicator values $I(x^1, x^2)$ using the normalized objective values f_i' and determine the maximum absolute indicator value $c = max_{x^1,x^2 \in P}|I(x^1, x^2)|$.
   2. $\forall x^1 \in P, F(x^1) = \sum_{x^2 \in P \setminus \{x^1\}} -e^{-I(\{x^2\},\{x^1\})/(c \cdot k)}$.
4: Enviromental selection: until the size of P does not exceed N, remove the individual with the smallest fitness value, and recalculate the fitness value of the remaining individuals.
5: Mating selection: Perform binary tournament selection with replacement on P in order to fill the temporary mating pool P'.
6: Variation: Apply recombination and mutation operators to the mating pool P' and add the resulting offspring to P.
7: **end while**

3.5 Multiple Objective Particle Swarm Optimization (MOPSO)

MOPSO [17] is an adapted version of Particle Swarm Optimization (PSO) to multi-objective optimization problems. MOPSO combines PSO [18] with the archiving strategy of PAES [19]. In this work, we use the version available in the EMOO repository [20].

The algorithm uses an external repository that stores non-dominated solutions in the swarm. The velocity and position of particles are updated using the standard equations of PSO but using a leader particle selected from the repository, instead of best neighbor as it is usual. The mechanism of leader selection is as follows: the fitness space is divided in hypercubic sectors, and one of the positions in the repository is randomly selected using a roulette algorithm that favors the sectors that are less populated. Therefore, particles are attracted by leaders located in the areas where fewer non-dominated positions have been found. The mechanism for inclusion in the repository ensures that it only stores non-dominated solutions. If the new solution dominates some of the solutions in the repository, those solutions are removed. The maximum number of particles in the repository is fixed, so when this limit is reached, before the insertion, a particle from the most-populated sector of the repository is removed. Upon completion of the specified number of iterations, the set of solutions in the repository is reported as the Pareto front.

Algorithm 5. MOPSO Pseudocode

```
1: Initialize the swarm
2: Calculate fitness for each particle, store fitness and position as PBestFitness and PBestPosition
3: Store the position and fitness of non-dominated particles in the Repository
4: while (not stopping criterion) do
5:    for Particle do
6:       Select Leader from the repository
7:       Update NewVelocity using standard equations and Leader as neighbor.
8:       Update NewPosition using standard equations of PSO
9:       Calculate NewFitness
10:      if NewFitness dominates PBestFitness then
11:         PBestFitness ← NewFitness and PBestPosition ← NewPosition
12:         if NewFitness is not dominated by solutions in the Repository then
13:            if Repository is full then
14:               Remove one solution
15:            end if
16:            Insert NewPosition and NewFitness in the Repository
17:            Remove from the Repository solutions dominated by the one just inserted
18:         end if
19:      end if
20:   end for
21: end while
```

3.6 Multi-Objective Cellular Genetic Algorithm (MOCell)

MOCell [21] is a cellular genetic algorithm (cGA). As other multi-objective metaheuristics, it includes an external archive to store the non-dominated solutions found so far. This archive is bounded and uses the crowding distance of NSGA-II to keep diversity in the Pareto front.

We have used here an asynchronous version of MOCell, similar to the one called aMOCell4 in [22], in which the cells are explored sequentially (asynchronously). The selection is based on taking an individual from the neighborhood of the current cell and another one chosen from the archive. After applying the genetic crossover and mutation operators, the new offspring is compared with the current one, replacing it if better; in the case of both solution be non-dominated, the worst individual in the neighborhood is replaced by the current one. In this two cases, the new individual is inserted into the archive.

Algorithm 6. MOCell Pseudocode

```
1: population ← initialize()
2: archive ← NULL
3: while (not stopping criterion) do
4:    for individual ← 1 to population.size() do
5:       neighbours ←getNeighborhood(population, position(individual));
6:       neighbours.add(position(individual));
7:       parent1 ←selection(neighbours);
8:       parent2 ←selection(archive);
9:       offspring←recombination(Pc, parent1, parent2);
10:      offspring←mutation(Pm, offspring);
11:      evaluateFitness(offspring);
12:      replacement(position(individual), offspring);
13:      insertIntoArchive(offspring);
14:   end for
15: end while
```

3.7 Non-dominated Sorting Differential Evolution (NSDEMO)

Differential Evolution (DE) [23,24,25] is an evolutionary algorithm introduced by Storn and Price in 1995. DE was designed to optimize (single-objective) problems over continuous domains. In this paper, we extend DE to solve multiobjective optimization problems. Like NSDE [26], our approach is a multiobjective differential evolution algorithm based on NSGA-II [4]. We call it NSDEMO: Non-dominated Sorting Differential Evolution for Multiobjective Optimization.

This algorithm replaces the crossover and mutation operators of the NSGA-II with the DE scheme. In particular, the DE/*rand*/1/*bin* strategy has been applied. The mutation is scaled using the factor F, while the crossover is controled with the parameter CR.

Algorithm 7. NSDEMO Pseudocode

```
1: initializeParameters
2: createPopulation
3: evaluate
4: assignRank
5: while (not stopping criterion) do
6:    for i = 0 to PopulationSize − 1 do
7:       selectThreeVectorsParents
8:       mutationDE and crossoverDE
9:       addChildToPopulation
10:      evaluate
11:   end for
12:   assignRankCrowding
13: end while
```

4 Proposed Parallel Approach

Multi-objective metaheuristic approaches in general, and evolutionary algorithms (MOEAs) in particular [3] are proven to be effective when solving multi-objective optimization problems, but they can be time and domain knowledge intensive when applied to solve real world instances. Several studies have been performed in order to reduce the resource expenditure when using MOEAs. These studies naturally lead to considering the MOEAs parallelization. In the parallel MOEA (pMOEA) island-based model [27] the population is divided into a number of independent subpopulations. Each subpopulation is associated to an island and a MOEA configuration is executed over each of them. Usually, each available processor constitutes an island. Each island evolves in isolation, but occasionally some solutions can be migrated between neighbour islands. Island-based models have shown good performance and scalability in many areas. Four basic island variants are seen to exist: all islands execute identical MOEAs/parameters (homogeneous), all islands execute different MOEAs/parameters (heterogeneous), each island evaluates different objective functions subsets, or each island represents a different region of the genotype/phenotype domains.

If we were able of finding a particular MOEA that clearly outperformed the other ones in solving a given MOP, the homogeneous island-based model using such MOEA would be the choice to consider. However, it is difficult to know a priori which MOEA is the most appropriate to solve a problem. If we consider that, when dealing with a real world problem, its objective functions can require a significant amount of computing time to be evaluated, it can be hard to choose the MOEA to be the basis of the homogeneous approach. As an alternative, the heterogeneous models allow to execute different MOEAs and/or parameters on the islands. By using heterogeneous models, the user avoids the selection of a specific MOEA to solve the problem. However, if some of the used MOEAs are not suitable to optimize the problem, the consequence can be a waste of resources.

The existence of a wide variety of MOEAs in the literature and the dependence on the problem domain and instance in the performance of the approaches hinders the user decision about the algorithm to be applied. For this reason, a promising approach appears in the application of hyperheuristics [28]. The underlying principle in using a hyperheuristic approach is that different algorithms have different strengths and weaknesses and it makes sense to combine them in an intelligent manner. A hyperheuristic solves the problem indirectly by recommending which solution method to apply at which stage of the solution process. One of the motivations is that the same hyperheuristic method can be applied to a wide range of problems. The goal is to raise the level of generality of decision support methodology perhaps at the expense of reduced - but still acceptable - solution quality when compared to tailor-made evolutionary approaches.

The proposed pMOEA model [29,30] breaks from the island-based model adding an adaptive property behaviour to it. Such property allows, by applying a hyperheuristic, to change in an automatic and dynamic way the MOEAs and/or parameters that are used in the islands along the pMOEA run. To the best of our knowledge, the application of hyperheuristics into parallel schemes aimed at multi-objective optimization is a novelty. The architecture of the new hybrid model is similar to the island-based model, i.e., it is constituted by a set of *slave islands* that evolves in isolation by applying an evolutionary algorithm to a given population. The number of islands and the different MOEAs to execute over the local populations are defined by the user. Also, as in the island-based model, a tunable migration scheme allows the exchange of solutions between neighbour islands. However, a new special island, the *master island*, is introduced into the scheme. It is in charge of maintaining the global solution achieved by the pMOEA and selecting the MOEA configurations that are executed on the slave islands. The *global solution* is obtained by selecting the non-dominated solutions from the ones locally achieved by the slave islands. Usually, it is not desirable to manage a global solution with unlimited size, so the NSGA-II crowding operator [4] is proposed as the way to limit the size of the global solution set.

In standard island models, only a global stop criterion is defined. However, in the proposed model, local stop criteria are also defined for the execution of the MOEAs on the islands. When a local stop criterion is reached, the island execution is stopped and the local results are sent to the master island. The master scores, according to a quality indicator, the different configurations defined by the user taking into account their obtained results. A configuration is a MOEA together with the set of parameters that define such MOEA, e.g. the mutation and crossover rate, the population size, the archive size, etc. Based on such score, the hyperheuristic is applied and the master selects the configuration that will continue executing on the idle island. If the new selected configuration is the same as the island current configuration, the local stop criterion is updated and the execution continues. Otherwise, the configuration is changed and the new selected MOEA begins its execution by randomly taking the initial population individuals from the current global solution. Finally, when the global stop criterion is reached, every island sends its local solution to the master and all the local solution sets are considered to generate the global final solution.

One crucial point for the correct operation of the model is the selection process performed by the hyperheuristic. Considering the results obtained through the executions, it is beneficial to grant more opportunities to the configurations with better expectations. The decision process must be light in order to avoid having idle processes. One possibility to predict the behaviour of the configurations in a fast way is to pay attention to the contribution [31] of every configuration to the global solution. In this work, the score of each configuration is calculated as the contribution metric of such configuration - considering the current global solution as the reference front - divided by the number of evaluations that it has performed. A probabilistic selection, based on the score of each configuration, is used to decide the next configuration to execute. It is important to note that the behaviour of the configurations can change along the execution. Moreover, the stochastic behaviour of the evolutionary approaches may lead to variations in the results achieved by each algorithm. Therefore, a non-promising configuration must have a low probability of being selected, but not a zero probability. Probabilistic selection methods are more conservative than elitist ones, tending to distribute the resources in a more uniform way. Probabilistic selection methods reduce the negative impact that a not accurate scoring method can introduce in the results. Even when the scoring method fails, some resources will be granted to good-behaved configurations, speeding up the optimization process. On the other hand, when the scoring method works correctly, some resources will be granted to non-promising configurations, slowing down the optimization. Preliminary studies shows that, in general, the usage of probabilistic methods is more suitable than the usage of elitist methods.

5 Experimental Evaluation

Initial experiments showed an irregular behaviour of different MOEAs when dealing with different MANET scenarios. In order to avoid the testing of many algorithms for solving each instance the adaptive model can been applied. Results for a mall MANET scenario demonstrate the validity of the approach. All the MOEAs presented in section 3 have been used inside the model. Tests have been run on a Debian GNU/Linux cluster of 8 Intel® Xeon^{TM} 3.20 Ghz bi-processor nodes with 1Gb RAM. The interconnection network is a Gigabit Ethernet. The compiler and MPI implementation used were *gcc 3.3* and MPICH *1.2.7*.

The new model has been compared with sequential MOEAs and with other standard pMOEAs. For each implemented MOEA a homogeneous scheme is considered: "*homo*-SPEA2", "*homo*-NSGA-II", "*homo*-IBEA", "*homo-adap*-IBEA", "*homo*-ESN", "*homo*-MOPSO", "*homo*-MOCell", and "*homo*-NSDEMO". Also, a heterogeneous scheme constituted by the eight implemented MOEAs is considered: "*heterogeneous*". The new proposal, labelled as "*8-adaptive*", also uses the eight different MOEAs but following the adaptive behaviour.

Each tested pMOEA is constituted by eight slave islands. The subpopulation size on each island has been fixed to 15 individuals, while the population size for every sequential execution has been fixed to 100 individuals. The maximum size of the external set for those algorithms maintaining an archive was fixed to 100 individuals in the sequential experiments and to 15 individuals in the parallel ones. The remaining parameterization of each MOEA was as follows:

- SPEA2: $p_m = 0.2$, $p_c = 0.9$
- NSGA-II: $p_m = 0.2$, $p_c = 0.9$
- IBEA, adaptive-IBEA: $p_m = 0.2$, $p_c = 0.9$, k = 0.002
- ESN: $\sigma = 0.1$
- MOPSO: $p_m = 0.2$, divisions in archive = 30
- MOCell: $p_m = 0.2$, $p_c = 0.9$
- NSDEMO: $F = 0.5, CR = 1.0$

For every algorithm not doing special emphasis on the mutation or crossover operators, a polynomial mutation [32] ($\eta_m = 20$) and a simulated binary crossover (SBX) [33] ($\eta_c = 20$) were applied. In every execution the same migration scheme is specified. It consists in an unrestricted topology where the migration is performed from a slave to a randomly selected partner. The migration probability has been fixed to 0.05 and the number of individuals to migrate was limited to 4 each time. The global stopping criterion for every execution was 25000 evaluations. The

Table 1. Average hypervolume achieved by the different pMOEAs

Parallel Model	Evaluations limit		
	5000	10000	25000
8-adaptive	0.723	0.735	0.742
heterogeneous	**0.711**	**0.721**	**0.729**
homo-SPEA2	0.713	**0.724**	0.738
homo-NSGA-II	**0.712**	**0.723**	0.739
homo-IBEA	0.715	**0.724**	0.733
homo-adap-IBEA	0.716	0.725	0.733
homo-ESN	**0.707**	**0.718**	**0.726**
homo-MOPSO	**0.682**	**0.683**	**0.685**
homo-MOCell	**0.690**	**0.702**	**0.712**
homo-NSDEMO	**0.713**	**0.720**	**0.727**

local stopping criterion in 8-adaptive executions was fixed to 15 generations. In all cases, the final solution was limited to 100 elements.

The first experiment compares the different aforementioned pMOEAs among them. For each type of execution, 30 repetitions have been performed and average values considered. In order to detect differences between the algorithms within short and long time ranges, three different number of individual evaluation limits have been considered: 5000, 10000 and 25000 evaluations. The computing time of a sequential execution with 25000 evaluations is approximately 50 hours. Table 1 shows the average hypervolume [34] achieved by each parallel model at the given limits. The hypervolume indicator makes possible to combine the quality information of convergence and diversity in a single value. The 8-adaptive configuration achieves the best results in every case. The dynamic mapping of the MOEAs into the islands allows to give more computational resources to the most suitable algorithms, thus improving the results of the heterogeneous model. Moreover, the simultaneous usage of different evolutionary algorithms makes possible to combine the benefits of each one, so that, the results of every homogeneous island-based model is improved.

In order to provide the results with confidence, the following statistical analysis has been performed [35,36]. First, a Kolmogorov-Smirnov test is performed in order to check whether the results follow a normal (gaussian) distribution. Every sample passes the normality test. The homogeneity of the variances for each pair of samples is ensured through the Levene test. Finally, the ANOVA test is passed to check the confidence levels. As our interest is focused in the new model, it has been statistically compared with the remaining pMOEAs. Table 1 shows data in bold when differences between such model and the new proposed model are significant. The new model achieves a better hypervolume in every case. In the case of 10000 evaluations, all differences are significant, except with *homo-adap-IBEA*, showing the good performance of the approach. Figure 1 presents, for each model, the average hypervolume achieved along the executions.

The second experiment analyzes the run-time behaviour of the sequential and adaptive models. The ideas presented in [37] were followed. Each MOEA, as well

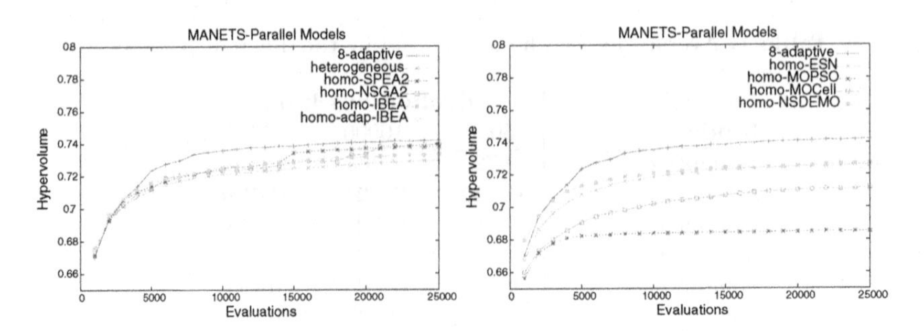

Fig. 1. Hypervolume achieved by the parallel models

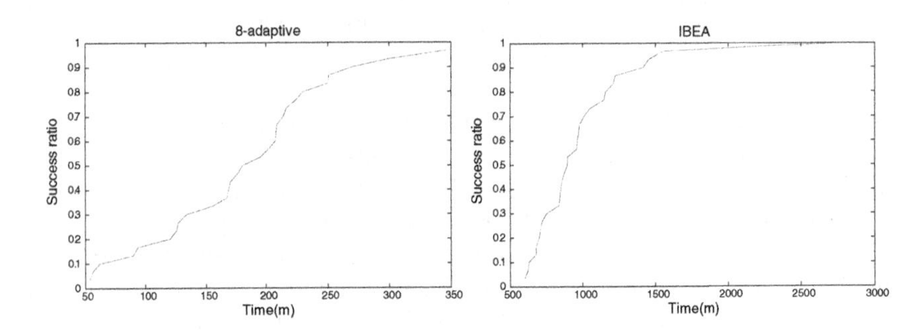

Fig. 2. Run length distributions

Table 2. Speedup of the proposed model and success ratio for sequential models

Sequential Models	8-adaptive Speedup	Success Ratio
SPEA2	7.25	80
NSGA-II	8.8	86.6
IBEA	5.57	100
adap-IBEA	5.09	86.6
ESN	9.75	83.3
MOPSO	-	0
MOCell	-	0
NSDEMO	11.59	36.6

as the new model, were executed using as finalization condition the achievement of a certain level of hypervolume quality: the 95% of the average achieved by the adaptive model in the first experiment. A second stopping criterion, consisting in executing a maximum number of 25000 evaluations was also considered. Figure 2 shows the run length distribution for the adaptive pMOEA and for the best behaved sequential MOEA (IBEA). For the remaining models a summary of the obtained information is presented in Table 2. For each sequential execution, the table shows the average speedup of the new model when the required quality is achieved, together with the success ratio, i.e. the probability of achieving the required hypervolume value, considering a maximum of 25000 evaluations. The

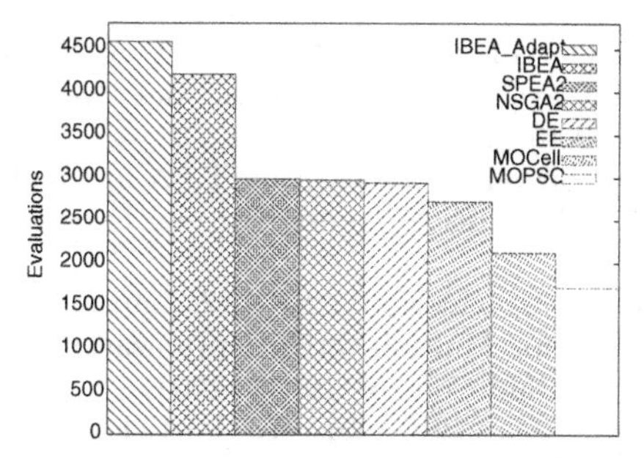

Fig. 3. Distribution of evaluations for the 8-adaptive model

parallel model obtains an almost linear speedup even when compared with the best MOEA, and a very high ratio of achieving the desired quality (96.6%).

Figure 3 shows, for the 30 executions of the 8-adaptive model, the average number of evaluations executed by each configuration. By comparing it with results in Table 2, it is clear that more computational resources are granted to the algorithms with best behaviour. IBEA and adap-IBEA, the best-behaved sequential algorithms, are the most used configurations, while MOPSO and MOCell, the worst-behaved sequential algorithms, are the least used configurations.

6 Conclusions and Future Work

An optimization approach for the broadcast operation in MANETs based on the DFCN protocol has been presented. In order to increase the level of generality of the proposed solution - so it can be applied to any MANET scenario - a new hybrid model, which adds an adaptive property to the well known island-based models by applying the operation principles of the hyperheuristics, was applied. The adaptive property allows to dynamically grant more computational resources to the most promising algorithms. Results achieved for a mall MANET scenario demonstrate the positive effect introduced by the hybridization. The new model provides high-quality solutions without forcing the user to have a prior knowledge about each MOEA behaviour when applied to any considered problem instance. The DFCN configurations obtained by the new parallel model clearly improve the ones obtained sequentially.

Future work is related to two different fields: the improvement of the broadcast in MANETs and the improvement of the designed optimization model. Other broadcast protocols can be considered to solve the same problem instances. In relation to the hybrid model, other MOEAs and even other kind of multi-objective optimization algorithms could be incorporated to the scheme. Some further studies concerning the model self-adaptation can be performed. In particular, other ways to measure the algorithms quality can be tested.

References

1. Macker, J., Corson, M.: Mobile Ad Hoc Networking and the IETF. ACM Mobile Computing and Communications Review 2(1) (1998)
2. Blum, C., Roli, A.: Metaheuristics in Combinatorial Optimization: Overview and Conceptual Comparison. ACM Computing Surveys 35(3), 268–308 (2003)

3. Coello, C.A.C., Lamont, G.B., Van Veldhuizen, D.A.: Evolutionary Algorithms for Solving Multi-Objective Problems (Genetic and Evolutionary Computation). Springer, New York (2006)
4. Deb, K., Pratap, A., Agarwal, S., Meyarivan, T.: A fast and elitist multiobjective genetic algorithm: NSGA-II. IEEE Transactions on Evolutionary Computation 6, 182–197 (2002)
5. Zitzler, E., Laumanns, M., Thiele, L.: SPEA2: Improving the Strength Pareto Evolutionary Algorithm for Multiobjective Optimization. Evolutionary Methods for Design, Optimization and Control, 19–26 (2002)
6. Reyes-Sierra, M., Coello, C.: Multi-Objective Particle Swarm Optimizers: A Survey of the State-of-the-Art. International Journal of Computational Intelligence Research 2(3), 287–308 (2006)
7. Hogie, L., Bouvry, P., Guinand, F.: An Overview of MANETs Simulation. Electronics Notes in Theorical Computer Science 150(1), 81–101 (2006)
8. Hogie, L.: Mobile Ad Hoc networks: modelling, simulation and broadcast-based applications. PhD thesis, Le Havre University and Luxembourg University (2007)
9. Williams, B., Camp, T.: Comparison of broadcasting techniques for mobile ad hoc networks. In: Proceedings of the ACM International Symposium on Mobile Ad Hoc Networking and Computing, pp. 194–205 (2002)
10. Hogie, L., Seredynski, M., Guinand, F., Bouvry, P.: A Bandwidth-Efficient Broadcasting Protocol for Mobile Multi-hop Ad hoc Networks. In: 5th International Conference on Networking (ICN 2006). IEEE, Los Alamitos (2006)
11. Alba, E., Dorronso, B., Luna, F., Nebro, A.J., Bouvry, P., Hogie, L.: A Cellular Multi-Objective Genetic Algorithm for Optimal Broadcasting Strategy in Metropolitan MANETs. Computer Communications 30(4), 685–697 (2007)
12. Alba, E., Cervantes, A., Gómez, J., Isasi, P., Jaraíz, M., León, C., Luque, C., Luna, F., Miranda, G., Nebro, A., Pérez, R., Segura, C.: Metaheuristic approaches for optimal broadcasting design in metropolitan mANETs. In: Moreno Díaz, R., Pichler, F., Quesada Arencibia, A. (eds.) EUROCAST 2007. LNCS, vol. 4739, pp. 755–763. Springer, Heidelberg (2007)
13. Bäck, T., Schwefel, H.: Evolutionary algorithms: Some very old strategies for optimization and adaptation. In: New Computing Techniques in Physics Research II: Proceedings of the Second International Workshop on Software Engineering, Artificial Intelligence, and Expert Systems for High Energy and Nuclear Physics, pp. 247–254 (1992)
14. Bäck, T., Rüdolph, G., Schwefel, H.: A survey of evolution strategies. In: Proceedings of the 4th International Conference on Genetic Algorithms, pp. 2–9 (1991)
15. Zitzler, E., Künzli, S.: Indicator-Based Selection in Multiobjective Search. In: Yao, X., Burke, E.K., Lozano, J.A., Smith, J., Merelo-Guervós, J.J., Bullinaria, J.A., Rowe, J.E., Tiňo, P., Kabán, A., Schwefel, H.-P. (eds.) PPSN 2004. LNCS, vol. 3242, pp. 832–842. Springer, Heidelberg (2004)
16. Zitzler, E., Thiele, L., Laumanns, M., Fonseca, C.M., Grunert da Fonseca, V.: Performance Assessment of Multiobjective Optimizers: An Analysis and Review. IEEE Transactions on Evolutionary Computation 7(2), 117–132 (2003)
17. Coello, C.A., Toscano, G., Salazar, M.: Handling multiple objectives with particle swarm optimization. IEEE Transactions on Evolutionary Computation 8(3), 256–279 (2004)
18. Kennedy, J., Eberhart, R., Shi, Y.: Swarm intelligence. Morgan Kaufmann Publishers, San Francisco (2001)
19. Knowles, J.D., Corne, D.W.: Approximating the nondominated front using the pareto archived evolution strategy. Evolutionary Computation 8(2), 149–172 (2000)

20. Coello, C.A., et al.: EMOO Repository, http://www.lania.mx/~ccoello/EMOO
21. Nebro, A.J., Durillo, J.J., Luna, F., Dorronsoro, B., Alba, E.: A cellular genetic algorithm for multiobjective optimization. In: Pelta, D.A., Krasnogor, N. (eds.) Proceedings of the Workshop on Nature Inspired Cooperative Strategies for Optimization (NICSO 2006), Granada, Spain, pp. 25–36 (2006)
22. Nebro, A.J., Durillo, J.J., Luna, F., Dorronsoro, B., Alba, E.: Design issues in a multiobjective cellular genetic algorithm. In: Obayashi, S., Deb, K., Poloni, C., Hiroyasu, T., Murata, T. (eds.) EMO 2007. LNCS, vol. 4403, pp. 126–140. Springer, Heidelberg (2007)
23. Price, K., Storn, R., Lampinen, J.A.: Differential Evolution: A Practical Approach to Global Optimization. Springer, Heidelberg (2006)
24. Storn, R., Price, K.: Differential Evolution - A Simple and Efficient Heuristic for Global Optimization over Continuous Spaces. J. of Global Optimization 11(4), 341–359 (1997)
25. Storn, R.: System design by constraint adaptation and Differential Evolution. IEEE Transactions on Evolutionary Computation 1(3), 22–34 (1999)
26. Iorio, A.W., Li, X.: Solving rotated multi-objective optimization problems using differential evolution. In: Webb, G.I., Yu, X. (eds.) AI 2004. LNCS (LNAI), vol. 3339, pp. 861–872. Springer, Heidelberg (2004)
27. Van Veldhuizen, D.A., Zydallis, J.B., Lamont, G.B.: Considerations in engineering parallel multiobjective evolutionary algorithms. IEEE Trans. Evolutionary Computation 7(2), 144–173 (2003)
28. Burke, E.K., Landa, J.D., Soubeiga, E.: Hyperheuristic Approaches for Multiobjective Optimisation. In: Metaheuristics International Conference, pp. 11.1–11.6 (2003)
29. León, C., Miranda, G., Segura, C.: Parallel Hyperheuristic: A Self-Adaptive Island-Based Model for Multi-Objective Optimization. In: Genetic and Evolutionary Computation Conference, pp. 757–758. ACM, New York (2008)
30. León, C., Miranda, G., Segura, C.: A Parallel Plugin-Based Framework for Multi-objective Optimization. In: International Symposium on Distributed Computing and Artificial Intelligence, vol. 50/2009, pp. 142–151. Springer, Heidelberg (2008)
31. Meunier, H., Talbi, E.G., Reininger, P.: A multiobjective genetic algorithm for radio network optimization. In: Congress on Evolutionary Computation (CEC 2000), La Jolla Marriott Hotel La Jolla, California, USA, pp. 317–324. IEEE Press, Los Alamitos (2000)
32. Deb, K., Goyal, M.: A combined genetic adaptive search (geneAS) for engineering design. Computer Science and Informatics 26(4), 30–45 (1996)
33. Deb, K., Agrawal, R.B.: Simulated binary crossover for continuous search space. Complex Systems 9, 115–148 (1995)
34. Zitzler, E., Thiele, L.: Multiobjective optimization using evolutionary algorithms - A comparative case study. In: Eiben, A.E., Bäck, T., Schoenauer, M., Schwefel, H.-P. (eds.) PPSN 1998. LNCS, vol. 1498, pp. 292–301. Springer, Heidelberg (1998)
35. Demšar, J.: Statistical comparisons of classifiers over multiple data sets. Journal of Machine Learning Research 7, 1–30 (2006)
36. Sheskin, D.: The handbook of parametric and nonparametric statistical procedures. CRC Press, Boca Raton (2003)
37. Hoos, H., Informatik, F., Hoos, H.H., Stutzle, T., Stutzle, T., Intellektik, F., Intellektik, F.: On the run-time behavior of stochastic local search algorithms for sat. In: Proceedings AAAI 1999, pp. 661–666 (1999)

Evolutionary Multiobjective Optimization for Dynamic Hospital Resource Management

Anke K. Hutzschenreuter[1,2], Peter A.N. Bosman[2], and Han La Poutré[1,2]

[1] Department Industrial Engineering & Innovation Sciences, Eindhoven University of Technology, The Netherlands
A.K.Hutzschenreuter@tue.nl

[2] Center for Mathematics and Computer Science, Amsterdam, The Netherlands
Peter.Bosman@cwi.nl, Han.La.Poutre@cwi.nl

Abstract. Allocating resources to hospital units is a major managerial issue as the relationship between resources, utilization and patient flow of different patient groups is complex. Furthermore, the problem is dynamic as patient arrival and treatment processes are stochastic. In this paper we present a strategy optimization approach where the parameters of different strategies are optimized using a multiobjective EDA. The strategies were designed such that they enable dynamic resource allocation with an offline EDA. Also, the solutions are understandable to health care professionals. We show that these techniques can be applied to this real-world problem. The results are compared to allocation strategies used in hospital practice.

1 Introduction

Today, many hospitals face great demands to reduce costs and improve quality of service, e.g. by reducing patient waiting times. In several European countries this is due to the introduction of a free market health care system, like in the Netherlands. In order to decrease costs, the occupancy rates of resources need to be increased. Increasing resource utilization, however, may lead to bottlenecks that cause the blocking of patient flows and thus increase patient waiting times. Therefore, the efficient allocation of resources is an important issue.

Hospital resource management is concerned with the efficient and effective deployment of resources, i.e. operating rooms and beds, when and where they are needed. In many hospitals, this is a major managerial issue, especially due to the complex relationship between resources, utilization and patient throughput for different patient groups[1]. Moreover, the problem is stochastic as resource usage at a hospital unit behaves like a stochastic process. Emergency patients arrive in urgent need for care, complications require patient transfers and the patients' length of stays are stochastic. Different patient treatment processes need to be considered that typically involve several hospital units. Often, resources (e.g. at the Intensive Care unit) are shared by multiple treatment processes. Thus, hospital resource management is a complex and highly dynamic problem.

M. Ehrgott et al. (Eds.): EMO 2009, LNCS 5467, pp. 320–334, 2009.

For the optimization of resource management three outcome measures are of interest to the hospital: patient throughput, i.e. the number of patients discharged from the hospital after treatment, resource costs and back-up capacity usage. In order to accommodate patients at the appropriate care level, a hospital unit may open an extra bed or transfer a patient temporarily to another unit until a bed becomes available. A well-designed hospital resource allocation features high patient throughput at low resource costs and back-up capacity usage. Previous work [2] showed that a trade-off is needed between these conflicting objectives.

Due to the stochastic patient processes and the actual patient flow being the result of resource availability, an analytical evaluation of a resource allocation is not feasible. Furthermore, changing the structure of the patient pathways or the underlying probability distributions is non-trivial in an analytical model. Therefore, the simulation tool described in [2] is important to be used for the evaluation of a resource allocation. Moreover, the decision space comprises allocations for each unit in a hospital. Due to the need of a complex simulation tool for evaluation, the huge decision space and multiple conflicting objectives, evolutionary algorithms (EAs) were chosen as solution technique, as they have been shown to be very powerful for multi-objective (MO) optimization [3,4,5].

For optimizing hospital resource management, we apply strategy (or policy) optimization, as advocated in [6]. Policies are parameterized functions that return an allocation decision for any given situation. The strategies' parameters are optimized using the EA. The advantage of using policies to solve stochastic dynamic optimization problems is that only one strategy has to be optimized that can be applied to a set of scenarios in the simulation. In cooperation with domain experts from the Catharina Hospital Eindhoven (CHE), the Netherlands, we designed strategies that enable dynamic resource allocations. The strategies can be easily understood by health care professionals which is important for the implementation and understanding in practice.

Thus, hospital resource management is a complex and dynamic problem that requires state-of-the-art techniques from dynamic MO research. Specifically, we combine strategy optimization with the SDR-AVS-MIDEA algorithm [7], an Estimation-of-Distribution (EDA) algorithm. The algorithm uses mixture distributions to stimulate the search for a broad Pareto-front and additionally contains techniques to prevent premature convergence (SDR-AVS). We demonstrate the applicability of these techniques to a real-world problem and their effectiveness.

Only few papers have addressed dynamic MO optimization, especially in stochastic environments. The approach presented in [5] is developed for seldom random changes of the environment and requires optimization from scratch if a change in the environment is detected. Our approach uses strategy optimization and therefore does not need to be re-optimized for each situation. Moreover, it can handle also frequent changes of the environment because the strategies describe what to do in any situation. In [4] the performance of the Non-dominated Sorting Genetic Algorithm version 2 (NSGA2) is evaluated for artificial objective functions. In our work, we use objective functions for a real-world application.

Work on hospital resource management can be found in the Operations Research and Operations Management literature. The models mainly focus on aggregated resource allocation policies, e.g. [1,8], or allocation policies for single units, e.g. [9,10]. Our approach allows for an in-depth analysis of allocation strategies also on the level of different hospital units. Furthermore, their work solely addresses static allocations whereas we consider also the optimization of dynamic resource allocation. The work in [11] provides theoretical results for hospital bed utilization. Our approach is more flexible and can easily be adopted to other settings. Moreover, earlier work considered hospital resource management as a single-objective optimization problem. In [12], the MO optimization problem is addressed. The model, however, is restricted to deterministic patient treatment processes. In our approach, we incorporate stochastic treatment processes that can be flexibly adjusted to other settings.

The remainder is organized as follows. First, we provide a model of the hospital domain and a description of the resource allocation problem in Section 2. Next, the allocation strategies and the algorithm used in our approach are presented in sections 3 and 4. The experiments are reported in Section 5. Finally, in Section 6 we provide our conclusions and an outlook on future work.

2 Simulation Model and Optimization Problem

2.1 Simulation Model

The simulation tool for hospital resource allocation is based on a case study at the CHE. The following features are included: patient characteristics influencing the patients' priority and pathway in the hospital and uncertainty related to the pathways. The model is described below. For a more detailed description, the reader is referred to [2].

Hospital care units. In general, a hospital can be divided into several, medically specialized, care units [13]. The units like nursing wards provide treatment and monitoring and are typically dedicated to a medical specialty such as cardiothoracic surgery (CTS). The operating room (OR) is typically shared by different specialties which are assigned time slots for performing surgical procedures (indicated by a prefix). The intensive care unit (ICU) is often divided into several subunits providing patient care and monitoring with different intensity. We distinguish between intensive care (IC), high care (HC) and medium care (MC). The post anesthesia care unit (PACU) is dedicated to patients recovering from anesthesia. The set of care units relevant for the simulation model is denoted by U with U={CTS-OR, IC, IC-HC, MC, CTS-HC, CTS-PACU, CTS ward, o}. o denotes the possible destinations of a patient's discharge from the hospital which are home or other care facilities, but also mortality.

Patient pathways and scheduling. We distinguish between scheduled patients (i.e. elective surgical patients from the waiting lists) and non-scheduled patients (i.e. emergency patients in urgent need for surgical and/or intensive

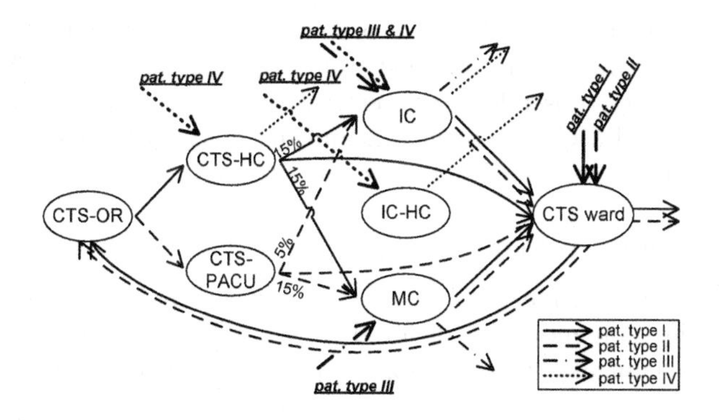

Fig. 1. Interference of CTS, other surgical and emergency patient pathways[1]

care). The set of patient groups is denoted by Θ. The Poisson arrival rate of patients is given as $\lambda_g, g \in \Theta$. We define a patient pathway of group $g \in \Theta$ as the sequence of actually required treatment operations and the respective length of stay (LoS). The patient process represents all possible pathways of patient type $g \in \Theta$ and is modeled by a probabilistic graph [14], $G^g = (V^g, A^g, P^g)$, where the set of nodes, $V^g \subset U$, represents the involved hospital units and the set of arcs, A^g, represents the possible adjacent treatment operations. The length of stay of a patient of group $g \in \Theta$ at hospital unit $u \in V^g$ is modeled as a random variable, LoS_u^g, that follows a probability distribution $P^{LoS_u^g}$. P^g is the set of conditional probability distributions defined on A^g with

$$P^g = \{Pr(v|u, g, t)|u \in V^g, (u, v) \in A^g, t \geq 0\} \text{ for } g \in \Theta. \tag{1}$$

$Pr(v|u, g, t)$ represents the probability that care provided at unit v is required given that a patient of group g has been admitted to unit u for t time units.

Resources are required to perform treatment operations at a hospital unit. Here, relevant resources are ORs and hospital beds. Often, hospital units operate autonomously which means that schedules and resources are managed locally by the units each applying their own (medical) priorities and preferences. The simulation model reflects the distributed decision making by representing each relevant unit by an agent. The policies for scheduling patient admissions and transfers implemented in the agent-based simulation tool were derived from the CHE case study. A detailed description is given in [2].

For the simulation we use four types of patient pathways (type I to IV) that were identified in the CHE case study for the CTS. Type I and II patients are CTS patients, for whom the immediate postoperative care is a priori indicated as CTS-HC and CTS-PACU, respectively. The type III pathway corresponds to the treatment process of emergency patients who arrive unexpectedly. The type IV patient path represents the inflow of other surgical patients in the system.

Figure 1 shows the four types of patient pathways[1]. The routing probabilities are indicated on the arrows where differing from 100%. Type III and IV pathways are restricted to their possible interference with type I and II patients at IC, IC-HC, CTS-HC and MC. The preceding and successive treatment steps are not considered because other dedicated resources are used.

2.2 Optimization Problem

In the following we consider a time horizon T with discrete time units t and n equidistant decision moments denoted by $t_i \in T' \subset T$ with $t_{i-1} < t_i$ for $i = 1, \ldots, n-1$. Typically, t would be in steps of hours and t_i would be in steps of days.

Decision variables & parameters. In the simulation model described above, we consider the number of allocated resources as free decision variables (i.e. control variables that impact the performance of the system). Formally, an allocation policy, $\pi(t_i) = (\pi_u(t_i), u \in U)$, determines the number of resources, $r_u(t)$, allocated to hospital unit u at time $t \in T$. Thus, we have that

$$r_u(t) = \pi_u(t_i)\ \forall u \in U, \forall t \in T : t \in [t_i, t_{i+1}),\ t_i \in T'. \tag{2}$$

The model parameters (i.e. the variables whose values characterize the problem instance) are listed below:

P^g: the conditional routing probability distribution of patient group $g \in \Theta$
$P^{LoS_u^g}$: the length of stay probability distribution for type $g \in \Theta$ at unit $u \in U$
λ_g: the (daily) arrival rate of patients of type $g \in \Theta$
r_u^{min}, r_u^{max}: the lower and upper bound for the resource capacity allocated to unit $u \in U$; the values are imposed by the layout of a hospital unit, the available equipment, staff and funds
c_u: the cost[2] for a resource at hospital unit $u \in U$; specifically, costs for the OR are only accounted for if allocated OR capacity remains unused due to cancelations of surgeries resulting from unavailable postoperative care beds[3].

Performance evaluation of resource allocations. In order to optimize resource allocation in hospitals a trade-off is needed between conflicting objectives, i.e. high patient throughput at low resource costs and back-up capacity usage. The outcome resulting from running the simulation applying allocation policy π is denoted by $F = F(\pi) = (F_0(\pi), F_1(\pi), F_2(\pi))$ with

[1] The actual patient routing may deviate from the medical indication depending on the available beds at the respective hospital care units. Patients may only be transferred to a higher care level than indicated. This gives more routing possibilities and thus makes the patient flows in Figure 1 more complex. The procedure is described in detail in [2].

[2] Costs for hospital resources relate to the daily costs for staff and materials and are expressed relative to the costs of a nursing ward bed.

[3] We assume that all fixed and variable costs for an operating room are covered by the surgical procedure that is to be performed. Therefore, only unused OR capacity is accounted for in the resource costs for the OR.

$F_0(\pi)$: the mean total throughput of patients under allocation π, defined as the number of patients discharged from the hospital after treatment.

$F_1(\pi)$: the mean total resource costs given by

$$F_1(\pi) = \sum_{u \in U \setminus \{CTS-OR\}} \sum_{t_i \in T'} c_u \cdot \pi_u(t_i) + c_{CTS-OR} \cdot uc_{CTS-OR}(\pi),$$

where $uc_{CTS-OR}(\pi)$ denotes the unused CTS-OR capacity due to canceled surgeries resulting from unavailable postoperative care beds given π.

$F_2(\pi)$: the mean total weighted back-up capacity usage under allocation π. The weighting factors correspond to the cost weights $c_u, u \in U$.

For optimizing resource management, $F_0(\pi)$ has to be maximized, while $F_1(\pi)$ and $F_2(\pi)$ have to be minimized. In the following we use as objective function

$$F'(\pi) = (-F_0(\pi), F_1(\pi), F_2(\pi)).$$

Optimization problem. The MO problem can thus be formulated as

$$\min F'(\pi) \tag{3}$$

where

$$\forall u \in U \, \forall t \in T : r_u(t) \in \mathbb{N} \cap [r_u^{min}, r_u^{max}]. \tag{4}$$

3 Strategies for Hospital Resource Management

As described in Section 2, hospital resource management is a highly stochastic and dynamic problem. In our approach, we use strategy optimization as advocated in [6]. Strategies are parameterized functions that return an allocation decision given the current situation. We thus have to optimize only one strategy that can be applied to a set of scenarios in the simulation because it describes what to do in any given situation. The strategies described below were developed in cooperation with domain experts from CHE. Therefore, the strategies can be easily understood by health care professionals which is important for the implementation and understanding under practical conditions.

In the following, the allocation strategies used in this study are described. Moreover, a mechanism for exchanging resources among units is described that enables the implementation of dynamic resource allocation in practice.

3.1 Static Resource Allocation Policies

Static allocation policies allocate a fixed number of resources to the different hospital units. We consider day-constant allocation policies, denoted by $\pi_u(t_i), u \in U$, given by $\pi_u(t_i) \equiv \mathfrak{r}_u \in \mathbb{N}$ for $t_i \in T'$. Day-constant policies are typically employed by hospitals and are also current practice at the CHE.

3.2 Dynamic Resource Allocation Strategy

A static allocation can do well in a relatively stable environment. This condition, however, does not hold in hospitals due to the stochastic patient treatment processes. Therefore, we consider dynamic strategies that return an allocation for the units in the network, given the current state of the units. This allows the resources (i.e. decision variables) to switch and track changes in the environment (i.e. the optimization problem) dynamically. Below, the state representation, the policy and its usage for dynamic resource allocation are described.

State description. The state at unit u at decision moment t_i, $s_u(t_i)$, is determined by the resource utilization rate at u, i.e. the ratio between the utilized capacity[4] at the start of day t_i and the resource capacity, $r_u(t_i^-)$, just before the adjustment at t_i, denoted by t_i^-. Formally, we have $s_u : T' \rightarrow \mathbb{R}_0^+$, $u \in U$, with

$$s_u(t_i) = \frac{\text{utilized capacity at unit } u \text{ at start of day } t_i}{r_u(t_i^-)}. \tag{5}$$

At the postoperative care units (CTS-PACU and CTS-HC) resources are available only for a couple of hours during the day. For these units the state at the start of day t_i defined in (5) may not be representative for the resource occupancy during the remainder of t_i, i.e. due to empty beds at the start of the day and canceled surgeries during the day. For these units, the *expected* resource utilization rate is used to determine $s_u(t_i)$. The *expected* utilized capacity for day t_i is calculated as the utilized capacity at time t_i minus the expected patient outflow plus the expected inflow (determined by the surgery scheme in the OR) for day t_i.

State-dependent allocation policy. A state-dependent allocation policy, denoted by $(\pi_u(t_i, s_u), t_i \in T', u \in U)$, is determined by five parameters: a base resource allocation, r_u^{base}, two adjustments, r_u^{decr} and r_u^{incr}, and two utilization thresholds, $\mathcal{UT}_u^{decr}$,$\mathcal{UT}_u^{incr}$ with $\mathcal{UT}_u^{decr} \leq \mathcal{UT}_u^{incr}$. We use an iterative step-function $\pi : T' \times \mathbb{R}_0^+ \rightarrow \mathbb{N}^{|U|}$ given as

$$\pi_u(t_i, s_u) = \begin{cases} \max\{r_u^{min}, r_u(t_i^-) - r_u^{decr}\} & \text{, if } s_u(t_i) < \mathcal{UT}_u^{decr} \\ r_u(t_i^-) & \text{, if } s_u(t_i) \in [\mathcal{UT}_u^{decr}, \mathcal{UT}_u^{incr}] \\ \min\{r_u^{max}, r_u(t_i^-) + r_u^{incr}\} & \text{, otherwise} \end{cases} \tag{6}$$

for $t_1, \ldots, t_{n-1}$ and

$$\pi_u(t_0, s_u) = r_u^{base}, \tag{7}$$

with $\pi_u(t_i, s_u) \in [r_u^{min}, r_u^{max}]$ $\forall t_i \in T', u \in U$. In (6) the current resources allocation, $r_u(t_i^-)$, is decreased by r_u^{decr} if the resource utilization rate is below the threshold $\mathcal{UT}_u^{decr}$. If the utilization rate is above $\mathcal{UT}_u^{incr}$, $r_u(t_i^-)$ is increased

[4] Note that due to the usage of back-up capacity the utilized capacity may exceed the allocated resources, thus s_u may be greater than 1.

by r_u^{incr}. Otherwise, the current allocation remains unchanged. Note that the policy specifies the allocation at the different units independently.

In the simulation the policy is applied at the start of every day after a warming-up period. Warming-up is necessary to avoid early convergence to minimal allocations due to the empty hospital in the start of a simulation run.

For the dynamic resource allocation problem (4) is changed to

$$\forall u \in U : r_u^{base} \in \mathbb{N} \cap [r_u^{min}, r_u^{max}], \tag{8}$$
$$\forall u \in U \, \forall t_i \in T' : s_u(t_i) \in \mathbb{R}_0^+, \tag{9}$$
$$\forall u \in U : r_u^{decr}, r_u^{incr} \in [0, 5] \tag{10}$$
$$\forall u \in U : \mathcal{UT}_u^{decr} \in [0, 1], \mathcal{UT}_u^{incr} \in [\mathcal{UT}_u^{decr}, \mathcal{UT}_u^{decr} + 1]. \tag{11}$$

As large adjustments are not desirable for hospital management, a maximal adjustment of 5 beds was chosen. Based on preliminary runs, a theoretical upper bound of 2 for $\mathcal{UT}_u^{incr}$ appeared to be more than sufficient.

3.3 Bed Exchange Mechanism for Dynamic Resource Allocation

In the state-dependent strategy described in Section 3.2, a large supply and stock of beds is assumed which enables the concurrent in- and decrease in resource capacity at the different units. In reality, however, bed availability is restricted by the available staff, in particular the number of personnel needed to per bed at a specific unit. Staff schedules need to be fixed at least several weeks in advance. The use of stand-by personnel is not common in the hospital domain. Therefore, a direct implementation of the policy described in Section 3.2 is often not practically feasible. To enable dynamic resource allocation in hospitals, we propose an exchange mechanism that is based on fixed personnel resources. The resources are exchanged among the hospital units to meet the current local need.

Here, $\pi_u(t_i, s_u)$ denotes the number of resources *required* by unit u at time t_i, determined by (6). The fixed personnel resources are determined by $r_u^{base}, u \in U$. The actual resource allocation, $r_u(t)$, is set by the mechanism below and not by (2).

We classify hospital units into three care levels, level 1 to level 3, based on the intensity of care and monitoring and the skill level of the personnel. Here, level 1 is the intensive care (IC), level 2 comprises the IC-HC, MC, CTS-HC and the CTS-PACU unit. Level 3 is the CTS ward, or shortly referred to as ward. From the application domain three rules arise for feasible bed exchanges:

R1: Due to staff training and physical requirements (i.e. access availability to the isolated electric power system in the hospital), beds can only be exchanged within the same or between adjacent care levels.

R2: Due to the staff assigned to a bed, shifting one bed from level l to level $l+1$ yields two beds at $l+1$ for $l = 1, 2$.

R3: Due to the personnel required to operate a bed, only an even number of beds can be shifted from level l to level $l-1$ for $l = 2, 3$ (i.e. the reverse of **R2**).

For the sake of reproducibility the mechanism and the method `shiftBeds` are described in detail in Algorithm 1. The number of resources available for exchange, $\mathcal{E}_l$, in care level l at time t_i is determined by

$$\mathcal{E}_l = \sum_{u \in \, level \, l} \max\{0, r_u(t_i^-) - \pi_u(t_i, s_u)\}, \, l = 1, 2, 3. \tag{12}$$

First, beds are shifted from level 1 to level 2. Then, level 2 beds are shifted to level 1 if necessary. Subsequently, beds of level 2 are exchanged within level 2. Finally, beds are exchanged between level 2 and 3. All exchanges are performed only if additional resources are required. The order of the care levels is based on the resource costs associated with the different units (given in Section 5.2). In future work, also other orderings of care levels will be considered. The order of units within a care level is determined randomly.

Through the mechanism, the implementation of (6) is extended with the above adjustments at time $t_i \in T'$, depending on the interaction with other units. This complex interaction mechanism answers to reality, however, it further complicates the optimization of resource management. Therefore, a state-of-the-art technique is needed for this optimization, which is described in Section 4.

4 EDA for Multi-objective Optimization

For the optimization of the dynamic and complex multi-objective resource allocation problem, we apply the SDR-AVS-MIDEA algorithm [7]. The algorithm was shown to be an efficient optimization technique for MO problems [7]. A brief outline of SDR-AVS-MIDEA is given in Section 4.1.

We use a strategy optimization approach with the policies defined in Section 3. The parameters of the strategies, specified in Section 5.1, are optimized using SDR-AVS-MIDEA. The fitness is determined using the simulation tool described in Section 2.1.

The optimization of the strategies is performed in an offline fashion. As the strategies are used online in the simulation, the anticipation of time-dependency effects [6], i.e. the impact of decisions taken now on the future, is implicitly included in the optimization of the strategies' parameters. Thus, MO techniques can be applied in a straightforward fashion to solve this dynamic problem. Since designing online MO appears to be rather hard, this approach yields an additional advantage. The policies proposed can be easily understood by health care professionals, so this approach is also practically implementable.

4.1 Outline of SDR-AVS-MIDEA

In this section, a brief outline of the evolutionary algorithm is given. For a detailed description the reader is referred to [7].

The algorithm divides the generated solutions into clusters of equal size that are kept separated in the objective space throughout a run. The use of clusters stimulates the search for a broad Pareto-front. New solutions are generated

Algorithm 1. Pseudo-code description of the bed exchange mechanism

Input: Set of hospital units, U, $\pi_u(t_i, s_u)\,\forall u \in U$, determined by (6), and $\mathcal{E}_l$, $l = 1, 2, 3$, determined by (12)
Result: $r_u(t_i)\,\forall u \in U$

```
for l = 1 to 3 do EX_l ← E_l;
/*exchange from level 1 to level 2 applying rules R1 and R2; if the
exchange results in more beds at unit u than required, a bed is
shifted to another unit of level 2 in need of additional resource
capacity                                                            */
while EX_1 > 0 and ∃u ∈ level 2 with π_u(t_i, s_u) > r_u(t_i^-) do
    if (π_u(t_i, s_u) − r_u(t_i^-)) is an even number then
        shiftBeds(1,2,IC,u,min{EX_1, (π_u(t_i, s_u) − r_u(t_i^-))/2})
    else
        shiftBeds(1,2,IC,u,min{EX_1, ⌈(π_u(t_i, s_u) − r_u(t_i^-))/2⌉});
        Find a v ∈ level 2, v ≠ u (if any) with π_v(t_i, s_v) > r_v(t_i^-) and
        shiftBeds(2,2,u,v,1);
/*exchange from level 2 to level 1 applying rules R1 and R3; if an
exchange from one unit is not feasible (R3), the exchange is
performed together with another unit of level 2 (if possible)       */
while EX_2 ≥ 2 and π_IC(t_i, s_IC) > r_IC(t_i^-) do
    Find v_1 ∈ U with π_v1(t_i, s_vj) < r_v1(t_i^-);
    if |π_u(t_i, s_u) − r_u(t_i^-)| is an even number then
        shiftBeds(2,1,v_1,IC,min{|π_v1(t_i, s_v1) − r_v1(t_i^-)|/2, π_IC(t_i, s_IC) −
        r_IC(t_i^-)});
    else
        Find v_2 ∈ U, v_2 ≠ v_1, with π_v2(t_i, s_v2) < r_v2(t_i^-) and
        shiftBeds(2,2,v_2,v_1,1) and shiftBeds(2,1,v_1,IC,min{⌈|π_v1(t_i, s_v1) −
        r_v1(t_i^-)|/2⌉, π_IC(t_i, s_IC) − r_IC(t_i^-)});
        if there is no such v_2 then
            shiftBeds(2,1,v_1,IC,min{⌊|π_v1(t_i, s_v1) − r_v1(t_i^-)|/2⌋, π_IC(t_i, s_IC) −
            r_IC(t_i^-)});
/*exchange within level 2 applying rule R1                          */
while EX_2 > 0 and ∃u ∈ level 2 with π_u(t_i, s_u) > r_u(t_i^-) do
    Find a v ∈ level 2, v ≠ u, with π_v(t_i, s_v) < r_v(t_i^-) and
    shiftBeds(2,2,v,u,min{|π_v(t_i, s_v) − r_v(t_i^-)|, π_u(t_i, s_u) − r_u(t_i^-)});
/*exchange from level 2 to level 3 applying rules R1 and R2         */
if EX_2 > 0 and π_ward(t_i, s_ward) > r_ward(t_i^-) then
    Find a v ∈ level 2, with π_v(t_i, s_v) < r_v(t_i^-) and
    shiftBeds(2,3,v,CTS ward,min{|π_v(t_i, s_v) − r_v(t_i^-)|, ⌊(π_ward(t_i, s_ward) −
    r_ward(t_i^-))/2⌋});
/*exchange from level 3 to level 2 applying rules R1 and R3         */
while EX_3 ≥ 2 and ∃u ∈ level 2 with π_u(t_i, s_u) > r_u(t_i^-) do
    shiftBeds(3,2,CTS ward,u,min{⌊|π_ward(t_i, s_ward) −
    r_ward(t_i^-)|/2⌋, π_u(t_i, s_u) − r_u(t_i^-)});
/*if no exchange is possible, the resource allocation remains
unchanged                                                           */
forall u ∈ U that were not yet considered do r_u(t_i) ← r_u(t_i^-);
```

Function. `shiftBeds`(*int* l^{from}, *int* l^{to}, *unit* u^{from}, *unit* u^{to}, *int* n)

$k = 1$;
if $l^{from} < l^{to}$ **then** $k = 2$;
else if $l^{from} > l^{to}$ **then** $k = 0.5$;
$r_{u^{to}}(t_i) \leftarrow r_{u^{to}}(t_i) + k \cdot n$, $r_{u^{from}}(t_i) \leftarrow r_{u^{from}}(t_i) - n$, $EX_{l^{from}} \leftarrow EX_{l^{from}} - n$;

according to the EDA principle. In each separate cluster a single normal distribution is used. The algorithm uses adaptive variance scaling (AVS) in combination with standard-deviation ratio (SDR) triggers to prevent premature convergence. This means that if the best fitness in a cluster is improved in one generation and the average improvement is more than one standard deviation away from the estimated mean of the distribution, then the variance of the estimated distribution is scaled up to increase the area of exploration. If, however, the improvements are obtained near the mean of the predicted distribution, then the variance is scaled down to allow for a faster convergence.

5 Experiments

5.1 Settings of SDR-AVS-MIDEA

The settings of the parameters in SDR-AVS-MIDEA are based on the guidelines reported in [7,15] with the percentile for truncation selection set to 0.3, $k = 4$ clusters. The guideline in [15] is used and results in a population size per cluster of 49 and 130 for the day-constant and the dynamic policies, respectively. The variance multiplier decreaser of AVS equals 0.9 and the SDR threshold is set to 1.0. As in [7], an elitist archive is maintained. To this end, the objective space is discretized in each objective with a discretization length of 10^{-3}. This provides sufficient granularity for the final Pareto-front approximations. We allowed 1600 generations for the different allocation policies.

In the EDA representation, the genes correspond to allocation policy parameters. For the day-constant policies described in Section 3.1, the genotype comprises the values for $\mathfrak{r}_u$, $u \in U$, with $\mathfrak{r}_u \in \mathbb{N} \cap [r_u^{min}, r_u^{max}]$. For the dynamic policies described in Section 3.2 and Section 3.3, a genotype is composed of values $r_u^{base} \in [r_u^{min}, r_u^{max}]$, $r_u^{decr}, r_u^{incr} \in [0, 5]$, $\mathcal{UT}_u^{decr} \in [0, 1]$, and $\mathcal{T}_u \in [0, 1]$ for $u \in U$. The parameter $\mathcal{T}_u$ is used to determine $\mathcal{UT}_u^{incr}$ by $\mathcal{UT}_u^{incr} = \mathcal{UT}_u^{decr} + \mathcal{T}_u$. The bounds, r_u^{min} and r_u^{max}, for the resource allocations were obtained from domain experts from CHE. These values are given in Table 1.

5.2 Settings of the Simulation Tool

Applying SDR-AVS-MIDEA to a real-world problem is associated with a large number of potential solutions using a complex simulation model. We run 10 simulation runs of 20 weeks including 8 weeks of warming-up to evaluate the allocation strategies. The warming-up period is not measured in the simulation outcomes. This setting results in a runtime of about 6 seconds per evaluation.

Table 1. Resource bounds, unit resource costs and benchmark day-constant policies obtained from CHE case study

	CTS-OR	CTS-HC	CTS-PACU	IC	IC-HC	MC	CTS ward
r_u^{min}	0	0	0	5	2	2	20
r_u^{max}	6	6	6	20	6	10	50
c_u	0.09	2	2	4	2	2	1
π_u^{CHE}	4	4	4	11	4	4	35

Table 2. Input parameters of patient pathways with LoS in hours (mean±stdev)

Type	Unit	LoS	P^g	Type	Unit	LoS	P^g
I	CTS-HC	15 ± 0	-	II	CTS-PACU	6 ± 0	-
	IC	48.48 ± 54	0.15		IC	42 ± 57.12	0.05
	MC	24.48 ± 38.52	0.15		MC	10.32 ± 22.08	0.15
	CTS ward	120 ± 22.08	0.7		CTS ward	120 ± 22.08	0.8
III	IC	89.48 ± 200.82	-	IV	IC-HC	34.94 ± 68.51	-

In a sensitivity analysis, the mean and variance of the relevant outcome values appear to be linear for increasing simulation run durations.

The settings for the patient pathways are based on the statistical data analysis conducted in the case study at the CHE. The relevant parameters of the different patient pathways introduced in Section 2.1 are given in Table 2. We use a Lognormal distribution for sampling patients' LoS. Arrivals of type III patients are Poisson with daily arrival rate $\lambda^{III} = 2$. Patients of type IV arrive daily in bulks between 2 and 4 patients. Costs for the different types of hospital resources related to the daily costs for staff and materials and are expressed in terms of relative costs of a nursing ward bed. The costs are given in Table 1. The OR costs account for the unused OR capacity due to cancelations of surgeries that result from unavailable postoperative care beds.

5.3 Results

One run of the EA takes approximately 10 hours for static strategy optimization and 30 hours for dynamic strategy optimization on a high-performance computer cluster. Specifically, we used 40 nodes running at speeds between 1.4Ghz and 2.2Ghz. We have run the EA for each strategy three times, yielding very comparable and stable results (within a strategy type).

In Figure 2 the results for the day-constant, the state-dependent strategies and the exchange mechanism are presented. In the application domain, the exact values for back-up capacity usage are of minor importance and a categorization of minimal (corresponding to $F_2 \in [0, 50)$), small ($F_2 \in [50, 100)$), medium, etc. is therefore sufficient for the representation of the optimization results. The results

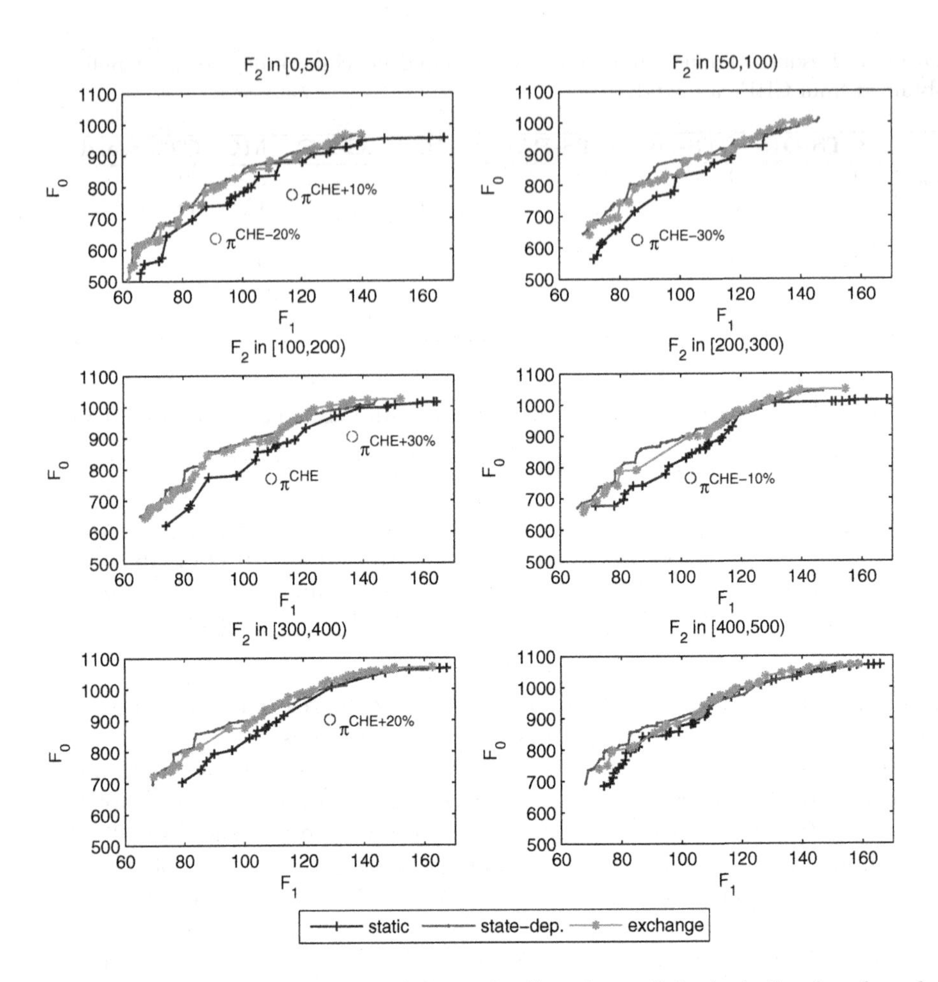

Fig. 2. Pareto-fronts for static and dynamic allocation policies including benchmarks from CHE

are confined to F_2-values below 500 as higher back-up capacity usage is not desirable for many hospitals. We depict the Pareto-fronts with respect to F_1 and F_0 for F_2-values in the predefined intervals. This in addition allows us better visibility of the results as opposed to 3D plots. To assess the performance of the policies the currently used day-constant resource allocation at the CHE and linearly scaled allocations are included as a benchmark for the relevant intervals. The currently employed policy of the CHE is denoted by π^{CHE} given in Table 1. Also, benchmarks determined by linearly scaled allocations are considered that are denoted by π^{CHE+i} with $\pi^{CHE+i} = (\lfloor \pi_u^{CHE} \cdot (1+i) + 0.5 \rfloor, u \in U, i = \pm 10\%, \pm 20\%, \pm 30\%)$.

The results in Figure 2 show that the benchmarks obtained from hospital practice are dominated by all policies proposed in this paper. Moreover, the dynamic resource allocation policies show higher performance compared to the static allocation policies. For F_2-values of above 300 and F_1-values higher than 120, the static and dynamic policies show similar performance. This can be explained by the small extent and frequency of allocation adjustments of the dynamic policies obtained for these F_1 and F_2 values. Since additional demand for care can be met by using back-up capacity, less allocation adjustments are necessary in these cases. The bed exchange policies show slightly lower performance compared to the state-dependent policies. The difference can be attributed to the interaction between the hospital units due to which required allocation adjustments cannot always be fully undertaken.

6 Conclusions

In this paper, multiobjective optimization for dynamic hospital resource management using evolutionary algorithms was addressed. We use a strategy optimization approach for which we designed policies that allow for the dynamic allocation of resources in hospital practice. Due to the complexity of the allocation strategies and the dynamic application domain, we used a state-of-the-art evolutionary MO technique, SDR-AVS-MIDEA. The fitness of the solutions was determined using a simulation tool developed for this application domain. Our results show that the benchmark allocations obtained from a case study could be considerably improved using the optimized strategies. Furthermore, we showed that policies that incorporate more dynamic resource allocations result in further improvements. These improvements are made possible by the design of the policy. SDR-AVS-MIDEA then is powerful enough to detect and exploit the additional possibilities. We demonstrated that proper design in combination with state-of-the-art EAs can make an important contribution and achieve an improvement for complex real-world dynamic MO problems as in hospital logistics.

By using our strategy types, we can circumvent the online MO optimization and use offline MO techniques to optimize the parameters of the strategies. Furthermore, our approach is feasible for stochastic fitness functions obtained from a simulation model.

In future work, we will develop allocation strategies that use more advanced anticipation models of the time-dependency effects. Furthermore, we will consider alternative orderings of care levels in the bed exchange mechanism. Also, we will further explore the settings of SDR-AVS-MIDEA in relation to the above extensions.

Acknowledgements

We thank Han Noot for his programming contributions, and Jan van Aarle, Ilona Blonk, Floor Haak and Dick Koning, for their expert feedback.

References

1. Harper, P.R., Shahani, A.K.: Modelling for the planning and management of bed capacities in hospitals. J. Oper. Res. Soc. 53, 11–18 (2002)
2. Hutzschenreuter, A.K., Bosman, P.A.N., Blonk-Altena, I., van Aarle, J., La Poutré, J.A.: Agent-based patient admission scheduling in hospitals. In: Proc. of 7th Int. Conf. on Autonomous Agents and Multiagent Systems (AAMAS 2008) - Industry and Applications Track, pp. 45–52 (2008)
3. Bosman, P.A.N., Thierens, D.: Multi–objective optimization with diversity preserving mixture–based iterated density estimation evolutionary algorithms. Int. J. Approx. Reason. 31, 259–289 (2002)
4. Bui, L.T., Abbass, H.A., Branke, J.: Multiobjective optimization for dynamic environments. In: Congress on Evol. Comp., vol. 3, pp. 2349–2356 (2005)
5. Farina, M., Deb, K., Amato, P.: Dynamic multiobjective optimization problems: test cases, approximations, and applications. IEEE T. Evol. Comp. 8, 425–442 (2004)
6. Bosman, P.A.N., La Poutré, J.A.: Learning and anticipation in online dynamic optimization with evolutionary algorithms: the stochastic case. In: GECCO 2007: Proc. of the 9th annual conference on Genetic and evol. comp., pp. 1165–1172 (2007)
7. Bosman, P.A.N., Thierens, D.: Adaptive variance scaling in continuous multi-objective estimation-of-distribution algorithms. In: GECCO 2007: Proc. of the 9th annual conference on Genetic and evol. comp., pp. 500–507 (2007)
8. Vissers, J.M.H.: Patient flow-based allocation of inpatient resources: A case study. E. J. Oper. Res. 105(2), 356–370 (1998)
9. Ridge, J.C., Jones, S.K., Nielsen, M.S., Shahani, A.K.: Capacity planning for intensive care units. E. J. Oper. Res. 105(2), 346–355 (1998)
10. Kim, S., Horowitz, I., Young, K.K., Buckley, T.A.: Flexible bed allocation and performance in the intensive care unit. J. Oper. Manag. 18(4), 427–443 (2000)
11. Kusters, R.J., Groot, P.M.A.: Modelling resource availability in general hospitals design and implementation of a decision support model. E. J. Oper. Res. 88(3), 428–445 (1996)
12. Blake, J.T., Carter, M.W.: A goal programming approach to strategic resource allocation in acute care hospitals. E. J. Oper. Res. 140(3), 541–561 (2002)
13. Vissers, J., Beech, R. (eds.): Health Operations Management: Patient flow logistics in health care. Health Management Series. Routledge (2005)
14. Bollobàs, B.: Graph Theory: An Introductory Course. Springer, Heidelberg (1979)
15. Bosman, P.A.N., Grahl, J., Thierens, D.: Enhancing the performance of maximum-likelihood gaussian edas using anticipated mean shift. In: Rudolph, G., Jansen, T., Lucas, S., Poloni, C., Beume, N. (eds.) PPSN 2008. LNCS, vol. 5199, pp. 133–143. Springer, Heidelberg (2008)

Multiobjective Decomposition of Positive Integer Matrix: Application to Radiotherapy

Thibaut Lust and Jacques Teghem

Faculté Polytechnique de Mons, Laboratory of Mathematics & Operational Research
9, rue de Houdain, 7000 Mons, Belgium
thibaut.lust@fpms.ac.be

Abstract. We consider the following problem: to decompose a positive integer matrix into a linear combination of binary matrices that respect the consecutive ones property. The positive integer matrix corresponds to fields giving the different radiation beams that a linear accelerator has to send throughout the body of a patient. Due to the inhomogeneous dose levels, leaves from a multi-leaf collimator are used between the accelerator and the body of the patient to block the radiations. The leaves positions can be represented by segments, that are binary matrices with the consecutive ones property. The aim is to find a decomposition that minimizes the irradiation time, and the setup-time to configure the multi-leaf collimator at each step of the decomposition. We propose for this NP-hard multiobjective problem a heuristic method, based on the Pareto local search method. Experimentations are carried out on different size instances and the results are reported. These first results are encouraging and are a good basis for the design of more elaborated methods.

1 Introduction

In this paper, we consider a problem dealing with the planning of an intensity modulated radiotherapy treatment (IRMT) to individual patients. The IRMT is usually composed of three phases [5]: the selection of beam angles through which radiation is delivered (geometry problem), the computation of an optimal intensity map for each selected beam angle (intensity problem) and the determination of a sequence of configurations of a multi-leaf collimator (realization problem). In this work, we only consider the realization problem, by taking into account three different objective. We present below the mathematical model of this problem.

Throughout we use the notation $[n] := \{1, 2, \cdots, n\}$ for positive integers n.

We consider a positive integer matrix A of size $m \times n$: $A = (a_{i,j})$ with $i \in [m]$ and $j \in [n]$. The matrix corresponds to fields giving the different radiation beams that a linear accelerator has to send throughout the body of a patient. The value $a_{i,j}$ of A gives the desired intensity that should be delivered to coordinate (i, j).

We have to decompose the matrix A into a set of segments. The segments correspond to the shape of a multi-leaf collimator (MLC) which is a system containing a collection of leaves that can be moved in parallel, in order to block the

M. Ehrgott et al. (Eds.): EMO 2009, LNCS 5467, pp. 335–349, 2009.

radiations (inhomogeneous dose levels are administrated: certain cancer targets receive a required amount of dose while functional organs are spared). Two types of leaves are used: left leaves that move from the left to the right and right leaves that move from the right to the left.

A segment can be represented by a special binary matrix of size $m \times n$ that describes the leaves positions. These matrices have to respect the consecutive ones property (C1), which means, in short, that the ones occur consecutively in a single block in each row (since we can only block the radiations with a left or a right leave).

A segment is noted $S = (s_{i,j})$ with $i \in [m]$ and $j \in [n]$. An example of a decomposition of a matrix A into segments is shown below.

$$\mathbf{A} = \begin{pmatrix} 4\,8\,3 \\ 5\,2\,1 \\ 5\,7\,2 \end{pmatrix} = 4\begin{pmatrix} 1\,1\,0 \\ 1\,0\,0 \\ 1\,1\,0 \end{pmatrix} + 2\begin{pmatrix} 0\,1\,1 \\ 0\,1\,0 \\ 0\,1\,1 \end{pmatrix} + 1\begin{pmatrix} 0\,1\,1 \\ 1\,0\,0 \\ 1\,1\,0 \end{pmatrix} + 1\begin{pmatrix} 0\,1\,0 \\ 0\,0\,1 \\ 0\,0\,0 \end{pmatrix}$$

The positions of the left and right leaves corresponding to a segment S are given by the l_i and r_i integers defined as follows:

$$0 \leq l_i < r_i \leq n+1 \quad (i \in [m])$$

$$s_{i,j} = \begin{cases} 1 \text{ if } l_i < j < r_i & (i \in [m], j \in [n]) \\ 0 \text{ otherwise.} \end{cases}$$

We denote the set of segments by $\mathcal{S}$.

For example, for the following segment:

$$\begin{pmatrix} 1\,1\,0 \\ 1\,0\,0 \\ 0\,0\,1 \end{pmatrix}$$

we have: $l_1 = 0, r_1 = 3; l_2 = 0, r_2 = 2$ and $l_3 = 2, r_3 = 4$.

It should be noted that in this work, a line full of zero is always represented by $l = 0$ and $r = 1$.

A feasible decomposition of A is a linear sum of segments and has the following form:

$$A = \sum_{k=1}^{K} u_k S^k \text{ with } u_k \in \mathbb{N}_0, S^k \in \mathcal{S}\ \forall k \in [K].$$

Two criteria are generally considered to evaluate the quality of a decomposition: the total irradiation time and the setup-time.

The total irradiation time, very important from a medical point of view, is the time during which a patient is irradiated. This criterion is proportional to the sum of the coefficients (decomposition time). The setup-time is the time to configure the MLC. This criterion is proportional to the number of segments (decomposition cardinality). It is important to minimize this criterion in order to reduce the time of the session and so the comfort of the patient.

We can formulate both objectives that is the decomposition time (DT) and the decomposition cardinality (DC) as follows:

$$\textbf{(DT)} \quad \min\left\{\sum_{k=1}^{K} u_k \;\middle|\; A=\sum_{k=1}^{K} u_k S^k, u_k \in \mathbb{N}_0, S^k \in \mathcal{S}\; \forall\, k \in [K]\right\}$$

$$\textbf{(DC)} \quad \min\left\{K \;\middle|\; A=\sum_{k=1}^{K} u_k S^k, u_k \in \mathbb{N}_0, S^k \in \mathcal{S}, \forall\, k \in [K]\right\}$$

Polynomial algorithms are known for the DT minimization [3,13]. Many optimal solutions can be found for this single-objective problem.

On the other hand, the DC minimization has been proved to be NP-Hard [2,4].

Some authors [2,7,11] optimize the objectives lexicographically: by first minimizing DT and then trying to reduce the DC while keeping the minimal DT. Taşkin *et al.* [14] and Wake *et al.* [15] recently considered both objectives at the same time, but by simply doing a linear sum of the objectives.

To be more realistic, we do not only consider in this paper constant times to move from one segment to the next. The variable setup-time is defined as follows:

$$(\mathbf{SU}_{var}) \quad \min\left\{\sum_{k=1}^{K-1} \mu(S^k, S^{k+1}) \;\middle|\; A=\sum_{k=1}^{K} u_k S^k, u_k \in \mathbb{N}_0, S^k \in \mathcal{S}, \forall\, k \in [K]\right\}$$

where μ is proportional to the time necessary to change the setup of the MLC from the configuration corresponding to S^k to the configuration corresponding to S^{k+1}. This objective is also known under the name overall leaf travel time [11].

The value μ between two segments S^k and S^{k+1} is computed as follows [13]:

$$\mu(S^k, S^{k+1}) = \max_{1\le i\le m} \max\left\{\left|l_i^{k+1} - l_i^k\right|, \left|r_i^{k+1} - r_i^k\right|\right\}$$

Once the segments are fixed, the minimization of the overall leaf travel time is equivalent to a search for a Hamiltonian *path* of minimal weight on the complete graph which has the segments as vertices and the weight function μ on the edges. The weight function μ has the property to be a metric [11].

This problem can be transformed to a TSP problem (Hamiltonian *cycle* of minimal weight) by adding a dummy vertex which has a distance of zero to all other vertices. However, with this transformation, the triangular inequality in Euclidean space no longer holds which makes the TSP problem a little bit harder to solve.

Since there is a positive correlation between DC and SU_{var}, the few authors that have considered SU_{var} tried to first minimize DC and then SU_{var} by generating the best sequence of segments. Kalinowski [11] uses a minimum spanning tree approximation to find the best sequence of segments, but it also possible to use an exact TSP algorithm, as done by Ehrgott *et al.* [6].

Siochi [13] considered both objectives at the same time, but through a linear sum.

In this paper, we will consider the three objectives (DT,DC,SU_{var}) simultaneously. The multiobjective formulation of the problem (P) considered is thus as follows:

$$
(P)\begin{cases}
\min z_1(x) & = \sum_{k=1}^{K} u_k \quad \textbf{(DT)} \\
\min z_2(x) & = K \quad \textbf{(DC)} \\
\min z_3(x) & = \sum_{k=1}^{K-1} \max_{1 \leq i \leq m} \max \left\{ \left| l_i^{k+1} - l_i^k \right|, \left| r_i^{k+1} - r_i^k \right| \right\} \quad (\mathbf{SU}_{var}) \\
\text{s.t} & A = \sum_{k=1}^{K} u_k S^k, u_k \in \mathbb{N}_0, S^k \in \mathcal{S}, \forall\, k \in [K]
\end{cases}
$$

We denote by X the feasible set in the decision space, defined by $X = \{x \in \{(u_k \in \mathbb{N}_0, S^k \in \mathcal{S})\}^K | A = \sum_{k=1}^{K} u_k S^k\}$. The feasible set in objective space is called Z and is defined by $Z = z(X) = \{(z_1(x), z_2(x), z_3(x)),\ \forall\, x \in X\} \subset \mathbb{Z}^3$.

To our knowledge, nobody has tried to find the efficient solutions, or even a good approximation of the efficient solutions of P. Our aim is to generate a good approximation of a minimal complete set [8] of the efficient solutions of P.

2 Pareto Local Search

The Pareto local search (PLS) method [1,12] is one of the simplest method for multiobjective optimization. This method is a purely local search algorithm, generalization in the multiobjective case of a basic metaheuristic: the hill-climbing method. The method does not require any objectives aggregation nor any numerical parameters, and is based on the notion of *Pareto local optimum set* [12].

The pseudo-code of the PLS method is given by the algorithm 1.

The method starts with a population P composed of potentially efficient solutions given by the initial population P_0, which is an input parameter of the method. Then, all the neighbors p' of each solution p of P are generated. If a neighbor p' is not weakly dominated by the current solution p, we try to add the solution p' to the approximation $\widehat{X_E}$ of the efficient set, which is updated with the procedure AddSolution. This procedure is not described in this work but simply consists of updating an approximation $\widehat{X_E}$ of the efficient set when a new solution p is added to $\widehat{X_E}$. This procedure has four parameters, the set $\widehat{X_E}$ to actualize, the new solution p, its evaluation $z(p)$ and an optional boolean variable called *Added* that returns $True$ if the new solution has been added and $False$ otherwise. If the solution p' has been added to $\widehat{X_E}$, the boolean variable *Added* is true and the solution p' is added to an auxiliary population P_a, which is updated also with the procedure AddSolution. Once all the neighbors of each solution of P have been generated, the algorithm starts again, with P equal to P_a, until $P = \emptyset$. The auxiliary population is used such that the neighborhood of each solution of the population P is explored, even if some solutions of P become dominated following the addition of a new solution to P. Thus, sometimes, neighbors are generated from a dominated solution.

Algorithm 1. PLS

```
Parameters ↓: An initial population P_0
Parameters ↑: An approximation X̂_E of the efficient set

--| Initialization of X̂_E and a population P with the initial population P_0
X̂_E ← P_0
P ← P_0
--| Initialization of an auxiliary population P_a
P_a ← ∅
while P ≠ ∅ do
   --| Generation of all the neighbors p' of each solution p ∈ P
   for all p ∈ P do
      for all p' ∈ N(p) do
         if z(p) ≰ z(p') then
            AddSolution(X̂_E ↕, p' ↓, z(p') ↓, Added ↑)
            if Added = true then
               AddSolution(P_a ↕, p' ↓, z(p') ↓)
   --| P is composed of the new potentially efficient solutions
   P ← P_a
   --| Reinitialization of P_a
   P_a ← ∅
```

3 Adaptation of PLS to the Multiobjective Decomposition Problem

3.1 Initial Population

Two initial solutions of good quality are generated and added to the initial population. The first solution is a good approximation of lexmin(DT,DC,SU_{var}) and the second one is a good approximation of lexmin(DT,SU_{var},DC). In both cases, we first minimize DT since polynomial algorithms are known for this problem. We can remark that a solution corresponding to lexmin(DT,SU_{var}) is also a solution corresponding to lexmin(DT,SU_{var},DC), since once SU_{var} is minimized, the DC value is set and can not be modified.

To approximate lexmin(DT,DC,SU_{var}), we first approximate lexmin(DT,DC) with the heuristic of Engel [7] which is efficient for this problem. We then apply a TSP heuristic to reduce the SU_{var} value of the solution. We use a very efficient heuristic for the TSP: the Lin and Kernighan heuristic implemented by Helsgaun (LKH) [10].

To approximate lexmin(DT,SU_{var}), we will adapt the heuristic of Engel since SU_{var} is correlated to the DC objective. The algorithm developed by Engel is a deterministic construction algorithm, that allows to find an optimal solution for the DT objective with a low DC value. He tackles this problem as follows: he removes different well-selected combinations of couples (u_k, S^k) from the current matrix until $A_{t+1} = 0$, with $A_{t+1} = A_t - uS$, where t represent the index of the step of the construction algorithm. He starts the algorithm with $A_0 = A$. A move consists thus of removing from the current matrix a segment multiplied by a certain coefficient.

At each step of the construction heuristic, the maximum coefficient (u_{max}) that can be used while ensuring that the optimal objective DT can be achieved is considered.

Engel has developed a theory to compute u_{max}. The coefficient u_{max} can be easily obtained in $O(mn^2)$. Using u_{max} allows to find very good results for the lexicographic problem (DT,DC) [6,7,11].

But once u_{max} has been defined, we also have to define which segment to use among all the segments that respect u_{max}. Kalinoswki [11] has developed a rule which gives slightly better results that the initial rule of Engel. The rule is as follows. If two zero columns are added to A, that is let:

$$a_{i,0} = a_{i,n+1} = 0 \ \ \forall i \in [m]$$

we can associate to A its difference matrix D of dimension $m \times (n+1)$:

$$d_{i,j} = a_{i,j} - a_{i,j-1} \ \ \forall i \in [m], j \in [n+1]$$

Now, we put

$$q(A) = \left|\{(i,j) \in [m] \times [n] : d_{i,j} \neq 0\}\right|,$$

and in the method of Kalinowski, we choose a segment S so that $q(A - uS)$ is minimized.

This method gives very good results for lexmin(DT,DC), but there is no theoretical evidence for that.

For lexmin(DT,SU$_{var}$), we keep the principle of the construction algorithm of Engel by removing well-selected combinations of couples (u_k, S^k) from the current matrix until $A(t+1) = 0$.

As in the Engel algorithm, it is worthwhile to take the maximal coefficient that allows to keep the minimal DT value, since the SU$_{var}$ objective is linked to the DC objective. For the definition of the segment that corresponds to u_{max}, we try three new rules. These three new rules are presented below.

For each line of the new segment to define, we have to choose between different intervals. If we sort out the intervals as follows: $\{(0,1), (0,2), \cdots, (0, n+1), (1,3), \cdots, (1, n+1), \cdots, (n-2, n), (n-2, n+1), (n-1, n+1)\}$, the first rule is to take the first feasible interval and the second rule is to take the last feasible interval. For these two rules, we expect to always move the leaves from the left to the right or to the right from the left to minimize the maximal distance between two consecutive segments. The third rule is to take the first feasible interval which is the closest to the preceding interval of the same line: the aim is to minimize the maximal distance between two consecutive segments. The first interval defined with this rule is the same than the one selected with the first rule. The results of the comparison between the different rules will be given at section 4.

3.2 Neighborhood

The neighborhood is the most important element of the PLS method. On the other hand, it is not trivial to produce neighbors from a feasible solution for the problem P. Removing one segment of the current solution is not enough. Removing two segments from the current solution requires to determine how to select both segments and how to recombine these two segments to produce a new feasible solution, ideally of better quality. Moreover, it will be difficult with this kind of technique to find neighbors with different DC values. Modifying the sequence of segments can improve the SU_{var} objective, but we can always apply a TSP algorithm at the end of the decomposition to improve this objective.

The neighborhood developed in this work is thus a bit complex. It works as follows:

1. Selection of a segment S that belongs to the current decomposition.
2. We modify a line i of S in the following way:

$$\begin{aligned} l_i &= l_i + (-1 \text{ or } 0 \text{ or } 1) \\ r_i &= r_i + (-1 \text{ or } 0 \text{ or } 1) \end{aligned}$$

3. We put S at the first place of the new decomposition.
4. We eventually modify the coefficient of this segment.
5. We construct a neighbor by adding the segments in the order of the current decomposition. If a segment is not feasible, we skip it. We adapt the coefficient of the segments added, which is equal to the minimum between the current coefficient of the segment and the maximal feasible coefficient.
6. The matrix that remains after adding all the possible segments is decomposed with the heuristic of Engel.
7. Once a decomposition is obtained, we optimize the SU_{var} objective by using a simple and fast TSP heuristic: the first improvement local search based on the 2-edges exchange moves [9].

This neighborhood requires the definition of many elements, but we will see that it is possible to explore many possibilities.

To illustrate the neighborhood, we show its functioning on the following example:

$$\mathbf{A} = \begin{pmatrix} 8 & 5 & 6 \\ 5 & 3 & 6 \end{pmatrix}$$

Ehrgott *et al.* [6] showed that for this example, DT and SU_{var} are contradictory, as well as DT and DC.

Let us consider that we start the neighborhood from the following solution, which minimizes the DT objective:

$$\mathbf{A} = \begin{pmatrix} 8 & 5 & 6 \\ 5 & 3 & 6 \end{pmatrix} = 3 \begin{pmatrix} 1 & 0 & 0 \\ 0 & 0 & 1 \end{pmatrix} + 1 \begin{pmatrix} 0 & 0 & 1 \\ 0 & 1 & 1 \end{pmatrix} + 3 \begin{pmatrix} 1 & 1 & 1 \\ 1 & 0 & 0 \end{pmatrix} + 2 \begin{pmatrix} 1 & 1 & 1 \\ 1 & 1 & 1 \end{pmatrix}$$

The DT value of this solution is optimal and equal to 9, DC is equal to 4 and the SU_{var} value is equal to (2+2+2)=6.

We apply the neighborhood to this solution:

1. We select the first segment:
$$\begin{pmatrix} 1\,0\,0 \\ 0\,0\,1 \end{pmatrix}$$
2. We select the second line. For this line, $l = 2$ and $r = 4$. We modify this line by putting $l = 1$, $r = 3$. We obtain this segment:
$$\begin{pmatrix} 1\,0\,0 \\ 0\,1\,0 \end{pmatrix}$$
3. We put this segment at the first place of the new decomposition.
4. The current coefficient of this segment is equal to 3. As for this segment, the maximal feasible coefficient that we can put is 3, we keep the coefficient equal to 3. The remaining matrix is:
$$\begin{pmatrix} 5\,5\,6 \\ 5\,0\,6 \end{pmatrix}$$
5. The first segment that we can consider is:
$$\begin{pmatrix} 0\,0\,1 \\ 0\,1\,1 \end{pmatrix}$$
but we can not add it.
The second segment is:
$$\begin{pmatrix} 1\,1\,1 \\ 1\,0\,0 \end{pmatrix}$$
We add it, with a coefficient equal to 3, that is its current coefficient. The remaining matrix is:
$$\begin{pmatrix} 2\,2\,3 \\ 2\,0\,6 \end{pmatrix}$$
The last segment in the initial decomposition is:
$$\begin{pmatrix} 1\,1\,1 \\ 1\,1\,1 \end{pmatrix}$$
but we can not add it.
6. The decomposition of the remaining matrix with the heuristic of Engel gives:
$$5\begin{pmatrix} 0\,0\,0 \\ 0\,0\,1 \end{pmatrix} + 2\begin{pmatrix} 1\,1\,1 \\ 1\,0\,0 \end{pmatrix} + 1\begin{pmatrix} 0\,0\,1 \\ 0\,0\,1 \end{pmatrix}$$
The decomposition obtained is thus:
$$3\begin{pmatrix} 1\,0\,0 \\ 0\,1\,0 \end{pmatrix} + 3\begin{pmatrix} 1\,1\,1 \\ 1\,0\,0 \end{pmatrix} + 5\begin{pmatrix} 0\,0\,0 \\ 0\,0\,1 \end{pmatrix} + 2\begin{pmatrix} 1\,1\,1 \\ 1\,0\,0 \end{pmatrix} + 1\begin{pmatrix} 0\,0\,1 \\ 0\,0\,1 \end{pmatrix}$$

But as we can see, it is possible to combine the second segment with the fourth one, to obtain this decomposition:

$$3\begin{pmatrix}1\,0\,0\\0\,1\,0\end{pmatrix}+5\begin{pmatrix}1\,1\,1\\1\,0\,0\end{pmatrix}+5\begin{pmatrix}0\,0\,0\\0\,0\,1\end{pmatrix}+1\begin{pmatrix}0\,0\,1\\0\,0\,1\end{pmatrix}$$

Therefore, each time we try to add a segment to a current decomposition, we check if we can combine this segment with other segments of the decomposition, in order to reduce the DC and SU_{var} values of the decomposition (this algorithm is not described in this work because of limited space).

The DT value of this solution is equal to 14, DC is equal to 4 and SU_{var} is equal to (2+3+1)=6.

7. The matrix of distances between the segments is as follows:

$$\begin{bmatrix} & S_1 & S_2 & S_3 & S_4\\ S_1 & 0 & 2 & 1 & 2\\ S_2 & 2 & 0 & 3 & 2\\ S_3 & 1 & 3 & 0 & 3\\ S_4 & 2 & 2 & 3 & 0\end{bmatrix}$$

We see that by doing a 2-edges exchange move ($S_3 + S_1 + S_2 + S_4$), we can improve the SU_{var} objective by one unit. We obtain the following neighbor:

$$5\begin{pmatrix}0\,0\,0\\0\,0\,1\end{pmatrix}+3\begin{pmatrix}1\,0\,0\\0\,1\,0\end{pmatrix}+5\begin{pmatrix}1\,1\,1\\1\,0\,0\end{pmatrix}+1\begin{pmatrix}0\,0\,1\\0\,0\,1\end{pmatrix}$$

The evaluation vector z of this solution is thus equal to (14,4,5).

We have thus obtained a new potentially efficient solution, and we can again apply the neighborhood from this solution:

1. We select the third segment:

$$\begin{pmatrix}1\,1\,1\\1\,0\,0\end{pmatrix}$$

2. We select the first line. For this line, $l = 0$ and $r = 4$. We modify this line by putting $r = 3$. We obtain this segment:

$$\begin{pmatrix}1\,1\,0\\1\,0\,0\end{pmatrix}$$

3. We put this segment at the first place of the new decomposition.
4. The current coefficient of this segment is equal to 5. As for this segment, the maximal feasible coefficient that we can put is 5, we keep the coefficient equal to 5. The remaining matrix is:

$$\begin{pmatrix}3\,0\,6\\0\,3\,6\end{pmatrix}$$

5. The first segment that we can consider is:

$$\begin{pmatrix} 0\,0\,0 \\ 0\,0\,1 \end{pmatrix}$$

We add it, with a coefficient equal to 5, that is its current coefficient. The remaining matrix is:

$$\begin{pmatrix} 3\,0\,6 \\ 0\,3\,1 \end{pmatrix}$$

The second segment is:

$$\begin{pmatrix} 1\,0\,0 \\ 0\,1\,0 \end{pmatrix}$$

We add it, with a coefficient equal to 3, that is its current coefficient. The remaining matrix is:

$$\begin{pmatrix} 0\,0\,6 \\ 0\,0\,1 \end{pmatrix}$$

The last segment in the initial decomposition is:

$$\begin{pmatrix} 0\,0\,1 \\ 0\,0\,1 \end{pmatrix}$$

We add it, with a coefficient equal to 1, that is its current coefficient.

6. The remaining matrix is:

$$\begin{pmatrix} 0\,0\,5 \\ 0\,0\,0 \end{pmatrix}$$

The decomposition of this matrix with the heuristic of Engel gives:

$$5\begin{pmatrix} 0\,0\,1 \\ 0\,0\,0 \end{pmatrix}$$

The decomposition obtained is thus:

$$5\begin{pmatrix} 1\,1\,0 \\ 1\,0\,0 \end{pmatrix} + 5\begin{pmatrix} 0\,0\,0 \\ 0\,0\,1 \end{pmatrix} + 3\begin{pmatrix} 1\,0\,0 \\ 0\,1\,0 \end{pmatrix} + 1\begin{pmatrix} 0\,0\,1 \\ 0\,0\,1 \end{pmatrix} + 5\begin{pmatrix} 0\,0\,1 \\ 0\,0\,0 \end{pmatrix}$$

But as we can see, it is possible to combine the second segment with the last one, and then the result of this combination with the fourth segment to obtain this decomposition:

$$5\begin{pmatrix} 1\,1\,0 \\ 1\,0\,0 \end{pmatrix} + 3\begin{pmatrix} 1\,0\,0 \\ 0\,1\,0 \end{pmatrix} + 6\begin{pmatrix} 0\,0\,1 \\ 0\,0\,1 \end{pmatrix}$$

with a SU_{var} value equal to (1+2)=3.

7. The matrix of distances between the segments is as follows:

$$\begin{bmatrix} & S_1 & S_2 & S_3 \\ S_1 & 0 & 1 & 2 \\ S_2 & 1 & 0 & 2 \\ S_3 & 2 & 2 & 0 \end{bmatrix}$$

   We see that it is impossible to improve the SU_{var} objective by doing 2-edges exchange moves.

   The DT value of this solution is equal to 14, and the DC and SU_{var} values are equal to 3 (which are the optimal values [6]).

Therefore, by applying two times the neighborhood from a solution that minimizes the DT objective, we have found a solution that minimizes the DC and SU_{var} values.

It should be noted that we have found one more potential non-dominated point for this problem: the point (10,4,4).

For this small problem, we have thus found three potential non-dominated points: (9,4,6), (10,4,4) and (14,3,3).

3.3 Final Optimization Step

For each potentially efficient solution found at the end of the PLS method, we apply the LKH heuristic, to eventually improve the SU_{var} value of the solutions.

4 Results

4.1 Rules for the Heuristic of Engel

We experiment here the different rules for the selection of the intervals in the heuristic of Engel. The rule “Min” is to take the first feasible interval which is the closest to the preceding interval, the rule “First” is to take the first feasible interval, the rule “Last” is to take the last feasible interval and the rule “Kali” is the rule developed by Kalinowski (see section 3.1).

To compare the four rules, we use the same instances of Engel, that is, matrix 15x15 with randomly generated elements (uniformly distributed) between zero and the parameter L. For each value of L we make the average on 1000 different matrices for three values: the DC objective, the SU_{var} objective and the SU_{var} optimized objective which is the value of SU_{var} obtained after optimization of the sequence of segments with the LKH heuristic.

The results are given at table 1, for L going from 3 to 16.

We remark that for the DC objective, the “Kali” rule is better than the others. But if we want to minimize the SU_{var} objective, we remark that the “Last” rule allows to obtain better results for the SU_{var} optimized objective. If we do not consider the optimization step, the rule “First” is the best for $L = 3$ and $L = 4$, the “Last” rule is the best for L going from 5 to 13 and the rule “Min” is the best for L going from 14 to 16.

Table 1. DC, SU_{var}, SU_{var} optimized values obtained by the different rules. A=15x15.

	DC				SU_{var}				SU_{var} optimized			
L	Min	First	Last	Kali	Min	First	Last	Kali	Min	First	Last	Kali
3	10.32	10.32	9.93	**9.72**	79.69	**69.57**	70.38	113.68	77.43	63.05	**61.23**	103.14
4	11.73	11.74	11.33	**10.94**	94.22	**83.74**	84.10	127.63	91.72	74.72	**71.63**	114.44
5	12.69	12.69	12.29	**11.76**	104.81	96.55	**95.11**	137.43	101.82	84.85	**80.34**	122.28
6	13.56	13.56	13.14	**12.50**	112.35	107.29	**105.53**	146.62	108.98	93.19	**88.62**	129.35
7	14.27	14.26	13.85	**13.12**	119.49	117.02	**114.22**	153.55	115.75	101.14	**95.83**	134.90
8	14.91	14.90	14.48	**13.71**	126.53	126.12	**122.94**	160.69	122.39	108.12	**102.30**	140.40
9	15.52	15.46	15.09	**14.20**	133.32	133.65	**129.75**	166.67	128.38	114.32	**107.66**	144.95
10	16.04	15.98	15.59	**14.69**	139.10	141.08	**137.05**	172.46	134.04	120.17	**113.93**	149.60
11	16.46	16.37	16.01	**15.06**	144.16	147.09	**142.57**	177.08	138.56	125.22	**118.31**	152.92
12	16.92	16.81	16.49	**15.46**	149.41	153.17	**148.31**	182.08	143.47	130.06	**122.71**	156.71
13	17.30	17.18	16.87	**15.81**	153.56	158.54	**153.42**	186.16	147.34	134.47	**126.85**	160.10
14	17.62	17.53	17.21	**16.13**	**157.19**	162.91	157.96	189.64	150.70	138.10	**130.82**	162.75
15	18.01	17.91	17.55	**16.52**	**162.00**	168.27	163.02	194.47	154.91	142.46	**135.04**	166.35
16	18.29	18.16	17.89	**16.79**	**166.22**	172.34	167.41	197.64	158.84	146.25	**138.53**	169.18

Therefore, the "Last" rule allows to considerably reduces the SU_{var} objective comparing to the rule of Kalinowski, even if this rule gives higher DC value on average. The running time of the heuristic of Engel with each of these rules is negligible.

4.2 Pareto Local Search

The initial population of PLS is updated with two solutions: the first one is a good approximation of lexmin(DT,DC,SU_{var}) (obtained with the "Kali" rule and LKH) and the second one is a good approximation of lexmin(DT,SU_{var},DC) (obtained with the "Last" rule and LKH). If there is one solution that dominates the other, the population will be composed of only one solution: the non-dominated one.

For the neighborhood, we adopt the following choices:

- We try all the possible segments for the segment that we put at the beginning of the new decomposition.
- Either we do not modify the segment or we modify it by trying all the possibilities of modification. We modify each line of the segment separately, by considering all the feasible possibilities for each line (equal to maximum 8).
- We try all the feasible coefficients for the segment that we put at the beginning of the decomposition.
- The remaining matrix is decomposed with the heuristic of Engel ("Last" rule).

If the number of segments of the current decomposition is equal to K and the maximal value of the matrix equal to L, a crude bound for the number of neighbors generated is equal to $KL(8m+1)$.

To experiment the method, we use the same instances than in section 4.1. As no state-of-the-art results are known for this multiobjective problem, we use very specific indicators to measure the quality of the approximations obtained.

At table 2, we report the number of potential non-dominated points found (NNDP), the number of phases of the PLS method before convergence and the running time in seconds (on a Intel Core 2 Duo T7500 2.2 GHz processor). We indicate for each indicator the minimum, maximum and mean value found. For each value of L we make the average on 20 different matrices.

Table 2. Indicators PLS(1)

	NNDP			Number of phases			Time(s)		
L	Mean	Min	Max	Mean	Min	Max	Mean	Min	Max
3	1.35	1	2	3.55	1	8	1.39	0.28	3.78
4	1.80	1	3	5.35	2	9	3.79	0.84	7.90
5	1.80	1	3	5.50	2	11	5.43	1.14	13.00
6	2.00	1	3	5.40	2	10	7.34	1.51	17.36
7	2.15	1	4	6.10	1	11	10.10	1.22	24.49
8	2.45	1	4	6.55	1	12	12.85	1.26	29.28
9	2.60	1	6	5.90	2	10	15.45	2.95	35.48
10	3.15	2	5	6.35	2	11	18.60	5.02	49.82
11	2.75	1	5	6.30	2	11	21.42	4.29	46.24
12	2.95	1	5	5.95	2	11	22.40	4.34	49.20
13	2.95	1	6	5.75	1	11	28.63	2.42	93.13
14	2.95	1	5	6.10	2	14	30.28	4.60	72.08
15	3.50	1	10	6.95	2	18	40.77	5.82	98.94
16	2.75	1	5	5.20	2	12	28.08	5.56	83.57

We remark that:

- The number of potential non-dominated points is not high: between 1 and 10, with a mean value between 1 and 4. The correlation between the objectives seems thus high for these instances.
- The mean number of phases is included between 3 and 7, what means that the neighborhood is efficient since at each phase improvements are realized. Please remind that if there is no new potential efficient solution generated during a phase, the PLS method stops since a Pareto local optimum set has been found.
- The mean running time is acceptable, between 1 and 41 seconds. But for some instances, the running time can be higher, until 99 seconds for an instance with $L = 15$.

To evaluate the quality of the results, we evaluate the improvement of DC by comparing the best value obtained with PLS to the value obtained by the heuristic of Engel with the "Kali" rule. We evaluate the improvement of SU_{var} by comparing the best SU_{var} value obtained with PLS to the SU_{var} value obtained with the heuristic of Engel with the "Kali" rule and LKH and to the value

Table 3. Indicators PLS(2)

	DT optimal			DT not necessary optimal		
	% DC	% SU_{var}		% DC	% SU_{var}	
L	Kali	Kali+LKH	Last+LKH	Kali	Kali+LKH	Last+LKH
3	0.00	44.35	12.49	0.45	44.52	12.77
4	0.42	40.11	8.12	0.42	40.11	8.12
5	0.83	37.90	12.20	0.83	37.90	12.20
6	0.38	33.93	7.98	0.38	33.93	7.98
7	0.00	31.18	7.74	0.00	31.24	7.83
8	0.33	33.32	7.79	0.33	33.32	7.79
9	0.00	31.06	7.97	0.00	31.23	8.17
10	0.33	28.68	6.75	0.33	28.98	7.12
11	0.31	26.11	5.61	0.31	26.47	6.06
12	0.00	25.30	6.81	0.00	25.30	6.81
13	0.31	25.36	6.11	0.31	26.25	7.19
14	0.29	24.69	5.70	0.29	24.78	5.81
15	0.29	24.44	6.90	0.29	25.76	8.53
16	0.00	22.59	6.65	0.00	22.89	7.00

obtained with the heuristic of Engel with the "Last" rule and LKH. Two cases are distinguished: initially we evaluate the improvements made if we keep the optimal value for DT, and secondly, we have no restrictions on the DT objective.

Results are shown at table 3 (mean values). We see that the improvements of the DC value are very small. Indeed, the heuristic of Engel with the "Kali" rule is known to give near-optimal results for lexmin(DT,DC) on random instances [11]. On the other hand, the improvements of the SU_{var} values are remarkable. Comparing to the values obtained by the heuristic of Engel with the "Kali" rule and LKH, we obtain improvements from 22 % to 44 %. Comparing to the heuristic of Engel with the "Last" rule and LKH (which is one of the initial solution of PLS), we obtain improvements from 5 % to 12 %.

We also see that allowing to deteriorate the DT value is only interesting for some instances. Both objectives DC and SU_{var} seem thus very correlated with the DT objective.

5 Conclusion and Discussion

We have presented in this paper first results for the multiobjective decomposition of positive integer matrices, within the framework of the radiotherapy treatment.

More experimentations on different types of instances (size and characteristics) and on real instances will have to be carried out to obtain more information about the correlation between the objectives, and to validate the approach.

But first results obtained with PLS are encouraging since the method allows to improve state-of-the-art results. This method could be the basis for a more elaborated method, like an evolutionary multiobjective algorithm where PLS would be used as an intensification operator.

Acknowledgments

We thank Céline Engelbeen for introducing this subject and for several discussions. T. Lust thanks the "Fonds National de la Recherche Scientifique" for a research fellow grant (Aspirant FNRS).

References

1. Angel, E., Bampis, E., Gourvès, L.: A dynasearch neighborhood for the bicriteria traveling salesman problem. In: Gandibleux, X., Sevaux, M., Sörensen, K., T'kindt, V. (eds.) Metaheuristics for Multiobjective Optimisation. Lecture Notes in Economics and Mathematical Systems, vol. 535, pp. 153–176. Springer, Berlin (2004)
2. Baatar, D., Hamacher, H., Ehrgott, M., Woeginger, G.: Decomposition of integer matrices and multileaf collimator sequencing. Discr. Appl. Math. 152, 6–34 (2005)
3. Bortfeld, T., Boyer, A., Kahler, D., Waldron, T.: X-ray filed compensation with multileaf collimators. Int. J. Radiat. Oncol. Biol. Phys. 28(3), 723–730 (1994)
4. Burkard, R.: Open problem session. In: Conference on Combinatorial Optimization, Oberwolfach, pp. 24–29 (November 2002)
5. Ehrgott, M., Güler, Ç., Hamacher, H.W., Shao, L.: Mathematical optimization in intensity modulated radiation therapy. 4OR: a Quarterly Journal of Operations Research 6(3), 162–199 (2008)
6. Ehrgott, M., Hamacher, H., Nußbaum, M.: Decomposition of matrices and static multileaf collimators: a survey. In: Alves, C., Pardalos, P., Vincente, L. (eds.) Optimization in medicine, pp. 27–48. Springer, Berlin (2007)
7. Engel, K.: A new algorithm for optimal multileaf collimator field segmentation. Discrete Appl. Math. 152(1-3), 35–51 (2005)
8. Hansen, P.: Bicriterion path problems. Lecture Notes in Economics and Mathematical Systems 177, 109–127 (1979)
9. Hansen, P., Mladenovic, N.: First vs. best improvement: An empirical study. Discrete Appl. Math. 154, 802–817 (2006)
10. Helsgaun, K.: An effective implementation of the lin-kernighan traveling salesman heuristic. European Journal of Operational Research 126, 106–130 (2000)
11. Kalinowski, T.: Realization of intensity modulated radiation fields using multileaf collimators. In: General Theory of Information Transfer and Combinatorics, pp. 1010–1055. Springer, Heidelberg (2006)
12. Paquete, L., Chiarandini, M., Stützle, T.: Pareto Local Optimum Sets in the Biobjective Traveling Salesman Problem: An Experimental Study. In: Gandibleux, X., Sevaux, M., Sörensen, K., T'kindt, V. (eds.) Metaheuristics for Multiobjective Optimisation. Lecture Notes in Economics and Mathematical Systems, vol. 535, pp. 177–199. Springer, Heidelberg (2004)
13. Siochi, R.A.C.: Minimizing static intensity modulation delivery time using an intensity solid paradigm. Int. J. Radiat. Oncol. Biol. Phys. 43, 671–680 (1999)
14. Taşkin, Z., Smith, J., Romeijn, H., Dempsey, J.: Optimal multileaf collimator leaf sequencing in IMRT treatment planning. Technical report, Department of Industrial and Systems Engineering, University of Florida (2007)
15. Wake, G.M.G.H., Boland, N., Jennings, L.S.: Mixed integer programming approaches to exact minimization of total treatment time in cancer radiotherapy using multileaf collimators. Computers and Operations Research 36(3), 795–810 (2009)

Comparison of MCDM and EMO Approaches in Wastewater Treatment Plant Design

Jussi Hakanen and Timo Aittokoski

Department of Mathematical Information Technology, P.O. Box 35 (Agora),
FI-40014 University of Jyväskylä, Finland
jussi.hakanen@jyu.fi, timo.aittokoski@jyu.fi

Abstract. This paper describes applying a new EMO algorithm for a real-world optimization problem arising from wastewater treatment. In addition, the results are compared to the ones obtained by applying the interactive multiobjective optimization tool IND-NIMBUS to the same problem. How the comparison should be made is not self-evident but we try to highlight the pros and cons of both the evolutionary multiobjective optimization and the multiple criteria decision making fields in the context of the wastewater treatment plant design problem considered.

1 Introduction

Many real-world optimization problems are multiobjective by their nature. In addition, sometimes the objective and the constraint function values originate from some numerical simulation procedure which is implementd as a (black-box) simulator. In that case, we can not usually make any assumptions of the behaviour of the functions (e.g., convexity, differentiability) to be utilized in optimization. Further, the simulation itself can be time consuming which means that obtaining the function values can take lots of time and it should be taken into account in optimization. Multiple conflicting objective functions can lead us to infinitely many optimal solutions called Pareto optimal solutions that are mathematically equivalent. The ultimate goal is that at the end of the optimization process, we need to have only one optimal solution which is best for the problem considered and is to be implemented in practice. Therefore, we need to have optimization tools that can handle multiple conlicting objectives, do not utilize any specific properties of the functions in the optimization problem and are able to produce the best solution for a decision maker (DM) to proceed with.

A traditional way to solve multiobjective optimization problems has been to apply the methods of multiple criteria decision making (MCDM) [5,17,26,27]. Recently, the evolutionary multiobjective optimization (EMO) approach [6,7,12,33] has been gaining ground and challenged the classical MCDM methods, especially, in solving engineering problems with multiple criteria. The methods of EMO are well suited for problems where little or no assumptions can be made of the objective or constraint functions because, for example, they do not utilize gradient information.

M. Ehrgott et al. (Eds.): EMO 2009, LNCS 5467, pp. 350–364, 2009.

The main difference between the MCDM and the EMO methods is that the former work with only one solution at a time while the latter operates on a set of solutions called a population. In other words, the MCDM methods produce one Pareto optimal solution at a time by, for example, solving a scalarized single objective problem that produces a Pareto optimal solution. On the other hand, the EMO approach tries to evolve a set of solutions so that they finally cover the whole Pareto front evenly and are as close as possible to the actual front.

The methods of multiobjective optimization can be classified, for example, according to the role of the DM in the solution process (see, for example, [17]). The classes are i) the methods that do not utilize any preference information, ii) a priori methods, iii) a posteriori methods, and iv) interactive methods. The methods in the first class do not utilize any preference information because, for example, the DM might not be available. A priori methods ask the preferences of the DM before the solution process, whereas a posteriori methods utilize the DM after the Pareto optimal solutions are computed. The last class consists of methods where the DM is able to input his/her preferences iteratively during the solution process. Note that the EMO methods fall into the class of a posteriori methods according to this classification. Recently, there has been efforts to combine the MCDM and the EMO approaches in order to utilize the strengths of both approaches (see, for example, [3]).

In this paper, we consider a real-world optimization problem arising from wastewater treatment plant (WWTP) design. The design problem has three conflicting objectives, it is based on numerical simulation of the wastewater treatment process and the objective function values come from a black-box process simulator which means that we do not have any gradients available. Thus, this problem is a typical real-world optimization problem. To solve the WWTP design problem we utilize both an MCDM and an EMO approach. To be more precise, we use an interactive multiobjective optimization tool IND-NIMBUS [18] and a new EMO algorithm introduced in [1]. In addition, we make a comparison of the results obtained, although the approaches are quite different in nature which makes the comparison difficult and it is not at all trivial how the comparison should be made. The issue of comparing the results of the MCDM and the EMO approaches is also very interesting in a more general setting but that is out of the scope of this paper.

Optimization of WWTP design by modelling and simulation usually involves comparisons of different process schemes or control strategies. The behaviour of the considered solutions is simulated, and the results are then compared to each other, usually in terms of investment or operational costs. The comparison can be done either by engineering judgement, as is usually the case or using an optimization algorithm (see e.g., [9,23]). However, using only one objective function instead of several individual criteria hides the interdependencies between different criteria and, thus, makes it difficult for the DM, who might be, e.g., a designer or a plant operator, to assess the true optimality of the solution. The DM may also have non-quantifiable priorities, such as operational stability and ease of operation, which may depend on many decision variables to be optimized.

Therefore, for a truly optimal design, the procedure must present the DM with solutions based on a multiobjective optimization approach, out of which (s)he can choose the best ones to be elaborated further.

The WWTP design has been previously considered by optimizing only one objective function, that in one way or another describes the costs of the process to be minimized (see, for example, [9,23]). So far, we haven't found any papers that deal with simulation-based multiobjective WWTP design. As the first attempt to tackle simulation-based multiobjective WWTP design, we solved the problem considered here in [10] by utilizing IND-NIMBUS. The interactive solution process of that study is briefly described in Section 3.3. In this paper, the new EMO algorithm has been applied to the same WWTP design problem and it is the first time it has been applied to any real-world optimization problem.

The rest of this paper is organized as follows. First in Section 2, we describe both the EMO and the MCDM approaches and, especially, the representative methods considered in this paper, namely the new EMO algorithm and IND-NIMBUS. Section 3 is devoted to the real-world application related to wastewater treatment plant design. We present solution processes and results for both the new EMO algorithm and IND-NIMBUS and compare the results obtained. Finally in Section 4, we make some conluding remarks about the study.

2 Possible Approaches

We consider multiobjective optimization problem of the form

$$\begin{aligned} &\text{minimize} \;\; \{f_1(\boldsymbol{x}), \ldots, f_k(\boldsymbol{x})\} \\ &\text{subject to } \boldsymbol{x} \in S, \end{aligned} \tag{1}$$

where k conflicting objective functions $f_i(\boldsymbol{x})$ are minimized simultaneously. The objective vectors $\boldsymbol{f}(\boldsymbol{x}) = (f_1(\boldsymbol{x}), \ldots, f_k(\boldsymbol{x}))^T$ belong to the objective space. The set S denotes the feasible region of the continuous decision (variable) vectors $\boldsymbol{x} \in \mathbb{R}^n$. The optimal solutions of problem (1) are called Pareto optimal solutions. A point $\boldsymbol{x}^* \in S$ is called a *Pareto optimal* solution if there does not exist another decision vector $\boldsymbol{x} \in S$ such that $f_i(\boldsymbol{x}) \leq f_i(\boldsymbol{x}^*)$ for all $i = 1, \ldots, k$ and $f_j(\boldsymbol{x}) < f_j(\boldsymbol{x}^*)$ for at least one j. The set of all Pareto optimal solutions is called the Pareto optimal set or the Pareto front. Note that there usually exists infinitely many Pareto optimal solutions that are all mathematically equivalent. Therefore, the best Pareto optimal solution depends on the problem in question and to select the best one, we need some additional information. The person who has the knowledge about the problem in question and can provide the additional information is the DM.

Next, we briefly introduce the methods used in this paper. In other words, we describe the new EMO algorithm and the interactive multiobjective optimization tool IND-NIMBUS.

2.1 EMO Approach

In this paper, we utilize a new, efficient EMO algorithm proposed in [1] for the first time to solve a real-world simulation-based problem. In contrast to scalarization based approaches, in EMO approaches, an approximation of the Pareto optimal set is created without any user intervention, which makes the basic idea of such approaches rather appealing; the DM is involved in the solution process only after the most time consuming computation is finished.

In [1], some difficulties of widely used dominance diversity preservation mechanism based EMO approaches were pointed out, such as convergence problems, deterioration of the population, difficulty to choose the proper population size and also problems with diversity maintenance. The algorithm presented in [1], based essentially on unrestricted population size, and on the point generation scheme of Differential Evolution [28], should overcome the previous issues, which are discussed more in detail in the following paragraphs.

Several current EMO approaches (e.g., [7,12,33]) are based on the concept of dominance, added with some mechanism to maintain a good diversity of the solutions in the objective space. With this approach, the population is usually ranked based on dominance, and naturally non-dominated solutions are considered better, and favored in reproduction. Further, if two solutions with the same rank must be ordered (typically while selecting members for the next generation), it is usually done using some mechanism to select the one which is located in a less crowded region of the objective space. In addition to the current population, also some external archive of good solutions may be maintained and used both for dominance and diversity assessment.

Convergence problems [11,16,25] of widely used EMO algorithms stem from the concept of dominance, in conjunction with the diversity preservation mechanisms used. If there is excess of non-dominated solutions to fit into the population, some of them must be pruned using a diversity preservation mechanism. In this case, some non-dominated solution located very near the Pareto optimal set may be replaced by some other non-dominated solution which improves diversity, but is at the same time located much farther from the Pareto optimal set. This behavior leads to oscillation in the solution quality, and prevents convergence to the Pareto optimal set.

If the population is deteriorated, in the history of all evaluated points during the optimization run, there exist solutions that dominate the ones in the current population. This behavior is closely related to the aforementioned convergence problems. When the population deteriorates, it is reasonable to say that some number of objective function evaluations has been wasted, and the population does not reflect the best possible solutions found during the optimization process.

Choosing the proper population size is not straightforward, because too few points cannot represent the characteristics of the Pareto optimal set properly. On the other hand, too many points may hinder the performance, because a large portion of the population may consist of dominated solutions, especially in the beginning of the optimization process when there are not enough non-dominated solutions found to fill the population. The population size is also related to the convergence problems, and deterioration. Obviously, the larger the population, the closer to the Pareto optimal set the algorithm may converge before the population starts to oscillate. This behavior was demonstrated in [1].

Diversity maintenance may also be a problematic issue. For example, the crowding distance of NSGA-II [7] (used also in other algorithms, e.g. [22,24]), actually works only in two dimensions, and may fail also in this case [13]. In [14], a viable way to implement the pruning needed in the diversity maintenance is presented. A relatively high computational cost may be considered as a drawback of this approach, but this in turn is probably not very significant if objective functions are costly.

In the new EMO algorithm, the focal point is to give up the idea of a fixed population size, and instead keep all the non-dominated solutions in the population. In this way, all problems discussed above are overcome, at least to some extent. The convergence to the Pareto optimal set is gained because the population cannot oscillate. There is no need to select the population size, and because of that in the beginning of the process the algorithm performs efficiently because the population is not filled to some predetermined size with bad quality dominated solutions. Also, the number of solutions in the population is increasing over time, thus having a better capability to capture the characteristics of the Pareto optimal set. The population of the proposed algorithm cannot deteriorate because it always contains all non-dominated solutions. Further, no explicit diversity preservation mechanism is needed. This may seem counter-intuitive, but it is essential to realize that neither NSGA-II nor DE-based EMO's (see, for example, [15,32]) strive to actively generate evenly spread solutions around the Pareto optimal set, rather they just select solutions located in less crowded regions to following populations. Thus, if all solutions are kept, also the diversity is maintained at least as well as, for example, in NSGA-II.

With the new EMO algorithm, all data about the behavior of objective functions gained during the optimization run is put in use, in contrast to several EMO approaches which exploit only information contained in the current population, and possibly in some rather small additional external archive. In other words, no objective function evaluations are wasted. This is highly important in solving real-world optimization problems that can be computationally demanding. Intuitively that approach feels advantageous, and numerical results presented in [1] comparing the performance of NSGA-II and the proposed algorithm also support this conception, as the proposed algorithm clearly outperformed the NSGA-II.

2.2 MCDM Approach

In this paper, our MCDM approach is based on the interactive multiobjective optimization method NIMBUS [17,19]. The reason behind this choice is that interactive methods have been found promising in solving real-world optimization problems with more than two objectives because they enable the DM to adjust his/her preferences while (s)he obtains new information during the interactive solution process and they are computationally efficient because they only need to generate few Pareto optimal solutions [17]. In addition, the WWTP design problem considered has already been solved with an implementation of the NIMBUS method [10].

The NIMBUS method is based on classification of the objective functions at the current Pareto optimal solution. The idea is to start with the so-called neutral compromise solution (NCS) that is a Pareto optimal solution approximately in the middle of the Pareto optimal set [31] and to generate a series of Pareto optimal solutions utilizing the preferences of the DM iteratively. The ultimate goal is to find the Pareto optimal solution that best satisfies the hopes of the DM (and to be implemented in practice).

In practice, the objective function values of the current Pareto optimal solution are shown to the DM and (s)he is asked to classify the objective functions f_i into the classes where the value of f_i

- should be improved as much as possible ($i \in I^{imp}$),
- should be improved until some specified aspiration level $\bar{z}_i$ ($i \in I^{asp}$),
- is satisfactory at the moment ($i \in I^{sat}$),
- can impair till some specified bound ϵ_i ($i \in I^{bound}$) and
- can change freely ($i \in I^{free}$).

According to the classification information provided by the DM, single objective subproblems are formed [19]. In other words, the multiobjective problem is scalarized. One of the subproblems is the standard NIMBUS subproblem and the other three subproblems are based on reference points [19]. The reference point-based subproblems come from the satisficing trade-off [21] and the GUESS [4] methods while one of them is based on achievement (scalarizing) functions [30]. Formulations of the subproblems can be found in [19].

New Pareto optimal solutions are obtained by solving the subproblems with a suitable single objective optimizer [19]. Note that the subproblems can be nonconvex depending on the objective and constraint functions of the original problem and, therefore, the solutions of the subproblems are Pareto optimal only when their global optima have been obtained. The new solutions obtained are shown to the DM and (s)he selects the best one to proceed with. The iterative solution process continues until the DM is satisfied, that is, when the DM want's not to improve any objective function. The other possibility to produce new Pareto optimal solutions in the NIMBUS method is to generate intermediate Pareto optimal solutions between any two Pareto optimal solutions already computed. Further information about the NIMBUS method can be found in [19].

To solve practical multiobjective optimization problems, an implementation of the NIMBUS method called IND-NIMBUS [18] has been developed (`http://ind-nimbus.it.jyu.fi`). IND-NIMBUS offers a graphical user-interface for the interaction between the DM and the NIMBUS method while it also provides graphical classification of the objective functions and several different visualizations to compare and evaluate the Pareto optimal solutions obtained during the interactive solution process. More information about IND-NIMBUS can be found in [18]. In [10], the WWTP design problem was solved with IND-NIMBUS as already mentioned.

3 Numerical Results

Next, we present the WWTP design problem that we solved with both the new EMO algorithm and IND-NIMBUS. In addition to presenting the results and describing the solution processes, we compare the results obtained.

3.1 On Wastewater Treatment Plant Design

Operational requirements of wastewater treatment plants, notably the effluent limits of nitrogen and phosphorus, are becoming more and more strict because of increased emphasis on environmental values. Consequently, more complex wastewater treatment processes are gaining ground. At the same time, the requirements for economical efficiency (for example, minimizing plant footprint and the consumption of chemicals and energy) as well as for operational reliability are also tightening. This makes the design of a WWTP a complex process involving trade-offs between a number of conflicting economical and operational criteria. Therefore, a simplified approach where all the aspects are gathered together, usually as estimated total costs, and optimized is not adequate anymore.

Mathematical modelling of WWTPs began gaining ground in the 1990s when experience on modelling solutions and computer power increased simultaneously. The overwhelming majority of modelling considers the activated sludge process (ASP), globally the most common method of wastewater treatment. Our example problem is also based on ASP. In this process, biomass (which is called activated sludge) suspended in the wastewater to be treated is cultivated and maintained in an aerated bioreactor. The wastewater is purified, i.e. organic carbon, nitrogen and phosphorus are removed, during its retention in the bioreactor. The bioreactor is followed by a clarifier basin, in which the biomass is separated by gravitational settling and returned to the bioreactor, and the treated wastewater is directed as overflow to futher treatment or to discharge. Excess activated sludge is removed from the process and treated separately. The schematical flow sheet of the process is presented in Figure 1.

The activated sludge models used are nonlinear differential equations, reflecting the nonlinear nature of microbial growth and solids separation. The activated

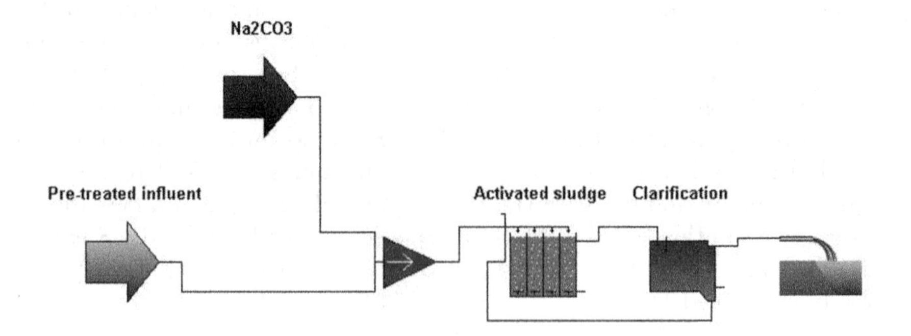

Fig. 1. A flowchart of the activated sludge process considered

sludge process was realized by using the commercial GPS-X process simulator (`http://www.hydromantis.com/software02.html`) that is developed especially for numerical simulation of wastewater treatment processes.

3.2 Problem Description

The process model used in this study describes a nitrifying activated-sludge process and the wastewater treated by the process corresponds to typical Finnish mechanically and chemically pre-treated municipal wastewater. The process performs nitrification, i.e. oxidation of ammonium nitrogen to nitrate nitrogen by autotrophic, slow-growing micro-organisms. The biochemical reactions involved consume a lot of oxygen and alkalinity. Oxygen is supplied by aeration compressors and alkalinity partly by influent wastewater, partly by adding chemicals, e.g. Na_2CO_3. Aeration consumes energy and chemicals cost money, so minimizing the need for aeration and alkalinity addition is important for the operational economy of the plant. Therefore, we consider the WWTP design problem with respect to three objective functions, namely the residual ammonium nitrogen concentration $[gN/m^3]$, the dose of alkalinity chemical $[m^3/d]$ and the consumption of energy by aeration $[kW]$. In what follows, we will denote them by N, A and E, respectively. The primary objective N is the most important one because it describes the quality of the treated wastewater and it should be kept at a sufficiently low level. The other two objective functions should be minimized.

As decision variables affecting the objectives selected, we consider the biomass concentration in the bioreactor (C_{MLSS}) $[kg/m^3]$, alkalinity chemical dosign rate $[m^3/d]$ and the O_2-concentration in the last section of the reactor $[g/m^3]$. C_{MLSS} should be kept as low as possible so that the secondary settler would not be overloaded in case of peak flows. The upper bounds used for the decision variables are 6.0, 500 and 2.5, respectively, while the lower bound for each is zero. Note that all these decision variables are continuous. As constraints of the optimization problem, we set restrictions for alkalinity of treated wastewater, that is, we require that the alkalinity remains between 1.5 and 2.0 moles/m^3. To summarize, we have three objective functions to be minimized, three continuous decision variables and one inequality constraint with both a lower and an upper bound. For a more detailed description of the process model, we refer to [10]. In order to get an idea of the simulation time, one simulation of the process considered, that is, one evaluation of the objective and the constraint functions, took about a couple of seconds with a standard PC. In other words, the problem is clearly more time consuming than standard test problems but not especially time consuming as a real-world optimization problem.

3.3 Solution Process with IND-NIMBUS

As already mentioned, this problem was solved with IND-NIMBUS in [10]. Next, we briefly describe the interactive solution process and the results presented there. The single objective subproblems produced by the NIMBUS method were

Table 1. Objective function values for the Pareto optimal solutions computed during the interactive solution process as well as the actions of the DM

	residual ammonium nitrogen concentration $[gN/m^3]$	alkalinity chemical dosing rate $[m^3/d]$	aeration energy consumption $[kW]$
best	0.03	0.45	308
worst	31.5	354	599
NCS	8.05	218	460
	$I^{asp}, \bar{z}_N = 1.0$	$I^{bound}, \epsilon_A = 330$	I^{sat}
2	3.52	286	490
3	**1.69**	**326**	**506**
4	4.90	298	477
	$I^{asp}, \bar{z}_N = 0.5$	I^{free}	$I^{bound}, \epsilon_E = 510$
5	1.11	336	515
6	**0.55**	**347**	**528**
	$I^{bound}, \epsilon_N = 1.0$	I^{free}	I^{imp}
7	9.36	246	448
8	30.2	7.23	308
9	0.90	333	519
	2 interm.	solutions between	5 & 6
10	***0.92***	***336***	***519***
11	0.72	332	524

solved by using the global Controlled Random Search (CRS) algorithm [2] because we don't know beforehand whether the problem is convex or not. In addition, CRS was given the maximum number of 400 objective function evaluations for each subproblem in order to not keep the DM waiting for too long. The DM involved in the solution process was an expert in the problem field in question.

At the beginning of the interactive solution process with IND-NIMBUS, approximations of the ideal (best) and the nadir (worst) objective vectors are computed. These bounds help the DM in classification because (s)he gets some idea of what kind of values are possible to achieve. In addition, the bounds are used by the NIMBUS method for scaling purposes. The approximations in our case were $(0.03, 0.45, 308)$ and $(31.5, 354, 599)$, respectively.

The solution process starts from the neutral compromise solution and it is the first Pareto optimal solution shown to the DM. In our case, NCS had the objective function values $(8.05, 218, 460)$. The value of N was not tolerable and indicated that, with these settings, the process would not work. Therefore, the DM made the first classification to improve the value of N until 1, and allowed the value of A to increase up to 330. The value of E was satisfactory at the moment. The resulting Pareto optimal solutions were $(3.52, 286, 490)$, $(1.69, 326, 506)$ and $(4.90, 298, 477)$. As can be seen, it was not possible to fully satisfy the classification specified.

Of the new solutions found, the second one seemed to be the most promising for the DM to continue with. The rest of the interactive solution process can be seen from Table 1 where the Pareto optimal solutions preferred in each step of the algorithm are shown as bold face and the final solution selected is shown both in bold face and italics. In addition, the actions of the DM (that is, classification or request for intermediate Pareto optimal solutions) are also shown.

At this point, the DM was satisfied and did not want to continue. As the final solution to be implemented, the DM could consider any one of the solutions 5, 9, 10, and 11 because the desired level of the main criterion, concentration of ammonium nitrogen in the effluent, is fulfilled. Since we must choose only one of them, the DM preferred number 10 ($N = 0.92gN/m^3$, $A = 336m^3/d$, $E = 519kW$) because it had the smallest biomass concentration in the bioreactor which makes it better for implementation.

3.4 Solution Process with the New EMO Algorithm

As already mentioned, here the new EMO algorithm is applied to a real-world multiobjective optimization problem for the first time. We set a budget for 5000 objective function evaluations to the new EMO algorithm because we want to compare the results to the ones obtained by IND-NIMBUS. Remember that with IND-NIMBUS, there was a limit of 400 evaluations for CRS in solving the single objective subproblems and there were altogether 14 such problems solved (ideal value computation for each objective and 11 Pareto optimal solutions) which equals to 5600 evaluations. That is the only termination criterion for the new algorithm because it has a varying population size. In addition, the parameters for DE-based point generation used were $CR = 0.1$ and $F = 0.8$.

As the WWTP design problem has one inequality constraint in addition to box constraints, we further enhanced the new EMO algorithm to handle also such constraints in this case. This was done simply by forbidding the solutions violating the constraints to enter the population, prior to selection of points to participate in the generation of the new trial point. Further, we concentrated on the most interesting part of the Pareto front where the primary objective function, the residual ammonium nitrogen concentration, obtains satisfying values (that is, values less than 2.0). The points with higher values of the primary objective function were also excluded from the population in the same way.

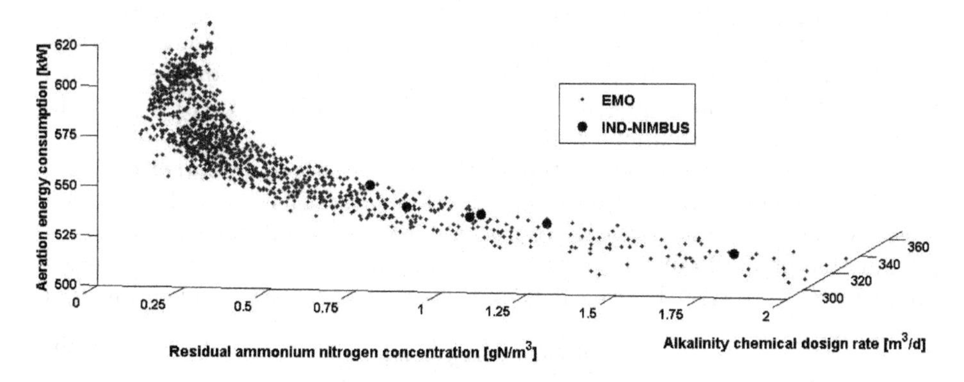

Fig. 2. An approximation of the Pareto front generated by the new EMO algorithm and Pareto optimal solutions produced by IND-NIMBUS

The results obtained with the new EMO algorithm can be seen in Figure 2. There the appromization of the Pareto front is illustrated as well as the six solutions obtained with IND-NIMBUS whose residual ammonium nitrogen concentration was less than 2.0. There are altogether 1271 points in the approximation shown in Figure 2. In other words, that is the final population size and the number of non-dominated solutions. As can be seen from Figure 2, the Pareto front for this problem seems to be more or less convex and well behaving. Further, it seems by looking at the picture that the new EMO algorithm has obtained as good solutions as IND-NIMBUS. Both the results of the new algorithm and IND-NIMBUS are more thoroughly analyzed in the next section where the comparison of the results is performed.

3.5 Comparison of the Results

A comparison between the results with IND-NIMBUS and the new EMO algorithm is not trivial. The former produced 14 Pareto optimal solutions of which one was selected as the final solution while the latter produced an approximation of the whole Pareto front but does not take into account the selection of the final solution. The idea behind our comparison in this paper is the following: if you have a fixed budget for function evaluations to be used in solving the WWTP design problem described here, is it better to use the new EMO algorithm or IND-NIMBUS? Note that in the case of two objective functions, it is usually better to use some EMO method instead of any interactive method.

Because the Pareto front is only three dimensional and it seems to be well behaving, it could be argued that the DM can choose the final solution by looking at Figure 2. In the case of a more complex three dimensional Pareto front or more than three objectives, the visualization is not reasonable, but some more sophisticated method (e.g., [8,20,29]) needs to be used to explore the complex or high dimensional approximated Pareto optimal set. In any case, the question to be answered is "are the solutions produced by different methods of equal accuracy"? We studied this by applying a local optimizer starting from the solutions produced by IND-NIMBUS. In addition, we selected solutions from the approximation produced by the new EMO algorithm that are closest to the ones produced by IND-NIMBUS and applied the same local optimizer starting from them. Then, we computed how large was the relative improvement gained with the local optimizer, in other words, how far from the actual Pareto front the above-mentioned solutions were.

To be more precise, the local optimizer that we used was `fminsearch` of MATLAB (with `MaxFunEvals`=1000 and `TolFun`=$1.0e^{-8}$) which is an implementation of the Nelder-Mead direct search algorithm. The nearest solutions from the approximation of the Pareto front were chosen so that for each solution $\hat{\boldsymbol{z}} \in \mathbb{R}^k$ produced by IND-NIMBUS we calculated the value of the achievement function

$$\max_{i=1,\ldots,k} \left[\frac{f_i(\boldsymbol{x}) - \hat{z}_i}{z_i^{nad} - z_i^{ideal}} \right] \tag{2}$$

for the whole final population and selected the one with the smallest value. The vectors $\boldsymbol{z}^{ideal}$ and $\boldsymbol{z}^{nad}$ are the approximations of the ideal and the nadir objective vectors, respectively. The relative improvement was calculated with respect to both the ideal and the nadir objective vectors. That is, we computed

$$\text{Imp}^{ideal} := \frac{||\boldsymbol{f}(\boldsymbol{x}) - \boldsymbol{f}(\tilde{\boldsymbol{x}})||}{||\boldsymbol{z}^{ideal}||} \text{ and } \text{Imp}^{nad} := \frac{||\boldsymbol{f}(\boldsymbol{x}) - \boldsymbol{f}(\tilde{\boldsymbol{x}})||}{||\boldsymbol{z}^{nad}||}, \quad (3)$$

where $\boldsymbol{x}$ is the solution produced by IND-NIMBUS/the new EMO algorithm and $\tilde{\boldsymbol{x}}$ is the solution found by `fminsearch`. To be more precise, $\tilde{\boldsymbol{x}}$ is obtained when function (2) is minimized with respect to the decision variable vector $\boldsymbol{x} \in S$. The norm $|| \cdot ||$ is the normal euclidean norm.

First of all, we considered only those solutions found by IND-NIMBUS that are in the interesting region of the Pareto optimal set, that is, the same region where we restricted the solutions to be in the case of the new EMO algorithm. There were six such solutions, numbered 3, 5, 6, 9, 10, and 11 in Table 1. Those solutions and the closest ones in the approximation of the Pareto front are shown in Table 2. It seems that the solutions are quite close to each other. Table 3 presents the relative improvements (3) computed for the solutions in Table 2. It can be seen, that the local optimizer was not able to improve four of the six solutions of the new EMO algorithm and the other two only very little which is not significant. On the other hand, the solutions produced by IND-NIMBUS could be improved a bit more, but also that improvement was not significant. To summarize, in this WWTP design problem both methods could produce solutions that are in practice in the true Pareto front.

Retrospectively, as the problem seems to be well behaving and almost convex (at least at a macroscopic level), we may think that IND-NIMBUS could have performed as well with less objective function evaluations using some local optimizer instead of CRS if this had been known beforehand. Unfortunately, with

Table 2. Corresponding solutions produced by IND-NIMBUS and the EMO algorithm

IND-NIMBUS	N	A	E	EMO	N	A	E
3	1.690	325.7	506.1		1.762	322.2	505.9
5	1.105	336.0	514.6		1.101	332.0	515.0
6	0.549	346.9	527.7		0.544	346.6	527.9
9	0.899	333.0	519.2		0.934	328.6	518.9
10	0.917	335.5	519.1		0.934	328.6	518.9
11	0.720	332.0	523.9		0.731	329.5	523.6

Table 3. Relative improvements

IND-NIMBUS	Imp^{ideal}	Imp^{nad}	EMO	Imp^{ideal}	Imp^{nad}
3	0.0093	0.0041		0.0	0.0
5	0.0060	0.0027		0.0	0.0
6	0.0007	0.0003		0.0	0.0
9	0.0011	0.0005		0.0001	4.5e-005
10	0.0006	0.0002		0.0001	4.5e-005
11	0.0007	0.0003		0.0	0.0

simulation based optimization problems, the convexity assumption may potentially be very dangerous to be made beforehand, without explicit understanding about the problem behavior.

On the other hand, in four of the six cases, the local search could not improve the results of the new EMO algorithm at all, suggesting that the EMO algorithm has converged to the true Pareto front. What our current analysis lacks, is the information about how early during the optimization run the EMO approach actually had converged to the true Pareto front. It is possible that, say, the last 3000 objective function evaluations have not improved the quality of the approximation, and thus, significant amount of evaluations could have been saved. Anyhow, it is worth to mention, that with the new EMO algorithm further evaluations are not wasted, because as a product of them, the population size grows bigger, and thus the approximation along the Pareto optimal set gets all the time more dense.

4 Conclusions

We have solved a real-world wastewater treatment plant design problem by using both an MCDM and an EMO approach. This was realized by using both an interactive multiobjective tool IND-NIMBUS and a new EMO algorithm based on the point generation scheme of differential evolution and an unrestricted population size to solve the problem. An attempt to compare the results obtained was also made. Comparing the results is not a trivial case because the MCDM and the EMO approaches have quite different goals in solving multiobjective optimization problems. The MCDM approach aims at producing one final Pareto optimal solution while the EMO approach generates an approximation of the whole Pareto front.

In the comparison, we proposed a question: if you have a fixed budget for function evaluations to be used in solving the WWTP design problem described here, is it better to use the new EMO algorithm or IND-NIMBUS? It seems that, at least in this case, it could be argued that it is more convenient to use the new EMO algorithm for the following reasons: the DM is involved in the solution process only after the approximation of the Pareto optimal set is generated, the solution quality of the EMO approach is in practice equal while achieved with less objective function evaluations, and finally, with the new EMO approach the DM gets more information, i.e., it is possible to gain understanding of the whole Pareto optimal set, instead of only certain regions of it, as is the case with the MCDM approach. However, with more complex and/or high dimensional Pareto fronts this argument may not be valid without further study.

Acknowledgements

The authors wish to thank Mr. Kristian Sahlstedt from Pöyry Environment Oy for providing his expertise with the process model and the simulator.

References

1. Aittokoski, T., Miettinen, K.: Efficient Evolutionary Method to Approximate the Pareto Optimal Set in Multiobjective Optimization. In: Proc. of EngOpt 2008, International Conference on Engineering Optimization, Rio de Janeiro (2008)
2. Ali, M.M., Storey, C.: Modified Controlled Random Search Algorithms. International Journal of Computer Mathematics 54, 229–235 (1994)
3. Branke, J., Deb, K., Miettinen, K., Slowinski, R. (eds.): Multiobjective Optimization: Interactive and Evolutionary Approaches. Springer, Heidelberg (2008)
4. Buchanan, J.T.: A Naiive Approach for Solving MCDM Problems: the GUESS Method. Journal of the Operational Research Society 48, 202–206 (1997)
5. Changkong, V., Haimes, Y.Y.: Multiobjective Decision Making: Theory and Methodology. Elsevier Science Publishing Co., Inc., Amsterdam (1983)
6. Deb, K.: Multi-Objective Optimization Using Evolutionary Algorithms. John Wiley & Sons, Ltd., Chichester (2001)
7. Deb, K., Pratap, A., Agarwal, S., Meyarivan, T.: A Fast and Elitist Multiobjective Genetic Algorithm: NSGA-II. IEEE Transactions in Evolutionary Computation 6, 182–197 (2002)
8. Eskelinen, P., Miettinen, K., Klamroth, K., Hakanen, J.: Pareto Navigator for Interactive Nonlinear Multiobjective Optimization. OR Spectrum (to appear)
9. Espírito-Santo, I., Fernandes, E., Araújo, M.M., Ferreira, E.C.: NEOS Server Usage in Wastewater Treatment Cost Minimization. In: Gervasi, O., Gavrilova, M.L., Kumar, V., Laganá, A., Lee, H.P., Mun, Y., Taniar, D., Tan, C.J.K. (eds.) ICCSA 2005. LNCS, vol. 3483, pp. 632–641. Springer, Heidelberg (2005)
10. Hakanen, J., Sahlstedt, K., Miettinen, K.: Simulation-Based Interactive Multiobjective Optimization in Wastewater Treatment. In: Proc. of EngOpt 2008, International Conference on Engineering Optimization, Rio de Janeiro (2008)
11. Hanne, T.: On the Convergence of Multiobjective Evolutionary Algorithms. European Journal of Operational Research 117, 553–564 (1999)
12. Knowles, J., Corne, D.: Approximating the Nondominated Front Using the Pareto Archived Evolution Strategy. Evolutionary Computation 8, 149–172 (2000)
13. Kukkonen, S., Deb, K.: Improved Pruning of Non-Dominated Solutions Based on Crowding Distance for Bi-Objective Problems. In: Proc. of 2006 IEEE Congress on Evolutionary Computation, Vancouver (2006)
14. Kukkonen, S., Deb, K.: A Fast and Effective Method for Pruning of Non-dominated Solutions in Many-Objective Problems. In: Runarsson, T.P., Beyer, H.-G., Burke, E.K., Merelo-Guervós, J.J., Whitley, L.D., Yao, X. (eds.) PPSN 2006. LNCS, vol. 4193, pp. 553–562. Springer, Heidelberg (2006)
15. Kukkonen, S., Lampinen, J.: GDE3: the Third Evolution Step of Generalized Differential Evolution. In: Proc. of IEEE Congress on Evolutionary Computation, Edinburgh, pp. 443–450 (2005)
16. Laumans, M., Thiele, L., Deb, K., Zitzler, E.: Combining Convergence and Diversity in Evolutionary Multi-Objective Optimization. Evolutionary Computation 10, 263–282 (2002)
17. Miettinen, K.: Nonlinear Multiobjective Optimization. Kluwer Academic Publishers, Boston (1999)
18. Miettinen, K.: IND-NIMBUS for Demanding Interactive Multiobjective Optimization. In: Trzaskalik, T. (ed.) Multiple Criteria Decision Making 2005, pp. 137–150. The Karol Adamiecki University of Economics in Katowice (2006)

19. Miettinen, K., Mäkelä, M.M.: Synchronous Approach in Interactive Multiobjective Optimization. European Journal of Operational Research 170, 909–922 (2006)
20. Monz, M., Küfer, K.H., Bortfeld, T.R., Thieke, C.: Pareto Navigation - Algorithmic Formulation of Interactive Multi-criteria IMRT Planning. Physics in Medicine and Biology 53, 985–998 (2008)
21. Nakayama, H., Sawaragi, Y.: Satisficing Trade-off Method for Multiobjective Programming. In: Grauer, M., Wierzbicki, A.P. (eds.) Interactive Decision Analysis, pp. 113–122. Springer, Heidelberg (1984)
22. Raquel, C.R., Naval Jr., P.C.: An Effective Use of Crowding Distance in Multiobjective Particle Swarm Optimization. In: Proc. of the Genetic and Evolutionary Computation (GECCO 2005), Washington DC, pp. 257–264 (2005)
23. Rivas, A., Irizar, I., Ayesa, E.: Model-Based Optimisation of Wastewater Treatment Plants Design. Environmental Modelling & Software 23, 435–450 (2008)
24. Robic, T., Filipic, B.: DEMO: Differential Evolution for Multiobjective Optimization. In: Coello Coello, C.A., Hernández Aguirre, A., Zitzler, E. (eds.) EMO 2005. LNCS, vol. 3410, pp. 520–533. Springer, Heidelberg (2005)
25. Rudolph, G., Agapie, A.: Convergence Properties of Some Multi-objective Evolutionary Algorithms. In: Proc. of IEEE Congress on Evolutionary Computation, pp. 1010–1016 (2000)
26. Sawaragi, Y., Nakayama, H., Tanino, T.: Theory of Multiobjective Optimization. Academic Press, Inc., London (1985)
27. Steuer, R.: Multiple Criteria Optimization: Theory, Computation and Applications. John Wiley & Sons, Inc., Chichester (1986)
28. Storn, R., Price, K.: Differential Evolution - a Simple and Efficient Heuristic for Global Optimization Over Continuous Spaces. Journal of Global Optimization 11, 341–359 (1997)
29. Trinkaus, H.L., Hanne, T.: knowCube: A Visual and Interactive Support for Multicriteria Decision Making. Computers & Operations Research 32, 1289–1309 (2005)
30. Wierzbicki, A.P.: A Mathematical Basis for Satisficing Decision Making. Mathematical Modelling 3, 391–405 (1982)
31. Wierzbicki, A.P.: Reference Point Approaches. In: Gal, T., Stewart, T.J., Hanne, T. (eds.) Multicriteria Decision Making: Advances in MCDM Models, Algorithms, Theory, and Applications, pp. 9-1–9-39. Kluwer Academic Publishers, Dordrecht (1999)
32. Zaharie, D.: Multi-objective Optimization with Adaptive Pareto Differential Evolution. In: Proc. of Symposium on Intelligent Systems and Applications (SIA 2003), Iasi (2003)
33. Zitzler, E., Laumanns, M., Thiele, L.: SPEA2: Improving the Strength Pareto Evolutionary Algorithm. Swiss Federal Institute of Technology, technical report TIK-Report 103 (2001)

A Trapezoidal Fuzzy Numbers-Based Approach for Aggregating Group Preferences and Ranking Decision Alternatives in MCDM

Alfonso Mateos and Antonio Jiménez

Artificial Intelligence Department, Technical University of Madrid (UPM)
Campus de Montegancedo S/N, Boadilla del Monte, 28660 Madrid, Spain
{amateos,ajimenez}@fi.upm.es

Abstract. In this paper we introduce an approach for solving multiattribute decision-making problems in which there are several decision-makers who individually and independently elicit their preferences. The preferences of each decision-maker are imprecise and represented by an imprecise additive multi-attribute utility function. We allow for incomplete information on the component utility functions and weights assessment, which leads to classes of utility functions and weight intervals, respectively. On the basis of this information, we introduce an approach for calculating the decision-maker group preferences using trapezoidal fuzzy numbers. The method consists of assigning trapezoidal fuzzy numbers to weights and component utilities and then, using an additive utility function to perform the evaluation process. The alternatives are then ranked by the trapezoidal fuzzy numbers representing them and the distances to some preset targets, i.e. the crisp maximum and minimum.

Keywords: Multicriteria Decision Making, Imprecision, Group Decision Making, Fuzzy Numbers.

1 Introduction

We consider a group decision-making problem in which there are several decision-makers (DMs) or groups of DMs located at different places that assess their preferences regarding a specified set of criteria and alternative performances. For reasons of time or space, these DMs may not have the chance to enter into a negotiation process in order to reach a consensus alternative. However, we assume that there is a desire to reach a consensus.

For example, this situation could match up with some e-democracy problems. E-democracy articulates political and democratic procedures involving citizens in societal decision making through the use of information and communication technologies, more recently focused on citizens' participation in discussion and deliberation on public matters rather than on electronic voting.

We assume that an objective hierarchy including all the relevant aspects related to the problem under consideration has been established, as has a set of n attributes associated with the lowest-level objectives, denoted by $X_1,\ldots,X_n$. Then, the performances of each

M. Ehrgott et al. (Eds.): EMO 2009, LNCS 5467, pp. 365–379, 2009.

alternative decision A^i in **A**, where $\mathbf{A} = \{A^1,\ldots,A^m\}$ is the available decision alternatives set, is defined by a vector $(x_1^i,\ldots,x_n^i)$, where x_j^i is the performance for attribute X_j.

We assume that there are d equally important DMs, whose preferences are assessed using a group decision support system, *Generic Multi-Attribute Analysis System* (GMAA)[1] [1,2]. The GMAA is a user-friendly decision support system that admits incomplete information concerning the quantification of the DM's preferences. This leads to classes of utility functions for the different attributes and imprecise attribute weights, i.e. a class of utility functions, denoted as $[u_i^{lL}(\bullet), u_i^{lU}(\bullet)]$, and a weight interval $[w_i^{lL}, w_i^{lU}]$ are available for each DM l, l=1,…,d, and each attribute X_i, where $L(U)$ means *L*ower (*U*pper).

Assuming the additive independence condition or an approximation [3,4,5], we have a global utility function for the DM l of the form

$$\mathbf{u}^l(A^j) = \sum_{i=1}^{n} w_i^l u_i^l\left(x_i^j\right),$$

where $w_i^l \in [w_i^{lL}, w_i^{lU}]$ and $u_i^l(\bullet) \in [u_i^{lL}(\bullet), u_i^{lU}(\bullet)]$ are the weight and utility intervals for attribute X_i, respectively.

For just one DM, the vector optimization problem can be stated as follows (provided that all the component utility functions are increasing):

$$\max \quad \mathbf{u}(A^j) = \left[\sum_{i=1}^{n} w_i^L u_i^L(x_i^L), \sum_{i=1}^{n} w_i^U u_i^U(x_i^U)\right]$$

$$\text{s.t.} \qquad A^j \in \mathbf{A}$$

This problem is solved in [6,7]. However, we consider a more generic decision-making situation involving d DMs whose preferences have to be aggregated.

The methodology proposed in this paper computes trapezoidal fuzzy numbers representing group preferences from weight and component utilities intervals elicited from each DM regarding the considered attributes. [8] proposed an analogous method in the statistics field, using density functions rather than membership functions. Then, decision alternatives are evaluated by means of an additive multiattribute utility function. The result of adding and subtracting trapezoidal numbers is a trapezoidal fuzzy number, but the product, quotient and inverse need not be a trapezoidal fuzzy number. [9] proposed a new approach for assigning distance between fuzzy numbers. They described a pseudo-metric on the set of fuzzy numbers and a metric on the set of trapezoidal fuzzy numbers. They used regular reducing functions and the *Hausdorf metric* to define the metric. Using this metric, we can approximate an arbitrary generalized left right number as a trapezoidal number. The resulting trapezoidal fuzzy numbers (representing overall utilities) are ranked on the basis of the distances between them and two preset targets: a minimum and a maximum utility. This distance overcomes general shortcomings such as the indiscriminative and counterintuitive

[1] http://www.dia.fi.upm.es/~ajimenez/GMAA

behaviour of several existing fuzzy rankings approaches, [10]. The above methodology is explained in detail in section 2.

In section 3, we give an example to illustrate the proposed methodology. In this example, a manufacturing company is looking to select a suitable city to set up a new factory. The results match up with the outcomes presented in [11]. Finally, some conclusions are outlined in section 4.

2 Methodology

The methodology we propose can be divided into three phases. In the first phase, the DMs' preferences are aggregated by means of trapezoidal fuzzy numbers. In the second phase, overall utilities for the alternatives under consideration are computed by means of a method that approximates the product of trapezoidal fuzzy numbers to a new trapezoidal fuzzy number. Finally, in the third phase, alternatives are ranked on the basis of the trapezoidal fuzzy numbers representing their overall utilities.

As the methodology we propose is based on trapezoidal fuzzy numbers, we first define this concept. We can define a trapezoidal fuzzy number as $\tilde{A} = (A_1, A_2, A_3, A_4)$. The membership function of this fuzzy number will be interpreted as follows:

$$\mu_{\tilde{A}}(x) = \begin{cases} \dfrac{x - A_1}{A_2 - A_1}, & \text{for} \quad A_1 \leq x \leq A_2 \\ 1, & \text{for} \quad A_2 \leq x \leq A_3 \\ \dfrac{A_4 - x}{A_4 - A_3}, & \text{for} \quad A_3 \leq x \leq A_4 \\ 0, & \text{otherwise} \end{cases}$$

and the α-cut interval for this shape is written as $\tilde{A}_\alpha = [(A_2\text{-}A_1)\alpha + A_1, -(A_4\text{-}A_3)\alpha + A_4]$ (see Fig. 1).

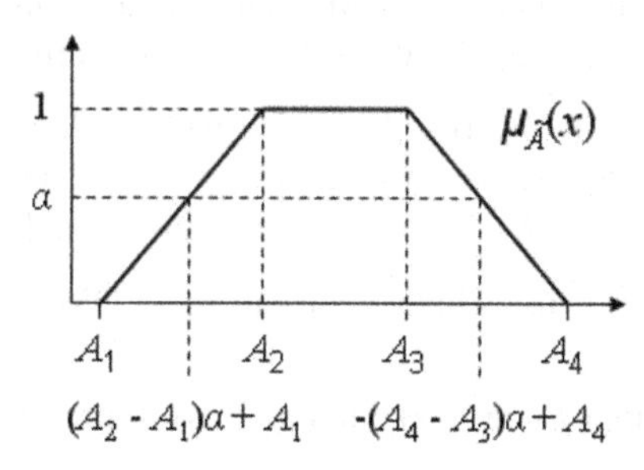

Fig. 1. Membership function and α-cut for a trapezoidal fuzzy number $\tilde{A}$

2.1 First Phase: Aggregation of the DMs' Preferences

Next, we show how to get a fuzzy number $\tilde{w}_i$ that represents group preferences from the respective weight intervals of each DM, $[w_i^{lL}, w_i^{lU}]$, l=1,…,d. The reasoning would be analogous for utility intervals.

Let us assume, without loss of generality, that $w_i^{1L} \leq w_i^{2L} \leq \cdots \leq w_i^{dL}$. The algorithm is divided into 10 steps. In the first one (step 0), variables j and N_j are initialized to 1. j represents the index of the DM that has the lowest lower end point of the weight interval for the considered iteration, and N_j is the number of weight intervals of DMs $j, j+1,\ldots,d$ whose intersection with the weight interval of the DM j is not empty. For instance, if j=3 and $N_3 = 4$, then the third DM is taken as the reference and the intersection among the weight intervals corresponding to the third, fourth, fifth and sixth DMs is not empty.

In steps 1 to 4, N_j is computed for the d DMs under consideration. In actual fact, in step 1, k, the index of the DM that has the highest lower end point of the weight interval for the considered iteration is initialized to j+1. In other words, j and k represent the first and the last DM considered in each iteration.

In step 2, the intersection of the weight intervals corresponding to DMs $j, j+1,\ldots,k$, denoted by I_j^k, is computed. In step 3, I_j^k is checked. If empty, then N_j is computed, and we either go back to step 1 to compute N_{j+1} or go to step 4 (N_j has been assessed for the d DMs under consideration); else ($I_j^k \neq \emptyset$), N_j is updated and k increased. In step 4, k is compared with the number of DMs: if it is lower than or equal to d, we go back to step 2, i.e. a cycle involving steps 2, 3 and 4 is formed until the above condition is satisfied; else ($k > d$) we go to step 5.

In step 5, once N_j have been computed for the d DMs under consideration, the DM (or DMs) whose weight interval intersects (non-empty) with more weight intervals corresponding to the remaining DMs is identified. In step 6 the number of DMs satisfying the above property is checked and, accordingly, we go to step 7 (just one DM) or 8 (more than one DM). Here, interval I_s (the interval whose values have possibility 1 of being selected as the weight of attribute i) is assessed.

In step 7, I_s is defined as the values of the intersection of weight intervals representing most DMs because these values are preferred by most DMs. In step 8, we denote the indexes corresponding to DMs with the same highest N_j by $s_1,\ldots,s_p$ (we can assume $s_1 \leq \ldots \leq s_p$). If $N_{s_1} = \ldots = N_{s_p} = 1$ then I_s is defined as those values that leave the same number of DMs to either side (these values can be considered as the median in statistics) because the median is a good statistic measure to define the group. However, if $N_{s_1} = \ldots = N_{s_p} \neq 1$, then I_s is defined as those values between the two most distant value intervals, which represent most DMs; $\lfloor x \rfloor$ denotes the integer side of the number x).

Finally, in step 9 the trapezoidal fuzzy number that represents group preferences is assessed.

Step 0. Let $j = 1$ and $N_j = 1$

Step 1. $k = j + 1$

Step 2. Compute $I_j^k = [I_j^{kL}, I_j^{kU}] = \bigcap_{l=j}^{k} [w_i^{lL}, w_i^{lU}]$.

Step 3. If $I_j^k = \emptyset$, then $N_j = k - j$

if $j = j + 1 < d$, then go to step 1
else $(j = j + 1 = d)$, $k = d + 1$ and go to step 4.
else $(I_j^k \neq \emptyset)$, $N_j = k - j + 1$ and $k = k + 1$.

Step 4. If $k \leq d$, then go to step 2; else, go to step 5.

Step 5. Calculate s given that $N_s = \max\{N_j, j=1,2,...,d\}$.

Step 6. If s is the only optimal solution in step 5, then go to step 7, else (it is not the only optimal solution), i.e. there are $s_1,\ldots,s_p$ (we can assume $s_1\leq\ldots\leq s_p$) optimal solutions with $N_{s_1} = \ldots = N_{s_p}$, go to step 8.

Step 7. Define $I_s = [I_s^L, I_s^U] = [I_s^{(s+N_s-1)L}, I_s^{(s+N_s-1)U}]$ and go to step 9.

Step 8. If $N_{s_1} = \ldots = N_{s_p} = 1$, then

if d is even, then $I_s = [I_s^L, I_s^U] = [w_i^{(d/2)U}, w_i^{(d/2+1)L}]$,
else (d is odd), $I_s = [I_s^L, I_s^U] = [w_i^{(\lfloor d/2 \rfloor+1)L}, w_i^{(\lfloor d/2 \rfloor+1)U}]$.
else, $I_s = [I_s^L, I_s^U] = [w_i^{(s_1+N_{s_1}-1)U}, w_i^{(s_p+N_{sp}-1)L}]$.

Step 9. Compute the trapezoidal fuzzy number that represents group preferences for the weight of attribute X_i, $\tilde{w}_i = (w_i^{1L}, I_s^L, I_s^U, \max_l w_i^{lU})$.

The approach for component utilities is analogous. Thus, the utility intervals for the different DMs $u_i^l(x_i^j) = [u_i^{lL}(x_i^j), u_i^{lU}(x_i^j)]$, $l=1,\ldots,d$, are aggregated as shown in the above algorithm for weights, leading to a trapezoidal fuzzy number $\tilde{u}_i(x_i^j)$.

2.2 Second Phase: Fuzzy Arithmetic to Evaluate Decision Alternatives

The evaluation of each alternative A^j will output the following fuzzy number

$$\tilde{u}(A^j) = \sum \tilde{w}_i \tilde{u}(x_i^j) \tag{1}$$

because the set of fuzzy numbers is closed under all arithmetic operations, [12,13]. However, it is well-known that the result of adding and subtracting trapezoidal fuzzy numbers is another trapezoidal fuzzy number. However, this is not the case for the product, where the resulting fuzzy number may be not trapezoidal [9,11].

Interval arithmetic is a popular way to do fuzzy arithmetic operations. This is possible because any α-cut of a fuzzy number is always an interval. Therefore, any fuzzy number can be represented as a series of intervals (one interval for each α -cut). The basics of interval arithmetic are given below. For any two intervals, $[a, b]$ and $[d, e]$, the arithmetic operations (we will only need addition and multiplication) are performed as follows:

- Addition: $[a, b] + [d, e] = [a+d, b+e]$;
- Product: $[a, b] \times [d, e] = [\min\{ad, ae, bd, be\}, \max\{ad, ae, bd, be\}]$.

Next, we give an example to illustrate the addition and multiplication of a trapezoidal fuzzy number: Let $\tilde{A} = (1, 5, 6, 9)$ and $\tilde{B} = (2, 3, 5, 8)$, then

$$\tilde{A} + \tilde{B} = (1, 5, 6, 9) + (2, 3, 5, 8) = (1+2, 5+3, 6+5, 9+8) = (3, 8, 11, 17).$$

The α-cuts of $\tilde{A}$ and $\tilde{B}$ are $\tilde{A}_\alpha = [4\alpha+1,-3\alpha+9]$ and $\tilde{B}_\alpha = [\alpha+2,-3\alpha+8]$ respectively. Since, for all α in [0,1], each element for each interval is positive, the multiplication of the α-cut intervals will be

$$\tilde{A}_\alpha \times \tilde{B}_\alpha = [(4\alpha+1)(\alpha+2), (-3\alpha+9)(-3\alpha+8)] = [4\alpha^2 + 9\alpha + 2, 9\alpha^2 - 51\alpha + 72]$$

If $\alpha = 0$, then $\tilde{A}_0 \times \tilde{B}_0 = [2, 72]$ and if $\alpha = 1$, then $\tilde{A}_1 \times \tilde{B}_1 = [4+9+2, 9-51+72] = [15, 30]$. The fuzzy number $\tilde{C} = \tilde{A} \times \tilde{B}$ is illustrated in Fig. 2. Fig. 2 shows that the resulting fuzzy number is not a trapezoidal fuzzy number.

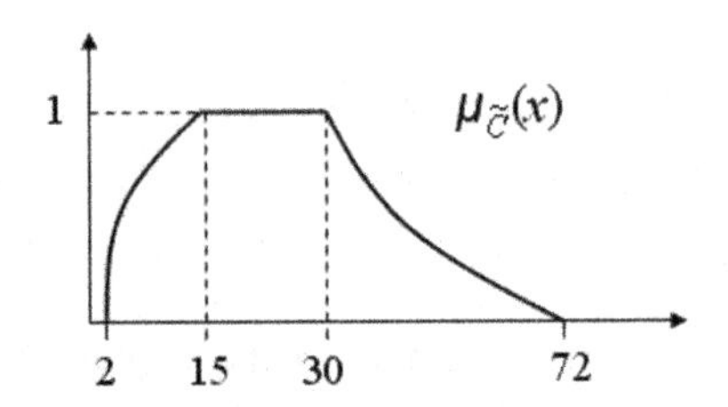

Fig. 2. Multiplication of trapezoidal fuzzy numbers $\tilde{C} = \tilde{A} \times \tilde{B}$

Decision alternatives will be ranked in the third phase of the methodology taking into account the above overall utilities expressed in terms of trapezoidal fuzzy numbers. The distance measures that will be used in this ranking process use any fuzzy numbers. From a computational point of view, though, it is recommendable to use trapezoidal fuzzy numbers. This is why we use an approximation of the product of two trapezoidal fuzzy numbers to a new trapezoidal fuzzy number.

The product of two trapezoidal fuzzy numbers, $\tilde{A}^i = (A_1^i, A_2^i, A_3^i, A_4^i)$ and $\tilde{A}^j = (A_1^j, A_2^j, A_3^j, A_4^j)$ is approximated by the trapezoidal fuzzy number $\tilde{A}^{ij} = (A_1^{ij}, A_2^{ij}, A_3^{ij}, A_4^{ij})$, as proposed in [9], where

$$\begin{aligned}
A_1^{ij} &= \frac{3}{2}\left(A_2^i - A_1^i\right)\left(A_2^j - A_1^j\right) + 2\left[\left(A_2^i - A_1^i\right)A_1^j + \left(A_2^j - A_1^j\right)A_1^i\right] + 3A_1^iA_1^j - 2A_2^iA_2^j,\\
A_2^{ij} &= A_2^iA_2^j,\\
A_3^{ij} &= A_3^iA_3^j,\\
A_4^{ij} &= \frac{3}{2}\left(A_4^i - A_3^i\right)\left(A_4^j - A_3^j\right) - 2\left[\left(A_4^i - A_3^i\right)A_3^j + \left(A_4^j - A_3^j\right)A_4^i\right] + 3A_4^iA_4^j - 2A_3^iA_3^j.
\end{aligned} \tag{2}$$

If we apply this method to the above example with $\tilde{A} = (1, 5, 6, 9)$ and $\tilde{B} = (2, 3, 5, 8)$, then we have $\tilde{C} = \tilde{A} \times \tilde{B} = (0, 15, 30, 67.5)$, as shown in Fig 3.

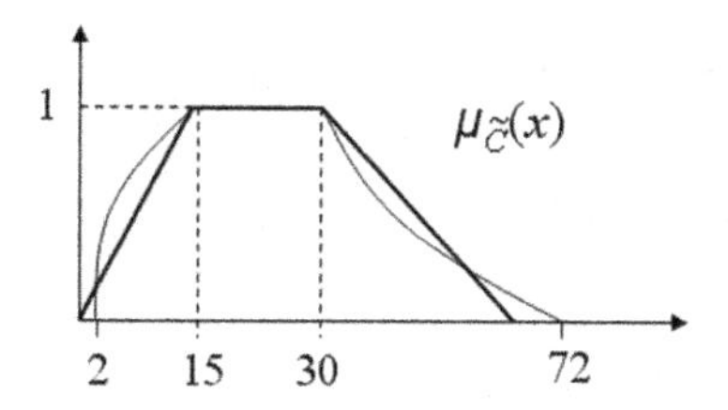

Fig. 3. Approximation of the multiplication of trapezoidal fuzzy numbers $\tilde{C} = \tilde{A} \times \tilde{B}$

In conclusion, an overall utility is computed for each decision alternative from expression (1) taking into account the above approximation for the product of trapezoidal fuzzy numbers.

2.3 Third Phase: Fuzzy Set Ranking

In this phase, we show how to rank the alternatives under consideration on the basis of the trapezoidal fuzzy numbers representing their overall utilities. Fuzzy set ranking methods can be divided into two classes, see [14]:

1. Methods that convert a fuzzy number into a crisp number by applying a mapping function F. Fuzzy numbers are then sorted by ranking crisp numbers produced by the mapping.
2. Methods that use fuzzy relations to compare pairs of fuzzy numbers, and then construct a relationship that gives the comparison a linguistic meaning.

Each methodology has its advantages and drawbacks. With respect to the first class, it has been argued that "by reducing the whole of our analysis to a single (crisp) number, we are loosing much of the information that we have purposely been keeping throughout our calculations" [15]. This methodology, on the other hand, produces a consistent ranking of all fuzzy sets considered (i.e. if $\tilde{A}$ is ranked greater than $\tilde{B}$, and $\tilde{B}$ is ranked greater than $\tilde{C}$, then $\tilde{A}$ will always be much greater than $\tilde{C}$). Also, there will always be a fuzzy set that is ranked as "best", "second best", "third best", and so on.

By keeping the comparison linguistic, the second class preserves the original fuzzy information of the problem. However, as [14] points out, "it may not always be possible to construct total ordering among all alternatives based on pairwise fuzzy preference relations". This means that even if $\tilde{A}$ is better than $\tilde{B}$ and $\tilde{B}$ is better than $\tilde{C}$, $\tilde{A}$ is not necessarily always better than $\tilde{C}$.

Literature review reveals that there are very many fuzzy set ranking methods. [16,17,18] offer a comprehensive survey of the available methods.

It is not the aim of this paper to prove which one is the best ranking method. Indeed, almost all the methods have pitfalls of some sort, such as being inconsistent with human intuition, indiscriminate and hard to interpret, [16,19,20]. However, the fuzzy ranking method suggested in [10] overcomes many of the problems intrinsic of the existing methods. Therefore, we use this method.

The distance measure between two interval numbers $\tilde{A} = [a_1, a_2]$ and $\tilde{B} = [b_1, b_2]$, [10] is

$$D(\tilde{A}, \tilde{B}) = \sqrt{\int_{-1/2}^{1/2}\int_{-1/2}^{1/2} \left\{ \left[\frac{a_1 + a_2}{2} + x(a_2 - a_1) \right] - \left[\frac{b_1 + b_2}{2} + y(b_2 - b_1) \right] \right\}^2 dxdy} = \\ = \sqrt{\left(\frac{a_1 + a_2}{2} - \frac{b_1 + b_2}{2} \right)^2 + \frac{1}{3}\left[\left(\frac{a_2 - a_1}{2} \right)^2 + \left(\frac{b_2 - b_1}{2} \right)^2 \right]}. \tag{3}$$

The integral in (3) shows that this distance takes into account every point in both intervals when computing the distance between two interval numbers. It is different from most existing distance measures for interval numbers, which often use only the lower and upper bound values ([21,22,23,24]).

[25] proposed a distance measure for intervals that also considers every point of the two intervals. Its general form, however, is too complicated, and the authors later restricted the measure to a particular case with a finite number of considered values for operational purposes.

As is well-known, a trapezoidal fuzzy number is denoted as $\tilde{A}^j = (A_1^j, A_2^j, A_3^j, A_4^j)$, and its α-cut is

$$\tilde{A}^j(\alpha) = [A^{jL}(\alpha), A^{jU}(\alpha)] = [A_1^j + (A_2^j - A_1^j)\alpha, A_4^j + (A_4^j - A_3^j)\alpha].$$

Using the distance measure for interval numbers defined in (3), a distance between two fuzzy numbers $\tilde{A}^i$ and $\tilde{A}^j$ can be defined as

$$D(\tilde{A}^i, \tilde{A}^j, f) = \sqrt{\frac{\int_0^1 \left\{ \begin{array}{l} \left(\frac{A^{iL}(\alpha) + A^{iU}(\alpha)}{2} - \frac{A^{jL}(\alpha) + A^{jU}(\alpha)}{2} \right)^2 + \\ + \frac{1}{3}\left[\left(\frac{A^{iU}(\alpha) - A^{iL}(\alpha)}{2} \right)^2 + \left(\frac{A^{jU}(\alpha) - A^{jL}(\alpha)}{2} \right)^2 \right] \end{array} \right\} f(\alpha)d\alpha}{\int_0^1 f(\alpha)d\alpha}}.$$

Here, f, which serves as a weighting function, is a continuous positive function defined on [0,1]. The distance is a weighted sum (integral) of the distances between two intervals at all α-cuts from 0 to 1. It is reasonable to choose f as an increasing function, which indicates that a greater weight is assigned to the distance between two intervals at a higher α level. For example, when the DM is risk-neutral, $f(\alpha) = \alpha$ seems to be reasonable. A risk-averse DM might want to put more weight on information at a higher α level by using other functions, such as $f(\alpha) = \alpha^2$ or a higher power of α. A constant ($f(\alpha)=1$), or even a decreasing function f, can be utilized for a risk-prone DM.

The method for ranking fuzzy numbers suggested in [10] is based on comparing the distance from fuzzy numbers to some preset targets: the crisp maximum and the crisp minimum. The idea is that a fuzzy number is ranked first if its distance to the crisp maximum is the smallest and its distance to the crisp minimum is the greatest. If only one of these conditions is satisfied, a fuzzy number might be outranked by the others depending upon the context of the problem (e.g., the attitude of the DM in a decision-making situation). If it is not known we propose ranking the alternatives according to the value of

$$R(\tilde{A}^i) = \frac{D(\tilde{A}^i, x, f)}{D_x} - \frac{D(\tilde{A}^i, y, f)}{D_y}$$

where $D_x = \max_j \{D(\tilde{A}^j, x, f)\}$, $D_y = \min_j \{D(\tilde{A}^j, y, f)\}$, $x = \min\{A_4^j, j = 1,\ldots,m\}$ and $y = \max\{A_1^j, j=1,\ldots,m\}$, where $\tilde{A}^j = (A_1^j, A_2^j, A_3^j, A_4^j)$. This ratio is always negative, and the smaller the distance between the fuzzy number ($\tilde{A}^i$) and the minimum value (x) and the greater the distance to the maximum value (y) is, the more negative the ratio is. Thus, the alternatives ranking will be based on this ratio, and the more negative the ratio is, the worse their ranking will be.

If the DM is risk-neutral ($f(\alpha)= \alpha$), then the distance between the fuzzy number $\tilde{A}^i$ and x is

$$D(\tilde{A}^i, x, \alpha) = \sqrt{\begin{aligned}&\left(\frac{A_2^i + A_3^i}{2} - x\right)^2 + \frac{1}{3}\left(\frac{A_2^i + A_3^i}{2} - x\right)\left[\left(A_4^i - A_3^i\right) - \left(A_2^i - A_1^i\right)\right] + \\ &+ \frac{2}{3}\left(\frac{A_3^i - A_2^i}{2}\right)^2 + \frac{1}{9}\left(\frac{A_3^i - A_2^i}{2}\right)\left[\left(A_4^i - A_3^i\right) + \left(A_2^i - A_1^i\right)\right] + \\ &+ \frac{1}{18}\left[\left(A_4^i - A_3^i\right)^2 + \left(A_2^i - A_1^i\right)^2 - \frac{1}{18}\left[\left(A_2^i - A_1^i\right)\right]\left(A_4^i - A_3^i\right)\right]\end{aligned}}.$$

If the DM is risk-prone ($f(\alpha)=1$), then the distance between the fuzzy number $\tilde{A}^i$ and x is

$$D(\tilde{A}^i, x, 1) = \sqrt{\begin{aligned}&\left(\frac{A_2^i + A_3^i}{2} - x\right)^2 + \frac{1}{2}\left(\frac{A_2^i + A_3^i}{2} - x\right)\left[\left(A_4^i - A_3^i\right) - \left(A_2^i - A_1^i\right)\right] + \\ &+ \frac{1}{3}\left(\frac{A_3^i - A_2^i}{2}\right)^2 + \frac{1}{6}\left(\frac{A_3^i - A_2^i}{2}\right)\left[\left(A_4^i - A_3^i\right) + \left(A_2^i - A_1^i\right)\right] + \\ &+ \frac{1}{9}\left[\left(A_4^i - A_3^i\right)^2 + \left(A_2^i - A_1^i\right)^2 - \frac{1}{9}\left[\left(A_2^i - A_1^i\right)\right]\left(A_4^i - A_3^i\right)\right]\end{aligned}}.$$

The assessment of the distance between the fuzzy number $\tilde{A}^i$ and y is analogous, except that x is replaced by y in the above expressions.

3 An Illustrative Example

In this section we adapt the numerical example used in [11] to the group decision-making situation and compare the solutions reached by Chou with the results by the proposed methodology.

A manufacturing company has to select a suitable city in which to set up a new factory. After preliminary screening, three candidate cities A^1, A^2 and A^3 are left for further evaluation. The company considers five attributes to select the best candidate: the investment conditions (X_1), expansion possibility (X_2), availability of required material (X_3), human resources (X_4), and weather conditions (X_5), respectively.

Imprecise attribute weights were elicited from three company partners (columns 2, 3 and 4 in Table 1), and the aggregation process (second phase of the proposed methodology) was carried out leading to the company weights (column 5 in Table 1) in terms of trapezoidal fuzzy numbers as follows.

Taking into account the weights intervals elicited from the three DMs for the ith attribute, we have three possibilities:

1. The intersection of the three weight intervals is empty (disjoint intervals).
2. Two weight intervals intersect and the third one is disjoint.
3. The intersection of the three intervals is not empty.

Table 1. Attribute weights for the three partners and aggregated preferences in terms of trapezoidal fuzzy numbers

Attribute	Partner 1	Partner 2	Partner 3	Company
X_1	[0.4,0.6]	[0.5,0.65]	[0.5,0.7]	(0.4,0.5,0.6,0.7)
X_2	[0.6,0.8]	[0.7,0.8]	[0.85,0.9]	(0.6,0.7,0.8,0.9)
X_3	[0.4,0.45]	[0.5,0.6]	[0.5,0.7]	(0.4,0.5,0.6,0.7)
X_4	[0.4,0.45]	[0.8,0.9]	[0.95,1.0]	(0.7,0.8,0.9,1.0)
X_5	[0.0,0.1]	[0.05,0.15]	[0.05,0.1]	(0.0,0.05,0.1,0.15)

The membership functions for the trapezoidal fuzzy numbers output as a consequence of the aggregation process are computed in a different way depending on the considered case.

Attribute X_4 matches up with the first case, see Table 1, and the membership function is (see Fig. 4a):

$$\mu_{\tilde{w}_i}(x) = \begin{cases} \dfrac{x-0.4}{0.8-0.4}, & \text{for} \quad 0.4 \le x \le 0.8 \\ 1, & \text{for} \quad 0.8 \le x \le 0.9 \\ \dfrac{1-x}{1-0.9}, & \text{for} \quad 0.9 \le x \le 1.0 \\ 0, & \text{otherwise} \end{cases}.$$

The reasoning behind this fuzzy number is: partners 1 and 3 have extreme weights, but are known to be looking for a consensus. Therefore, it is reasonable to assume that they will agree on some of the weights between their respective intervals. But, we also know that the weights for partner 2 are in this region, and it is reasonable to assume that these will be highly possible weights. So, from these weights, there will be a linearly decreasing possibility down to the value zero at the lower and upper end points of the intervals provided by partners 1 and 3, 0.4 and 1.0, respectively.

Attributes X_2 and X_3 match up with the second case, see Table 1. The membership function in attribute X_2 is as follows (see Fig. 4b):

$$\mu_{\tilde{w}_i}(x) = \begin{cases} \dfrac{x-0.6}{0.7-0.6}, & \text{for} \quad 0.6 \le x \le 0.7 \\ 1, & \text{for} \quad 0.7 \le x \le 0.8 \\ \dfrac{0.9-x}{0.9-0.8}, & \text{for} \quad 0.8 \le x \le 0.9 \\ 0, & \text{otherwise} \end{cases}.$$

Weights for partners 1 and 2 intersect in the interval [0.7, 0.8], and both are disjoint with the third weight interval (partner 3). As partners 1 and 2 agree within an interval, it is logical to think that this interval represents what would be a consensus in a negotiation process. Also, as they constitute a majority, the greatest possibility (equal to 1) should be assigned to [0.7, 0.8]. From this interval, there will be a linearly decreasing possibility down to the value zero at the lower and upper end points of the intervals provided by partners 1 and 3, w_i^{1L} and w_i^{3U}, respectively.

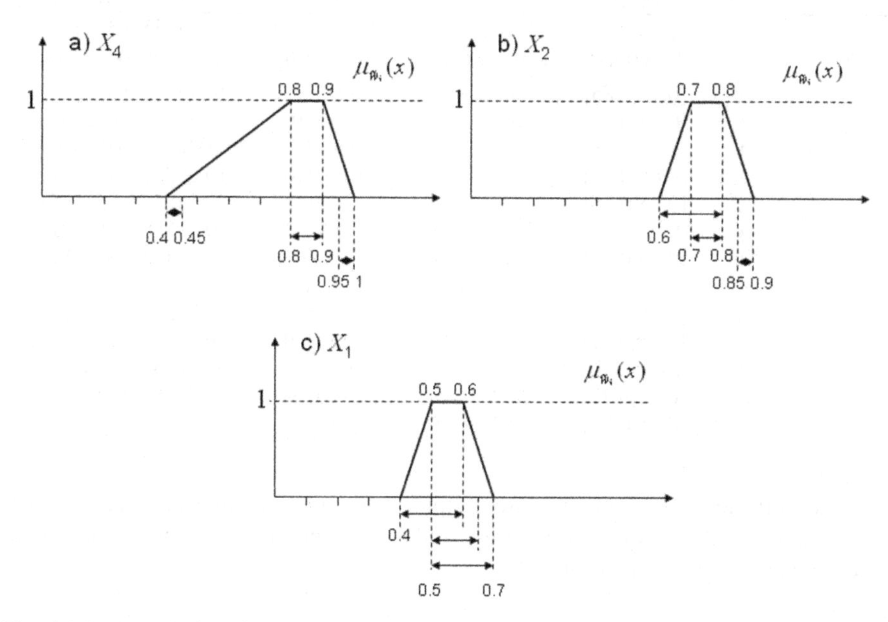

Fig. 4. Membership functions for the trapezoidal fuzzy numbers representing group preferences for attributes X_1, X_2, and X_4

Finally, attributes X_1 and X_5 match up with the third case, see Table 1. The membership function in attribute X_1 is as follows (see Fig. 4c):

$$\mu_{\tilde{w}_i}(x) = \begin{cases} \dfrac{x-0.4}{0.5-0.4}, & \text{for} \quad 0.4 \le x \le 0.5 \\ 1, & \text{for} \quad 0.5 \le x \le 0.6 \\ \dfrac{0.7-x}{0.7-0.6}, & \text{for} \quad 0.6 \le x \le 0.7 \\ 0, & \text{otherwise} \end{cases}.$$

Now, the intersection of the three weight intervals is not empty. Consequently, the highest possibility (equal to 1) should be assigned to this intersection interval [0.5, 0.6]. From this interval, there will be a linearly decreasing possibility down to the value zero at the lower and upper end points of the intervals provided by partners 1 and 3, 0.4 and 0.7, respectively.

The partners also assessed component utilities (preferences concerning city performances) in each attribute (columns 3, 4 and 5 in Table 2). They were then aggregated into trapezoidal fuzzy numbers (column 6 in Table 2) representing the company preferences.

Table 2. Partner preferences concerning the city performances for the different attributes and aggregated preferences in terms of trapezoidal fuzzy numbers

Attribute	City	Partner 1	Partner 2	Partner 3	Company
	A^1	[0.2,0.3]	[0.25,0.0.3]	[0.31,0.35]	(0.2,0.25,0.3,0.35)
X_1	A^2	[0.8,0.83]	[0.84,0.9]	[0.85,1.0]	(0.8,0.85,0.9,1.0)
	A^3	[0.75,0.85]	[0.8,0.86]	[0.88,0.9]	(0.75,0.8,0.85,0.9)
	A^1	[0.75,0.78]	[0.79,0.85]	[0.8,0.9]	(0.75,0.8,0.85,0.9)
X_2	A^2	[0.8,0.9]	[0.85,0.91]	[0.95,1.0]	(0.8,0.85,0.9,1.0)
	A^3	[0.65,0.68]	[0.7,0.75]	[0.7,0.8]	(0.65,0.7,0.75,0.8)
	A^1	[0.35,0.55]	[0.4,0.6]	[0.62,0.65]	(0.35,0.45,0.55,0.65)
X_3	A^2	[0.75,0.0.78]	[0.8,0.85]	[0.88,0.9]	(0.75,0.8,0.85,0.9)
	A^3	[0.65,0.67]	[0.7,0.75]	[0.77,0.8]	(0.65,0.7,0.75,0.8)
	A^1	[0.8,0.9]	[0.85,0.9]	[0.95,1.0]	(0.8,0.85,0.9,1.0)
X_4	A^2	[0.75,0.77]	[0.78,0.85]	[0.8,0.9]	(0.75,0.8,0.85,0.9)
	A^3	[0.75,0.77]	[0.78,0.85]	[0.8,0.9]	(0.75,0.8,0.85,0.9)
	A^1	[0.35,0.4]	[0.45,0.55]	[0.6,0.65]	(0.35,0.45,0.55,0.65)
X_5	A^2	[0.75,0.77]	[0.78,0.85]	[0.8,0.9]	(0.75,0.8,0.85,0.9)
	A^3	[0.75,0.77]	[0.78,0.85]	[0.8,0.9]	(0.75,0.8,0.85,0.9)

Overall utilities for each city are output using (1), taking into account the approximation of the product of two trapezoidal fuzzy numbers to a new trapezoidal fuzzy number, are shown in Table 3.

Table 3. Overall utilities for the three candidade cities

Overall utilities	Trapezoidal fuzzy numbers
$\tilde{u}\,(A^1)$	(1.2175,1.5275,1.875,2.295)
$\tilde{u}\,(A^2)$	(1.61625,2.02,2.45,2.98125)
$\tilde{u}\,(A^3)$	(1.46625,1.84,2.24,2.66625)

Thus, the problem is now to rank the three trapezoidal fuzzy numbers using their distances to the minimum and maximum utility, x and y, see Table 4.

Table 4. Distances to the minimum and maximum utilities

	$f(\alpha)=\alpha$			$f(\alpha)=1$		
Utilities	$D(\bullet, x, \alpha)$	$D(\bullet, y, \alpha)$	$R(\bullet)$	$D(\bullet, x, 1)$	$D(\bullet, y, 1)$	$R(\bullet)$
$\tilde{u}\,(A^1)$	0.55495	1.28362	-1.12510	0.32877	1.63615	-2.37808
$\tilde{u}\,(A^2)$	1.08021	0.78325	0	1.20823	0.63615	0
$\tilde{u}\,(A^3)$	0.87200	0.96990	-0.43013	0.78277	0.94592	-0.88430

Looking at Table 4 we find that the rankings provided by $D(\bullet, x, \alpha)$ and $D(\bullet, y, \alpha)$ under a risk-neutral perspective ($f(\alpha)=\alpha$) match up. A^2 is the best, whereas A^1 is the worst alternative regarding the distance to the minimum utility (x). The same applies to the distances to the maximum utility (y), A^2 is again the best and A^1 the worst alternative. Thus, ratio $R(\bullet)$ for each alternative does not need to be used. On the other hand, the ranking output from a risk-prone perspective ($f(\alpha)=1$) matches up with the above, as shown in Table 4. In both cases, the ranking output on the basis of the ratio would be

$$A^2 \succ A^3 \succ A^1$$

as expected. Note that the results for this example match up with the outcomes presented in [11], where the same example was considered but a different methodology was applied.

4 Conclusions

In many real decision-making problems there is more than one DM, and the preferences for each one are modelled by an additive utility function. To make the job easier for DMs, they are allowed to provide ranges instead single values in response to the questions analysed in the methods for quantifying DM preferences. Thus, for each DM and attribute, we have a class of utility functions and a weight interval. However, for reasons of time or space, these DMs may not have the chance to enter into a negotiation process for the purpose of reaching a consensus alternative.

In this paper, we have introduced a possibility for aggregating the individual preferences of several DMs to get group preferences, assuming that there is a desire to reach a consensus. Using logic reasoning, we show that group weights and utilities can be considered as trapezoidal fuzzy numbers. The result is that the overall group global utility for each alternative is a trapezoidal fuzzy number, calculated as a sum and product of trapezoidal fuzzy numbers.

The problem then is to rank trapezoidal fuzzy numbers, for which purpose we have proposed using the distances between them and two preset targets, the minimum and the maximum utility. For this purpose, we have considered a notion of distance measure between two interval numbers based on an integral. This takes into account every point in both intervals, and α-cuts for the considered trapezoidal fuzzy numbers. This way, we can consider risk-prone, risk-neutral and risk-averse DMs.

The proposed methodology has been illustrated using an example taken from the literature: the selection of a suitable city to set up a new factory We reached the same conclusions.

Acknowledgments. This paper was supported by the Madrid Regional Government project S-0505/TIC/0230 and the Spanish Ministry of Education and Science project TIN2008-06796-C04-02.

References

1. Jiménez, A., Ríos-Insua, S., Mateos, A.: A Decision Support System for Multiattribute Utility Evaluation Based on Imprecise Assignment. Dec. Support Syst. 36, 65–79 (2003)
2. Jiménez, A., Ríos-Insua, S., Mateos, A.: A Generic Multi-Attribute Analysis System. Comput. Opl. Res. 33, 1081–1101 (2006)
3. Keeney, R.L., Raiffa, H.: Decisions with Multiple Objectives. Cambridge University Press, Cambridge (1993)
4. Raiffa, H.: The Art and Science of Negotiation. Harvard Univ. Press, Cambridge (1982)
5. Stewart, T.J.: Robustness of Additive Value Function Method in MCDM. J. Multi-Criteria Decision Anal. 5, 301–309 (1996)
6. Mateos, A., Jiménez, A., Ríos Insua, S.: A Multiattribute Solving Dominance and Potential Optimality in Imprecise Multi-Attribute Additive Problems. Reliab. Eng. Syst. Safety 79, 253–262 (2003)
7. Mateos, A., Ríos-Insua, S., Jiménez, A.: Dominance, Potential Optimality and Alternative Ranking in Imprecise Multi-Attribute Model. J. Opl. Res. Soc. 58, 326–336 (2007)
8. Mateos, A., Jiménez, A., Ríos-Insua, S.: Monte Carlo Simulation Techniques for Group Decision-Making with Incomplete Information. Eur. J. Oper. Res. 174, 1842–1864 (2006)
9. Abbasbandy, S., Amirfakhrian, M.: The Nearest Trapezoidal Form of a Generalized Left Right Fuzzy Number. Int. J. Approx. Reas. 43, 166–178 (2006)
10. Tran, L., Duckstein, L.: Comparison of Fuzzy Numbers Using a Fuzzy Distance Measure. Fuzzy Sets Syst. 130, 331–341 (2002)
11. Chou, C.C.: The representation of multiplication operation on fuzzy numbers and application to solving fuzzy multiple criteria decision making problems. In: Yang, Q., Webb, G. (eds.) PRICAI 2006. LNCS (LNAI), vol. 4099, pp. 161–169. Springer, Heidelberg (2006)
12. Kaufman, A., Gupta, M.M.: Introduction to Fuzzy Arithmetic. Van Nostrand Reinold Company Inc., New York (1985)
13. Klir, G.J., Yuan, B.: Fuzzy Sets and Fuzzy Logic: Theory and Applications. Prentice Hall, New Jersey (1995)
14. Yuan, Y.: Criteria for Evaluating Fuzzy Ranking Methods. Fuzzy Sets Syst. 44, 139–157 (1991)
15. Freeling, A.N.S.: Fuzzy Sets and Decision Analysis. IEEE Trans. Syst. Man Cybernetics 6, 341–354 (1980)
16. Bortoland, G., Degani, R.: A Review of Some Methods for Ranking Fuzzy Subsets. Fuzzy Sets Syst. 15, 1–19 (1985)
17. Wang, X., Kerre, E.E.: Reasonable Properties for the Ordering of Fuzzy Quantities (I). Fuzzy Sets Syst. 118, 375–385 (2001a)
18. Wang, X., Kerre, E.E.: Reasonable Properties for the Ordering of Fuzzy Quantities (II). Fuzzy Sets Syst. 118, 387–405 (2001b)
19. Chen, S.J., Hwang, C.L.: Fuzzy Multiple Attribute Decision Making. Springer, Berlin (1992)
20. Zhu, Q., Lee, E.S.: Comparison and Ranking of Fuzzy Numbers. In: Kacprzyk, J., Fedrizi, M. (eds.) Fuzzy Regression Analysis, pp. 21–44. Omnitech Press, Warsaw and Physica-Verlag, Heidelberg (1992)

21. Bárdossy, A., Hagaman, R., Duckstein, L., Bogardi, I.: Fuzzy Least Squares Regression: Theory and Applications. In: Kacprzyk, J., Fedrizi, M. (eds.) Fuzzy Regression Analysis, pp. 181–193. Omnitech Press, Warsaw and Physica-Verlag, Heidelberg (1992)
22. Diamond, P.: Fuzzy Least Squares. Inf. Sci. 46, 141–157 (1988)
23. Diamond, P., Hörner, R.: Extended Fuzzy Linear Models and Least Squares Estimates. Comput. Math. Appl. 33, 15–32 (1997)
24. Diamond, P., Tanaka, H.: Fuzzy Regression Analysis. In: Slowinski, R. (ed.) Fuzzy Sets in Decision Analysis, Operations Research and Statistics, pp. 349–387. Kluwer Academic Publishers, Boston (1998)
25. Bertoluzza, C., Corral, N., Salas, A.: On a new Class of Distances Between Fuzzy Numbers. Mathware Soft Comput. 2, 71–84 (1995)

Multicriteria Relational Clustering: The Case of Binary Outranking Matrices

Yves De Smet and Stefan Eppe

CoDE – SMG,
Université Libre de Bruxelles, Brussels, Belgium (ULB)
{yves.de.smet,stefan.eppe}@ulb.ac.be

Abstract. In this paper we address the question of multicriteria relational clustering. We develop a method that builds clusters and relations between these clusters on the basis of a binary outranking matrix. An extension of the k-means algorithm is presented and tested on artificial data sets. Finally, an example about the clustering of journals from the field of "Operations Research and Management Science" is illustrated.

1 Introduction

In the field of multicriteria decision aid, one usually distinguishes three main problems [4]. The first one can be formulated as identifying a subset of alternatives that are considered to be the best or the more interesting ones. This is referred to as the choice problem. The second one is related to the ranking of the actions from the best to the worst one. Finally, in sorting problems, the decision maker focuses on the assignment of alternatives to pre-defined ordered categories [3,5,6].

Recently some researchers have addressed the question of detecting clusters in multicriteria contexts. In an earlier paper, De Smet et al. [2] have developed an extension of the traditional k-means algorithm to tackle this problem. The main contribution of their work relies on the definition of a distance measure that is based on binary preference relations between the alternatives. Intuitively, two actions will be considered as similar if they are indifferent, incomparable, being preferred or preferred to the same alternatives. In this way, the intrinsic nature of the multicriteria problem is taken into account in the k-means algorithm.

In this article, we present an extension of the aforementioned work. Our aim is not only to detect clusters in a multicriteria context but also to identify relations between these clusters. This last issue was not treated in [2]. Typically we are looking for information such as "two clusters A and B are incomparable while they are both preferred to a third cluster C". As mentioned in Cailloux et al. [1], this is referred to as *relational multicriteria clustering.*

The paper is organized in three main sections. Section 2 is dedicated to the model and the algorithm. Tests are performed on artificial data sets in Section 3. Finally an illustrative example is proposed in Section 4. The idea is to apply the algorithm on a binary outranking matrix obtained from the comparison of 60

M. Ehrgott et al. (Eds.): EMO 2009, LNCS 5467, pp. 380–392, 2009.

journals from the "Operations Research & Management Science" category of the 2006 edition of the Thomson Scientific's Journal Citation Report.

2 The Model

Let us consider a set of actions, noted $A = \{a_1, a_2, \dots a_n\}$, and a binary outranking relation S defined on A. For any pair of actions a_i, $a_j \in A$, $a_i S a_j$ expresses that "a_i is at least as good as a_j" ([4]). This aggregated relation has been built from the comparison of every couple of actions with respect to a set $G = \{g_1, g_2, \dots, g_m\}$ of m criteria using a given multicriteria outranking method.

The outranking relation S induces a preference structure (as defined in [4]) with the following relations: Preference (P), Indifference (I), Incomparability (J). Formally, we have:

$$\begin{cases} a_i P a_j \Leftrightarrow a_i S a_j,\ a_j \neg S a_i \\ a_i I a_j \ \Leftrightarrow a_i S a_j,\ a_j S a_i \\ a_i J a_j \Leftrightarrow a_i \neg S a_j,\ a_j \neg S a_i \end{cases}$$

The aim of the method is to aggregate the provided preference information into a relational partition of A, noted (P_k, S_{P_k}) and where

- $P_k = \{C_1, C_2, \dots, C_k\}$ is a partition of A into k clusters
- $S_{P_k} \subset P_k \times P_k$ is an antisymmetric binary relation on P_k

Similarly $C_i S_{P_k} C_j$ expresses that "cluster C_i is at least as good as cluster C_j". Naturally, we can assume that S_{P_k} is antisymmetric.

Let us remark that the number of possible relational partitions (P_k, S_{P_k}) is given by the Sterling number of the second kind, multiplied by the number of possible permutations inside the clusters relations, thus $3^{\frac{k(k-1)}{2}} \cdot S(n,k)$.

As already mentioned, our goal is to apply an extension of the widely used k-means clustering method that integrates the multicriteria nature of the provided input data. In order to achieve this, we will have to define:

- how to build the centroid of a cluster,
- a multicriteria *distance* between actions and cluster's centroids, and
- a *fitness* measure that allows to quantitatively evaluate a given relational partition

As the definition of the distance is a key element for applying an adapted k-means clustering algorithm, we will use the distance defined in [2], which integrates this information by using a particular representation for each action, whether real or fictitious, called the profile. More formally,

Definition 1. *[2] The profile $Q(a_i)$ of action $a_i \in A$ is defined as being a 4-uple $\langle\, J(a_i)\,,\ P^+(a_i)\,,\ P^-(a_i)\,,\ I(a_i)\,\rangle$, where:*

$$\begin{cases} J(a_i) \ \ \ = \{a_j \in A \mid a_i J a_j\} = Q_1(a_i) \\ P^+(a_i) = \{a_j \in A \mid a_i P a_j\} = Q_2(a_i) \\ P^-(a_i) = \{a_j \in A \mid a_j P a_i\} = Q_3(a_i) \\ I(a_i) \ \ \ \ = \{a_j \in A \mid a_i I a_j\} = Q_4(a_i) \end{cases}$$

In other words, we consider a profile of a given action a_i as a partition of A into 4 subsets denoted by $J(a_i)$, $P^+(a_i)$, $P^-(a_i)$ and $I(a_i)$, each of which contains the actions $a_j \in A$ satisfying respectively the relations a_iJa_j, a_iPa_j, a_jPa_i, a_iIa_j.

In what follows we will consider that two actions will be similar if they are indifferent, preferred or being preferred by the same actions. Put in another way, "two actions will be as close as their profiles are alike".

Definition 2. *[2] Let $Q(a)$ be the profile of $a \in A$, the distance between two actions a_i, $a_j \in A$ is defined as follows:*

$$d\left(Q(a_i), Q(a_j)\right) = 1 - \frac{1}{n}\sum_{l=1}^{4} |Q_l(a_i) \cap Q_l(a_j)| \ ,$$

where $n = |A|$. Considering a relational partition (P_k, S_{P_k}), we would like to characterize a representative element of each cluster that we will call a centroid.

Definition 3. *Based on the relational partition (P_k, S_{P_k}), let c_i be the centroid of cluster $C_i \in P_k$. The profile of c_i is defined as follows:*

$$Q(c_i) = \left\langle \bigcup_{j \in \mathcal{E}_J(C_i)} C_j, \bigcup_{j \in \mathcal{E}_{P^+}(C_i)} C_j, \bigcup_{j \in \mathcal{E}_{P^-}(C_i)} C_j, \bigcup_{j \in \mathcal{E}_I(C_i)} C_j \right\rangle$$

where the following subsets of cluster indices associated to a given cluster C_i have been defined:

$$\begin{cases} \mathcal{E}_J(C_i) &= \{ j \in \{1,..,k\} \mid C_j \in P_k,\ C_i \neg S_{P_k} C_j,\ C_j \neg S_{P_k} C_i \} \\ \mathcal{E}_{P^+}(C_i) &= \{ j \in \{1,..,k\} \mid C_j \in P_k,\ C_i S_{P_k} C_j,\ C_j \neg S_{P_k} C_i \} \\ \mathcal{E}_{P^-}(C_i) &= \{ j \in \{1,..,k\} \mid C_j \in P_k,\ C_i \neg S_{P_k} C_j,\ C_j S_{P_k} C_i \} \\ \mathcal{E}_I(C_i) &= \{ j \in \{1,..,k\} \mid C_j \in P_k,\ C_i S_{P_k} C_j,\ C_j S_{P_k} C_i \} \end{cases}$$

Let us stress that this definition differs from the one given in [2].

In order to evaluate the quality of a relational partition (P_k, S_{P_k}), we compute the average distance of all actions to the centroid of the cluster they are assigned to. The fitness measure, denoted by f, will be the complementary value.

Definition 4. *Let A be a set of actions, S be a binary outranking relation on A and (P_k, S_{P_k}) be a relational partition, the fitness f associated to that relational partition in regard to A and S is defined as follows:*

$$f\left(A, S, P_k, S_{P_k}\right) = 1 - \frac{1}{n}\sum_{r=1}^{k}\sum_{a_i \in C_r} \mathrm{d}\left(Q(a_i), Q(c_r)\right) \ ,$$

where $n = |A|$. Although this fitness measure only looks like an internal homogeneity indicator, as it tries to minimize the distance between all actions and their associated centroid, it is not. Indeed, since the fitness is based on a distance which, in turn, is calculated by using profiles, the relational connection between the clusters are implicitly included in this measure (since they are used for the definition of the centroid's profiles).

This can be shown in the following way, by explicitly introducing the definition of the distance into the fitness definition:

$$f(A, S, P_k, S_{P_k}) = 1 - \frac{1}{n}\sum_{r=1}^{k}\sum_{a_i\in C_r}\left[1 - \frac{1}{n}\sum_{l=1}^{4}|Q_l(a_i)\cap Q_l(c_r)|\right]$$
$$= \frac{1}{n^2}\sum_{r=1}^{k}\sum_{a_i\in C_r}\sum_{l=1}^{4}|Q_l(a_i)\cap Q_l(c_r)|$$

In addition to this scalar fitness measure, a richer indicator can be defined in order to obtain a fine grained evaluation.

Definition 5. *Let A be a set of actions, S be a binary outranking relation on A and (P_k, S_{P_k}) be a relational partition. The matrix of fitness indices Φ of that relational partition in regard to A and S is a $k \times k$ matrix, which elements φ_{cr}, $\forall c, r \in \{1, .., k\}$ are defined by*

$$\varphi_{rq} = \frac{1}{|C_r| \cdot |C_q|}\sum_{a_i\in C_r}\left|Q_{l^*_{rq}}(a_i)\cap C_q\right|.$$

where l^*_{rq} is the index related to the existing relation between C_r and C_q.

Intuitively, φ_{rq} expresses the fitness at the relational level by providing a quantitative evaluation of what proportion of cluster r's actions have the same relation to cluster q's actions, as it is expressed by the calculated relation (denoted $l^*_{rq} \in \{1, .., 4\}$) between the clusters r and q. For instance let us assume that C_r is incomparable to C_q. As a consequence $l^*_{rq} = 1$, φ_{rq} will quantify in which way the actions belonging to C_q are present or not in $J(a_i)$ for every a_i of C_r.

We finally gathered all necessary tools to implement the adapted k-means algorithm described below.

The Binary RMCC Clustering Algorithm

`Inputs:` $A = \{a_1, a_2, \ldots, a_n\}, S, k$
$\forall a_i \in A$ compute $Q(a_i)$
Randomly initialize the relational partition (P_k, S_{P_k}), such that
$\forall C_h \in P_k, |C_h| > 0$
`repeat`
  `foreach` $C_j \in P_k$ `do`
    Compute $Q(c_j)$
  `endfor`
  `foreach` $a_i \in A$ `do`
    Assign a_i to C_j such that
    $\mathrm{d}(Q(a_i), Q(c_j)) \leq \mathrm{d}(Q(a_i), Q(c_m))\ \forall C_j, C_m \in P_k$
  `endfor`
  Given P_k, determine S_{P_k} in order to maximize f
`until` The relational partition no longer changes

If two or more centroids are at the same minimal distance from the considered action, it is arbitrarily assigned to the cluster with the smallest index. Additionally at each step we ensure that at least one element belongs to every cluster. Without this condition a cluster could possibly disappear during the execution.

Once a given partition P_k has been determined, the relation between each pair of clusters (C_i, C_j) has to be calculated. We choose the relation so as to maximize the fitness f of the relational partition.

$$S^*_{P_k} = \underset{S_{P_k}}{\text{argmax}}\, f\left(A, S, P_k, S_{P_k}\right)$$

As a consequence, $3^{\frac{k(k-1)}{2}}$ tests have to be performed each time a new partition P_k is considered. Because this would rapidly become a bottleneck when increasing the number of clusters (for $k = 6$ already $\sim 10^8$), an equivalent voting procedure has been adopted.

Proposition 1. *For a given set (A, S, P_k), applying the hereafter described voting procedure to determine the cluster's relations is equivalent to maximizing the fitness function $f(A, S, P_k, S_{P_k})$.*

For each pair of clusters (C_r, C_q), their relation $l^*_{rq} \in \{1, .., 4\}$ will be determined such that

$$l^*_{rq} = \begin{cases} \underset{l_{rq}\in\{1,..,3\}}{\text{argmax}} \left(\sum\limits_{a_i\in C_r} \sum\limits_{a_j \in C_q} \left| Q_{l_{rq}}(a_i) \cap a_j \right| \right) & , \textit{for } r \neq q \\ 4 & , \textit{for } r = q \end{cases}$$

Proof. Intuitively, the voting procedure defined above will consist in comparing the n_r actions of cluster C_r with the n_q actions of cluster C_q. The goal being to show that the fitness is maximal with these calculated relations, we will create a centroid profile $Q(c_r)$ for each cluster $C_r \in P_k$ in the following way $\forall r \in \{1, .., k\}$:

$$Q_l(c_r) = \bigcup_{q\in \mathcal{J}_l(c_r)} C_q \,,$$

where $\mathcal{J}_l(c_r) = \left\{ q \in \{1, .., k\} \mid l^*_{rq} = l \right\}, \forall l \in \{1, .., 4\}$.

Armed with these centroid's profiles it can be shown that the given voting procedure maximizes the fitness f. The proof is done by rewriting the definition of the fitness:

$$f\left(A, S, P_k, S_{P_k}\right) = 1 - \tfrac{1}{n} \sum_{r=1}^{k} \sum_{a_i \in C_r} \mathrm{d}\left(Q(a_i), Q(c_r)\right)$$

$$= \tfrac{1}{n^2} \sum_{r=1}^{k} \sum_{a_i\in C_r} \sum_{l=1}^{4} \left| Q_l(a_i) \cap Q_l(c_r) \right|$$

$$= \tfrac{1}{n^2} \sum_{r=1}^{k} \sum_{a_i\in C_r} \sum_{l=1}^{4} \left| Q_l(a_i) \cap \left(\bigcup_{q\in\mathcal{J}_l(c_r)} C_q \right) \right|$$

$$= \tfrac{1}{n^2} \sum_{r=1}^{k} \sum_{a_i\in C_r} \sum_{l=1}^{4} \left| \bigcup_{q\in\mathcal{J}_l(c_r)} \left[Q_l(a_i) \cap C_q \right] \right|$$

$$= \frac{1}{n^2} \sum_{r=1}^{k} \sum_{a_i \in C_r} \sum_{l=1}^{4} \sum_{C_q \in P_k} \sum_{a_j \in C_q} \delta_{[l=l^*_{rq}]} \left|Q_l(a_i) \cap a_j\right|$$

$$= \frac{1}{n^2} \sum_{r=1}^{k} \sum_{a_i \in C_r} \sum_{q=1}^{k} \sum_{a_j \in C_q} \left|Q_{l^*_{rq}}(a_i) \cap a_j\right|$$

$$= \frac{1}{n^2} \sum_{r=1}^{k} \sum_{q=1}^{k} \left[\sum_{a_i \in C_r} \sum_{a_j \in C_q} \left|Q_{l^*_{rq}}(a_i) \cap a_j\right| \right]$$

The proposition naturally follows from the last expression and the definition of l^*_{rq} (i.e. the way a cluster relation is elected), since $\forall r, q \in \{1, .., k\}, \forall l \in \{1, .., 4\}$

$$\sum_{a_i \in C_r} \sum_{a_j \in C_q} \left|Q_{l^*_{rq}}(a_i) \cap a_j\right| \geqslant \sum_{a_i \in C_r} \sum_{a_j \in C_q} \left|Q_l(a_i) \cap a_j\right| .$$

3 Empirical Tests

Validation tests have been carried out on artificial data sets in order to analyze the sensitivity of the adapted k-means algorithm with respect to several parameters. The generated data sets will be briefly described in the first part of this section. We will thereafter focus on the actual results and the conclusions we may draw from them.

Five relational models have been considered, each having distinctive features. We briefly present them in Table 1.

We will now use our relational model $\tilde{S}$ as defined in Table 1 in order to build a reference outranking matrix, denoted $\tilde{\Sigma}$, and whose actual form will also depend on the number of actions n_j, $j \in \{1, .., k\}$ that will be assigned to each cluster j. The number of actions of the artificial data set will be $n = \sum_{j=1}^{k} n_j$. The reference outranking matrix is computed by transforming each element $\tilde{s}_{ij}$ of the relational matrix $\tilde{S}$ into a sub-matrix of dimension $n_i \times n_j$, with all its elements being set to the value of $\tilde{s}_{ij}$. The assembly of all these sub-matrices finally yields the reference outranking matrix.

For instance, for the case 1 and a set of cluster cardinals $\{n_1, n_2, n_3\} = \{2, 3, 1\}$, the relation matrix $\tilde{S}_1$ would extend to the 6×6 reference outranking matrix

$$\tilde{\Sigma}_1 = \begin{bmatrix} 1\,1\,1\,1\,1\,1 \\ 1\,1\,1\,1\,1\,1 \\ 0\,0\,1\,1\,1\,0 \\ 0\,0\,1\,1\,1\,0 \\ 0\,0\,1\,1\,1\,0 \\ 0\,0\,0\,0\,0\,1 \end{bmatrix} .$$

$\tilde{\Sigma}$ represents an ideal case, where the algorithm performs best. Unfortunately, this is an utopian configuration. In order to represent more complex data structures, we will have to consider ways of applying perturbations. This will be done in two ways, described hereafter.

Table 1. Relational models used for validation

Ref	Representation	Relation matrix
1	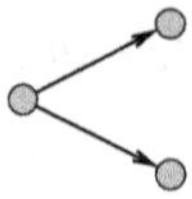	$\tilde{S}_1 = \begin{bmatrix} 1\,1\,1 \\ 0\,1\,0 \\ 0\,0\,1 \end{bmatrix}$
2	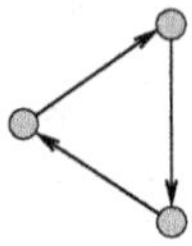	$\tilde{S}_2 = \begin{bmatrix} 1\,1\,0 \\ 0\,1\,1 \\ 1\,0\,1 \end{bmatrix}$
3	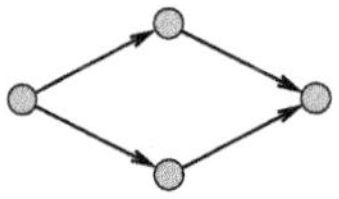	$\tilde{S}_3 = \begin{bmatrix} 1\,1\,1\,0 \\ 0\,1\,0\,1 \\ 0\,0\,1\,1 \\ 0\,0\,0\,1 \end{bmatrix}$
4	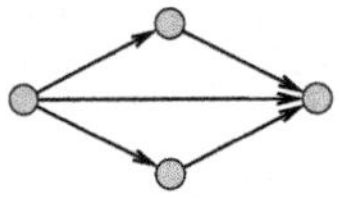	$\tilde{S}_4 = \begin{bmatrix} 1\,1\,1\,1 \\ 0\,1\,0\,1 \\ 0\,0\,1\,1 \\ 0\,0\,0\,1 \end{bmatrix}$
5	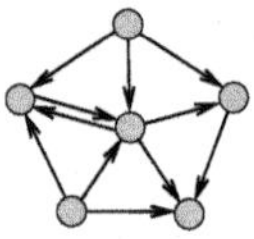	$\tilde{S}_5 = \begin{bmatrix} 1\,1\,0\,0\,1\,1 \\ 0\,1\,1\,0\,0\,0 \\ 0\,0\,1\,0\,0\,0 \\ 0\,0\,1\,1\,1\,1 \\ 0\,0\,0\,0\,1\,1 \\ 0\,1\,1\,0\,1\,1 \end{bmatrix}$

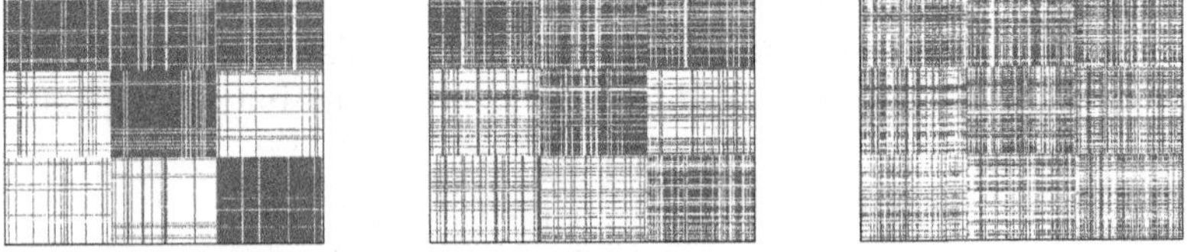

Fig. 1. Examples of $\tilde{S}_T$ with respectively ν_{perm} =20%, 40% and 75% of relational cluster-action perturbations applied on case 1

First, we introduce what we will call *relational cluster-action perturbation.* This type of perturbation consists in randomly altering the relation between a given action $a_i \in A$ and all actions of a given cluster C_j. The result of performing this perturbation with several magnitudes is shown on Figure 1, where the matrices have been represented by a two-color raster, the values 0 and 1 being respectively shown as white and gray pixels. The parameter of this perturbation is a normalized perturbation factor which is defined as $\nu_{perm} = \frac{n_{perm}}{k \cdot n}$, where n_{perm} is the number of permutations that will be applied.

In addition to the *relational cluster-action perturbation*, we also artificially introduce *noise* into the reference outranking matrix $\tilde{\Sigma}$. This is done by randomly picking pairs of actions and randomly change their relation. Similarly to the previous perturbation, a normalized noise factor $\nu_{noise} = \frac{2 \cdot n_{noise}}{n \cdot (n-1)}$ (where n_{noise} is the number of pairs of actions that alterations are applied to) is defined. The application of noise on a reference outranking matrix is shown at Figure 2.

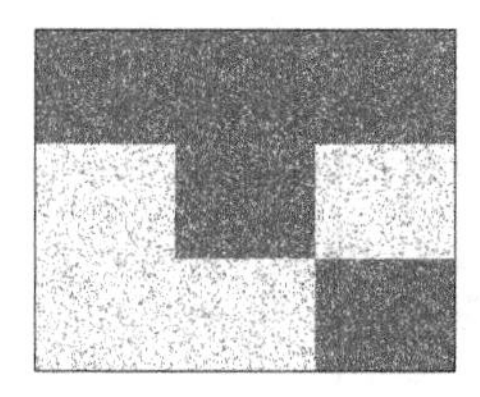
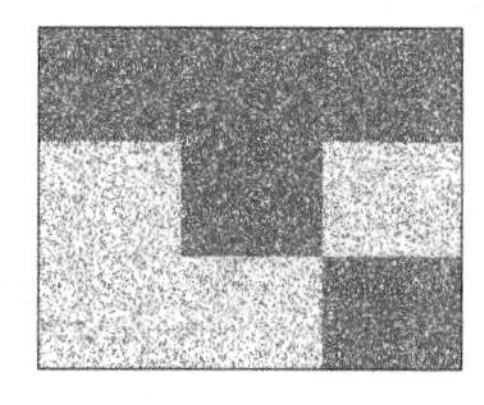
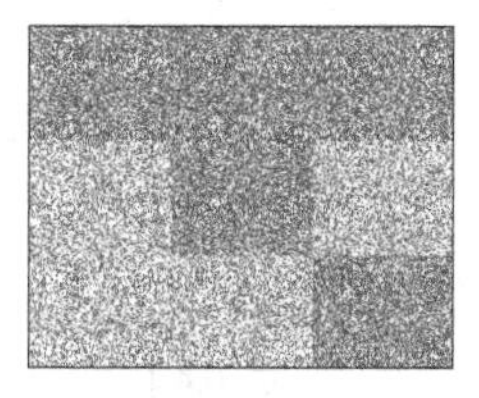

Fig. 2. Examples of $\tilde{S}_T$ with 20%, 40% and 75% of noise applied on case 1

Table 2 summarizes the different parameters that we will play with to generate artificial data sets.

Table 2. Parameters for artificial data sets

Parameter	Symbol
Number of actions per cluster	$n = \{n_1, \ldots, n_k\}$
Relational cluster-action perturbation	$\nu_{perm} \in [0, 1]$
Noise	$\nu_{noise} \in [0, 1]$

As the k-means algorithm badly suffers from its dependency on random initialization, it has been repeated until no new improvements appeared during 10 consecutive executions. The best result is kept and returned as the result of the algorithm.

Results

In the following, only the results obtained for the reference case 3 are presented in some detail, because the results for all five studied cases (defined in Table 1) have shown to be very similar.

The results presented in Table 3 are averaged outcomes of the algorithm repeated 20 times. These results are graphically shown on Figure 3.

Table 3. Results for reference case 3 ($n = 50$)

Noise\Pert	0,00	0,05	0,10	0,15	0,20	0,25
0,00	1,00	0,93	0,85	0,83	0,86	0,71
0,05	0,96	0,92	0,88	0,79	0,74	0,79
0,10	0,94	0,93	0,82	0,81	0,75	0,67
0,15	0,90	0,74	0,79	0,67	0,71	0,64
0,20	0,88	0,83	0,79	0,72	0,68	0,69
0,25	0,82	0,81	0,74	0,70	0,63	0,61

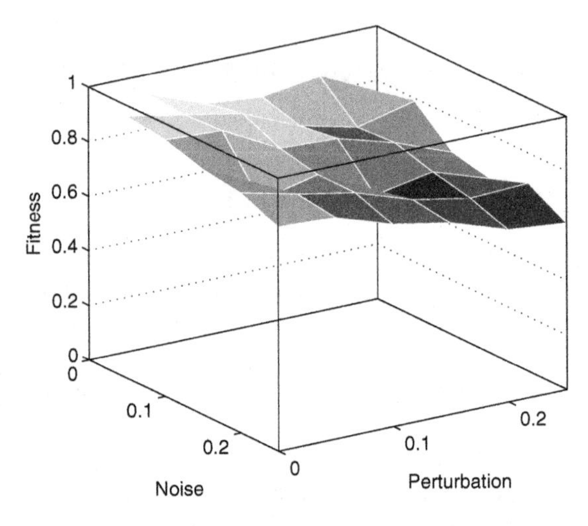

Fig. 3. Fitness depending on applied perturbation and noise. Results for reference case 3 ($n = 50$).

When compared to the results for other reference cases, it is striking to notice that it does not depend on the underlying relational structure nor the relative weights of the different clusters. This behavior may be explained by the fact that the fitness measure, directly depending on the average distance of all actions to their respectively assigned cluster's centroid, "only" counts the number of differences between each action's profile and its cluster's centroid profile, regardless on the type of difference that arises.

A second type of tests that has been carried out on the above defined artificial data sets is the alteration of the relative cluster sizes. As in the first series, the clusters of all underlying data sets had the same number of elements, this equilibrium has been modified in order to observe the influence of that parameter. For each graph of Figure 4, the number of elements in each cluster is indicated in the legend.

As can easily be retrieved from Figure 4, clusters of very different sizes are handled in a satisfying way with respect to the fitness, presenting no noticeable degradation of the result's fitness.

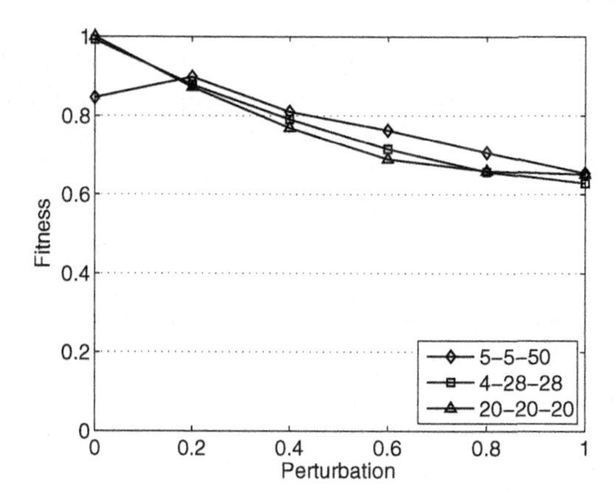

Fig. 4. Variation of the results under the influence of altered cluster size proportions for the reference case 1

4 Illustrative Example

Let us now consider an application of the presented algorithm to the practical case of scientific journal ranking. The data used for this example is taken from the 2006 edition of Thomson Scientific's Journal Citation Reports (JCR), from which only journals gathered in the subject category of "Operations Research & Management Science" have been extracted.

The goal is to partition the set of selected journals into a given number of clusters and to elicit their relations in regard to different criteria. Amongst all cited criteria leading to the JCR's ranking of journals, we will select two, which an outranking rule will be applied on, in order to obtain a pairwise outranking relation on all selected journals. The number of considered criteria has been limited to two in order to allow an easy representation of the journals in two dimensions. Not too much emphasis will be put on the motivations for the choices made, since their primary purpose is to furnish illustrative input data.

The used (and transformed) basic information for each journal are, for a given JCR year y:

- The impact factor IF_y, denoted as g_1
- The total cites divided by the number of published articles $\frac{TC_y}{A_y}$, as the second criterion g_2

In order to derive the outranking matrix we will use an extension of the dominance relation. More formally:

$$aSb \Leftrightarrow \forall h \in \{1, 2\}, \, g_h(a) \geq g_h(b) - \sigma_h \, ,$$

where σ_h is a threshold that is set to be a fifth of criterion's h measured amplitude.

Results

Applying the algorithm to the data set in Table 4 and setting the number of clusters to $k = 5$, yields the results presented in Figure 5. This relational partition has a quite good fitness of $f = 87,5\%$. It is noticeable that 4 regular clusters with perfect transitive outranking relations are obtained. A fifth, "pathological" cluster, only holds two journals (i.e. "INFOR" and "INT J FLEX MANUF SYS"), which are special because of the very low number A_y of published articles. Indeed, a very small value of A_y makes the criteria g_2 artificially explode, without having any actual meaning. Although the relation to the worst of the regular cluster may be questionable, this result allows to isolate and draw attention to these two special journals.

Figure 5 also integrates the elicited outranking relations, represented as arrows, between the clusters. The direction of these arrows express the preference

Table 4. The data set used for illustrating the proposed method

Journal	TC_y	A_y	$\frac{TC_y}{A_y}$	IF_y	Journal	TC_y	A_y	$\frac{TC_y}{A_y}$	IF_y
ANN OPER RES	1833	115	15.9	0.544	MANAGE SCI	12110	133	91.1	1.931
APPL STOCH MODEL BUS	161	34	4.7	0.342	MATH METHOD OPER RES	357	61	5.9	0.400
ASIA PAC J OPER RES	96	32	3.0	0.258	MATH OPER RES	1908	55	34.7	0.875
COMPUT OPER RES	2402	221	10.9	1.147	MATH PROGRAM	3644	52	70.1	1.475
COMPUT OPTIM APPL	641	57	11.2	0.851	MIL OPER RES	55	17	3.2	0.241
CONCURRENT ENG-RES A	233	25	9.3	0.482	NAV RES LOG	1511	73	20.7	0.548
DECIS SUPPORT SYST	1645	160	10.3	1.119	NETW SPAT ECON	126	20	6.3	0.514
DISCRETE EVENT DYN S	200	19	10.5	0.545	NETWORKS	1257	57	22.1	0.609
ENG OPTIMIZ	368	53	6.9	0.571	OMEGA-INT J MANAGE S	1124	62	18.1	1.327
EUR J OPER RES	11003	838	13.1	1.096	OPER RES	6135	82	74.8	1.467
EXPERT SYST APPL	1254	222	5.6	1.177	OPER RES LETT	1195	112	10.7	0.517
IIE TRANS	2146	85	25.2	0.797	OPTIM CONTR APPL MET	164	24	6.8	0.735
INFOR	283	3	94.3	0.275	OPTIM ENG	132	17	7.8	0.711
INFORMS J COMPUT	624	57	10.9	0.907	OPTIM METHOD SOFTW	427	57	7.5	0.554
INT J COMPUT INTEG M	290	64	4.5	0.297	OPTIMIZATION	470	45	10.4	0.408
INT J FLEX MANUF SYS	255	4	63.8	0.452	OR SPECTRUM	351	38	9.2	0.562
INT J INF TECH DECIS	145	40	3.6	0.718	PROBAB ENG INFORM SC	266	37	7.2	0.577
INT J PROD ECON	2601	212	12.3	0.995	PROD OPER MANAG	770	51	15.1	2.123
INT J PROD RES	4281	278	15.4	0.56	PROD PLAN CONTROL	689	59	11.7	0.561
INT J SYST SCI	954	85	11.2	0.492	QUAL RELIAB ENG INT	323	66	4.9	0.500
INT J TECHNOL MANAGE	531	87	6.1	0.356	QUEUEING SYST	894	54	16.6	0.851
INTERFACES	953	42	22.7	0.575	RAIRO-OPER RES	118	30	3.9	0.088
J GLOBAL OPTIM	1237	111	11.1	0.813	RELIAB ENG SYST SAFE	1774	159	11.2	1.004
J IND MANAG OPTIM	55	54	1.0	0.722	SAFETY SCI	629	54	11.6	0.427
J OPER MANAG	1971	82	24.0	1.851	SYST CONTROL LETT	3376	95	35.5	1.634
J OPER RES SOC	2873	159	18.1	0.784	TECHNOVATION	798	60	13.3	1.004
J OPER RES SOC JPN	190	34	5.6	0.200	TRANSPORT RES B-METH	1801	67	26.9	1.948
J OPTIMIZ THEORY APP	2655	135	19.7	0.688	TRANSPORT RES E-LOG	350	47	7.4	1.000
J QUAL TECHNOL	1432	26	55.1	1.184	TRANSPORT SCI	1613	35	46.1	1.427
J SCHEDULING	331	31	10.7	1.000					

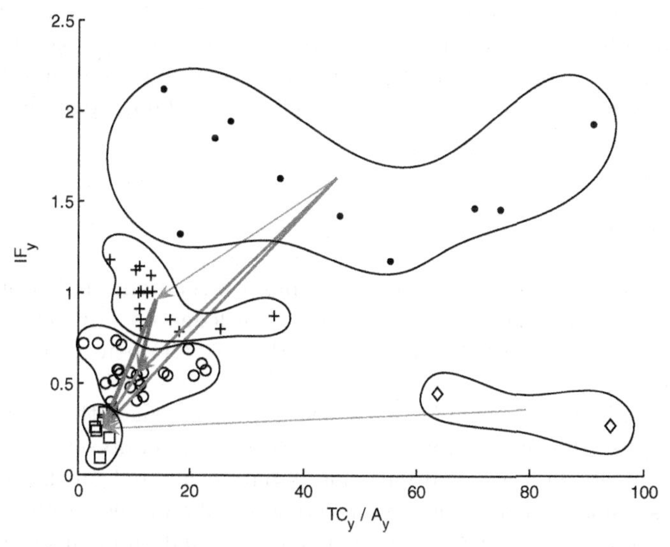

Fig. 5. Computed clusters and their relations with the proposed algorithm

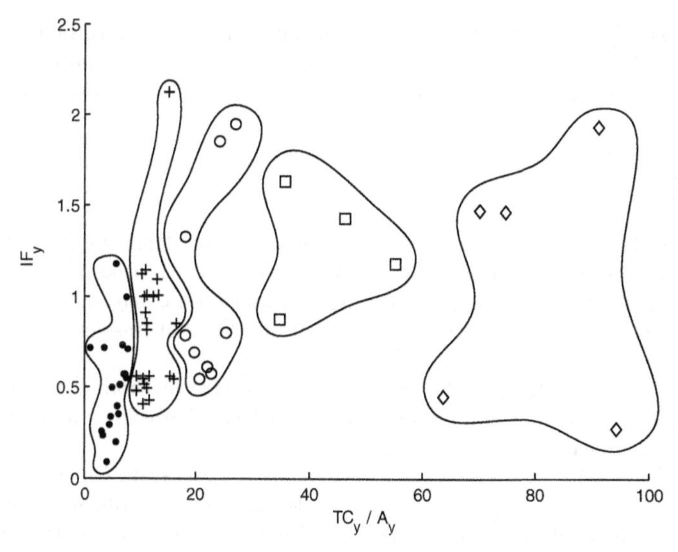

Fig. 6. The classical k-means algorithm applied with an euclidean distance on the data set $(TC_y/A_y, IF_y)$

between each pair of clusters. Additionally, the matrix of fitness indices is given by

$$\Phi = \begin{bmatrix} 0{,}88 & 0{,}81 & 1{,}00 & 1{,}00 & 1{,}00 \\ 0{,}81 & 0{,}87 & 0{,}92 & 1{,}00 & 0{,}54 \\ 1{,}00 & 0{,}92 & 0{,}65 & 0{,}99 & 1{,}00 \\ 1{,}00 & 1{,}00 & 0{,}99 & 0{,}18 & 0{,}67 \\ 1{,}00 & 0{,}54 & 1{,}00 & 0{,}67 & 0{,}50 \end{bmatrix}.$$

This result can be compared with the one obtained with the classical k-means method when an euclidean metric is directly applied on the prepared data set $(TC_y/A_y, IF_y)$. See Figure 6. Our aim here is to stress that significant differences might exist between the two outputs. Nevertheless, we acknowledge that we have not refined the application of the k-means algorithm.

5 Conclusion

In this paper we have presented a model to build a relational partition on the basis of a binary outranking matrix. To our knowledge no previous work has addressed that kind of question and we hope that it will open many directions for future research.

For the sake of simplicity we have applied a k-means algorithm to solve the problem. Of course the use of refined meta-heuristics could improve the obtained results. Up to now, the relational partition is only evaluated on the basis of a unique fitness value which integrate the homogeneity of the groups and the coherence of the clusters relations. Obviously the evaluation of such a partition involves different conflicting criteria and give rise to multi-objective optimization problems.

From a methodological point of view several extensions have to be considered. Among them, the extension to valued outranking matrices or the development of a weighted distance that will differentiate the four preference relations.

References

1. Cailloux, O., Lamboray, C., Nemery, P.: A taxonomy of clustering procedures. In: Proceedings of the 66th Meeting of the EWG on MCDA, Marrakech, Maroc (2007)
2. De Smet, Y., Montano Guzmán, L.: Towards multicriteria clustering: An extension of the k-means algorithm. European Journal of Operational Research 158, 390–398 (2004)
3. Nemery, P., Lamboray, C.: FlowSort: a flow-based sorting method with limiting and central profiles. TOP (Journal of the Spanish Society of Statistics and Operations Research) (to appear)
4. Vincke, P.: Multicriteria Decision-aid. John Wiley & Sons Ltd., Chichester (1992)
5. Yu, W.: Aide multicritère à la décision dans le cadre de la probl'ematique du tri: m'ethodes et applications, PhD thesis, LAMSADE, Universit'e Paris Dauphine, Paris (1992)
6. Zopounidis, C., Doumpos, M.: Multicriteria classification and sorting methods. A literature review. European Journal of Operational Research 138, 229–246 (2002)

Towards the Early Diagnosis of Alzheimer's Disease via a Multicriteria Classification Model

Amaury T. Brasil Filho, Plácido R. Pinheiro, and André L.V. Coelho

Graduate Program in Applied Informatics, University of Fortaleza,
Av. Washington Soares 1321, J30, Fortaleza CE, Brazil
abrasil@gmail.com, {placido,acoelho}@unifor.br

Abstract. The very early detection of Alzheimer's disease (AD) has been deeply investigated in numerous studies in the past years. These studies have demonstrated that the pathology usually arises decades before the clinical diagnosis is effectively made, and so a reliable identification of AD in its earliest stages is one of the major challenges clinicians and researchers face nowadays. In the present study, we introduce a new approach developed upon a specific Multicriteria Decision Aid (MCDA) classification method to assist in the early AD diagnosis process. The MCDA method is centered on the concept of prototypes, that is, alternatives that serve as class representatives related to a given problem, and has its performance index very dependent upon the choice of values of some control parameters. In such regard, two techniques, one based on ELECTRE IV methodology and the other on a customized genetic algorithm, are employed in order to select the prototypes and calibrate the control parameters automatically. Moreover, a new database has been designed taking as reference both the functional and cognitive recommendations of the Scientific Department of Cognitive Neurology and Aging of the Brazilian Academy of Neurology and a neuropsychological battery of exams made available by the well-known Consortium to Establish a Registry for Alzheimer's Disease (CERAD). Various experiments have been performed over this database in a manner as to either fine-tune the components of the MCDA model or to compare its performance level with that exhibited by other state-of-the-art classification algorithms.

1 Introduction

The Alzheimer's disease (AD) is a progressive and degenerative disease of the brain which causes a serious impairment over its two main activities: thinking and memory. According to Celsis [5], AD is the most common form of dementia among the elderly population, comprising up to 75% of all dementia cases. AD causes a gradual loss of intellectual abilities with deterioration in cognition, function, and behavior, affecting many aspects of an individual life.

This way, with the decline of the normal functioning over the nervous and other bodily systems, and with the natural behavioral and personality changes, the identification of what constitutes abnormal impairment becomes a hard task. Davidoff [6] argues that the problem over the AD diagnosis is not only related to

M. Ehrgott et al. (Eds.): EMO 2009, LNCS 5467, pp. 393–406, 2009.

the current level of understanding of the disease, but also to the comprehension of the normal process involving the patients age. For the author, there are yet no consistent established set of values for what would be a normal level of impairment in the elderly. To overcome these difficulties, some authors [2,9,17] have demonstrated that the AD first symptoms appears relatively early in life, and it evolves during lifetime. This fact raises the chances of identifying the pathology decades before a clinical diagnosis of dementia can be made.

In the present study, a Multicriteria Decision Analysis (MCDA) classification approach, which is developed upon the method recently proposed by Goletsis et al. [14] (referred to hereafter as *gMCDA classifier*), is employed towards the effective early diagnosis of Alzheimer's disease. The gMCDA classifier makes use of the concept of prototypes, that is, special alternatives representing the classes of a problem, and has associated with itself some control parameters related to the expert's preference modeling process. As some of the experiments reported here reveal, the appropriate selection of prototypes as well as the calibration of control parameters are key issues to leverage the gMCDA classifier's performance. This way, our approach combines two complementary techniques, one based on ELECTRE IV methodology [25] and the other on a customized genetic algorithm [8], in order to select the best prototypes and effectively calibrate the control parameters, respectively.

Trying to detect potential patients with AD as early as possible, many studies [3,4,20,22] have investigated potential tests and exams that, through a functional and cognitive analysis, may help the early AD detection. In this context, to evaluate the effectiveness of our MCDA classification approach in the early AD detection, we have developed a special-purpose AD-related database by following the recommendations of the Scientific Department of Cognitive Neurology and Aging of the Brazilian Academy of Neurology [23] and by making use of a neuropsychological battery of exams made available by the well-known Consortium to Establish a Registry for Alzheimer's Disease (CERAD) [12]. Various experiments have been performed over this database in a manner as to either fine-tune the components of the MCDA model or to compare its performance level with that exhibited by other state-of-the-art classification algorithms.

The rest of the paper is organized as follows. The next section presents an overview of some related work concerning the themes of AD and MCDA classification. The third section outlines the main conceptual ingredients of the gMCDA classifier and the methodologies that were employed to solve the prototype and parameter selection tasks. The fourth section provide details of the database we have designed while the fifth section is dedicated to discuss the AD classification experiments we have conducted so far over the new database. Finally, the last section concludes the paper and brings remarks on future work.

2 Related Work

A classification problem refers to the assignment of a group of alternatives to a set of predefined classes, also known as categories. During the last decades these problems have been tackled using a high variety of statistical and machine learning techniques. Recently, the area of Multicriteria Decision Aid (MCDA) [10,25] has also brought new methodologies and techniques to solve these problems.

The main difference between the MCDA classification methods and others coming from related disciplines, as artificial neural networks (ANN), Bayesian models, rule-based models, decision trees, etc. [29], lies in the way that the MCDA methods incorporate the decision maker's preferences into the categorization process. In the ANN field, for instance, the work of French et al. [13] performs a comparison between an ANN model and a linear discriminant analysis (LDA) algorithm to classify and to stage the degree of dementia. The results demonstrated that the ANN algorithm clearly outperformed the LDA one in terms of classification accuracy, highlighting the utility of using ANN for group classification of patients with AD and staging dementia severity using neuropsychological data.

Figueiredo et al. [11] present an algorithm that classifies individuals into four different groups (i.e., clinically diagnosed groups of elderly normal, demented, AD, and vascular dementia subjects). The classification is performed after the analysis of computer tomography image data from brain and using an optimal interpolative neural network. Another classification work related to dementia disorders among the elderly [30] uses a nave credal classifier to address two different classification problems: discrimination between demented and control patients, and the assignment from among the different types of dementia. The dataset was developed from a set of measures collected among of a series of computerized tests (tasks), which assess some cognitive faculties of the patient.

Sandip et al. [27] realize the AD classification based on a molecular test that evaluates characteristic changes in the concentrations of signaling proteins in the blood, generating a detectable disease-specific molecular phenotype. By this way, through a molecular biomarker in blood plasma, the model classifies the patients into AD or non-AD and identifies those presymptomatic individuals with mild cognitive impairment which will eventually convert to Alzheimer's disease.

In the MCDA field, a decision making model has been recently proposed by Castro and Pinheiro [3,4] to assist the specialist in the early diagnosis of the Alzheimer's disease. Differently from our approach, this model uses the Macbeth software [7] to construct the judgement matrices and the value scales for each fundamental point of view (FPV) already defined. Each patient's information is judged by the decision maker for each FPV; then the Macbeth software generates the value scales that will be used in the final judgment of the patient's diagnosis. Instead of providing the classification itself, this sort of model gives the possibilities of a patient acquiring or not a certain type of dementia in the future.

3 Multicriteria Decision Analysis

Zopounidis and Doumpos [31] define that the decision making problems, according to their nature, the policy of the decision maker, and the overall objective of the decision, may require the choice, ranking, or the assignment of the considered alternatives into predefined classes.

The practical approach that concerns the classification problems motivated researches in developing different methods and mathematical models to solve these problems trying to achieve the highest classification rate. A substantial overview on MCDA methods can be found in [1,19,21] where the authors address the definitions and the problems that are involved in the decision making process.

These methods have been successfully applied to real world problems. The major difficulty during their employment, however, is that, in order to produce models that comply with the decision maker's expectations, a set of control parameters, such as threshold variables, weights, coefficients, etc., needs to be properly set in advance, which turns out to be a hard task to be dealt with. Some authors, like Belacel [1] and Jacquet-Lagréze & Siskos [16], have already provided some alternatives to counter this sort of drawback, although their solutions seem to be rather specific to the contexts that were investigated and yet no general recipes are available to be deployed in all methods and circumstances.

As pointed out by Zopounidis and Doumpos [31], the great majority of works conducted on the MCDA classification theme has focused on the development of novel MCDA classification methods, not giving much emphasis on characterizing and comparing their distinctive problems. Likewise, the authors also advocate that future research on this field should consider a more deep investigation into some important practical issues, such as the analysis of the interdependencies of the control parameters of the algorithms, the statistical validation of the generated models, the analysis of performance over large data sets, and the establishment of links between MCDA classifier models and those coming from related disciplines, such as Pattern Recognition, Machine Learning, and Data Mining [29].

In this context, we have developed an approach developed upon a specific MCDA classification model, which is also composed of two complementary techniques: one responsible for eliciting the values of the classifier's control parameters and the other in charge of selecting the best prototypes from the dataset in accordance with the decision maker's preferences. The gMCDA classifier and the associated techniques are detailed in the sequel.

3.1 MCDA Classifier

The MCDA classification method that we have chosen to cope with the AD classification problem was proposed by Goletsis et al. [14]. Like PROAFTN [1], this method makes use of the concept of prototypes, that is, special alternatives to serve as references against which new alternatives are compared (matched) with. One distinctive aspect of this scheme with respect to other MCDA-based ones is that it presents less control parameters to be adjusted (only some thresholds and criteria weights). In what follows, we provide further details of the formulation behind the gMCDA classifier.

Analytically, the model can be defined as in the following way. Let A be the finite set of alternatives; F the set of n features (in the nominal sorting problem, they are also known as criteria), with $n \geq 1$; w_j the weight of a specific criterion, with $\sum_j w_j = 1$; C the set of categories of a problem, where $C = \{C^1, C^2, \ldots, C^K\}$ and $K > 1$, and $B^h = \{b_p^h | 1, \ldots, L^h\}$ and $\{h = 1, \ldots, K\}$ the set of prototypes of the category C^h, where b_p^h represents the pth prototype of the category C^h and L^h the number of prototypes of this category. Each alternative in A and B is characterized by a feature vector $\overline{g}$ containing its feature values for all n criteria in the F set. Each alternative is compared with each prototype b_p^h under each criterion j.

As described by Goletsis et al. [14], during this comparison, the first element to be computed is the Similarity Index (SI), denoted as $SI_j(a, b_p^h)$. This index is calculated for each criterion, and its purpose is to model the criteria into a five-zone similarity index. In order to compute this index, two thresholds must be specified.

The first threshold that needs to be specified is the similarity threshold, q_j, which represents the maximum allowed criterion difference between the alternatives and the prototypes, i.e. $|g_j(a) - g_j(b_p^h)|$. Using this, the alternatives can be judged as similar under a specific criterion.

The second threshold used in the calculus of SI_j is the dissimilarity threshold, p_j, representing the minimum allowed criterion difference between an alternative a and prototype b_p^h. This threshold needs to be defined in order to consider the alternatives totally dissimilar.

The similarity index SI_j is computed as described below:

$$SI_j(a, b_p^h) = \begin{cases} 1, & \text{if } |g_j(a) - g_j(b_p^h)| \leq q_j \\ \left(\frac{|g_j(a) - g_j(b_p^h)| - p_j}{q_j - p_j}\right), & \text{if } q_j < |g_j(a) - g_j(b_p^h)| < p_j \\ 0, & \text{if } |g_j(a) - g_j(b_p^h)| \geq p_j \end{cases} \quad (1)$$

After the computation of the similarity index, the next step is to compute the Concordance Index (CI). This index indicates the overall similarity concordance of an alternative a with a prototype b_p^h. This index is computed as follows:

$$CI(a, b_p^h) = \sum_j w_j \, SI_j(a, b_p^h). \quad (2)$$

Each alternative will have its CI computed for all prototypes of all classes.

The next step is the computation of the Membership Degree (MD) of an alternative a to a category h. The membership degree applies the best CI among the alternatives to the category h. MD is computed as follows:

$$MD(a, C^h) = arg\ max\{CI(a, b_1^h), \ldots, CI(a, b_{Lh}^h)\}. \quad (3)$$

Finally, the last step is the assignment of the alternative a to a category $C(a)$ with the maximum MD calculated to all the groups of prototypes. The formula is given below:

$$C(a) = arg\ max_h\, MD(a, C^h). \quad (4)$$

The gMCDA classifier, as presented above, was first applied in the ischemic beat classification problem [14]. Aiming to overcome the necessity of manually tuning the parameters, Goletsis et al. employed a genetic algorithm instance devised for such a purpose. In this paper, we have followed the same strategy while dealing with the classification of AD individuals (see Subsection 3.3). As an improvement, our approach also stipulates that the prototypes should be selected through a complementary MCDA technique, which is discussed in the next subsection.

3.2 ELECTRE IV

One of the complementary techniques applied cojointly with the gMCDA classifier tackles the problem of prototype selection. This technique is also based on the MCDA principles, but conversely is based on the concept of sorting of alternatives and criteria.

According to Zopounidis and Doumpos [31], the indirect techniques are widely used for developing sorting models that employ the outranking concept. To apply this technique, the decision analyst specifies the parameters based on an interactive inquiry process with the decision maker. This process ensures that the decision maker preferences will be correctly captured in the model.

Differently from other similar algorithms [18], the ELECTRE IV method [26] does not require the specification of a weight value for each criterion. Conversely, the decision analyst chooses the criterion that it wants to work with and then ELECTRE IV combines them to give birth to the outranking relations. This approach avoids the problem of trying to quantify how important a criterion is. Each criterion can be either defined as a benefit or cost criterion. When the decision analyst considers a cost criterion, the lower the criterion value, the higher its merit; the converse is true for a benefit criterion.

To employ this method to rank the alternatives of a class, the decision analyst should define only the preference and indifference thresholds for each criterion. Specifically in our MCDA approach, the ELECTRE IV method will assume the role of the indirect technique responsible for the prototype selection activity.

Basically, the ELECTRE IV method can be divided into five stages: 1) criteria selection; 2) calculus of the relative thresholds; 3) construction of weak and strong outranking relations; 4) construction of the downward and upward ranks; and 5) elicitation of the final rank.

The first step to employ the ELECTRE IV algorithm is to select the criteria that will be used during the ranking process. The second stage is the determination of the relative thresholds. This phase basically sets the relation of two alternatives under some criterion. It can be defined that two alternatives are indifferent, strictly preferred, or weakly preferred over a criterion k. After that, it is necessary to construct the weak and strong outranking relations for every pair of alternatives [28]. At this point, an alternative i will either strongly or weakly outrank an alternative j based on several restrictions that compares the relative ranks and the thresholds defined [28]. The next step determines the strengths, weaknesses and the qualification of each alternative, and, based on these numbers, defines the downward and upward ranks. Finally, the final rank is set using the mean of the upward and downward ranks.

3.3 Genetic Algorithm

Genetic algorithms (GAs) comprise the class of evolutionary algorithms that uses a specific vocabulary borrowed from natural genetics [8]. The data structures representing the individuals (genotypes) of the population are often called chromosomes; these are one-chromosome (haploid) individuals encoding potential solutions to a problem. In standard GAs, the individuals are represented as strings of bits. Each unit of a chromosome is termed a gene, located in a certain place in the chromosome called locus. The different values a gene can assume are the alleles. The problem to be solved is captured in an objective (fitness) function that allows evaluating the adequacy of any potential solution.

As each chromosome corresponds to the encoded value of a candidate solution, it has to be decoded into an appropriate form for evaluation and is then assigned a fitness value according to the objective. For each chromosome is assigned a probability of reproduction, so that its likelihood of being selected is proportional to its fitness relative to the other chromosomes in the population. If the fitness of each chromosome is a strictly positive number to be maximized, selection is traditionally performed via an algorithm called Roulette Wheel selection [8]. The assigned probabilities of reproduction result in the generation of a population of chromosomes probabilistically selected from the current population. The selected chromosomes will generate offspring via the use of probabilistic genetic operators, namely, crossover (recombination of gene blocks) and mutation (perturbation through genetic variation) each one associated with a specific rate. Each new generation contains a higher proportion of the characteristics of the previous generation good members, providing a good possibility to converge to an optimal solution of the problem.

According to [8], GAs have successfully been applied to a wide variety of problems, including those which are hard to be solved by other methods. In the MCDA field, their application primarily concerns the task of control parameter optimization [15,14], the same investigated in this paper.

4 Diagnosis of Alzheimer's Disease

The early diagnosis of Alzheimer's disease can bring benefits to the patients and their families. With the constant development of drug therapies and therapeutic advances, an early treatment process can be highly advantageous. The families can feel the benefits as they prepare for the patient management, raising their quality of life.

According to Davidoff [6], the difficulty to assert if a patient has AD or any senile dementia is not just related to the current level of the understanding of this disease, but also to the relatively poorly comprehended aspects that concerns the elderly. This difficulty is also attested in [24], wherein the authors highlight that, despite the high incidence of dementia in the elderly population, doctors fail to detect them in 21 to 72% of the cases.

As mentioned before, this study seeks to assist the decision maker (clinician) in the early AD diagnosis. To achieve this objective, we have manually designed a specific dataset of cases taking as reference the neuropsychological battery of CERAD standardized assessments and the Brazilian consensus of cognitive and functional evaluation. These are discussed in the following two subsections.

4.1 CERAD

The Consortium to Establish a Registry for Alzheimer's Disease (CERAD) was founded in 1986 after the Health Research Extension Act of 1985 with a specific focus on issues of diagnosis and diagnostic standardization [12]. At that time, besides the fact that there had been an increasing interest over the illness, there was no uniform guideline over some issues, like diagnostic criteria, testing methods, and classifications of the disease severity, that could be followed. CERAD is a distinctive collaborative initiative to attend to this need.

CERAD has developed some standardized assessment instruments from different manifestations of Alzheimer's disease: clinical neuropsychology; neuropathology; behavior rating scale for dementia; family history interviews; and assessment of service needs. In this way, the CERAD battery improved the ability of specialists and researchers to describe and correlate clinical, neuropsychological, and neuropathologic aspects of AD.

4.2 The Novel Dataset

In order to provide a way to detect the presence of AD as soon as possible, we have followed the recommendations of the Scientific Department of Cognitive Neurology and Aging of the Brazilian Academy of Neurology [23] while crafting our dataset of cases. This consensus specifies the recommendations over the clinical diagnosis of AD through a functional and cognitive perspective, and therefore the database was designed by following the strategy of correlating clinical and neuropsychological assessments of CERAD with recommendations provided by the Brazilian consensus.

In particular, the language evaluation exams allow for both a quantitative and qualitative diagnosis, showing the profile of the linguistic disorder [23]. For the Brazilian consensus, the Boston Naming Test is one of the recommended tests that can be applied to break down the language aspects of a patient. This way, the first criterion (attribute) considered in the dataset relates to the amount of right answers given by each patient.

According to [23], the dementia diagnosis should be established in a clinical exam, documented as the Mini-Mental State Examination. To comply with the consensus, we turned this assessment into the second criterion associated with each case. This criterion reflects the sum of answers correctly assigned by each patient.

The third AD cognitive criterion designates a set of cognitive skills related to social relationships and that guarantee a proper, responsible, and effective conduct of the patient [23]. Among the tests available in CERAD battery, we have used the Verbal Fluency exam. This test requests the patient to verbalize the highest number of animals as possible during a certain period of time. The criterion is defined by the number of items mentioned in a minute, excluding the repeated ones.

One of the main characteristics of AD is the impairment of memory. The Brazilian consensus stresses the importance of the memory evaluation and suggests the memorization of lists of words as an exam that can be applied to detect any sort of brain impairment during the early stages of the disease. This exam asks the patient to remember a ten-word list after a short period of time to evaluate the status of the short-term memory. The CERAD assessment applies three lists of ten words, so the database criterion we have devised specifies the overall number of words that were remembered by the patient.

The last criterion introduced relates to the concept of constructional ability. This CERAD assessment provides a non-verbal measure of the patient's mental health through the manipulation of geometric figures. The criterion denotes the number of elements correctly-assigned by the patient.

Besides the fact that the neuropsychological assessments available in the CERAD battery of exams were applied to more than 5,000 patients, only 119 cases could be effectively used in our experiments. This number was achieved after cross-correlating the neuropsychological and clinical assessments in order to certify whether the patient had effectively developed AD or not. By these means, the resulting dataset encompasses 5 criteria and 119 alternatives (cases).

5 Experiments

In this section, we provide details of the experiments we have conducted so far over the database introduced previously. First, we concentrate on the prototype selection and control parameter calibration tasks conducted, respectively, by the ELECTRE IV and GA engines. Then, we report on the results we have achieved while applying the overall MCDA approach over the AD database, presenting a comparison with some state-of-the-art classifiers.

5.1 ELECTRE IV Engine

As already mentioned, the ELECTRE IV method [26] has been applied to assist in the prototype selection task through an indirect perspective. In such case, the decision analyst is responsible for providing the system with his/her preferences, which are effectively captured through the preference and indifference parameters (thresholds) associated with ELECTRE IV, so that the method can sort the alternatives.

Since the alternatives are ranked, the number of prototypes chosen while conducting the experiments was 7% of the original dataset. From that number, the prototypes were separated into their classes respecting their original distribution in the dataset. It is interesting to note that the application of the ELECTRE IV

Table 1. Criteria preference and indifference thresholds

Criteria	Description	+p	+q	-p	-q
C1	Boston Naming Test	0.9	0.39	-0.9	-0.39
C2	Mini-Mental State Examination	0.9	0.39	-0.9	-0.39
C3	Verbal Fluency	1.1	0.45	-1.1	-0.45
C4	Word List	1.1	0.45	-1.1	-0.45
C5	Constructional Praxis	0.9	0.35	-0.9	-0.35

method can vary depending on the type of dataset that is under consideration. In cases where the problem presents more than two classes, the ELECTRE IV should be applied for each class separately, sorting the best alternatives of each class. This occurs because the classification problems often present conflicting criteria.

When applied to our AD dataset, as it only presents two categories,the ELECTRE IV engine needs to be applied only once to sort the patients from the most probable of not having Alzheimer to those most probable of manifesting the disease. In our experiments, we have ranked the patients from the non-AD to the AD category. For this purpose, we have established the same preference and indifference thresholds for all criteria, as they are all benefit criteria and have the same numerical ranges. For this dataset, all criteria were considered as relevant, so we have avoided discarding any attribute. Table 1 shows the preference and indifference values that were elicited for each criterion from the decision maker (clinician).

5.2 The Genetic Algorithm Engine

According to our approach, after the best prototypes are selected by the ELECTRE IV engine, a customized GA is then employed in order to automatically estimate the gMCDA classifier's control parameters (thresholds). The GA components [8] have been configured as follows: a population of 50 individuals (which initially is randomly generated) is evolved at each generation; the Roulette Wheel operator is used to select individuals to reproduce; individuals are recombined through a single-point crossover and the offspring is mutated according to a uniform distribution over the parameters' ranges; the crossover and mutation rates are 80% and 15%, respectively; and the stop criterion adopted is to go through 500 generations of evolution.

To experiment with the GA, we have randomly generated 10 pairs of stratified training/test datasets from the original database, allocating 80% of the samples for training and the remaining for test. After the training phase, the best chromosome (configuration of thresholds) discovery is applied to the test data.

6 Classification Results

In order to assess the potentials of the whole MCDA approach, we have decided to compare the gMCDA classifier assisted with the ELECTRE IV and GA

Table 2. Accuracy over the Alzheimer test data of the gMCDA classifier acting alone for randomly-chosen sets of prototypes and arbitrary parameter values

	1	2	3	4	5	6	7	8	9	10
1	58.82%	56.3%	64.71%	58.82%	55.46%	58.82%	65.55%	68.07%	53.78%	63.87%
2	64.7%	60.5%	63.87%	71.43%	64.71%	62.19%	71.43%	65.55%	67.23%	68.07%
3	64.7%	60.5%	63.87%	71.43%	64.71%	62.19%	71.43%	65.55%	68.07%	68.07%
4	73.1%	64.7%	67.22%	74.79%	60.5%	72.27%	71.43%	71.43%	71.43%	77.31%
5	73.1%	64.7%	67.22%	75.63%	62.19%	72.27%	71.43%	70.59%	70.59%	77.31%
6	73.1%	64.7%	66.39%	74.79%	60.5%	72.27%	70.59%	70.59%	70.59%	77.31%
7	71.43%	64.7%	67.22%	74.79%	61.34%	72.27%	70.59%	70.59%	70.59%	76.47%
8	66.39%	62.19%	67.22%	73.95%	60.5%	68.07%	73.95%	66.39%	69.75%	72.27%
9	64.7%	64.7%	68.08%	79.83%	65.55%	72.27%	69.75%	70.59%	64.71%	68.91%
10	63.02%	60.5%	63.87%	71.43%	64.71%	62.19%	71.43%	64.71%	67.23%	68.07%

Table 3. Performance of the gMCDA classifier assisted with the ELECTRE IV and GA engines applied to the 10 test sets

1	2	3	4	5	6	7	8	9	10	Mean	S.D.
91.66%	84%	83.33%	95.83%	91.66%	95.83%	87.5%	95.83%	91.66%	85.71%	90.28%	4.88

Table 4. Performance measures for the AD diagnosis

Classification Algorithm	Classification Rate (%)
J48	75.63%
NBTree	84.033%
OneR	82.352%
NaiveBayes	75.63%
MCDA Classification Model	90.28%

engines with the gMCDA classifier acting alone. For such a purpose, 10 different groups of prototypes were randomly selected from the AD dataset and 10 different sets of control parameter values were arbitrarily chosen. In this respect, Table 2 shows the performance levels produced by the simple gMCDA classifier when varying both the sets of prototypes and parameters. It is easily noticeable that the gMCDA classifier shows high sensitivity to the choice of both prototypes and cut-off threshold values. As mentioned earlier, the choice of the prototypes and control parameters seems indeed to be a key issue to be properly dealt with in order to leverage the classifier's performance.

Table 3 shows the accuracy rates achieved by the augmented gMCDA classifier over the same 10 test data partitions. By contrasting these results with those shown in Table 2, it is possible to observe that, for some sets of threshold values, the MCDA classification model could have its performance improved by more than 20%, taking the mean results over the 10 sets of random prototypes. Moreover, in some runs, the gMCDA classifier's performance could increase for as high as 33%.

Finally, to provide a flavor of comparison with other classification algorithms, we have resorted to some well-known classification models available in the WEKA workbench [29]. Table 4 brings the average accuracy levels achieved with each contestant model over the 10 derived datasets. The performance level achieved by the gMCDA classifier was superior to those achieved by the other models. It should be emphasized that for each of the four additional classifiers we performed some preliminary experiments in order to manually calibrate its associated control parameters. However, we can not guarantee that the sets of parameters effectively obtained were in fact the optimal ones at all.

From the results discussed above, one can conclude that the ELECTRE IV and GA engines have demonstrated good potential in solving the prototype and parameter selection problems. After the decision analyst sets the preferential information for the AD problem, the first engine selects the prototypes independently of the classifier's parameters. One relevant factor that relates to the application of this engine is the fact that the decision analyst can, depending on the problem, reduce the number of criteria used for the prototype selection and classification purposes. We particulary feel that this strategy is interesting to be employed in situations that the decision analyst has a vast knowledge over the domain that is under consideration. After the application of the ELECTRE IV, the prototypes are employed whenever the GA runs over the derived training dataset. By this means, the control parameters of the gMCDA classifier could have their values set in consonance with the prototypes already selected by the ELECTRE IV engine, thus leveraging the overall classification rate.

7 Conclusion and Future Work

The continuous growth of the elderly population in the last years has led to a high increase in the prevalence of different types of dementia. Among these, the most frequently-diagnosed one is the Alzheimer's disease [6]. For this reason, and also due to the fact that the effectiveness of clinical treatments depends very much on the current stage of the disease, the early diagnosis of AD has been a goal recently pursued by several initiatives.

Different from other studies over the AD, in this paper, our purpose was to assess the performance achieved by an extended version of a recently-proposed MCDA classification model [14]. In this context, the employment of the ELECTRE IV algorithm revealed that the prototype selection task really exerts an important role over the MCDA classification process. Along with the ELECTRE IV, a GA engine was deployed to assist the in the automatic calibration of the control parameter values (weights and thresholds) associated with the gMCDA classifier.

For this assessment, we have developed a new database of cases taking as reference the CERAD neuropsychological battery of assessments. This battery was chosen because it complies with the recommendations of the Brazilian Academy of Neurology and has been used by several other studies. Overall, the devised MCDA approach could achieve satisfactory levels of accuracy while classifying the patients in the conducted experiments, leveraging the performance of the gM-CDA classifier as proposed by Goletsis et al. [14]. The average performance level achieved with the augmented classifier compares favorably with those achieved with other well-known classifiers [29].

As future work, we feel that it is possible to elicit novel criteria through the correlation of the Brazilian consensus with other batteries of assessments. Likewise, the integration of the whole model developed here with other related MCDA models, such as the one developed in [3,4], could be a promissing strategy to pursue in order to better cope with the early AD diagnosis.

References

1. Belacel, N.: Multicriteria assignment method PROAFTN: Methodology and medical applications. European Journal of Operational Research 125, 175–183 (2000)
2. Braak, H., Braak, E.: Frequency of stages of Alzheimer-related lesions in different age categories. Neurobiology of Aging 18, 351–357 (1997)
3. Castro, A.K., Pinheiro, P.R., Pinheiro, M.C.: Applying a decision making model in the early diagnosis of alzheimer's disease. In: Yao, J., Lingras, P., Wu, W.-Z., Szczuka, M.S., Cercone, N.J., Ślęzak, D. (eds.) RSKT 2007. LNCS, vol. 4481, pp. 149–156. Springer, Heidelberg (2007)
4. Castro, A.K., Pinheiro, P.R., Pinheiro, M.C.: A multicriteria model applied in the diagnosis of alzheimer's disease. In: Wang, G., Li, T., Grzymala-Busse, J.W., Miao, D., Skowron, A., Yao, Y. (eds.) RSKT 2008. LNCS, vol. 5009, pp. 612–619. Springer, Heidelberg (2008)
5. Celsis, P.: Age-related cognitive decline, mild cognitive impairment or preclinical alzheimer's disease? Annals of Medicine 32, 6–14 (2000)
6. Davidoff, A.D.: Issues in the clinical diagnosis of alzheimer's disease. American Journal of Alzheimer's Disease and Other Dementias 1(1), 9–15 (1986)
7. Bana e Costa, C.A., Corte, J.M., Vasnick, J.C.: Macbeth. LSE-OR Working Paper (2003)
8. Eiben, A.E., Smith, J.E.: Introduction to Evolutionary Computing. Springer, Heidelberg (2003)
9. Elias, M.F., Beiser, A., Wolf, P., Au, R., White, R.F., D'Agostino, R.B.: The preclinical phase of Alzheimer disease: A 22-year prospective study of the Framinghan cohort. Arch. Neurol. 57, 808–813 (2000)
10. Figueira, J., Mousseau, V., Roy, B.: ELECTRE methods. Multiple Criteria Decision Analysis: State of the Arts Surveys, 133–162 (2005)
11. Figueiredo, R.J.P., et al.: Neural-network-based classification of cognitively normal, demented, alzheimer disease and vascular dementia from single photon emission with computed tomography image data from brain. Medical Sciences 92, 5530–5534 (1995)
12. Fillenbaum, G.G., et al.: Consortium to establish a registry for Alzheimer's disease (CERAD): The first twenty years. Alzheimer's & Dementia 4, 96–109 (2008)
13. French, B.M., Dawnson, M.R.W., Dobbs, A.R.: Classification and staging of dementia of the alzheimer type: A comparison between neural networks and linear discriminant analysis. Arch. Neurol. 54(8), 1001–1009 (1997)
14. Goletsis, Y., Papaloukas, C., Fotiadis, D., Likas, A., Michalis, L.: Automated ischemic beat classification using genetic algorithms and multicriteria decision analysis. IEEE Transactions on Biomedical Engineering 51(10), 1717–1725 (2004)

15. Gouvenir, H.A., Erel, E.: Multicriteria inventory classification using a genetic algorithm. European Journal of Operations Research 105(1), 29–37 (1998)
16. Jacquet-Lagréze, E., Siskos, J.: Preference disaggregation: Twenty years of MCDA experience. European Journal of Operational Research 130, 233–245 (2001)
17. Kawas, C.H., et al.: Visual memory predicts Alzheimer's disease more than a decade before diagnosis. Neurology 60(7), 1089–1093 (2003)
18. Keeney, R.L., Raiffa, H.: Decisions with Multiple Objectives: Preferences and Value Trade-Offs. Cambridge University Press, Cambridge (1993)
19. Massaglia, M., Ostanello, A.: N-TOMIC: A decision support for multicriteria segmentation problems. Lecture Notes in Economics and Mathematics Systems, vol. 356, pp. 167–174 (1991)
20. Mortimer, J.A., et al.: Very early detection of Alzheimer neuropathology and the role of brain reserve in modifying its clinical expression. Journal of Geriatric Psychiatry and Neurology 18(4), 218–223 (2005)
21. Mousseau, V., Slowinski, R., Zielniewicz, P.: ELECTRE TRI 2.0a: Methodological guide and user's documentation. Technical report, Universite de Paris-Dauphine (1999)
22. Nestor, P.J., Scheltens, P., Hodges, J.R.: Advances in the early detection of alzheimer's disease. Neurodegeneration 5, S34–S41 (2004)
23. Nitrini, R., Caramelli, P., Bottino, C.M., Damasceno, B.P., Brucki, S.M., Anghinah, R.: Diagnóstico de doença de Alzheimer no Brasil: Avaliação cognitiva e funcional. Arquivos de Neuro-Psiquiatria 63, 713–719 (2005)
24. Pinholt, E.M., Kroenke, K., Hanley, J.F., Kussman, M.J., Twyman, P.L., Carpenter, J.L.: Functional assessment of the elderly: A comparison of standard instruments with clinical judgment. Arch. Intern. Med. 147, 484–488 (1987)
25. Roy, B.: Multicriteria Methodology for Decision Aiding. Kluwer Academic Publishers, Dordrecht (1996)
26. Roy, B., Hugonard, B.: Ranking of suburban line extension projects on the paris metro system by a multicriteria method. Transportation Research 16, 301–312 (1982)
27. Sandip, R., et al.: Classification and prediction of clinical alzheimer's diagnosis based on plasma signaling proteins. Nature Medicine 13, 1359–1362 (2007)
28. Ukkusuri, S.V., Karoonsoontawong, A., Kockelman, K.M.: Congestion Pricing Technologies: A Comparative Evaluation in New Transportation Research Progress. Nova Science Publishers (2007)
29. Witten, I.H., Frank, E.: Data Mining: Practical Machine Learning Tools and Techniques, 2nd edn. Morgan Kaufmann, San Francisco (2005)
30. Zaffalon, M., Wesnes, K., Petrini, O.: Reliable diagnoses of dementia by the naive credal classifier inferred from incomplete cognitive data. Artificial Intelligence in Medicine 29, 61–79 (2003)
31. Zopounidis, C., Doumpos, M.: Multicriteria classification and sorting methods: A literature review. European Journal of Operational Research 138(2), 229–246 (2002)

Many-Objective Optimization by Space Partitioning and Adaptive ϵ-Ranking on MNK-Landscapes

Hernán Aguirre[1,2] and Kiyoshi Tanaka[2]

[1] Fiber-Nanotech Young Researcher Empowerment Program
[2] Shinshu University, Faculty of Engineering,
4-17-1 Wakasato, Nagano, 380-8553 Japan
{ahernan,ktanaka}@shinshu-u.ac.jp

Abstract. This work proposes a method to search effectively on *many*-objective problems by instantaneously partitioning the objective space into subspaces and performing one generation of the evolutionary search in each subspace. The proposed method uses a partition strategy to define a schedule of subspace sampling, so that different regions of objective space could be emphasized at different generations. In addition, it uses an adaptive ϵ-ranking procedure to re-rank solutions in each subspace, giving selective advantage to some of the solutions initially ranked highest in the whole objective space. Adaptation works to keep the actual number of highest ranked solutions in each subspace close to a desired number. The performance of the proposed method is verified on MNK-Landscapes. Experimental results show that convergence and diversity of the solutions found can improve remarkably on $4 \leq M \leq 10$ objectives.

1 Introduction

Multiobjective evolutionary algorithms (MOEAs) [1,2] optimize simultaneously two or more objective functions, aiming to find a set of trade-off solutions in a single run of the algorithm. Most state of the art MOEAs use Pareto dominance within the selection procedure of the algorithm to rank solutions. Selection based on Pareto dominance is thought to be effective for problems with convex and non-convex fronts and has been successfully applied in two and three objectives problems.

Recently, there is a growing interest on applying MOEAs to solve *many*-objective optimization problems, i.e. problems with four or more objectives. However, current research reveals that the number of Pareto non-dominated solutions gets substantially larger as we increase the number of objectives of the problem [3,4]. Hence, ranking by Pareto dominance becomes coarser and too many solutions are assigned the same rank. This affects the effectiveness of selection, severely deteriorating the performance of MOEAs [5,6,7].

In this work, we propose a method to search on many-objective problems by instantaneously partitioning the objective space into subspaces and performing one generation of the evolutionary search in each subspace. Partitioning of the

M. Ehrgott et al. (Eds.): EMO 2009, LNCS 5467, pp. 407–422, 2009.

objective space into subspaces aims to instantaneously emphasize the search within smaller regions of objective space. The proposed method uses a partition strategy to define a schedule of subspace sampling, so that different regions could be emphasized at different generations. In addition, it uses an adaptive ϵ-ranking procedure to re-rank solutions in each subspace, giving selective advantage to some of the solutions initially ranked highest in the whole objective space, so that selection can put more emphasizes in exploitation. Adaptation in the re-ranking procedure works to keep the actual number of highest ranked solutions in each subspace close to a desired number. The combination of space partitioning, partitioning strategy, and adaptive ϵ-ranking allows to perform an effective search aiming to improve convergence and diversity of solutions on many-objective problems.

In this paper, we implement the proposed method using NSGA-II's framework [8]. We test the proposed method on MNK-Landscapes [3,4] with $4 \leq M \leq 10$ objectives, $N = 100$ bits, and $0 \leq K \leq 50$ epistatic interactions. Experimental results show that convergence and diversity of the solutions found can improve remarkably on $4 \leq M \leq 10$ objectives for all K.

2 Multiobjective Optimization Concepts and Definitions

Let us consider, without loss of generality, a maximization multiobjective problem with M objectives:

$$maximize \;\; \boldsymbol{f}(\boldsymbol{x}) = (f_1(\boldsymbol{x}), f_2(\boldsymbol{x}), \cdots, f_M(\boldsymbol{x})) \tag{1}$$

where $\boldsymbol{x} \in \mathcal{X}$ is a solution vector in the solution space $\mathcal{X}$, and $f_1, f_2, \cdots, f_M$ the M objective functions to be optimized.

Definition 1 (Objective space ϕ). *The objective space of the problem is determined by the set $\phi = \{f_1, f_2, \cdots, f_M\}$ of the M objective functions to be optimized.*

One dimensional comparison and Pareto optimality are two popular methods used to decide what solution to choose from a set of solutions. Yu [9] showed that these two methods are extreme cases in the entire domain of domination structures and that there are infinity valid methods lying between them, which suitability depends on how much information is known on the decision maker's preferences. Within the EMO community, these other domination structures are also known as relaxed forms of Pareto dominance and one method to implement them is ϵ-dominance [10]. Pareto dominance and ϵ-dominance concepts are of special relevance to this work and are defined as follows.

Definition 2 (Pareto dominance). *A solution $\boldsymbol{x}$ is said to Pareto dominate other solution $\boldsymbol{y}$ in the objective space ϕ if the two following conditions are satisfied:*

$$\begin{array}{ll} \forall f_m \in \phi & f_m(\boldsymbol{x}) \geq f_m(\boldsymbol{y}) \;\wedge \\ \exists f_m \in \phi & f_m(\boldsymbol{x}) > f_m(\boldsymbol{y}). \end{array} \tag{2}$$

Here, $\boldsymbol{x}$ dominates $\boldsymbol{y}$ is denoted by $\boldsymbol{f}(\boldsymbol{x}) \succeq \boldsymbol{f}(\boldsymbol{y})$.

Definition 3 (ϵ-dominance). *A solution $\boldsymbol{x}$ is said to ϵ-dominate other solution $\boldsymbol{y}$ in the objective space ϕ if the two following conditions are satisfied:*

$$\begin{array}{ll} \forall f_m \in \phi & (1+\epsilon) f_m(\boldsymbol{x}) \geq f_m(\boldsymbol{y}) \ \wedge \\ \exists f_m \in \phi & (1+\epsilon) f_m(\boldsymbol{x}) > f_m(\boldsymbol{y}). \end{array} \tag{3}$$

where $\epsilon > 0.0$. Here, $\boldsymbol{x}$ ϵ-dominates $\boldsymbol{y}$ is denoted by $\boldsymbol{f}(\boldsymbol{x}) \succeq^{\epsilon} \boldsymbol{f}(\boldsymbol{y})$.

Other important concepts we use to describe our algorithm are defined as follows.

Definition 4 (Subspace ψ). *A subspace ψ of ϕ is a lower dimension space that includes some of the functions in ϕ, i.e. $\psi \subset \phi$.*

Definition 5 (Non-overlapping subspaces). *Two subspaces $\psi_1 \subset \phi$ and $\psi_2 \subset \phi$ are said to be non-overlapping if they have no common objectives, i.e. $\psi_1 \cap \psi_2 = \emptyset$.*

Definition 6 (Space partition Ψ_{N_S}). *An space ϕ is said to be partitioned into N_S subspaces, denoted as Ψ_{N_S}, if all subspaces are non-overlapping and no objective function in ϕ is left unassigned to a subspace, i.e. $\Psi_{N_S} = \{\psi_1, \psi_2, \cdots, \psi_{N_S} \mid \psi_1 \cap \psi_2 \cdots \cap \psi_{N_S} = \emptyset \wedge \psi_1 \cup \psi_2 \cdots \cup \psi_{N_S} = \phi\}$.*

Definition 7 (Subspace ϵ-dominance). *A solution $\boldsymbol{x}$ is said to ϵ-dominate other solution $\boldsymbol{y}$ in the subspace ψ if:*

$$\begin{array}{ll} \forall f_m \in \psi & (1+\epsilon) f_m(\boldsymbol{x}) \geq f_m(\boldsymbol{y}) \ \wedge \\ \exists f_m \in \psi & (1+\epsilon) f_m(\boldsymbol{x}) > f_m(\boldsymbol{y}). \end{array} \tag{4}$$

Here, $\boldsymbol{x}$ ϵ-dominates $\boldsymbol{y}$ in the subspace ψ is denoted by $\boldsymbol{f}(\boldsymbol{x}) \succeq^{\epsilon}_{\psi} \boldsymbol{f}(\boldsymbol{y})$.

3 Method

3.1 Concept

In this section, we describe the proposed method to search on many-objective problems by space partitioning and adaptive ϵ-ranking. In the following, we call this method ϵ Ranking Multiobjective Optimizer (ϵR-EMO). The goal of ϵR-EMO is to find a set of solutions with good properties of convergence and diversity. To achieve its goal, ϵR-EMO first ranks solutions by Pareto dominance calculated in the whole objective space. Then, it instantaneously partitions the objective space into subspaces, re-ranks solutions for each subspace using an adaptive subspace-ϵ-ranking procedure, and performs one generation of the evolutionary search within each subspace. During the next cycle of the algorithm, parents and offspring from all subspaces will be joined together so that they will be ranked again in the whole objective space.

By partitioning the objective space into subspaces, we aim to instantaneously emphasize the search within smaller regions of objective space. At each generation, we don't search in all possible subspaces. Instead, we define a schedule of subspace sampling by using a partition strategy. Re-ranking of solutions by the adaptive subspace-ϵ-ranking aims to give selective advantage to some of the usually too many solutions assigned highest rank in a many-objective subspace, so that selection can put more emphasizes in exploitation. Adaptation in the re-ranking procedure works to keep the actual number of highest ranked solutions in each subspace close to a desired number. The combination of space partitioning, partitioning strategy, and adaptive subspace-ϵ-ranking, aims to effectively search on many-objective problems.

In the following we first explain the general flow of the proposed method using NSGA-II's framework [8] and then explain in detail its distinctive features.

Procedure 1. ϵR-EMO

Input: N_S, number of subspaces at each generation. α, desired number of solutions with highest rank in each subspace (as a fraction of the entire parent population)
Output: $\mathcal{F}_1$, set of Pareto non-dominated solutions

1: $\mathcal{P} \leftarrow \emptyset$, $\mathcal{Q} \leftarrow random$ // initialize parent $\mathcal{P}$ and offspring $\mathcal{Q}$ populations
2: $\epsilon_1, \epsilon_2, \cdots, \epsilon_{N_S} \leftarrow 0.0$
3: **repeat**
4: evaluation($\mathcal{Q}, \phi$) // $\phi = \{f_1, f_2, \cdots, f_M\}$
5: $\mathcal{F} \leftarrow$ non-domination-sorting($\mathcal{P} \cup \mathcal{Q}$) // $\mathcal{F} = \{\mathcal{F}_i\}$ $(i = 1, 2 \cdots, N_F)$
6: crowding-distance($\mathcal{F}$)
7: $\Psi_{N_S} \leftarrow$ subspace-partition(ϕ, N_S) // $\Psi_{N_S} = \{\psi_1, \psi_2, \cdots, \psi_{N_S}\}$
8: $P \leftarrow \emptyset$, $Q \leftarrow \emptyset$
9: **for** $s = 1$ to N_S **do**
10: $\mathcal{F}^{\epsilon_s} \leftarrow$ subspace-ϵ-ranking (ψ_s, ϵ_s, $\mathcal{F}$) // $\mathcal{F}^{\epsilon_s} = \{\mathcal{F}_j^{\epsilon_s}\}$ $(j = 1, 2 \cdots, N_F^{\epsilon_s})$
11: $\epsilon_s \leftarrow$ adaptation(ϵ_s, α, $|\mathcal{F}_1^{\epsilon_s}|$) // adapt ϵ_s for the next generation
12: $\mathcal{P}_s \leftarrow$ truncation($\mathcal{F}^{\epsilon_s}$) // $|\mathcal{P}_s| = |\mathcal{P}|/N_S$, $|\mathcal{F}^{\epsilon_s}| = |\mathcal{P}| + |\mathcal{Q}|$
13: $\mathcal{Q}_s \leftarrow$ recombination and mutation($\mathcal{P}_s$) // $|\mathcal{Q}_s| = |\mathcal{Q}|/N_S$
14: $\mathcal{P} \leftarrow \mathcal{P} \cup \mathcal{P}_s$, $\mathcal{Q} \leftarrow \mathcal{Q} \cup \mathcal{Q}_s$
15: **end for**
16: **until** termination criterion is met
17: **return** $\mathcal{F}_1$

3.2 ϵR-EMO

ϵR-EMO implemented in NSGA-II's framework [8] is illustrated in Procedure 1. See that solutions are evaluated in all M objectives $\phi = \{f_1, f_2, \cdots, f_M\}$, ranked based on Pareto dominance, and assigned a crowding measure using non-domination sorting and crowding distance procedures [8], respectively (lines 4-6). After this initial ranking, solutions are classified in sets of non-dominated solutions $\mathcal{F} = \{\mathcal{F}_i\}$ $(i = 1, 2, \cdots, N_F)$. Next, the objective space ϕ is partitioned into N_S non-overlapping subspaces $\Psi_{N_S} = \{\psi_1, \psi_2, \cdots, \psi_{N_S}\}$ (line 7). Then, for each subspace ψ_s, solutions $\mathcal{F}$ are re-ranked and re-classified in $\mathcal{F}^{\epsilon_s} = \{\mathcal{F}_j^{\epsilon_s}\}$ $(i = 1, 2, \cdots, N_F^{\epsilon_s})$ using a subspace-ϵ-ranking procedure, where $N_F^{\epsilon_s} \geq N_F$, creating a finer grained ranking of solutions (line 10). Subspace-ϵ-ranking uses parameter ϵ_s to control the number of highest ranked individuals $|\mathcal{F}_1^{\epsilon_s}|$. Parameter ϵ_s is adapted every generation (line 11) to keep $|\mathcal{F}_1^{\epsilon_s}|$ close to a desired number. The parent population in each subspace $\mathcal{P}_s$ is obtained by truncating $\mathcal{F}^{\epsilon_s}$ based on rank and crowding distance (line 12). That is, groups of solutions $\mathcal{F}_j^{\epsilon_s}$ are assigned iteratively to $\mathcal{P}_s$ by rank order, starting with $\mathcal{F}_1^{\epsilon_s}$. If $\mathcal{F}_j^{\epsilon_s}$ overfills $\mathcal{P}_s$, crowding distance calculated in the whole space ϕ is used to choose the required number o solutions. Mating for recombination is carried out by binary tournament, where winners are decided by rank in the subspace ψ_s breaking ties by crowding distance in ϕ.

3.3 Subspace Partitioning

In our approach, we partition the M dimensional space $\phi = \{f_1, f_2, \cdots, f_M\}$ into N_S non-overlapping subspaces $\Psi_{N_S} = \{\psi_1, \psi_2, \cdots, \psi_{N_S}\}$. All subspaces have the same dimension $M_S = M/N_S$ in case $r = (M \mod N_S)$ is zero. Otherwise, r of the N_S subspaces have dimension $M_S = \lfloor M/N_S \rfloor + 1$ and the rest $M_S = \lfloor M/N_S \rfloor$. The number of all possible ways to partition ϕ into subspaces of dimension M_S is very large. In our approach, we don't explicitly search in all possible subspaces at each generation. Instead, we set N_S to a small value and define a schedule of subspace sampling by using a partitioning strategy. We investigate three strategies to partition ϕ. Namely, *random*, *shift*, and *fixed* partition strategies.

Random strategy randomly assigns objectives $f_i \in \phi$ to subspaces $\psi_s \in \Psi_{N_S}$. With this strategy, any possible M_S dimensional subspace of ϕ could be formed. However, it does not seek to correlate the s-th subspace ψ_s from generation t to the next.

Shift strategy, at the first generation, assigns deterministically objectives $f_i \in \phi$ to subspaces $\psi_s \in \Psi_{N_S}$, so that objectives assigned to a given ψ_s are ordered by objective index i. Then, in subsequent generations, the objective with highest index in the s-th subspace is shifted to the $((s+1) \mod N_S)$-th subspace, $\forall \psi_s \in \Psi_{N_S}$. This strategy correlates the s-th subspace from generation t to the next. In fact, subspace ψ_s at generation t overlaps with ψ_s at generation $t-1$ in all but one objective. However, not all possible M_S dimensional subspaces of ϕ could be formed.

Fixed strategy assigns deterministically objectives $f_i \in \phi$ to subspaces $\psi_s \in \Psi_{N_S}$ and keep the same assignment throughout the generations. With this strategy only N_S subspaces of ϕ could be formed.

These strategies would allow us to verify the impact of subspace sampling on the quality of solutions and the effect of subspace correlation from one generation to the next on the adaptation of ϵ for subspace-ϵ-ranking.

3.4 Adaptation of ϵ

In our method, solutions are re-ranked in each subspace by using a subspace-ϵ-ranking procedure in which the number of solutions assigned highest rank depends on the value set to ϵ (≥ 0) and on the instantaneous distribution of solutions in objective space (see below). Although it is difficult to tell in advance exactly how many solutions will be assigned highest rank for a given value of ϵ, we know that larger values of ϵ decrease the number of highest ranked solutions and vice versa. The algorithm takes advantage of this correlation to adapt ϵ at each generation in order to keep the actual number of highest ranked solutions close to a desired number [12]. The desired number of highest ranked solution in each subspace is specified by $\alpha \times |\mathcal{P}|$, where α is a parameter in the range [0.0,1.0] set by the user and $|\mathcal{P}|$ is the size of the entire parent population.

In our method, instead of using one ϵ for all subspaces, we adapt one ϵ_s for each one of the N_S subspaces ψ_s. Note that the actual combination of objectives that define ψ_s change with time, depending on the partition strategy. So, adaptation of ϵ_s reacts to the characteristics of the different instantaneous subspaces (actual combinations of f_i) assigned to ψ_s. Since the dimension of the subspace is strongly correlated to the value of ϵ that renders the desired number of highest ranked solutions $\alpha \times |P|$, when the space ϕ is partitioned we make sure that the dimension of the subspace ψ_s remains the same throughout the generations.

3.5 Subspace-ϵ-Ranking

Subspace-ϵ-ranking fine grains ranking of solutions initially ranked by Pareto dominance in the objective space ϕ, using a randomized ϵ-sampling procedure in the subspace $\psi \subset \phi$ that favors a good distribution of solutions based on dominance regions wider than conventional Pareto dominance. Subspace-ϵ-ranking extends ϵ-ranking [13], where ϵ-sampling acts on ϕ instead of ψ. In the following, we first explain ϵ-sampling and then subspace-ϵ-ranking.

ϵ-sampling assumes that there is a set of equally ranked solutions from which a subset should be chosen to give them selective advantage in order to proceed further with the evolutionary search. That is, ϵ-sampling acts as a decision making procedure, not to find a final solution, but to help selection of the evolutionary algorithm. Hence, the sampling heuristic must reflect criteria that favor an effective search. Here, the sample of solutions to be given selective advantage are obtained with the following criteria,

- Extreme solutions are always part of the sample.
- Each (not extreme) sampled solution is the sole sampled representative of its area of influence. The area of influence of the sampled solutions is determined by a domination region wider than Pareto dominance, i.e. ϵ-dominance.
- Sampling of (not extreme) solutions follows a random schedule.

The first criterion tries to push the search towards the optimum values of each fitness function, aiming to find non-dominated solutions in a wide area of objective space. The second criterion assures that only one solution in a given zone of objective space is given higher rank, trying to distribute the search effort more or less uniformly among the different zones represented in the actual population. The third criterion dynamically establishes the zones that are represented in the sample. Also, in the case that there are several solutions within each zone, it increases the likelihood that the sampled solutions that will be given higher rank are different from one generation to the next, increasing the possibility of exploring wider areas of objective and variable space.

Procedure 2 illustrates ϵ-sampling algorithm. Let us denote $\mathcal{A}$ the set of solutions that have been assigned the same rank based on conventional Pareto dominance, for example by applying non-domination sorting [8]. ϵ-sampling returns the sampled solutions $\mathcal{S} \subset \mathcal{A}$ that will be given selective advantage as well as the set of solutions $\mathcal{D}^\epsilon$ to be demoted. See that extreme solutions are the first to be assigned to the sample $\mathcal{S}$ (lines 1,2). Then, one by one, solutions are randomly chosen and included in $\mathcal{S}$ (lines 4-6), whereas solutions that lie in the wider domination region of the randomly picked solution are assigned to $\mathcal{D}^\epsilon$ (lines 7,8). Note that subspace ϵ-dominance $\boldsymbol{f}(z) \succeq^\epsilon_\psi \boldsymbol{f}(y)$ is used. Fig. 1 (a) illustrates the application of ϵ-sampling on the set of solutions $\mathcal{A} = \mathcal{F}_1$. The numbers close to the solutions represents the random schedule in which solutions are sampled (0 means extreme solutions, which are all selected at once).

Procedure 2. ϵ-sampling (ψ, ϵ, $\mathcal{A}$, $\mathcal{S}$, $\mathcal{D}^\epsilon$)

Input: Subspace ψ, ϵ-dominance factor ϵ and a set of solutions $\mathcal{A}$
Output: $\mathcal{S}$ and $\mathcal{D}^\epsilon$ ($\mathcal{S} \cup \mathcal{D}^\epsilon = \mathcal{A}$). $\mathcal{S}$ contains the sample of solutions from $\mathcal{A}$, whereas $\mathcal{D}^\epsilon$ contains ϵ-dominated solutions in ψ

1: $\mathcal{X} \leftarrow \{x \in \mathcal{A} \mid f_m(x) = \max(f_m(\cdot)), \forall f_m \in \psi\}$
2: $\mathcal{S} \leftarrow \mathcal{X}$, $\mathcal{A} \leftarrow \mathcal{A} \setminus \mathcal{X}$, $\mathcal{D}^\epsilon \leftarrow \emptyset$
3: **while** $\mathcal{A} \neq \emptyset$ **do**
4: $\quad r \leftarrow rand()$ // $1 \leq r \leq |\mathcal{A}|$
5: $\quad z \leftarrow r$-th solution $\in \mathcal{A}$
6: $\quad \mathcal{S} \leftarrow \mathcal{S} \cup \{z\}$
7: $\quad \mathcal{Y} \leftarrow \{y \in \mathcal{A} \mid \boldsymbol{f}(z) \succeq^\epsilon_\psi \boldsymbol{f}(y), z \neq y\}$
8: $\quad \mathcal{D}^\epsilon \leftarrow \mathcal{D}^\epsilon \cup \mathcal{Y}$
9: $\quad \mathcal{A} \leftarrow \mathcal{A} \setminus \{\{z\} \cup \mathcal{Y}\}$
10: **end while**
11: **return**

The ϵ-sampling procedure works on a set of equally ranked solutions, however within a population there could be several sets of such solutions (each set with a different rank). Here, we explain subspace-ϵ-ranking to re-rank all possible sets of equally ranked solutions using ϵ-sampling.

Subspace-ϵ-ranking is applied at each generation for each subspace after non-domination sorting to reclassify the sets $\mathcal{F}_i$ ($i = 1, \cdots, N_F$). Procedure 3 describes subspace-ϵ-ranking algorithm. See that the reclassified sets $\mathcal{F}^{\epsilon_s}_j$ ($j = 1, \cdots, N^{\epsilon_s}_F$) now contains only the sample of solutions $\mathcal{S} \subset \mathcal{F}_i$ found by ϵ-sampling (lines 8,9). Also, see that solutions $\mathcal{D}^\epsilon$, which are not part of the sample (line 8) are demoted by joining them with solutions of a lower ranked set in the next iteration of the loop (line 4). Thus, $\mathcal{F}^{\epsilon_s}_1$ contains some of the solutions initially ranked first, but $\mathcal{F}^{\epsilon_s}_j$, $j > 1$, can contain solutions that initially were assigned to sets with different ranks. This gives chance to lateral diversity present

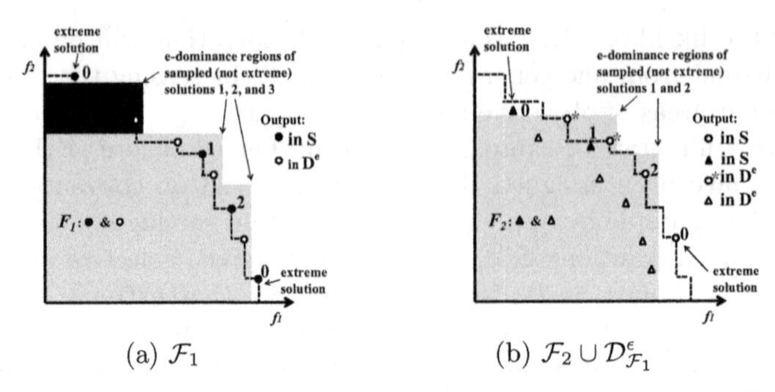

Fig. 1. ϵ-sampling on (a) the set of solutions initially ranked first and (b) on the set of solutions initially ranked second joined with solutions demoted from the first set

in the initial ranking of solutions and can punish highly crowded solutions even if they are initially ranked first by conventional Pareto dominance.

Procedure 3. subspace-ϵ-ranking (ψ_s, ϵ_s, $\mathcal{F}$, $\mathcal{F}^{\epsilon_s}$)

Input: Subspace ψ_s, ϵ-dominance factor ϵ_s and solutions $\mathcal{F}$ classified in fronts $\mathcal{F}_i$ $(i = 1, \cdots, N_F)$ by non-domination sorting
Output: $\mathcal{F}^{\epsilon_s}$, solutions re-classified in groups $\mathcal{F}_j^{\epsilon_s}$ $(j = 1, \cdots, N_F^{\epsilon_s})$

1: $\mathcal{D}^{\epsilon} \leftarrow \emptyset$, $i \leftarrow 1$, $j \leftarrow 1$
2: **repeat**
3:     **if** $i \leq N_F$ **then**
4:         $\mathcal{A} \leftarrow \mathcal{F}_i \cup \mathcal{D}^{\epsilon}$, $i \leftarrow i + 1$
5:     **else**
6:         $\mathcal{A} \leftarrow \mathcal{D}^{\epsilon}$
7:     **end if**
8:     ϵ-sampling(ψ, ϵ_s, $\mathcal{A}$, $\mathcal{S}$, $\mathcal{D}^{\epsilon}$)
9:     $\mathcal{F}_j^{\epsilon_s} \leftarrow \mathcal{S}$, $j \leftarrow j + 1$
10: **until** $\mathcal{D}^{\epsilon} = \emptyset$
11: **return**

Fig. 1 illustrates ϵ-ranking calling on ϵ-sampling to re-rank the set $\mathcal{F}_1$ of solutions initially ranked first and the set $\mathcal{F}_2$ of solutions ranked second joined with the demoted solutions $\mathcal{D}^{\epsilon}$ from $\mathcal{F}_1$. The example illustrates the application of ϵ-sampling to a 2 dimensional objective space ϕ. When ϵ-sampling is applied to a subspace $\psi \subset \phi$, the non-dominated solutions in ϕ projected in the subspace ψ (assuming a 2 dimensional subspace) would look similar to Fig. 1 **(b)**. ϵ-sampling will be applied to all projected solutions.

4 Test Problems, Performance Measures and Parameters

4.1 Multiobjective MNK-Landscapes

In this work we test the performance of the algorithms on multiobjective MNK-Landscapes. A multiobjective MNK-Landscape [3,4] is defined as a vector function mapping binary strings into real numbers $\boldsymbol{f}(\cdot) = (f_1(\cdot), f_2(\cdot), \cdots, f_M(\cdot))$:

$\mathcal{B}^N \rightarrow \Re^M$, where M is the number of objectives, $f_i(\cdot)$ is the i-th objective function, $\mathcal{B} = \{0, 1\}$, and N is the bit string length. $\boldsymbol{K} = \{K_1, \cdots, K_M\}$ is a set of integers where K_i ($i = 1, 2, \cdots, M$) is the number of bits in the string that epistatically interact with each bit in the i-th landscape. Each $f_i(\cdot)$ can be expressed as an average of N functions as follows

$$f_i(\boldsymbol{x}) = \frac{1}{N} \sum_{j=1}^{N} f_{i,j}(x_j, z_1^{(i,j)}, z_2^{(i,j)}, \cdots, z_{K_i}^{(i,j)}) \tag{5}$$

where $f_{i,j} : \mathcal{B}^{K_i+1} \rightarrow \Re$ gives the fitness contribution of bit x_j to $f_i(\cdot)$, and $z_1^{(i,j)}, z_2^{(i,j)}, \cdots, z_{K_i}^{(i,j)}$ are the K_i bits interacting with bit x_j in the string $\boldsymbol{x}$. The fitness contribution $f_{i,j}$ of bit x_j is a number between [0.0, 1.0] drawn from a uniform distribution. Thus, each $f_i(\cdot)$ is a non-linear function of $\boldsymbol{x}$ expressed by a Kauffman's NK-Landscape model of epistatic interactions [11]. In addition, it is also possible to arrange the epistatic pattern between bit x_j and the K_i other interacting bits. That is, the distribution $D_i = \{random, nearest\ neighbor\}$ of K_i bits among N. Thus, M, N, $\boldsymbol{K} = \{K_1, K_2, \cdots, K_M\}$, and $\boldsymbol{D} = \{D_1, D_2, \cdots, D_M\}$, completely specify a multiobjective MNK-Landscape.

4.2 Performance Measures

In this work, we use the hypervolume $\mathcal{H}$ and coverage $\mathcal{C}$ measures [14] to evaluate and compare the performance of the algorithms. The measure $\mathcal{H}$ calculates the volume of the M-dimensional region in objective space enclosed by a set of non-dominated solutions and a dominated reference point. Let $\mathcal{A}$ be a set of non-dominated solutions. The hypervolume of $\mathcal{A}$ can be expressed as

$$\mathcal{H}(\mathcal{A}) = \cup_{i=1}^{|\mathcal{A}|} (\mathcal{V}_i - \cap_{j=1}^{i-1} \mathcal{V}_i \mathcal{V}_j) \tag{6}$$

where $\mathcal{V}_i$ is the hypervolume rendered by the point $\boldsymbol{x}_i \in \mathcal{A}$ and the reference point. In this work, the reference point is set to $[0.0, \cdots, 0.0]$. Given two sets of non-dominated solutions $\mathcal{A}$ and $\mathcal{B}$, if $\mathcal{H}(\mathcal{A}) > \mathcal{H}(\mathcal{B})$ then set $\mathcal{A}$ can be considered better on convergence and/or diversity of solutions. To calculate $\mathcal{H}$, we use Fonseca et al. [15] algorithm, which significantly reduces computational time.

The coverage $\mathcal{C}$ measure [14] provides complementary information on convergence. Let us denote $\mathcal{A}$ and $\mathcal{B}$ the sets of non-dominated solutions found by two algorithms. $\mathcal{C}(\mathcal{A}, \mathcal{B})$ gives the fraction of solutions in $\mathcal{B}$ that are dominated at least by one solution in $\mathcal{A}$. More formally,

$$\mathcal{C}(\mathcal{A}, \mathcal{B}) = \frac{|\{\boldsymbol{b} \in \mathcal{B} | \exists \boldsymbol{a} \in \mathcal{A} : \boldsymbol{f}(\boldsymbol{a}) \succeq \boldsymbol{f}(\boldsymbol{b})\}|}{|\mathcal{B}|}. \tag{7}$$

$\mathcal{C}(\mathcal{A}, \mathcal{B}) = 1.0$ indicates that all solutions in $\mathcal{B}$ are dominated by solutions in $\mathcal{A}$, whereas $\mathcal{C}(\mathcal{A}, \mathcal{B}) = 0.0$ indicates that no solution in $\mathcal{B}$ is dominated by solutions in $\mathcal{A}$. Since usually $\mathcal{C}(\mathcal{A}, \mathcal{B}) + \mathcal{C}(\mathcal{B}, \mathcal{A}) \neq 1.0$, both $\mathcal{C}(\mathcal{A}, \mathcal{B})$ and $\mathcal{C}(\mathcal{B}, \mathcal{A})$ are required to understand the degree to which solutions of one set dominate solutions of the other set.

4.3 Parameters

In this work, we test the performance of the algorithm on MNK-Landscapes with $4 \leq M \leq 10$ objectives, $N = 100$ bits, number of epistatic interactions $K = \{0, 1, 3, 5, 10, 15, 25, 35, 50\}$ ($K_1, \cdots, K_M = K$), and *random* epistatic patterns among bits for all objectives ($D_1, \cdots, D_M = random$). Results presented below show the average performance of the algorithms on 50 different problems randomly generated for each combination of M, N and K. In the plots, error bars show 95% confidence intervals on the mean.

In the following sections we analyze results by ϵR-EMO, comparing them with results by conventional NSGA-II. The algorithms use parent and offspring populations of size $|\mathcal{P}| = |\mathcal{Q}| = 100$, two point crossover for recombination with rate $p_c = 0.6$, and bit flipping mutation with rate $p_m = 1/N$ per bit. The number of evaluations is set to 3×10^5. We study the performance of ϵR-EMO setting the number of subspaces to $N_S = \{1, 2\}$, varying the parameter α. For $N_S = 1$ (no subspace partitioning, $\Psi_{N_S} = \{\psi_1 = \phi\}$), we set $\alpha = \{1.0, 0.7, 0.5, 0.3\}$ so that the desired number of solutions with highest rank after subspace-ϵ-ranking is $\alpha \times |\mathcal{P}| = \{100, 70, 50, 30\}$, respectively. For $N_S = 2$ subspaces (subspace partition $\Psi_{N_S} = \{\psi_1, \psi_2\}$), we set $\alpha = \{0.5, 0.35, 0.25, 0.15\}$ so that $\alpha \times |\mathcal{P}| = \{50, 35, 25, 15\}$, respectively, in each of the two subspaces.

5 Experimental Results and Discussion

5.1 Performance by ϵR-EMO with No Objective Space Partitioning

In this section, we discuss the performance of ϵR-EMO when no objective space partitioning is considered ($N_S = 1$), setting the fraction between desired number of highest ranked individuals and population size to $\alpha = \alpha^*$ that achieves maximum hypervolume $\mathcal{H}$. Fig. 2 (a) shows the average ratio $\frac{\mathcal{H}(E)}{\mathcal{H}(N)}$, where E and N denote the set of solutions found by ϵR-EMO and conventional NSGA-II, respectively. Thus, a ratio greater than 1.0 indicates better $\mathcal{H}$ by ϵR-EMO than conventional NSGA-II. As a reference, we include a horizontal line to represent the $\mathcal{H}(N)$ values normalized to 1.0. From this figure, we can see that ϵR-EMO can significantly improve $\mathcal{H}$ on $4 \leq M \leq 10$ objectives problems, for all values of K (up to 27% improvement). Note that improvements on $\mathcal{H}$ become larger as we increase the number of objectives M from 4 to 6, whereas improvements on $\mathcal{H}$ are similarly high for $8 \leq M \leq 10$. Due to space limitations, we include results for $M = \{4, 6, 8, 10\}$ only and not for $M = \{5, 7, 9\}$.

Improvements on $\mathcal{H}$ can be due to solutions with better convergence, better diversity, or both. To complement the analyzes of results on $\mathcal{H}$ we also present results using the $\mathcal{C}$ measure. Fig. 2 (b) shows the average $\mathcal{C}$ values between conventional NSGA-II and ϵR-EMO set with α^*. From this figure, we can be see that $\mathcal{C}(N, E)$ is close to 0.0 for most K and M. This indicates that there are almost no solutions by conventional NSGA-II that dominate solutions by ϵR-EMO. On the other hand, the values of $\mathcal{C}(E, N)$ are very high for 4 objectives (in the range 0.60-0.85) and reduce gradually as we increase M up to 10 objectives (in the

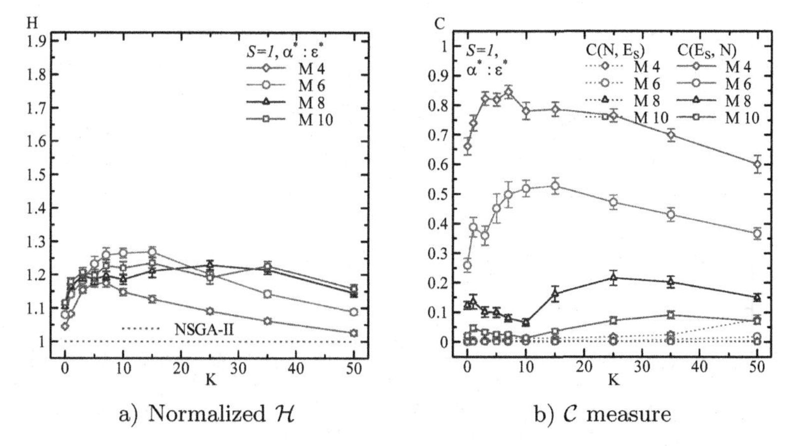

a) Normalized $\mathcal{H}$ b) $\mathcal{C}$ measure

Fig. 2. Normalized $\mathcal{H}$ and $\mathcal{C}$ between NSGA-II and ϵR-EMO when no objective space division ($S = 1$) is considered. ϵR-EMO is set to α^* that achieves maximum $\mathcal{H}(E)$.

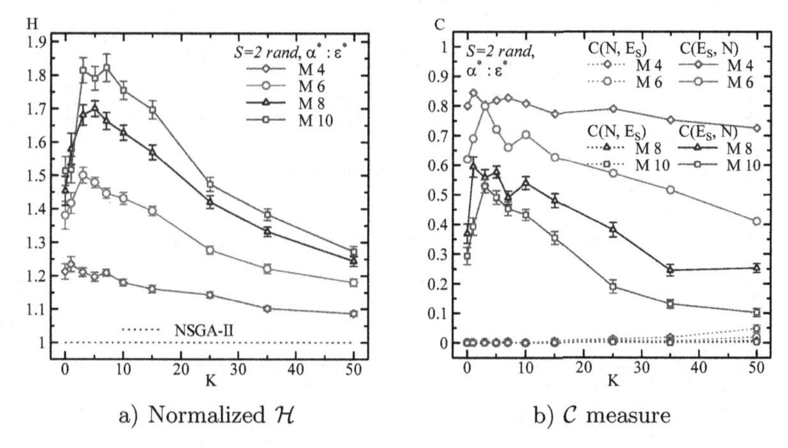

a) Normalized $\mathcal{H}$ b) $\mathcal{C}$ measure

Fig. 3. Results by NSGA-II and ϵR-EMO partitioning the objective space in two subspaces ($S = 2$) using *random* strategy and setting α^* that achieves maximum $\mathcal{H}(E)$

range 0.01-0.08). This suggests that a better convergence of solutions contributes to the increases of $\mathcal{H}$ by ϵR-EMO on $M = 4$ problems. As we increase M, gains on diversity gradually become more significant than gains on convergence as the reason for the significant improvement of $\mathcal{H}$ on $6 \leq M \leq 10$.

5.2 Performance by ϵR-EMO with Objective Space Partitioning

In this section we analyze the performance of ϵR-EMO partitioning instantaneously the objective space into two subspaces using the *random*, *shift*, and *fixed* partition strategies introduced in section 3.

First, we show results by the *random* partitioning strategy in Fig. 3. Looking at Fig. 3 (a) and comparing with Fig. 2 (a), we can see that ranking on subspaces using a *random* strategy leads to a remarkable improvement on $\mathcal{H}$ for all values

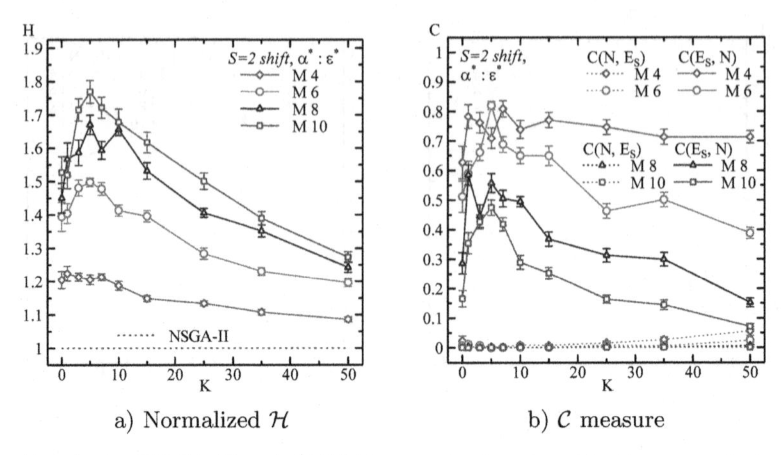

a) Normalized $\mathcal{H}$ b) $\mathcal{C}$ measure

Fig. 4. Results by NSGA-II and ϵR-EMO partitioning the objective space in two subspaces ($S = 2$) using *shift* strategy and setting α^* that achieves maximum $\mathcal{H}(E)$

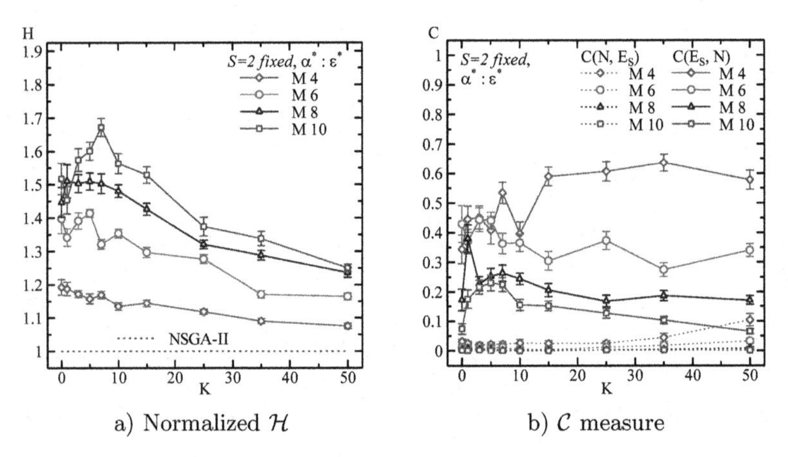

a) Normalized $\mathcal{H}$ b) $\mathcal{C}$ measure

Fig. 5. Results by NSGA-II and ϵR-EMO partitioning the objective space in two subspaces ($S = 2$) using *fixed* strategy and setting α^* that achieves maximum $\mathcal{H}(E)$

of M and K (up to 82.5% improvement). Note that the increase on $\mathcal{H}$ gets bigger with the number of objectives M. Looking at at Fig. 3 (b) and comparing with Fig. 2 (b), we can see that $\mathcal{C}(E, N)$ also increases for any value of K and M, whereas $\mathcal{C}(N, E)$ remains close to zero. That is, convergence also improves substantially.

Next, we discuss results by the *shift* partition strategy shown in Fig. 4 (a) and (b). From these figures note that the *shift* strategy also leads to a remarkable improvement on $\mathcal{H}$ and $\mathcal{C}$. Comparing the *shift* and *random* strategies, the latter leads to slightly better results than the former especially for $M \geq 8$ and $K \leq 15$. As mentioned above, the *random* strategy can sample any possible subspace of ϕ, whereas the *shift* strategy can sample most but not all subspaces of ϕ. The number of subspaces unable to sample the *fixed* strategy increase with the dimension of the objective space, especially if we keep constant the number of subspaces. Better results by the *random* strategy suggests that sampling all possible subspaces becomes relevant as the number of objectives increase.

Results by the *fixed* partition strategy are shown in Fig. 5 (a) and (b). See that the *fixed* strategy leads to smaller $\mathcal{H}$ and $\mathcal{C}(E, N)$ than the *random* and *shift* partition strategies. Comparing to ϵR-EMO with no subspace partitioning, the *fixed* strategy leads to higher $\mathcal{H}$ on all M but with smaller $\mathcal{C}(E, N)$ on $M \leq 6$. The *fixed* strategy only explores N_S of all possible subspaces of ϕ and it is seems not an appropriate strategy to achieve best performance on both convergence and diversity of solutions.

Finally, note that as we increase K (non-linearity of the problem) improvements on both $\mathcal{H}$ and $\mathcal{C}$ reduce regardless of the partition strategy. This suggests that in addition to better ranking strategies, we should also look into ways to improve recombination and mutation to achieve better performance on highly non-linear problems.

5.3 Analysis of α

In this section we analyze the parameter α that determines the desired number of highest ranked solution in each subspace. As an example, Fig. 6 shows $\mathcal{H}$ and $\mathcal{C}$ results achieved by different settings of α on $M = 8$ objectives landscapes partitioning the objective space in $N_S = 2$ subspaces using *shift* strategy. From this figure, see that $\alpha \geq 0.25$ (at least 50%of the parent population in each subspace is given highest rank) leads to high performance, whereas results by $\alpha =$ 0.15 are clearly lower on $M = 8$ objectives landscapes. Analyzing performance by $\alpha \geq 0.25$, see that setting α to 0.35 or 0.25 leads to best performance for most K, both on $\mathcal{H}$ and $\mathcal{C}$. However, see that setting α to 0.5 could give highest performance especially on small K. Although there is not an absolute winner among $\alpha \geq 0.25$ values, it is important to note that subspace partitioning's

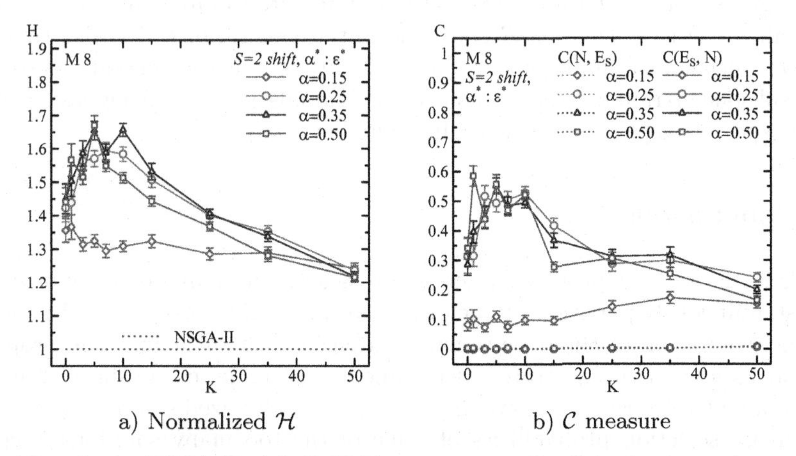

a) Normalized $\mathcal{H}$ b) $\mathcal{C}$ measure

Fig. 6. Results by NSGA-II and ϵR-EMO partitioning the objective space in two subspaces ($S = 2$) using *shift* strategy and varying α on $M = 8$ objectives

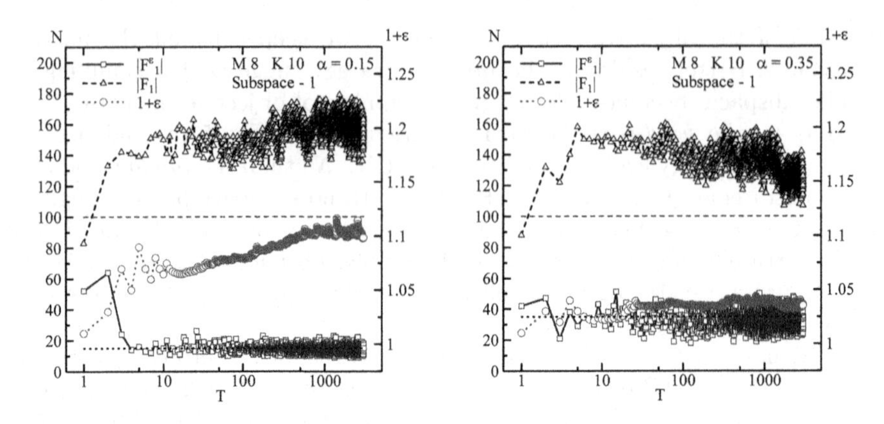

Fig. 7. Adaptation of ϵ in ϵR-EMO for $\alpha = 0.15$ and $\alpha = 0.35$ on $M = 8$ objectives and $K = 10$ epistatic bits

lower bound performance ($\min \mathcal{H} \wedge \min \mathcal{C}(E, N), \forall \alpha \in \{0.5, 0.35, 0.25\}$) is by far better than the performance by no subspace partitioning (see Fig. 6 and compare with Fig. 2). Analyzing our data for other values of M, in general, we see that performance by $\alpha = 0.25$ is better than 0.35 when the number of objectives decrease to $M = 6$ and $M = 4$; whereas performance by 0.5 and 0.35 is better than 0.25 when we increase M to 10 objectives. As a rule of thumb, when the space is partitioned into 2 subspaces, $\alpha = 0.25$ works well on $M = 4$ and $M = 6$, $\alpha = 0.35$ on $M = 8$, and $\alpha = 0.5$ on $M = 10$.

Fig. 7 illustrates adaptation of ϵ_s for $\alpha = 0.15$ and $\alpha = 0.35$ for one of the two subspaces (the adaptation trend in the other subspace is similar) in a $M = 8$ and $K = 10$ landscape. The horizontal dashed line at $N = 100$ indicates the size of the overall parent population $|P|$ and the horizontal dotted line the desired number of individuals $\alpha \times |\mathcal{P}|$ with highest rank in the subspace. From these figures note that the number of non-dominated individuals $|\mathcal{F}_1|$ (considering all objectives) exceeds $\mathcal{P}$ since the initial generations. See also that the adaptive mechanism appropriately varies ϵ_s throughout the generations so that after subspace-ϵ-ranking the number of individuals $|\mathcal{F}_1^{\epsilon_s}|$ with highest rank in the subspace is kept around the desired number $\alpha \times |\mathcal{P}|$.

6 Conclusions

In this work, we have proposed a method to search on many-objective problems by instantaneously partitioning the objective space into subspaces and performing one generation of the evolutionary search in each subspace. The proposed method uses a partition strategy to define the schedule of subspace sampling and an adaptive re-ranking method that uses a randomized sampling procedure to increase selection probabilities of some of the too many solutions assigned highest rank in a many-objective subspace. We tested the performance of the proposed method on MNK-Landscapes with $4 \leq M \leq 10$ objectives, $N = 100$

bits and $0 \leq K \leq 50$ epistatic interactions, showing that both convergence and diversity of the obtained solutions can improve remarkably on problems with $4 \leq M \leq 10$ objectives for any level of epistatic interactions K. We also showed that uniformly sampling all possible subspaces throughout the generations leads to better performance.

As future works, we would like to study the effects of larger population sizes and more than two subspaces. Also, we should compare the proposed method with other approaches for many-objective optimization.

Acknowledgment

This study was performed through Special Coordination Funds for Promoting Science and Technology of the Ministry of Education, Culture, Sports, Science and Technology, of the Japanese Government.

References

1. Deb, K.: Multi-Objective Optimization using Evolutionary Algorithms. John Wiley & Sons, Chichester (2001)
2. Coello, C., Van Veldhuizen, D., Lamont, G.: Evolutionary Algorithms for Solving Multi-Objective Problems. Kluwer Academic Publishers, Boston (2002)
3. Aguirre, H., Tanaka, K.: Insights on Properties of Multiobjective MNK-Landscapes. In: Proc. 2004 IEEE Congress on Evolutionary Computation, pp. 196–203. IEEE Service Center, Los Alamitos (2004)
4. Aguirre, H., Tanaka, K.: Working Principles, Behavior, and Performance of MOEAs on MNK-Landscapes. European Journal of Operational Research 181(3), 1670–1690 (2007)
5. Purshouse, R., Fleming, P.: Conflict, harmony, and independence: Relationships in evolutionary multi-criterion optimisation. In: Fonseca, C.M., Fleming, P.J., Zitzler, E., Deb, K., Thiele, L. (eds.) EMO 2003. LNCS, vol. 2632, pp. 16–30. Springer, Heidelberg (2003)
6. Aguirre, H., Tanaka, K.: Selection, drift, recombination, and mutation in multiobjective evolutionary algorithms on scalable MNK-landscapes. In: Coello Coello, C.A., Hernández Aguirre, A., Zitzler, E. (eds.) EMO 2005. LNCS, vol. 3410, pp. 355–369. Springer, Heidelberg (2005)
7. Hughes, E.J.: Evolutionary Many-Objective Optimization: Many Once or One Many? In: Proc. 2005 IEEE Congress on Evolutionary Computation, vol. 1, pp. 222–227. IEEE Service Center, Los Alamitos (2005)
8. Deb, K., Agrawal, S., Pratap, A., Meyarivan, T.: A Fast Elitist Non-Dominated Sorting Genetic Algorithm for Multi-Objective Optimization: NSGA-II, KanGAL report 200001 (2000)
9. Yu, P.L.: Cone Convexity, Cone Extreme Points, and Nondominated Solutions in Decision Problems with Multiobjectives. Journal of Optimization Theory and Applications 14(3), 319–377 (1974)
10. Laumanns, M., Thiele, L., Deb, K., Zitzler, E.: Combining Convergence and Diversity in Evolutionary Multi-objective Optimization. Evolutionary Computation 10(3), 263–282 (2002)

11. Kauffman, S.A.: The Origins of Order: Self-Organization and Selection in Evolution. Oxford University Press, Oxford (1993)
12. Aguirre, H., Tanaka, K.: Adaptive ϵ-Ranking on MNK-Landscapes. In: Proc. 2009 IEEE Symposium on Computational Intelligence in Multicriteria Decision Making (2009)
13. Aguirre, H., Tanaka, K.: Robust Optimization by ϵ-Ranking On High Dimensional Objective Spaces. In: Li, X., Kirley, M., Zhang, M., Green, D., Ciesielski, V., Abass, H.A., Michalewicz, Z., Hendtlass, T., Deb, K., Tan, K.C., Branke, J., Shi, Y. (eds.) SEAL 2008. LNCS, vol. 5361, pp. 421–431. Springer, Heidelberg (2008)
14. Zitzler, E.: Evolutionary Algorithms for Multiobjective Optimization: Methods and Applications, PhD thesis, Swiss Federal Institute of Technology, Zurich (1999)
15. Fonseca, C., Paquete, L., López-Ibáñez, M.: An Improved Dimension-sweep Algorithm for the Hypervolume Indicator. In: Proc. 2006 IEEE Congress on Evolutionary Computation, pp. 1157–1163. IEEE Service Center, Los Alamitos (2006)

Online Objective Reduction to Deal with Many-Objective Problems

Antonio López Jaimes, Carlos A. Coello Coello*, and Jesús E. Urías Barrientos

CINVESTAV-IPN (Evolutionary Computation Group)
Departamento de Computación
Av. IPN No. 2508, Col. San Pedro Zacatenco
México, D.F. 07360, México
tonio.jaimes@gmail.com, ccoello@cs.cinvestav.mx, eduardo.uriasb@gmail.com

Abstract. In this paper, we propose and analyze two schemes to integrate an objective reduction technique into a multi-objective evolutionary algorithm (MOEA) in order to cope with many-objective problems. One scheme reduces periodically the number objectives during the search until the required objective subset size has been reached and, towards the end of the search, the original objective set is used again. The second approach is a more conservative scheme that alternately uses the reduced and the entire set of objectives to carry out the search. Besides improving computational efficiency by removing some objectives, the experimental results showed that both objective reduction schemes also considerably improve the convergence of a MOEA in many-objective problems.

Keywords: Many-objective optimization, dimensionality reduction, objective reduction.

1 Introduction

Since the first implementation of a Multi-objective Evolutionary Algorithm (MOEA) in the mid 1980s [1], a wide variety of new MOEAs have been proposed, gradually improving in both their effectiveness and efficiency to solve multi-objective problems (MOPs) [2]. However, most of these algorithms have been evaluated and applied to problems with only two or three objectives, in spite of the fact that many real-world problems have more than three objectives [3,4].

Recent experimental [5,6,7] and analytical [8,9] studies have shown that MOEAs based on Pareto optimality [10] scale poorly in MOPs with a high number of objectives (4 or more). Although this limitation seems to affect only to Pareto-based MOEAs, optimization problems with a large number of objectives (also known as many-objective problems) introduce some difficulties common to any other multi-objective optimizer. Three of the most serious difficulties due to high dimensionality are the following:

---

* The second author is also affiliated to the UMI LAFMIA 3175 CNRS at CINVESTAV-IPN.

M. Ehrgott et al. (Eds.): EMO 2009, LNCS 5467, pp. 423–437, 2009.

1. *Deterioration of the Search Ability.* One of the reasons for this problem is that the proportion of nondominated solutions (*i.e.*, equally good solutions) in a population increases rapidly with the number of objectives [11]. According to Bentley *et al.* [12] the number of nondominated k-dimensional vectors on a set of size n is $O(\ln^{k-1} n)$. As a consequence, in a many-objective problem, the selection of solutions is carried out almost at random or guided by diversity criteria. In fact, Mostaghim and Schmeck [13] have shown that a random search optimizer achieves better results than the NSGA-II [14] in a problem with 10 objectives.
2. *Dimensionality of the Pareto front.* Due to the 'curse of dimensionality', the number of points required to represent accurately a Pareto front increases exponentially with the number of objectives. Formally, the number of points necessary to represent a Pareto front with k objectives and resolution r is given by kr^{k-1} (*e.g.*, see [15]). This poses a challenge both to the data structures to efficiently manage that number of points and to the density estimators to achieve an even distribution of the solutions along the Pareto front.
3. *Visualization of the Pareto front.* Clearly, with more than three objectives is not possible to plot the Pareto front as usual. This is a serious problem since visualization plays a key role for a proper decision making. Parallel coordinates [16] and self-organizing maps [17] are some of the methods proposed to ease the decision making in high dimensional problems. However, more research in this field is required.

Currently, there are mainly two approaches to solve many-objective problems, namely:

1. Adopt or propose an optimality relation that yields a solution ordering finer than that yielded by Pareto optimality. Among these alternative relations we can find k-optimality [11], preference order ranking [18], and a method that controls the dominance area [19].
2. Reduce the number of objectives of the problem during the search process or, *a posteriori*, during the decision making process [20,21,22]. The main goal of this kind of reduction techniques is to identify the redundant objectives (or redundant to some degree) in order to discard them. A redundant objective is one that can be removed without changing the dominance relation[1] induced by the original objective set.

In the current paper we propose to incorporate an objective reduction method into a Pareto-based MOEA in order to cope with many-objective problems. By selecting a computationally efficient objective reduction method we can expect that the resulting MOEA improves its efficiency, since a smaller number of objective functions are evaluated. While this may be true, the omission of some objective implies some loss of information that could be important to converge to the real Pareto front. On the other hand, this omission can be useful to cope with

[1] The dominance relation induced by a given set F of objectives is defined by $\preceq_F = \{(\boldsymbol{x}, \boldsymbol{y}) | \forall f_i \in F : f_i(\boldsymbol{x}) \leq f_i(\boldsymbol{y})\}$.

the deterioration of the search ability of Pareto-based MOEAs in many-objective problems. With this in mind we propose two schemes to integrate an efficient reduction method into a MOEA in such a way that the resulting MOEA can be useful even in problems with inexpensive objective functions. Additionally, one of the goals of this work is to investigate if an objective reduction method represents a benefit or a damage to the search ability. The results show that the proposed schemes improve the computational efficiency of a common MOEA even in problems with low computational-cost functions. More important, the experiments show that the reduction techniques employed also improve the search ability of the MOEA. Therefore, the benefit of reducing the objective set is greater than the negative effect caused by the loss of information. In [23] is also incorporated an objective reduction method into a MOEA, however the objective in that work is to improve the efficiency of hypervolume-based MOEAs which have exponential complexity in the number of objectives.

The remainder of this paper has the following structure. Section 2 presents the objective reduction technique selected to be incorporated into a MOEA. In Section 3 we describe two schemes to incorporate the reduction method during the search. The assessment of the proposed reduction schemes is presented in Section 4. Finally, in Section 5 we draw some conclusions about the proposed reduction schemes, as well as some possible paths for future research.

2 An Objective Reduction Technique Based On Correlation

The success of an objective reduction method during the search mainly depends on the balance between the overhead incurred by the reduction method itself, and the time saved by omitting some objective function evaluations. For this reason, an efficient reduction method is more likely beneficial in a wide variety of problems. In the following, we shortly describe three objective reduction methods recently proposed in the specialized literature.

Saxena and Deb [20] proposed a method for reducing the number of objectives based on principal component analysis. This method consists of an iterative scheme where the nondominated set obtained by the NSGA-II [14] is analyzed in order to gradually obtain a smaller objective set. The time complexity of each iteration[2] of this algorithm is $O(ms^2 + s^3) + O(gm^2 s)$, where the second term corresponds to NSGA-II's complexity, s is the number of objectives, m is the size of the nondominated set, and g the number of generations for each run of NSGA-II.

Brockhoff and Zitzler [21] proposed two greedy algorithms to reduce the number of objectives. One of them finds a minimum objective subset that yields a given error δ (degree of change of the dominance relation). The other algorithm finds a k-sized objective subset with the minimum possible error. Both algorithms use the ϵ-dominance relation to measure the change of the dominance

[2] The total number of iterations depends on a threshold cut parameter and on the particular nondominated sets generated by the NSGA-II.

relation when objectives are discarded. The time complexity for these algorithms is $O(\min\{m^2s^3, m^4s^2\})$ and $O(m^2s^3)$, respectively.

Similar to Brockhoff and Zitzler, López Jaimes *et al.* [22] proposed two schemes to reduce the number of objectives. The first algorithm is intended to determine a minimum subset of objectives that yields the minimum possible error, while the second one finds a subset of objectives of a given size that yields the minimum error. These algorithms are based on a feature selection technique which uses correlation between nondominated vectors to estimate the conflict between each pair of objectives. The complexity of both algorithms is $O(ms^2)$.

Since López Jaimes *et al.*'s algorithms have a lower time complexity, they are suitable to be integrated into a MOEA since the chances that their computational time savings overcome their overhead are larger than those of the other methods described here. However, in this study we have only chosen the algorithm that finds a subset of objectives of a given size.

2.1 Details of the Selected Objective Reductions Method

The algorithm that finds a k-sized objective subset (KOSSA) uses a correlation matrix to estimate the conflict between each pair of objectives. This matrix is computed using the nondominated set generated by some MOEA. A negative correlation between a pair of objectives means that one objective increases while the other decreases and vice versa. This way, we could interpret that the more negative the correlation between two objectives, the more conflict between them.

Since the interest is in the negative correlation, we use $1 - \rho(f_1, f_2) \in [0, 2]$ to measure the degree of negative correlation (where $\rho(f_1, f_2)$ is the correlation between objectives f_1 and f_2). Thus, a value of 2 indicates that objectives f_1 and f_2 are completely negatively correlated (totally in conflict) and a result of zero indicates that the objectives are completely positively correlated (without any conflict).

The central part of the objective reduction algorithm is divided in three steps:

1. Divide the objective set into homogeneous neighborhoods of size q around each objective. The conflict between objectives takes the role of the distance. That is, the more conflict between two objectives, the more distant they are in the "conflict" space. Figure 1(a) shows only two neighborhoods of a hypothetical situation with eight objectives and $q = 2$.
2. Select the most compact neighborhood. That is, the neighborhood with the minimum distance to its q-th nearest-neighbor. Figure 1(b) shows the farthest neighbor for each of the two neighborhoods. In the example, the neighborhood on the left is the most compact one.
3. Retain the center of that neighborhood and discard its q neighbors. In this process, the distance to the q-th neighbor can be thought of as the error committed by removing the q objectives (see Figure 1(c)).

The pseudocode of the reduction algorithm, KOSSA, is presented in Figure 2. In this pseudocode each entry, $r_{i,j}$, of the correlation matrix represents the conflict between objective f_i and f_j. In particular, $r_{i,q}$ denotes the conflict between objective f_i and its q-th nearest-neighbor.

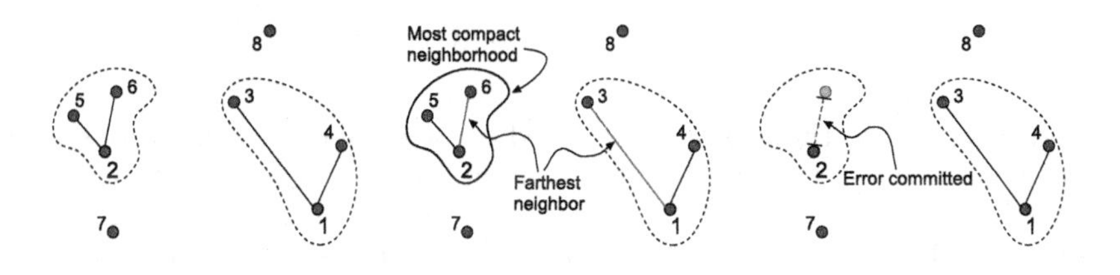

(a) Divide the objective set into neighborhoods around each objective.

(b) Select the most compact neighborhood.

(c) Retain the center and remove the neighbors.

Fig. 1. Basic strategy of the objective reduction method employed

Input: Nondominated set A.
Initial objective set $F = \{f_i, i = 1, ..., s\}$.
Number of neighbors $q \leq |F| - k$.
Size of the desired objective subset, k.
Step 0: Compute the correlation matrix using A.
Step 1: $F' \leftarrow F$.
Step 2: Find objective f_i^{min} which corresponds to
$r_{i,q}^{min} \leftarrow \min_{f_i \in F'} \{r_{i,q}\}$.
Step 3: Retain f_i^{min} and discard its q neighbors from F'.
Let $error \leftarrow r_{i,q}^{min}$.
Step 4: **If** $q > |F'| - k$ **then** $q \leftarrow |F'| - k$.
Step 5: **If** $|F'| = k$ **then** go to **Step 8** to stop.
Compute again $r_{i,q}^{min} \leftarrow \min_{f_i \in F'} \{r_{i,q}\}$.
Step 6: **While** $r_{i,q}^{min} > error$ and $q > 1$ **do**:
$q \leftarrow q - 1$.
$r_{i,q}^{min} \leftarrow \min_{f_i \in F'} \{r_{i,q}\}$.
Step 7: Go to **Step 2**.
Step 8: Return set F' as the reduced objective set.

Fig. 2. Pseudocode of the objective reduction algorithm KOSSA

3 Integration Schemes of the Objective Reduction Method into a MOEA

When some objectives are discarded from the original problem some information is being lost. The magnitude of this loss depends on the degree of redundancy among the objectives.

In any case, we have to balance the benefit of discarding some objectives along with the computational cost of the reduction algorithm. Two benefits are clear from removing some objectives, namely: *i*) the avoidance of the computation of some possible computational expensive objective functions, and *ii*) the speedup in execution of the MOEA, specially if its complexity time largely depends on the number of objectives.

Next, we will describe two schmes to incorporate the KOSSA method into a MOEA. First, we propose a simple scheme where the objective set is reduced successively during most of the search and only towards the end of the search all the objectives are integrated. This scheme is divided in three stages:

1. In the first stage the MOEA is executed for a number of generations using all the objectives. The MOEA obtains an initial approximation of the Pareto front which will be the first input of the objective reduction method, KOSSA.
2. The second stage is the main stage of the scheme where the objective set is gradually reduced through several generations. In this stage, every certain number of generations KOSSA is executed to reduce the objective set and then the execution of the MOEA is resumed. This process is repeated until the desired objective set size has been reached.
3. In the last stage all the objectives are taken up again to obtain the final approximation of the Pareto front.

The detailed scheme with successive reductions is described in Algorithm 1, where P denotes the best population obtained so far by the MOEA.

Algorithm 1. Pseudocode of the successive reduction scheme.

Input:
- R: Number of reductions during the search.
- k: Size of the minimum objective set allowed.
- G_{max}: Total number of generations.
- G_{pre}: Generations before the reduction stage.
- G_{post}: Generations after the reduction stage.

```
1: G ← G_pre;    F' ← F
2: k' ← ⌈(|F| − k)/R⌉   ▷ Number of objectives discarded per reduction.
3: for r ← 1 until R + 2 do
4:     for g ← 1 until G do
5:         MOEA(P, F')
6:     if r ≠ R + 2 then
           ▷ Reduce the current objective set F'.
7:         if r ≤ R then
8:             F' ← KOSSA(P, F', |F'| − k')
9:             G ← (G_max − G_pre − G_post)/R
10:        else
               ▷ Integrate all the objectives at the end of the search.
11:            F' ← F
12:            G ← G_post
```

In the current implementation of this scheme we decided to schedule the reduction phases equally distributed during the reduction stage. However, other schedules are possible, for instance the number of generations for the next reduction can be shortened each time, since the population converges faster after each reduction. A similar decision can be made with regard to the number of objectives discarded on each reduction. Currently, the same number of objectives is removed at each reduction as it can be seen in the third statement of Algorithm 1.

Although this scheme has the advantage (computationally speaking) of omitting the evaluation of many objectives during most of the search, it is possible that the loss of information diminishes the MOEA's convergence ability. Therefore, we also proposed a less aggressive scheme which integrates the entire objective set periodically during the search to counterbalance the loss of information. As in the scheme described previously, this mixed scheme starts the search using the whole objective set for some generations. However, it alternates the reduction process with the integration of the original objectives during the remainder of the search. Algorithm 2 presents the details of the mixed scheme.

Algorithm 2. Pseudocode of the mixed reduction scheme.

Input:

R: Number of reductions during the search.
k: Size of the minimum objective set allowed.
G_{max}: Total number of generations.
G_{pre}: Generations before the reduction stage.
p_{red}: Percentage of generations using the reduced objective set.

$p_{int} \leftarrow 1 - p_{red}$.
$G_{red} \leftarrow p_{red} \times (G_{max} - G_{pre})/R$
$G_{int} \leftarrow p_{int} \times (G_{max} - G_{pre})/R$
$G \leftarrow G_{pre}$
$F' \leftarrow F$
$k' \leftarrow \lceil (|F| - k)/R \rceil$ ▷ Number of objectives discarded per reduction.

for $r \leftarrow 1$ until $2R + 1$ **do**
  for $g \leftarrow 1$ until G **do**
    MOEA(P, F')

  if $r \neq 2R + 1$ **then**
    ▷ Reduce the current objective set F'.
    if $r \bmod 2 = 1$ **then**
      $F' \leftarrow$ KOSSA$(P, F', |F'| - k')$
      $G \leftarrow G_{red}$
    else
      ▷ Integrate all the objectives for the next generations.
      $F' \leftarrow F$
      $G \leftarrow G_{int}$

---

4 Assessment of the Objective Reduction Schemes Coupled with a MOEA

In order to evaluate the performance of the schemes presented in the previous section we chose the NSGA-II as a testbed. As we mention in previous sections, the worth of using an objective reduction method depends on its computational cost, the time complexity of the MOEA (specially if it depends on the number of objectives), the computational cost of the objective functions, and on the effect caused by the removal of objectives.

In order to investigate the effect of these factors, we carried out two types of experiments. The first group of experiments attempts to provide an overall assessment of all those factors in order to determine if the reduction method is advantageous. To do so, instead of using the number of evaluations as a stopping criterion, we use the real computational time instead. By doing so, we can decide if the overall benefits of the reduction method are greater than its possible damages. In the second group of experiments we want to investigate if a reduction method increases or decreases the number of generations required to reach a certain quality of the approximation set produced.

In both types of experiments we compare the NSGA-II equipped with the reduction method (REDGA) against the original NSGA-II. The following problems were adopted in all the experiments: the 0/1 multi-objective knapsack problem with 200 items, and a variation, proposed in [24], of the well-known problem DTLZ2 (denoted here by DTLZ2$_{BZ}$) with 30 variables. All the runs were executed in a single-core computer with a 2.13 GHz CPU.

In the first group of experiments the results were evaluated using the additive ϵ-Indicator [25], which is defined as

$$I_{\epsilon+}(A,B) = \inf_{\epsilon \in \mathbb{R}} \{\forall \boldsymbol{z}^2 \in B \ \exists \boldsymbol{z}^1 \in A : \boldsymbol{z}^1 \succeq_{\epsilon+} \boldsymbol{z}^2\}$$

for two nondominated sets A and B, where $\boldsymbol{z}^1 \succeq_{\epsilon+} \boldsymbol{z}^2$ iff $\forall i : z_i^1 \leq \epsilon + z_i^2$, for a given ϵ. In other words, $I_{\epsilon+}(A,B)$ is the minimum value such that aggregated to any objective vector in B, then $A \succeq B$. In general, $I_{\epsilon+}(A,B) \neq I_{\epsilon+}(B,A)$ so we have to compute both values. The smaller $I_{\epsilon+}(A,B)$ and larger $I_{\epsilon+}(B,A)$, the better A over B.

4.1 Overall Assessment of the Reduction Schemes

In these experiments we used four instances for each of the two test problems employed with 4, 6, 8 and 10 objectives. For each number of objectives we fixed the following time windows: 2, 4, 6 and 10 seconds. For all the 30 runs and problems we used a population of 300 individuals. For NSGA-II we employed a crossover probability of 0.9 and a mutation probability of $1/N$ (N is the number of variables). In the knapsack problem we used a binary representation with a mutation probability of $1/n$ (n is the length of the chromosome).

In order to study the successive reduction scheme we reduced in all cases the objective set until a size of $k = 3$ and the percentage of generations before and after the reduction stage was fixed to 20% and 5%, respectively. Here, we studied two scenarios: one that reduces all the required objectives in one reduction (REDGA-S-1), while the other one uses, among all possible number of reductions, an intermediate number of reductions considering a final set of size $k = 3$ (REDGA-S-m). That is, for 6, 8 and 10 objectives were used 2, 3, and 4 reductions, respectively. In the mixed reduction scheme we only used an intermediate number of reductions for every number of objectives (REDGA-X-m), and

Table 1. Results of the reduction schemes with respect to the ϵ-Indicator in the DTLZ2$_{BZ}$ problem using a fixed-time stopping criterion

DTLZ2$_{BZ}$ with 4 objectives					
$I_{\epsilon+}(A,B)$	**REDGA-S-1**	**REDGA-S-m**	**REDGA-X-m**	**NSGA-II**	Average
REDGA-S-1	-	-	-	0.04450	**0.04450**
REDGA-S-m	-	-	-	-	
REDGA-X-m	-	-	-	-	
NSGA-II	0.06469	-	-	-	0.06469
Average	**0.06469**			0.04450	

DTLZ2$_{BZ}$ with 6 objectives					
$I_{\epsilon+}(A,B)$	**REDGA-S-1**	**REDGA-S-m**	**REDGA-X-m**	**NSGA-II**	Average
REDGA-S-1	-	0.05961	0.05723	0.06019	0.05901
REDGA-S-m	0.05259	-	0.05085	0.05849	**0.05398**
REDGA-X-m	0.05850	0.05614	-	0.05421	0.05628
NSGA-II	0.07447	0.07711	0.07972	-	0.07710
Average	0.06185	**0.06429**	0.06260	0.05763	

DTLZ2$_{BZ}$ with 8 objectives					
$I_{\epsilon+}(A,B)$	**REDGA-S-1**	**REDGA-S-m**	**REDGA-X-m**	**NSGA-II**	Average
REDGA-S-1	-	0.08583	0.07711	0.07179	0.07824
REDGA-S-m	0.06905	-	0.08195	0.06341	**0.07147**
REDGA-X-m	0.07386	0.08171	-	0.06944	0.07500
NSGA-II	0.09882	0.10616	0.11782	-	0.10760
Average	0.08058	0.09123	**0.09229**	0.06821	

DTLZ2$_{BZ}$ with 10 objectives					
$I_{\epsilon+}(A,B)$	**REDGA-S-1**	**REDGA-S-m**	**REDGA-X-m**	**NSGA-II**	Average
REDGA-S-1	-	0.09108	0.09182	0.08316	0.08869
REDGA-S-m	0.06916	-	0.07072	0.07926	**0.07305**
REDGA-X-m	0.07998	0.08554	-	0.06840	0.07797
NSGA-II	0.11608	0.12159	0.11480	-	0.11749
Average	0.08841	**0.09940**	0.09245	0.07694	

the other parameters were $k = 3$, $p_{red} = 0.85$ and 20% of the total generations were accomplished before the reduction stage. The results of the ϵ-Indicator for these scenarios on problem DTLZ2$_{BZ}$ are presented in Table 1. Since for four objectives REDGA-S-m and REDGA-X-m are equivalent to the REDGA-S-1 scheme, we only show the results of this scheme against NSGA-II.

As we can clearly see in Table 1, all the reduction schemes perform better than NSGA-II for every number of objectives. Besides, the advantage of the reduction schemes over the NSGA-II increases with the number of objectives. On the other hand, except for 8 objectives, the scheme REDGA-S-m achieved better results than the REDGA-X-m which is the second best in this comparison. This means that the strategy of integrating all the objectives periodically did not improve the performance of the reduction scheme. As somewhat expected, the REDGA-S-1 scheme did not obtain results as good as the other reduction schemes. A possible explanation is that, in spite of the fact that REDGA-S-1 carries out

Table 2. Results of the reduction schemes with respect to the ϵ-indicator in the 0/1 Knapsack problem using a fixed-time stopping criterion

Knapsack with 4 objectives					
$I_{\epsilon+}(A,B)$	**REDGA-S-1**	**REDGA-S-m**	**REDGA-X-m**	**NSGA-II**	Average
REDGA-S-1	-	-	-	205	**205**
REDGA-S-m	-	-	-	-	
REDGA-X-m	-	-	-	-	
NSGA-II	241	-	-	-	241
Average	**241**			205	
Knapsack with 6 objectives					
$I_{\epsilon+}(A,B)$	**REDGA-S-1**	**REDGA-S-m**	**REDGA-X-m**	**NSGA-II**	Average
REDGA-S-1	-	408	264	318	**330.0**
REDGA-S-m	371	-	269	352	330.7
REDGA-X-m	372	403	-	306	360.3
NSGA-II	448	414	378	-	413.3
Average	397.0	**408.3**	303.7	325.3	
Knapsack with 8 objectives					
$I_{\epsilon+}(A,B)$	**REDGA-S-1**	**REDGA-S-m**	**REDGA-X-m**	**NSGA-II**	Average
REDGA-S-1	-	646	478	505	543.0
REDGA-S-m	457	-	323	290	**356.7**
REDGA-X-m	441	465	-	345	417.0
NSGA-II	564	472	438	-	491.3
Average	487.3	**527.7**	413.0	380.0	
Knapsack with 10 objectives					
$I_{\epsilon+}(A,B)$	**REDGA-S-1**	**REDGA-S-m**	**REDGA-X-m**	**NSGA-II**	Average
REDGA-S-1	-	455	424	423	434.0
REDGA-S-m	503	-	411	376	**430.0**
REDGA-X-m	760	667	-	493	640.0
NSGA-II	533	455	522	-	503.3
Average	**598.7**	525.7	452.3	430.7	

more evaluations than the other schemes in the given time, this advantage is not enough to counteract the negative effect caused by the loss of information. In this sense, the REDGA-S-m scenario represents a better tradeoff between these factors.

As in the previous problem, NSGA-II was the worst algorithm in the 0/1 knapsack problem regarding the ϵ-Indicator (see Table 2). Nonetheless, the REDGA-S-1 scheme presented a better performance than in DTLZ2$_{BZ}$, *i.e.*, with 4 objectives it was the second best and with 10 it was the best scheme. The reason is that knapsack's objective functions are more computationally expensive than those of the problem DTLZ2$_{BZ}$. This allowed that REDGA-S-1 could perform many more generations than any other scheme. This is a clear example that the balance between the computational cost of the objective functions and the overhead of the reduction scheme plays an important role on the success of the reduction scheme. Furthermore, it acts as a guide to decide what type of reduction scheme to choose. If the objective functions are expensive then it may be convenient to use an aggressive scheme such as REDGA-S-1; otherwise, the REDGA-S-m could be more appropriate.

4.2 Effect of the Reduction Schemes on MOEA's Search Ability

In order to investigate how a reduction scheme affects the MOEA's convergence ability we compare the reduction schemes using the number of generations as the stopping criterion. In these experiments we used a population of 300 individuals for every number of objectives, and all the algorithms were executed for 200 generations (60 000 evaluations). In this experiment we adopt only DTLZ2$_{BZ}$ since convergence can be easily measured given that the nondominated vectors of its true Pareto front have the property $D = \sum_{i=1}^{s} f_i^2 = 1$, where s is the number of objectives. The distribution of the values of D for each algorithm are shown in Figure 3. The horizontal axis represents the D values obtained by each algorithm and the vertical axis denotes the frequency of a given D value. As well as in other

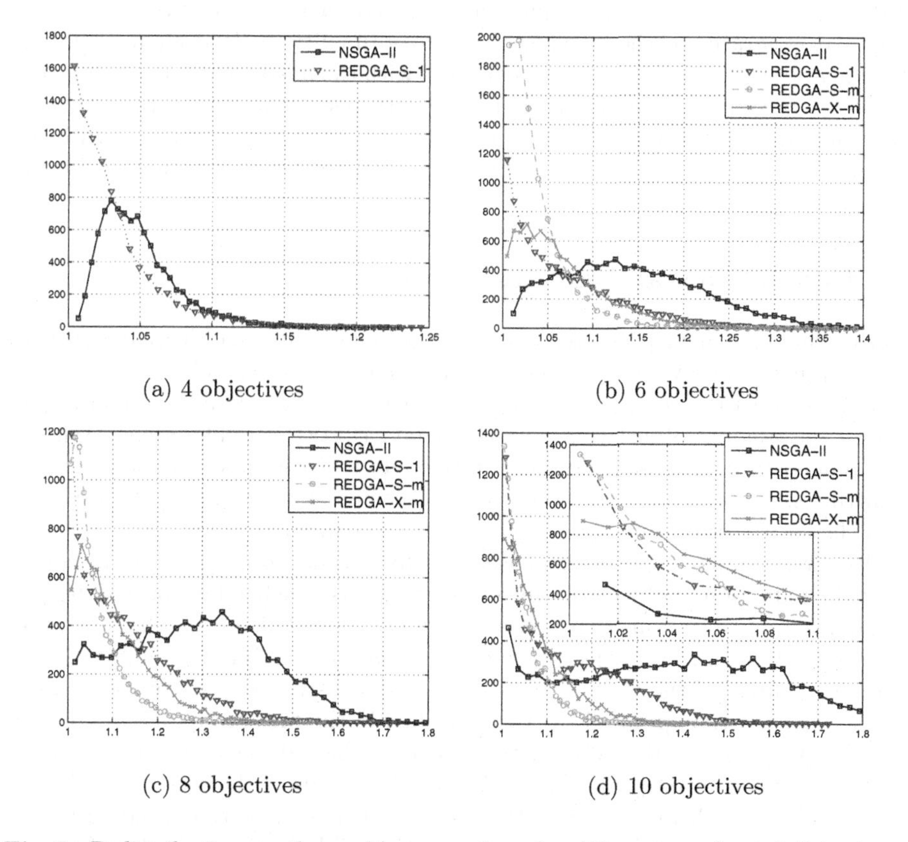

(a) 4 objectives (b) 6 objectives

(c) 8 objectives (d) 10 objectives

Fig. 3. D distribution on the problem DTLZ2$_{BZ}$ for different number of objectives. $D = 1$ corresponds to the true Pareto front.

Table 3. Results of the reduction schemes with respect to the value D in the DTLZ2$_{BZ}$ problem using a fixed-generations stopping criterion

Obj		REDGA-S-1	REDGA-S-m	REDGA-X-m	NSGA-II
4	**Average**	**1.0305**	-	-	1.0488
	Std. Dev.	0.0289	-	-	**0.0266**
6	**Average**	1.0672	**1.0358**	1.0649	1.1445
	Std. Dev.	0.0609	**0.0334**	0.0496	0.0799
8	**Average**	1.1276	**1.0607**	1.1040	1.2863
	Std. Dev.	0.1113	**0.0561**	0.0805	0.1559
10	**Average**	1.1402	**1.0501**	1.0786	1.3787
	Std. Dev.	0.1234	**0.0487**	0.0690	0.2218

studies [13,6], Figure 3 shows that the performance of NSGA-II decays as the number of objectives increases. In addition, all the reduction schemes perform better than NSGA-II in all cases. This means that the reduction schemes, besides reducing execution time also help Pareto-based MOEAs to recover the search ability deteriorated by the inability of Pareto optimality to discriminate solutions in many-objective problems. In concordance with the fixed-time experiments, the REDGA-S-m achieves the best convergence with respect to the average D value presented in Table 3. Like all the algorithms, its convergence decreases with the number of objectives. However REDGA-S-m is the scheme less affected by the number of objectives.

5 Conclusions and Future Work

In this paper, we have presented two schemes to integrate an objective reduction method into a MOEA. One of these schemes reduces successively the number of objectives until the required size has been reached and only at the final generations the original objective set is used again (REDGA-S). The second scheme is intended to counterbalance the negative effect of the loss of information by omitting some objectives (REDGA-X). This scheme uses alternately the reduced and the entire set of objectives to carry out the search.

The first group of experiments based on a fixed-time stopping criterion showed that the reduction of objectives during the search is beneficial in spite of the loss of information since it also saves computational time. This means that the overhead introduced by the objective reduction method was small enough to speed up the execution of the MOEA even with the inexpensive objective functions used in the study. Although in all the cases studied in the first group of experiments the MOEA coupled with the reduction scheme achieved better results than the MOEA alone, we have to carefully select the parameters of the reduction scheme. There is an equilibrium point in the number of objectives that need to be removed in order to achieve the best tradeoff possible between the benefits and damages obtained by the reduction scheme. To illustrate this, it is sufficient to consider that, although the REDGA-S scheme with only one reduction is the one that saves more time per generation, it did not present as good performance as a less aggressive configuration such as the REDGA-S-m. On the other hand, the periodic incorporation of the entire objective set did not improve the performance of the successive reduction scheme, which is simpler.

One important finding is that a reduction scheme besides reducing the execution time of a MOEA also helps to remedy the limitation of Pareto optimality for dealing with problems having a large number of objectives. The results showed that all the reduction schemes studied outperformed the original MOEA even when a stopping criterion based on a fixed number of generations was used. This shreds light into the usefulness of objective reduction schemes since they bring advantages both in efficiency and effectiveness.

As part of our future work, we want to study the performance of the objective reduction methods in problems with less conflict among their objectives. We would expect that in those problems the benefit of using a reduction method would be greater since the loss of information is smaller than in many-objective conflicting problems. Given their encouraging results, it would be interesting to compare the reduction schemes proposed against methods that have shown good performance in many-objective problems.

Acknowledgments

The first author acknowledges support from CONACyT and from CINVESTAV-IPN to pursue graduate studies in computer science.

References

1. Schaffer, J.D.: Multiple Objective Optimization with Vector Evaluated Genetic Algorithms. In: Genetic Algorithms and their Applications: Proceedings of the First International Conference on Genetic Algorithms, pp. 93–100. Lawrence Erlbaum, Mahwah (1985)
2. Coello Coello, C.A., Lamont, G.B., Van Veldhuizen, D.A.: Evolutionary Algorithms for Solving Multi-Objective Problems, 2nd edn. Springer, Heidelberg (2007)
3. Fonseca, C.M., Fleming, P.J.: Multiobjective Optimization and Multiple Constraint Handling with Evolutionary Algorithms—Part II: A Application Example. IEEE Transactions on Systems, Man, and Cybernetics, Part A: Systems and Humans 28(1), 38–47 (1998)
4. Hughes, E.J.: Radar waveform optimisation as a many-objective application benchmark. In: Obayashi, S., Deb, K., Poloni, C., Hiroyasu, T., Murata, T. (eds.) EMO 2007. LNCS, vol. 4403, pp. 700–714. Springer, Heidelberg (2007)
5. Hughes, E.J.: Evolutionary Many-Objective Optimisation: Many Once or One Many? In: 2005 IEEE Congress on Evolutionary Computation (CEC 2005), Edinburgh, Scotland, vol. 1, pp. 222–227. IEEE Service Center, Los Alamitos (2005)
6. Wagner, T., Beume, N., Naujoks, B.: Pareto-, aggregation-, and indicator-based methods in many-objective optimization. In: Obayashi, S., Deb, K., Poloni, C., Hiroyasu, T., Murata, T. (eds.) EMO 2007. LNCS, vol. 4403, pp. 742–756. Springer, Heidelberg (2007)

7. Praditwong, K., Yao, X.: How well do multi-objective evolutionary algorithms scale to large problems. In: IEEE Congress on Evolutionary Computation, CEC 2007, pp. 3959–3966 (September 2007)
8. Teytaud, O.: How entropy-theorems can show that approximating high-dim pareto-fronts is too hard. In: Bridging the Gap between Theory and Practice - Workshop PPSN-BTP, International Conference on Parallel Problem Solving from Nature PPSN (2006)
9. Knowles, J., Corne, D.: Quantifying the effects of objective space dimension in evolutionary multiobjective optimization. In: Obayashi, S., Deb, K., Poloni, C., Hiroyasu, T., Murata, T. (eds.) EMO 2007. LNCS, vol. 4403, pp. 757–771. Springer, Heidelberg (2007)
10. Pareto, V.: Cours D'Economie Politique. F. Rouge (1896)
11. Farina, M., Amato, P.: On the Optimal Solution Definition for Many-criteria Optimization Problems. In: Proceedings of the NAFIPS-FLINT International Conference 2002, Piscataway, New Jersey, pp. 233–238. IEEE Service Center, Los Alamitos (2002)
12. Bentley, J.L., Kung, H.T., Schkolnick, M., Thompson, C.D.: On the average number of maxima in a set of vectors and applications. J. ACM 25(4), 536–543 (1978)
13. Mostaghim, S., Schmeck, H.: Distance based ranking in many-objective particle swarm optimization. In: Rudolph, G., Jansen, T., Lucas, S., Poloni, C., Beume, N. (eds.) PPSN 2008. LNCS, vol. 5199, pp. 753–762. Springer, Heidelberg (2008)
14. Deb, K., Agrawal, S., Pratab, A., Meyarivan, T.: A Fast Elitist Non-Dominated Sorting Genetic Algorithm for Multi-Objective Optimization: NSGA-II. In: Deb, K., Rudolph, G., Lutton, E., Merelo, J.J., Schoenauer, M., Schwefel, H.-P., Yao, X. (eds.) PPSN 2000. LNCS, vol. 1917, pp. 849–858. Springer, Heidelberg (2000)
15. Sen, P., Yang, J.: Multiple criteria decision support in engineering design. Springer, London (1998)
16. Wegman, E.J.: Hyperdimensional data analysis using parallel coordinates. Journal of the American Statistical Association 85, 664–675 (1990)
17. Obayashi, S., Sasaki, D.: Visualization and data mining of pareto solutions using self-organizing map. In: Fonseca, C.M., Fleming, P.J., Zitzler, E., Deb, K., Thiele, L. (eds.) EMO 2003. LNCS, vol. 2632, pp. 796–809. Springer, Heidelberg (2003)
18. di Pierro, F.: Many-Objective Evolutionary Algorithms and Applications to Water Resources Engineering. PhD thesis, School of Engineering, Computer Science and Mathematics, UK (August 2006)
19. Sato, H., Aguirre, H.E., Tanaka, K.: Controlling dominance area of solutions and its impact on the performance of mOEAs. In: Obayashi, S., Deb, K., Poloni, C., Hiroyasu, T., Murata, T. (eds.) EMO 2007. LNCS, vol. 4403, pp. 5–20. Springer, Heidelberg (2007)
20. Deb, K., Saxena, D.K.: Searching for Pareto-optimal solutions through dimensionality reduction for certain large-dimensional multi-objective optimization problems. In: 2006 IEEE Congress on Evolutionary Computation (CEC 2006), Vancouver, BC, Canada, pp. 3353–3360. IEEE, Los Alamitos (2006)
21. Brockhoff, D., Zitzler, E.: Are All Objectives Necessary? On Dimensionality Reduction in Evolutionary Multiobjective Optimization. In: Runarsson, T.P., Beyer, H.-G., Burke, E.K., Merelo-Guervós, J.J., Whitley, L.D., Yao, X. (eds.) PPSN 2006. LNCS, vol. 4193, pp. 533–542. Springer, Heidelberg (2006)
22. López Jaimes, A., Coello Coello, C.A., Chakraborty, D.: Objective Reduction Using a Feature Selection Technique. In: 2008 Genetic and Evolutionary Computation Conference (GECCO 2008), Atlanta, USA, pp. 674–680. ACM, New York (2008)

23. Brockhoff, D., Zitzler, E.: Improving Hypervolume-based Multiobjective Evolutionary Algorithms by Using Objective Reduction Methods. In: 2007 IEEE Congress on Evolutionary Computation (CEC 2007), Singapore, pp. 2086–2093. IEEE Press, Los Alamitos (2007)
24. Brockhoff, D., Zitzler, E.: Offline and Online Objective Reduction in Evolutionary Multiobjective Optimization Based on Objective Conflicts. TIK Report 269, Institut für Technische Informatik und Kommunikationsnetze, ETH Zürich (April 2007)
25. Zitzler, E., Thiele, L., Laumanns, M., Fonseca, C.M., da Fonseca, V.G.: Performance Assessment of Multiobjective Optimizers: An Analysis and Review. IEEE Transactions on Evolutionary Computation 7(2), 117–132 (2003)

Adaptation of Scalarizing Functions in MOEA/D: An Adaptive Scalarizing Function-Based Multiobjective Evolutionary Algorithm

Hisao Ishibuchi, Yuji Sakane, Noritaka Tsukamoto, and Yusuke Nojima

Department of Computer Science and Intelligent Systems, Graduate School of Engineering,
Osaka Prefecture University, 1-1 Gakuen-cho, Naka-ku, Sakai, Osaka 599-8531, Japan
{hisaoi,nojima}@cs.osakafu-u.ac.jp,
{sakane,nori}@ci.cs.osakafu-u.ac.jp
http://www.ie.osakafu-u.ac.jp/~hisaoi/ci_lab_e

Abstract. It is well-known that multiobjective problems with many objectives are difficult for Pareto dominance-based algorithms such as NSGA-II and SPEA. This is because almost all individuals in a population are non-dominated with each other in the presence of many objectives. In such a population, the Pareto dominance relation can generate no strong selection pressure toward the Pareto front. This leads to poor search ability of Pareto dominance-based algorithms for many-objective problems. Recently it has been reported that better results can be obtained for many-objective problems by the use of scalarizing functions. The weighted sum usually works well in scalarizing function-based algorithms when the Pareto front is convex. However, we need other functions such as the weighted Tchebycheff when the Pareto front is non-convex. In this paper, we propose an idea of automatically choosing between the weighted sum and the weighted Tchebycheff for each individual in each generation. The characteristic feature of the proposed idea is to use the weighted Tchebycheff only when it is needed for individuals along non-convex regions of the Pareto front. The weighted sum is used for the other individuals in each generation. The proposed idea is combined with a high-performance scalarizing function-based algorithm called MOEA/D (multiobjective evolutionary algorithm based on decomposition) of Zhang and Li (2007). Effectiveness of the proposed idea is demonstrated through computational experiments on modified multiobjective knapsack problems with non-convex Pareto fronts.

1 Introduction

Evolutionary multiobjective optimization (EMO) has been and will continue to be one of the most active research areas in the field of evolutionary computation. EMO algorithms have been successfully applied to various multiobjective optimization problems [3], [4]. Almost all well-known EMO algorithms such as NSGA-II [5] and SPEA [26] use the Pareto dominance relation together with a crowding measure for the fitness evaluation of each individual. In this sense, we call them Pareto dominance-based algorithms. Pareto dominance-based algorithms usually work very well for finding Pareto-optimal or near Pareto-optimal solutions along the Pareto front of

M. Ehrgott et al. (Eds.): EMO 2009, LNCS 5467, pp. 438–452, 2009.

an optimization problem with two or three objectives. The search ability of those algorithms is, however, severely deteriorated by the increase in the number of objectives as pointed out in the literature (See [18], [22] for early studies, and [12], [13] for a short review on evolutionary many-objective optimization studies). This is because almost all individuals in a population become non-dominated with each other under many objectives. When all individuals in a population are non-dominated, the Pareto dominance relation cannot generate any selection pressure toward the Pareto front. In this case, the fitness evaluation of each individual is based on only a crowding measure in each Pareto dominance-based algorithm. Thus their search ability is severely deteriorated by the increase in the number of objectives. It has been demonstrated in the literature [7]-[10], [16], [17] that better results can be obtained by the use of scalarizing functions for many-objective problems with more than three objectives.

Whereas Pareto dominance-based algorithms usually work well on two-objective and three-objective problems, it is not always the case in their applications to combinatorial optimization problems [6]. For example, it has been demonstrated in some studies [16], [17], [23] that Pareto dominance-based algorithms are not good at finding Pareto-optimal or near Pareto-optimal solutions around the edges of the Pareto front of a two-objective knapsack problem. A possible reason for this observation is that the Pareto dominance relation has a too large effect on the fitness evaluation for the two-objective knapsack problem especially in the early stage of evolution. This means that a crowding measure does not have a sufficient effect on the fitness evaluation for increasing the diversity of solutions. As a result, good solutions are obtained in a small region around the center of the Pareto front. It has been reported in the above-mentioned studies [16], [17] that much better results in terms of the diversity of solutions are obtained by scalarizing function-based algorithms.

As we have already explained, scalarizing function-based algorithms have several advantages such as the scalability to many-objective problems and high search ability for combinatorial optimization problems. Moreover, scalarizing function-based fitness evaluation needs much less computation load than Pareto dominance-based one especially for many-objective problems. That is, computational efficiency is another advantage of scalarizing function-based algorithms.

One promising approach for improving the search ability of EMO algorithms is the hybridization with local search [11], [14]-[17], [24]. Local search in such a hybrid EMO algorithm usually uses a scalarizing function (e.g., the weighted sum in multiobjective genetic local search [11] and an augmented achievement scalarizing function in a recent hybrid EMO algorithm [24]). It is easy to incorporate local search into scalarizing function-based algorithms whereas the implementation of local search in Pareto dominance-based algorithms is not always easy. High compatibility with local search is also another advantage of scalarizing function-based algorithms.

One important issue in the implementation of scalarizing function-based algorithms is the choice of an appropriate scalarizing function. For example, both the weighted sum and the weighted Tchebycheff were implemented and compared in a scalarizing function-based algorithm called MOEA/D (multiobjective evolutionary algorithm based on decomposition [25]). It is well-known that the weighted sum cannot appropriately handle multiobjective problems with non-convex Pareto fronts whereas the weighted Tchebycheff can handle them. Moreover, the weighted Tchebycheff has nice theoretical properties [20]. Thus the weighted Tchebycheff seems to be a better choice in scalarizing function-based algorithms than the weighted sum. It was, however, clearly demonstrated in [25] that better results were obtained from the weighted sum than the weighted Tchebycheff in the application of MOEA/D to multiobjective knapsack problems. This observation may suggest the use of the weighted sum for multiobjective problems with convex Pareto fronts whereas its use is not appropriate for other problems with non-convex Pareto fronts.

From these discussions, the shape of the Pareto front seems to be a good indicator for choosing an appropriate scalarizing function. One difficulty in using this indicator is that the shape of the Pareto front is often unknown before we find a large number of Pareto-optimal solutions. Another possible difficulty is related to the use of the same scalarizing function for all individuals in every generation.

When the current population is far from the Pareto front, we can use the weighted sum even if its shape is non-convex. This situation is illustrated by a population of small squares in the bottom-left area of each plot in Fig. 1. This figure shows experimental results of MOEA/D with the weighted Tchebycheff on modified two-objective 500-item knapsack problems with non-convex Pareto fronts. In each plot, three populations at the first, 20th and 500th generations are shown by squares, triangles and circles, respectively. We can see that the population at the first generation is far from the Pareto front. Thus the weighted sum can be used in the early stage of evolution (e.g., in the first 20 generations). When the current population is close to the non-convex Pareto front (e.g., after the 20th generation), the use of the weighted Tchebycheff seems to be appropriate for all solutions in Fig. 1 (a). On the other hand, we need the weighted Tchebycheff only for solutions around the non-convex region (A) of the Pareto front in Fig. 1 (b). The weighted sum can be used throughout the execution of MOEA/D for the other solutions around the two convex regions (B and C) of the Pareto front in Fig. 1 (b). Motivated by these discussions, we propose an idea of automatically choosing between the weighted sum and the weighted Tchebycheff.

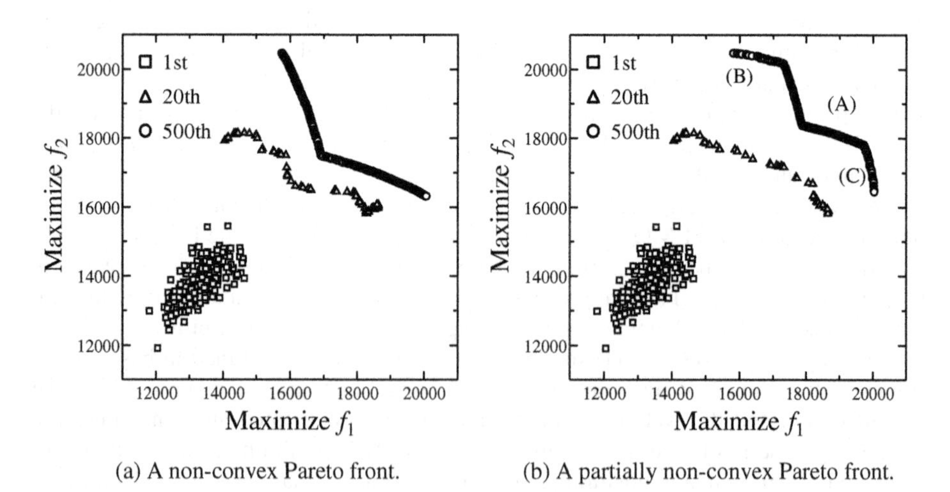

(a) A non-convex Pareto front.

(b) A partially non-convex Pareto front.

Fig. 1. Experimental results of MOEA/D [25] with the weighted Tchebycheff on modified two-objective 500-item knapsack problems with non-convex Pareto fronts

This paper is organized as follows. In Section 2, we explain the two scalarizing functions (i.e., the weighted sum and the weighted Tchebycheff) used in MOEA/D. In Section 3, we explain MOEA/D. After that, we compare the two scalarizing functions with each other in Section 4 through computational experiments on multiobjective 500-item knapsack problems with two, four and six objectives. In Section 5, we explain how we can generate multiobjective knapsack problems with non-convex Pareto fronts by modifying the original ones in Zitzler and Thiele [26]. For example, we can generate the two knapsack problems in Fig. 1. In Section 6, we explain our idea for automatically choosing between the weighted sum and the weighted Tchebycheff for the use in MOEA/D. We examine the effectiveness of the proposed idea through computational experiments on modified multiobjective knapsack problems with non-convex Pareto fronts in Section 7. Finally we conclude this paper in Section 8.

2 Scalarizing Functions

Let us consider the following m-objective maximization problem:

$$\text{Maximize } \boldsymbol{f}(\boldsymbol{x}) = (f_1(\boldsymbol{x}),\ f_2(\boldsymbol{x}),\ ...,\ f_m(\boldsymbol{x}))\,, \tag{1}$$

where $\boldsymbol{f}(\boldsymbol{x})$ is the m-dimensional objective vector, $f_i(\boldsymbol{x})$ is the i-th objective to be maximized, and $\boldsymbol{x}$ is the decision vector. Since we use multiobjective knapsack problems in our computational experiments, we explain our idea for the multiobjective maximization problem in (1).

One of well-known and frequently-used scalarizing functions is the weighted sum with a non-negative weight vector $\boldsymbol{\lambda} = (\lambda_1, \lambda_2, ..., \lambda_m)$:

$$g^{WS}(\boldsymbol{x} \mid \boldsymbol{\lambda}) = \lambda_1 \cdot f_1(\boldsymbol{x}) + \lambda_2 \cdot f_2(\boldsymbol{x}) + \ ... \ + \lambda_m \cdot f_m(\boldsymbol{x})\,, \tag{2}$$

where λ_i is a non-negative weight for the i-th objective $f_i(\boldsymbol{x})$. We assume that the weight vector $\boldsymbol{\lambda}$ satisfies $\lambda_1 + \lambda_2 + ... + \lambda_m = 1$ and $\lambda_i \geq 0$ for $i = 1, 2, ..., m$.

Another well-known scalarizing function is the weighted Tchebycheff. Let Ω be a set of solutions. A reference point $z^* = (z_1^*, z_2^*, ..., z_m^*)$ in the objective space can be specified by the best objective value of each objective in Ω as

$$z_i^* = \max\{f_i(\boldsymbol{x}) \mid \boldsymbol{x} \in \Omega\},\ i = 1, 2, ..., m\,. \tag{3}$$

When Ω is the set of all feasible solutions, z^* is called the ideal vector.

The weighted Tchebycheff measures the distance from the reference point z^* to a solution $\boldsymbol{x}$ in the objective space as follows:

$$g^{TE}(\boldsymbol{x} \mid \boldsymbol{\lambda}, z^*) = \max_{i=1,2,...,m} \{\lambda_i \cdot \mid z_i^* - f_i(\boldsymbol{x}) \mid\}\,. \tag{4}$$

As in Zhang and Li [25], we specified the reference point z^* for each generation in our computational experiments as follows:

$$z_i^* = 1.1 \cdot \max\{ f_i(\boldsymbol{x}) \mid \boldsymbol{x} \in \Omega(t)\}, \; i = 1, 2, ..., m, \tag{5}$$

where $\Omega(t)$ is the secondary population at the t-th generation of MOEA/D [25]. The reference point z^* is updated whenever the best value in (5) is updated.

3 MOEA/D of Zhang and Li (2007)

MOEA/D (multiobjective evolutionary algorithm based on decomposition) was proposed by Zhang and Li [25] as a high-performance scalarizing function-based algorithm. We use MOEA/D because it is a simple but very powerful EMO algorithm. MOEA/D has a number of advantages over Pareto dominance-based algorithms such as the scalability to many-objective problems, high performance for combinatorial optimization problems, computational efficiency of fitness evaluation, and high compatibility with local search.

The main characteristic of MOEA/D is that a multiobjective problem is handled as a collection of a large number of single-objective problems. Each single-objective problem has a scalarizing function with a different weight vector. Each weight vector has a single individual in each population. This idea is similar to a cellular EMO algorithm of Murata et al. [21] where a different weight vector was assigned to each cell. In both algorithms, each individual in the current population was governed by a scalarizing function with a different weight vector. The decomposition of an original multiobjective problem into a number of multiobjective problems was also discussed in other studies (e.g., cone separation by Branke et al. [1]). High performance of MOEA/D has already been demonstrated for well-known test problems such as the ZDT and DTLZ series and knapsack problems in [25], flowshop scheduling problems in [2], and some difficult problems with complicated Pareto fronts [19].

MOEA/D uses a prespecified number of uniformly distributed weight vectors satisfying the following two conditions:

$$\lambda_1 + \lambda_2 + ... + \lambda_m = 1, \tag{6}$$

$$\lambda_i \in \left\{0, \frac{1}{H}, \frac{2}{H}, ..., \frac{H}{H}\right\}, \; i = 1, 2, ..., m, \tag{7}$$

where H is a user-definable positive integer. The number of weight vectors is calculated as $N = {}_{H+m-1}C_{m-1}$ [25]. For example, we have 101 weight vectors by specifying H as $H = 100$ for a two-objective problem: $\boldsymbol{\lambda}$ = (0, 1), (0.01, 0.99), ..., (1, 0).

Let us denote the generated N weight vectors as $\{\boldsymbol{\lambda}^1, \boldsymbol{\lambda}^2, ..., \boldsymbol{\lambda}^N\}$. Each weight vector $\boldsymbol{\lambda}^k$ has the nearest T weight vectors (including $\boldsymbol{\lambda}^k$ itself) as its neighbors where T is a user-definable positive integer. We denote the T neighbors of $\boldsymbol{\lambda}^k$ by $B(\boldsymbol{\lambda}^k)$, which can be viewed as the neighborhood of size T for the weight vector $\boldsymbol{\lambda}^k$. The distance between two weight vectors is measured by the standard Euclidean distance.

The same neighborhood structure is used for individuals since each weight vector has a single individual. Let us denote the individual associated with the weight vector $\boldsymbol{\lambda}^k$ by $\boldsymbol{x}^k$. Then we denote the T individuals associated with the T weight vectors in $B(\boldsymbol{\lambda}^k)$ by $B(\boldsymbol{x}^k)$, which is referred to as the neighborhood of $\boldsymbol{x}^k$. We also call the T individuals in $B(\boldsymbol{x}^k)$ as the neighbors of $\boldsymbol{x}^k$. In MOEA/D, genetic operations for each individual are locally performed among its neighbors as in cellular algorithms.

Now we have N weight vectors. We also have the T neighbors in $B(\boldsymbol{\lambda}^k)$ for each weight vector $\boldsymbol{\lambda}^k$, $k=1,2,...,N$. As in standard evolutionary algorithms, the first step of MOEA/D is to generate an initial population. It should be noted that the population size is the same as the number of the weight vectors (i.e., N). We randomly generate an initial individual for each weight vector. Next we generate an offspring for each weight vector by selection, crossover and mutation. When an offspring is to be generated for the weight vector $\boldsymbol{\lambda}^k$, a couple of parents are randomly selected among the T neighbors of $\boldsymbol{x}^k$ in $B(\boldsymbol{x}^k)$. Then an offspring is generated by crossover and mutation. Let us denote the generated offspring by $\boldsymbol{y}^k$. If the offspring $\boldsymbol{y}^k$ is better than the current individual $\boldsymbol{x}^k$, $\boldsymbol{x}^k$ is replaced with $\boldsymbol{y}^k$. The two individuals $\boldsymbol{x}^k$ and $\boldsymbol{y}^k$ are compared with each other by the scalarizing function with the weight vector $\boldsymbol{\lambda}^k$ (i.e., the weighted sum or the weighted Tchebycheff). The newly generated offspring $\boldsymbol{y}^k$ is also compared with all neighbors in $B(\boldsymbol{x}^k)$. This comparison is performed using the weight vector of each neighbor. If $\boldsymbol{y}^k$ is better than some neighbors, they are replaced with $\boldsymbol{y}^k$. The genetic operations (i.e., selection, crossover, mutation) and the comparison of the newly generated offspring with all neighbors in $B(\boldsymbol{x}^k)$ are performed for each individual $\boldsymbol{x}^k$ (i.e., $k=1,2,...,N$) in the current population. We used the total number of examined solutions as the stopping condition in our computational experiments.

MOEA/D has a secondary population (i.e., an archive population) for storing non-dominated solutions. The secondary population is updated by newly generated offspring throughout the execution of MOEA/D. The secondary population is used just for storing non-dominated solutions. That is, no individual in the secondary population is used in the genetic operations for generating new offspring. This means that the secondary population has no effect on the search behavior of MOEA/D. The use of non-dominated solutions in the secondary population usually significantly improves the search ability of EMO algorithms (e.g., [11], [26]). In MOEA/D, the "replace-if-better" strategy is used for all individuals. This replacement strategy can be viewed as a kind of elitism. Thus MOEA/D has high search ability without utilizing non-dominated solutions in the secondary population as parents in the genetic operations for generating new offspring.

4 Comparison of the Two Scalarizing Functions in MOEA/D

In this section, we compare the weighted sum and the weighted Tchebycheff with each other through computational experiments. We used the two-objective and four-objective knapsack problems in Zitzler and Thiele [26]. We also generated a six-objective 500-item knapsack problem. We denote these test problems by the number of objectives and the number of items as the 2-500, 4-500 and 6-500 problems. In our computational experiments, we used the following parameter specifications:

Population size (which is the same as the number of weight vectors):
200 (2-500), 220 (4-500), and 252 (6-500),
Parameter H for generating weight vectors:
199 (2-500), 9 (4-500), and 5 (6-500),
Stopping condition (i.e., the total number of examined solutions):
100,000 (2-500), 150,000 (4-500), and 200,000 (6-500),
Crossover probability: 0.8 (Uniform crossover),
Mutation probability: 1/500 (Bit-flip mutation),
Neighborhood size T (i.e., the number of neighbors): 10.

Our computational experiments were performed in the same manner as in Zhang and Li [25]. For example, we used the same greedy repair method as in [25]. We show experimental results of a single run of MOEA/D with the weighted sum in Fig. 2 (a) and the weighted Tchebycheff in Fig. 2 (b). In each plot, we show all individuals in the current and secondary populations at each generation.

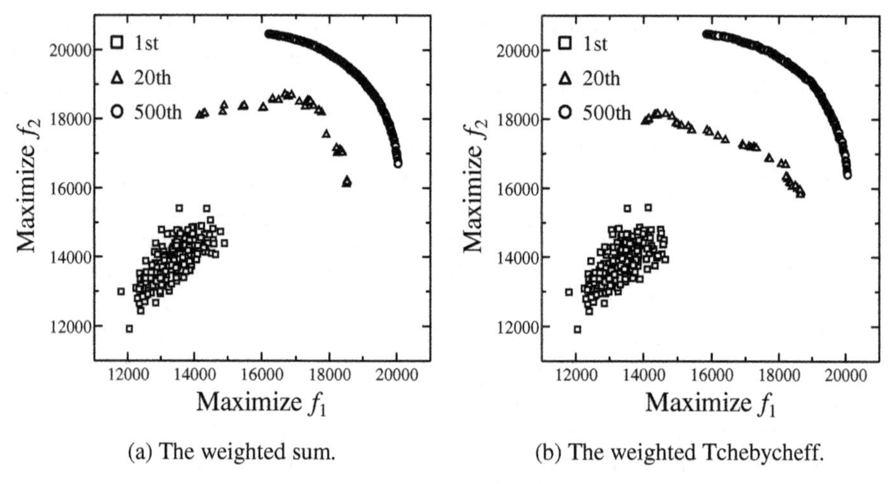

(a) The weighted sum. (b) The weighted Tchebycheff.

Fig. 2. Experimental results of a single run of MOEA/D with the weighted sum and the weighted Tchebycheff on the 2-500 knapsack problem of Zitzler and Thiele [26]

In Fig. 2, we cannot observe any clear difference in the final results at the 500th generation between Fig. 2 (a) with the weighted sum and Fig. 2 (b) with the weighted Tchebycheff. There exists, however, a clear difference in the intermediate results at the 20th generation between the two plots in Fig. 2. That is, we can see that the faster convergence speed was achieved by the weighted sum in Fig. 2 (a) than the weighted Tchebycheff in Fig. 2 (b) in the application of MOEA/D to the 2-500 test problem.

In Figs. 3-5, we compare the two scalarizing functions with each other using the hypervolume measure. Each figure shows the histogram of the values of the hypervolume measure obtained by 100 runs of MOEA/D with each scalarizing function. Whereas there is no clear difference between the two scalarizing functions for the 2-500 problem in Fig. 3 (also see Fig. 2), much better results were obtained by the weighted sum for the 4-500 problem in Fig. 4 and for the 6-500 problem in Fig. 5.

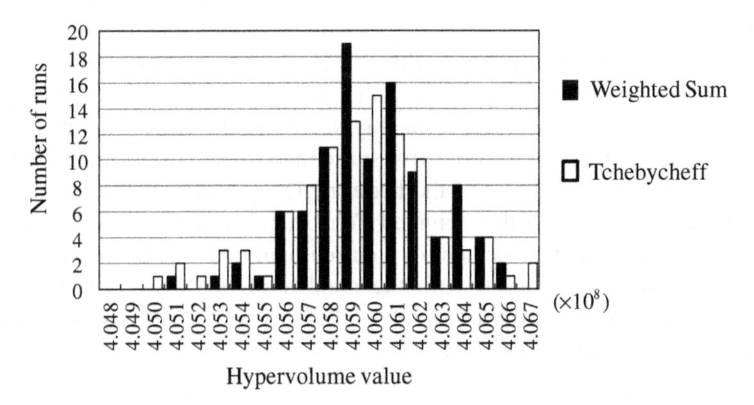

Fig. 3. Distribution of 100 values of the hypervolume measure (2-500 problem)

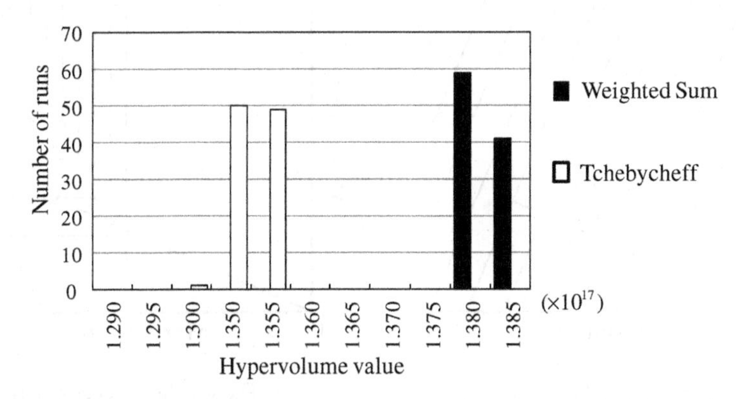

Fig. 4. Distribution of 100 values of the hypervolume measure (4-500 problem)

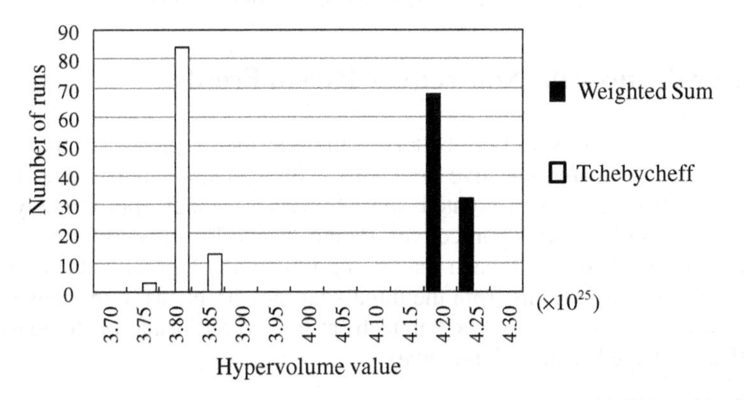

Fig. 5. Distribution of 100 values of the hypervolume measure (6-500 problem)

The deterioration in the performance of the weighted Tchebycheff by the increase in the number of objectives can be explained by the shape of contour lines of each scalarizing function. In the case of the weighted sum, a contour line is a line, plane or hyper-plane (see Fig. 6 (a) where the weight vector is assumed to be (0.7, 0.3)). The objective space is divided into two subspaces by a contour line. One subspace is a better region than the contour line while the other is a worse region. If a current solution is on the contour line, it is replaced with a newly generated offspring only when the offspring is in the better region. Roughly speaking, the size of the better region can be viewed as being 1/2 of the objective space independent of the number of objectives. On the other hand, the size of the better region is $(1/2)^m$ of the m-dimensional objective space in the case of the weighted Tchebycheff (see Fig. 6 (b)). This means that the probability of the replacement of a current individual with a newly generated offspring exponentially decreases with the number of objectives. As a result, the search ability of MOEA/D with the weighted Tchebycheff function was deteriorated by the increase in the number of objectives in Figs. 3-5.

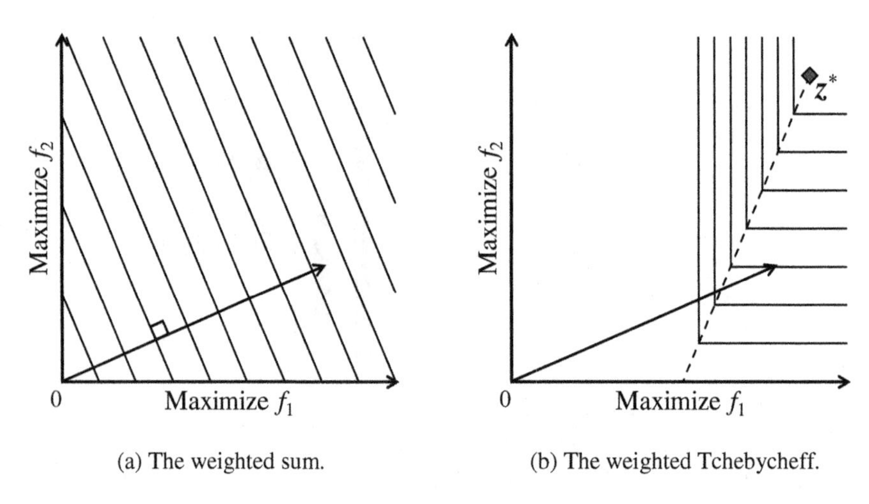

(a) The weighted sum.

(b) The weighted Tchebycheff.

Fig. 6. Contour lines of each scalarizing function

5 Test Problems with Non-convex Pareto Fronts

As shown in the previous section, better results are often obtained by the weighted sum especially for many-objective problems with convex Pareto fronts than the weighted Tchebycheff. The weighted sum, however, cannot appropriately handle multiobjective problems with non-convex Pareto fronts. In order to further examine the behavior of MOEA/D with each scalarizing function, we generated test problems with non-convex Pareto fronts from the three knapsack problems in the previous section. Our idea is to pull the convex Pareto front toward the inside of the feasible region so that its shape becomes non-convex.

Let G be a point in the feasible region of the objective space (see Fig. 7). We denote the location of the point G in the objective space by $(g_1, g_2, ..., g_m)$. Using this point, the value of each objective is modified for each individual $\boldsymbol{x}$ as

$$f_i(\boldsymbol{x}) := f_i(\boldsymbol{x}) - \min\{0, \min_{j=1,2,...,m} \{\alpha \cdot (f_j(\boldsymbol{x}) - g_j)\}\},\ i = 1, 2, ..., m, \tag{8}$$

where α is a user-definable positive parameter. By the modification in (8), all individuals dominating the point G are pulled toward the inside of the feasible region (i.e., toward the point G). Other individuals are not modified. The parameter α specifies the amount of the modification. In Fig. 7, the value of α is specified as $\alpha = 0.6$.

One important issue in (8) is the specification of the point G. We generated two types of test problems: Type C and Type D. In Type C, the point G is specified so that G is completely dominated by the Pareto front (see Fig. 7 (a)). As a result, the entire Pareto front becomes non-convex. On the other hand, the point G is specified so that G is dominated by a half of the Pareto front (see Fig. 7 (b)) in Type D. As a result, only the center region of the Pareto front becomes non-convex.

When the Pareto front of an original multiobjective problem is unknown, we can roughly estimate its range by applying an EMO algorithm to the problem and/or independently optimizing each objective. Then we can specify the point G based on the estimation of the range of the Pareto front. Due to the page limitation, we only report experimental results on the two test problems in Fig. 7 while we generated other test problems with non-convex Pareto fronts from the 4-500 and 6-500 problems.

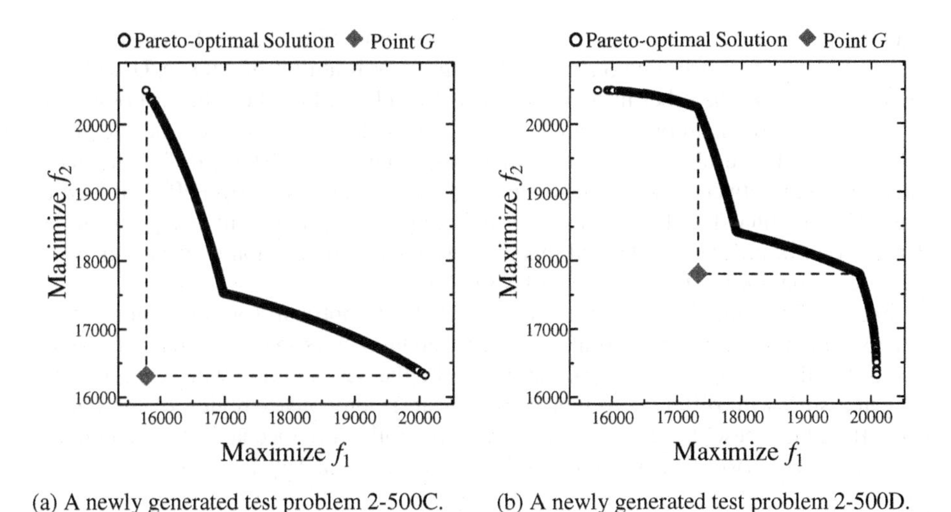

(a) A newly generated test problem 2-500C. (b) A newly generated test problem 2-500D.

Fig. 7. Two test problems generated from the 2-500 knapsack problem (α=0.6)

In the same manner as in the previous section, we applied MOEA/D to the 2-500C problem in Fig. 7 (a). Experimental results are shown in Fig. 8. Fig. 8 (a) clearly demonstrates that the weighted sum cannot appropriately handle non-convex Pareto

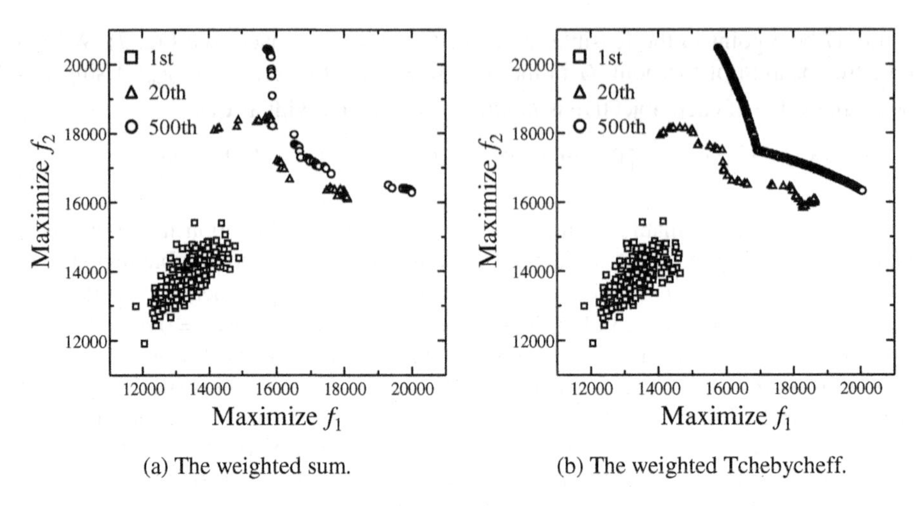

(a) The weighted sum.

(b) The weighted Tchebycheff.

Fig. 8. Experimental results of MOEA/D on the 2-500C test problem

fronts. All individuals in the main population converged to the two edges of the Pareto front whereas some other solutions were stored in the secondary population.

6 Our Idea for Automatically Choosing a Scalarizing Function

Our experimental results on many-objective problems with convex Pareto fronts in Fig. 4 and Fig. 5 clearly demonstrated the high search ability of MOEA/D with the weighted sum. On the other hand, experimental results in Fig. 8 on the two-objective problem with the non-convex Pareto front showed the advantage of the weighted Tchebycheff. These observations motivated us to devise a mechanism for automatically choosing between the weighted sum and the weighted Tchebycheff. Our idea is to use the weighted Tchebycheff only for individuals along non-convex regions of the Pareto front. In order to implement this idea, we have to detect non-convex regions of the Pareto front during the execution of MOEA/D.

Since the weighted sum cannot find any Pareto-optimal solutions along non-convex regions of the Pareto front, different weight vectors may have the same individual. In other words, a single individual can be the best solution with respect to the weighted sum with different weight vectors near a non-convex region of the Pareto front. Based on these discussions, we propose the following mechanism for automatically choosing between the weighted sum and the weighted Tchebycheff.

Proposed Idea: If a current individual $\boldsymbol{x}^k$ and at least K neighbors in its neighborhood $B(\boldsymbol{x}^k)$ have the same objective vector, we use the weighted Tchebycheff together with the weight vector $\boldsymbol{\lambda}^k$. Otherwise we use the weighted sum together with $\boldsymbol{\lambda}^k$.

In the proposed idea, K is a user-definable integer. In an extreme case of $K = 0$, the weighted Tchebycheff is always used. In the other extreme case with an infinitely large value of K, the weighted sum is always used.

7 Effectiveness of Our Idea

We applied MOEA/D with the proposed idea to the test problems in Fig. 7. We used four values 1, 2, 4, 9 as the threshold parameter K. In Fig. 9 and Fig. 10, we show experimental results on the 2-500C problem in Fig. 7 (a) and the 2-500D problem in Fig. 7 (b), respectively. We can see from Fig. 9 and Fig. 10 that the proposed idea works well for finding Pareto-optimal or near Pareto-optimal solutions along the non-convex regions of the Pareto fronts. The effect of the specification of the threshold parameter K is also clearly demonstrated in Fig. 9 and Fig. 10.

We also examined the performance of the proposed idea for the 2-500, 4-500, 6-500 and their C versions using the hypervolume measure as in Fig. 5. The weighted sum and the weighted Tchebycheff worked well for many objectives and non-convex Pareto fronts, respectively. Intermediate results between these two scalarizing functions were almost always obtained from the proposed idea.

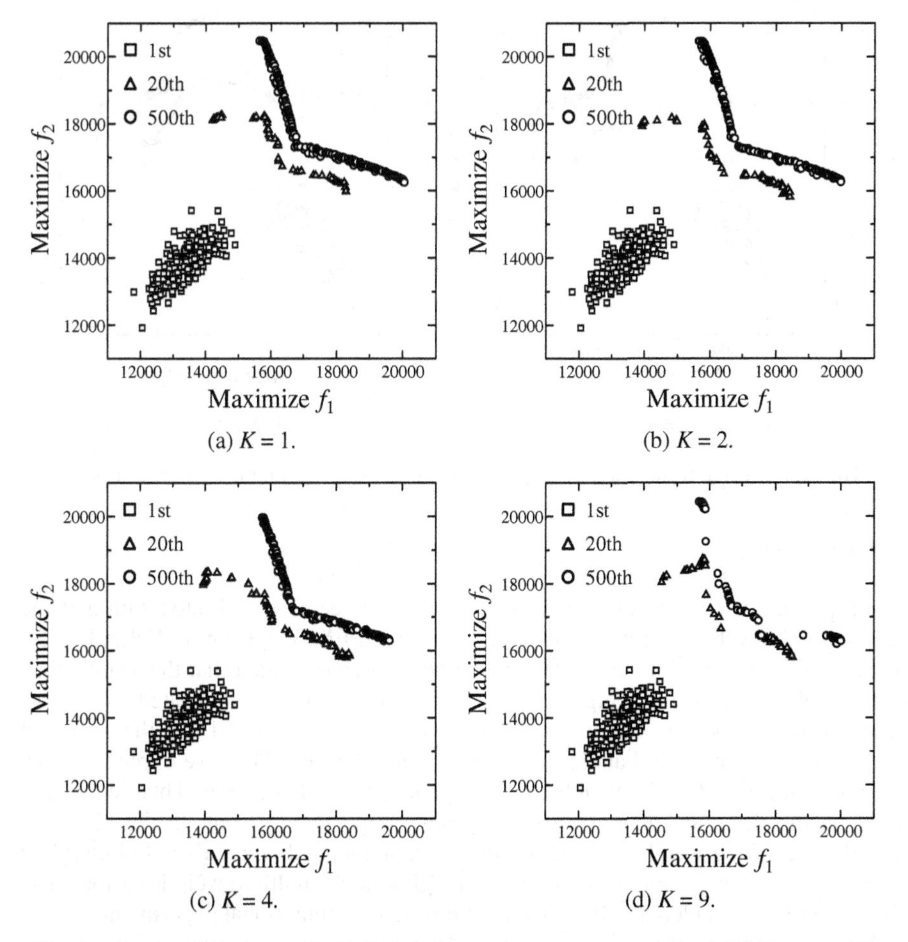

Fig. 9. Experimental results of MOEA/D with the proposed idea on the 2-500C test problem

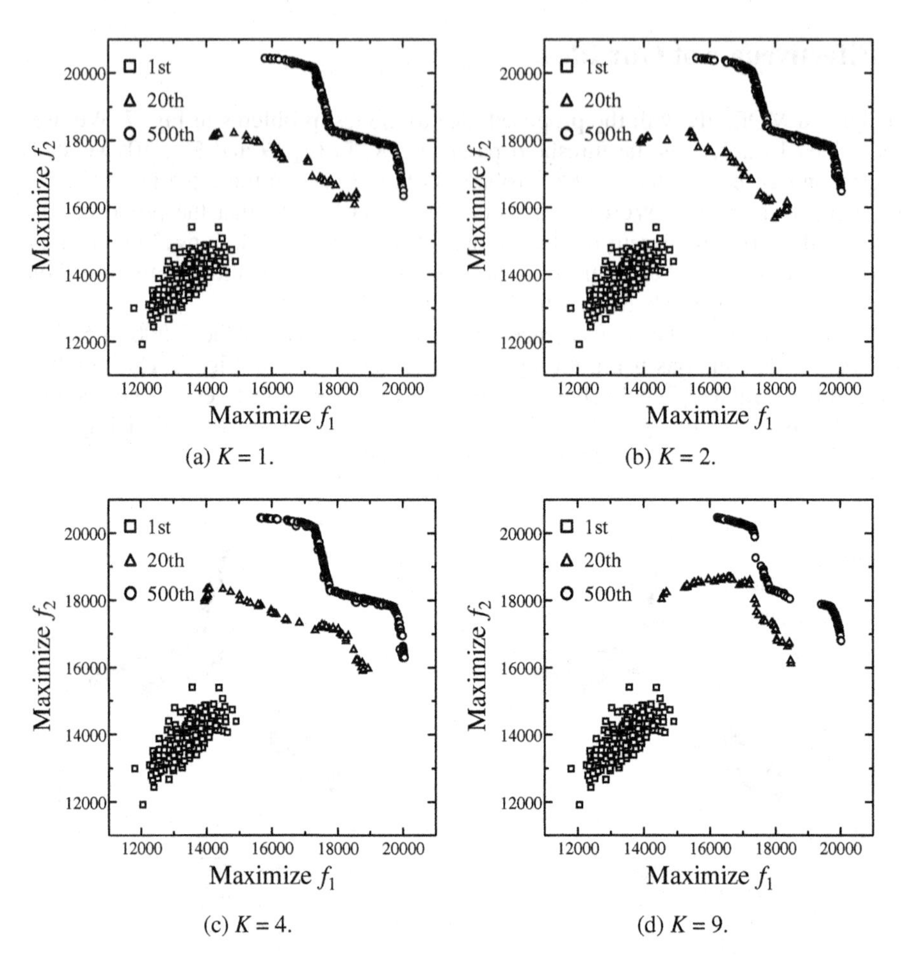

Fig. 10. Experimental results of MOEA/D with the proposed idea on the 2-500D test problem

8 Conclusions

We proposed an idea of automatically choosing between the weighted sum and the weighted Tchebycheff in a scalarizing function-based EMO algorithm MOEA/D [25]. To explain the motivation behind the proposal, we demonstrated that the search ability of MOEA/D with the weighted Tchebycheff is severely deteriorated by the increase in the number of objectives. We also visually demonstrated that the weighted sum cannot appropriately handle non-convex Pareto fronts. Then we explained how we can decide whether the weighted Tchebycheff is needed or not. This decision is based on the detection of non-convex regions of the Pareto front. In the proposed idea, the use of the weighted Tchebycheff is recommended when a single individual has the best weighted sum value among its neighbors for multiple weight vectors. The effectiveness of this idea was demonstrated through computational experiments.

We presented our idea in a very simple form. So it has a number of issues to be further discussed in future studies. One issue is the location of the reference point z^*

for the weighted Tchebycheff (we specified it by (5) in this paper). When only a part of the Pareto front is non-convex as in the 2-500D problem in Fig. 7 (b), it may be a good idea to locate the reference point z^* close to the non-convex region in order to concentrate the multiobjective search by the weighted Tchebycheff on the non-convex region. Another issue is the frequency of the change between the weighted sum and the weighted Tchebycheff. It seems that too frequent changes may have a negative effect on the search of MOEA/D. One critical issue is the possibility that the two scalarizing functions drive the current individual to totally different areas in the objective space even when they have the same weight vector. In this case, frequent changes may severely deteriorate the search ability of MOEA/D with our idea.

One possible remedy for such a negative effect of changing between the two scalarizing functions is to calibrate them so that they drive the current individual to the same area when they have the same weight vector. Another possible remedy is to use both the weighted sum and the weighted Tchebycheff as an integrated function. Actually we examined the use of the augmented weighted Tchebycheff in MOEA/D in our preliminary computational experiments. We observed high sensitivity of the search ability of MOEA/D with the augmented weighted Tchebycheff to the specification of the weight value for the augmented term. Different problems needed different specifications of the weight value. The point is how to adjust the weight for each problem. The proposed idea for detecting non-convex regions can be used in such an adjustment mechanism. The proposed idea can be also used for automatically specifying an appropriate objective function for each solution during the execution of local search in hybrid EMO algorithms (i.e., multiobjective memetic algorithms: MOMAs).

This work was partially supported by Grant-in-Aid for Scientific Research for Scientific Research (B) (20300084).

References

1. Branke, J., Schmeck, H., Deb, K., Reddy, M.: Parallelizing Multi-Objective Evolutionary Algorithms: Cone Separation. In: Proc. of 2004 Congress on Evolutionary Computation, pp. 1952–1957 (2004)
2. Chang, P.C., Chen, S.H., Zhang, Q., Lin, J.L.: MOEA/D for Flowshop Scheduling Problems. In: Proc. of 2008 Congress on Evolutionary Computation, pp. 1433–1438 (2008)
3. Coello, C.A.C.: Evolutionary Multi-Objective Optimization: A Historical View of the Field. IEEE Computational Intelligence Magazine 1, 28–36 (2006)
4. Deb, K.: Multi-Objective Optimization Using Evolutionary Algorithms. John Wiley & Sons, Chichester (2001)
5. Deb, K., Pratap, A., Agarwal, S., Meyarivan, T.: A Fast and Elitist Multiobjective Genetic Algorithm: NSGA-II. IEEE Trans. on Evolutionary Computation 6, 182–197 (2002)
6. Gandibleux, X., Ehrgott, M.: 1984-2004 – 20 Years of Multiobjective Metaheuristics. But What About the Solution of Combinatorial Problems with Multiple Objectives? In: Coello Coello, C.A., Hernández Aguirre, A., Zitzler, E. (eds.) EMO 2005. LNCS, vol. 3410, pp. 33–46. Springer, Heidelberg (2005)
7. Hughes, E.J.: Multiple Single Objective Sampling. In: Proc. of 2003 Congress on Evolutionary Computation, pp. 2678–2684 (2003)
8. Hughes, E.J.: Evolutionary Many-Objective Optimization: Many Once or One Many? In: Proc. of 2005 Congress on Evolutionary Computation, pp. 222–227 (2005)

9. Hughes, E.J.: MSOPS-II: A General-Purpose Many-Objective Optimiser. In: Proc. of 2007 Congress on Evolutionary Computation, pp. 3944–3951 (2007)
10. Ishibuchi, H., Doi, T., Nojima, Y.: Incorporation of Scalarizing Fitness Functions into Evolutionary Multiobjective Optimization Algorithms. In: Runarsson, T.P., Beyer, H.-G., Burke, E.K., Merelo-Guervós, J.J., Whitley, L.D., Yao, X. (eds.) PPSN 2006. LNCS, vol. 4193, pp. 493–502. Springer, Heidelberg (2006)
11. Ishibuchi, H., Murata, T.: A Multi-Objective Genetic Local Search Algorithm and Its Application to Flowshop Scheduling. IEEE Trans. on Systems, Man, and Cybernetics - Part C: Applications and Reviews 28, 392–403 (1998)
12. Ishibuchi, H., Tsukamoto, N., Hitotsuyanagi, Y., Nojima, Y.: Effectiveness of Scalability Improvement Attempts on the Performance of NSGA-II for Many-Objective Problems. In: Proc. of 2008 Genetic and Evolutionary Computation Conference, pp. 649–656 (2008)
13. Ishibuchi, H., Tsukamoto, N., Nojima, Y.: Evolutionary Many-Objective Optimization: A Short Review. In: Proc. of 2008 Congress on Evolutionary Computation, pp. 2424–2431 (2008)
14. Ishibuchi, H., Yoshida, T., Murata, T.: Balance between Genetic Search and Local Search in Memetic Algorithms for Multiobjective Permutation Flowshop Scheduling. IEEE Trans. on Evolutionary Computation 7, 204–223 (2003)
15. Jaszkiewicz, A.: Genetic Local Search for Multi-Objective Combinatorial Optimization. European Journal of Operational Research 137, 50–71 (2002)
16. Jaszkiewicz, A.: On the Performance of Multiple-Objective Genetic Local Search on the 0/1 Knapsack Problem - A Comparative Experiment. IEEE Trans. on Evolutionary Computation 6, 402–412 (2002)
17. Jaszkiewicz, A.: On the Computational Efficiency of Multiple Objective Metaheuristics: The Knapsack Problem Case Study. European Journal of Operational Research 158, 418–433 (2004)
18. Khara, V., Yao, X., Deb, K.: Performance Scaling of Multi-objective Evolutionary Algorithms. In: Fonseca, C.M., Fleming, P.J., Zitzler, E., Deb, K., Thiele, L. (eds.) EMO 2003. LNCS, vol. 2632, pp. 376–390. Springer, Heidelberg (2003)
19. Li, H., Zhang, Q.: Multiobjective Optimization Problems with Complicated Pareto Sets, MOEA/D and NSGA-II. IEEE Trans. on Evolutionary Computation (in press); available from IEEE Xplore
20. Miettinen, K.: Nonlinear Multiobjective Optimization. Kluwer Academic Publishers, Boston (1998)
21. Murata, T., Ishibuchi, H., Gen, M.: Specification of Genetic Search Directions in Cellular Multi-objective Genetic Algorithms. In: Zitzler, E., Deb, K., Thiele, L., Coello Coello, C.A., Corne, D.W. (eds.) EMO 2001. LNCS, vol. 1993, pp. 82–95. Springer, Heidelberg (2001)
22. Purshouse, R.C., Fleming, P.J.: Evolutionary Many-Objective Optimization: An Exploratory Analysis. In: Proc. of 2003 Congress on Evolutionary Computation, pp. 2066–2073 (2003)
23. Sato, H., Aguirre, H.E., Tanaka, K.: Local Dominance and Local Recombination in MOEAs on 0/1 Multiobjective Knapsack Problems. European Journal of Operational Research 181, 1708–1723 (2007)
24. Sindhya, K., Deb, K., Miettinen, K.: A Local Search Based Evolutionary Multi-Objective Optimization Approach for Fast and Accurate Convergence. In: Rudolph, G., Jansen, T., Lucas, S., Poloni, C., Beume, N. (eds.) PPSN 2008. LNCS, vol. 5199, pp. 815–824. Springer, Heidelberg (2008)
25. Zhang, Q., Li, H.: MOEA/D: A Multiobjective Evolutionary Algorithm Based on Decomposition. IEEE Trans. on Evolutionary Computation 11, 712–731 (2007)
26. Zitzler, E., Thiele, L.: Multiobjective Evolutionary Algorithms: A Comparative Case Study and the Strength Pareto Approach. IEEE Trans. on Evolutionary Computation 3, 257–271 (1999)

Combining Aggregation with Pareto Optimization: A Case Study in Evolutionary Molecular Design

Johannes W. Kruisselbrink[1], Michael T.M. Emmerich[1], Thomas Bäck[1], Andreas Bender[2], Ad P. IJzerman[2], and Eelke van der Horst[2]

[1] LIACS, Leiden University, Niels Bohrweg 1, Leiden, NL
[2] LACDR, Leiden University, Einsteinweg 55, Leiden, NL

Abstract. This paper is motivated by problem scenarios in automated drug design. It discusses a modeling approach for design optimization problems with many criteria that can be partitioned into objectives and fuzzy constraints. The purpose of this remodeling is to transform the original criteria such that, when using them in an evolutionary search method, a good view on the trade-off between the different objectives and the satisfaction of constraints is obtained.

Instead of reducing a many objective problem to a single-objective problem, it is proposed to reduce it to a multi-objective optimization problem with a low number of objectives, for which the visualization of the Pareto front is still possible and the size of a high-resolution approximation set is affordable. For design problems where it is reasonable to combine certain objectives and/or constraints into logical groups by means of desirability indexes, this method will yield good trade-off results with reduced computational effort. The proposed methodology is evaluated in a case-study on automated drug design where we aim to find molecular structures that could serve as estrogen receptor antagonists.

1 Introduction

Using Pareto optimization for multi-objective optimization problems becomes problematic when the number of objectives increases. There are at least two reasons for this. Firstly, for dimensions higher than three or four the visualization of the Pareto front, and thus of the trade-off between attainable non-dominated solutions, becomes increasingly difficult. Secondly, the memory space needed for maintaining a high resolution approximation set of the Pareto front can grow exponentially with the number of objectives.

One possibility in such cases is to retreat to an aggregation approach and thereby to renounce the advantages of Pareto front visualization and exploration, such as getting intuition about the nature of the trade-offs and conflicts and the possibility to generate a diverse set of alternatives that can then be discussed by the design experts on the basis of more subjective criteria.

This paper suggests a middle ground between Pareto optimization with many objectives and aggregation to a single objective problem. The idea is to apply partial aggregation of the objectives in order to recast a many objectives

M. Ehrgott et al. (Eds.): EMO 2009, LNCS 5467, pp. 453–467, 2009.

optimization problem into an optimization problem with a moderate number of objectives. This approach is suitable for cases where the objectives can be grouped into different categories and aggregates can be build for each category. In particular we are interested in problems for which a large number of objectives relate to so-called fuzzy constraints.

Fuzzy constraints arise in decision models where the constraint boundary is not defined in a precise way. However, there is often a region where a constraint is fully satisfied and a region where a constraint is strictly violated. In the latter case, a solution will be clearly inferior to any solution with non-violated constraints, regardless of its objective function values. In the 'gray area', where the violation of a constraint is debatable, the degree of violation can serve as an objective to be minimized.

In order to normalize the objective space, both objectives and fuzzy constraints can be transformed into desirability functions ([6,5,17]). They are an intuitive way for the decision maker for indicating how desirable certain objective values are, using the same scale for each objective. Moreover, they can be used, as we will discuss later, to model fuzzy constraints. Several desirability functions can be aggregated by their product. Desirability has been considered earlier in the context of evolutionary multi-objective optimization by Trautmann and Mehnen in a theoretical study [14] and industrial application [15]. They are used to integrate a-priori preferences of the decision maker, allowing to focus on interesting parts of the Pareto front. In that respect they are closely related to Guided Dominance methods by Branke and Deb [2].

In contrast to the previous mentioned approaches, we propose the use of desirability functions to model fuzzy constraints and reduce the number of objectives in many-objective optimization problems. For this, we group objectives and constraints in logically separable groups, among which the set of fuzzy constraints forms one group, using desirability indexes. Then we perform Pareto optimization on the lower dimensional, normalized objective space the aggregated desirability indexes give rise to. Using the proposed remodeling method in a multi-objective evolutionary search scheme will reduce the computational effort that is due to maintaining the non-dominated solution set and it will yield a good view on the trade-offs.

In a case study, an automated drug design problem, the aim is to find estrogen receptor antagonists using three activity prediction models. This problem has a large number of criteria which can be partitioned into fuzzy or soft constraints, such as the solubility of a candidate molecule, and typically a small set of objectives which in our case are the maximization of predicted activities on certain targeted cells. This study extends the work presented in [8,13].

The paper starts with a short definition of the general form of the multi-objective optimization problems that we consider in this paper in Section 2. In Section 3 the concept of desirability indexes and related work is reviewed. Section 4 discusses ways to combine desirability indexes with Pareto optimization. Section 5 provides preliminaries for the case study. Then, in Section 6 we present the results of the proposed approaches on the drug design problem. The article ends with a summary and outlook in Section 7.

2 Problem Class Description

We consider multi-objective design optimization problems consisting of a small number of objectives and a number of fuzzy or soft constraints:

$$\begin{array}{ll} \min f_i(x), & i = 1, \ldots, N \\ s.t.\ \ g_j(x) \gtrsim 0, & j = 1, \ldots M \end{array} \tag{1}$$

Here, x can be of any domain D and $\gtrsim$ is a fuzzified version of $\geq$. It has the linguistic interpretation $g_j(x)$ is essentially bigger than 0 (see [7]). To express this linguistic variable we have to define a membership function, which in the setting of the constraints of (1) will be a one-sided desirability function for maximization. We postpone the discussion of details to section 3.

The aim is not only to provide one optimal solution but rather to provide a broad variety of solutions as design options which are both different in the search space as well as representing different trade-offs in the objective space. Moreover, we do not necessarily want to reject solutions that do not strictly violate constraints, but rather give insight in the trade-off between the objectives and the degree of constraint satisfaction.

For many design optimization problems, the aim as we have stated here is more realistic than those usually considered in optimization approaches. Diversity is a necessary criterion because for most design optimization problems, the mathematical methods of expressing fitness functions are limited, and having a high fitness score does not necessarily guarantee high performance. It is therefore desirable to provide a diverse set of possible candidate solutions.

The desire also to show solutions that are very good with respect to the objective functions but which do not satisfy the constraints occurs because in design problems one should be aware of the fact that there is a designer involved who will evaluate these results. In many cases, a designer is also interested in these high quality solutions, although they do not comply to (all of) the constraints. This can happen, for example, when the designer might think he/she will be able to fix such constraint violations or when the performance gain outweighs the constraint violations (soft constraints). This work emphasizes on the second case, where the boundary between constraint violation and satisfaction is fluent.

3 Multiobjective Optimization and Desirability Functions

This section will review two general solution concepts for the discussed class of multiobjective optimization (that we will combine later): Pareto optimization and desirability functions.

3.1 Pareto Optimization

A common way to solve problems with multiple conflicting objectives is to search for the set of all non-dominated solutions, also called the Pareto front. In cases where the full Pareto front cannot be obtained, an approximation set of it can be computed, which ideally should consist only of non-dominated solutions and should cover the entire Pareto front with a high resolution.

This Pareto front (approximation) can be used by the decision maker or systems designer to get an intuition about the nature of the conflict and the trade-off between various objectives. Based on this he/she can come to an informed decision on what good compromise solutions could be.

However, for optimization techniques like evolutionary algorithms, this approach is considered to be limited to a small number of objectives, as the number of solutions needed to approximate the Pareto front with a high resolution can grow exponentially with the number of objectives. For example, given a continuous problem with m objectives, in the non-degenerate case the Pareto front is a $m-1$ dimensional manifold and we need $\prod_{i=1}^{m-1}(w_i/\epsilon)$ points to capture a Pareto front [9]. Here we assume that the ith objective has a range of w_i and ϵ is the desired maximal distance between points in the Pareto front approximation which determines the resolution. In addition, for a high number of objectives it is difficult to get a visual impression of the Pareto front shape which limits the benefit of the approach as a way to guide the intuition of the decision maker.

In conclusion; a full spread Pareto front approximation is useful, but in many cases only feasible for less than 5 objectives.

3.2 Desirability Indexes

The concept of desirability was first introduced by Harrington [6] in the scope of industrial quality management. Harrington considered the optimization of a process or a product with respect to a set of design variables $X = x_1, \ldots, x_n$. The behavior of the process or product was assumed to be mapped to a number of quality criteria $Y = y_1, \ldots, y_m$ with $y_i = f_i(X)$ which denote its qualitative properties. The goal was to reach certain levels of quality, which could either be to *minimize* or *maximize* the performance level or to *reach a specified target.*

In order to allow for a better comparison between the differently scaled quality criteria with respect to their desired levels Harrington proposed to map the quality criteria to the open unit interval (0; 1), where a value close to zero stands for 'poor quality' whereas a value close to 1 stands for 'high quality'. This mapping was based on target values (desired values) for the non-normalized quality levels, and the functions yielding such a mapping were named desirability functions. Desirability functions of the Harrington type have a signature:

$$d_i(y_i) : \mathbb{R} \to (0, 1) \, , \, i = 1, \ldots, m \qquad (2)$$

Harrington distinguished between one-sided desirability functions and two-sided desirability functions. The former are used to model objectives to be maximized or minimized and are represented by monotonuous functions. The two-sided desirability functions reflect the cases where a target value for the quality levels has to be approached as close as possible. The general form of the desirability functions of Harrington is that of an exponential function. We refer to [6] for the mathematical description of this mapping for both the one-sided and the two-sided quality criteria.

An extension to the work of Harrington was later proposed by Derringer [5]. Derringer proposed to allow to express also for which values an objective is fully satisfactory (setting the value of the desirability to its maximum) or when a fuzzy constraint is strictly violated (setting the value of the desirability index to 0). This idea is sketched in Fig. 1.

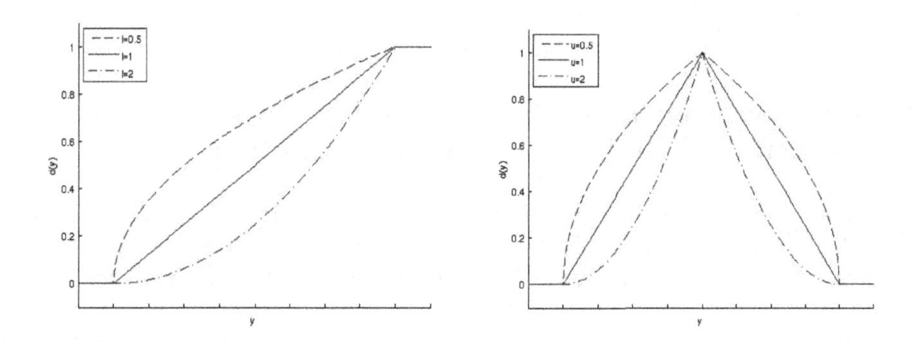

Fig. 1. Derringer desirability functions

A useful property of desirability functions is that they can easily be aggregated into one single quality value, named a desirability index, which can be used directly for optimization. Harrington used the geometric mean for aggregation:

$$\max_{x_1,\ldots,x_n} D(x_i,\ldots,x_n) = \sqrt[m]{\prod_{i=1}^{m} d_i(f_i(x_1,\ldots,x_n))} \tag{3}$$

For multi-objective optimization, this yields an easy aggregation of all objectives into one objective function. For the two-sided desirability functions, optimal solutions are also Pareto optimal with respect to the original problem [17].

Aggregation is not a new concept in multi-objective optimization and dates back to [12]. However the proposed desirability indexes allow for a more controlled way of aggregating different objective functions than, for instance, the weighted sum approach. As opposed to the latter it is also possible to obtain points in concave parts of the Pareto front and because of the product used in the aggregation it tends to focus more on the central part of a Pareto front where good compromises are located. Some of the classical utility function classes also use product forms, but they lack normalization and the flexibility in using one- and two-sided functions that will be essential in the methods we will propose.

Also it is much more efficient for convergence to a single non-dominated solution as the linear order allows for a more focused search. However, the downside of using a desirability index is that it yields a much smaller spread of the solutions with respect to the Pareto front than would be obtained by applying Pareto optimization with e.g. an ϵ-archive.

In conclusion; aggregating the objectives into one objective function by means of desirability functions and a desirability index is a good way of reducing the number of objectives and is especially useful when many objectives make Pareto optimization impossible. However, it may focus too much on particular areas of the Pareto front.

As a compromise between scalarization and many-objective optimization we will propose to remodel the many-objective optimization problem to a multi-objective problem with a low number of objectives.

4 Incorporating Desirability Functions in Fuzzy Constrained Multi-Objective Optimization

In this section we turn back to the problem class description in section 2 and we will propose two approaches which will remodel the objectives and fuzzy constraints into desirability functions and combine Pareto optimization with desirability indexes. By combining Pareto optimization with desirability indexes, a method will be created which can visualize the trade-off between the different objectives very well without requiring large population sizes.

4.1 Remodeling the Objectives and Fuzzy Constraints

For both approaches, we first remodel the original problem to a normalized problem by mapping the objective functions and the constraints to $[0, 1]$ desirability intervals which need to be maximized. When we consider to have objective functions that need to be minimized, this yields a transformation for the objective functions f_i, $i = 1, \ldots, M$ that looks like this:

$$\hat{f}_i(x) = \exp\left(d_i \cdot (f_i^* - f_i(x))\right) \rightarrow \max \text{ , } i = 1, \ldots, M \tag{4}$$

Here, f^* denotes the fitness value of the global minimum or an estimation of it which should be smaller than the real global minimum. The parameter d_i can be used for scaling purposes. Note that also a different mapping might be used, but this mapping has all the desired properies: It increases monotonously as f_i approaches the minimum, it obtains values in the $[0, 1]$-interval, and its smoothness can be easily controlled by means of the parameter d_i.

The constraints that are usually considered are of (or can be reduced to) one of these forms:

$$A_j \lesssim g_j(x) \text{ or } g_j(x) \lesssim B_j \text{ or } A_j \lesssim g_j(x) \lesssim B_j \tag{5}$$

As noted, the relation $A \lesssim B$ reads A is essentially smaller than B (fuzzy constraints). Similar to the flexible programming approach suggested by Inuiguchi et al. [7] in the context of mathematical programming, we will remodel the constraints of (5) to become more of a quality score function (an objective or, in the terminology of flexible progamming a *fuzzy goal*). Instead of fuzzy membership functions [7], we make use of Derringer and Suich's [5] type of desirability functions, often simply referred to as Derringer desirability functions. We split up the constraints into three types of regions: rejection regions, gray areas, and acceptance regions.

Using this principle, we model the constraints with Derringer desirability functions which map each constraints to an interval $[0,1]$. Doing so will yield 0 for strictly violated constraints, 1 for fully satisfied constraints, and the gray area yielding values between 0 and 1 indicating a degree of satisfaction on a linear scale. The leftmost case in (5) translates to:

$$d_j(x) = \begin{cases} 0 & , g_j < \mathrm{LB}_j \\ \left(\frac{g_j(x)-\mathrm{LB}_j}{A_j-\mathrm{LB}_j}\right)^{l_j} & , \mathrm{LB}_j <= g_j(x) < A_j \\ 1 & , g_j(x) >= A_j \end{cases} \tag{6}$$

Here, LB_j denotes the absolute lower rejection bound determining the rejection region, the area between LB_j and A_j denotes the gray area, and everything beyond A_j is accepted. The parameter l_j is used to control the shape of the curve that maps the gray region to the interval $[0,1]$. Having $l_j = 1$ yields a linear curve, $l_j > 1$ generates a convex curve that will be relatively mild on the solutions in the gray area, and $l_j < 1$ yields a convex shape which will be more strict on the solutions in the gray area.

Using the same principles we obtain for the second case of (5):

$$d_j(x) = \begin{cases} 1 & , g_j(x) <= B_j \\ \left(\frac{g_j(x)-\mathrm{UB}_j}{B_j-\mathrm{UB}_j}\right)^{u_j} & , B_j < g_j(x) <= \mathrm{UB}_j \\ 0 & , g_j(X) > \mathrm{UB}_j \end{cases} \tag{7}$$

Here UB_j is the upper rejection bound beyond which everything is rejected, and u_j is the minimization counterpart of l_j.

For the third case we use a slight adaptation of the two-sided Derringer desirability functions in order to map values that are in the acceptance region to a value of 1. This yields the following two sided desirability function:

$$d_j(x) = \begin{cases} 0 & , g_j < \mathrm{LB}_j \\ \left(\frac{g_j(x)-\mathrm{LB}_j}{A_j-\mathrm{LB}_j}\right)^{l_j} & , \mathrm{LB}_j <= g_j(x) < A_j \\ 1 & , A_j <= g_j(x) <= B_j \\ \left(\frac{g_j(x)-\mathrm{UB}_j}{B_j-\mathrm{UB}_j}\right)^{u_j} & , B_j < g_j(x) <= \mathrm{UB}_j \\ 0 & , g_j(X) > \mathrm{UB}_j \end{cases} \tag{8}$$

Figure 1 illustrates the mapping of the original constraint function $g_j(x)$ to a Deringer desirability function.

Having mapped all objectives and constraints to desirability functions, they can now be combined in any logical kind of way into desirability indexes. Moreover, where we want a more specific view on the trade-off between different groups of objectives and/or constraints, we can choose to keep them separate and perform Pareto optimization on those groups as separate groups. Of the multiple schemes that can be thought of we will propose two methods for doing so.

4.2 Approach 1

Every objective function f_i is mapped to a function $\hat{f}_i$ using (4) and all objective functions are combined into one objective function by taking the product:

$$f_{\text{objectives}} = \hat{f}_1 \cdot \hat{f}_2 \cdot \ldots \cdot \hat{f}_N \tag{9}$$

The constraints are modeled by desirability functions (we suggest to take $u_j = l_j = \frac{1}{2}$ in order to be mild on the constraints) aggregated in a second objective function:

$$f_{\text{constraints}} = \hat{g}_1 \cdot \hat{g}_2 \cdot \ldots \cdot \hat{g}_M \tag{10}$$

This yields a two objective optimization problem instead of the original $M \times N$ objectives (we also count the converted fuzzy constraints) while still preserving the possibility to get a view on a part of the trade-offs of the multi-objective optimization problem.

4.3 Approach 2

The objectives are kept separate and only the constraints are combined using desirability indexes as in approach 1. This yields an $N + 1$ Pareto optimization problem. It will allow for getting a better view on the trade-offs between the different objectives, but will also be computationally more challenging and is only possible when at least the number of objectives is low (between 2 and 4). Color can be used to visualize the value of the fourth dimension ($f_{\text{constraints}}$).

5 Evolutionary Algorithms in Molecular Design

This research is based on a problem that originates from the application of evolutionary algorithms for automated drug design. In this section we will briefly describe the problem that will be used as a case study for the proposed approaches; the search for Estrogen Receptor Antagonists.

Evolutionary Algorithms for automated drug design are increasingly gaining on practical applicability. This can be attributed to an increasing quality of methods to predict the biological and pharmacological properties of molecules, as well as increased knowledge in the evolutionary algorithms field on how to deal with the complex search spaces induced by the graph-like structures of molecules. Successful recent approaches are described in [19,8,16,13].

Although the use of computational techniques is taking over in all parts of the drug design process, in the end it is the intuitive knowledge of the medicinal chemist that is necessary in order to distinguish the promising candidate molecules from the 'obviously' bad solutions. For the medicinal chemist, such a distinction is very easily made, but based on intuition and heuristics which are very difficult to capture in any kind of objective function.

For automated drug design methods it is very important to realize this as it changes the objective of the search. The objective should not be to obtain only a single, optimal solution, corresponding to a narrowly defined class of molecular structures, but rather to provide a broad set of solutions which can be used as lead components in the experimental follow-up. In practice, it is well-known that drugs often fail during research and development stages. Providing multiple different starting points increases the chance of finding a class of compounds that will succeed into more advanced drug development stages.

5.1 Test Case: Search for Estrogen Antagonists

As test case we consider the search for estrogen receptor antagonists. As objective functions we use three different activity prediction models (support vector machine models) which predict the quality of molecules with respect to being an antagonist for the estrogen receptor, based on molecular similarity [1] to a set of steroidal as well as non-steroidal reference compounds:

- f_1: activity prediction of an SVM based on ECFP6 fingerprints [18]
- f_2: activity prediction of an SVM based on AlogP2 Estate Counts [18]
- f_3: activity prediction of an SVM based on MDL [18]

As the output values of the activity models are values between 0 and 1 (1 = active), we are lucky that the objective functions do not have to mapped to the interval $[0, 1]$ in order to be merge-able into a desirability index.

5.2 Constraints: Drug-Likeness and the Lipinski Rules

Besides the fact that molecules should be active against a certain target, very important boundary conditions are that they should have reasonable structures and they should be usable as drugs (drug-like [10]). When applying evolutionary search for automated molecular design a major problem is that there is a tendency to generate molecular structures that look very weird and very wrong to the medicinal chemist. In order to limit the search only to solutions that can be considered as feasible drug molecules, a number of constraints can be used which can indicate the feasibility of a candidate molecule.

There are many possible ways that can be used to get an indication of the feasibility of candidate molecular structures. The way that we have adopted in our approach is to use fuzzy bounds for a number of relatively easy calculatable properties together with boundary values which are chemically reasonable (based on the drug-likeness properties suggested by Lipinski [11]):

Descriptor	LB	A	B	UB	Descriptor	LB	A	B	UB
Num H-acceptors	0	1	6	10	Num H-donors	0	1	3	5
Molecular solubility	-6	-4	∞	∞	Molecular weight	150	250	450	600
ALogP	0	1	4	5	Minimized energy	0	0	80	150

6 Experiments, Results and Discussion

The two proposed approaches are tested together with one other approach (that will be referred to as approach 3) which is the aggregation of all objectives and constraints into one desirability index as objective function that we will use for benchmarking:

$$f = \hat{f}_1 \cdot \hat{f}_2 \cdot \ldots \cdot \hat{f}_N \cdot \hat{g}_1 \cdot \hat{g}_2 \cdot \ldots \cdot \hat{g}_M \qquad (11)$$

The **experimental setup** is as follows: The three approaches will be used in an adapted version of the evolutionary algorithm that was presented in [13]. The modifications are that for approach 1 and approach 2, a NSGA-II [4] selection scheme is used, and approach 3 implements a single objective selection mechanism. The genetic operators for the graph-like structures of the molecules are the same as used in [13]. Furthermore, the following **parameter settings** are adopted: parent population size: 80, number of offspring: 120, number of generations: 1000 (in compliance with previous studies). For each approach we run the algorithm for 4 runs.

6.1 Results and Discussion

The most illustrative results can be found in the 4D plots of Fig. 2, which visualize the obtained Pareto front approximations of the three approaches.

Approach 1 yields a very broad range of different design options, approach 2 shows a slightly less broad Pareto front approximation, and approach 3 yields solutions that lie very close together in the objective space. Although this is something that is to be expected, the difference between the single objective aggregation and the combination of aggregation and Pareto optimization can be noted as remarkable. Also, the difference between approach 1 and approach 2 seems much smaller than the difference between approach 2 and approach 3.

Hence, it can be argued that the gain in diversity (seen from the objectives) is much higher going from one to two objectives than when going from two to four objectives. These results support the main argument of this paper that Pareto optimization is good, but with a further increasing number of objectives, the upkeep becomes more difficult while the gain in diversity is only marginal.

Another noteworthy observation is that in all four runs, approach three converged to solutions that scored very high on the constraints (almost all solutions in the final set had a constraint value of 1 or close to 1). From the search space, this can be explained by the fact that many molecular structures can be constructed which comply to the constraint and that the constraint scores are much easier to optimize than the other objectives. This results in a rapid convergence to feasible solutions which closes down the road to less feasible solutions that have better objective function values.

This is opposed to approach 1 and 2, where a broad set of trade-off solutions is provided. Moreover, when looking at run three of approach 2, solutions were found that were very good on the objectives but which did not perform well on the constraints. Interestingly, when we look at the cluster of solutions in Fig. 3,

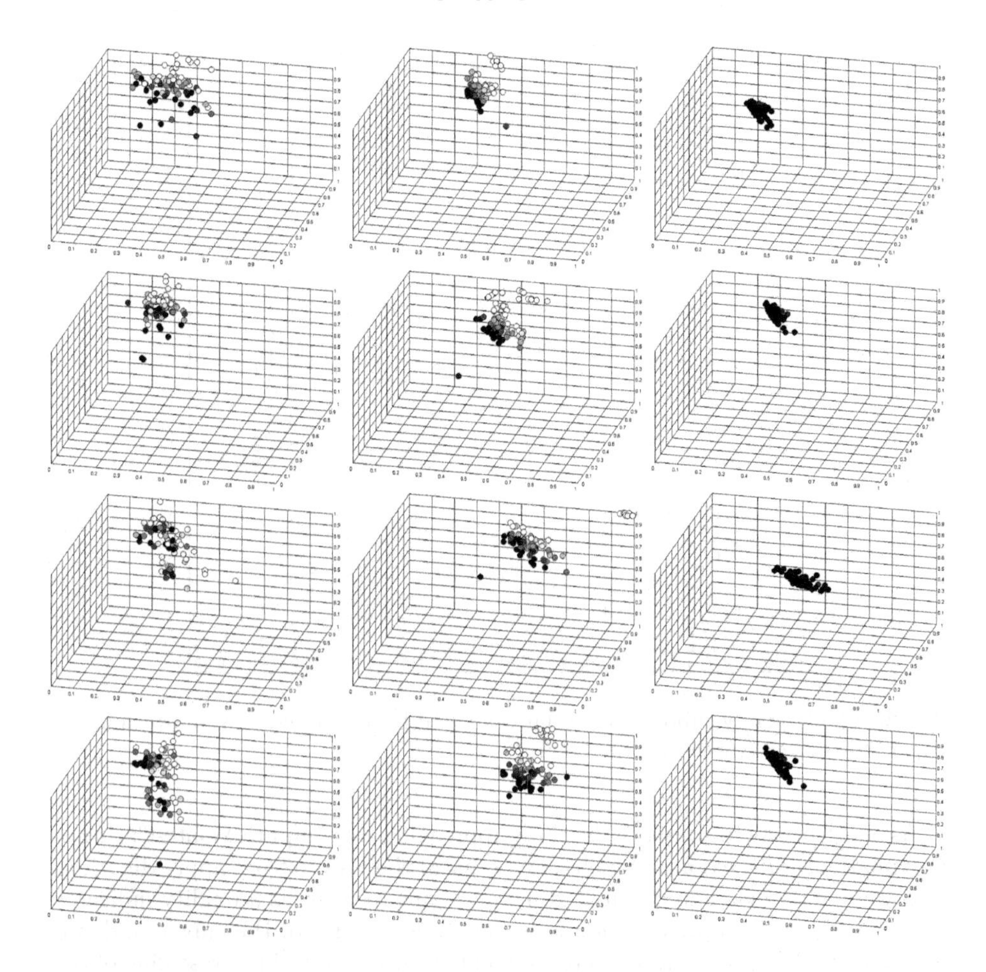

Fig. 2. Resulting Pareto front approximations after 1000 generations of runs of the three methods. The left column shows the runs of approach 1, the middle column show the runs of approach 2, and the right column shows the runs of approach 3. The three objective functions (f_1, f_2, and f_3) are plotted on the x, y and z axis. The fourth dimension, the constraint scores computed with (10), is visualized by the grayscale (scaling from white = strictly violated to black = accepted).

we recognize the first solution of this subset to be the well known and marketed drug Tamoxifen which is an Estrogen Receptor Antagonist used against breast cancer. This is one of those molecules that we surely want to find, but which would not have been found with the single objective aggregation because it simply does not comply to the (heuristic based) constraints. Especially in the field of molecular design, there is no clear way to distinguish good molecules from bad molecules, and we should account for these cases were high quality solutions that do not comply to the usual heuristics can still be very promising. We state that this also holds for many other design optimization problems.

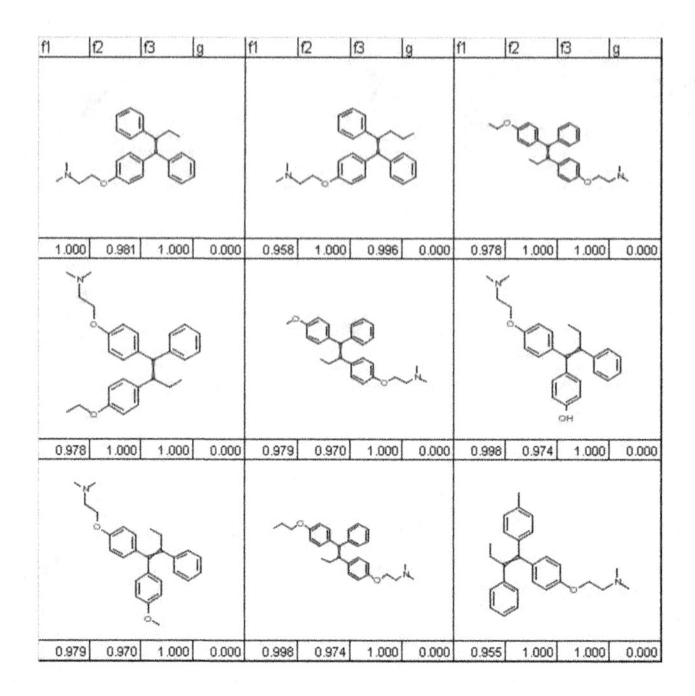

Fig. 3. A high quality subset of the solutions found by approach 2 in run 3 with in the upper left the very succesful drug known as Tamoxifen

As another way to compare the three approaches we compute Monte-Carlo approximations for the dominated hypervolume of the Pareto front approximations that were obtained. The dominated hypervolume measure is a quality indicator for a Pareto front approximation that measures the size of the subspace dominated by it with respect to a reference point. A nice result of the rescaling of the objectives and constraints to the interval $[0, 1]$ that is the hyperspace of possible solutions is now encapsulated in the domain $[0, 1]^d$ (with in our case $d = 4$) which provides us with a natural choice for a reference point in the lower left corner of that cube, and with an integration region $[0, 1]^4$ for Monte Carlo integration of the hypervolume. The latter was computed with a uniform sampling of $N = 10000$ sample points $\mathbf{q}_1, \ldots, \mathbf{q}_N \in [0, 1]^4$ using:

$$\mathcal{H}(P) = \frac{1}{N} \sum_{i=1}^{N} I(\mathbf{q}_i) \; , \; I(\mathbf{q}) = \begin{cases} 1, & \text{if } \mathbf{q} \text{ is dominated by P} \\ 0, & \text{otherwise} \end{cases} \tag{12}$$

Using this approximation of the dominated hypervolume on the results of the three approaches yields the results as shown in Table 1. We obtain that the mean values of approach 1 and 2 are better than the one of approach 3. Moreover approach 2 outperforms approach 3. However, a significant result (p=0.06, two-sided t-test) was only obtained when comparing approach 2 to approach 3, in favor of approach 2. For stronger claims, more test runs are needed.

Table 1. The dominated Hypervolume approximations

	Approach 1	Approach 2	Approach 3
Run 1	0.3975	0.3343	0.2088
Run 2	0.3271	0.3505	0.3155
Run 3	0.3223	0.5365	0.2855
Run 4	0.3004	0.5225	0.3560
Mean	0.3368	0.4360	0.29145
Variance	0.0018	0.0117	0.00387

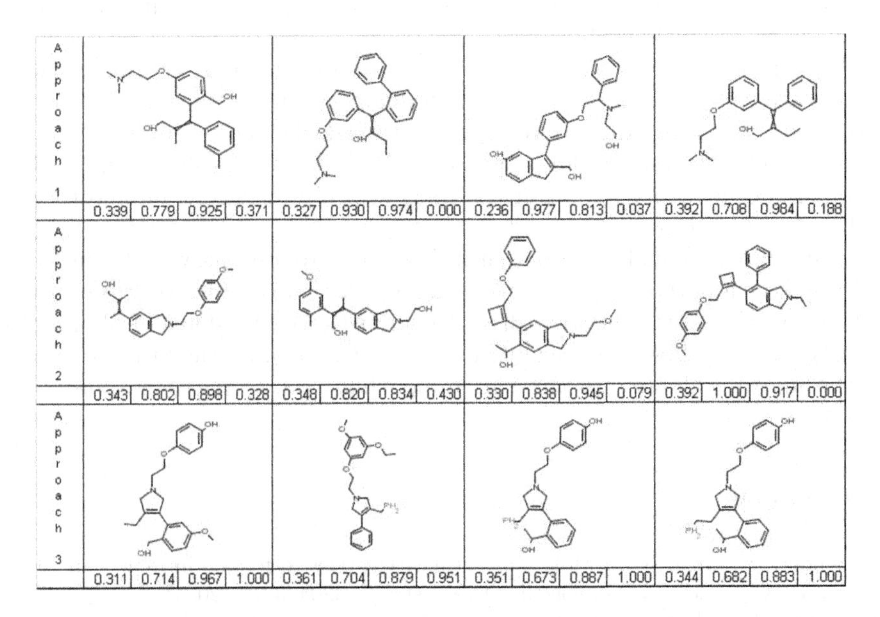

Fig. 4. Random subsets (size 4) from Pareto fronts obtained in run 1 of each approach

Finally, to get an impression about the obtained diversity in the search space, Fig. 4 shows 5 random solutions from the first run of each of the approaches. From this we can conclude, similarly to what is observed in the objective space, that approach 1 yields the most diverse solutions, and that the diversity decreases when reducing the number of objectives. However, the diversity in the search space is much less notable than it is in the objective space. Even with the results of approach 1, the solutions still seem to look very similar. This argues for a more active control of diversity (by means of niching techniques), which is not only focusing on maintaining diversity in the objective space, but also controls diversity in the search space.

7 Summary and Outlook

We have presented a method of converting a multi-objective optimization problem with fuzzy constraints into a many objective optimization problem and proposed two approaches to aggregate the many objectives into an optimization problem with only few objectives. The transformation of fuzzy constraints to desirability functions which can be treated as objectives is a straightforward and simple way of implementing fuzzy constraints, but it generates a many objective optimization problem. Combining the different objectives into logically seperable groups by means of desirability indexes is thereafter a good way to reduce the number of objectives. A nice side-effect of this rescaling is that the objective space is mapped to a hypercube bounded on the interval $[0,1]^d$. In order to still obtain relatively good approximations of the Pareto front, the number of objectives should not be reduced to one, but it pays off to reduce the many objectives to two or four objectives.

This approach is designed for design optimization problems where fuzzy constraints are very common and where high quality solutions that do not comply to the constraints can still be interesting. Although the approach was motivated by the application in molecular design, we believe that this problem scenario can be found in other problem domains, such as design problems in automotive and civil engineering, too.

For evolutionary molecular design we find the proposed way of dealing with multi-objective optimization problems with fuzzy constraints very promising, and we will also test it on other drug design tasks. As the shape of desirability indexes does not allow to model all kinds of preferences, for the future it will be worthwhile to evaluate also alternative modeling and aggregation approaches, such as those given in [3].

References

1. Bender, A., Glen, R.C.: Molecular similarity: a key technique in molecular informatics. Organic and Biomolecular Chemistry 2, 3204–3218 (2004)
2. Branke, J., Deb, K.: Integrating user-preferences into evolutionary multi-objective optimization. In: Jin, Y. (ed.) Knowledge integration into evolutionary multiobjective optimization, Berlin, pp. 461–478 (2004)
3. Cvetkovic, D., Parmee, I.C.: Preferences and their application in evolutionary multiobjective optimization. IEEE TEC (6), 1, 42–57 (2002)
4. Deb, K., Pratap, A., Agarwal, S., Meyarivan, T.: A fast and elitist multiobjective genetic algorithm: NSGA-II. IEEE Transactions on Evolutionary Computation 6, 182–197 (2002)
5. Derringer, G., Suich, R.: Simultaneous optimization of several response variables. Journal of Quality Technology 12, 214–219 (1980)
6. Harrington, E.C.: The desirability function. Industrial Quality Control 21, 494–498 (1965)
7. Inuiguchi, M., Ichihashi, H., Tanaka, H.: Fuzzy programming: A survey of Recent Developments. Stochastic versus Fuzzy Approaches to Multiobjective Mathematical Programming under Uncertainty, 45–68 (1990)
8. Lameijer, E.-W., Kok, J.N., Bäck, T., IJzerman, A.P.: The molecule evoluator: An interactive evolutionary algorithm for the design of drug-like molecules. Journal of Chemical Information and Modeling 46(2), 545–552 (2006)

9. Laumanns, M., Thiele, L., Zitzler, E.: Archiving with guaranteed convergence and diversity in multi-objective optimization. In: Proceedings of the Genetic and Evolutionary Computation Conference, pp. 439–447 (2002)
10. Li, Q., Bender, A., Pei, J., Lai, L.: A large descriptor set and a probabilistic kernel-based classifier significantly improve druglikeness classification. Journal of Chemical Information and Modeling 47(5), 1776–1786 (2007)
11. Lipinski, C., Lombardo, F., Dominy, B., Feeney, P.: Experimental and computational approaches to estimate solubility and permeability in drug discovery and developments settings. Advanced Drug Delivery Reviews 46(1-3), 3–26 (2001)
12. Keeney, R.L., Raiffa, H.: Decision with Multiple Objectives. Wiley, NY (1976)
13. Kruisselbrink, J.W., Bäck, T., van der Horst, E., IJzerman, A.P.: Evolutionary algorithms for automated drug design towards target molecule properties. In: Proceedings of the Conference on Genetic and Evolutionary Computation, pp. 1555–1562 (2008)
14. Trautmann, H., Mehnen, J.: A method for including a-priori preferences in multicriteria optimization. TR 49/2005, University of Dortmund, Germany, SFB 475 (2005)
15. Trautmann, H., Mehnen, J.: Integration of Expert's Preferences in Pareto Optimization by Desirability Function Techniques. In: CIRP ICME 2006, Ischia, Italy, pp. 293–298 (2006)
16. Nicolaou, C.A., Pattichis, C.S.: Multi-objective de novo drug design using evolutionary graphs. Chemistry Central Journal 2008 2(suppl. 1), 7 (2007)
17. Trautmann, H., Weihs, C.: Pareto-Optimality and Desirability Indices. Department of Statistics, University of Dortmund (2004)
18. SciTegic, Inc. 10188 Telesis Court, Suite 100, San Diego, CA 92121, USA., http://www.scitegic.com/products_services/pipeline_pilot.htm
19. Sharma, B., Parmee, I.C., Whittaker, M., Sedwell, A.: Drug discovery: exploring the utility of cluster oriented genetic algorithms in virtual library design. In: Congress on Evolutionary Computation, pp. 668–675 (2005)

Many-Objective Optimization for Knapsack Problems Using Correlation-Based Weighted Sum Approach

Tadahiko Murata[1,2] and Akinori Taki[3]

[1] Department of Informatics, Kansai University, 2-1-1, Ryozenji, Takatsuki, Osaka 569-1073, Japan
[2] Policy Grid Computing Laboratory, Institute for Socionetwork Strategies, Kansai University, 3-3-35, Yamate, Suita, Osaka 564-8680, Japan
[3] Graduate School of Informatics, Kansai University, 2-1-1, Ryozenji, Takatsuki, Osaka 569-1073, Japan
{murata,ataki}@pglab.kansai-u.acjp

Abstract. In this paper, we examine the effectiveness of an EMO (Evolutionary Multi-criterion Optimization) algorithm using a correlation based weighted sum for many objective optimization problems. Recently many EMO algorithms are proposed for various multi-objective problems. However, it is known that the convergence performance to the Pareto-frontier becomes weak in approaches using archives for non-dominated solutions since the size of archives becomes large as the number of objectives becomes large. In this paper, we show the effectiveness of using a correlation information between objectives to construct groups of objectives. Our simulation results show that while an archive-based approach, such as NSGA-II, produces a set of non-dominated solutions with better objective values in each objective, the correlation-based weighted sum approach can produce better compromise solutions that has averagely better objective values in every objective.

Keywords: Many objective optimization, problem correlation, weighted sum approach, archive-based approach.

1 Introduction

In this decade, researches in EMO (Evolutionary Multi-criterion Optimization) algorithms have mainly focused on multi-objective optimization problems with a few number of objectives. The better performance of an archive-based (or Pareto-based) approach[1] such as NSGA-II [1] or SPEA [2] are reported in those problems. Recently it is reported that the performance of the archive-based approach is degraded when the number of objectives increases [3-5]. This is because that the performance of converging to the Pareto front becomes weak in those archive based approach when they are applied to many objective optimization problems. Wagner et al. [5] showed that MSOPS [6] and SMS-EMOA [7] are better than NSGA-II or SPEA2.

[1] The term "archive-based approach" is used for the approaches that keep non-dominated solutions obtained during their search to produce following generations. The definition of "Pareto" is given in Section 2.

M. Ehrgott et al. (Eds.): EMO 2009, LNCS 5467, pp. 468–480, 2009.

In this paper, we combine a weighted-sum approach[2] with an archive-based approach in order to compensate the archive-based approach for the weakness in their searching pressure toward Pareto optimal solutions of many-objective optimization problems. By aggregating several objectives into an objective using weight values, we reduce the number of objectives for the archive-based approach. Correlations between two objectives are used to aggregate some objectives into an objective. We examine the influence of correlations among objectives to aggregate objectives into one objective.

Our simulation results show that while an archive-based approach, such as NSGA-II, produces non-dominated solutions that have better objective values in some objectives, our proposed approach can produce better compromise solutions that has averagely better objective values in every objective.

The following section reveals challenges that arise in many objective optimization problems. Then Section 3 shows our proposed approach that employs the correlation information between objectives. Simulation results are provided to compare our proposed approach with NSGA-II. Finally Section 4 summarizes this paper and show some further challenges in the proposed algorithm.

2 Challenge in Many Objective Optimization

In general, a k-objective maximization problem is written as follows:

$$\text{Maximize } \mathbf{f}(\mathbf{x}) = (f_1(\mathbf{x}), f_2(\mathbf{x}), \ldots, f_k(\mathbf{x})), \tag{1}$$

$$\text{subject to } \mathbf{x} \in \mathbf{X}, \tag{2}$$

where $\mathbf{f}(\mathbf{x})$ is the k-dimensional objective vector that consists of k objectives to be maximized, and $\mathbf{x}$ is the decision vector in the feasible region of the decision space $\mathbf{X}$.

One aim in multi-objective optimization problems is to search for the set of Pareto optimal solutions. The Pareto optimality can be explained using two feasible solutions $\mathbf{y} \in \mathbf{X}$ and $\mathbf{z} \in \mathbf{X}$ of the k-objective maximization problem in (1) and (2). If the following conditions are satisfied, $\mathbf{z}$ is said to dominate $\mathbf{y}$:

$$\text{Maximize } \forall i, f_i(\mathbf{y}) \le f_i(\mathbf{z}) \text{ and } \exists i, f_i(\mathbf{y}) < f_i(\mathbf{z}). \tag{3}$$

That is, at least one among k objectives, $\mathbf{z}$ is superior to $\mathbf{y}$.

When $\mathbf{y}$ is not dominated by any other feasible solutions in $\mathbf{X}$, the solution $\mathbf{y}$ can be referred to as a Pareto-optimal solution of the k-objective maximization problem in (1) and (2). The set of all Pareto optimal solutions forms the tradeoff surface in the objective space. EMO algorithms are designed to search for a set of well-distributed non-dominated solutions that approximates the entire set of Pareto optimal solutions. They are expected to converge to the set of Pareto optimal solutions as soon as possible, and simultaneously keep well-distributed solutions on the set.

[2] The term "weighted-sum approach" is used for the approaches that uses the aggregated sum of objective values using weight values for the fitness function.

Currently NSGA-II [1] and SPEA [2] are well-known EMO algorithms that are frequently used for various multi-objective optimization problems. They have common features: fitness evaluation based on Pareto-dominance, diversity maintenance, and the utilization of non-dominated solutions as elite individuals. EMO algorithms usually work very well on two- or three-objective problems. However, their search ability becomes worse when the number of objectives increases. We show a reason why the search ability of the archive-based approach using knapsack problems. The purpose of multi-objective knapsack problems in Zitzler and Thiele [2] is to maximize values of multiple knapsacks under their weight limitation.

As Ishibuchi et al. [8] examined to see the performance of an archive-based approach, NSGA-II, we also employ NSGA-II and apply it to multi-objective 500-item knapsack problems with two and four-objectives. Our two-objective and four-objective problems are the same ones in [2]. We also generated a ten-objective problem in the same manner as in [2]. In our experiments we used the following parameter settings:

Population size: 100,
Archive size: 100,
Crossover probability: 1.0 (uniform crossover),
Mutation probability: 0.1 (for each individual),
Stopping conditions: 10 000 generations.

In NSGA-II, we employed $(\mu+\lambda)$-ES generation update. We specified μ and λ as $\mu=\lambda=100$. That is, the best 100 individuals are selected from the current and offspring populations with 200 individuals in total.

Fig. 1 shows the average number of non-dominated solutions among the 200 individuals in the current and offspring populations in each generation over 10 runs. 2-500, 4-500 and 10-500 show the results obtained for two-objective 500-item knapsack problem, four-objective 500-item, and ten-objective 500-item, respectively. Similar figures were also depicted in the literatures (e.g., [8,9]).

In Fig. 1, we can see that the number of non-dominated solutions increases through the running of NSGA-II. We can also see that the number of non-dominated solutions increases as we increase the number of objectives from two to ten. Although it is obvious that the number of non-dominated solutions increases when the number of objectives increases, archive-based approaches may have a difficulty in their generation update. Since we employed $(\mu+\lambda)$-ES generation update in NSGA-II, when the number of the non-dominated solutions exceeds 100, the best 100 individuals are selected only from those non-dominated solutions using the secondary criterion (i.e., crowding distance). As a result, the generation update is done with respect to the secondary criterion. This means that NSGA-II mainly tries to increase the diversity of individuals rather than the convergence to the set of Pareto optimum solutions. As we can see for the ten-objective knapsack problem, the number of non-dominated solutions exceeds 100 in early stages of NSGA-II's search. This means that the search in the ten-objective problem is mainly done to increase the diversity.

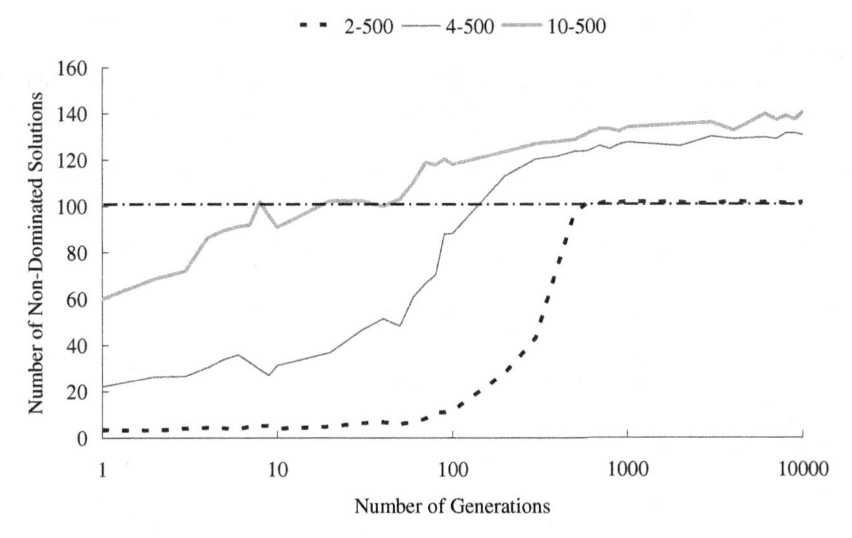

Fig. 1. Average number of non-dominated solutions among the 200 individuals in the current and offspring population in each generation of NSGA-II

In order to see more detail solution updates in NSGA-II, we examined the sum of all objectives and the sum of the range of objective values over all objectives in the current population that are shown in [8]. The sum of all objectives is described as follows:

$$\mathrm{sum}(\mathbf{x}) = \sum_{\mathrm{i=1}}^{\mathrm{k}} f_i(\mathbf{x})\,. \tag{4}$$

This index can be regarded as a measure of the convergence to the set of Pareto solutions. Since the knapsack problems are maximization problems, the larger, the better. The maximum value of the sum of all objectives was calculated over all individuals in the current population in each generation during the execution of NSGA-II.

$$\mathrm{Sum}(\Psi) = \max_{\mathbf{x}\in\Psi} \mathrm{sum}(\mathbf{x})\,, \tag{5}$$

where Ψ denotes the current population.

The other measure is the sum of the range of objective values over all objectives in the current population:

$$\mathrm{Range}(\Psi) = \sum_{i=1}^{k} [\max_{\mathbf{x}\in\Psi} f_i(\mathbf{x}) - \min_{\mathbf{x}\in\Psi} f_i(\mathbf{x})]\,. \tag{6}$$

Since $\mathrm{Range}(\Psi)$ indicates the difference between the maximum and the minimum for each objectives in the current population, the larger value means the large variety (or gap) in the population. Since an aim of EMO algorithms is to show non-dominated solutions as many as possible, the larger variety is better. However, the larger range does not directly mean the large variety because the range becomes larger when the EMO algorithm finds several solutions that optimize only one objective with the sacrifice of the other objectives. The score of the range should be examined with the score of the sum.

For the same experimental results in Fig. 1, the sum of all objectives in (5) and the range in (6) are shown in Figs. 2 and 3, respectively. The experimental results shown in Figs. 2 and 3 are normalized by the average value of each measure at the first generation over 10 runs. Thus, the value of each measure is always 100 at the initial generation.

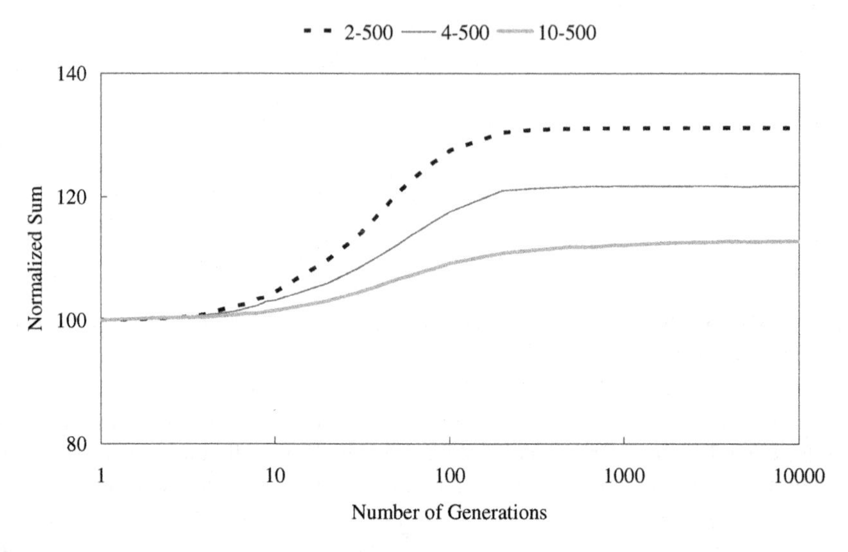

Fig. 2. Sum of all objectives. Experimental results are normalized using the result at the first generation for each knapsack problem.

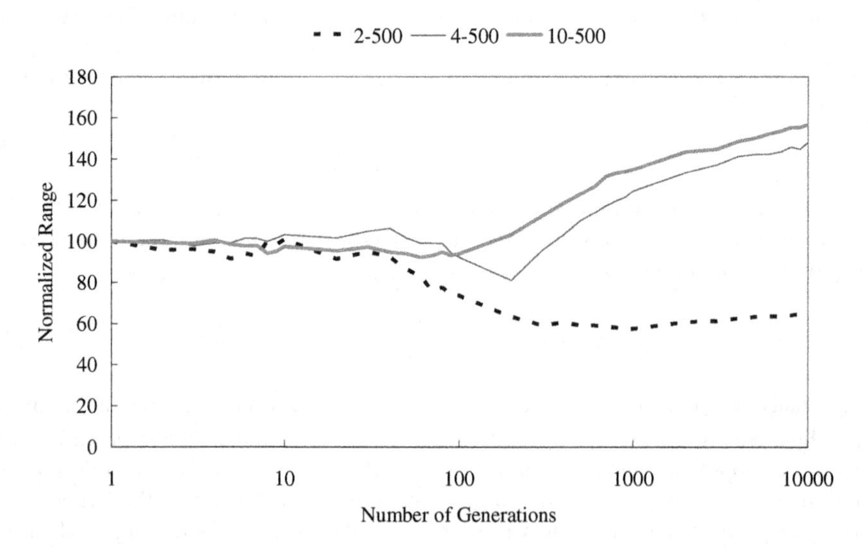

Fig. 3. Range of objective values. Experimental results are normalized as in Fig. 2.

From Fig. 2, the sum of all objectives was first increased through multi-objective optimization by NSGA-II. However, it started to stagnate after a certain number of generations. On the other hand, Fig. 3 shows that the range was not increased in the early stage, but started to increase in the latter stage.

When comparing Fig. 3 with Fig. 1, we can see that the range measure in Fig. 3 started to increase after the number of non-dominated solutions exceeded 100 in Fig. 1. This is because that after the number of non-dominated solutions was larger than 100, the secondary criterion (i.e., crowding distance) had a large effect on the fitness evaluation in NSGA-II. On the other hand, the sum of all objectives could increase while the number of non-dominated solutions is smaller than 100. This is because that the primary criterion had a dominant effect to approach the set of Pareto optimum solutions. Through these experiments, we can see the weakness of archive-based approach when the number of objectives increases.

3 Correlation-Based Weighted Sum Approach

In this paper, we combine a correlation-based weighted sum approach with an archive-based approach in order to cope with the challenges of many objective optimization problems. As we see in the previous section, archive-based approaches have difficulties when the number of objectives is many (more than three). In this paper, we propose a method to reduce the number of objectives using weighted values.

Weighted sum approaches was firstly introduced to EMO algorithms by Murata and Ishibuchi [10,11] for multi-objective flowshop scheduling problems. They randomly generate weight vectors for selecting parent individuals from a population. In order to have weight vectors proportionally for the object space, they proposed a weight vector generation method in a cellular genetic algorithm [12]. Although the weighted sum approach is known to have a convergence problem in multiobjective optimization problems with a concave set of Pareto solutions, this approach attracts attentions again among researchers in EMO community who tackle many objective problems. Recently Hughes [6,13] proposed a multiple single objective pareto sampling method (MSOPS) where multiple weight vectors are generated according to an utopia point that is defined as the minimum achieved objective value in each objective dimension.

Although the above mentioned weighted sum approaches employ various weight vectors to evaluate each individual in a population, the correlation between objectives are not considered. In this paper, we try to employ the correlation information between objectives to combine several correlated objectives into one group. However, it should be noted that the generation update is done in the manner of an archive-based approach, that is NSGA-II in this paper. The correlation information is employed to merge several objectives into one group. The aim of aggregating several objectives is to reduce the number of objectives in the archive-based approach that does not work when the number of objectives becomes large. We describe the procedure of the proposed algorithm in the following subsection, and show experimental results on a ten-objective knapsack problem and a forty-objective problem.

3.1 Correlation-Based Weighted Sum Approach

As we see from the computational experiments in Section 2, the archive-based approach such as NSGA-II has its advantage in finding a set of non-dominated solutions with the wide range. Each solution in the set of non-dominated solutions obtained by the archive-based approach may have a good performance in one or a few objectives. Therefore we combine several objectives that correlate each other, then make several groups of objectives. We propose the following method to combine objectives.

Step 1: Determine the number of groups, K, and the number of generations when the combination of objectives is held.

Step 2: Execute an EMO algorithm until the generation specified in Step 1.

Step 3: Using the solutions that are archived during the execution of the EMO algorithm, calculate correlation between each of two objectives.

Step 4: Find the combination of objectives that maximize the average value of the following measure:

$$performance(\Omega_k) = \frac{\sum_{i,j \in \Omega_k} corr(f_i, f_j)}{|\Omega_k| - 1}, \quad k = 1,2,...,K, \quad \text{if } |\Omega_k| \geq 2, \tag{7}$$

$$performance(\Omega_k) = 1, \ k = 1,2,...,K, \text{ if } |\Omega_k| = 1, \tag{8}$$

where $corr(f_i, f_j)$ is the correlation between the objectives f_i and f_j, Ω_k is a set of objectives of k-th group, $|\Omega_k|$ is the cardinality of the set Ω_k. Then, calculate the average of $performance(\Omega_k)$ over K groups. It is noted that the correlation $corr(f_i, f_j)$ is calculated by the non-dominated solutions that are registered in the archives until the prespecified generation.

Step 4.1: Generate a specified number of groups by selecting objectives randomly.

Step 4.2: Calculate the average of $performance(\Omega_k)$ over K groups.

Step 4.3: Modify objectives in the groups by exchanging one objective in a group to another randomly selected group.

Step 4.4: Calculate the average of $performance(\Omega_k)$ over K groups. If the calculated average becomes better than that of the previous groups, accept the modification of groups in Step 4.3. Then go to Step 4.3. If it becomes worse, go back to the previous combination of groups. Then go to Step 4.3. Repeat Steps 4.3 and 4.4 until no further improvement is found.

Step 4.5: Move to Step 4.1 until the prespecified stopping condition is satisfied, terminate the procedure.

Step 5: Using the groups determined in Step 4, merge objectives with a weight vector as follows:

$$f_{\Omega_k}(\mathbf{x}) = \frac{1}{|\Omega_k|} f_{\Omega_k^1}(\mathbf{x}) + \frac{1}{|\Omega_k|} f_{\Omega_k^2}(\mathbf{x}) + \ldots + \frac{1}{|\Omega_k|} f_{\Omega_k^{|\Omega_k|}}(\mathbf{x}), \; k = 1, 2, \ldots, K, \quad (9)$$

where $f_{\Omega_k}(\mathbf{x})$ is the fitness value of the solution $\mathbf{x}$ for the group Ω_k, $f_{\Omega_k^i}(\mathbf{x})$ ($i = 1, 2, \ldots, |\Omega_k|$) is the objective value of $\mathbf{x}$ for the i-th objective in the group Ω_k.

Step 6: With K merged objective functions $f_{\Omega_k}(\mathbf{x})$ ($k = 1, 2, \ldots, K$), execute the EMO algorithm to search for a set of non-dominated solutions.

It is noted that K, the number of groups, is better to be smaller than four or three, because the larger number of K is not suitable for the archive-based approach.

3.2 10-Objective Knapsack Problem

We applied the proposed algorithm in Subsection 3.1 to a 10-objective knapsack problem. The problem is basically generated in the same manner as in [2]. After generating the problem, we slightly modify the problem in order to give a positive correlation among some objectives. Our 10-objective knapsack problem is to maximize the value of 10 knapsacks. When certain items have larger values in some knapsacks, those items should be selected for those knapsacks. We add a value α to items 1 through 250 of Knapsacks 1 through 5, and a value α to items 251 through 500 of Knapsacks 6 through 10. In this way, the problem has the positive correlation among first five knapsacks, and among the latter five knapsacks. We defined $\alpha = 5$ to generate the problem. The other parameter values in the proposed algorithm are follows:

The number of groups: $K = 2$,
Sampling non-dominated solutions for calculating the correlation: every 5 generation,
The number of generation when the correlations are calculated: 500th generation,
Stopping condition for determining the groups in Step 4.5: 100,
Stopping condition for EMO algorithm: 2 000 generations.

Figs. 4 and 5 show each of the non-dominated solutions found by the original NSGA-II and the NSGA-II with the proposed correlation-based weighted sum approach. The horizontal axis shows the ID numbers of ten knapsacks, and the vertical axis shows the values of knapsacks. Each line in these figure describes each obtained non-dominated solution with the value in each knapsack.

When we merge several objectives at the 500th generation using correlation information, Knapsacks 1 through 5, and Knapsacks 6 through 10 are accurately divided into two groups. That is, the proposed correlation method could find the appropriate set of knapsacks that have the same modified items. From Fig. 4, we can see that the original NSGA-II could find the non-dominated solutions in a wide range. On the other hand, the NSGA-II with the proposed method in Fig. 5 could find the narrow width with respect to the range, however, the obtained non-dominated solutions locates in the better region of the solutions obtained by the original NSGA-II. That is, the solutions obtained by the NSGA-II have high values in some knapsacks but they have low values in the other knapsacks. However, the solutions obtained by the NSGA-II with the proposed method can keep the better value in every knapsack.

As we discussed in Section 2, the NSGA-II tried to maximize the variety in non-dominated solutions after the number of solutions becomes larger than its population. Therefore the solution obtained by the NSGA-II has a tendency to maximize a few objectives with the sacrifice of maximizing the other objectives. On the other hand, the NSGA-II with the proposed approach tries to maximize two combined objectives after aggregating objectives using weight values.

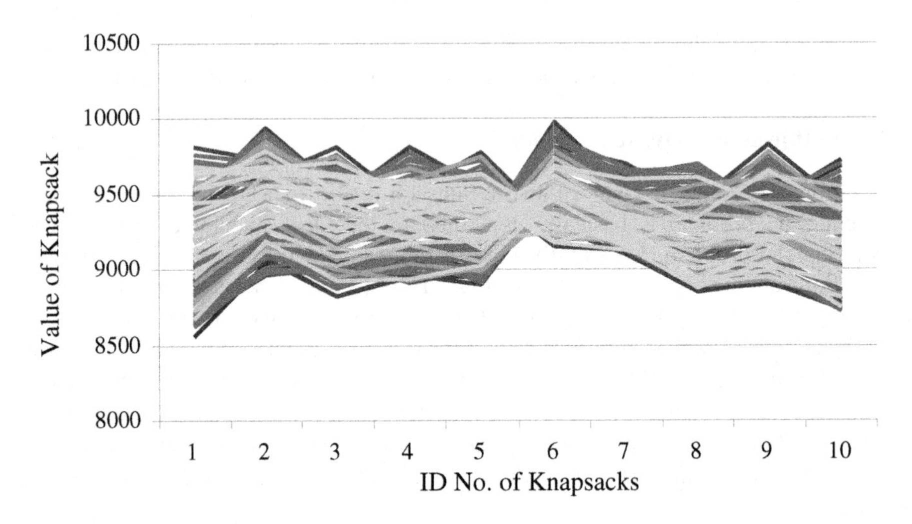

Fig. 4. Values of knapsacks of each non-dominated solutions obtained by the original NSGA-II

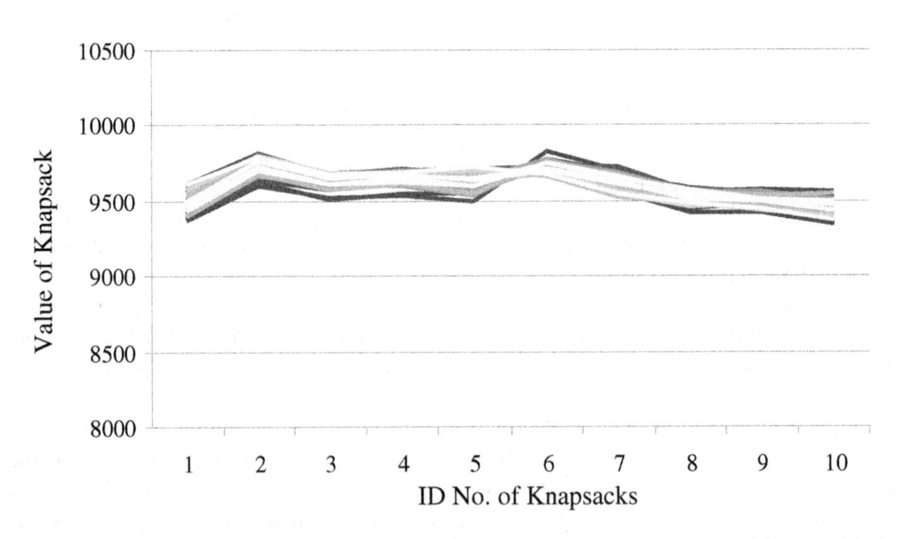

Fig. 5. Values of knapsacks of each non-dominated solutions obtained by the NSGA-II with the proposed correlation-based weighted sum approach

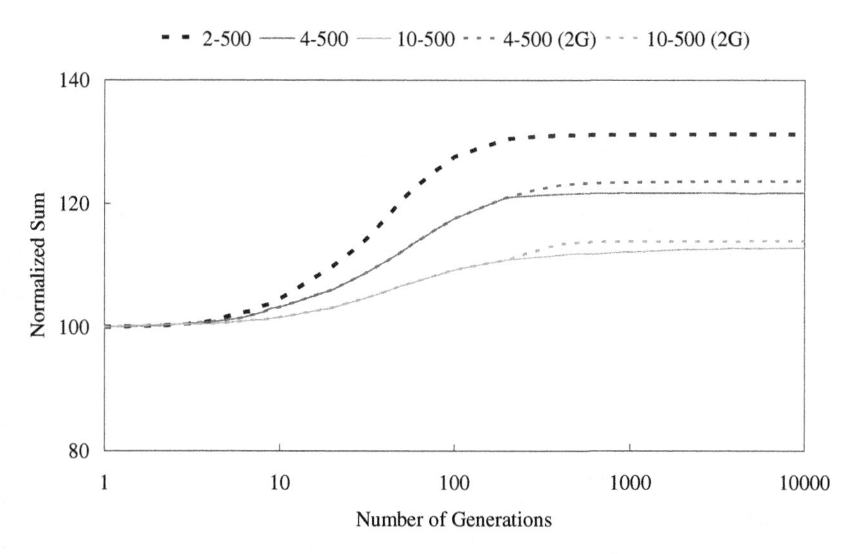

Fig. 6. Normalized sum obtained by NSGA-II with the proposed method to compare with the original NSGA-II

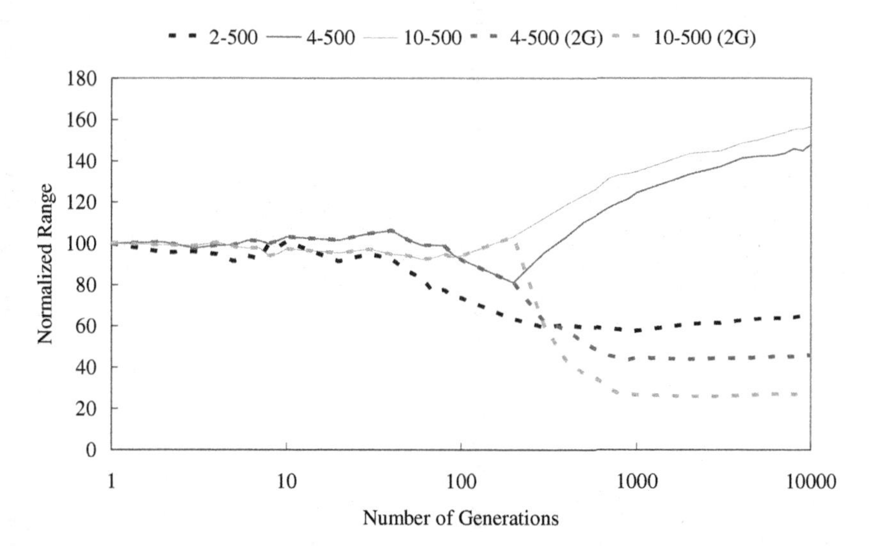

Fig. 7. Normalized range obtained by NSGA-II with the proposed method to compare with the original NSGA-II

The characteristics of the NSGA-II with the proposed method can be seen in Figs. 6 and 7. These figures show the average sum of all objectives and the average range of objective values over ten trials as in Figs. 2 and 3 in Section 2. We applied the NSGA-II with the proposed method to the same problem in Figs. 2 and 3. We merge objectives in NSGA-II at the 250th generation. In Fig. 6, the notation (2G)

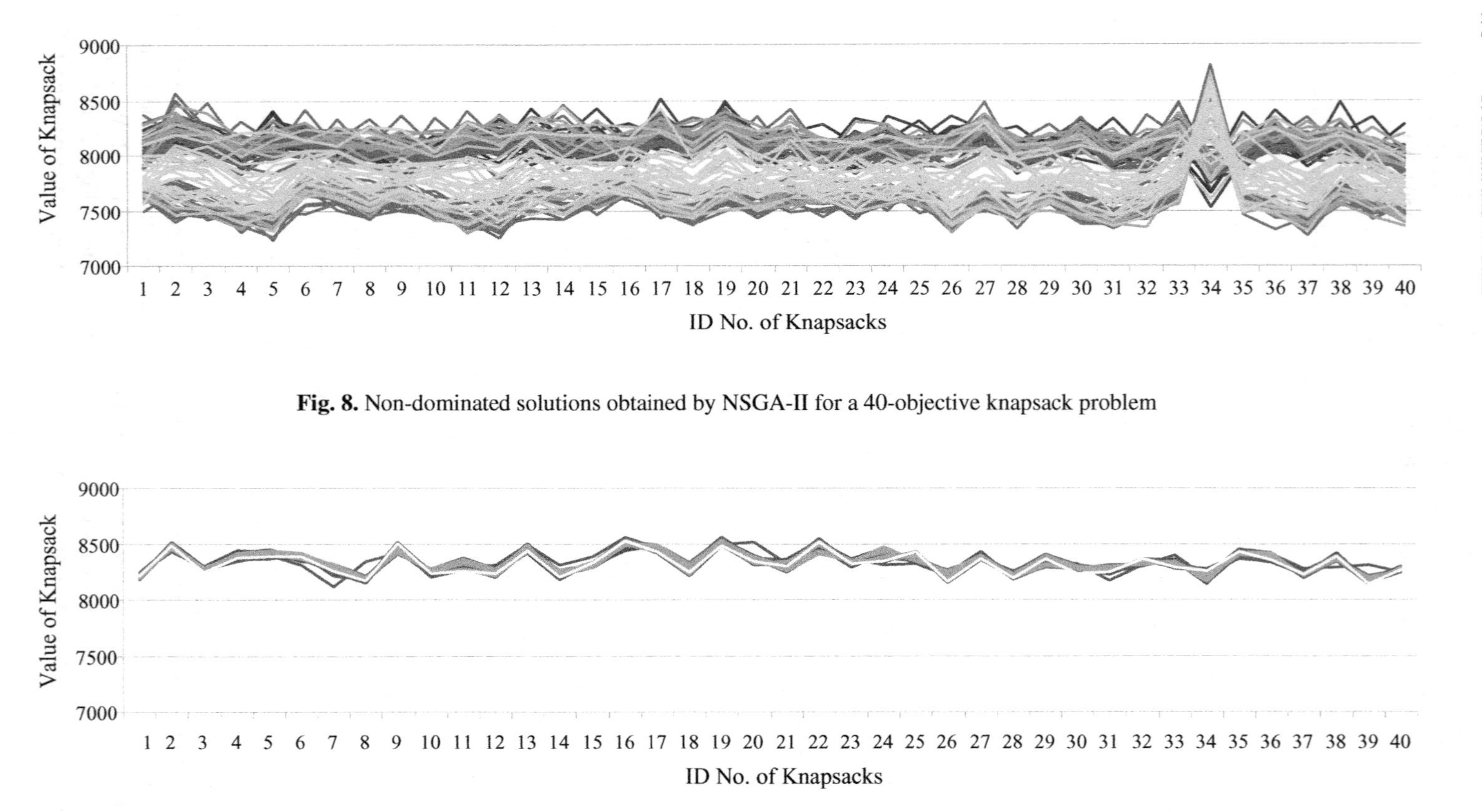

Fig. 8. Non-dominated solutions obtained by NSGA-II for a 40-objective knapsack problem

Fig. 9. Non-dominated solutions obtained by NSGA-II with the proposed method for a 40-objective knapsack problem shows the scores obtained by NSGA-II with the proposed method. 2-500, 4-500, and 10-500 are the same scores in Fig. 2. Fig. 6 clearly shows that the proposed method can improve the normalized sum after 250th generation.

Fig. 7 shows that the range of the solutions obtained by NSGA-II with the proposed method becomes small although the range of the original NSGA-II becomes large. Although the larger range is better in the sense of the variety in non-dominated solutions, the smaller sum means that the obtained solutions tends to maximize a few objectives with the sacrifice of the others. From Figs. 6 and 7, we can see that the NSGA-II with the proposed method can find the better set of solutions although its range becomes smaller.

3.3 40-Objective Knapsack Problem

We generated a 1000-item problem with 40 objectives in the same manner as in [2]. We did not give any bias to the problem in this problem as we did it to a 10-objective problem in Subsection 3.2. As was in Figs. 4 and 5, Figs. 8 and 9 show that each of non-dominated solutions obtained by the original NSGA-II and the NSGA-II with the proposed method. From these figures, we can also see in Fig. 8 that the original NSGA-II could find the set of non-dominated solutions with a wide range and solutions that have a good value in only some objectives. On the other hand, as we can see from Fig. 9, the NSGA-II with the proposed method found the set of solutions with a better value almost in every objective. As we can see from the results on the ten-objective problem, we can also observe that the NSGA-II with the proposed method can find better non-dominated solutions than the original NSGA-II.

4 Conclusion and Further Challenges

In this paper, we proposed a correlation-based weighted sum approach in order to cope with the challenges of many objective optimization problems. As we see through computational experiments, EMO algorithms that keep non-dominated solutions in its archive would have problems since the number of non-dominated solutions easily exceeds the size of the archive. After exceeding the archive, EMO algorithms may lost the selection pressure toward the Pareto solutions, but improve the variety of the set of non-dominated solutions. One way to avoid this tendency in archive-based approaches is to increase the size of archive, though, it may be difficult to have enough size of archive when the number of objectives increases more.

The proposed algorithm tries to reduce the number of objectives using weighted sum approach. Although the variety of a set of non-dominated solutions may be lost in the EMO algorithm that employs our proposed method, the better set of solutions. The simulation results for a 40-objective problem clearly show that the original NSGA-II produces many extreme solutions that have better performance, however, only in one or a few objectives. Those solutions have worse performance in the other objectives. The extreme solutions could be found by sacrificing the other objectives. On the other hand, the EMO algorithm with the proposed method could find better solutions that do not compensate other objectives.

Although the proposed method can improve the performance of NSGA-II, still there are challenges in the method. Since several parameters are introduced to the algorithm, that is, the number of groups, and the timing for making groups, they should be determined in advance. In order to apply the proposed method to EMO algorithms with less efforts, these parameters should be automatically determined. We should improve the method in these points.

Acknowledgments. We appreciate the effective comments of referees to improve this work. This work was supported in part by the Japan MEXT (Ministry of Education, Culture, Sports, Science and Technology) under Collaboration with Local Communities Project for Private Universities starting 2005.

References

1. Deb, K., Pratap, S., Agarwal, S., Meyarivan, T.: A fast and elitist multiobjective genetic algorithm: NSGA-II. IEEE Transaction on Evolutionary Computation 6(2), 182–197 (2002)
2. Zitzler, E., Thiele, L.: Multiobjective evolutionary algorithms: A comparative case study and strength Pareto approach. IEEE Transaction on Evolutionary Computation 3(4), 257–271 (1999)
3. Hughes, E.J.: Evolutionary many-objective optimization: Many once or one many? In: Proc. of 2005 IEEE Congress on Evolutionary Computation, pp. 222–227 (2005)
4. Ishibuchi, H., Nojima, Y., Doi, T.: Comparison between single-objective and multi-objective genetic algorithms: Performance comparison and performance measures. In: Proc. of 2006 Congress on Evolutionary Computation, pp. 3959–3966 (2006)
5. Wagner, T., Beume, N., Naujoks, B.: Pareto-, aggregation-, and indicator-based methods in many-objective optimization. In: Obayashi, S., Deb, K., Poloni, C., Hiroyasu, T., Murata, T. (eds.) EMO 2007. LNCS, vol. 4403, pp. 742–756. Springer, Heidelberg (2007)
6. Hughes, E.J.: Multiple single objective pareto sampling. In: Proc. of 2003 Congress on Evolutionary Computation, pp. 2678–2684 (2003)
7. Emmerich, M., Beume, N., Naujoks, B.: An EMO algorithm using the hypervolume measure as selection criterion. In: Coello Coello, C.A., Hernández Aguirre, A., Zitzler, E. (eds.) EMO 2005. LNCS, vol. 3410, pp. 62–76. Springer, Heidelberg (2005)
8. Ishibuchi, H., Tsukamoto, N., Nojima, Y.: Evolutionary many-objective optimization. In: Proc. of 3rd Int'l. Workshop on Genetic and Evolving Fuzzy Systems, pp. 47–52 (2008)
9. Sato, H., Aguirre, H.E., Tanaka, K.: Controlling dominance area of solutions and its impact on the performance of mOEAs. In: Obayashi, S., Deb, K., Poloni, C., Hiroyasu, T., Murata, T. (eds.) EMO 2007. LNCS, vol. 4403, pp. 5–20. Springer, Heidelberg (2007)
10. Murata, T., Ishibuchi, H., Tanaka, H.: Multi-objective genetic algorithm and its application to flowshop scheduling. Computer and Industrial Engineering Journal 30(4), 957–968 (1996)
11. Ishibuchi, H., Murata, T.: A multi-objective genetic local search algorithm and its application to flowshop scheduling. IEEE Trans. on Systems, Man, and Cybernetics, Part C 28(5), 601–618 (1999)
12. Murata, T., Ishibuchi, H., Gen, M.: Specification of genetic search directions in cellular multi-objective genetic algorithms. In: Zitzler, E., Deb, K., Thiele, L., Coello Coello, C.A., Corne, D.W. (eds.) EMO 2001. LNCS, vol. 1993, pp. 82–95. Springer, Heidelberg (2001)
13. Hughes, E.J.: MSOP-II: A general-purpose many-objective optimiser. In: Proc. of 2007 Congress on Evolutionary Computation, pp. 3944–3951 (2007)

An Elitist GRASP Metaheuristic for the Multi-objective Quadratic Assignment Problem

Hui Li and Dario Landa-Silva

Automated Scheduling Optimisation and Planning Research Group
School of Computer Science, The University of Nottingham, United Kingdom
{hzl,jds}@cs.nott.ac.uk

Abstract. We propose an elitist Greedy Randomized Adaptive Search Procedure (GRASP) metaheuristic algorithm, called mGRASP/MH, for approximating the Pareto-optimal front in the multi-objective quadratic assignment problem (mQAP). The proposed algorithm is characterized by three features: elite greedy randomized construction, adaptation of search directions and cooperation between solutions. The approach builds starting solutions in a greedy fashion by using problem-specific information and elite solutions found previously. Also, mGRASP/MH maintains a population of solutions, each associated with a search direction (i.e. weight vector). These search directions are adaptively changed during the search. Moreover, a cooperation mechanism is also implemented between the solutions found by different local search procedures in mGRASP/MH. Our experiments show that mGRASP/MH performs better or similarly to several other state-of-the-art multi-objective metaheuristic algorithms when solving benchmark mQAP instances.

1 Introduction

The quadratic assignment problem (QAP) models many real-world optimization problems in diverse areas such as operations research, economics, etc. One of its major applications is facility location, where a set of facilities should be assigned to different locations. The objective is to find an assignment of all facilities to all locations, such that the total cost is minimized. The QAP is a **NP**-hard combinatorial optimization problem [1]. So, there is no known exact algorithm for solving the QAP in polynomial time. Recently, the multi-objective QAP (mQAP) has been investigated by researchers in the multi-objective optimization community [2,3]. Unlike the single-objective QAP, the mQAP involves multiple types of flows between any two facilities.

Over the last decades, research on multi-objective metaheuristics, such as evolutionary algorithms, simulated annealing, and tabu search, has attracted a lot of attention from the scientific community. A majority of these algorithms use either Pareto dominance or weighting method for fitness assignment. For example, two representative Pareto-based evolutionary multi-objective (EMO) algorithms - NSGA2 [4] and SPEA2 [5] rank the members of the population by comparing them in terms of Pareto domination while MOEA/D [6] defines the

M. Ehrgott et al. (Eds.): EMO 2009, LNCS 5467, pp. 481–494, 2009.

fitness of individuals by using weighted functions. To find a well-distributed set of solutions, some strategies, such as estimating the density of non-dominated solutions and maintaining a set of uniform weights, have been used to maintain the diversity of population in these algorithms.

It is well-known that well-designed genetic operators play an important role in improving the performance of evolutionary algorithms. The proximate optimality principle (POP) [7] assumes that good solutions share some similarities in the decision space. This principle holds for many real-world problems. Based on this principle, Zhang and Sun [8] proposed a genetic operator, called guided mutation, to sample solutions in promising areas of the search space. This is achieved by modifying the elite solutions found previously and then using global information from a probabilistic model. The combination of guided mutation with iterated local search produced competitive results for solving the QAP in [8].

GRASP [9] is one of the most successful metaheuristics for combinatorial optimization. It is a multi-start local search approach. In each iteration of GRASP, two procedures are involved: greedy randomized construction of starting solutions and a local search procedure. A multi-objective version of GRASP was proposed in [10] to handle multi-objective knapsack problem. In that algorithm, each solution is improved along a certain direction by local search. However, the local optima obtained in different iterations do not interact with each other. As shown in [6] and [11], cooperation between solutions with similar search directions and the adaptive change of these search directions is beneficial. In this paper, we propose an elitist multi-objective GRASP metaheuristic called mGRASP/MH. We assess the performance of mGRASP/MH by applying it to a number of benchmark mQAP instances and comparing its performance to that of some existing multi-objective algorithms.

The remainder of this paper is organized as follows. Section 2 formulates the mQAP and discusses fast local search for this problem. Section 3 discusses some important issues of the basic GRASP algorithm for single objective optimization. Section 4 presents the proposed mGRASP/MH for the mQAP. Experimental results are presented and discussed in Section 5 while Section 6 concludes the paper.

2 The Multi-objective Quadratic Assignment Problem

2.1 Mathematical Formulation

Given a location matrix $A = \{a_{ij}\}_{n\times n}$ and flow matrices $B^k = \{b^k_{rs}\}_{n\times n\times m}, k = 1,\ldots,m$, the mQAP is to minimize the following objective functions simultaneously:

$$C(\pi) = \{C^1(\pi),\ldots,C^m(\pi)\}, \pi \in \Omega \quad (1)$$

with

$$C^k(\pi) = \sum_{i=1}^{n}\sum_{j=1}^{n} a_{ij} b^k_{\pi_i \pi_j}, k = 1,\ldots,m \quad (2)$$

where

- n is the number of locations/facilities, m is the number of objectives (i.e. types of flows), $\pi = (\pi_1, \ldots, \pi_n)$ is a permutation of $L = \{1, \ldots, n\}$, Ω is the set of all permutations, $C(\pi)$ is a vector of m objective functions $C^k(\pi), i = 1, \ldots, m$.
- a_{ij} is the distance between locations i and j, and $b^k_{\pi_i \pi_j}$ is the k-th flow between facilities π_i and π_j.

In the case of conflicting objectives, there is no solution π^* which is optimal for all objective functions $C^k(\pi), k = 1, \ldots, m$. Instead, the optimal solution π^* to the mQAP in (1) is often defined as the trade-off solution in terms of Pareto optimality. Assume u and v are objective vectors, u is said to *dominate* v if and only if $u_k \leq v_k$ for all $k = 1, \ldots, m$, and $\exists s \in \{1, \ldots, m\}$, $u_s < v_s$. A solution π^* is said to be Pareto-optimal to (1) if $C(\pi^*)$ is not dominated by $C(\pi)$ for any $\pi \in \Omega$. The *Pareto-optimal front* (POF) is the set of objective vectors of all Pareto-optimal solutions.

In the mathematical programming community, multi-objective optimization problems are often tackled using some form of weighted sum method that combines multiple objective functions into a single scalar function as follows:

$$f(\pi|\lambda) = \sum_{k=1}^{m} \lambda_k \cdot C^k(\pi) \tag{3}$$

where $\lambda = (\lambda_1, \ldots, \lambda_m)^T$ is the weight vector with $\lambda_k \geq 0, k = 1, \ldots, m$ and $\sum_{k=1}^{m} \lambda_k = 1$. Each component of λ can be regarded as the preference w.r.t each objective. The global minima of $f(\pi)$ in (3) is also Pareto-optimal to the mQAP in (1). By minimizing the scalar functions (3) with appropriate weight vectors, a good approximation of the POF is likely to be obtained. However, the weighted sum method cannot solve the multi-objective optimization problems with non-convex POF. Despite this, the weighed sum method has been successfully applied to solve many multi-objective combinatorial optimization problems.

2.2 Fast Local Search

Local search based on 2-opt operator has been widely used to tackle some permutation-based combinatorial optimization problems. In the QAP, the neighborhood of the current solution consists of all solutions obtained by exchanging the positions of two elements in its permutation [12] (i.e., 2-opt swap). Since all elements in the new solution, except the exchanged ones, remain the same, the computation of the objective function value for neighboring solutions can be done quickly by considering only those exchanged elements. In the case of the mQAP, the computation of the function values of neighboring solutions is very similar. Assume that i and j are two positions exchanged in permutation π, the difference $\Delta(\pi, k, i, j)$ of function values regarding the k-th flow before and after exchanging elements i and j can be stated as:

$$\begin{aligned}\Delta(\pi, k, i, j) = &(a_{jj} - a_{ii})(b^k_{\pi_i\pi_i} - b^k_{\pi_j\pi_j}) + \\ &(a_{ji} - a_{ij})(b^k_{\pi_i\pi_j} - b^k_{\pi_j\pi_i}) + \\ &\textstyle\sum_{s=1,s\neq i,j}^{n}((a_{sj} - a_{si})(b^k_{\pi_s\pi_i} - b^k_{\pi_s\pi_j}) + \\ &(a_{js} - a_{is})(b^k_{\pi_i\pi_s} - b^k_{\pi_j\pi_s}))\end{aligned} \tag{4}$$

When A and $B^k, k = 1, \ldots, m$, are symmetric,

$$\Delta(\pi, k, i, j) = 2 \sum_{s=1,s\neq i,j}^{n} (a_{sj} - a_{si})(b^k_{\pi_s\pi_i} - b^k_{\pi_s\pi_j}) \tag{5}$$

Then, the function value of the neighboring solution $\bar{\pi}$ after swapping the elements i and j is

$$C^k(\bar{\pi}) = C^k(\pi) + \Delta(\pi, k, i, j), k = 1, \ldots, m. \tag{6}$$

The computational complexity in (6) is only $O(n)$, which is much less than the complexity of evaluating $C(\bar{\pi})$ in (1) (i.e. $O(n^2)$).

3 Greedy Randomized Adaptive Search Procedure

GRASP is a multi-start metaheuristic algorithm, which repeatedly improves starting solutions by local search. At each iteration of GRASP, a greedy randomized constructive procedure and a local search procedure are involved. The best local optimum collected over all local searches is retained and returned as the final solution of GRASP.

3.1 Greedy Randomized Construction

A greedy randomized construction procedure for building starting solutions is shown in Fig. 1. Initially, a partial solution S is set as an empty set. Then, the greedy function values of all unselected components in E are evaluated. To make better contribution to the partial solution S, a restricted candidate list (RCL) is formed by the components with low g values in E. One of the commonly-used strategies to determine RCL is to select the elements with g values between

$$[g^{min}, g^{min} + \alpha \times (g^{max} - g^{min})],$$

where $g^{min} = \min\{g(e)|e \in \text{E}\}$ and $g^{max} = \max\{g(e)|e \in \text{E}\}$. Here, $\alpha \in [0, 1]$ is a parameter to balance the greediness and randomness of the partial solution S. When $\alpha = 0$, only the component with the minimal g value will be selected. This component should make the biggest contribution to the partial solution. On the contrary, when $\alpha = 1$, all candidate components in E have equal chance to be selected. That is, the construction procedure will pick unselected components randomly. In practice, α is set to be either fixed or adaptive.

```
1  begin
2      S := ∅ and E := {all components of solution}.
3      while E is not empty do
4          foreach e in E do compute greedy function value g(e);
5          Define RCL as the set of elements in E with low g values;
6          Select an element ē ∈ RCL randomly;
7          Add ē to partial solution (i.e., S := S ∪ {ē});
8          Remove ē from E (i.e., E := E\{ē}).
9      end
10 end
```

Fig. 1. Greedy Randomized Constructive Procedure of GRASP

3.2 Local Search Procedure

Following the construction step, local search is applied to improve starting solutions. Two basic strategies - first improvement and best improvement, are often considered to accept local search moves. In first improvement, the first neighbor with better objective function value examined is accepted as the new current solution. In contrast, best improvement examines all neighbors and accepts the best one as the new current solution. More sophisticated local search methods with good global search ability, such as simulated annealing and tabu search, have also been suggested to improve the starting solutions in GRASP [13].

4 The Proposed mGRASP/MH Algorithm

4.1 Motivation

In [10], a GRASP algorithm, denoted mGRASP here, was developed to tackle the multi-objective knapsack problem. Like single-objective GRASP algorithms, mGRASP uses a greedy randomized construction step and a local search step. At each iteration, a weighted sum function is defined as the utility function for selecting greedy elements in the construction step and accepting better neighbors in the local search step.

To find a diverse set of Pareto-optimal solutions, mGRASP uses multiple distinct weight vectors evenly spread. According to the experimental setting reported in [10], up to one thousand weight vectors are used in one thousand iterations of mGRASP. Note that each iteration of mGRASP is independent from the other iterations. As shown in [6,11], the adaptation of finite weight vectors and the cooperation between solutions with similar weight vectors could benefit the diversity and convergence in multi-objective search. These strategies can be easily used in mGRASP.

Inspired by the POP principle, the guided mutation operator generates solution in a different way to greedy randomized construction [8]. This operator uses the global information in a probabilistic model to disturb the elite solutions

```
1  Algorithm 1: mGRASP/MH
   input  : N: population size, α: balance factor between greediness and
            randomness, β: proportion of components from elite solution
   output: NDS: the set of all non-dominated solutions
2  Initialize P = {π^(1), ..., π^(N)} and W = {λ^(1), ..., λ^(N)}.
3  begin
4      repeat
5          foreach i ∈ {1, ..., N} do
6              Step 1: Generate greedy solution π based on λ^(i) and π^(i);
7              Step 2: Apply local search on π to produce π' and update NDS;
8              Step 3: Replace the worse members in P with π';
9              Step 4: Modify the search direction λ^i adaptively.
10         end
11     until stopping condition is satisfied ;
12 end
```

Fig. 2. Framework of mGRASP/MH

found during the search. This idea has not yet been used in multi-objective algorithms. Then, we improve the performance of mGRASP by constructing promising starting solutions based on elite solutions.

4.2 mGRASP/MH for the QAP

We propose an elitist multi-objective GRASP metaheuristic in this paper, called mGRASP/MH. At each iteration, a population $P = \{\pi^{(1)}, \ldots, \pi^{(N)}\}$ of solutions and a set of corresponding weight vectors $W = \{\lambda^{(1)}, \ldots, \lambda^{(N)}\}$ are maintained. The framework of mGRASP/MH is shown in Fig. 2. The four main steps in lines 6-9 are involved in the main loop of mGRASP/MH. In the following, each of these steps is detailed.

Step 1: Elitist-based Greedy Construction. Unlike the greedy randomized construction algorithm in Fig. 1, the construction algorithm shown in Fig. 3 uses not only problem-specific greedy information but also the elite solution $\pi^{(i)}$ found in the previous local search. Parameter α is used to balance the greediness and the randomness of the partial solution. The parameter $\beta \in [0, 1]$ is used to control the proportion of components copied from the elite solution $\pi^{(i)}$. n_0 is the number of elements copied from $\pi^{(i)}$. ϕ is a random order of locations. L' denotes the set of locations assigned. In lines 4-6, n_0 components in $\pi^{(i)}$ are directly copied into a new solution π. Line 7 calculates the cost of the partial solution containing the components only from elite solutions. LOC and FAC in line 8 are the set of locations and facilities unassigned.

The ground set E is composed of all unassigned (location, facility) pairs. For each pair, the growth in cost is computed in lines 11-13. The associated g value is obtained in line 14. In line 16, RCL is formed by selecting a set of (location,

```
1  Algorithm 2: ElitistGreedyConstruction(α, β, λ^(i), π^(i))
   input  : λ^(i): current weight vector, π^(i): elite solution
   output: π: greedy randomized elite solution
2  begin
3      Set φ = {φ_1, ..., φ_n} to be a random permutation of L = {1, ..., n},
       n_0 = ⌊β × n⌋, and L' = ∅;
4      for c = 1 to n_0 do
5          π_{φ_c} = π^(i)_{φ_c}; L' = L' ∪ {φ_c};
6      end
7      for k = 1 to m do C^k = Σ_{i∈L'} Σ_{j∈L'} a_ij · b^k_{π_i π_j};
8      Set LOC = {φ_{n_0+1}, ..., φ_n} and FAC = {π^(i)_{φ_{n_0+1}}, ..., π^(i)_{φ_n}};
9      while FAC is not empty do
10         foreach (lc, fc) ∈ M = LOC × FAC do
11             for k = 1 to m do
12                 Δ(lc, fc, k) = Σ_{p∈L'} a_{p,lc} b^k_{π_p,fc} + Σ_{q∈L'} a_{lc,q} b^k_{fc,π_q} + a_{lc,lc} b^k_{fc,fc}
13             end
14             g(lc, fc) = Σ_{k=1}^{m} λ^(i)_k · (C^k + Δ(lc, fc, k));
15         end
16         RCL = {(lc, fc) | g^min ≤ g(lc, fc) ≤ g^min + α(g^max − g^min)};
17         Randomly select a pair (lc', fc') from RCL and set π_{lc'} = fc' and
           L' = L' ∪ {lc'};
18         for k = 1 to m do C^k = C^k + Δ(lc', fc', k);
19         Set LOC = LOC\{lc'} and FAC = FAC\{fc'}.
20     end
21     return π;
22 end
```

Fig. 3. Elitist-based Greedy Construction Procedure for the mQAP

facility) pairs with the g values between $[g^{min}, g^{min} + \alpha(g^{max} - g^{min})]$, where $g^{min} = \min\{g(lc, fc)|(lc, fc) \in M\}$ and $g^{max} = \max\{g(lc, fc)|(lc, fc) \in M\}$. One pair (lc', fc') of (location, facility) is randomly selected from RCL and updates the partial solution in line 17. In line 18, the total cost of partial solution with the pair selected in the previous step is computed. Line 19 removes lc' and fc' from the sets of unassigned locations and facilities respectively. This procedure is repeated until the set FAC is empty. Finally, a complete solution is returned.

Step 2: Local Search. After constructing an elite greedy solution, a local search procedure is triggered and guided by the weighted sum function with $\lambda^{(i)}$ in (3). In mGRASP/MH, 2-opt local search with first improvement is used for the mQAP. Each local search procedure is terminated if there is no solution in its neighborhood with better fitness. Since all members of the population have different weight vectors (i.e. search directions), the set of all local optima found for all search directions is likely to cover the POF reasonably well. The set NDS

is updated when a successful local move is made. On the one hand, the current solution is added to *NDS* if it is not dominated by any member of *NDS*. On the other hand, any members of *NDS* dominated by the current solution are removed from this set.

Step 3: Selection. As discussed in [11,6], optimal solutions obtained with similar weight vectors should be similar in the objective space and decision space. Cooperation between solutions with similar weighted sum functions can be very helpful for finding good approximations to the POF. Therefore, the local optima obtained in **Step 2** is very likely to be better than the solutions in the population with similar weight vectors. In this paper, we compare π with all $\pi^{(i)} \in P, i = 1, \ldots, N$. If $f(\pi|\lambda^{(i)}) < f(\pi^{(i)}|\lambda^{(i)})$, then set $\pi^{(i)} = \pi$. In mGRASP, solutions found in different iterations do not interact.

Step 4: Modification of Search Direction. Ideally, finding the optimal solutions of all weighted sum functions leads to a good approximation of the POF. However, this is impossible in mGRASP/MH since a population of fixed size is used. In [11], we have suggested an adaptive mechanism to tune the weight vector of each solution according to the locations of some solutions previously examined. In this mechanism, the non-dominated neighboring solution π' that is nearest to $\pi^{(i)}$ is identified. For each objective k, if $C^k(\pi') < C^k(\pi^{(i)})$, then decrease $\lambda_k^{(i)}$ by δ (> 0); otherwise, increase by δ. If $\lambda_k^{(i)}$ exceeds the bounds, then use the nearest bound to replace it. As a result, the optimal solution of the weighted sun function with the modified weight vector should be moved away from π' in the objective space. In such a way, the sparse part of POF can be explored more intelligently and efficiently. In this paper, we use this strategy in a slightly different manner. Each search direction is modified with a probability.

5 Computational Experiments

5.1 Performance Assessment

To quantitatively evaluate the non-dominated solutions found by each algorithm, we use both the generational distance (GD) metric and the inverted generational distance (IGD) metric. Assume S is the final set of non-dominated solutions found by multi-objective algorithm and S^* is a set of reference solutions, either the true POF or a very good approximation. The GD metric measures the average distance from S to S^*, while the IGD metric measures the average distance from S^* to S [14]. These two metrics can be formulated as follows: $\text{GD}(S, S^*) = \frac{1}{|S|}\sum_{u \in S} \min\{dist(u,v)|v \in S^*\}$ and $\text{IGD}(S^*, S) = \frac{1}{|S^*|}\sum_{u \in S^*} \min\{dist(u,v)|v \in S\}$, where $dist(u,v)$ is the Euclidean distance between two objective vectors. The smaller the GD or IGD values, the better quality of the set S. In this paper, the reference set for each instance is formed by collecting all non-dominated solutions found by five algorithms in 20 runs.

5.2 Experimental Settings

We used a set of 18 benchmark mQAP instances to test the performance of mGRASP/MH. These test instances were generated by Knowles [15] and are available at *http://dbkgroup.org/knowles/mQAP/*. The correlation values between flow matrices of these test instances are shown in Table 1.

Table 1. Correlations between the flows of the 18 benchmark mQAP test instances

Instance	$c(B^1, B^2)$	Instance	$c(B^1, B^2)$	Instance	$c(B^1, B^2)$,$c(B^1, B^3)$
KC10-2fl-1uni	0	KC20-2fl-1uni	0	KC30-3fl-1uni	(0, 0)
KC10-2fl-2uni	0.8	KC20-2fl-2uni	0.7	KC30-3fl-2uni	(0.4,0.4)
KC10-2fl-3uni	-0.8	KC20-2fl-3uni	-0.7	KC30-3fl-3uni	(-0.4, -0.4)
KC10-2fl-1rl	0	KC20-2fl-1rl	0	KC30-3fl-1rl	(0.4, 0)
KC10-2fl-2rl	0.7	KC20-2fl-2rl	0.4	KC30-3fl-2rl	(0.7, -0.5)
KC10-2fl-3rl	-0.7	KC20-2fl-3rl	-0.4	KC30-3fl-3rl	(-0.4, -0.4)

We compared mGRASP/MH to mGRASP and to three state-of-the-art EMO algorithms - MOEA/D, NSGA2, and SPEA2. In MOEA/D, the mQAP is converted into a number of single objective subproblems. These subproblems are optimized by an evolutionary algorithm simultaneously. The best solutions to all subproblems found so far are retained in its population. The distribution of these solutions is controlled by the diversity of weight vectors. Each offspring solution in MOEA/D is improved by local search. In both NSGA2 and SPEA2, the non-dominated solutions found so far have priority to survive in the population. The diversity of these non-dominated solutions is maintained by estimating their density. In this paper, we use cycle crossover [16] and mutation based on the 2-opt swap for the MOEA/D, NSGA2, and SPEA2 algorithms.

In both mGRASP and mGRASP/MH, α is set to 0.1. Parameter β is set to 0.5. That is, half of the components in elite solutions are copied to the construction procedure of mGRASP/MH. The population size (N) in mGRASP/MH is 50 for all instances. The δ value for changing weight is 0.01. The population size in NSGA2, SPEA2, and MOEA/D is 100. In MOEA/D, the neighborhood size of each subproblem is 20 for all test instances.

We run each algorithm on each instance 20 times. All algorithms are coded in C++ and executed on a PC with CPU (Intel (R) Core (TM) 2, 1.86GHZ) and RAM (2GB). Every algorithm uses the same computational time for the same test instance. The computational times used for the instances with 10, 20, and 30 locations are set to 10, 20, and 30 seconds, respectively.

5.3 Discussions of Results

The mean GD and IGD values found by the five algorithms are summarized in Table 2 and Table 3. It is evident that mGRASP/MH and MOEA/D clearly outperform the other three algorithms on all test instances. Among the five algorithms, NSGA2 and SPEA2 show the worst performance with respect to

Table 2. The mean GD values of non-dominated solutions found in 20 runs

Instance	mGRASP/MH	mGRASP	MOEA/D	NSGA2	SPEA2
KC10-2fl-1uni	0	592	1730	4462	6152
KC10-2fl-2uni	5305	0	5490	11800	13845
KC10-2fl-3uni	0	1	111	1357	2893
KC10-2fl-1rl	0	1129	22132	236966	321468
KC10-2fl-2rl	22086	16300	34471	157128	151661
KC10-2fl-3rl	0	1129	14979	244293	285310
KC20-2fl-1uni	9225	21758	11269	48813	53635
KC20-2fl-2uni	9138	58660	16364	65180	61904
KC20-2fl-3uni	3758	6966	4934	22133	29537
KC20-2fl-1rl	580688	2069384	509229	2996725	2565999
KC20-2fl-2rl	205812	1124948	155082	1372892	1117776
KC20-2fl-3rl	168651	476440	145244	1194632	1251489
KC30-3fl-1uni	41072	55178	18945	132735	163554
KC30-3fl-2uni	64156	111067	26085	153182	156566
KC30-3fl-3uni	30308	36855	14684	94685	123557
KC30-3fl-1rl	1302906	2491688	302268	3264761	3667731
KC30-3fl-2rl	877695	1931606	297531	3038431	3281139
KC30-3fl-3rl	917218	1427153	313038	3450839	3880325

Table 3. The mean IGD values of non-dominated solutions found in 20 runs

Instance	mGRASP/MH	mGRASP	MOEA/D	NSGA2	SPEA2
KC10-2fl-1uni	7	460	2211	6590	7795
KC10-2fl-2uni	4715	0	4915	11284	13196
KC10-2fl-3uni	0	6	147	2393	4387
KC10-2fl-1rl	266	3555	45512	318513	382993
KC10-2fl-2rl	8414	10460	128988	212026	226922
KC10-2fl-3rl	14	2403	37239	300822	357818
KC20-2fl-1uni	8509	21360	12058	53492	58575
KC20-2fl-2uni	10500	58830	16987	66425	64604
KC20-2fl-3uni	3526	6677	4878	35764	44289
KC20-2fl-1rl	467232	1980738	433020	2914559	2623621
KC20-2fl-2rl	280650	1259521	192956	1895681	1520627
KC20-2fl-3rl	205030	653760	153859	1594337	1534329
KC30-3fl-1uni	38396	54552	20578	141325	167422
KC30-3fl-2uni	63583	110308	26415	161061	163284
KC30-3fl-3uni	29342	36927	16133	127932	154106
KC30-3fl-1rl	1519861	3350333	474028	5962007	7018525
KC30-3fl-2rl	1062987	2837723	421962	4538068	4986717
KC30-3fl-3rl	974208	1658247	395310	4072323	4503648

minimizing the GD and IGD values. The main reason for this might be that no local search is used to improve offspring solutions in these two approaches.

The non-dominated solutions found by all five algorithms after 20 runs on the four 2-objective instances with zero correlation between flow matrices are

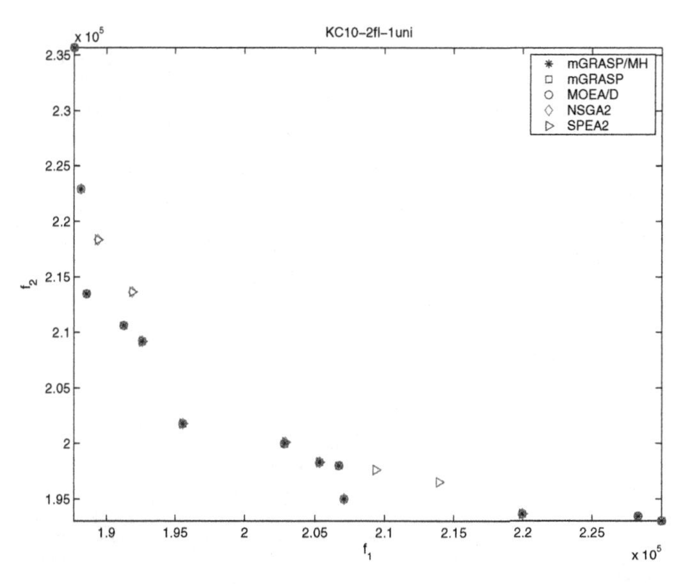

Fig. 4. Non-dominated solutions found by mGRASP/MH, mGRASP, MOEA/D, NSGA2, and SPEA2 on KC10-2fl-1uni in 20 runs

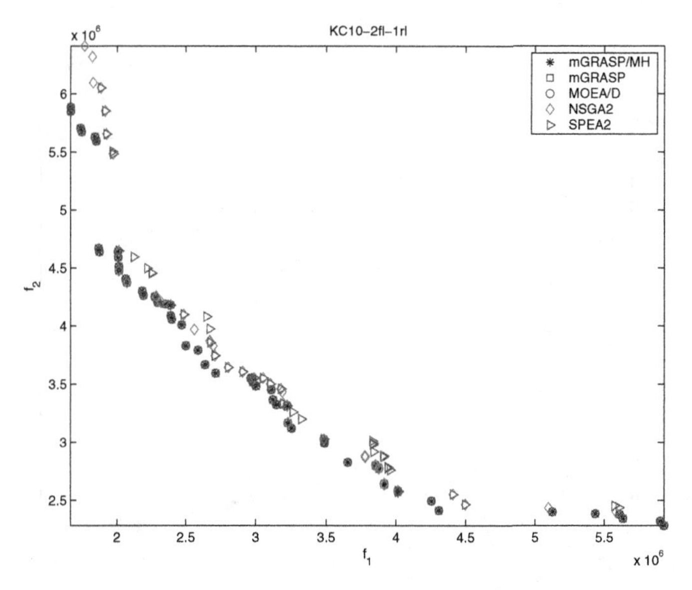

Fig. 5. Non-dominated solutions found by mGRASP/MH, mGRASP, MOEA/D, NSGA2, and SPEA2 on KC10-2fl-1rl in 20 runs

plotted in Figs. 4-7. It can be observed from Fig. 4 that all five algorithms find almost the same set of non-dominated solutions on instance KC10-2fl-1uni. The results in Fig. 6 and Fig. 7 show that both mGRASP/MH and MOEA/D clearly perform better than mGRASP on KC20-2fl-1uni and KC20-2fl-1rl. Figs. 5-7

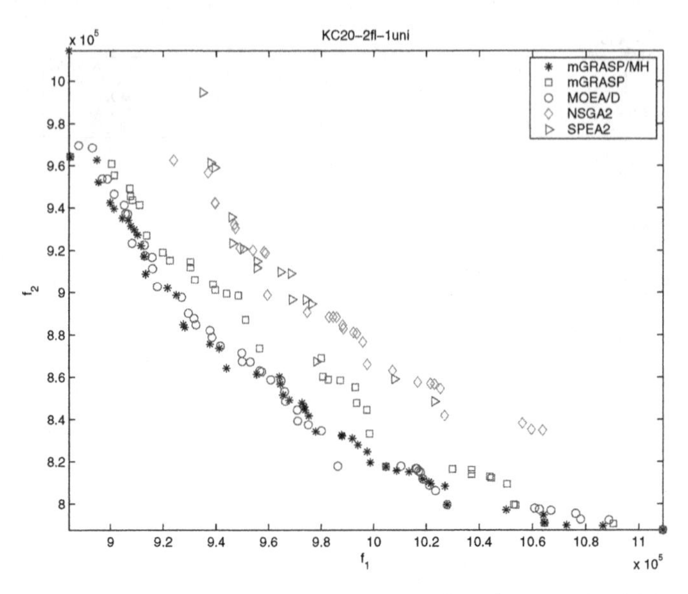

Fig. 6. Non-dominated solutions found by mGRASP/MH, mGRASP, MOEA/D, NSGA2, and SPEA2 on KC20-2fl-1uni in 20 runs

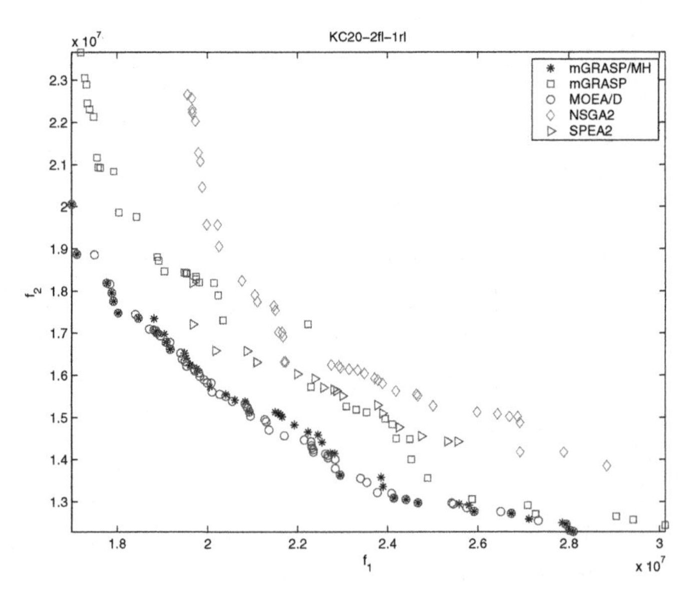

Fig. 7. Non-dominated solutions found by mGRASP/MH, mGRASP, MOEA/D, NSGA2, and SPEA2 on KC20-2fl-1rl in 20 runs

show that three local search-based metaheuristics - mGRASP/MH, mGRASP, and MOEA/D, find better solutions than two Pareto-based EMO algorithms - NSGA2 and SPEA2 on three instances - KC10-2fl-1rl, KC20-2fl-1uni, and KC20-2fl-1rl.

Results in Table 2 and Table 3 on six 3-objective test instances show that both mGRASP/MH and MOEA/D perform better than the other three algorithms in terms of the GD and IGD metrics. It can also be seen that mGRASP/MH finds the worse GD and IGD values than MOEA/D on these instances. It is easy to understand the reason behind the worse performance of mGRASP/MH. The greedy randomized construction procedure in mGRASP/MH has higher computational complexity than the crossover and mutation operators used in MOEA/D. Within the restricted computational time, MOEA/D could improve more new solutions by local search. This is also part of the reason that mGRASP/MH performs better than mGRASP. The former only builds half of the starting solution by greedy randomized construction procedure. Therefore, mGRASP/MH needs less time in the construction of a starting solution.

6 Conclusions

We proposed an elitist GRASP metaheuristic algorithm called mGRASP/MH to tackle the mQAP (multi-objective quadratic assignment problem). In the proposed approach, elitist-based greedy randomized construction, cooperation between solutions, and weight-vector adaptations are used to accelerate convergence and diversify the search. Our experimental results show that mGRASP/MH is competitive with MOEA/D and outperforms mGRASP and two Pareto-based EMO algorithms - NSGA2 and SPEA2 on the benchmark problem instances considered here. It has also been shown that the multi-objective metaheuristic algorithms using local search perform better than those without local search for the mQAP.

In this paper, the construction of starting solutions copies parts or components from elite solutions. Under the framework of mGRASP/MH, it is very easy to use other advanced techniques, such as guided mutation [8], cooperative strategy [17,18] and path relinking [19]. Complex memory structure for storing historical information from the search, probability distributions in guided mutation, all these should benefit the global search ability of mGRASP/MH. The cooperation between solutions obtained by different local search procedures can be implemented by considering path relinking [19] or tabu mechanisms [18]. These are some of our future research directions.

Acknowledgement. The work was funded by the UK's EPSRC, under grant EP/E019781/1. The authors would like to thank anonymous reviewers for their helpful and constructive comments.

References

1. Garey, M.R., Johnson, D.S.: Computers and Intractability: A Guide to the Theory of NP-Completeness. Series of Books in the Mathematical Sciences. W. H. Freeman, New York (1979)
2. Knowles, J.D., Corne, D.W.: Towards landscape analyses to inform the design of hybrid local search for the multiobjective quadratic assignment problem. In: Soft Computing Systems - Design, Management and Applications (HIS 2002), pp. 271–279 (2002)

3. Paquete, L., Stützle, T.: A study of stochastic local search algorithms for the biobjective QAP with correlated flow matrices. European Journal of Operational Research 169(3), 943–959 (2006)
4. Deb, K., Agrawal, S., Pratap, A., Meyarivan, T.: A fast and elitist multiobjective genetic algorithm: NSGA-II. IEEE Trans. on Evolutionary Computation 6(2), 182–197 (2002)
5. Zitzler, E., Laumanns, M., Thiele, L.: SPEA2: Improving the strength pareto evolutionary algorithm for multiobjective optimization. In: EUROGEN 2001 - Evolutionary Methods for Design, Optimisation and Control with Applications to Industrial Problems, Athens, Greece, pp. 95–100 (2001)
6. Zhang, Q., Li, H.: MOEA/D: A multiobjective evolutionary algorithm based on decomposition. IEEE Trans. on Evolutionary Computation 11(6), 712–731 (2007)
7. Glover, F., Laguna, M.: Tabu Search. Kluwer Academic Publishers, Dordrecht (1998)
8. Zhang, Q., Sun, J.: Iterated local search with guided mutation. In: The 2006 IEEE Congress on Evolutionary Computation (CEC 2006), Vancouver (2006)
9. Feo, T.A., Resende, M.G.C.: A probabilistic heuristic for a computationally difficult set covering problem. Operations Research Letters 8, 67–71 (1989)
10. Vianna, D.S., Arroyo, J.E.C.: A GRASP algorithm for the multi-objective knapsack problem. In: The First International Conference on Quantitative Evaluation of Systems. IEEE Computer Society, Los Alamitos (2004)
11. Li, H., Landa-Silva, J.D.: Evolutionary multi-objective simulated annealing with adaptive and competitive search direction. In: The 2008 IEEE Congress on Evolutionary Computation (CEC 2008), Hong Kong, pp. 3310–3317. IEEE Press, Los Alamitos (2008)
12. Taillard, É.D.: Robust taboo search for the quadratic assignment problem. Parallel Computing 17(4-5), 443–455 (1991)
13. Resende, M.G.C.: Metaheuristic hybridization with GRASP. In: Chen, Z.L., Raghavan, S. (eds.) Tutorials in Operations Research, INFORMS (2008)
14. Czyzak, P., Jaszkiewicz, A.: Pareto-simulated annealinga metaheuristic technique for multi-objective combinatorial optimization. J. Multi-Criteria Decis. Anal. 7(1), 34–47 (1998)
15. Knowles, J.D., Corne, D.W.: Instance generators and test suites for the multiobjective quadratic assignment problem. In: Fonseca, C.M., Fleming, P.J., Zitzler, E., Deb, K., Thiele, L. (eds.) EMO 2003. LNCS, vol. 2632, pp. 295–310. Springer, Heidelberg (2003)
16. Oliver, I.M., Smith, D.J., Holland, J.R.C.: A study of permutation crossover operators on the traveling salesman problem. In: Proceedings of the Second International Conference on Genetic Algorithms on Genetic algorithms and their application, Hillsdale, NJ, USA, pp. 224–230. L. Erlbaum Associates Inc., Mahwah (1987)
17. Burke, E.K., Landa-Silva, J.D.: The influence of the fitness evaluation method on the performance of multiobjective search algorithms. European Journal of Operational Research 169(3), 875–897 (2006)
18. Landa-Silva, J.D., Burke, E.K.: Asynchronous cooperative local search for the office space allocation problem. INFORMS Journal on Computing 19(4), 575–587 (2007)
19. Oliveira, C.A.S., Pardalos, P.M., Resende, M.G.C.: GRASP with path-relinking for the quadratic assignment problem. In: Ribeiro, C.C., Martins, S.L. (eds.) WEA 2004. LNCS, vol. 3059, pp. 356–368. Springer, Heidelberg (2004)

Multi-Objective Particle Swarm Optimizers: An Experimental Comparison

Juan J. Durillo[1], José García-Nieto[1], Antonio J. Nebro[1], Carlos A. Coello Coello[2,*], Francisco Luna[1], and Enrique Alba[1]

[1] Department of Computer Science, University of Málaga, Spain
{durillo,jnieto,antonio,flv,eat}@lcc.uma.es
[2] Department of Computer Science, CINVESTAV-IPN, Mexico
ccoello@cs.cinvestav.mx

Abstract. Particle Swarm Optimization (PSO) has received increasing attention in the optimization research community since its first appearance in the mid-1990s. Regarding multi-objective optimization, a considerable number of algorithms based on Multi-Objective Particle Swarm Optimizers (MOPSOs) can be found in the specialized literature. Unfortunately, no experimental comparisons have been made in order to clarify which MOPSO version shows the best performance. In this paper, we use a benchmark composed of three well-known problem families (ZDT, DTLZ, and WFG) with the aim of analyzing the search capabilities of six representative state-of-the-art MOPSOs, namely, NSPSO, SigmaMOPSO, OMOPSO, AMOPSO, MOPSOpd, and CLMOPSO. We additionally propose a new MOPSO algorithm, called SMPSO, characterized by including a velocity constraint mechanism, obtaining promising results where the rest perform inadequately.

Keywords: Particle Swarm Optimization, Multi-Objective Optimization, Comparative Study.

1 Introduction

The relative simplicity and competitive performance of the Particle Swam Optimization (PSO) [11] algorithm as a single-objective optimizer have favored the use of this bio-inspired technique when dealing with many real-word optimization problems [17]. A considerable number of these optimization problems require the optimization of more than one objective at the same time which are in conflict with respect to each other. These properties, along with the fact that PSO is a population-based metaheuristic, have made it a natural candidate to be extended for multi-objective optimization. Since the first proposed Multi-Objective Particle Swarm Optimizer (MOPSO) developed by Moore and Chapman in 1999 [15], more than thirty different MOPSOs have been reported in the specialized literature. Reyes and Coello [17] carried out a survey of the existing MOPSOs, providing a complete taxonomy of such algorithms. In that

* The fourth author is also affiliated to the UMI 1375 CNRS.

M. Ehrgott et al. (Eds.): EMO 2009, LNCS 5467, pp. 495–509, 2009.

work, the authors considered as the main features of all existing MOPSOs the following ones: the existence of an external archive of non-dominated solutions, the selection strategy of non-dominated solutions as leaders for guiding the swarm, the neighborhood topology, and the existence or not of a mutation operator.

In this work, we are interested in analyzing six representative state-of-the-art MOPSOs in order to provide hints about their search capabilities. Five of them were selected from Reyes and Coello's survey, namely: NSPSO [14], SigmaMOPSO [16], OMOPSO [18], AMOPSO [19], and MOPSOpd [1]. An approach not covered in the survey is also compared: MOCLPSO [9].

With the aim of assessing the performance of these algorithms, we have used three benchmarks of multi-objective functions covering a broad range of problems with different features (concave, convex, disconnected, deceptive, etc.). These benchmarks include the test suites Zitzler-Deb-Thiele (ZDT) [20], the Deb-Thiele-Laumanns-Zitzler (DTLZ) problem family [5], and the Walking-Fish-Group (WFG) test problems [10]. The experimental methodology we have followed consists of computing a pre-fixed number of function evaluations and then comparing the obtained results by considering three different quality indicators: additive unary epsilon [13], spread [4], and hypervolume [21]. The results of our study reveal that many MOPSOs have difficulties when facing some multi frontal problems. We analyze this issue and propose a new algorithm, called SMPSO, which incorporates a velocity constraint mechanism. We find that SMPSO shows a promising behavior on those problems where the other algorithms fail.

The remainder of this paper is organized as follows. Section 2 includes basic background about PSO and MOPSO algorithms. In Section 3, we briefly review the studied approaches focusing on their main features. Section 4 is devoted to the experimentation, including the parameter settings and the methodology adopted in the statistical tests. In Section 5, we analyze the obtained results regarding the three quality indicators mentioned before. The results are discussed in Section 6, where a new MOPSO based on a constraint velocity mechanism is introduced. Finally, Section 7 contains the conclusions and some possible paths for future work.

2 PSO Background

PSO is a population-based metaheuristic inspired on the social behavior of birds within a flock. In a PSO algorithm each potential solution to the problem is called a *particle* and the population of solutions is called a *swarm*. The way in which PSO updates the particle $\boldsymbol{x}_i$ at the generation t is through the formula:

$$\boldsymbol{x}_i(t) = \boldsymbol{x}_i(t-1) + \boldsymbol{v}_i(t) \quad (1)$$

where the factor $\boldsymbol{v}_i(t)$ is known as velocity and it is given by

$$\boldsymbol{v}_i(t) = w * \boldsymbol{v}_i(t-1) + C1 * r1 * (\boldsymbol{x}_{pbest_i} - \boldsymbol{x}_i) + C2 * r2 * (\boldsymbol{x}_{gbest_i} - \boldsymbol{x}_i) \quad (2)$$

In this formula, $\boldsymbol{x}_{pbest_i}$ is the best solution that $\boldsymbol{x}_i$ has viewed, $\boldsymbol{x}_{gbest_i}$ is the best particle (also known as the *leader*) that the entire swarm has viewed, w is

Algorithm 1. Pseudocode of a general PSO algorithm.

```
1: initializeSwarm()
2: locateLeader()
3: generation = 0
4: while generation < maxGenerations do
5:    for each particle do
6:       updatePosition() // flight (Formulas 1 and 2)
7:       evaluation()
8:       updatePbest()
9:    end for
10:   updateLeader()
11:   generation ++
12: end while
```

the inertia weight of the particle and controls the trade-off between global and local experience, $r1$ and $r2$ are two uniformly distributed random numbers in the range $[0, 1]$, and $C1$ and $C2$ are specific parameters which control the effect of the personal and global best particles.

Algorithm 1 describes the pseudo-code of a general single-objective PSO. The algorithm starts by initializing the swarm (Line 1), which includes both the positions and velocities of the particles. The corresponding *pbest* of each particle is initialized, as well as the leader (Line 2). Then, during a maximum number of iterations, each particle *flies* through the search space updating its position (Line 6), it is evaluated (Line 7), and its *pbest* is also calculated (Lines 6-8). At the end of each iteration, the leader is updated. As commented before, the leader can be the *gbest* particle in the swarm. However, it can be a different particle depending on the *social structure* of the swarm (i.e., the topology of the neighborhood of each particle) [12].

To apply a PSO algorithm in multi-objective optimization the previous scheme has to be modified to cope with the fact that the solution of a problem with multiple objectives is not a single one but a set of non-dominated solutions. Issues that have to be considered are [17]:

1. How to select the set of particles to be used as leaders?
2. How to retain the non-dominated solutions found during the search?
3. How to maintain diversity in the swarm in order to avoid convergence to a single solution?

The pseudo-code of a general MOPSO is included in Algorithm 2. After initializing the swarm (Line 1), the typical approach is to use an external archive to store the leaders, which are taken from the non-dominated particles in the swarm. After initializating the leaders archive (Line 2), some quality measure has to be calculated (Line 3) for all the leaders to select usually one leader for each particle of the swarm. In the main loop of the algorithm, the flight of each particle is performed after a leader has been selected (Lines 7-8) and, optionally, a mutation or *turbulence* operator can be applied (Line 9); then, the particle is evaluated and its corresponding *pbest* is updated (Lines 10-11). After each iteration, the set of leaders is updated and the quality measure is calculated again (Lines 13-14). After the termination condition, the archive is returned as the

Algorithm 2. Pseudocode of a general MOPSO algorithm.

```
1: initializeSwarm()
2: initializeLeadersArchive()
3: determineLeadersQuality()
4: generation = 0
5: while generation < maxGenerations do
6:    for each particle do
7:       selectLeader()
8:       updatePosition() // flight (Formulas. 1 and 2)
9:       mutation()
10:      evaluation()
11:      updatePbest()
12:   end for
13:   updateLeadersArchive()
14:   determineLeadersQuality()
15:   generation ++
16: end while
17: returnArchive()
```

result of the search. For further details about the operations contained in the MOPSO pseudocode, please refer to [17].

3 Studied Approaches

The studied approaches we have considered in this work can be classified as *Pareto-based* MOPSOs [17]. The basic idea, commonly found in all these algorithms, is to select as leaders the particles that are non-dominated with respect to the swarm. However, this leader selection scheme can be slightly different depending on the additional information each algorithm includes on its own mechanism (e.g., information provided by a density estimator). We summarize next the main features of the considered MOPSOs:

- **Non-dominated Sorting PSO:** NSPSO [14] incorporates the main mechanisms of NSGA-II [4] to a PSO algorithm. In this approach, once a particle has updated its position, instead of comparing the new position only against the *pbest* position of the particle, all the *pbest* positions of the swarm and all the new positions recently obtained are combined in just one set (given a total of $2N$ solutions, where N is the size of the swarm). Then, NSPSO selects the best solutions among them to conform the next swarm (by means of a non-dominated sorting). This approach also selects the leaders randomly from the leaders set (stored in an external archive) among the best of them, based on two different mechanisms: a niche count and a nearest neighbor density estimator. This approach uses a mutation operator that is applied at each iteration step only to the particle with the smallest density estimator value.
- **SigmaMOPSO:** In SigmaMOPSO [16], a sigma value is assigned to each particle of the swarm and of an external archive. Then, a given particle of the swarm selects as its leader to the particle of the external archive with the closest sigma value. The use of the sigma values makes the selection pressure of PSO even higher, which may cause premature convergence in some cases.

To avoid this, a turbulence operator is used, which is applied on the decision variable space.
- **Optimized MOPSO:** The main features of OMOPSO [18] include the use of the crowding distance of NSGA-II to filter out leader solutions and the combination of two mutation operators to accelerate the convergence of the swarm. The original OMOPSO algorithm makes use of the concept of ϵ-dominance to limit the number of solutions produced by the algorithm. We consider here a variant discarding the use of ϵ-dominance, being the leaders archive the result of the execution of the technique.
- **Another MOPSO:** AMOPSO [19] uses the concept of Pareto dominance to determine the flight direction of a particle. The authors adopt clustering techniques to divide the population of particles into several swarms. This aims at providing a better distribution of solutions in the decision variable space. Each sub-swarm has its own set of leaders (non-dominated particles). In each sub-swarm, a PSO algorithm is executed (leaders are randomly chosen) and, at some point, the different sub-swarms exchange information: the leaders of each swarm are migrated to a different swarm in order to variate the selection pressure. Also, this approach does not use an external archive since elitism in this case is an emergent process derived from the migration of leaders.
- **Pareto Dominance MOPSO:** in MOPSOpd [1], the authors propose methods based exclusively on Pareto dominance for selecting leaders from a non-dominated external archive. Three different selection techniques are presented: one technique that explicitly promotes diversity (called Rounds by the authors), one technique that explicitly promotes convergence (called Random), and finally one technique that is a weighted probabilistic method (called Prob) reaching a compromise between Random and Rounds. Additionally, MOPSOpd uses a turbulence factor that is added to the position of the particles with certain probability; we have used the same operator applied in SigmaMOPSO.
- **Comprehensive Learning MOPSO:** MOCLPSO [9] incorporates a Pareto dominance mechanism to the CLPSO algorithm for selecting leaders from non-dominated external archive. In this approach, a crowding distance method is used to estimate the density of the solutions once the external archive reaches its maximum allowable size. The distance values of all the archive members are calculated and sorted from large to small. The first $Nmax$ (maximum size of archive) members are kept whereas the remaining ones are deleted from the archive. The leaders are randomly chosen from this external archive of non-dominated solutions. In MOCLPSO, no perturbation methods are applied to keep the diversity through the evolution steps.

4 Experimentation

In this section, we detail the parameter settings we have used, as well as the methodology followed in the experiments.

The benchmarking MOPs chosen to evaluate the six MOPSOs have been the aforementioned ZDT [20], DTLZ [5], and WFG [10] test suites, leading to a total number of 21 problems. The two latter families of MOPs have been used with their bi-objective formulation. For assessing the performance of the algorithms, we have considered three quality indicators: additive unary epsilon indicator ($I^1_{\epsilon+}$) [13], spread (Δ) [4], and hypervolume (HV) [21]. The two first indicators measure, respectively, the convergence and the diversity of the resulting Pareto fronts, while the last one measures both convergence and diversity.

All the algorithms have been implemented using jMetal [7], a Java-based framework for developing metaheuristics for solving multi-objective optimization problems.

4.1 Parameterization

We have chosen a common subset of parameter settings which are the same to all the algorithms. Thus, the size of the swarm and the leader archive, when applicable, is fixed to 100 particles, and the stopping condition is always to perform 250 iterations (yielding a total of 25,000 function evaluations). If we consider NSPSO, for example, the swarm size and the number of iterations used in [14] is 200 and 100, respectively. Our approach has been to establish common settings in order to make a fair comparison, keeping the rest of the parameters according to the papers where the algorithms were originally described.

The parameter settings are summarized in Table 1. For those particular parameters that have not been explained, please see the references for further details.

4.2 Methodology

To assess the search capabilities of the algorithms, we have made 100 independent runs of each experiment, and we have obtained the median, $\tilde{x}$, and interquartile range, IQR, as measures of location (or central tendency) and statistical dispersion, respectively. Since we are dealing with stochastic algorithms and we want to provide the results with statistical confidence, the following statistical analysis has been performed in all this work [6]. Firstly, a Kolmogorov-Smirnov test is applied in order to check whether the values of the results follow a normal (Gaussian) distribution or not. If the distribution is normal, the Levene test checks for the homogeneity of the variances. If samples have equal variance (positive Levene test), an ANOVA test is done; otherwise a Welch test is performed. For non-Gaussian distributions, the non-parametric Kruskal-Wallis test is used to compare the medians of the algorithms. We always consider a confidence level of 95% (i.e., significance level of 5% or p-value below 0.05) in the statistical tests. Successful tests are marked with '+' symbols in the last column in all the tables containing the results; conversely, '-' means that no statistical confidence was found (p-value > 0.05). The best result for each problem has a gray colored background. For the sake of a better understanding of the results, we have also used a clearer grey background to indicate the second best result.

Table 1. Parameterization

Common parameters	
Swarm size	100 Particles
Iterations	250
NSPSO [14]	
Variant	CD (Crowding distance)
C_1, C_2	2.0
w	Decreased from 1.0 to 0.4
SigmaMOPSO [16]	
Archive size	100
C_1, C_2	2.0
w	0.4
Mutation	$newPosition = position + rand(0.0, 1.0) * position$
Mutation probability	0.05
OMOPSO [18]	
Archive size	100
C_1, C_2	$rand(1.5, 2.0)$
w	$rand(0.1, 0.5)$
Mutation	uniform + non-uniform + no mutation
Mutation probability	Each mutation is applied to 1/3 of the swarm
AMOPSO [19]	
Number of subswarms	5
C_1, C_2	2.0
w	0.4
MOPSOpd [1]	
Archive Size	100
C_1, C_2	1.0
w	0.5
Mutation	$newPosition = position + rand(0.0, 1.0) * position$
Mutation probability	0.05
Selection method	Rounds
MOCLPSO [9]	
Archive Size	100
C_1, C_2	*N/A*
w	0.9 to 0.2

To further analyze the results statistically, we have also included a post-hoc testing phase which allows for a multiple comparison of samples [8]. We have used the `multcompare` function provided by Matlab© for that purpose.

5 Computational Results

This section is devoted to evaluating and analyzing the results of the experiments. We start by discussing the values obtained after applying the $I^1_{\epsilon+}$ quality indicator, which are contained in Table 2. We can observe that OMOPSO clearly outperforms the rest of MOPSOs according to this indicator, achieving the lowest (best) values in 13 out of the 21 problems composing the benchmark. It also obtains six second best values. The next best performing algorithms are SigmaMOPSO, MOPSOpd, and AMOPSO, which get similar numbers of best and second best results. Thus, we can claim that OMOPSO produces solution sets having better convergence to the Pareto fronts in most of the benchmark problems considered in our study. All the results have statistical significance, as it can be seen in the last column, where only '+ ' symbols are found.

The values obtained after applying the Δ quality indicator are included in Table 3. We can observe again that OMOPSO is clearly the best performing

Table 2. Median and interquartile range of the $I^1_{\epsilon+}$ quality indicator

Problem	NSPSO $\bar{x}_{IQR}$	SigmaMOPSO $\bar{x}_{IQR}$	OMOPSO $\bar{x}_{IQR}$	AMOPSO $\bar{x}_{IQR}$	MOPSOpd $\bar{x}_{IQR}$	MOCLPSO $\bar{x}_{IQR}$	
ZDT1	$4.57e-1_{3.7e-1}$	$3.07e-2_{2.6e-2}$	$6.36e-3_{5.1e-4}$	$2.41e-1_{8.0e-2}$	$6.75e-2_{1.6e-2}$	$3.74e-1_{8.8e-2}$	+
ZDT2	$1.54e+0_{8.5e-1}$	$1.00e+0_{0.0e+0}$	$6.19e-3_{5.4e-4}$	$6.33e-1_{8.3e-1}$	$1.00e+0_{8.9e-1}$	$6.45e-1_{1.4e-1}$	+
ZDT3	$9.14e-1_{4.1e-1}$	$9.75e-1_{8.3e-1}$	$1.32e-2_{7.7e-3}$	$7.30e-1_{3.5e-1}$	$1.66e-1_{1.1e-1}$	$5.97e-1_{2.0e-1}$	+
ZDT4	$4.14e+1_{1.6e+1}$	$8.30e+0_{6.8e+0}$	$5.79e+0_{4.3e+0}$	$1.21e+1_{7.6e+0}$	$4.23e+0_{2.1e+0}$	$1.71e+1_{1.3e+1}$	+
ZDT6	$1.81e-1_{3.2e-1}$	$5.91e-3_{1.1e-3}$	$4.65e-3_{4.2e-4}$	$1.69e-1_{6.0e-2}$	$1.21e-1_{7.0e-2}$	$3.38e+0_{3.8e-1}$	+
DTLZ1	$2.30e+1_{8.0e+0}$	$2.54e+1_{1.3e+1}$	$1.92e+1_{1.1e+1}$	$8.46e+0_{1.9e+1}$	$1.72e+1_{1.1e+1}$	$2.12e+1_{8.0e+0}$	+
DTLZ2	$4.41e-2_{6.5e-2}$	$1.13e-1_{9.1e-2}$	$6.72e-3_{9.1e-4}$	$1.25e-1_{3.9e-2}$	$9.26e-2_{5.1e-2}$	$3.95e-2_{3.8e-2}$	+
DTLZ3	$1.04e+2_{6.2e+1}$	$1.79e+2_{7.5e+1}$	$8.86e+1_{9.5e+1}$	$4.41e+1_{9.0e+1}$	$1.23e+2_{6.5e+1}$	$2.37e+2_{5.7e+1}$	+
DTLZ4	$8.91e-2_{5.9e-2}$	$3.00e-1_{4.5e-2}$	$3.18e-2_{1.0e-2}$	$2.20e-1_{1.1e-1}$	$6.33e-2_{3.0e-2}$	$2.56e-2_{8.6e-3}$	+
DTLZ5	$3.92e-2_{3.6e-2}$	$1.11e-1_{9.8e-2}$	$6.62e-3_{8.9e-4}$	$1.22e-1_{4.3e-2}$	$9.10e-2_{4.0e-2}$	$3.31e-2_{3.0e-2}$	+
DTLZ6	$1.47e+0_{7.9e-1}$	$1.00e+0_{2.9e-1}$	$5.36e-3_{4.8e-4}$	$1.75e-1_{9.1e-1}$	$1.57e+0_{1.3e+0}$	$4.77e+0_{3.2e-1}$	+
DTLZ7	$1.33e+0_{1.4e+0}$	$1.27e+0_{2.7e-2}$	$7.13e-3_{6.8e-4}$	$3.00e-1_{1.9e-1}$	$1.65e-1_{1.1e-1}$	$4.94e-1_{1.0e-1}$	+
WFG1	$1.36e+0_{7.7e-2}$	$1.00e+0_{9.3e-2}$	$1.35e+0_{4.9e-2}$	$1.53e+0_{3.0e-2}$	$1.10e+0_{2.0e-1}$	$1.31e+0_{5.1e-2}$	+
WFG2	$1.67e-2_{5.5e-3}$	$4.87e-2_{3.6e-2}$	$1.04e-2_{1.7e-3}$	$3.57e-1_{1.8e-1}$	$7.24e-2_{2.1e-2}$	$5.96e-2_{3.7e-2}$	+
WFG3	$2.00e+0_{5.3e-4}$	$2.00e+0_{4.2e-3}$	$2.00e+0_{1.6e-5}$	$2.10e+0_{1.2e-1}$	$2.00e+0_{4.5e-5}$	$2.12e+0_{2.0e-1}$	+
WFG4	$1.09e-1_{1.8e-2}$	$6.06e-2_{2.7e-2}$	$5.98e-2_{1.5e-2}$	$3.21e-1_{8.1e-2}$	$5.57e-2_{1.8e-2}$	$8.04e-2_{2.4e-2}$	+
WFG5	$8.34e-2_{2.0e-2}$	$6.36e-2_{1.2e-3}$	$6.37e-2_{9.0e-4}$	$6.24e-1_{3.3e-1}$	$3.24e-1_{3.5e-1}$	$2.57e-1_{2.2e-1}$	+
WFG6	$1.04e-1_{6.6e-2}$	$5.60e-1_{3.8e-1}$	$1.79e-2_{2.5e-3}$	$4.63e-1_{1.3e-1}$	$3.30e-1_{2.6e-1}$	$2.40e-1_{2.3e-1}$	+
WFG7	$4.05e+2_{6.1e+3}$	$5.75e+2_{1.8e+2}$	$1.94e+2_{1.7e+3}$	$3.77e+1_{1.5e+1}$	$6.16e+1_{1.1e+1}$	$2.44e+1_{3.4e+1}$	+
WFG8	$5.24e-1_{9.2e-2}$	$5.66e-1_{1.9e-1}$	$5.06e-1_{3.4e-2}$	$8.30e-1_{1.2e-1}$	$5.39e-1_{2.3e-2}$	$7.70e-1_{6.0e-2}$	+
WFG9	$6.38e-2_{2.0e-2}$	$2.89e-2_{1.7e-3}$	$2.95e-2_{2.5e-3}$	$3.25e-1_{2.5e-1}$	$1.11e-1_{4.6e-2}$	$1.49e-1_{2.1e-1}$	+

algorithm, yielding the lowest (best) values in 16 out of the 21 problems. Considering the next algorithms according to the best and second best indicator values, we find SigmaMOPSO, NSPSO, and MOCLPSO. AMOPSO is the worst performer according to the Δ indicator, not achieving any best nor second best result.

Table 3. Median and interquartile range of the Δ quality indicator

Problem	NSPSO $\bar{x}_{IQR}$	SigmaMOPSO $\bar{x}_{IQR}$	OMOPSO $\bar{x}_{IQR}$	AMOPSO $\bar{x}_{IQR}$	MOPSOpd $\bar{x}_{IQR}$	MOCLPSO $\bar{x}_{IQR}$	
ZDT1	$7.19e-1_{1.0e-1}$	$4.11e-1_{3.9e-1}$	$1.00e-1_{1.4e-2}$	$9.57e-1_{1.6e-1}$	$6.03e-1_{1.1e-1}$	$7.70e-1_{6.4e-2}$	+
ZDT2	$9.82e-1_{9.4e-2}$	$1.00e+0_{0.0e+0}$	$9.45e-2_{1.8e-2}$	$1.00e+0_{6.0e-2}$	$1.00e+0_{2.8e-1}$	$8.03e-1_{7.4e-2}$	+
ZDT3	$8.17e-1_{9.7e-2}$	$1.09e+0_{3.6e-1}$	$7.35e-1_{5.2e-2}$	$9.00e-1_{1.5e-1}$	$8.59e-1_{6.7e-2}$	$8.85e-1_{5.7e-2}$	+
ZDT4	$9.53e-1_{8.0e-2}$	$1.00e+0_{3.3e-3}$	$8.78e-1_{5.2e-2}$	$1.03e+0_{2.5e-2}$	$1.00e+0_{2.4e-2}$	$9.32e-1_{8.2e-2}$	+
ZDT6	$1.39e+0_{6.6e-2}$	$2.89e-1_{3.6e-1}$	$8.78e-2_{1.2e+0}$	$1.12e+0_{1.5e-1}$	$1.20e+0_{2.7e-1}$	$9.67e-1_{4.1e-2}$	+
DTLZ1	$8.38e-1_{1.2e-1}$	$1.14e+0_{1.7e-1}$	$7.77e-1_{1.1e-1}$	$1.13e+0_{2.6e-1}$	$8.72e-1_{2.0e-1}$	$7.90e-1_{7.2e-2}$	+
DTLZ2	$6.02e-1_{1.5e-1}$	$1.01e+0_{1.4e-1}$	$1.81e-1_{2.3e-2}$	$1.15e+0_{1.8e-1}$	$1.21e+0_{8.6e-2}$	$7.92e-1_{8.7e-2}$	+
DTLZ3	$9.31e-1_{2.0e-1}$	$1.23e+0_{1.6e-1}$	$7.90e-1_{1.1e-1}$	$1.09e+0_{4.3e-1}$	$8.55e-1_{1.3e-1}$	$7.69e-1_{8.5e-2}$	+
DTLZ4	$7.17e-1_{1.7e-1}$	$1.41e+0_{8.0e-1}$	$6.77e-1_{7.9e-2}$	$1.46e+0_{2.7e-1}$	$1.10e+0_{9.2e-2}$	$7.33e-1_{5.3e-2}$	+
DTLZ5	$5.99e-1_{9.3e-2}$	$1.00e+0_{1.7e-1}$	$1.77e-1_{2.6e-2}$	$1.16e+0_{1.9e-1}$	$1.21e+0_{9.3e-2}$	$7.89e-1_{8.9e-2}$	+
DTLZ6	$8.18e-1_{4.0e-1}$	$1.28e+0_{1.0e+0}$	$1.18e-1_{1.7e-2}$	$1.23e+0_{4.4e-1}$	$8.35e-1_{1.5e-1}$	$8.04e-1_{7.2e-2}$	+
DTLZ7	$9.08e-1_{1.6e-1}$	$7.96e-1_{2.4e-1}$	$5.21e-1_{6.8e-3}$	$1.02e+0_{2.4e-1}$	$7.95e-1_{1.3e-1}$	$8.51e-1_{7.0e-2}$	+
WFG1	$1.14e+0_{5.5e-2}$	$7.50e-1_{1.2e-1}$	$1.17e+0_{6.0e-2}$	$1.30e+0_{3.9e-2}$	$1.16e+0_{7.8e-2}$	$1.12e+0_{4.2e-2}$	+
WFG2	$8.65e-1_{9.0e-2}$	$9.61e-1_{8.5e-2}$	$7.64e-1_{5.5e-3}$	$9.94e-1_{1.9e-1}$	$1.22e+0_{7.0e-2}$	$1.11e+0_{5.8e-2}$	+
WFG3	$5.00e-1_{2.6e-2}$	$4.96e-1_{2.5e-2}$	$3.78e-1_{8.7e-3}$	$1.20e+0_{8.7e-2}$	$1.19e+0_{1.3e-1}$	$9.04e-1_{6.2e-2}$	+
WFG4	$6.25e-1_{5.0e-2}$	$5.01e-1_{7.7e-2}$	$5.06e-1_{6.3e-2}$	$1.14e+0_{1.3e-1}$	$4.83e-1_{4.4e-2}$	$6.18e-1_{4.9e-2}$	+
WFG5	$3.59e-1_{4.5e-2}$	$1.44e-1_{2.0e-2}$	$1.44e-1_{2.0e-2}$	$1.03e+0_{1.7e-1}$	$1.13e+0_{2.3e-1}$	$8.06e-1_{9.7e-2}$	+
WFG6	$5.98e-1_{8.1e-2}$	$6.34e-1_{2.1e-1}$	$1.63e-1_{2.5e-2}$	$1.09e+0_{1.7e-1}$	$1.23e+0_{7.0e-2}$	$8.32e-1_{7.6e-2}$	+
WFG7	$3.71e+1_{5.8e+2}$	$4.07e+1_{5.5e+2}$	$1.59e+1_{2.1e+2}$	$1.13e+0_{1.3e+1}$	$1.31e+0_{7.1e+2}$	$9.13e+1_{8.7e+2}$	+
WFG8	$7.19e-1_{8.4e-2}$	$9.08e-1_{1.7e-1}$	$7.93e-1_{8.8e-2}$	$1.02e+0_{1.4e-1}$	$8.68e-1_{6.6e-2}$	$7.88e-1_{5.3e-2}$	+
WFG9	$5.07e-1_{1.3e-1}$	$2.22e-1_{2.6e-2}$	$2.24e-1_{2.7e-2}$	$1.19e+0_{1.5e-1}$	$7.54e-1_{5.2e-2}$	$7.29e-1_{6.3e-2}$	+

After applying a quality indicator that measures convergence and another one that measures diversity, the *HV* indicator should confirm the previous results. The *HV* values, included in Table 4, show that OMOPSO generates solution sets with the highest (best) values in 15 out of the 21 problems. Thus, we can state that according to the parameterization, quality indicators, and benchmark problems considered in this work, OMOPSO is clearly the most salient technique among the six considered in our study.

Table 4. Median and interquartile range of the HV quality indicator

Problem	NSPSO $\bar{x}_{IQR}$	SigmaMOPSO $\bar{x}_{IQR}$	OMOPSO $\bar{x}_{IQR}$	AMOPSO $\bar{x}_{IQR}$	MOPSOpd $\bar{x}_{IQR}$	MOCLPSO $\bar{x}_{IQR}$	
ZDT1	$1.54e-1_{2.4e-1}$	$6.54e-1_{8.3e-3}$	$6.61e-1_{1.5e-4}$	$3.81e-1_{9.3e-2}$	$5.94e-1_{1.7e-2}$	$3.28e-1_{4.6e-2}$	+
ZDT2	–	–	$3.28e-1_{2.5e-4}$	$4.10e-2_{1.9e-1}$	$0.00e+0_{2.6e-1}$	$6.54e-2_{3.7e-2}$	+
ZDT3	$1.12e-1_{1.2e-1}$	$3.21e-1_{2.3e-1}$	$5.10e-1_{3.8e-3}$	$2.45e-1_{1.1e-1}$	$4.38e-1_{7.2e-2}$	$2.55e-1_{3.2e-2}$	+
ZDT4	–	–	–	–	–	–	-
ZDT6	$3.09e-1_{1.3e-1}$	$4.01e-1_{3.1e-4}$	$4.01e-1_{1.5e-4}$	$2.31e-1_{4.1e-2}$	$3.50e-1_{5.7e-2}$	–	+
DTLZ1	–	–	–	–	–	–	–
DTLZ2	$1.64e-1_{5.9e-2}$	$1.64e-1_{2.1e-2}$	$2.10e-1_{4.5e-4}$	$1.23e-1_{2.4e-2}$	$1.78e-1_{2.5e-2}$	$2.01e-1_{2.3e-3}$	+
DTLZ3	–	–	–	–	–	–	-
DTLZ4	$1.37e-1_{5.1e-2}$	–	$1.96e-1_{6.1e-3}$	$7.62e-2_{9.8e-2}$	$1.90e-1_{9.8e-3}$	$1.96e-1_{4.0e-3}$	+
DTLZ5	$1.71e-1_{3.5e-2}$	$1.65e-1_{2.3e-2}$	$2.11e-1_{5.4e-4}$	$1.22e-1_{2.9e-2}$	$1.77e-1_{2.0e-2}$	$2.01e-1_{2.1e-3}$	+
DTLZ6	–	–	$2.12e-1_{4.4e-5}$	$8.77e-2_{1.5e-1}$	–	–	+
DTLZ7	$1.59e-2_{9.7e-2}$	$2.18e-1_{1.7e-2}$	$3.34e-1_{3.2e-4}$	$2.00e-1_{7.1e-2}$	$2.53e-1_{5.5e-2}$	$1.01e-1_{1.3e-2}$	+
WFG1	$8.98e-2_{8.3e-3}$	$1.21e-1_{2.2e-3}$	$1.04e-1_{1.0e-2}$	$6.22e-2_{7.4e-3}$	$1.69e-1_{7.2e-2}$	$1.01e-1_{5.1e-3}$	+
WFG2	$5.61e-1_{2.5e-3}$	$5.60e-1_{1.7e-3}$	$5.64e-1_{1.0e-4}$	$4.68e-1_{3.9e-2}$	$5.57e-1_{3.6e-3}$	$5.60e-1_{1.8e-3}$	+
WFG3	$4.40e-1_{3.3e-4}$	$4.38e-1_{8.0e-4}$	$4.42e-1_{5.4e-5}$	$4.04e-1_{1.2e-2}$	$4.27e-1_{1.8e-2}$	$4.30e-1_{1.3e-2}$	+
WFG4	$1.78e-1_{7.0e-3}$	$2.00e-1_{1.6e-3}$	$2.02e-1_{1.6e-3}$	$1.27e-1_{1.2e-2}$	$2.07e-1_{1.3e-3}$	$2.00e-1_{2.3e-3}$	+
WFG5	$1.96e-1_{2.8e-4}$	$1.96e-1_{8.8e-5}$	$1.96e-1_{6.3e-5}$	$1.60e-1_{1.7e-2}$	$1.68e-1_{5.9e-2}$	$1.90e-1_{1.9e-3}$	+
WFG6	$1.75e-1_{2.6e-2}$	$1.90e-1_{1.9e-2}$	$2.09e-1_{3.5e-4}$	$9.88e-2_{2.8e-2}$	$1.60e-1_{4.7e-2}$	$2.01e-1_{1.9e-3}$	+
WFG7	$2.03e+1_{2.7e+3}$	$2.02e+1_{1.1e+3}$	$2.09e+1_{1.7e+4}$	$1.14e+1_{1.4e+2}$	$9.49e+2_{4.2e+2}$	$2.01e+1_{2.7e+3}$	+
WFG8	$1.07e-1_{8.7e-3}$	$1.33e-1_{4.2e-3}$	$1.26e-1_{3.0e-3}$	$6.08e-2_{1.9e-2}$	$1.41e-1_{3.0e-3}$	$1.33e-1_{1.9e-3}$	+
WFG9	$2.24e-1_{6.1e-3}$	$2.34e-1_{4.1e-4}$	$2.34e-1_{6.6e-4}$	$1.87e-1_{1.1e-2}$	$2.29e-1_{4.7e-3}$	$2.30e-1_{1.1e-3}$	+

The results corresponding to problems ZDT4, DTLZ1, and DTLZ3 deserve additional comments. We have used the '–' symbol in Table 4 to indicate those experiments in which the HV value is equal to 0, meaning that the solution sets obtained by the algorithms are outside the limits of the Pareto front; when applying the HV indicator these solutions are not taken into account, because otherwise the obtained results would be unreliable. In the case of the three aforementioned problems, none of the six algorithms is able to achieve a HV greater than 0 over the 100 independent runs. We can also see that other problems are difficult to solve by some techniques, e.g., ZDT2 and DTLZ6. The statistical tests indicate that the results of the Δ and HV indicators have statistical confidence. To provide further statistical information, we show in Table 5 those problems for which no statistical differences appear between OMOPSO and the rest of algorithms considering the three quality indicators. It can be observed that statistical differences exist for most of the pair-wise comparisons.

Table 5. Non-successful statistical tests between OMOPSO and the rest of the algorithms

	$I^1_{\epsilon+}$	Δ	HV
AMOPSO	- DTLZ3 -	- - -	- - -
MOCLPSO	- DTLZ1, DTLZ4 -	ZDT6 DTLZ1, DTLZ3 WFG8	- DTLZ4 WFG1, WFG4
MOPSOpd	ZDT4 DTLZ1, DTLZ3 WFG3, WFG4	- - WFG1, WFG4	- - -
NSPSO	- DTLZ3 WFG1, WFG8	- DTLZ4 -	- - -
SigmaMOPSO	- - WFG4, WFG5, WFG9	ZDT6 - WFG4, WFG5, WFG9	- - WFG5, WFG9

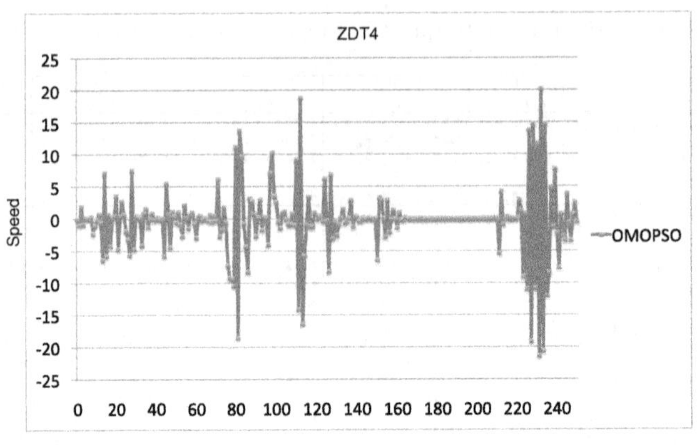

Fig. 1. Tracing the velocity of the second variable of OMOPSO when solving ZDT4

6 Discussion

The conclusion drawn from the analysis of the results in the previous section is that OMOPSO performs the best in our study. In this section, we carry out the same experiments but using OMOPSO and NSGA-II in order to put the results of the first one in context. Such a comparison will allow us to know how competitive OMOPSO is. Before that, we investigate why OMOPSO (as well as the rest of MOPSOs) is unable to solve the ZDT4, DTLZ1, and DTLZ3 problems. If we consider ZDT4, it is a well-known problem characterized by having many local optima (it is a multifrontal problem). We have traced the velocity of the second variable in the first particle in OMOPSO when facing the solution of ZDT4 (the second variable takes values in the interval $[-5, +5]$, which provides a better illustration of the following analysis than using the first variable, which ranges in $[0, 1]$). The obtained values after the 250 iterations are depicted in Fig. 1. We can observe that the velocity values suffer a kind of erratic behavior in some points of the execution, alternating very high with very low values. Let us note that the limits of the second variable in ZDT4 are $[-5, +5]$, and the velocity takes values higher than ± 20. The consequence is that this particle is moving to its extreme values continuously, so it is not contributing to guide the search.

To find out whether this is one of the reasons making OMOPSO unable to solve multi frontal MOPs, we have modified it by including a velocity constraint mechanism, similar to the one proposed in [2]. In addition, the accumulated velocity of each variable j (in each particle) is also bounded by means of the following equation:

$$v_{i,j}(t) = \begin{cases} delta_j & \text{if } v_{i,j}(t) > delta_j \\ -delta_j & \text{if } v_{i,j}(t) \leq -delta_j \\ v_{i,j}(t) & \text{otherwise} \end{cases} \quad (3)$$

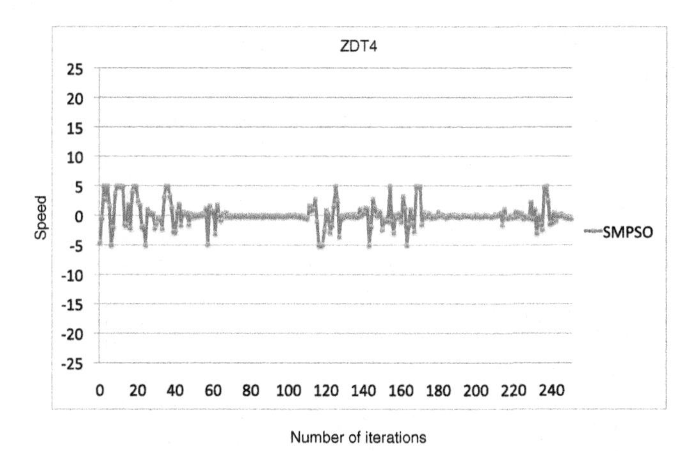

Fig. 2. Tracing the velocity of the second variable of SMPSO when solving ZDT4

where

$$delta_j = \frac{(upper\_limit_j - lower\_limit_j)}{2} \tag{4}$$

This way, we can ensure an effective new position calculation. We have called the resulting algorithm SMPSO (Speed-constrained Multi-objective PSO). In Fig. 2 we show again the velocity of the particle representing the second parameter of ZDT4. We can observe that the erratic movements of the velocity have vanished, so the particle is taking values inside the bounds of the variable and thus it is moving along different regions of the search space. To evaluate the effect of the changes in SMPSO, we have included this algorithm in the comparison between OMOPSO and NSGA-II. We have solved all the problems again, following the same methodology. The parameter settings of NSGA-II are: the population size is 100 individuals, we have used SBX and polynomial mutation [3] as operators for crossover and mutation operators, respectively, and the distribution indexes for both operators are $\eta_c = 20$ and $\eta_m = 20$, respectively. The crossover probability is $p_c = 0.9$ and the mutation probability is $p_m = 1/L$, where L is the number of decision variables.

In Table 6, we include the median and interquartile range of NSGA-II, O-MOPSO, and SMPSO corresponding to the $I^1_{\epsilon+}$ quality indicator. We observe that SMPSO yields the best values in 11 out of the 12 problems comprising the ZDT and DTLZ benchmarks. If we focus on the WFG problems, the lowest (best) metric values are shared between OMOPSO (six problems) and NSGA-II (three problems), while SMPSO obtains the second lowest values in 8 out of the 9 WFG problems. These results indicate first, that OMOPSO is competitive when compared against NSGA-II concerning convergence and, second, that the velocity constraint mechanism included in SMPSO improves globally the behavior of OMOPSO considering all the benchmark problems.

Table 6. NSGA-II vs OMOPSO vs SMPSO: Median and interquartile range of the $I^1_{\epsilon+}$ quality indicator

Problem	NSGA-II $\tilde{x}_{IQR}$	OMOPSO $\tilde{x}_{IQR}$	SMPSO $\tilde{x}_{IQR}$	
ZDT1	$1.37e-2_{3.0e-3}$	$6.36e-3_{5.1e-4}$	$5.78e-3_{3.8e-4}$	+
ZDT2	$1.28e-2_{2.3e-3}$	$6.19e-3_{5.4e-4}$	$5.66e-3_{3.0e-4}$	+
ZDT3	$8.13e-3_{1.9e-3}$	$1.32e-2_{7.7e-3}$	$6.09e-3_{1.3e-3}$	+
ZDT4	$1.49e-2_{3.0e-3}$	$5.79e+0_{4.3e+0}$	$7.93e-3_{1.4e-3}$	+
ZDT6	$1.47e-2_{2.8e-3}$	$4.65e-3_{4.2e-4}$	$4.87e-3_{4.8e-4}$	+
DTLZ1	$7.13e-3_{1.6e-3}$	$1.92e+1_{1.1e+1}$	$3.73e-3_{5.4e-4}$	+
DTLZ2	$1.11e-2_{2.7e-3}$	$6.72e-3_{9.1e-4}$	$5.81e-3_{6.0e-4}$	+
DTLZ3	$1.04e+0_{1.2e+0}$	$8.86e+1_{9.5e+1}$	$6.57e-3_{1.0e-2}$	+
DTLZ4	$1.13e-2_{9.9e-1}$	$3.18e-2_{1.0e-2}$	$6.54e-3_{8.8e-4}$	+
DTLZ5	$1.05e-2_{2.5e-3}$	$6.62e-3_{8.9e-4}$	$5.77e-3_{6.1e-4}$	+
DTLZ6	$4.39e-2_{3.4e-2}$	$5.36e-3_{4.8e-4}$	$5.22e-3_{4.4e-4}$	+
DTLZ7	$1.04e-2_{2.8e-3}$	$7.13e-3_{6.8e-4}$	$5.46e-3_{4.3e-4}$	+
WFG1	$3.52e-1_{4.6e-1}$	$1.35e+0_{4.9e-2}$	$1.34e+0_{4.6e-2}$	+
WFG2	$7.10e-1_{7.0e-1}$	$1.04e-2_{1.7e-3}$	$1.40e-2_{3.4e-3}$	+
WFG3	$2.00e+0_{5.8e-4}$	$2.00e+0_{1.6e-5}$	$2.00e+0_{3.9e-4}$	+
WFG4	$3.26e-2_{6.7e-3}$	$5.98e-2_{1.5e-2}$	$6.46e-2_{6.0e-3}$	+
WFG5	$8.41e-2_{8.3e-3}$	$6.37e-2_{9.0e-4}$	$6.40e-2_{2.0e-3}$	+
WFG6	$4.14e-2_{1.6e-2}$	$1.79e-2_{2.5e-3}$	$2.56e-2_{3.8e-3}$	+
WFG7	$3.47e+2_{8.1e+3}$	$1.94e+2_{1.7e+3}$	$2.67e+2_{3.8e+3}$	+
WFG8	$3.38e-1_{2.3e-1}$	$5.06e-1_{3.4e-2}$	$4.32e-1_{7.8e-2}$	+
WFG9	$3.73e-2_{7.5e-3}$	$2.95e-2_{2.5e-3}$	$3.15e-2_{3.3e-3}$	+

Table 7. NSGA-II vs OMOPSO vs SMPSO: Median and interquartile range of the Δ quality indicator

Problem	NSGA-II $\tilde{x}_{IQR}$	OMOPSO $\tilde{x}_{IQR}$	SMPSO $\tilde{x}_{IQR}$	
ZDT1	$3.70e-1_{4.2e-2}$	$1.00e-1_{1.4e-2}$	$8.66e-2_{1.6e-2}$	+
ZDT2	$3.81e-1_{4.7e-2}$	$9.45e-2_{1.8e-2}$	$7.46e-2_{1.5e-2}$	+
ZDT3	$7.47e-1_{1.8e-2}$	$7.35e-1_{5.2e-2}$	$7.17e-1_{1.7e-2}$	+
ZDT4	$4.02e-1_{5.8e-2}$	$8.78e-1_{5.2e-2}$	$1.53e-1_{2.2e-2}$	+
ZDT6	$3.56e-1_{3.6e-2}$	$8.78e-2_{1.2e+0}$	$7.28e-1_{1.2e+0}$	+
DTLZ1	$4.03e-1_{6.1e-2}$	$7.77e-1_{1.1e-1}$	$1.14e-1_{1.8e-2}$	+
DTLZ2	$3.84e-1_{3.8e-2}$	$1.81e-1_{2.3e-2}$	$1.59e-1_{2.3e-2}$	+
DTLZ3	$9.53e-1_{1.6e-1}$	$7.90e-1_{1.1e-1}$	$1.98e-1_{3.3e-1}$	+
DTLZ4	$3.95e-1_{6.4e-1}$	$6.77e-1_{7.9e-2}$	$1.70e-1_{2.5e-2}$	+
DTLZ5	$3.79e-1_{4.0e-2}$	$1.77e-1_{2.6e-2}$	$1.58e-1_{2.2e-2}$	+
DTLZ6	$8.64e-1_{3.0e-1}$	$1.18e-1_{1.7e-2}$	$1.14e-1_{2.1e-2}$	+
DTLZ7	$6.23e-1_{2.5e-2}$	$5.21e-1_{6.8e-3}$	$5.20e-1_{2.0e-3}$	+
WFG1	$7.18e-1_{5.4e-2}$	$1.17e+0_{6.0e-2}$	$1.12e+0_{5.0e-2}$	+
WFG2	$7.93e-1_{1.7e-2}$	$7.64e-1_{5.5e-3}$	$8.26e-1_{3.5e-2}$	+
WFG3	$6.12e-1_{3.6e-2}$	$3.78e-1_{8.7e-3}$	$3.84e-1_{6.4e-3}$	+
WFG4	$3.79e-1_{3.9e-2}$	$5.06e-1_{6.3e-2}$	$5.51e-1_{7.0e-2}$	+
WFG5	$4.13e-1_{5.1e-2}$	$1.44e-1_{2.0e-2}$	$1.50e-1_{2.8e-2}$	+
WFG7	$3.79e+1_{4.6e+2}$	$1.59e+1_{2.1e+2}$	$2.44e+1_{3.1e+2}$	+
WFG6	$3.90e-1_{4.2e-2}$	$1.63e-1_{2.5e-2}$	$2.47e-1_{4.1e-2}$	+
WFG8	$6.45e-1_{5.5e-2}$	$7.93e-1_{8.8e-2}$	$8.08e-1_{5.4e-2}$	+
WFG9	$3.96e-1_{4.1e-2}$	$2.24e-1_{2.7e-2}$	$2.46e-1_{2.8e-2}$	+

The values obtained when applying the Δ and HV indicators are included in Tables 7 and 8, respectively. We can observe that we can practically draw the same conclusions obtained from the $I^1_{\epsilon+}$ indicator, i.e., the algorithms obtain the lowest values in the same problems according to the convergence and diversity indicators. In all the experiments included in this section all the statistical tests are significant, which actually grounds our claims. If we focus in the HV and in those problems in which OMOPSO obtained a value of 0 (ZDT4, DTLZ1, and DTLZ3), we see that the velocity constraint mechanism added to SMPSO allows it to successfully solve them. NSGA-II also outperforms OMOPSO in this sense, only presenting difficulties in DTLZ3.

Table 8. NSGA-II vs OMOPSO vs SMPSO: Median and interquartile range of the *HV* quality indicator

Problem	NSGA-II $\bar{x}_{IQR}$	OMOPSO $\bar{x}_{IQR}$	SMPSO $\bar{x}_{IQR}$	
ZDT1	$6.59e-1_{4.4e-4}$	$6.61e-1_{1.5e-4}$	$6.62e-1_{1.5e-4}$	+
ZDT2	$3.26e-1_{4.3e-4}$	$3.28e-1_{2.5e-4}$	$3.28e-1_{1.1e-4}$	+
ZDT3	$5.15e-1_{2.3e-4}$	$5.10e-1_{3.8e-3}$	$5.15e-1_{5.1e-4}$	+
ZDT4	$6.56e-1_{4.5e-3}$	–	$6.61e-1_{3.8e-4}$	+
ZDT6	$3.88e-1_{2.3e-3}$	$4.01e-1_{1.5e-4}$	$4.01e-1_{1.0e-4}$	+
DTLZ1	$4.88e-1_{5.5e-3}$	–	$4.94e-1_{3.4e-4}$	+
DTLZ2	$2.11e-1_{3.1e-4}$	$2.10e-1_{4.5e-4}$	$2.12e-1_{2.3e-4}$	+
DTLZ3	–	–	$2.12e-1_{2.8e-3}$	+
DTLZ4	$2.09e-1_{2.1e-1}$	$1.96e-1_{6.1e-3}$	$2.09e-1_{3.3e-4}$	+
DTLZ5	$2.11e-1_{3.5e-4}$	$2.11e-1_{5.4e-4}$	$2.12e-1_{2.1e-4}$	+
DTLZ6	$1.75e-1_{3.6e-2}$	$2.12e-1_{4.4e-5}$	$2.12e-1_{4.8e-5}$	+
DTLZ7	$3.33e-1_{2.1e-4}$	$3.34e-1_{3.2e-4}$	$3.34e-1_{7.3e-5}$	+
WFG1	$5.23e-1_{1.3e-1}$	$1.04e-1_{1.0e-2}$	$9.70e-2_{5.3e-3}$	+
WFG2	$5.61e-1_{2.8e-3}$	$5.64e-1_{1.0e-4}$	$5.62e-1_{5.7e-4}$	+
WFG3	$4.41e-1_{3.2e-4}$	$4.42e-1_{5.4e-5}$	$4.41e-1_{1.1e-4}$	+
WFG4	$2.17e-1_{4.9e-4}$	$2.02e-1_{1.6e-3}$	$1.96e-1_{2.0e-3}$	+
WFG5	$1.95e-1_{3.6e-4}$	$1.96e-1_{6.3e-5}$	$1.96e-1_{5.8e-5}$	+
WFG6	$2.03e-1_{9.0e-3}$	$2.09e-1_{3.5e-4}$	$2.05e-1_{1.1e-3}$	+
WFG7	$2.09e+1_{3.3e+4}$	$2.09e+1_{1.7e+4}$	$2.06e+1_{8.2e+4}$	+
WFG8	$1.47e-1_{2.1e-3}$	$1.26e-1_{3.0e-3}$	$1.40e-1_{1.9e-3}$	+
WFG9	$2.37e-1_{1.7e-3}$	$2.34e-1_{6.6e-4}$	$2.33e-1_{4.1e-4}$	+

Table 9 contains those problems for which no statistical confidence exist considering the three algorithms and the three quality indicators. The results of OMOPSO against NSGA-II are significant in all the problems but DTLZ3 with respect to the Δ indicator. Concerning SMPSO, there a few cases where the results are not significant, but they do not alter the analysis carried out.

Table 9. Non-successful statistical tests among NSGA-II, OMOPSO, and SMPSO

$I^1_{\epsilon+}$	OMOPSO	SMPSO
NSGA-II		WFG3, WFG8
OMOPSO	N/A	ZDT6, DTLZ6, WFG1, WFG4

Δ	OMOPSO	SMPSO
NSGA-II	DTLZ3	WFG2
OMOPSO	N/A	ZDT6, DTLZ6

HV	OMOPSO	SMPSO
NSGA-II		ZDT6
OMOPSO	N/A	DTLZ6, DTLZ7, WFG8

We can summarize this section by stating that OMOPSO, the most salient of the six MOPSOs studied in this work, is a competitive algorithm when compared with NSGA-II, and we have shown that its search capabilities can be improved by including a velocity constraint mechanism. However, although SMPSO outperforms both NSGA-II and OMOPSO in the ZDT and DTLZ problems, it does not achieve the best result in the WFG benchmark. This indicates that more research has to be done. It is also necessary to consider a broader set of problems as well as studying in more depth the effect of modulating the speed in a MOPSO.

7 Conclusions and Further Work

We have evaluated six MOPSO algorithms over a set of three well-known benchmark problems by using three different quality indicators. For each experiment, 100 independent runs have been carried out, and statistical tests have been applied to know more about the confidence of the obtained results. In the context of the problems analyzed, the experimentation methodology, and the parameter settings used, we can state that OMOPSO is clearly the most salient of the six compared algorithms. The results have also shown that all the algorithms are unable to find accurate Pareto fronts for three multi frontal problems. We have studied this issue and we have proposed the use of a velocity constraint mechanism to enhance the search capability in order to solve these problems. The resulting algorithm, SMPSO, shows significant improvements when compared with respect to OMOPSO and NSGA-II. As part of our future work, we plan to study the convergence speed of MOPSO algorithms in order to determine whether they are faster than other multi-objective evolutionary algorithms in reaching the Pareto front of a problem.

Acknowledgments. This work has been partially funded by the "Consejería de Innovación, Ciencia y Empresa", Junta de Andalucía under contract P07-TIC-03044 DIRICOM project, `http://diricom.lcc.uma.es`. Juan J. Durillo is supported by grant AP-2006-03349 from the Spanish Ministry of Education and Science. Francisco Luna acknowledges support from the Spanish Ministry of Education and Science and FEDER under contract TIN2005-08818-C04-01 (the OPLINK project).

References

1. Álvarez-Benítez, J.E., Everson, R.M., Fieldsend, J.E.: A MOPSO Algorithm Based Exclusively on Pareto Dominance Concepts. In: Coello Coello, C.A., Hernández Aguirre, A., Zitzler, E. (eds.) EMO 2005. LNCS, vol. 3410, pp. 459–473. Springer, Heidelberg (2005)
2. Clerc, M., Kennedy, J.: The particle swarm - explosion, stability, and convergence in a multidimensional complex space. IEEE Transactions on Evolutionary Computation 6(1), 58–73 (2002)
3. Deb, K.: Multi-Objective Optimization Using Evolutionary Algorithms. John Wiley & Sons, Chichester (2001)
4. Deb, K., Pratap, A., Agarwal, S., Meyarivan, T.: A fast and elitist multiobjective genetic algorithm: NSGA-II. IEEE Transactions on Evolutionary Computation 6(2), 182–197 (2002)
5. Deb, K., Thiele, L., Laumanns, M., Zitzler, E.: Scalable Test Problems for Evolutionary Multiobjective Optimization. In: Abraham, A., Jain, L., Goldberg, R. (eds.) Evolutionary Multiobjective Optimization. Theoretical Advances and Applications, pp. 105–145. Springer, Heidelberg (2005)

6. Demšar, J.: Statistical comparisons of classifiers over multiple data sets. J. Mach. Learn. Res. 7, 1–30 (2006)
7. Durillo, J.J., Nebro, A.J., Luna, F., Dorronsoro, B., Alba, E.: jMetal: A Java Framework for Developing Multi-Objective Optimization Metaheuristics. Technical Report ITI-2006-10, Departamento de Lenguajes y Ciencias de la Computación, University of Málaga, E.T.S.I. Informática, Campus de Teatinos (December 2006)
8. Hochberg, Y., Tamhane, A.C.: Multiple Comparison Procedures. Wiley, Chichester (1987)
9. Huang, V.L., Suganthan, P.N., Liang, J.J.: Comprehensive learning particle swarm optimizer for solving multiobjective optimization problems. Int. J. Intell. Syst. 21(2), 209–226 (2006)
10. Huband, S., Hingston, P., Barone, L., While, L.: A Review of Multiobjective Test Problems and a Scalable Test Problem Toolkit. IEEE Transactions on Evolutionary Computation 10(5), 477–506 (2006)
11. Kennedy, J., Eberhart, R.: Particle swarm optimization. In: Fourth IEEE International Conference on Neural Networks, pp. 1942–1948 (1995)
12. Kennedy, J., Eberhart, R.C.: Swarm Intelligence. Morgan Kaufmann Publishers, San Francisco (2001)
13. Knowles, J., Thiele, L., Zitzler, E.: A Tutorial on the Performance Assessment of Stochastic Multiobjective Optimizers. Technical Report 214, Computer Engineering and Networks Laboratory (TIK), ETH Zurich (2006)
14. Li, X.: A Non-dominated Sorting Particle Swarm Optimizer for Multiobjective Optimization. In: Cantú-Paz, E., Foster, J.A., Deb, K., Davis, L., Roy, R., O'Reilly, U.-M., Beyer, H.-G., Kendall, G., Wilson, S.W., Harman, M., Wegener, J., Dasgupta, D., Potter, M.A., Schultz, A., Dowsland, K.A., Jonoska, N., Miller, J., Standish, R.K. (eds.) GECCO 2003. LNCS, vol. 2723, pp. 37–48. Springer, Heidelberg (2003)
15. Moore, J., Chapman, R.: Application of particle swarm to multiobjective optimization. Technical report, Department of Computer Science and Software Engineering, Auburn University (1999)
16. Mostaghim, S., Teich, J.: Strategies for finding good local guides in multi-objective particle swarm optimization (MOPSO). In: Proceedings of the IEEE Swarm Intelligence Symposium, SIS 2003, pp. 26–33 (2003)
17. Reyes-Sierra, M., Coello, C.: Multi-Objective Particle Swarm Optimizers: A Survey of the State-of-the-Art. International Journal of Computational Intelligence Research 2(3), 287–308 (2006)
18. Reyes Sierra, M., Coello Coello, C.A.: Improving PSO-based multi-objective optimization using crowding, mutation and ϵ-dominance. In: Coello Coello, C.A., Hernández Aguirre, A., Zitzler, E. (eds.) EMO 2005. LNCS, vol. 3410, pp. 505–519. Springer, Heidelberg (2005)
19. Toscano, G., Coello, C.: Using Clustering Techniques to Improve the Performance of a Multi-objective Particle Swarm Optimizer. In: Deb, K., et al. (eds.) GECCO 2004. LNCS, vol. 3102, pp. 225–237. Springer, Heidelberg (2004)
20. Zitzler, E., Deb, K., Thiele, L.: Comparison of Multiobjective Evolutionary Algorithms: Empirical Results. Evolutionary Computation 8(2), 173–195 (2000)
21. Zitzler, E., Thiele, L.: Multiobjective Evolutionary Algorithms: A Comparative Case Study and the Strength Pareto Approach. IEEE Transactions on Evolutionary Computation 3(4), 257–271 (1999)

Adapting to the Habitat: On the Integration of Local Search into the Predator-Prey Model

Christian Grimme, Joachim Lepping, and Alexander Papaspyrou

Robotics Research Institute - Section Information Technology,
TU Dortmund University, 44221 Dortmund, Germany
{christian.grimme,joachim.lepping,alexander.papaspyrou}@udo.edu

Abstract. Traditionally, Predator-Prey Models—although providing a more nature-oriented approach to multi-objective optimization than many other standard Evolutionary Multi-Objective Algorithms—suffer from inherent diversity loss for non-convex problems. Still, the approach to peg single objectives to a predator allows a very simple algorithmic design. The building-block configuration of the predators offers potent means for fine-tuning and tackling multi-objective problems in a problem-specific way. In the work at hand, we propose the integration of local search heuristics into the classic model approach in order to overcome the unsatisfactory behavior for the aforementioned problem class. Our results show that, introducing a gradient-based local search mechanism to the system, deficiencies with respect to diversity loss can be highly ameliorated while keeping the beneficial properties of the Predator-Prey Model.

Keywords. Predator-Prey Model, Evolutionary Multi-Objective Algorithm, Local Search.

1 Introduction

Over the last years, the original Predator-Prey Model (PPM) by Laumanns [1] has been repeatedly revisited as an alternative approach to multi-objective optimization. In contrast to established and well working algorithms like NSGA-II [2], SMS-EMOA [3], or SPEA [4], which mainly rely on a dominance based selection operator, the PPM approach mimics aspects of the natural interplay of predators and prey. In this process, each predator targets a single objective, and it is expected that the joint influence of all predating individuals affects the prey population in such a way that good trade-off solutions survive.

Studies have shown, however, that the traditional PPM does not behave ideally in all cases [5]: although the strong influence of single-objective selection guarantees the convergence towards the feasible region of the search space, it often leads to a loss of locally[1] optimal or intermediate, but Pareto-optimal solutions.

[1] In the sense of single-objective optimization.

M. Ehrgott et al. (Eds.): EMO 2009, LNCS 5467, pp. 510–524, 2009.

A promising direction of research towards the improvement of this model is the building block-wise design of variation operators. Grimme and Lepping [6] have proposed a framework to create problem-specific composite operators by coupling a single operator to a predator individual. With this approach, the emergence induced by each predator's influence on the prey leads to an adjustable common influence of all operators. This principle was successfully used to solve simple test problems and even a multi-objective scheduling problem [7].

Still, the combination of predator-bound variation operators—although working well for convex problems—does not prevent the model-inherent diversity loss in the non-convex case. We therefore propose the integration of a local search mechanism into the until now passive prey individuals in order to tackle this important issue of the PPM. This inclusion induces more independence and, to a certain extent, makes the prey immune to the effect of predator action. In order to direct the individuals to a feasible region in search space, we apply a gradient-based single-step local search process to each prey, assuming good-natured test problems and approximated gradients[2].

The paper is organized as follows: Section 2 gives a very brief introduction to the PPM and critically reviews some problems. Next, the local search mechanism is described in Section 3 and the integration of local search into the PPM is explained in Section 4. In Section 5 the proposed extension is evaluated on the basis of common test problems and performance measures. Eventually, Section 6 summarizes the results and points out directions for future research.

2 The Predator-Prey Model

The interplay of predators and prey is a well known paradigm from biology and has first been applied to multi-objective optimization by Laumanns and colleagues [1]. The basic idea is to expose spatially distributed prey, which represent solutions for a multi-objective problem, to the evolutionary influence of predators that pursue only a single objective. In that process, the emergence of all predators is expected to force the prey population towards good trade-off solutions.

2.1 Abstract View

We assume a modified [6] version of Laumanns' early approach, where predators are constructed according to a building block model in order to fulfill specific tasks, each of which proportionally adjustable, on the spatially distributed prey population which is located on a toroidal grid[3].

Predator building blocks are basically the movement (mostly random walk) behavior, selection (a single objective), and reproduction (a variation operator).

[2] This is the case for most real-world applications.

[3] Note that the distribution on the toroidal grid does not necessarily map to the individuals distribution in search space. In fact, in most realizations the initialization of prey is random in search space, while their distribution on the toroidal grid is independently random.

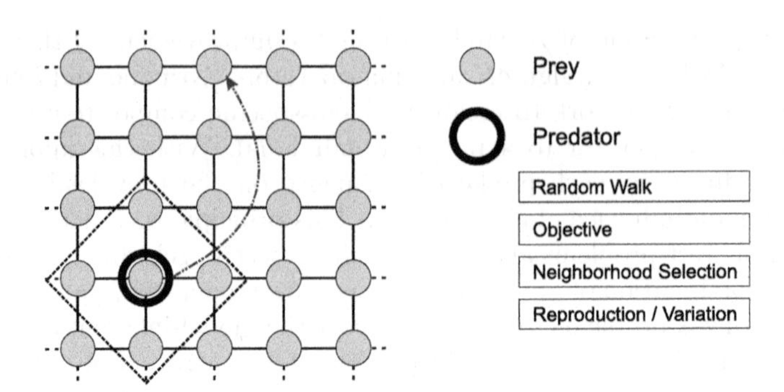

Fig. 1. Schematic depiction of the Predator-Prey Model. The passive prey represent solutions for a given multi-objective optimization problem and are spatially distributed. The predators roam over the population and hunt concerning a single objective.

The combination of these blocks defines a single predator individual. In this way, it is possible to construct predators with specific characteristics, that tackle a certain aspect of the whole considered multi-objective problem. While the movement building block causes change in the area of influence of a single predator, the evolutionary process of the prey population can be steered regarding the predator's objective using individual reproduction methods (e.g. a predator can be configured to select regarding a first objective, move always 5 steps, and use simple Gaussian mutation as variation operator for a steady-state reproduction mechanism).

This approach simplifies the construction of a fine-tuned algorithm which allows the integration of expert knowledge via the variation operators (see Figure 1). The general PPM algorithm consists of the following consecutive steps:

Movement. A predator performs a movement from its current position on the toroidal grid. This is usually done randomly in order to visit each location equally often on the long term. The movement is often limited to a small number of steps to locally and temporally restrict a predator's influence on the prey.

Selection. A predator spans—according to a given neighborhood function—a selection neighborhood as subpopulation of the spatially distributed individuals and evaluates the contained prey individuals regarding its objective. Eventually, following an elitist scheme, the worst prey is eliminated from the selection neighborhood.

Reproduction. A predator applies its reproduction building block on the remaining prey from the previous step and repopulates the empty location with newly created individuals (in the simple reproduction scheme, only one descendant is created and inserted into the population). Common variation operators in this context are Gaussian mutation or Simplex recombination, but—depending on the considered problem—specially tailored variation operators could also be applied, see Grimme et al. [7].

The here applied PPM instance is detailed in Section 5.1, while a detailed description of the general PPM concept is given by Grimme et al. [5]. This work also provides comprehensive insight into the effects of building blocks and resulting population dynamics.

2.2 Criticism

Although previous works [6,5,7] generally confirm the applicability of the modular PPM, the test problems that this approach has been applied to are of simple nature and defined by convex functions. For more complex test problems, especially those basing on non-convex functions, the PPM fails to approximate diverse Pareto sets.

Considering the results of these studies, deficiencies in performance stem from the strong single-objective selection mechanism that leaves almost no room for the survival of Pareto-optimal solutions: Such individuals are typically suboptimal with respect to a single objective. This leads—under the influence of mutation only—to relatively stable prey subpopulations, each of which representing the extremal solution of one objective. The additional application of recombination operators can ameliorate this tendency of 'extremal drift' and produce solutions within the convex hull, but does not address other problem classes.

A usual criticism to the PPM is the amount of function evaluations needed to evolve the prey population to a good solution. This amount often strongly exceeds the standard value of $30,000$ function evaluations which is set as bound for state-of-the-art approaches. This behavior is due to the mainly used steady-state reproduction mechanisms as well as a slow emergence of the predators influence in the beginning of the evolutionary process for an analysis [5].

Summarizing, this analysis shows that, in order to foster the improvement of the PPM, besides the traditional mechanisms such as predator behavior and variation operators, other mechanisms within the model have to be considered as extension points.

3 A Simple Gradient-Based Local Search Mechanism

The inclusion of gradient-based, deterministic local search mechanisms into multi-objective optimization tasks has a long tradition which is rooted in an important inherent property of multi-objective problems: near the Pareto-set the gradients of a solution are almost contradictory. Ester and colleagues [8,9], for example, used this property for their deterministic and stochastic global multi-objective optimization methods back in the 1980ies. In EMO research, the potential of considering gradient information for local search also has been discovered: Brown and Smith [10] review the basic theoretical principles of gradient-directed multi-objective search while Bosman and de Jong [11] already try to efficiently combine several local search techniques inside a MOEA. Experiments towards the hybridization of globally good performing evolutionary approaches for multi-objective optimization and local search are also conducted by Harada and colleagues [12], who propose a Pareto-descent method, extend it for constraint handling, and compare it to other gradient methods. Shukla [13] focuses on the efficiency of such methods for unconstrained test problems and their effect on the computational complexity of the host algorithm. Recently, Schuetze and colleagues [14] proposed a so-called 'hill-climber with sidestep' that also relies on gradient information and successfully integrated it into NSGA-II.

In this paper, we propose and evaluate the integration of such techniques into the PPM by allowing prey individuals to conduct a local search themselves, using online approximated gradient-related information for determining the descent direction for the point in search space they represent.

Preliminaries. For the local optimization task we consider a single point $\boldsymbol{x} \in S \subseteq \mathbb{R}^n$ in decision space. Further, we consider M objectives that define the objective space $O \subseteq \mathbb{R}^M$.

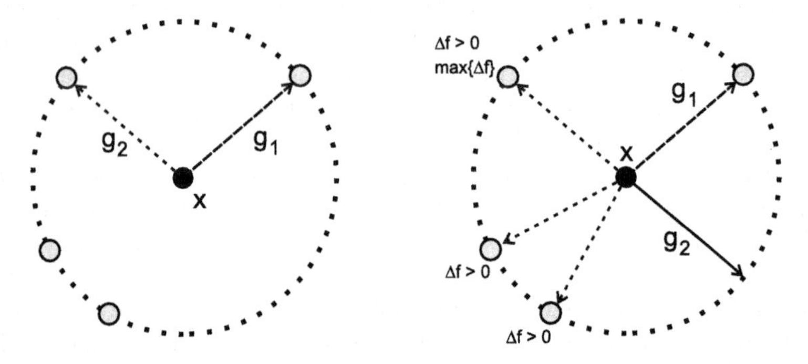

Fig. 2. Sketch of the simple sampling mechanism used to approximate gradient directions for all objectives: Random points on a hypersphere surface are used to select the direction with maximum negative slope. If no point with negative slope is detected, use the counter direction of maximum positive slope.

3.1 Approximation of Gradients

The finite difference method yields approximative information on the slope directions for a given point $\boldsymbol{x}$ within its nearest environment. In order to find a good approximation, several probes have to be taken around $\boldsymbol{x}$. Generally, with a given sampling radius $r \in \mathbb{R}$ (usually $0 < r \ll 1$), a point $\boldsymbol{q} = (q_1, \ldots, q_i, \ldots, q_n)$ for probing is located on the surface of an n-dimensional hypersphere and can be determined uniformly distributed using Equation 1.

$$q_i = x_i + r \cdot \sin(\lambda_i) \cdot \prod_{j=i}^{n} \cos \lambda_j \tag{1}$$

$$\text{with } \lambda_1 = \frac{\pi}{2},\ \lambda_2 = \mathcal{U}(0, 2\pi), \text{ and } \lambda_i = \mathcal{U}(-\frac{\pi}{2}, \frac{\pi}{2}),\ i = 1, \ldots, n$$

Figure 2 shows the sampling with four randomly distributed points on the surface of a two-dimensional hypersphere (circle). Using the sampled points, the direction of maximum negative slope is selected as gradient approximation. If the sampled points do not yield any negative slope, the inverse direction of maximum positive slope is used. For a sufficiently small sampling radius r, this direction yields a descent direction with increasing probability.

3.2 Determining the Pareto-Descent Direction

With the approximated gradient for each objective at hand, three cone types can be constructed:

Descent cones, which promise a benefit for all objectives and thus are preferable in order to reach the Pareto set;
Contradictory cones, which favor some objectives over others; and
Ascent cones, which lead to a deterioration of all objectives.

Depending on the position of the point $\boldsymbol{x}$ relative to the Pareto set, the cones are of different size, see Figure 3: If $\boldsymbol{x}$ is far from the Pareto set, the descent cone is large. Otherwise, the descent cone is rather small while the contradictory cone is large. Altogether with the gradient information, these properties are used to perform a local search step.

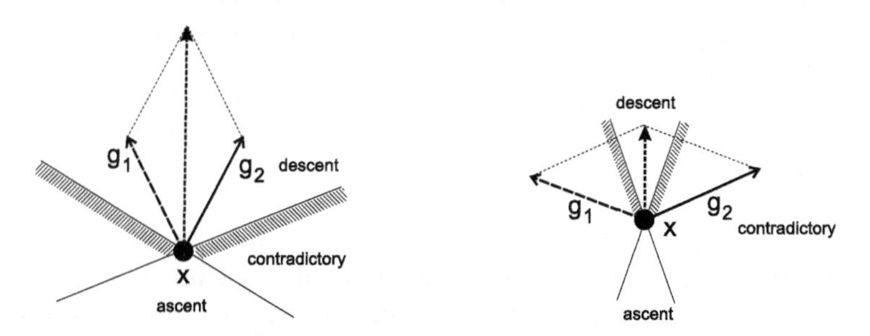

Fig. 3. Depiction of the applied local search based on approximated gradient directions for all objectives (here with two dimensions): if descent gradients point approximately to the same direction, the solution is far from the (local) Pareto set and the area of overall beneficial descent is large. Otherwise, the solution is close to the (local) Pareto-set and the beneficial area is small.

The search step length and direction is determined by combining the normalized gradient vectors as shown in Equation 2.

$$\boldsymbol{x}' = \boldsymbol{x} + \boldsymbol{z},\ \boldsymbol{z} = \frac{1}{\sqrt{M}} \sum_{i=1}^{M} \omega_i \cdot \boldsymbol{g}_i \tag{2}$$

$$\text{with } \boldsymbol{z}, \boldsymbol{g}_i \in \mathbb{R}^n \text{ and } \omega_i = \mathcal{U}(1, M),\ \sum_{i=1}^{M} \omega_i = M$$

After combining the gradient vectors, the resulting step is normalized with respect to the dimensions to avoid large steps for $M > 2$ objectives. Simultaneously, each gradient vector is weighted with a uniformly distributed value to increase or reduce its influence in the heading direction. If point $\boldsymbol{x}$ is far from the Pareto set, the resulting direction $\boldsymbol{z}$ will not leave the descent cone. Close to the Pareto set, however, the weights will more frequently result in a step direction towards the contradictory cone. This is beneficial in order to favor diversity for solutions close to the Pareto set. Obviously, a parameter for switching between descent and diversity steps is not necessary: The local search mechanism automatically adjusts to the situation using gradient information inherent to the population.

4 Hybridizing the Predator Prey Model

Following the building block approach, the natural way of integrating local search to the Predator-Prey Model is the prey itself, as it represents the only other entity in the PPM which is generally enabled to act. Until now, all other PPM models assume a completely passive prey individual, having the only purpose of representing a solution of the evaluated problem. As such, it is a sensible choice to let the prey to optimize itself within its habitat. This also reflects natural evolution and allows for self-immunization against the effects of the environment to a certain extent.

Overall, it is expected that the integration of local search mechanisms into the prey leads to the following benefits for the model dynamics by exploiting its inherent properties:

1. It stabilizes the population in terms of preserving and advancing locally optimal solutions by seeking solutions that are globally superior. This is achieved without strong selection pressure of a single-objective selection as normally induced by predator individuals.
2. It allows for improvements on a very fine-grained level, not only with respect to convergence towards locally optimal solutions, but also from the diversity point of view when having reached those: Due to the large area covered by the contradictory cone, individuals tend to move along the local Pareto front, inducing a diverse set of solutions, see Section 3.
3. It exploits the inherent parallelism of the PPM by enabling fully parallelizable local optimization without disturbing or impairing the global algorithm execution.

The predators, in turn, ensure that the prey individuals overall converge into the direction of globally optimal solutions. This is achieved by the pressure which is induced by the interplay of selection and variation: while the selection operator provides a global comparison of local solutions and only keeps a certain variety alive within the population, the variation operator[4] incorporates innovation and exploration effects into the population.

[4] Which is still coupled to the predator individual, as suggested in the standard model.

5 Evaluation

In order to assess the performance of the local search-extended PPM, we conducted a comparative evaluation between the classic PPM and our new proposal. Our main focus therefore lies on the effect of integrating support mechanisms into the traditional model in order to show the expected influence on the system's dynamics. As such, a comparison with other EMOAs is beyond the scope of this work and not intended by giving comparison results.

5.1 Setup

For both, the traditional model (PPM-T) and the local-search-enhanced model (PPM-L) we consider the same basic parameter settings. Generally we assume a toroidal grid as population structure as graph $G = (V, E)$, where $v \in V$ are vertices and $e \in E$ are edges along which predators can move. The movement of a predator is set as uniform random walk given by Equation 3. Starting at vertex v a neighboring target vertex v_t is reached with the same probability.

$$walk(v)(v_t) = \begin{cases} \frac{1}{d(v)} & \text{if } (v, v_t) \in E \\ 0 & \text{otherwise.} \end{cases} \tag{3}$$

A neighborhood $N(v, rad)$ can be generally described by Equation 4 with a starting point $N(v, rad = 0) = v$ and neighborhood radius rad.

$$N(v, rad) = \bigcup_{(v,\nu) \in E} N(\nu, rad - 1) \tag{4}$$

The here applied selection and reproduction neighborhood is constructed according to the neighborhood function given in Equation 5. In this simple case we always consider a von-Neumann neighborhood with $rad = 1$ as also depicted in Figure 1.

$$nbh(v_t, v) = \begin{cases} 1 & \text{if } v \in N(v_t, 1), \\ 0 & \text{otherwise.} \end{cases} \tag{5}$$

For PPM-T we applied Gaussian mutation with fixed mutation step size and Simplex recombination as variation operators. We manually fine-tuned the parameters to achieve good results for a fair comparison. Moreover, we varied the population size between 900 and 1600 individuals[5]. For PPM-L we used Gaussian mutation as the only predator-based variation operator. In addition to this, we enabled parametrized local search using a sampling radius $0.1 \leq r \leq 0.2$ and varied the number of sample points between 3 and 5. As population size, we always used 225 individuals on a 15×15 toroidal grid. For both setups, we limited the number of function evaluations to a maximum of 200,000, which is a standard value for the predator prey approach.

[5] Derived from a 30×30 and 40×40 torus grid size, respectively.

Table 1. Test problems used for the comparison of the traditional PPM and the extended model with local search

Name	Test Problem	Initialization	Constraints
Multisphere [1]	$f_1(x) = x_1^2 + x_2^2$ $f_2(x) = (x_1 - 2)^2 + x_2^2$	$x \in [-10, 10]^2$	none
Kursawe [15]	$f_1(x) = \sum_{i=1}^{n-1} -10 \cdot exp(-0.2 \cdot \sqrt{x_i^2 + x_{i+1}^2})$ $f_2(x) = \sum_{i=1}^{n} \lvert x_i\rvert^{0.8} + 5 \cdot \sin^3(x_i)$	$x \in [2, 4]^2$	none
ZDT3 [16]	$f_1(x) = x_1$ $f_2(x) = g(x) \cdot h(x)$ $g(x) = 1 + \frac{9}{n-1} \cdot \sum_{i=2}^{n} x_i^2$ $h(x) = \left[2 - \sqrt{\frac{x_1}{g(x)}} - \frac{x_1}{g(x)} \cdot \sin(10 \cdot \pi x_1)\right]$	$x_1 \in [0.1, 1]$ $x_i \in [0, 1]$ $i = 2 \dots 10$	$x_1 \in [0.1, 1]$ $x_i \in [0, 1]$ $i = 2 \dots 10$
Viennet [17]	$f_1(x) = 0.5 \cdot (x_1^2 + x_2^2) + \sin(x_1^2 + x_2^2)$ $f_2(x) = \frac{(3x_1 - 2x_2 + 4)^2}{8} - \frac{(x_1 - x_2 + 1)^2}{27} + 15$ $f_3(x) = \frac{1}{x_1^2 + x_2^2 + 1} - 1.1 \cdot \exp(-x_1^2 - x_2^2)$	$x \in [-3, 3]^2$	none

As a benchmark for the performance of both PPM variants, we used a selection of multicriteria test problems as listed in Table 1. Our test suite consists of two- and three-objective test problems with different properties that pose varying challenges to EMOAs. There, we feature two—one convex, one non-convex—problems with a single Pareto front, one problem with multiple disconnected local Pareto fronts, and one problem with three objectives. Reference figures of the test problems' true Pareto-front are given in the Appendix.

For assessing the quality of the gained results, we applied two standard measures for quantifying convergence and diversity, namely the Generational Distance (GD) [18] and the normalized Hypervolume (HV) [19]. Equation 6 explains the calculation of the Generational Distance which the mean distance of an approximated solution from true Pareto-front P_{true}. Basically, the minimum euclidean distance for all approximated solution points to the true front is considered and normalized by the number of exact solutions.

$$GD = \frac{\sum_{i=1}^{n} d_i}{|P_{true}|} \tag{6}$$

The Hypervolume metric denotes the volume enclosed by a reference point $p_r \in \mathbb{R}^M$ and the approximated non-dominated solutions $\mathcal{L} = \{\ell \in \mathbb{R}^M\}$. Let Q_{ℓ,p_r} be the Hypervolume enclosed by a solution ℓ with p_r perpendicular to the coordinate axes. The complete Hypervolume is then given by: $HV := \bigcup_{\ell \in \mathcal{L}} Q_{\ell,p_r}$.

5.2 Discussion of Results

Following, we analyze the behavior of both models and discuss their impact on the results of the test problems. Although each MOO-problem was tested in 30

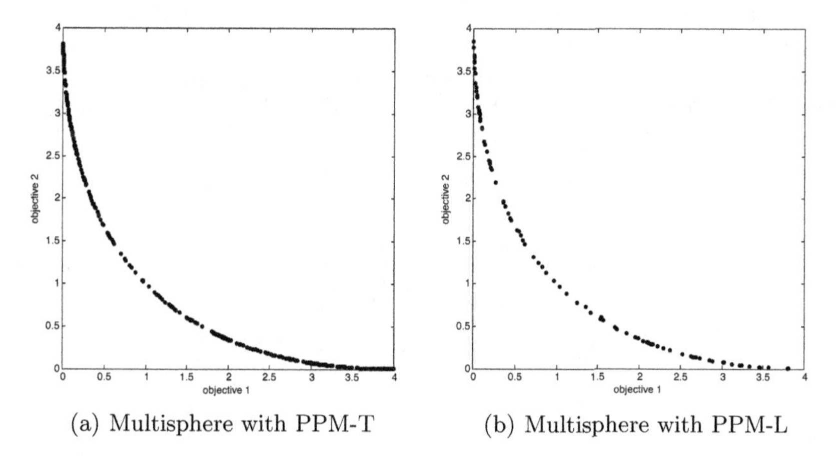

(a) Multisphere with PPM-T (b) Multisphere with PPM-L

Fig. 4. Depiction of exemplary results for both approaches on the Multisphere test problem

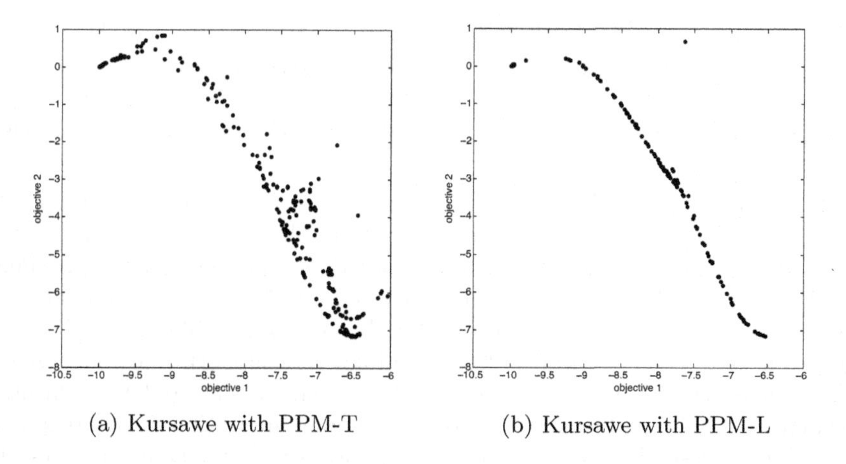

(a) Kursawe with PPM-T (b) Kursawe with PPM-L

Fig. 5. Depiction of exemplary results for both approaches on the Kursawe test problem

experiments, the displayed Pareto-fronts are exemplary depictions of a single trail. Intentionally, we did not apply a Pareto filter for the values shown in the figures, as to allow a broader interpretation of the algorithms' performance.

On the very simple Multisphere test problem, the PPM-T is known to perform well, due to the specifically adapted Simplex recombination operator which fosters convergence to the true Pareto front. This fact is also reflected by Figure 4, where a very diverse Pareto front can be seen. On the other hand, although reaching the best HV value, the obtained solutions are not very stable (see Table 2). The PPM-L, in turn, reaches almost competitive, however far more stable results for HV.

Our second test problem, Kursawe, poses a much higher challenge for PPM-T, as the Pareto front dissolves into convex and concave parts, see Figure 5(a). Especially, the concave part of the Pareto front is poorly approximated. With the integration of local search, PPM-L offers a much better approximation for both parts of the Pareto front, see Figure 5(b), which is also reflected by the mean HV and GD values as shown in Table 2.

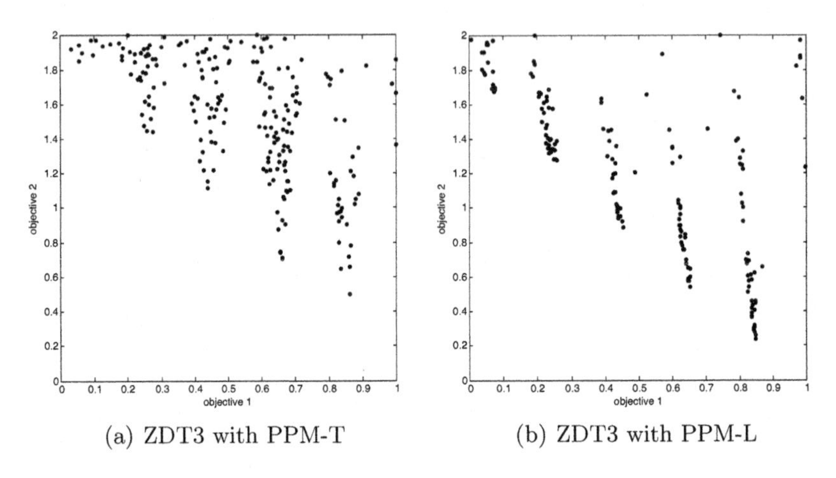

(a) ZDT3 with PPM-T (b) ZDT3 with PPM-L

Fig. 6. Depiction of exemplary results for both approaches on the ZDT3 test problem

For ZDT3, the PPM-T shows a stagnation at the local Pareto fronts, see Figure 6(a). This effect probably originates from the fixed step length of the Gaussian mutation. Due to the properties of the test problem, the approximation of the global Pareto fronts is increasingly improbable and degenerates to a random search. The local search in PPM-L, in turn, is able to stabilize the solutions near to the local Pareto fronts and actively supports the convergence towards them, see Figure 6(b). The global influence of the predators through their mutation building block still introduces enough innovation to ensure global convergence. These observations are also supported by the GD and HV results in Table 2.

Finally, the Viennet test problem shows that the local search mechanism is successfully transferable to problems with more than two objectives: Here again, the diversity of the solutions gained from PPM-L is comparable with PPM-T, as shown by HV and GD results, see Table 2. However, as shown in Figure 7, PPM-L is able to approximate the solution set parts more complete than PPM-T.

Overall, the extension of PPM with local search is advantageous for all evaluated test problems with respect to diversity and stability of their respective solution: The HV measure shows constantly better performance for PPM-L, while the standard deviation of the results is very small. In addition to that, problems with multiple local Pareto fronts, as exemplarily shown with ZDT3, seem to benefit from PPM-L with respect to convergence; this is backed by the GD measure. For a detailed list of results, see Table 2.

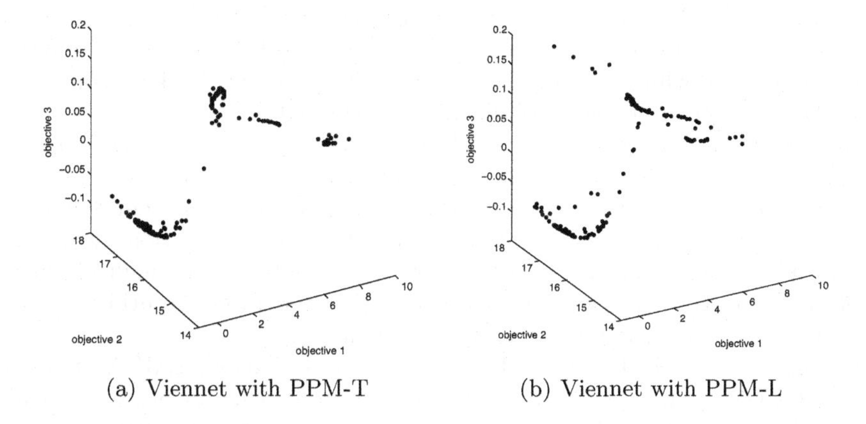

(a) Viennet with PPM-T (b) Viennet with PPM-L

Fig. 7. Depiction of exemplary results for both approaches on the Viennet test problem

Table 2. Numerical results for PPM-T and PPM-L on the four considered test problems. For performance evaluation the Hypervolume (HV) and Generational Distance (GD) measures were applied (statistical values from 30 experiments).

		HV		GD	
		PPM-T	PPM-L	PPM-T	PPM-L
Multisphere	best	0.831197	0.829462	1.59898e-4	3.49630e-4
	mean	0.726294	**0.828406**	**2.10974e-4**	3.77680e-4
	std	0.076993	1.32709e-4	1.27843e-5	2.96064e-6
Kursawe	best	0.764955	0.769532	5.65940e-5	1.27920e-4
	mean	0.761400	**0.767443**	4.76703e-4	**2.00163e-4**
	std	0.001330	2.48120e-4	6.81601e-5	1.00732e-5
ZDT3	best	0.435715	0.492830	0.007884	0.81808e-4
	mean	0.401769	**0.486205**	0.013962	**0.001481**
	std	0.008062	9.64324e-4	0.002432	9.37421e-5
Viennet	best	0.821630	0.831471	0.002081	0.001900
	mean	0.800788	**0.819030**	0.004751	**0.004607**
	std	0.004586	0.001931	0.000823	9.07811e-4

6 Conclusion and Future Work

In the presented work, the traditional Predator-Prey Model has been extended with a local search mechanism in order to stabilize the system's dynamics and to foster diversity and convergence towards the global Pareto front. To this end, the gradient-based search was integrated into the hitherto passive prey as an additional building block. This mechanism enables the prey to reach good solutions by walking along a cooperative descent cone as long as it is far from the local Pareto front, while favoring contradictory directions around local solutions. Herewith, a better preservation of diversity is achieved.

Both the traditional and the new PPM have been compared using four well known test problems to demonstrate the benefits of local search integration. We were able to show that the expected stabilizing effect to the system dynamics could be achieved. Moreover, the results show that the integration yields significant advantages regarding diversity. For test problems with multiple local Pareto fronts convergence was also improved; this illustrates the utility gained by blending local search and global optimization.

A comparison of PPM with local search to other well known EMOAs was not part of this work, but is to be conducted in future research. Obviously, there are many function evaluations necessary to reach a good solution set. Although the population size for the new variant of the PPM has been reduced significantly (from 1600 prey individuals to only 225 individuals), emergence effects still seem to take a long time to show effect in the whole population. This aspect of model dynamics has to be investigated further.

Another area that leaves room for further improvement is the optimization of the local search mechanism: Although it can be externalized due to the parallel nature of the PPM, its performance depends strongly on the approximation accuracy of the gradients used for cooperative descent. Here, advanced mechanisms are to be developed, too.

References

1. Laumanns, M., Rudolph, G., Schwefel, H.P.: A Spatial Predator-Prey Approach to Multi-objective Optimization: A Preliminary Study. In: Eiben, A.E., Bäck, T., Schoenauer, M., Schwefel, H.-P. (eds.) PPSN 1998. LNCS, vol. 1498, pp. 241–249. Springer, Heidelberg (1998)
2. Deb, K., Agrawal, S., Pratab, A., Meyarivan, T.: A Fast Elitist Non-Dominated Sorting Genetic Algorithm for Multi-Objective Optimization: NSGA-II. In: Deb, K., Rudolph, G., Lutton, E., Merelo, J.J., Schoenauer, M., Schwefel, H.-P., Yao, X. (eds.) PPSN 2000. LNCS, vol. 1917, pp. 849–858. Springer, Heidelberg (2000)
3. Beume, N., Naujoks, B., Emmerich, M.: SMS-EMOA: Multiobjective selection based on dominated hypervolume. European Journal of Operational Research 181, 1653–1669 (2007)
4. Zitzler, E., Laumanns, M., Thiele, L.: SPEA2: Improving the Strength Pareto Evolutionary Algorithm. Technical Report 103, Computer Engineering and Communication Networks Lab (TIK), Swiss Federal Institute of Technology (ETH), Zürich (2001)
5. Grimme, C., Lepping, J., Papaspyrou, A.: Exploring the Behavior of Building Blocks for Multi-Objective Variation Operator Design using Predator-Prey Dynamics. In: Thierens, D., others (eds.) Proceedings of the Genetic and Evolutionary Computation Conference (GECCO 2007), London, pp. 805–812. ACM Press, New York (2007)

6. Grimme, C., Lepping, J.: Designing Multi-objective Variation Operators Using a Predator-Prey Approach. In: Obayashi, S., Deb, K., Poloni, C., Hiroyasu, T., Murata, T. (eds.) EMO 2007. LNCS, vol. 4403, pp. 21–35. Springer, Heidelberg (2007)
7. Grimme, C., Lepping, J., Papaspyrou, A.: The parallel predator-prey model: A step towards practical application. In: Rudolph, G., et al. (eds.) PPSN 2008. LNCS, vol. 5199, pp. 681–690. Springer, Heidelberg (2008)
8. Ester, J.: Systemanalyse und mehrkriterielle Entscheidung. VEB Verlag Technik, Berlin (1987)
9. Peschel, M.: Ingenieurtechnische Entscheidungen. Modellbildung und Steuerung mit Hilfe der Polyoptimierung. VEB Verlag Technik, Berlin (1980)
10. Brown, M., Smith, R.E.: Directed Multi-Objective Optimization. International Journal of Computers, Systems and Signals 6, 3–17 (2005)
11. Bosman, P.A.N., de Jong, E.D.: Combining Gradient Techniques for Numerical Multi-Objective Evolutionary Optimization. In: GECCO 2006: Proceedings of the 8th Annual Conference on Genetic and Evolutionary Computation, pp. 627–634. ACM, New York (2006)
12. Harada, K., Sakuma, J., Kobayashi, S.: Local Search for Multiobjective Function Optimization: Pareto Descent Method. In: GECCO 2006: Proceedings of the 8th Annual Conference on Genetic and Evolutionary Computation, pp. 659–666. ACM, New York (2006)
13. Shukla, P.K.: On Gradient Based Local Search Methods in Unconstrained Evolutionary Multi-objective Optimization. In: Obayashi, S., Deb, K., Poloni, C., Hiroyasu, T., Murata, T. (eds.) EMO 2007. LNCS, vol. 4403, pp. 96–110. Springer, Heidelberg (2007)
14. Schuetze, O., Sanchez, G., Coello, C.A.C.: A New Memetic Strategy for the Numerical Treatment of Multi-Objective Optimization Problems. In: Proceedings of the 10th Annual Conference on Genetic and Evolutionary Computation, pp. 705–712. ACM Press, New York (2008)
15. Kursawe, F.: A Variant of Evolution Strategies for Vector Optimization. In: Schwefel, H.-P., Männer, R. (eds.) PPSN 1990. LNCS, vol. 496, pp. 193–197. Springer, Heidelberg (1991)
16. Zitzler, E., Deb, K., Thiele, L.: Comparison of Multiobjective Evolutionary Algorithms: Empirical Results. Evolutionary Computation 8(2), 173–195 (2000)
17. Viennet, R., Fontiex, C., Marc, I.: Multicriteria Optimization Using a Genetic Algorithm for Determining a Pareto Set. Journal of Systems Science 27(2), 255–260 (1996)
18. Veldhuizen, D.V., Lamont, G.: Multiobjective Evolutionary Algorithm Research: A History and Analysis. Technical Report Dept. Elec. Comput. Eng. Air Force TR-98-03, Air Force Inst. Technol. (1998)
19. Zitzler, E., Thiele, L.: Multiobjective Evolutionary Algorithms: A Comparative Case Study and the Strength Pareto Approach. IEEE Transactions on Evolutionary Computation 3(4), 257–271 (1999)

Test Problem References

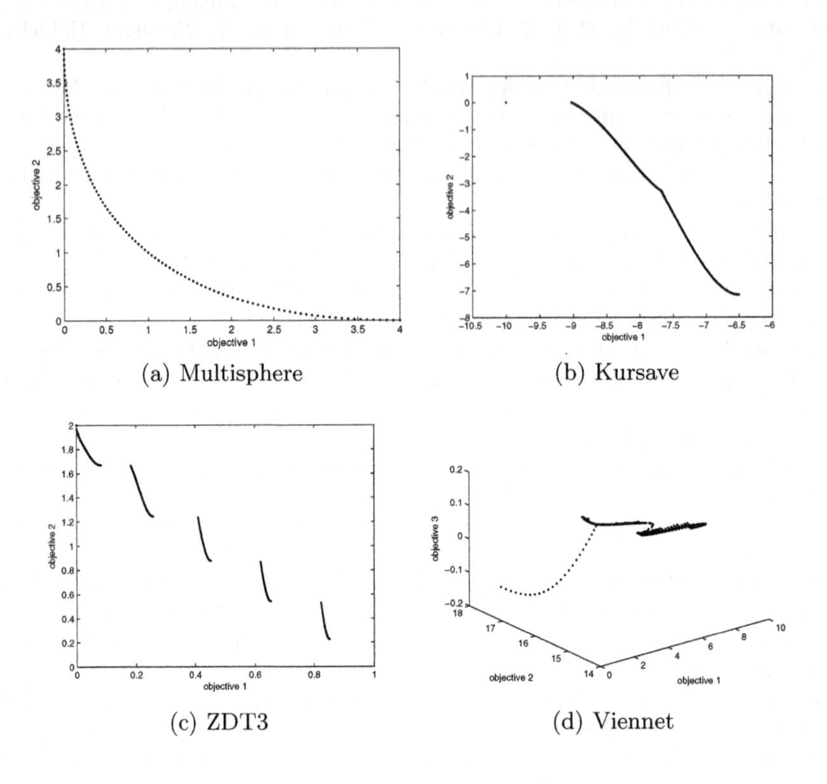

(a) Multisphere

(b) Kursave

(c) ZDT3

(d) Viennet

Multiobjective Distinct Candidates Optimization (MODCO): A Cluster-Forming Differential Evolution Algorithm

Peter Dueholm Justesen[1] and Rasmus K. Ursem[2]

[1] Department of Computer Science
University of Aarhus,
Åbogade 34, DK-8200 Århus N, Denmark
juste@daimi.au.dk

[2] Department of Fluid Mechanics, R&T
Grundfos Management A/S
Poul Due Jensens Vej 7, DK-8850 Bjerringbro, Denmark

Abstract. Traditionally, Multiobjective Evolutionary Algorithms (MOEAs) aim at approximating the entire true pareto-front of their input problems. However, the actual number of solutions with different trade-offs between objectives in a resulting pareto-front is often too large to be applicable in practice. The new field Multiobjective Distinct Candidates Optimization (MODCO) research is concerned with the optimization of a low and user-defined number of clearly distinct candidates. This dramatically decreases the amount of post-processing needed in the decision making process of which solution to actually implement, as described in our related technical repport "Multiobjective Distinct Candidates Optimization (MODCO): A new Branch of Multiobjective Optimization Research" [9].

In this paper, we introduce the first algorithm designed for the challenges of MODCO; providing a given number of distinct solutions as close as possible to the true pareto-front. The algorithm is using subpopulations to enforce clusters of solutions, in such a way that the number of clusters formed can be set directly. The algorithm is based on the Differential Evolution for Multiobjective Optimization (DEMO) algorithm versions, but is exchanging the crowding/density measure with two alternating secondary fitness measures. Applying these measures ensures that subpopulations are attracted towards knee regions while also making them repel each other if they get too close to one another. This way subpopulations traverse different parts of the objective space while forming clusters each returning a single distinct solution.

Keywords: Multiobjective Optimization, Evolutionary Algorithms, Differential Evolution, Distinct Candidates, Subpopulations, Clustering, NSGA-II, SPEA2, DEMO, MODCO.

1 Introduction

Multiobjective optimization (MO) is the discipline of finding the optimal set of solutions of problems having several, usually conflicting objectives. In traditional

M. Ehrgott et al. (Eds.): EMO 2009, LNCS 5467, pp. 525–539, 2009.

algorithmic research on MO, algorithms are evaluated on the following three goals: The algorithm's ability to (1) find a set of solutions as close to the true pareto-front as possible, (2) ensure an even distribution of solutions on the pareto-front, and (3) have a high spread of solutions, i.e., to find extreme solutions for each of the involved objectives. In short, closeness, distribution, and spread. Although these goals are desirable from a theoretical point-of-view, they are not from a practical point-of-view. Post-processing several hundred alternative solutions returned by traditional MO algorithms is often impossible because time and money may disallow further analysis of many solutions using advanced simulations or physical prototypes.

To address this issue and a number of other practically motivated challenges, we introduce Multiobjective Distinct Candidates Optimization (MODCO) as a new branch of MO research. In short, MODCO defines three alternative goals to MO research. The ideal MODCO algorithm aims at (1) closeness to true pareto front, (2) return a user-defined number of distinct solutions, and (3) return solutions in knee regions. In short, (1) closeness, (2) global distinctiveness, and (3) local multiobjective optimality. Whereas this paper introduces the first MODCO algorithm, a more elaborate argumentation for the soundness of MODCO and why this should be considered a new branch of MO research is given in our related technical repport [9].

In this paper, we propose the Cluster-Forming Differential Evolution algorithm (CFDE), using subpopulations subject to Differential Evolution (DE). It features a user-defined number of candidate solutions (K_{NC}), a user-defined performance distinctiveness (K_{PD}), and the ability to converge to knee regions, see [9] for further details. Hence, this first MODCO algorithm does not allow user-defined design distinctiveness (K_{DD}) or incorporation of simulator accuracy (K_{SA}). In short, performance distinctiveness (K_{PD}), design distinctiveness (K_{DD}) and category distinctiveness (K_{CD}) allow the user to set how similar or different the few returned solutions should be wrt. performance (objective space), design (search space), and the MODCO problem's category functions (see [9]).

2 Cluster-Forming Differential Evolution

Differential Evolution (DE) was suggested by Price and Storn [4], and was initially designed for single-objective optimization. However, with the growing interest in solving multiobjective problems, it has been successfully adopted in multiobjective evolutionary algorithms. Recently, Robič and Filipič combined both NSGA-II and SPEA2 selection with the Differential Evolution scheme for solution reproduction. These alternate versions were named DEMO^{NSII} and DEMO^{SP2} and have been shown to outperform the more classic multiobjective genetic algorithms NSGA-II and SPEA2 [1,2,3,5,6]. The Cluster-Forming Differential Evolution (CFDE) algorithm is based on the DEMO-scheme, but is modified for cluster forming using subpopulations and two specialized secondary fitness measures.

2.1 Data Structures and Notation

The population in the CFDE algorithm is stored in a vector of vectors, with each entry in the vector defining a subpopulation. For the rest of this article, we assume without loss of generality that the user-defined number of distinct candidates K_{NC} divides the population size N, i.e. that $N \bmod K_{NC} == 0$. We further use a centroid vector of M-dimensional points, and a temporary offspring vector for migration.

In pseudocode, we let P_i refer to subpopulation i of the full population P, whereas $x_{i,j}$ refers to the individual at the j'th position in subpopulation i. For global ranking and truncation, we denote the current generation population P_t, and subpopulations $P_{i,t}$.

We access cluster centroids using C_i to denote the M-dimensional point in objective space corresponding to the centroid of subpopulation i, and $C_{i,m}$ to access the m'th entry in this. Lastly, we let $f(x)$ denote the objective vector of individual x, and $f_m(x)$ to denote the m'th entry of this.

2.2 Main Algorithm

Pseudocode of the main algorithm can be seen in Algorithm 1, where we let $minDist(C_i)$ denote the function returning the minimum distance from centroid C_i to the nearest other centroid. Further, the calculation of σ is problem-dependent, as explained in [9], and further demonstrated later in this paper.

First, the CFDE algorithm performs global mating with replacement, as in usual DE. However, it stores the incomparable offspring in a temporary offspring

Algorithm 1. Cluster-Forming Differential Evolution

Require: Population size N, K_{NC}, K_{PD}
Ensure: K_{NC} different non-dominated individuals.
1: Initialize K_{NC} subpopulations with N/K_{NC} random individuals in each
2: **while** Halting criterion has not been met **do**
3:   Perform global DE-based mating - store incomparable offspring
4:   Calculate subpopulation centroids C_i
5:   Migrate incomparable offspring to nearest subpopulation wrt. centroid
6:   **for** All $P_i \in P$ **do**
7:     **if** $minDist(C_i) < \sigma$ **then**
8:       Assign nearest other centroid distance to each individual $x_{i,j} \in P_i$
9:     **else**
10:       Assign knee utility function value to each individual $x_{i,j} \in P_i$
11:     **end if**
12:   **end for**
13:   Assign final fitness wrt. global pareto rank, then secondary fitness
14:   Truncate subpopulations wrt. final fitness
15: **end while**
16: Return K_{NC} solutions, by returning the non-dominated solution closest to the subpopulation centroid from each subpopulation.

vector, until it can be determined, which subpopulation they should belong to. From the parent part (see figure 1) of the subpopulations, a centroid for each is then calculated. Following this, the incomparable offspring are migrated to the subpopulations with the nearest centroid.

At this point, the CFDE algorithm determines which of the two secondary fitness measures to use for each subpopulation. The secondary fitness measure is each individual's distance to the nearest other centroid, if the subpopulations centroid is too close to its nearest neighboring centroid. In case the centroid of the subpopulation is sufficiently far away from its neighbours, the secondary fitness measure is Branke et al.'s utility function, favoring individuals in knee regions [8].

Still, CFDE maintain focus on convergence towards the true pareto-front, so it then assigns to each individual a *global* pareto rank using the NSGA-II non-dominated sorting. This is used for assigning each individual a final fitness, such that the final fitness incorporates both rank and secondary fitness measure in a total order.

Finally each subpopulation is truncated to the original size of N/K_{NC} using the truncation mechanism of NSGA-II. Here, the main differences are, that truncation is done locally in subpopulations, and that the subpopulations are truncated using one of the two secondary fitness measures, which is incorporated in the final fitness. Hence, some subpopulations may be truncated using distance, and others using the knee utility function. This way subpopulations may be attracted to different knee regions, while forming clusters during the evolutionary process. As we return only one solution from each subpopulation, we get the wanted number of distinct solutions returned.

2.3 Subpopulation Based Differential Evolution

For subpopulation based DE, we will use the DE scheme with replacement, such that parents are replaced if they are dominated by their offspring, and offspring are discarded if they are dominated by their parents. However, the subpopulation approach makes it necessary to determine which subpopulation a newly created offspring should belong to, in case the offspring is incomparable, such that it neither dominates or is dominated by its parent.

In case an offspring is incomparable to its parent, it is stored in a temporary vector in order to enable migration after we have calculated subpopulation centroids. This ensures that incomparable offspring will always belong to the subpopulation they are closest to, and that the parent part of the subpopulation remains fixed. This is illustrated in figure 1, which shows the population after migration. Note that we denote the parent part as the N/K_{NC} first entries in the subpopulation vector, even though some of these entries may have been overwritten by dominating offspring candidates. This is natural, as these offspring contribute to mating as soon as they have replaced their parents.

The subpopulation based Differential Evolution algorithm is based on the *rand/1/bin* standard DE scheme, and is depicted in Algorithm 2. As usual, we

Algorithm 2. Subpopulation based Differential Evolution

Require: Parent $x_{i,j}$, crossover factor CF, scaling factor F.
Ensure: Candidate $c_{i,j}$.
1: Randomly select three individuals x_{i_1,j_1}, x_{i_2,j_2}, x_{i_3,j_3} from P, where $x_{i,j}$, x_{i_1,j_1}, x_{i_2,j_2} and x_{i_3,j_3} are pairwise different.
2: Calculate candidate $c_{i,j}$ as $c = x_{i_1,j_1} + F \cdot (x_{i_2,j_2} - x_{i_3,j_3})$.
3: Modify candidate $c_{i,j}$ by binary crossover with the parent $x_{i,j}$ using CF.

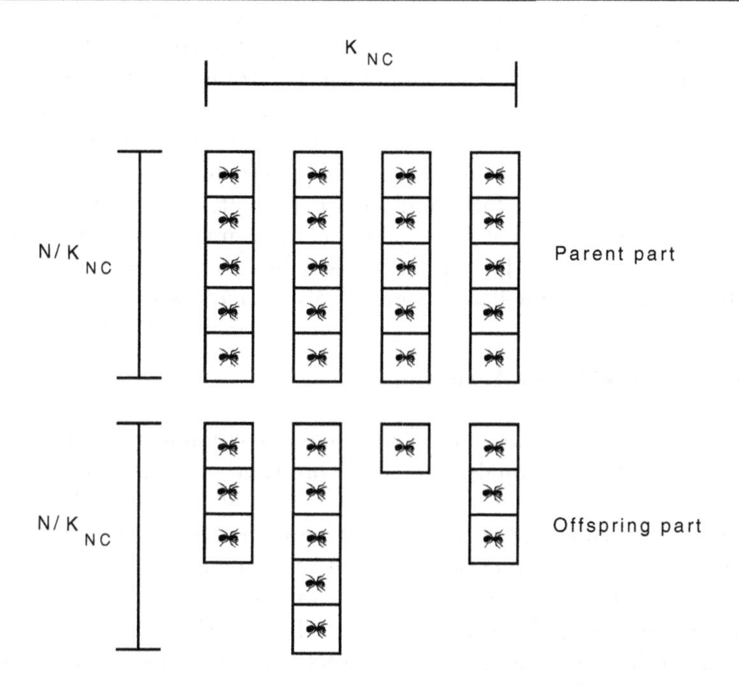

Fig. 1. Population after migration

perform the algorithm for all individuals[1] $x_{i,j} \in P$, creating one offspring per parent, which is treated according to the cases described above.

2.4 Calculating Subpopulation Centroids

The simple subpopulation centroid calculation is a measure of the center of the current elite part of the subpopulation, namely the parent part, as depicted in figure 1 and described above. This part consists of the elite from the previous generation, mixed with the offspring, which replaced their parents. These individuals are the current best, and therefore we can use this fixed size part of the subpopulation to define a centroid. For each subpopulation P_i, we calculate the centroid $C_i = [C_{i,1}, C_{i,2}...C_{i,M}]$ as the average point of the elite in objective space:

[1] Which at this point is only the parent part, see figure 1.

$$C_{i,m} = \frac{\sum_{j=1}^{N/K_{NC}} f_m(x_{i,j})}{N/K_{NC}}, m = 1..M. \quad (1)$$

2.5 Migration

All incomparable offspring are at this point in a temporary offspring vector. To migrate these offspring, the algorithm iterates through the offspring vector, and move each individual to the subpopulation, whose centroid is the closest to the individual in objective space. More formally, this is done by appending the individual to the subpopulation, while removing it from the offspring vector. Note, that we do not allow duplicates within subpopulations.

After this migration, the full population is present, which is depicted in figure 1, but it may be noted that some subpopulations may increase their offspring part size beyond N/K_{NC}. This, however, is a theoretical possibility, and most often subpopulation sizes will vary from N/K_{NC} to $2N/K_{NC}$.

2.6 Secondary Fitness Assignment

Having calculated the subpopulation centroids, we can now use this to assign a secondary fitness to individuals, depending on whether subpopulations are too close to each other or not, wrt σ. So we may assign different secondary measures to subpopulations. However, this is no problem with regard to truncation, which also happens within subpopulations, so if two individuals are not judged by the same secondary measure, they will not be compared during truncation.

Incomparable offspring are at this point distributed to subpopulations, as seen in figure 1. Therefore we can go through the full subpopulations, and assign to each individual a secondary fitness. The measures proposed are the utility-function proposed by Branke et al. [8], which focus search on knee regions, and a new centroid distance measure, which makes subpopulations repel each other.

Using Utility Function. The utility function proposed in [8], is intended to discover knee regions by calculating an average fitness value for a large number of randomly sampled weight vectors. If this average fitness is good, the individual is more likely to reside in a knee-region. Knee-regions are characterized by the fact that a small improvement in one objective, will result in large deterioration in another objective.

The utility function takes only one argument, *precision*, denoting the number of sample weight vectors to apply. Let λ denote the weight vector of dimension M, with $\sum_m \lambda_m = 1$. Then we calculate the secondary fitness SF with precision *precision* of each individual $x \in P_i$ as:

$$SF(x) = \frac{\sum_{p=1}^{precision} \lambda_p \cdot f(x)}{precision} \quad (2)$$

Using Centroid Distance. The centroid distance is the distance in objective space, from the individual to the nearest other subpopulation centroid. Now, let $dist(x, y)$ denote the distance in objective space between point x and point y, each of dimension M. Further, let $min(S)$ denote the minimal element of the set S. For the subpopulation P_i, we may then assign to each individual $x \in P_i$ a secondary fitness SF as:

$$SF(x) = min(\{dist(f(x), C_j), j = 1..K_{NC}, j \neq i\}) \quad (3)$$

2.7 Assigning Global Pareto Rank

The global pareto rank is intended to guide the search towards the best individuals wrt. dominance. Two very popular ways of assigning a pareto rank are based on the non-dominated sorting from NSGA-II or the strength pareto approach of SPEA2. These ranks can be viewed as a raw fitness, and are traditionally complemented by a crowding/density measure. Therefore, it is straight forward to exchange the original measures with the secondary fitnesses introduced above. We show how to do this using the approach of NSGA-II, but it is also possible to use the SPEA2 scheme for ranking and truncating.

The elitism mechanism in NSGA-II is based on non-dominated sorting of the current population P_t. For each individual $\in P_t$, non-dominated sorting assigns a non-dominated rank equal to the non-dominated front label, which is used to group the individuals into fronts. Here, the first front[2] consists of the populations non-dominated solutions, the next front consists of solutions dominated only by the first front and so on. Different ways of performing non-dominated sorting are described in [2].

NSGA-II further uses a crowding measure intended to be maximized, such that the higher the value, the better the individual. Assuming a maximization problem, we are also interested in maximizing our secondary fitness, such that we may directly replace the crowding measure with our assigned secondary fitness. This way we can utilize the truncation mechanism from NSGA-II directly, since we now have the two values needed for sorting, namely a rank to be minimized and a secondary fitness to be maximized. Note that for minimization objectives, we need to multiply the corresponding weights of Branke's knee utility function with -1, whereas the centroid distance always is to be maximized.

2.8 Truncation

We can now easily apply the truncation mechanism of NSGA-II, with the major difference being that the procedure is now performed on each subpopulation $P_{i,t} \in P_t$ in generation t instead of on the full population. Truncation ensures that the size of $P_{i,t+1}$ is exactly N/K_{NC}.

After ranking, as many fronts as possible are accommodated in the next generation subpopulation $P_{i,t+1}$, where we include the lowest ranked fronts first. The

[2] Where all individuals are assigned rank 1.

last front to be inserted is sorted wrt. the secondary fitness measure, and only the best individuals wrt. this are chosen for inclusion. This truncation mechanism favors rank first, and a good secondary fitness value next.

3 Experimental Setup and Results

In this section, we demonstrate the CFDE algorithm on 3 kinds of problems; the 2D ZDT problems [2,1], the knee problems of Branke et al. [8], and all non-constrained 3D DTLZ problems [7], all with problem settings as suggested in the respective papers. We do this with respect to the MODCO goals:

- Convergence performance (MODCO goal 1)
- Global distinctiveness (MODCO goal 2)
- User-defined performance distinctiveness (MODCO goal 2)
- Local multiobjective optimality (MODCO goal 3)

Using this taxonomy, we address the different issues of CFDE usage according to the MODCO goals. First, we check CFDE convergence against the DEMO versions, which have demonstrated good performance on many problems [5,6]. Next, we want to demonstrate convergence to K_{NC} clusters, how we may change solution diversity by setting K_{PD}, and finally that CFDE are able to locate knees.

For reference, we here provide an overview of the parameters used for the experiments. For all experiments performed with the DEMO algorithm versions, we have used a population size $N = 100$, and a DE setting with $F = 0.5$ and $CF = 0.3$, as used in [5,6]. Number of generations used by DEMO for the test problems is listed in Table 1.

Table 1. Settings for DEMO^{NSII} and DEMO^{SP2}

Problem	ZDT1	ZDT2	ZDT3	ZDT4	ZDT6	DO2DK	DEB2DK	DEB3DK
Generations	250	250	250	500	500	200	200	200
Problem	DTLZ1	DTLZ2	DTLZ3	DTLZ4	DTLZ5	DTLZ6	DTLZ7	
Generations	300	300	500	500	200	500	200	

In table 2 we see the settings for the CFDE runs. As we will see, even with a lower number of generations, CFDE is very competitive to the DEMO versions on almost the entire test suite. For all runs except one, we have used $N = 100$, and the same DE settings as for DEMO; $F = 0.5$ and $CF = 0.3$. The exception is DTLZ3, were we have used a population size of 160 individuals, to match $K_{NC} = 8$. This way, we use approximately as many evaluations on this problem as DEMO, as we perform 300 generations for CFDE[3]. We demonstrate the use of K_{PD} only on knee problems, as this is only relevant for such problems. For ZDT and DTLZ problems, we therefore always set $\sigma = \infty$, effectively disabling knee search. For DTLZ problems, we have used a higher K_{NC} to ensure that we find both extreme and intermediate trade-offs.

[3] 160 individuals $\cdot$ 300 iterations = 48000 evaluations for CFDE vs. 50000 for DEMO.

Table 2. Settings for CFDE

Problem	ZDT1	ZDT2	ZDT3	ZDT4	ZDT6	DO2DK	DEB2DK	DEB3DK
Generations	100	100	100	500	500	200	200	200
K_{NC}	5	5	5	5	5	5	4	5
Problem	DTLZ1	DTLZ2	DTLZ3	DTLZ4	DTLZ5	DTLZ6	DTLZ7	
Generations	300	300	300	200	200	500	200	
K_{NC}	10	10	8	10	10	10	4	

3.1 Convergence Performance

To deal with the different cardinality of more standard MOEAs and the CFDE algorithm, we use the universal notion of dominance. Here, we compare the CFDE algorithm against DEMO^{NSII} and DEMO^{SP2}. One algorithmic argument for MODCO is that the low number of returned solutions allows a more focused search because MODCO does not aim at an even distribution, see [9] for details. Consequently, a MODCO algorithm should be able to return solutions closer to the true pareto front.

We wish to investigate in which extent the returned solutions from the CFDE algorithm dominate the most similar solutions from the returned population of the competing MOEAs, where similarity is measured as distance in objective space. This way we see if the CFDE approach is competitive to simply picking K_{NC} solutions from the resulting populations of the DEMO versions.

For all results below, these are generated using the NSGA-II version of global ranking and truncation in the CFDE algorithm. We have used 20 runs for both the DEMO versions and for CFDE on each problem. For each generated population of CFDE, we have compared each resulting individual to its most similar counterpart from each of the DEMO populations. This gives a percentage of the amount of dominating, dominated and incomparable individuals CFDE was able to produce and is independent of K_{NC} and K_{PD}. Using $K_{NC} = 5$ we thereby get $5 \cdot 20 \cdot 20 = 2000$ comparisons for the problem, while $K_{NC} = 10$ yields 4000 comparisons.

As can be seen from Table 3, CFDE is performing extremely well on the 2D ZDT problems and on the more simple knee problems. It is clear, that the

Table 3. CFDE versus DEMO^{NSII} and DEMO^{SP2}

CFDE vs. DEMO(NSII)	ZDT1	ZDT2	ZDT3	ZDT4	ZDT6	DO2DK	DEB2DK	DEB3DK
Dominates	0	0	0.0995	0.7700	0	0.4680	0.8718	0.4455
Dominated	0	0	0	0	0	0	0	0
Incomparable	1	1	0.9005	0.2300	1	0.5320	0.1282	0.5545
CFDE vs. DEMO(SP2)								
Dominates	0	0	0.0810	0.6800	0.0100	0.2265	0.9081	0.3710
Dominated	0	0	0.0005	0	0.0400	0.0030	0	0
Incomparable	1	1	0.9185	0.3200	0.9500	0.7705	0.0919	0.6290

compared DEMO individuals almost never dominate the CFDE individuals, while CFDE often produce individuals dominating their DEMO counterpart. For the more simple problems ZDT1 and ZDT2, all algorithms find optimal solutions and are therefore incomparable. For ZDT3, CFDE seems to outperform both DEMO versions, even with only 100 iterations. This may be caused by the more focused search around subpopulation centroids performed by CFDE during the run. Even for the much more difficult problem ZDT4 with many local fronts, CFDE seems to dominate again maybe caused by the more focused search not spending effort on equally distributing individuals. However, for ZDT6 with a low density of solutions near the true pareto-front, DEMO shows equal performance to CFDE.

Table 4. CFDE versus DEMO^{NSII} and DEMO^{SP2}

CFDE vs. DEMO(NSII)	DTLZ1	DTLZ2	DTLZ3	DTLZ4	DTLZ5	DTLZ6	DTLZ7
Dominates	0.1432	0.3085	0.3112	0.3210	0.3580	0	0.0493
Dominated	0	0	0	0	0	0.2000	0
Incomparable	0.8568	0.6915	0.6888	0.6790	0.6420	0.8000	0.9507
CFDE vs. DEMO(SP2)							
Dominates	0.1740	0.1600	0.2960	0.3765	0.3255	0.0000	0.0781
Dominated	0	0	0.0300	0	0	0.1750	0
Incomparable	0.8260	0.8400	0.6740	0.6235	0.6745	0.8250	0.9219

In Table 4, it is clear that on DTLZ1 and DTLZ2, CFDE outperforms both DEMO version with an equal number of evaluations. This is however more clear for DEMO^{NSII} than for DEMO^{SP2}. For DTLZ3, CFDE outperforms both DEMO versions, using a higher populations size, but with a lower number of iterations resulting in a similar number of evaluations as noted before. For this problem, CFDE needed a higher diversity than normal to perform well due to the heavy complexity of the problem. On DTLZ4, CFDE outperforms DEMO with less than half the number of iterations, most likely due to the optimal front being a flat curve easy to attain. This also goes for DTLZ5, but with an equal number of evaluations performed. Only on DTLZ6, the harder version of DTLZ5, DEMO outperforms CFDE. This problem is also hard for more traditional MOEAs as noted in [7]. For DTLZ7, we have used $K_{NC} = 4$, as there are 4 optimal planes, which are all being attained by CFDE which on this problem demonstrates only a slightly superior performance.

Overall, CFDE outperforms or has same performance as the two DEMO versions on all test problems except DTLZ6. So CFDE appears competitive to the DEMO versions, even if we have only compared the most similar solutions. More elaborate measures for MOEAs with different result cardinality is future works.

3.2 Global Distinctiveness

Global distinctiveness is achieved by the CFDE algorithm using the centroid distance to repel subpopulations. Figure 2 and 3 displays the returned results of 20 runs of the CFDE algorithm on ZDT1 and ZDT3, using $K_{NC} = 5$ and $K_{NC} = 10$. As mentioned above, we here set $\sigma = \infty$. Similar robust convergence is seen for the other test problems, i.e., CFDE found roughly the same set of distinct candidates in repeated runs.

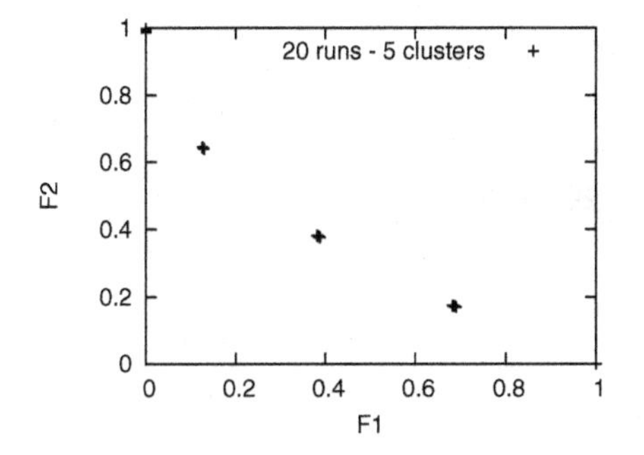

Fig. 2. ZDT1 plot - 20 runs - 5 clusters

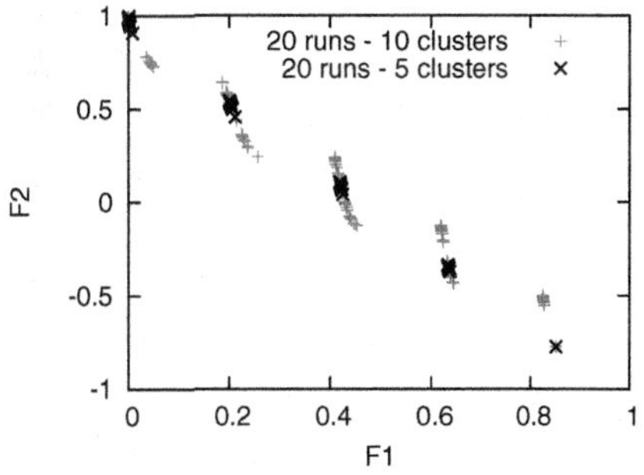

Fig. 3. ZDT3 plot - 20 runs

As can be seen in figure 2, all of the 20 runs of CFDE returned similar distinct solutions. In figure 3, we see that using $K_{NC} = 5$ ensures a result returned from each of the 5 patches of the true pareto-front and again we see only a small variation. However, using $K_{NC} = 10$ makes the returned results be much more spread in the 20 runs, since there are now more clusters to be formed than there are discontinuous patches. As can be observed from the density, solutions will here most often seek the most outer part of the patches making the returned solutions as distinct as possible.

3.3 User-Defined Performance Distinctiveness

The MODCO parameter $K_{PD} \in [0, 1]$ allows the DM to set how distinct the returned solutions should be. A low value corresponds to a low distinctiveness and a high value to a high distinctiveness. See [9] for further information.

To demonstrate the effects of changing K_{PD}, we have chosen DEB3DK as an illustrative example. DEB3DK is interesting with its 3 objectives and single knee, since this allow us to visualize the effect of altering the balance between knee search and subpopulation repelling in 3D. We have used DEB3DK with only one knee, i.e. $K = 1$.

We first demonstrate the calculation of σ used in the CFDE algorithm. First, we will assume settings $K_{PD} = 1$ and $K_{NC} = 5$. For DEB3DK, we may use reference points $z^{**} = (0, 0, 0)$ and $z^I = (8, 8, 8)$, which span the interesting part of the objective space. Then we can calculate:

$$\sigma = K_{PD}/K_{NC} \cdot ||z^{**} - z^I|| = 1/5 \cdot \sqrt{192} \approx 2.77 \tag{4}$$

In this setting, subpopulations will repel each other if they get within a distance of 2.77 of each other's centroids. Setting $K_{PD} = 1$ should ensure maximum

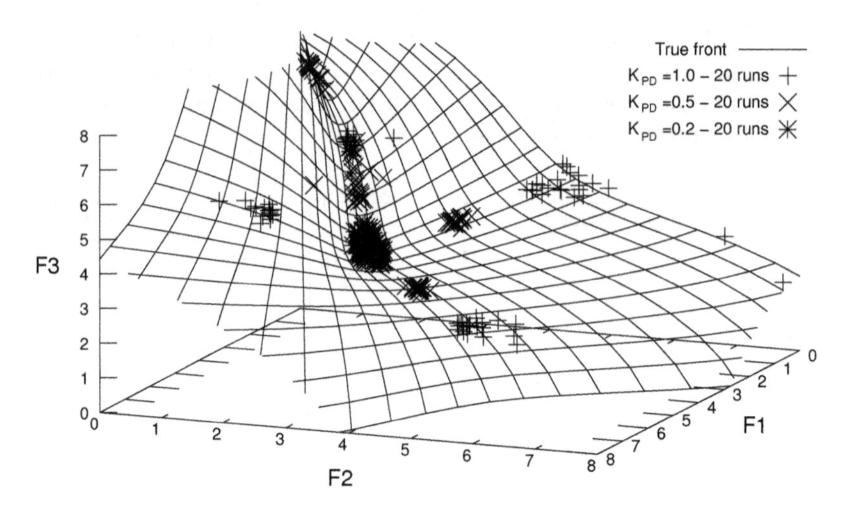

Fig. 4. DEB3DK plot - investigating user defined performance distinctiveness

global distinctiveness, such that we get clusters uniformly spread across the objective space spanned by the reference points. Contrary, setting K_{PD} close to zero[4] enables clusters to get closer to each other while searching for knees.

Figure 4 illustrates the results of setting K_{PD} to 0.2, 0.5 and 1.0. For $K_{PD} = 1$, the 5 clusters are equidistant around the single knee, where one cluster is placed. The four clusters not in the knee are repelled from each other as they reach a distance of 2.77 between centroids, as was demonstrated in the example calculation above. Interestingly, the four clusters not in the knee are located in the partial knees on the lines forming a cross. For $K_{PD} = 0.5$, we always hit the knee with one cluster. Further, it can be seen that some solutions has found other knee regions, crawling towards the one in the middle, but these are still not allowed to get too close to each other. Setting $K_{PD} = 0.2$ results in all clusters getting very close to the single knee region. Overall, it is clear that increasing K_{PD} indeed makes clusters repel each other more.

3.4 Local Multiobjective Optimality

Figure 4 and 5 illustrates the knee problems DO2DK and DEB2DK, with the resulting CFDE individuals of 20 runs. In the DO2DK problem, we set $K = 4$ and $s = 1.0$, such that we have exactly the same settings as has been used for creating the results illustrated in figure 4 in [8]. For the 20 runs depicted in figure 4, it may be noticed, that the density of solutions near knee regions is very high. When using $K_{NC} = 5$, CFDE finds the 4 knee regions very precisely, while one cluster typically hits an outer solution, or is caught in-between knee regions. For DO2DK, we have used $K_{PD} = 0.75$ corresponding to $\sigma = 1.5$. This way we keep clusters separated, while still allowing for knees to be found.

[4] Note that setting $K_{PD} = 0$ allows for subpopulations to overlap.

For the DEB2DK problem, we have used $K = 4$ to replicate the results illustrated in figure 5 in [8]. In figure 5, we again see that for the 20 runs the density of solutions near knees are very high. Here, we have to set K_{NC} to be equal to the number of knees, and it is clear that all knees are discovered in all runs. Here, we have used $K_{PD} = 0.2$ corresponding to $\sigma = 0.5$. This is low, so the centroid distance assignment is rarely used. Hence, subpopulations converge to knees, and as long as $\sigma > 0$, the clusters formed will not overlap.

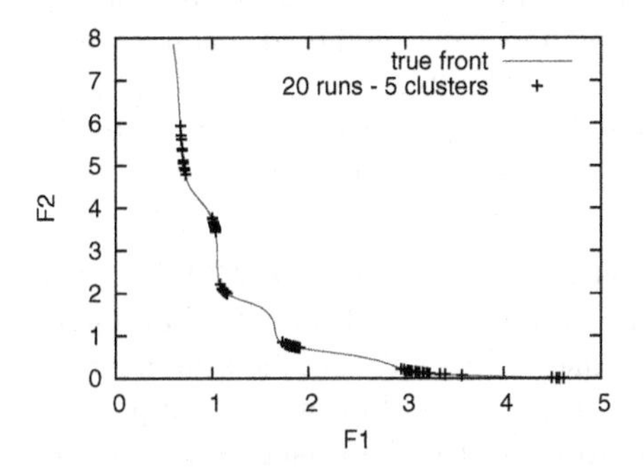

Fig. 5. DO2DK plot - 20 runs - 5 clusters - 4 knees

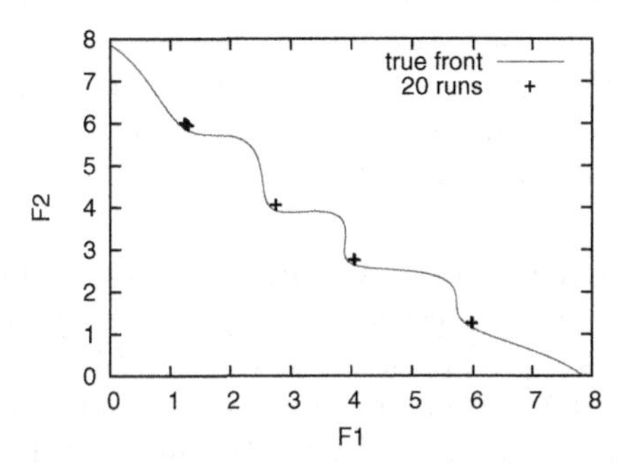

Fig. 6. DEB2DK plot - 20 runs - 4 clusters - 4 knees

From the figures above, and further figure 4, it is clear, that CFDE is indeed able to locate knee regions. Further, it has been demonstrated how to balance the search using K_{PD} resulting in different σ values.

3.5 Further Aspects

In this section we discuss two further aspects of the CFDE algorithm, namely initialization in two dimensions, and an approach for constrained problems.

Initialization. When initializing the first population, it is possible to make an easy initial clustering for problems with 2 objectives. We simply sort the initial random population based on the individuals first objective, before these are inserted into subpopulations. This way, less effort is spend on forming clusters during start of the run, enhancing efficiency. The approach has been used in all tests on 2D problems. The approach is only applicable for 2-dimensional problems, as for $M > 2$, there can be established no total order of individuals based on their objective vector.

Constrained Problems. Constrained problems require further comparison when applying selection. We recommend the GDE3 approach for such problems [10], such that individuals with no constraint violation will always be preferred over individuals who do violate constraints. If two individuals both violate constraints, the most dominant one in *constraint space* is favoured. This will guide the search towards non-violated individuals, not loosing the property of being able to guide the search across constrained areas in objective space.

4 Conclusion

In this paper, we introduce the first MODCO algorithm called CFDE. Using subpopulation based Differential Evolution, the proposed CFDE algorithm allows the DM to specify the number of solutions returned, and how distinct they should be. The algorithm forms the desired number of clusters during the run using using two alternating secondary fitness measures assigned locally. These guides the search according to the goals of MODCO [9].

To conclude on this paper, we will again use the taxonomy presented in the start of the results section, Section 3, according to the MODCO goals:

Convergence. CFDE clearly outperforms DEMO^{NSII} and DEMO^{SP2} when comparing distinct individuals on a broad test suite with ZDT, DTLZ and knee based problems. The reason for this is most likely the more focused search giving better convergence by allowing individuals to get very close within subpopulations. We give this as the most likely reason, as the approach resembles local search based exploitation having previously demonstrated good results e.g. in [11].

Global Distinctiveness. Regarding global distinctiveness, CFDE shows robust behavior by converging to the same set of distinct candidates on many independent runs. On test problems with no knees it was shown, that returned results of CFDE do try to achieve maximum global distinctiveness, but that this is naturally also depending on the number of subpopulations in use.

User-Defined Distinctiveness. Further, our parameter investigations on K_{PD} shows that different settings of K_{PD} allows a user-controlled performance distinctiveness. Here in the form of distance-based performance distinctiveness [9].

Local Multiobjective Optimality. As for local multiobjective optimality, our experiments have shown that the CFDE algorithm is able to locate knees even for changing parameter settings. This is important because we typically have no a priori knowledge on the structure of the problem's objective space.

Overall, the CFDE algorithm seems to comply very well with the goals of MODCO, and seems to be a very good alternative to post processing the results of more traditional MOEAs in order to decrease cardinality.

References

1. Deb, K., Pratap, A., Agarwal, S., Meyarivan, T.: A fast and elitist multiobjective genetic algorithm: NSGA-II. IEEE transactions on Evolutionary Computation 6, 182–197 (2002)

2. Deb, K.: Multi-Objective Optimization using Evolutionary Algorithms. Wiley, Chichester (2002)
3. Zitzler, E., Laumanns, M., Thiele, L.: SPEA2: Improving the strength pareto evolutionary algorithm. Technical Report 103, Computer Engineering and Networks Laboratory (TIK), Swiss Federal Institute of Technology (ETH) Zurich, Gloriastrasse 35, CH-8092 Zurich, Switzerland (2001)
4. Price, K.V., Storn, R.: Differential Evolution - a simple evolution strategy for fast optimization. Dr. Dobb's journal 22, 18–24 (1997)
5. Robič, T., Filipič, B.: DEMO: Differential Evolution for Multiobjective Optimization. In: Coello Coello, C.A., Hernández Aguirre, A., Zitzler, E. (eds.) EMO 2005. LNCS, vol. 3410, pp. 520–533. Springer, Heidelberg (2005)
6. Robič, T., Filipič, B.: Differential Evolution versus Genetic Algorithms in Multiobjective Optimization. In: Obayashi, S., Deb, K., Poloni, C., Hiroyasu, T., Murata, T. (eds.) EMO 2007. LNCS, vol. 4403, pp. 257–271. Springer, Heidelberg (2007)
7. Deb, K., Thiele, L., Laumanns, M., Zitzler, E.: Scalable Test Problems for Evolutionary Multi-Objective Optimization. In: Proceedings of the 2002 Congress on Evolutionary Computation, CEC 2002, pp. 825–830 (2002)
8. Branke, J., Deb, K., Dierolf, H., Osswald, M.: Finding Knees in Multi-objective Optimization. In: Yao, X., Burke, E.K., Lozano, J.A., Smith, J., Merelo-Guervós, J.J., Bullinaria, J.A., Rowe, J.E., Tiňo, P., Kabán, A., Schwefel, H.-P. (eds.) PPSN 2004. LNCS, vol. 3242, pp. 722–731. Springer, Heidelberg (2004)
9. Ursem, R.K., Justesen, P.D.: Multiobjective Distinct Candidates Optimization (MODCO) – A new Branch of Multiobjective Optimization Research. Technical Report no. 2008-01, Dept. of Computer Science, University of Aarhus (2008), Download: `http://www.daimi.au.dk/~ursem/publications/Ursem_EMO2009_MODCO.pdf`
10. Kukkonen, S., Lampinen, J.: GDE3: The third Evolution Step of Generalized differential evolution. In: Proceedings of the 2005 Congress on Evolutionary Computation, CEC 2005, pp. 443–450 (2005)
11. Karthis, S., Deb, K., Miettinen, K.: A Local Search Based Evolutionary Multiobjective Optimization for Fast and Accurate Convergence. In: Rudolph, G., Jansen, T., Lucas, S., Poloni, C., Beume, N. (eds.) PPSN 2008. LNCS, vol. 5199, pp. 815–824. Springer, Heidelberg (2008)

An Evolutionary Algorithm to Estimate the Nadir Point in MOLP

Maria João Alves and João Paulo Costa

Faculty of Economics of University of Coimbra / INESC Coimbra
Av. Dias da Silva, 165, 3004-512 Coimbra, Portugal
mjalves@fe.uc.pt, jpaulo@fe.uc.pt

Abstract. In this paper we propose an evolutionary algorithm to estimate the minimum (*nadir*) objective values over the efficient set in multiple objective linear programming problems (MOLP). Nadir values provide valuable information for characterizing the ranges of the objective function values over the efficient set. However, they are very hard to compute in the general case. The proposed algorithm uses a population of weight vectors with particular characteristics, which are then used as parameters in the optimization of weighted-sums of the objective functions. The population evolves through a process of selection, recombination and mutation. The algorithm has been tested on a number of random MOLP problems for which the nadir point is known. A result comparison with an exact method is shown and discussed.

Keywords: Evolutionary multi-objective optimization, nadir point, multi-objective linear programming.

1 Introduction

Consider the multiple objective linear programming (MOLP) problem with k objective functions (criteria), n decision variables and m constraints:

$$\begin{array}{ll} \max & z_1(x) = c^1 x \\ \max & z_2(x) = c^2 x \\ \ldots & \\ \max & z_k(x) = c^k x \\ \text{s.t.} & x \in S = \{x \in \Re^n \mid Ax \le b,\ x \ge 0\} \end{array} \tag{1}$$

where A is a $m \times n$ matrix, $b \in \Re^m$ and $c^i \in \Re^n$ (i=1,…,k). Assume that S is non-empty and bounded.

Let E denote the set of all efficient (nondominated or Pareto optimal) solutions of the MOLP problem. Further, let z^* be the *ideal* point whose components z_i^*, i=1,…,k, are the maximum objective values over the efficient set E, and let z_i^{nad}, i=1,…,k, be the minimum objective values over E, which form the so-called *nadir* point z^{nad}. Below we denote by $\tilde{z}^{nad}$ an estimate of z^{nad}.

M. Ehrgott et al. (Eds.): EMO 2009, LNCS 5467, pp. 540–553, 2009.

The ideal and the nadir values together provide valuable information for characterizing the ranges of the objective values over the efficient set. While the ideal values are easy to obtain by maximizing each objective function individually in the feasible region S, the nadir values are very hard to determine in the general case (except in the bi-objective case) because the efficient set is not known explicitly and it is non-convex, even in MOLP. Due to these difficulties, the minimum column values of the *payoff* table have been often considered as estimates of the nadir values. In a *payoff* table the ith row is an objective vector resulting from maximizing the ith objective individually, and we assume that all row vectors are nondominated. Thus, the minimum objective values in the *payoff* table do not underestimate, but may overestimate, the nadir values. Computational experiments have demonstrated that the discrepancies between the minima in the *payoff* table and the minima over the efficient set can be very large ([1], [2]).

Research on computing the nadir point has not been vast. Authors have recognized that computing the nadir objective values is an important but difficult task. A few dedicated approaches have been proposed, which either aim at finding the exact nadir values ([1], [2], [3]) or to approximate them using heuristic algorithms ([4], [5], [6]). Finding each component of the nadir point is a particular instance of the problem of optimizing a function over the efficient solution set, which has been addressed by several authors. An overview of existing algorithms for this global optimization problem is reported in [7]. Although these approaches are theoretically able to compute the nadir values, they present rather complex algorithms.

Among the more recent algorithms, there are the evolutionary approaches proposed by Deb et al. [6] to estimate the nadir point for linear and nonlinear multi-objective problems. The authors propose some modifications on the evolutionary multi-objective optimization procedure NSGA-II [8] to focus its search on extreme solutions of the nondominated solution set.

Also, an exact method for computing the nadir values in MOLP has been recently developed by the authors of this paper [2]. This method was applied to a set of MOLP problems whose results are used to test the performance of the evolutionary algorithm we propose herein.

In this paper we propose a new evolutionary algorithm for MOLP problems which aims at estimating all the components of the nadir point in one run. It uses a population of weight vectors with particular characteristics having the purpose of finding good estimates of the nadir objective values. The weight vectors are in turn used as parameters in the optimization of weighted-sums of the objective functions, so the computed solutions to the MOLP problem (1) are guaranteed to be efficient. Reproduction and mutation are applied to the individuals (weight vectors) and a particular fitness function and a selection procedure are employed to update the population for the next generation. This evolutionary algorithm is described in Section 2. Section 3 reports the computational experience, firstly considering a three-objective linear problem addressed in [6] and then a set of 60 problems with 3 to 6 objective functions used in [2]. A result comparison with the exact method [2] is provided. The paper finishes with some concluding remarks in Section 4.

2 The Evolutionary Algorithm

It is known that the nadir objective values of a MOLP problem can be found among its basic efficient solutions. Moreover, the nadir value of the objective z_i is provided by an efficient solution of the sub-problem with (k-1) objectives obtained by dropping $z_i(x)$ from the MOLP problem (**1**) (see [3]). Consequently, the ith component of the nadir point can be found among the efficient solutions of (**1**) that optimize the weighted-sum program (**2**) with weight vectors $w \geq 0$ such that $w_i = 0$ (or nearly zero).

$$\max \sum_{j=1}^{k} w_j z_j(x) \qquad (2)$$
$$\text{s.t. } x \in S$$

The weights used in (**2**) must be strictly positive to ensure that only efficient solutions are computed. So, zero components of w are replaced with ε, a very small positive scalar but computationally significant (e.g. 0.0001). Below, a component of w equal to ε is referred to as a *zero*-component.

The basic principle of the proposed algorithm is to use weight vectors with at least one *zero*-component to constitute the populations of the evolutionary process. The individuals are of type $w=(w_1, w_2, \ldots, w_k)$ with $w_j \in [\varepsilon, 1]$ for all j=1,…,k and $w_i = \varepsilon$ for at least one i. They are coded as real vectors.

2.1 Initial Population

Define N as the population size, which is considered to be a multiple of k. To form the initial population P_0, the first k individuals are defined deterministically while a randomized scheme is used to generate the other (N-k) individuals.

The first k individuals are extreme weight vectors, which lead to efficient solutions that optimize individually each objective function. They are: $w^{(1)}=(1, \varepsilon, \varepsilon, \ldots, \varepsilon)$, $w^{(2)} = (\varepsilon, 1, \varepsilon, \ldots, \varepsilon)$, $w^{(3)} =(\varepsilon, \varepsilon, 1, \ldots, \varepsilon)$, …, $w^{(k)} =(\varepsilon, \varepsilon, \varepsilon, \ldots, 1)$. The remaining individuals are generated as follows. Create (N-k) vectors w in which each w_j is a random number between ε and 1. Then, split this set of vectors into k equal-sized groups and set $w_1=\varepsilon$ in the vectors from the first group, $w_2=\varepsilon$ in the vectors from the second group, and so on. Hence, the initial population has a configuration like the one shown in figure 1, where grey cells represent *zero*-components and empty cells represent randomly generated values.

For each $w \in P_0$ the weighted-sum (**2**) is optimized, producing a set of efficient solutions to the MOLP problem. Let E_0 designate the set of objective vectors $z=z(x)$ of the efficient solutions obtained in this step. These vectors are used to initialize the estimate of the nadir point. So, $\tilde{z}_i^{nad} = \min\{z'_i \mid z' \in E_0\}$, $i = 1,\ldots,k$.

As E_0 includes the nondominated vectors that maximize each objective function individually (generated by the first k individuals of the population), then the initial approximation for the nadir point is at least as good as the one obtained from the payoff table.

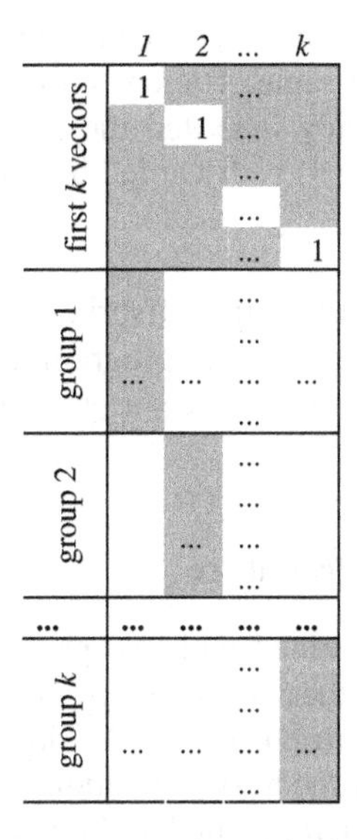

Fig. 1. Configuration of the initial population

2.2 Recombination and Mutation

A random mate selection is employed, in which N pairs of individuals are randomly chosen from the current population P_t (where t denotes the generation index).

For each pair w^{p1} and w^{p2} of parents selected for reproduction a number r from 0 to 1 is drawn at random. The child w^c of w^{p1} and w^{p2} is generated as follows. First, set w^c to be the convex combination of w^{p1} and w^{p2} defined by r, i.e., $w_j^c = rw_j^{p1} + (1-r)w_j^{p2}$ for all $j=1,\dots,k$. Then, assign the *zero*-component(s) in the following way: for each $i \in \{1,\dots,k\}$ such that $w_i^{p1} = \varepsilon$ or $w_i^{p2} = \varepsilon$ draw at random $a_i \in]0,1]$; if ($a_i \leq r$ and $w_i^{p1} = \varepsilon$) or ($a_i > r$ and $w_i^{p2} = \varepsilon$) then set $w_i^c = \varepsilon$. As a result of this process, a *zero*-component common to both parents is always inherited by the child.

The mutation operator is next applied to the offspring considering a probability p_m (typically, 0.01) of mutating each gene. Mutation of a gene consists of replacing that gene with a random number between ε and 1.

After recombination and mutation, a set of N offspring weight vectors is obtained. However, some of them may be considered non-eligible for integrating the offspring population. A w^c is discarded when

(a) it has no *zero*-component,
(b) it has k or (k-1) *zero*-components, or
(c) it results from a pure convex combination of parents that correspond to the same efficient solution of the MOLP problem.

Weight vectors with none or all components equal to ε are not considered for natural reasons. In addition, a vector with all components but one equal to ε would act (in principle) like one of the first k individuals from the initial population, leading to a nondominated objective vector in the payoff table. Finally, an offspring weight vector whose parents correspond to the same efficient solution and is a pure convex combination of their parents (i.e, it only inherited the *zero*-components common to both parents and did not suffer mutation) is discarded because it leads to the same efficient solution. Actually, the region of weight vectors associated with one efficient solution is convex.

The remaining children constitute the offspring population P^c with $N^c \leq N$ individuals. For each $w^c \in P^c$ the weighted-sum program (**2**) is optimized, yielding an efficient solution x^c whose objective vector is $z^c = z(x^c)$. Then, $\tilde{z}_i^{nad}$ is updated with z_i^c for all $i \in \{1,\ldots,k\}$ such that $z_i^c < \tilde{z}_i^{nad}$.

2.3 Fitness and the Selection Procedure

The offspring population P^c is joined with the parent population P_t in an auxiliary pool P^{aux} (of size $N^c + N$) and a fitness value is assigned to each individual in this pool. A selection procedure is then carried out on P^{aux} in order to choose the individuals that constitute the population for the next generation.

The fitness assignment to an individual w both incorporates a quality measure, which results from the differences between the values of $\tilde{z}^{nad}$ and of the nondominated objective values corresponding to w, and incorporates density information in order to preserve diversity. The latter aims at ensuring that the algorithm searches for *good* approximation values in all the dimensions $i = 1,\ldots,k$ of the nadir point, not giving privilege to some i in relation to others.

Consider a weight vector $w^a \in P^{aux}$ and let z^a be the corresponding nondominated objective vector obtained from the optimization step. Then, its quality measure θ^a is defined as:

$$\theta^a = \max_{i=1,\ldots,k} \left\{ \frac{\tilde{z}_i^{nad} - z_i^a}{z_i^* - \tilde{z}_i^{nad}} \right\} \tag{3}$$

Since $\tilde{z}^{nad}$ has been updated for all nondominated solutions already computed, then $\tilde{z}_i^{nad} - z_i^a \leq 0$ for every z^a and every i. For a given z^a, picking the *maximum* difference for i=1,…,k means evaluating the solution by the objective function that is closer to the respective component of the nadir approximation vector. The denominator is intended to normalize the ranges. Hence, a solution with the current minimum value in any objective is assigned the maximum value of the quality measure, that is $\theta^a = 0$, whereas the others are assigned negative values. The measure θ^a (and then the fitness value) is to be maximized.

Although the quality measure provides a rank of the individuals, this rank does not take into account that there may be many solutions close to the nadir point in the same component(s), while only a few solutions are close to the nadir point in other components. This fact could lead to a low rate of survival of individuals associated with the latter solutions in relation to the former ones. The evolutionary process could thus favour some objective functions (in the sense of obtaining a good approximation of the corresponding nadir values) over the others. Therefore, additional information is incorporated to define the fitness of w^a.

Let $i^a \in \{1,\ldots,k\}$ be the/a *max-ratio index* in (**3**) for the individual w^a, that is, i^a is the/a component index such that $\theta^a = (\tilde{z}^{nad}_{i^a} - z^a_{i^a})/(z^*_{i^a} - \tilde{z}^{nad}_{i^a})$. Further let $COUNT\,(i^a)$ be the number of individuals in P^{aux} for which the max-ratio index is equal to i^a, i.e., $COUNT\,(i^a) = |\{\, w' \in P^{aux} : i' = i^a \}|$, where $|\,.\,|$ denotes the cardinality of a set. An individual will become less powerful with the increase of the cardinality of its subset.

Since $\theta^a \leq 0$ and it is to be maximized, in the following expression of $F(w^a)$ that defines the fitness value of w^a, the $COUNT(i^a)$ factor weakens θ^a (except when $\theta^a = 0$ which is only attained by the *best* individuals).

$$F(w^a) = \theta^a \times COUNT\,(i^a) \tag{4}$$

Then, the environmental selection procedure chooses N elements of P^{aux} to constitute the population of the next generation, P_{t+1}, in the following way:

First, k individuals are chosen according to an elitist selection mechanism. These are individuals that have yielded the current estimate of the nadir point, one individual for each component i=1,…k of the nadir point. Hence, for each i=1,…k, the first individual w^a such that $F(w^a)$=0 and $i^a = i$ is selected. Next, N-k binary tournaments are performed in P^{aux} to select the other individuals that form P_{t+1}. In each binary tournament, the individual with higher fitness is chosen. During the whole selection procedure, once an individual is copied to P_{t+1} it is also deleted from P^{aux}.

2.4 Steps of the Algorithm

The complete algorithm can be synthesized as follows:

Input: N (population size)
T (maximum number of generations)
p_m (probability of mutation)

Output: $\tilde{z}^{nad}$ (estimate of the nadir point)

Step 1. *Initialization*: Generate an initial population P_0 of weight vectors $w \in \Re^k_+$ such that the first k vectors are extreme weight vectors (which enable to compute the payoff table) and the other N-k vectors are randomly generated with at least one *zero*-component. For each $w \in P_0$ optimize the weighted-sum program (**2**), and then initialize $\tilde{z}^{nad}$. Set t =0.

Step 2. *Offspring generation*: Randomly select N pairs of individuals from P_t for reproduction. Apply recombination to each pair to generate an offspring and then apply the mutation operator with probability p_m. Discard offspring that do not meet some criteria (*cf.* Section 2.2) forming an offspring population P^c with $N^c \leq N$ elements. Optimize the program (**2**) for each $w^c \in P^c$ and update $\tilde{z}^{nad}$.

Step 3. *Fitness assignment and environment selection*: Calculate the fitness value for all individuals in P_t and P^c (*cf.* Section 2.3) and join these individuals in an auxiliary pool P^{aux}. *Elitism*: select k *best* individuals of P^{aux} with respect to each component of the nadir point to introduce into the next population P_{t+1}. Then, perform N-k binary tournaments in P^{aux} to fill up P_{t+1}. Each selected individual is copied to P_{t+1} and deleted from P^{aux}.

Increment the generation counter, $t = t$+1. If $t = T$ then stop, else go to Step 2.

3 Computational Experience

The algorithm was implemented in Delphi 2007 for Windows and tested in a computer CPU T9500, 2.6 GHz with 4 GB of RAM. In this section we report the results of two experiments. The first experiment concerns a 3-objective linear problem tested in [6] (this is the only linear problem addressed in [6] as the proposed approaches are devoted to general multi-objective programming problems, linear and non-linear). In the second experiment, we applied the algorithm to 60 MOLP problems with 3 to 6 objective functions for which the true nadir points are known because these problems have already been analysed with the exact method in [2]. We have chosen this set of problems because the quality of the results returned by the evolutionary algorithm can be better evaluated by comparing them with the true nadir values (which are available in the URL referenced in [2]). We have also run the exact method on the computer referred to above in order to compare computing times. Therefore, all the times presented herein were recorded in the same machine.

3.1 A Three-Objective Linear Problem

Consider the following three-objective linear problem, which was used by Deb et al. [6] and had been previously described in [5]. The nadir point of this problem is z^{nad} = (0, 0, 0) and none of the nadir values is in the payoff table [5].

$$\begin{array}{ll} \max & z_1 = 11x_2 + 11x_3 + 12x_4 + 9x_5 + 9x_6 - 9x_7 \\ \max & z_2 = 11x_1 + 11x_3 + 9x_4 + 12x_5 + 9x_6 - 9x_7 \\ \max & z_3 = 11x_1 + 11x_2 + 9x_4 + 9x_5 + 12x_6 + 12x_7 \\ \text{s.t.} & \sum_{i=1}^{7} x_i = 1, \quad x_i \geq 0, \ i = 1, 2, \ldots, 7 \end{array} \tag{5}$$

Deb et al. [6] applied three NSGA-II approaches (NSGA-II, extremized crowded NSGA-II and worst crowded NSGA-II) to this problem considering a standard parameter setting, in which the population size was set to 100. A repair mechanism was used to make every solution a feasible one. The authors notice that the extremized crowded NSGA-II is able to find both minimum and maximum objective values corresponding to the efficient solutions. That approach is able to provide a stable and correct estimate of the nadir point. On the other hand, the worst crowded NSGA-II is unable to estimate the nadir point reliably.

We applied our evolutionary algorithm to this problem, setting p_m=0.01 and considering two population sizes, 10 and 18. Note that, while NSGA-II approaches work with feasible solutions to the MOLP problem, the algorithm proposed herein works with efficient solutions, which justifies setting the population size to a much smaller number. We have analysed the quality of the nadir point estimate with the generation number in 30 runs of the algorithm. Since the algorithm provides very often the true nadir point to this problem, we report in figure 2 the number of completely successful runs out of the 30 runs, i.e. the percentage of runs in which all the nadir components were correctly estimated. For N=10, one out of the 30 runs was not able to reach all the nadir components until the 20th generation (it failed to reach one nadir value). For N=18, all the 30 runs reached all the nadir components in 5 or less generations.

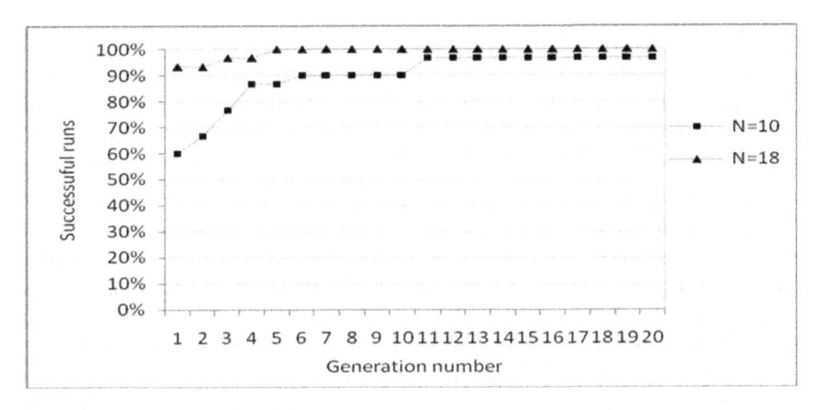

Fig. 2. Percentage of the 30 runs that yielded the true nadir point in problem (5)

It should be noticed that, although the algorithm can find the nadir point without difficulty, this is not an efficient algorithm to solve this so small problem. In fact, we present it only for illustration purposes, because the nadir point of this problem can be computed very easily and quickly by an exact method such as [2].

3.2 Test Problems

The problem set used in this experiment to test the evolutionary algorithm consists of MOLP problems randomly generated with k = 3, 4, 5 and 6 objective functions and the following numbers of decision variables (n) × constraints (m): 60×30, 80×40 and 100×50. For each combination of k with (n×m), five instances (with indexes a, b, c, d and e) were generated, in a total of 60 test problems. The coefficients of the objective functions (c_{ij}, $i = 1,\ldots k$, $j = 1,\ldots,n$), the technical coefficients (a_{ij}, $i = 1,\ldots m$, $j = 1,\ldots,n$) and the right-hand sides of the constraints (b_i, $i = 1,\ldots m$) are integer numbers randomly generated in the following ranges (where p means probability): .

$$\begin{cases} -100 \le c_{ij} \le -1 & p=0.2 \\ 0 \le c_{ij} \le 100 & p=0.8 \end{cases} \qquad \begin{cases} -100 \le a_{ij} \le -1 & p=0.1 \\ a_{ij} = 0 & p=0.1 \\ 1 \le a_{ij} \le 100 & p=0.8 \end{cases} \qquad 100 \le b_i \le \sum_{j=1}^{n} a_{ij}$$

These problems are available at http://www4.fe.uc.pt/mjalves .

We have first conducted some preliminary tests in order to set the parameter values of the evolutionary algorithm: N (population size), T (number of generations) and p_m (probability of mutation). It has sounded satisfactory to consider p_m=0.01 and define N and T as linear functions of k for this problem set. Accordingly, the parameters were defined as N=6k and T=10k. We suppose, however, that these parameter values may not suffice for large values of k and perhaps a quadratic function of k could be more adequate for those cases.

For each problem, 30 runs of the evolutionary algorithm with different random seeds were carried out. Below we report on results of two studies.

The first study is designed to analyse the effect on the performance of the evolutionary algorithm of increasing the number of objective functions or the size of the problem ($n \times m$). For this study we only use the instances with index a. The problems $a\_3\_60 \times 30$, $a\_4\_60 \times 30$, $a\_5\_60 \times 30$, $a\_6\_60 \times 30$ and further $a\_7\_60 \times 30$ aim to analyse the effect of incrementing k keeping equal the numbers of variables and constraints (the last problem was added specifically for this study as it does not belong to the original set). Next, fixing k=5 and varying ($n \times m$) we consider the sequence $a\_5\_60 \times 30$, $a\_5\_80 \times 40$, $a\_5\_100 \times 50$ in order to analyse the effect of increasing the size of the problem keeping equal the number of objective functions. The results of these problems are shown in Table 1, which includes a column for each objective function presenting the minimum value in the payoff table, the true nadir value and the best, worst and median estimate values obtained from the 30-runs experiment.

The computing time required by the evolutionary algorithm to estimate the nadir point in this series of problems varies from 0.5 seconds (in $a\_3\_60 \times 30$) to 3.7 seconds (in $a\_7\_60 \times 30$), increasing by about 1 second for each objective function that is added to the problem and keeping equal the other dimensions of the problem. On the other side, the computing time with the exact method is initially low (0.2 seconds in $a\_3\_60 \times 30$) but increases very quickly: for a given k it can be ten times higher than for k-1. Its maximum is about 1600 seconds in $a\_7\_60 \times 30$. In case that k is fixed to 5 and $n \times m$ is increased the differences between the computing times with the evolutionary algorithm and the times with the exact method are not so significant. It is worthwhile to note that a relevant conclusion from the previous experience with the exact method states that the computational effort increases significantly with the number of objectives, rather than with the number of variables and constraints [2].

Concerning the quality of the solution, we observed that the true nadir point of the instance $a\_3\_60 \times 30$ was obtained in 28 out of the 30 runs of the algorithm. However, one run takes more time than executing the exact method. In the instance with 4 objective functions, all the true nadir values were also found several times. In the other instances with k = 5 to 7, the number of true nadir values found by the algorithm at least once varied from 3 to 5. For all instances used in this experiment only one nadir value is in the payoff table, so the other 34 nadir values are lower than the corresponding payoff minima. Note that we consider payoff tables constructed with nondominated vectors, so they do not underestimate the nadir values. The worst estimate produced for each value in this 30-runs experiment improved the payoff minimum for 30 out of these 34 values. One of the remaining 4 values is also the median estimate (for z_4 in $a\_5\_60 \times 30$), while the other three worst values were produced in only 1 or 2 runs.

The second study considers all the 60 test problems. We first consider the performance measure proposed in [6] for cases in which the nadir point is known. This measure consists of computing the normalized Euclidean distance (D) between the true nadir point and the estimated nadir point.

$$D = \sqrt{\sum_{i=1}^{k} \left(\frac{\tilde{z}_i^{nad} - z_i^{nad}}{z_i^{*} - z_i^{nad}} \right)^2} \quad (6)$$

For each instance we calculate the measure D by averaging the Ds values obtained from the 30 runs of the algorithm. The same rule is applied to calculate the computing time spent by the evolutionary algorithm in each instance. Further, *Avg D* and *Avg time* are computed for each group $k\_n \times m$. They are given by the average values of D and *time*, respectively, of the five instances a to e from each group.

Graphs of Figure 3 display the *Avg D* values. As can be seen in this figure, the average normalized Euclidean distance between the true nadir point and the estimated nadir point increases with k (which in part is explained by the mathematical formula of D) but is small in all cases: it range from 0.007 (in group 3_60×30) to 0.179 (in group 6_100×50).

Table 1. Results of the instances *a_k_*60×30 with k=3,...,7, *a*_5_80×40 and *a*_5_100×50

		z_1	z_2	z_3	z_4	z_5	z_6	z_7
	Min payoff	*59.93*	*434.66*	*103.81*				
	True nadir	**-162.79**	**321.21**	**64.68**				
*a*_3_60×30	*Best estimate*	**-162.79**	**321.21**	**64.68**				
	Worst estimate	-34.31	384.79	**64.68**				
	Median estimate	**-162.79**	**321.21**	**64.68**				
	Min payoff	*59.93*	*303.21*	*103.81*	***226.04***			
	True nadir	**-162.79**	**277.89**	**64.68**	**226.04**			
*a*_4_60×30	*Best estimate*	**-162.79**	**277.89**	**64.68**	**226.04**			
	Worst estimate	*59.93*	*303.21*	94.47	**226.04**			
	Median estimate	**-162.79**	297.60	**64.68**	**226.04**			
	Min payoff	*59.93*	*303.21*	*103.81*	*226.04*	*188.51*		
	True nadir	**-359.98**	**227.19**	**64.68**	**-39.78**	**-59.36**		
*a*_5_60×30	*Best estimate*	**-359.98**	**227.19**	**64.68**	75.01	**-59.36**		
	Worst estimate	-162.79	297.60	94.47	*226.04*	-51.44		
	Median estimate	-332.83	283.05	78.97	*226.04*	**-59.36**		
	Min payoff	*59.93*	*303.21*	*53.46*	*226.04*	*188.51*	*435.26*	
	True nadir	**-359.98**	**192.75**	**30.22**	**-49.17**	**-59.36**	**3.50**	
*a*_6_60×30	*Best estimate*	**-359.98**	**192.75**	**30.22**	**-49.17**	**-59.36**	90.27	
	Worst estimate	-57.94	203.71	40.24	193.82	175.81	211.84	
	Median estimate	-250.48	200.07	**30.22**	**-49.17**	-55.40	107.60	
	Min payoff	*59.93*	*303.21*	*53.46*	*226.04*	*188.51*	*-208.34*	*467.77*
	True nadir	**-515.47**	**192.75**	**30.22**	**-135.54**	**-59.36**	**-546.68**	**-26.25**
*a*_7_60×30	*Best estimate*	**-515.47**	200.07	**30.22**	-49.17	**-59.36**	-538.54	1.72
	Worst estimate	-369.02	278.65	*53.46*	87.16	41.01	-468.82	72.12
	Median estimate	-507.82	200.07	**30.22**	66.06	**-59.36**	-538.54	1.72
	Min payoff	*-187.37*	*154.09*	*-546.20*	*14.69*	*5.21*		
	True nadir	**-430.86**	**-161.40**	**-623.38**	**-140.72**	**-59.79**		
*a*_5_80×40	*Best estimate*	-428.53	**-161.40**	**-623.38**	**-140.72**	-24.30		
	Worst estimate	-330.14	16.57	**-623.38**	-68.46	1.06		
	Median estimate	-383.62	-159.50	**-623.38**	**-140.72**	-24.30		
	Min payoff	*939.16*	*164.93*	*184.17*	*1906.06*	*319.05*		
	True nadir	**-495.68**	**-374.06**	**-16.25**	**574.26**	**-90.83**		
*a*_5_100×50	*Best estimate*	-493.10	**374.06**	**-16.25**	**574.26**	**-90.83**		
	Worst estimate	-394.56	-237.90	101.03	1126.06	101.17		
	Median estimate	-470.49	-297.97	33.32	1037.10	-90.83		

Figure 4 shows the average computational times (*Avg time*) spent by the evolutionary algorithm and by the exact method. A logarithmic scale is used to display the times (in seconds) due to the large differences between the exact method times and the evolutionary algorithm times in problems with 5 and 6 objective functions. Observing this figure we can realize that for problems with 3 objective functions and up to 100 variables and 50 constraints the exact method proposed in [2] is faster than the evolutionary algorithm, besides ensuring that the true nadir point is yielded. On the other side, for problems with 5 and 6 objective functions the differences in times are very significant. An estimate of the nadir point for a problem with 6 objective functions, 100 variables and 50 constraints can be obtained in about 1% of the time taken by the exact method.

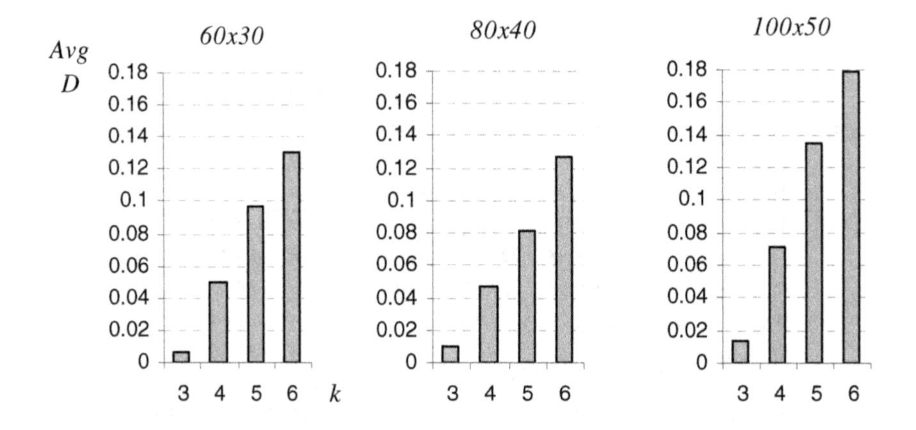

Fig. 3. Average measure *D* for each group

Although the results presented seem to indicate that the evolutionary algorithm can produce good estimates for this set of problems, we believe that, besides *D*, there are other performance measures that can be useful to assess the quality of an estimated nadir point.

First of all, note that the normalization factors may be critical because large ranges of $z_i^* - z_i^{nad}$ may lead to a low discrimination of $\left(\tilde{z}_i^{nad} - z_i^{nad}\right)/\left(z_i^* - z_i^{nad}\right)$ and give an incorrect perception of the quality of the estimate. So, we propose to first compute, for each component, a normalized difference (d_i) between the estimate and the nadir value, which uses the minimum of the payoff table instead of the ideal value to define the normalization factor:

$$d_i = \begin{cases} \dfrac{\tilde{z}_i^{nad} - z_i^{nad}}{z_i^{payoff} - z_i^{nad}} & \text{if } z_i^{payoff} \neq z_i^{nad} \\ 0 & \text{if } z_i^{payoff} = z_i^{nad} \end{cases} \qquad i = 1,\dots,k \tag{7}$$

where z_i^{payoff} is the minimum of the objective *i* in the payoff table.

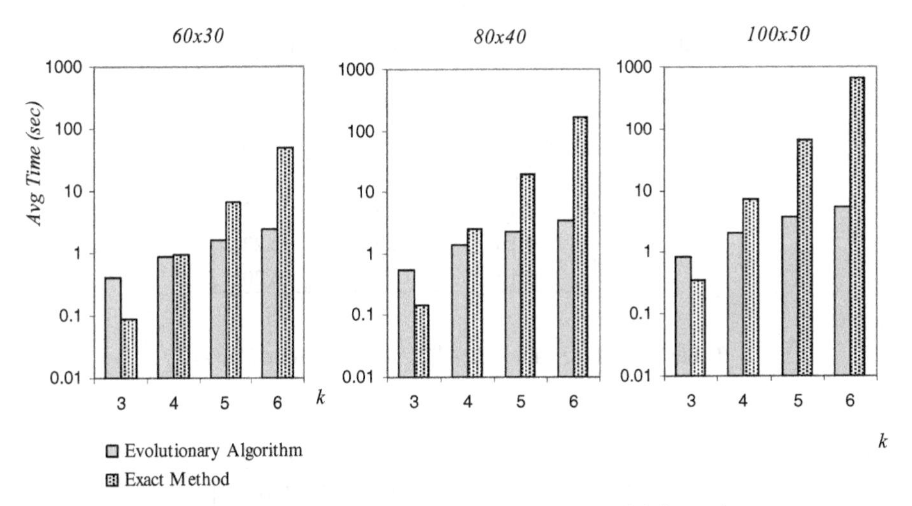

Fig. 4. Average computing times (in a logarithmic scale) for each group

The d_i values can range from 0 to 1, where 1 is the worst possible value which is attained when the estimate is equal to the payoff minimum and this is not the true nadir value. We now consider two performance measures based on d_i. The first one, in (8), computes the average normalized difference d_i for i=1,…,k. The second one, in (9), evaluates a point estimate by its worst component by computing the maximum difference d_i.

$$\bar{d} = \frac{1}{k}\sum_{i=1}^{k} d_i \tag{8}$$

$$d^{\max} = \max_{i=1,\ldots,k} d_i \tag{9}$$

As before, the average values of $\bar{d}$ and $d^{\max}$ for each group of equal sized problems (k_n×m) have been computed. These values are shown in tables 2 and 3 and are plotted in graphs of figure 5. As can be seen, apart from the problems with 3 objective functions and the group 4_60×30 the values of these performance measures do not show a sharp tendency of increasing with the size of the problem or the number of objective functions. Values of $\bar{d}$ indicate that on average an estimate value differs from the true nadir value about 20% of the range from the nadir value to the minimum in the payoff table; the $d^{\max}$ measure indicates that the worst component-wise estimate may be in the middle of the range or closer to the minimum in the payoff table (when $d^{\max}$ >0.5). We should however note that, while the normalization factors that use the ideal values ($z_i^* - z_i^{nad}$) may lead to a low discrimination of the ratios used in D, the normalization factors used herein ($z_i^{payoff} - z_i^{nad}$) may produce the opposite effect when there is a small difference between a nadir value and the corresponding minimum in the payoff table. This can lead to a large d_i (and, consequently, large $d^{\max}$ or $\bar{d}$) even when the estimate is, in absolute value, close to the true nadir value.

Table 2. Average values of $\bar{d}$ for each group *k_n×m*

		$n \times m$		
		60x30	80x40	100x50
k	3	0.02	0.07	0.10
	4	0.11	0.20	0.17
	5	0.18	0.20	0.22
	6	0.16	0.23	0.20

Table 3. Average values of $d^{\max}$ for each group *k_n×m*

		$n \times m$		
		60x30	80x40	100x50
k	3	0.07	0.20	0.24
	4	0.36	0.57	0.45
	5	0.51	0.58	0.58
	6	0.56	0.71	0.52

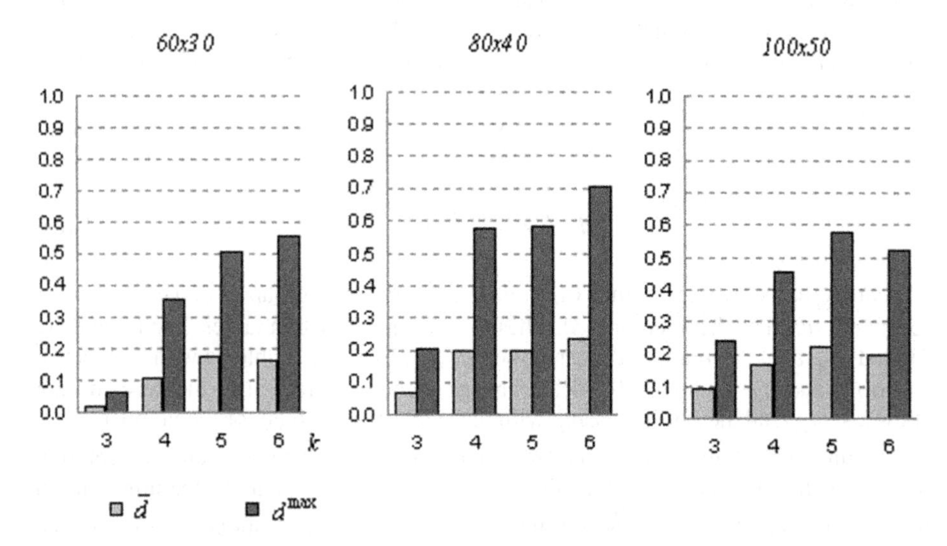

Fig. 5. Average values of $\bar{d}$ and $d^{\max}$ for each group

At last, we have analysed the number of true nadir values found in at least one run of this 30-runs experiment, disregarding the cases where the nadir values are certainly found because they are in the payoff table. In this set of problems, the total number of nadir objective values is 270 (45, 60, 75 and 90 respectively for problems with 3, 4, 5 and 6 objective functions) and 237 of them are not in the payoff table (i.e., 26, 51, 70 and 90 respectively for 3 to 6-objective problems). The algorithm was able to find

76% of the *hidden* nadir values (180 out of the 237 exact values) at least once. The disaggregated percentages are 100%, 88%, 76% and 62%, respectively for the problems with 3, 4, 5 and 6 objectives.

4 Concluding Remarks

In this paper we have presented an evolutionary algorithm for estimating the nadir point in MOLP problems. This algorithm has been compared with an exact method that we have previously developed. A computational experiment was conducted using 60 MOLP problems with 3 to 6 objective functions, 60, 80 and 100 variables, and 30, 40 and 50 constraints.

The evolutionary algorithm sounds to be efficient in problems with 4 or more objective functions as it was able to find a good approximation of the nadir point in a short time. In problems with 3 objective functions and up to 100 variables and 50 constraints the exact method was faster, besides ensuring that the true nadir point is computed. We have computed three performance measures. The first one was proposed by Deb. et al. [6] and the other two are new metrics that use the minima of the payoff table instead of the ideal values to define normalization factors. All of them assume that the true nadir values are known, and present advantages and disadvantages.

The evolutionary algorithm seems promising but further tests must be carried out using larger problems with more objective functions. Clearly, other performance measures must be studied for the cases where the true nadir point is not known.

References

1. Isermann, H., Steuer, R.E.: Computational Experience Concerning Payoff Tables and Minimum Criterion Values Over the Efficient Set. European Journal of Operational Research 33, 91–97 (1987)
2. Alves, M.J., Costa, J.P.: An Exact Method for Computing the Nadir Values in Multiple Objective Linear Programming. European Journal of Operational Research (2008) (accepted for publication), Results available at: `http://www4.fe.uc.pt/mjalves`
3. Ehrgott, M., Tenfelde-Podehl, D.: Computation of Ideal and Nadir Values and Implications for their use in MCDM Methods. European Journal of Operational Research 151, 119–139 (2003)
4. Dessouky, M.I., Ghiassi, M., Davis, W.J.: Estimates of the Minimum Nondominated Criterion Values in Multiple-Criteria Decision-Making. Engineering Costs and Production Economics 10, 95–104 (1986)
5. Korhonen, P., Salo, S., Steuer, R.E.: A Heuristic for Estimating Nadir Criterion Values in Multiple Objective Linear Programming. Operations Research 45, 751–757 (1997)
6. Deb, K., Chaudhuri, S., Miettinen, K.: Towards Estimating Nadir Objective Vector using Evolutionary Approaches. In: Proceedings of the 8th annual conference on Genetic and evolutionary computation, pp. 643–650. ACM Press, New York (2006)
7. Yamamoto, Y.: Optimization Over the Efficient Set: Overview. Journal of Global Optimization 22, 285–317 (2002)
8. Deb, K., Agrawal, S., Pratap, A., Meyarivan, T.: A Fast and Elitist Multi-Objective Genetic Algorithm: NSGA-II. IEEE Transactions on Evolutionary Computation 6, 182–197 (2002)

Interactive Evolutionary Multiobjective Optimization Using Robust Ordinal Regression

Jürgen Branke[1], Salvatore Greco[2], Roman Słowiński[3,4], and Piotr Zielniewicz[3]

[1] Warwick Business School, University of Warwick, Coventry, CV4 7AL, UK
juergen.branke@wbs.ac.uk
[2] Faculty of Economics, University of Catania, 95129 Catania, Italy
salgreco@unict.it
[3] Institute of Computing Science, Poznan University of Technology, 60-965 Poznan, Poland
{roman.slowinski,piotr.zielniewicz}@cs.put.poznan.pl
[4] Systems Research Institute, Polish Academy of Sciences, 01-447 Warsaw, Poland

Abstract. This paper proposes the Necessary-preference-enhanced Evolutionary Multiobjective Optimizer (NEMO), a combination of an evolutionary multiobjective optimization method, NSGA-II, and an interactive multiobjective optimization method, GRIP. In the course of NEMO, the decision maker is able to introduce preference information in a holistic way, by simply comparing some pairs of solutions and specifying which solution is preferred, or comparing intensities of preferences between pairs of solutions. From this information, the set of all compatible value functions is derived using GRIP, and a properly modified version of NSGA-II is then used to search for a representative set of all Pareto-optimal solutions compatible with this set of derived value functions. As we show, this allows to focus the search on the region most preferred by the decision maker, and thereby speeds up convergence.

1 Introduction

Most of past research on evolutionary multiobjective optimization (EMO) attempts to approximate the complete Pareto-optimal front by a set of well-distributed representatives of Pareto-optimal solutions. The underlying reasoning is that in the absence of any preference information, all Pareto-optimal solutions have to be considered equivalent.

On the other hand, in most practical applications, the decision maker (DM) is eventually interested in only a small subset of good solutions, or even a single most preferred solution. In order to come up with such a result, it is necessary to involve the DM. This is the underlying idea of another multiobjective optimization paradigm: interactive multiobjective optimization (IMO). IMO deals with the identification of the most preferred solution by means of a systematic dialogue with the DM. Only recently, the scientific community has discovered the great potential of combining the two paradigms (for a recent survey, see

M. Ehrgott et al. (Eds.): EMO 2009, LNCS 5467, pp. 554–568, 2009.

[14]). From the point of view of EMO, involving the DM in an interactive procedure will allow to focus the search on the area of the Pareto front which is most relevant to the DM. This, in turn, may allow to find preferred solutions faster. In particular, in the case of many objectives, EMO has difficulties, because the number of Pareto-optimal solutions becomes huge, and Pareto-optimality is not sufficiently discriminative to guide the search into better regions. Integrating user's preferences promises to alleviate these problems, allowing to converge faster to the preferred region of the Pareto-optimal front.

This paper combines NSGA-II [3], a widely used EMO technique, with an IMO methodology from multiple criteria decision aiding (MCDA), originally conceived to deal with a limited number of alternatives. This methodology relies on the Robust Ordinal Regression approach recently implemented in the two methods, UTA^{GMS} [10] and GRIP [7]. In these methods, the user is presented with a small set of alternatives and can state his/her preferences by specifying a holistic preference of one alternative over another, or comparing intensities of preferences between pairs of alternatives. The user can also compare intensities of preferences with respect to single criteria. Robust ordinal regression then identifies the whole set of additive value functions (also called utility functions) compatible with the preference information given by the user. This permits to compare any pair of alternatives x and y in a simple and intuitive way, as follows:

- x is necessarily at least as good as y, if this is true for all compatible value functions,
- x is possibly at least as good as y, if this is true for at least one compatible value function.

The interactive EMO method we are proposing, called NEMO (Necessary-preference-enhanced Evolutionary Multiobjective Optimization), takes the information about necessary preferences into account during optimization, focusing search on the most promising parts of the Pareto-optimal front. More specifically, robust ordinal regression based on information obtained through interaction with the user determines the set of all compatible value functions, and an EMO procedure searches for all non-dominated solutions with respect to all compatible value functions in parallel. In the context of EMO, the alternatives considered in GRIP are solutions of a current population.

We believe that the integration of GRIP into EMO is particularly promising for two reasons:

1. The preference information required by GRIP is very basic and easy to provide by the DM. All that is asked for is to compare two solutions, and to reveal whether one is preferred over the other. Additionally, the DM can compare the intensity of preference between pairs of solutions.
2. The resulting set of compatible value functions implicitly reveals also an appropriate scaling of the criteria, an issue that is largely ignored by the EMO community so far.

The paper is organized as follows. The next section provides a brief overview of existing EMO/IMO hybrids. Section 3 describes the basic concepts of robust ordinal regression, presenting UTA^{GMS} and GRIP. Then, Section 4 presents the basic ideas of our method, NEMO. Preliminary empirical results are reported in Section 5. The paper concludes with a summary and some ideas for future research.

2 Interactive Evolutionary Multiobjective Optimization

There are various ways in which user preferences can be incorporated into EMO. Furthermore, there are many IMO techniques, and most of them are suitable for combination with EMO.

A form of preference information often used is a reference point, and various ways to guide the search towards a user-specified reference point have been proposed. Perhaps the earliest such approach has been presented in [8], which gives a higher priority to objectives in which the goal is not fulfilled. [5] suggests to use the distance from the reference point as a secondary criterion following the Pareto ranking. [22] uses an indicator-based evolutionary algorithm, and an achievement scalarizing function to modify the indicator and force the algorithm to focus on the more interesting part of the Pareto front.

In the guided MOEA proposed in [2], the user is allowed to specify preferences in the form of maximally acceptable trade-offs like "one unit improvement in objective i is worth at most a_{ji} units in objective j". The basic idea is to modify the dominance criterion accordingly, so that it reflects the specified maximally acceptable trade-offs.

[4] proposes an interactive decision support system called I-MODE that implements an interactive procedure built over a number of existing EMO and classical decision making methods. The main idea of the interactive procedure is to allow the DM to interactively focus on interesting region(s) of the Pareto front. The DM has options to use several tools for generation of potentially Pareto-optimal solutions concentrated in the desired regions. For example, he/she may use weighted sum approach, utility function based approach, Tchebycheff function approach or trade-off information. The preference information is then used by an EMO to generate new solutions in the most interesting regions.

There are several additional papers that integrate EMO and IMO, but due to space constraints, we refer the interested reader to two recent reviews [14,1]. Instead, in the following, we shall restrict our attention to three papers that perhaps come closest to what we propose in this paper, namely [11], [19], and [13].

[11] suggests a procedure which asks the user to rank a few alternatives, and from this derives constraints for linear weighting of the objectives consistent with the given ordering. Then, these are used within an EMO to check whether there is a feasible linear weighting such that solution x is preferable to solution y. If this is not the case, it is clear that y is preferred to x. The approach differs from ours in two important aspects: first, the interaction with the user is only prior to EMO, while our approach interacts with the user during optimization. Second, the utility function model is only a linear weighting of the objectives, while we consider general additive value functions.

The interactive evolutionary algorithm proposed by [19] allows the user to provide preference information about pairs of solutions during the run. Based on this information, the authors compute the "most compatible" weighted sum of objectives (i.e., a linear achievement scalarizing function) by means of linear programming, and use this as single substitute objective for some generations of the evolutionary algorithm. This concept presented in this paper is truly interactive, as preference information is collected during the run. However, as it reduces the preference information to a single linear weighting of the objectives, the power of EMO, which is capable of simultaneously searching for multiple solutions with different trade-offs, is not exploited. Furthermore, since only partial preference information is available, there is no guarantee that the weight vector obtained by solving the linear programming model defines the DM's value function, even if the value function has the form of a weighted sum (naturally, the bias may become even more significant when the DM's preferences cannot be modeled with a linear function).

The method of [13] is based on the Pareto memetic algorithm (PMA). The original PMA samples the set of scalarizing functions drawing a random weight vector for each single iteration and using this for selection and local search. In the proposed interactive version, preference information from pairwise comparisons of solutions is used to reduce the set of possible weight vectors. While this approach is more flexible in terms of the considered value function model, and changes the value function from generation to generation, it still does not make explicit use of the EMO's capability to search for multiple solutions in parallel.

Furthermore, all of the methods discussed above require a pre-defined scaling of the objectives, while we propose a new way that allows to automatically and continuously adjust the scaling of the objectives to the most likely user preferences given the information gathered so far.

3 Robust Ordinal Regression

In MCDA, the *preference information* may be either direct or indirect, depending whether it specifies directly values of some parameters used in the *preference model* (e.g., trade-off weights, aspiration levels, discrimination thresholds, etc.) or whether it specifies some examples of holistic judgments from which compatible values of the preference model parameters are induced. Eliciting direct preference information from the DM can be counterproductive in real-world decision making situations because of a high cognitive effort required. Eliciting indirect preferences is less demanding in terms of cognitive effort. Indirect preference information is mainly used in the ordinal regression paradigm. According to this paradigm, a holistic preference information on a subset of some reference or training solutions is known first and then a preference model compatible with the information is built and applied to the whole set of solutions in order to rank them.

The *ordinal regression* paradigm emphasizes the discovery of intentions as an interpretation of actions rather than as a priori position, which was called by

March the posterior rationality [16]. It has been known for at least fifty years in the field of multidimensional analysis. It is also concordant with the induction principle used in machine learning. This paradigm has been applied within the two main MCDA approaches: those using a value function as preference model [21,18,12,20], and those using an outranking relation as preference model [15,17]. This paradigm has also been used since mid nineties' in MCDA methods involving a new, third family of preference models - a set of dominance decision rules induced from rough approximations of holistic preference relations [9].

Recently, the ordinal regression paradigm has been revisited with the aim of considering the whole set of value functions compatible with the preference information provided by the DM, instead of a single compatible value function used in UTA-like methods [12,20]. This extension, called *robust ordinal regression*, has been implemented in a method called UTA^{GMS} [10], and further generalized in another method called GRIP [7]. UTA^{GMS} and GRIP are not revealing to the DM one compatible value function, but they are using the whole set of compatible (general, not piecewise-linear only) additive value functions to set up a necessary weak preference relation and a possible weak preference relation in the whole set of considered solutions.

3.1 Concepts: Definitions and Notation

We are considering a multiple criteria decision problem where a finite set of solutions $A = \{x, \ldots, y, \ldots w, \ldots\}$ is evaluated on a family $F = \{g_1, g_2, \ldots, g_n\}$ of n criteria. Let $I = \{1, 2, \ldots, n\}$ denote the set of criteria indices. We assume, without loss of generality, that the smaller $g_i(x)$, the better solution x on criterion g_i, for all $i \in I$, $x \in A$. A DM is willing to rank the solutions of A from the best to the worst, according to his/her preferences. The ranking can be complete or partial, depending on the preference information provided by the DM and on the way of exploiting this information.

Such a decision-making problem statement is called *multiple criteria ranking problem*. It is known that the only information coming out from the formulation of this problem is the dominance ranking. For any pair of solutions $x, y \in A$, one of the four situations may arise in the dominance ranking: x is preferred to y (x dominates y but y does not dominate x), y is preferred to x (y dominates x but x does not dominate y), x is indifferent to y (x and y dominate each other), or x is incomparable to y (neither x dominates y nor y dominates x). Usually, the dominance ranking is very poor, i.e., the most frequent situation is x incomparable to y, in particular if the number of objectives is high.

In order to enrich the dominance ranking, the DM has to provide preference information which is used to construct an aggregation model making the solutions more comparable. Such an aggregation model is called preference model. It induces a preference structure on set A, whose proper exploitation permits to work out a ranking proposed to the DM.

In what follows, the evaluation of each solution $x \in A$ on each criterion $g_i \in F$ will be denoted by $g_i(x)$.

Let G_i denote the value set (scale) of criterion g_i, $i \in I$. Consequently,

$$G = G_1 \times G_2 \times \ldots \times G_n$$

represents the evaluation space. From a pragmatic point of view, it is reasonable to assume that $G_i \subseteq \mathbb{R}$, for $i = 1, \ldots, m$. More specifically, we will assume that the value space on each criterion g_i is bounded, such that $G_i = [\alpha_i, \beta_i]$, where α_i, β_i, $\alpha_i < \beta_i$ are the worst and the best (finite) evaluations, respectively. Thus, $g_i \ : \ A \rightarrow G_i$, $i \in I$. Therefore, each solution $x \in A$ is associated with an evaluation solution denoted by $\underline{g}(x) = (g_1(x), g_2(x), \ldots, g_n(x)) \in G$. For notational simplicity, we will also write x_i instead of $g_i(x)$, so $\underline{g}(x) = (x_1, x_2, \ldots, x_n) \in G$.

We consider a weak preference relation $\succeq$ on A which means, for each pair of solutions $x, y \in A$,

$$x \succeq y \Leftrightarrow \text{“}x \text{ is at least as good as } y\text{”}.$$

This weak preference relation can be decomposed into its asymmetric and symmetric parts, as follows,

1) $x \succ y \equiv [x \succeq y \text{ and } not(y \succeq x)] \Leftrightarrow$ “x is preferred to y”, and
2) $x \sim y \equiv [x \succeq y \text{ and } y \succeq x] \Leftrightarrow$ “x is indifferent to y”.

3.2 The Ordinal Regression Method for Learning the Whole Set of Compatible Value Functions

The additive value function considered in ordinal regression is defined on A such that for each $\underline{g}(x) \in G$,

$$U(\underline{g}(x)) = \sum_{i=1}^{n} u_i(g_i(x_i)), \quad (1)$$

where, u_i are non-increasing marginal value functions, $u_i : G_i \rightarrow \mathbb{R}$, $i \in I$. For the sake of simplicity, we shall write (1) as follows,

$$U(x) = \sum_{i=1}^{n} u_i(x). \quad (2)$$

Recently, two new methods, UTA^{GMS} [10] and GRIP [7], have generalized the classical ordinal regression approach of the UTA method [12] in several aspects:

- taking into account all additive value functions (1) compatible with the preference information, while UTA is using only one such function,
- considering marginal value functions of (1) as general non-decreasing functions, and not piecewise-linear, as in UTA,
- asking the DM for a ranking of reference solutions which is not necessarily complete (just pairwise comparisons),

- taking into account additional preference information about intensity of preference, expressed both comprehensively and with respect to a single criterion,
- avoiding the use of the exogenous, and not neutral for the result, parameter ε in the modeling of strict preference between solutions.

UTA^{GMS} produces two rankings on the set of solutions A, such that for any pair of solutions $a, b \in A$:

- in the *necessary* ranking, a is ranked at least as good as b if and only if, $U(a) \geq U(b)$ for *all* value functions compatible with the preference information,
- in the *possible* ranking, a is ranked at least as good as b if and only if, $U(a) \geq U(b)$ for *at least one* value function compatible with the preference information.

GRIP produces four more necessary and possible rankings on the set of solutions A, and on the set of pairs of solutions, $A \times A$.

The necessary ranking can be considered as robust with respect to the preference information. Such robustness of the necessary ranking refers to the fact that any pair of solutions is ranked in the same way whatever the additive value function compatible with the preference information. Indeed, when no preference information is given, the necessary ranking boils down to the weak dominance relation (i.e., a is necessarily at least as good as b, if $g_i(a) \leq g_i(b)$ for all $g_i \in F$), and the possible ranking is a complete relation. Every new pairwise comparison of reference solutions, for which the dominance relation does not hold, is enriching the necessary ranking and it is impoverishing the possible ranking, so that they converge with the growth of the preference information.

Moreover, such an approach has another feature which is very appealing in the context of MOO. It stems from the fact that it gives space for interactivity with the DM. Presentation of the necessary ranking, resulting from a preference information provided by the DM, is a good support for generating reactions from the DM. Namely, he/she could wish to enrich the ranking or to contradict a part of it. Such a reaction can be integrated in the preference information considered in the next calculation stage.

The idea of considering the whole set of compatible value functions was originally introduced in UTA^{GMS}. GRIP (Generalized Regression with Intensities of Preference) can be seen as an extension of UTA^{GMS} permitting to take into account additional preference information in form of comparisons of intensities of preference between some pairs of reference solutions. For solutions $x, y, w, z \in A$, these comparisons are expressed in two possible ways (not exclusive): (i) comprehensively, on all criteria, like "x is preferred to y at least as much as w is preferred to z"; and, (ii) partially, on each criterion, like "x is preferred to y at least as much as w is preferred to z, on criterion $g_i \in F$ ". In the following, we shall use GRIP.

3.3 The Preference Information Provided by the Decision Maker

The DM is expected to provide the following preference information in the dialogue stage of the procedure:

- A partial preorder $\succsim$ on A^R whose meaning is: for some $x, y \in A^R$

$$x \succsim y \Leftrightarrow \text{“}x \text{ is at least as good as } y\text{”}.$$

  Moreover, $\succ$ (preference) is the asymmetric part of $\succsim$, and $\sim$ (indifference) is its symmetric part.
- A partial preorder $\succsim^*$ on $A^R \times A^R$, whose meaning is: for some $x, y, w, z \in A^R$,

$$(x, y) \succsim^* (w, z) \Leftrightarrow \text{“}x \text{ is preferred to } y \text{ at least as much as } w \text{ is preferred to } z\text{”}.$$

  Also in this case, $\succ^*$ is the asymmetric part of $\succsim^*$, and $\sim^*$ is its symmetric part.
- A partial preorder $\succsim_i^*$ on $A^R \times A^R$, whose meaning is: for some $x, y, w, z \in A^R$, $(x, y) \succsim_i^* (w, z) \Leftrightarrow$ “x is preferred to y at least as much as w is preferred to z” on criterion g_i, $i \in I$.

In the following, we also consider the weak preference relation $\succsim_i$ being a complete preorder whose meaning is: for all $x, y \in A$,

$$x \succsim_i y \quad \Leftrightarrow \quad \text{“}x \text{ is at least as good as } y\text{” on criterion } g_i, \quad i \in I.$$

Weak preference relations $\succsim_i$, $i \in I$, are not provided by the DM, but they are obtained directly from the evaluation of solutions x and y on criteria g_i, i.e., $x \succsim_i y \Leftrightarrow g_i(x) \leq g_i(y)$, $i \in I$.

3.4 Linear Programming Constraints

In this subsection, we present a set of constraints that interprets the preference information in terms of conditions on the compatible value functions.

To be compatible with the provided preference information, the value function $U : A \to [0, 1]$ should satisfy the following constraints corresponding to the DM's preference information:

a) $U(w) > U(z)$ if $w \succ z$
b) $U(w) = U(z)$ if $w \sim z$
c) $U(w) - U(z) > U(x) - U(y)$ if $(w, z) \succ^* (x, y)$
d) $U(w) - U(z) = U(x) - U(y)$ if $(w, z) \sim^* (x, y)$
e) $u_i(w) \geq u_i(z)$ if $w \succsim_i z$, $i \in I$
f) $u_i(w) - u_i(z) > u_i(x) - u_i(y)$ if $(w, z) \succ_i^* (x, y)$, $i \in I$
g) $u_i(w) - u_i(z) = u_i(x) - u_i(y)$ if $(w, z) \sim_i^* (x, y)$, $i \in I$

Moreover, the following normalization constraints should also be taken into account:

h) $u_i(x_i^*) = 0$, where x_i^* is such that $x_i^* = \max\{g_i(x) : x \in A\}$;
i) $\sum_{i \in I} u_i(y_i^*) = 1$, where y_i^* is such that $y_i^* = \min\{g_i(x) : x \in A\}$.

For computational details, the reader is referred to [7].

3.5 The Most Representative Value Function

The robust ordinal regression builds a set of additive value functions compatible with preference information provided by the DM and results in two rankings, necessary and possible. Such rankings answer to robustness concerns, since they provide, in general, "more robust" conclusions than a ranking made by an arbitrarily chosen compatible value function. However, in some decision-making situations, it may be desirable to give a score to different solutions, and despite the interest of the rankings provided, some users would like to see, and they indeed need, to know the "most representative" value function among all the compatible ones This allows assigning a score to each solution. Recently, a methodology to identify the "most representative" function in GRIP without losing the advantage of taking into account all compatible value functions has been proposed in [6]. The idea is to select among all compatible value functions the most discriminant value function for consecutive solutions in the necessary ranking, i.e., that value function which maximizes the difference of scores between solutions related by preference in the necessary ranking. To break ties, one can wish to minimize the difference of scores between solutions not related by preference in the necessary ranking. This can be achieved using the following procedure:

1. Determine the necessary preference relations in the considered set of solutions.
2. For all pairs of solutions (a, b), such that a is necessarily preferred to b, add the following constraints to the linear programming constraints of GRIP: $U(a) \geq U(b) + \varepsilon$.
3. Maximize the objective function ε.
4. Add the constraint $\varepsilon = \varepsilon^*$, with ε^* being the resulting maximal ε from point 3), to the linear programming constraints of point 2).
5. For all pairs of solutions (a, b), such that neither a is necessarily preferred to b nor b is necessarily preferred to a, add the following constraints to the linear programming constraints of GRIP and to the constraints considered in above point 4): $U(a) - U(b) \leq \delta$ and $U(b) - U(a) \leq \delta$.
6. Minimize δ.

This procedure maximizes the minimal difference between values of solutions for which the necessary preference holds. If there is more than one such value function, the above procedure selects the most representative compatible value function giving the largest minimal difference between values of solutions for which the necessary preference holds, and the smallest maximal difference between values of solutions for which the possible preference holds.

Notice that the concept of the "most representative" value function thus defined is still based on the necessary and possible preference relations, which remain crucial for GRIP. In a sense, it gives the most faithful representation of these necessary and possible preference relations. Notice also that the above procedure can be simplified by joint maximization of $M\varepsilon - \delta$ where M is a "big value".

In the following, we will use the most representative value function for continuously adapting the scaling of the objectives in a non-linear way.

4 Necessary-Preference-Enhanced Evolutionary Multiobjective Optimization – NEMO

Our main idea is to integrate the concept of GRIP into an EMO approach, in particular NSGA-II [3]. NSGA-II is one of today's most prominent and most successful EMO algorithms. It ranks individuals according to two criteria.

The primary criterion is the so-called dominance-based ranking. This method ranks individuals by iteratively determining the non-dominated solutions in the population (non-dominated front), assigning those individuals the next best rank and removing them from the population. The result is a partial ordering, favoring individuals closer to the Pareto-optimal front.

As secondary criterion, individuals which have the same dominance-rank (primary criterion) are sorted according to crowding distance, which is defined as the sum of distances between a solution's neighbors on either side in each dimension of the objective space. Individuals with a large crowding distance are preferred, as they are in a less crowded region of the objective space, and favoring them aims at preserving diversity in the population.

In our approach, we will

1. Replace the dominance-based ranking by the *necessary* ranking. The necessary ranking is calculated analogously to the dominance-based ranking, but taking into account the preference information by the user through the necessary preference relations. More specifically, first put in the best rank those solutions which have no competitor which would be necessarily preferred, remove them from the population, etc.
2. Replace the crowding-distance by a distance calculated taking into account the multidimensional scaling given by the "most representative value function" among the whole set of compatible value functions (see sub-section 3.5). While in NSGA-II the crowding distance is calculated in the space of objective functions, in NEMO it is calculated in the space of marginal value functions which are components of the "most representative" value function. Given a solution x, its crowding distance is calculated according to the following formula:

$$\text{Crowding distance}(x) = \sum_{i=1}^{n} \left|u_i(y^i) - u_i(z^i)\right| - \left|\sum_{i=1}^{n} \left[U(y^i) - U(z^i)\right]\right|,$$

where U is the "most representative value function", u_i are its marginal value functions, and y^i and z^i are left and right neighbors of x in dimension of marginal value u_i. Remark that for a given n, we can have up to $2n$ different neighbors of x in all dimensions, due to non-univocal selection of solutions with equal marginal values. In fact, we select the neighbors such as to diversify them as much as possible.

Algorithm 1. Basic NEMO

```
Generate initial solutions randomly
Elicit user preferences {Present to user a pair of solutions and ask for a preference
information}
Determine necessary ranking {Will replace dominance ranking in NSGA-II}
Determine secondary ranking {Order solutions within a front, based on crowding
distance measured in terms of the "most representative value function"}
repeat
   Mating selection and offspring generation
   if Time to ask DM then
      Elicit user preferences
   end if
   Determine necessary ranking
   Determine secondary ranking
   Environmental selection
until Stopping criterion met
Return all preferred solutions according to necessary ranking
```

Preferences are elicited by asking the DM to compare some pairs of solutions, and specify a preference relation between them. This is done during the run of the NSGA-II.

The overall algorithm is outlined in Algorithm 1. Although the general procedure is rather straightforward, there are several issues that need to be considered:

1. How many pairs of solutions are shown to the DM, and when? Here, we decide to ask for one preference relation every k generations, i.e., every x generations, NSGA-II is stopped, and the user is asked to provide preference information about one given pair of individuals.
2. Which pairs of solutions shall be shown to the DM for comparison? Here, we randomly pick a small set of non-dominated solutions (according to the necessary ranking). This also prevents the user from specifying inconsistent information.

5 Experimental Results

An empirical evaluation of interactive EMO methods is difficult, because the test environment has to include a model of the user behavior. For testing, we use the simple 30-dimensional ZDT1 test function. We assume that our artificial user makes decisions with respect to a simple predefined value function $U(x) = -(0.6f_1(x) + 0.4f_2(x))$. This function is unknown to NEMO, but is used to simulate user's comparisons of solutions when preferences are elicited. In every k-th generation, NEMO randomly selects two individuals from the non-dominated solutions according to the necessary ranking, and receives as feedback the solution preferred by the DM according to the predefined value function. In particular, only pairwise comparisons of solutions are considered here, while intensities of preferences between pairs of solutions are not (yet) considered. The population size has been set to 32.

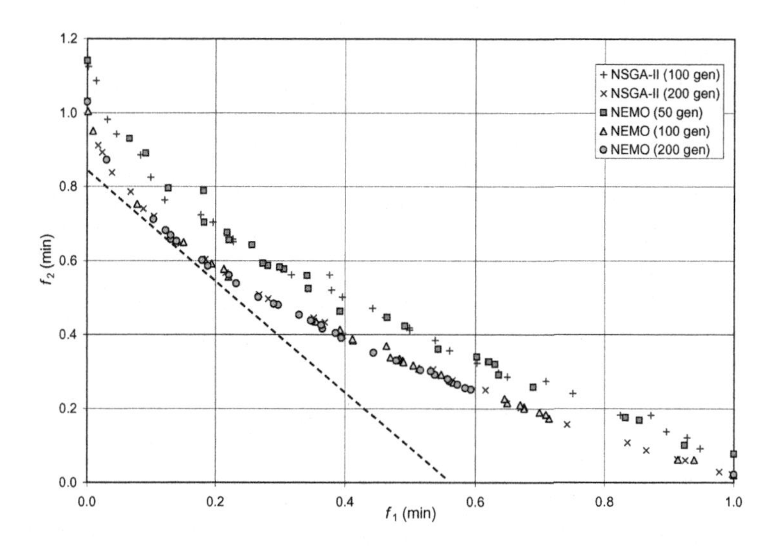

Fig. 1. Results of NEMO and NSGA-II on ZDT1 after 50, 100 and 200 generations, with preference elicitation every 20 generations. The dashed line indicates the artificial user's value function.

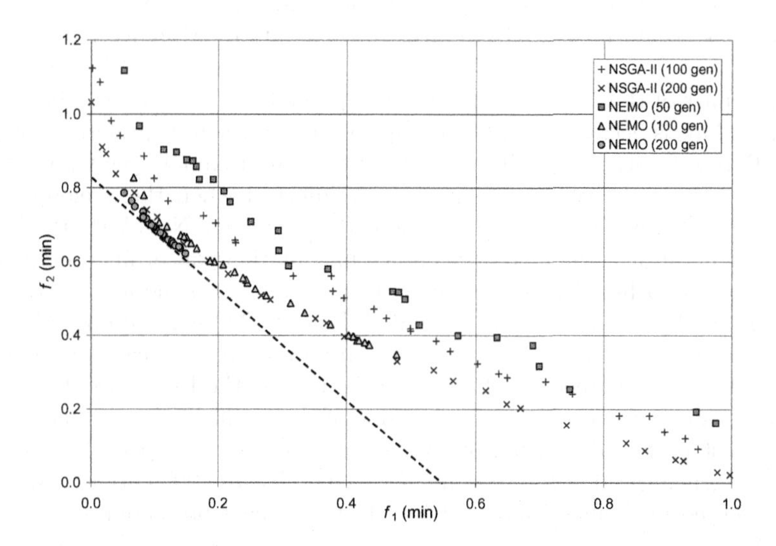

Fig. 2. Results of NEMO and NSGA-II on ZDT1 after 50, 100 and 200 generations, with preference elicitation in every generation. The dashed line indicates the artificial user's value function.

Figure 1 shows results of NEMO and NSGA-II after 50, 100 and 200 generations, when the preference information concerning one pairwise comparison is gathered every 20 generations. As can be seen, NEMO converges faster than

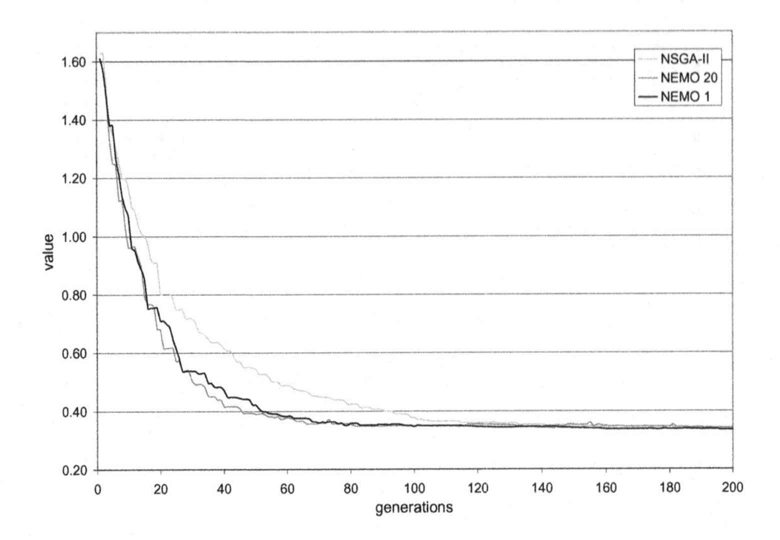

Fig. 3. Value of the most preferred solutions in successive generations of NEMO and NSGA-II

NSGA-II. After 50 generations, the solutions obtained by NEMO are as good as the solutions obtained by NSGA-II after 100 generations. Moreover, in the course of generations, the population of solutions obtained by NEMO is narrowed to a smaller part of the Pareto front than the population of solutions obtained by NSGA-II. This is because NEMO concentrates on on the user-preferred solutions on the Pareto front, while NSGA-II attempts to approximate the whole front.

The tendency observed in Figure 1 is reinforced when the preferences are gathered more often. Figure 2 shows results of NEMO and NSGA-II after 50, 100 and 200 generations, when the preference information concerning one pairwise comparison is gathered in every generation. After 100 generations NEMO is reaching equally good solutions as NSGA-II after 200 generations. Moreover, due to the richer preference information than in the previous case, the solutions obtained by NEMO are focused on a smaller part of the Pareto front.

Figure 3 shows the evolution of the value of the artificial user's value function for the most preferred solution in successive generations. It permits to observe the convergence speed of NEMO and NSGA-II. "NEMO 20" corresponds to the case presented in Figure 1, and "NEMO 1" to the case presented in Figure 2.

6 Conclusion

We presented an interactive EMO method called NEMO. It combines the advantages of the well known EMO method NSGA-II with an MCDA method GRIP enabling the user interaction based on robust ordinal regression. The main advantages of the proposed methodology are the following:

1. It models the user's preferences in terms of very general value functions,
2. It requires a preference information expressed in a simple and intuitive way (comparisons of solutions or comparisons of intensities of preferences),
3. It considers all value functions compatible with the user's preferences, with the goal to generate a representative approximation of all Pareto-optimal solutions compatible with any of these value functions,
4. With respect to crowding distance, it permits to calculate distances in utility space, rather than objective space, thereby alleviating the need of scaling the objectives.

Preliminary empirical results show that the proposed NEMO method works as expected and is able to converge faster to the user-preferred solutions than NSGA-II without taking user preferences into account.

Clearly, a more thorough empirical analysis on a variety of test functions and value functions is necessary. Also, we are currently elaborating and extending the approach in various directions. In particular, we are implementing improved interaction mechanisms, with adaptive methods to determine when a DM should be asked for preference information, and what individuals to present for comparison. We will also extend the current interaction to allow additional preference information to be incorporated. Apart from the above mentioned intensities of preferences, we plan to integrate into GRIP maximum/minimum trade-off information, e.g., one unit improvement in objective f_1 is worth at most w units worsening in objective f_2.

Finally, we plan to elaborate a slightly different approach: instead of calculating the necessary preference relation in the population of solutions, we could look for solutions that are the best for at least one compatible value function. The expected advantages of this new approach are speeding up of calculations and of the convergence to the most interesting part of the Pareto front.

References

1. Branke, J.: Consideration of partial user preferences in evolutionary multiobjective optimization. In: Branke, J., Deb, K., Miettinen, K., Słowiński, R. (eds.) Multiobjective Optimization - Interactive and Evolutionary Approaches. LNCS, vol. 5252, pp. 157–178. Springer, Heidelberg (2008)
2. Branke, J., Kaußler, T., Schmeck, H.: Guidance in evolutionary multi-objective optimization. Advances in Engineering Software 32, 499–507 (2001)
3. Deb, K., Agrawal, S., Pratap, A., Meyarivan, T.: A fast and elitist multi-objective genetic algorithm: NSGA-II. IEEE Transactions on Evolutionary Computation 6(2), 182–197 (2002)
4. Deb, K., Chaudhuri, S.: I-MODE: An interactive multi-objective optimization and decision-making using evolutionary methods. Technical Report KanGAL Report No. 2007003, Indian Institute of Technology Kanpur (2007)
5. Deb, K., Sundar, J., Udaya Bhaskara Rao, N., Chaudhuri, S.: Reference point based multi-objective optimization using evolutionary algorithms. International Journal of Computational Intelligence Research 2(3), 273–286 (2006)

6. Figueira, J., Greco, S., Słowiński, R.: Identifying the "most representative" value function among all compatible value functions in the GRIP method. In: 68th Meeting of the EURO Working Group on MCDA, Chania, October 2-3 (2008)
7. Figueira, J., Greco, S., Słowiński, R.: Building a set of additive value functions representing a reference preorder and intensities of preference: GRIP method. European Journal of Operational Research 195(2), 460–486 (2009)
8. Fonseca, C.M., Fleming, P.J.: Genetic algorithms for multiobjective optimization: Formulation, discussion, and generalization. In: Proceedings of the Fifth International Conference on Genetic Algorithms, pp. 416–423 (1993)
9. Greco, S., Matarazzo, B., Słowiński, R.: Rough sets theory for multicriteria decision analysis. European Journal of Operational Research 129(1), 1–47 (2001)
10. Greco, S., Mousseau, V., Słowiński, R.: Ordinal regression revisited: Multiple criteria ranking with a set of additive value functions. European Journal of Operational Research 191(2), 415–435 (2008)
11. Greenwood, G.W., Hu, X.S., D'Ambrosio, J.G.: Fitness functions for multiple objective optimization problems: combining preferences with Pareto rankings. In: Belew, R.K., Vose, M.D. (eds.) Foundations of Genetic Algorithms, pp. 437–455. Morgan Kaufmann, San Francisco (1997)
12. Jacquet-Lagrèze, E., Siskos, Y.: Assessing a set of additive utility functions for multicriteria decision making: The UTA method. European Journal of Operational Research 10, 151–164 (1982)
13. Jaszkiewicz, A.: Interactive multiobjective optimization with the Pareto memetic algorithm. Foundations of Computing and Decision Sciences 32(1), 15–32 (2007)
14. Jaszkiewicz, A., Branke, J.: Interactive multiobjective evolutionary algorithms. In: Branke, J., Deb, K., Miettinen, K., Słowiński, R. (eds.) Multiobjective Optimization - Interactive and Evolutionary Approaches. LNCS, vol. 5252, pp. 179–193. Springer, Heidelberg (2008)
15. Kiss, L., Martel, J.M., Nadeau, R.: ELECCALC - an interactive software for modelling the decision maker's preferences. Decision Support Systems 12, 757–777 (1994)
16. March, J.: Bounded rationality, ambiguity and the engineering of choice. Bell Journal of Economics 9, 587–608 (1978)
17. Mousseau, V., Słowiński, R.: Inferring an ELECTRE TRI model from assignment examples. Journal of Global Optimization 12, 157–174 (1998)
18. Pekelman, D., Sen, S.K.: Mathematical programming models for the determination of attribute weights. Management Science 20, 1217–1229 (1974)
19. Phelps, S., Köksalan, M.: An interactive evolutionary metaheuristic for multiobjective combinatorial optimization. Management Science 49(12), 1726–1738 (2003)
20. Siskos, Y., Grigoroudis, V., Matsatsinis, N.: UTA methods. In: Figueira, F., Greco, S., Ehrgott, M. (eds.) Multiple Criteria Decision Analysis: State of the Art Surveys, pp. 297–343. Springer, Heidelberg (2005)
21. Srinivasan, V., Shocker, A.D.: Estimating the weights for multiple attributes in a composite criterion using pairwise judgments. Psychometrika 38, 473–493 (1973)
22. Thiele, L., Miettinen, K., Korhonen, P.J., Molina, J.: A preference-based interactive evolutionary algorithm for multiobjective optimization. Working papers w-412, Helsinki School of Economics, Helsinki (2007)

A Hybrid Integrated Multi-Objective Optimization Procedure for Estimating Nadir Point

Kalyanmoy Deb[1,⋆], Kaisa Miettinen[2], and Deepak Sharma[1]

[1] Department of Mechanical Engineering
Indian Institute of Technology Kanpur, PIN 208016, India
{deb,dsharma}@iitk.ac.in
[2] Department of Mathematical Information Technology
P.O. Box 35 (Agora), FI-40014
University of Jyväskylä, Finland
kaisa.miettinen@jyu.fi

Abstract. A nadir point is constructed by the worst objective values of the solutions of the entire Pareto-optimal set. Along with the ideal point, the nadir point provides the range of objective values within which all Pareto-optimal solutions must lie. Thus, a nadir point is an important point to researchers and practitioners interested in multi-objective optimization. Besides, if the nadir point can be computed relatively quickly, it can be used to normalize objectives in many multi-criterion decision making tasks. Importantly, estimating the nadir point is a challenging and unsolved computing problem in case of more than two objectives. In this paper, we revise a previously proposed serial application of an EMO and a local search method and suggest an integrated approach for finding the nadir point. A local search procedure based on the solution of a bi-level achievement scalarizing function is employed to extreme solutions in stabilized populations in an EMO procedure. Simulation results on a number of problems demonstrate the viability and working of the proposed procedure.

1 Introduction

A *nadir* point signifies, in principle, opposite to that meant by an *ideal* point, in the context of multi-objective optimization. An ideal point is an M-dimensional objective vector (where M is the number of objectives) constructed with best feasible objective values and is a comparatively easy to compute. For minimization problems, in principle, this calls for solving M single-objective minimization problems and collecting each optimal objective values to form the ideal point. On the other hand, a nadir point is constructed with the worst objective values of Pareto-optimal solutions. In minimization problems, this task is different from

⋆ Also Finland Distinguished Professor, Helsinki School of Economics, PO Box 1210, FIN-00101, Helsinki, Finland, (Kalyanmoy.Deb@hse.fi).

M. Ehrgott et al. (Eds.): EMO 2009, LNCS 5467, pp. 569–583, 2009.

simply maximizing M objective functions one at a time. This is because the search of the worst value of an objective must be restricted within the Pareto-optimal solutions. This is the reason why the estimation of nadir point has been found to be a complex task [13,11] and there does not exist any provable algorithm for the task, even for linear multi-objective optimization problems having three or more objectives.

With the advent of efficient evolutionary optimization procedures for multi-objective optimization, some attention has been made in the recent past in developing procedures for estimating the nadir point. Simplistic ideas, such as finding a set of Pareto-optimal solutions by an EMO procedure and then choosing the extreme solutions for estimating the nadir point, to more sophisticated ideas, such as replacing the focus of EMO to find a wide-spreaded set of solutions on the entire Pareto-optimal front to find only the critical extreme Pareto-optimal points [4,17], are suggested. Most of these EMO methodologies have shown to find an approximation of the nadir point, rather than to estimate the exact nadir point. Recent studies [6,5] suggested a two-step serial procedure of employing a modified NSGA-II procedure to identify extreme near Pareto-optimal solutions and then a local search procedure to converge to the true extreme Pareto-optimal points.

In this study, we suggest and simulate a hybrid integrated approach in which a local search procedure is used within the modified NSGA-II algorithm sparingly to achieve the nadir point estimation task. We restrict our discussions for real-parameter optimization problems, but the concept can very well be used for other types of optimization problems. The suggested local search procedure is based on utilizing a reference point based approach, a so-called achievement scalarizing function [18] which is widely used in the MCDM field. Using this scalarized function, any point in the objective space can be projected on the Pareto optimal front and the scalarizing function does not need any artificial information like weights [15]. In the procedure proposed, the achievement scalarizing function is used in a bi-level manner to guarantee getting reliable enough information about extreme values in the Pareto optimal front for estimating the nadir point. Based on a statistical analysis of the performance of the NSGA-II procedure, the execution of the local search event is decided dynamically at every generation. Both NSGA-II and local search procedures are terminated using statistical performance criteria. Simulation results on a number of test problems and three engineering problems are presented to demonstrate the efficacy of the proposed procedure.

2 Nadir Objective Vector

We consider multi-objective optimization problems involving M conflicting objectives ($f_i : \mathcal{S} \rightarrow \mathbf{R}$) as functions of decision variables $\mathbf{x}$:

$$\begin{aligned} &\text{minimize } \{f_1(\mathbf{x}), f_2(\mathbf{x}), \ldots, f_M(\mathbf{x})\}, \\ &\text{subject to } \mathbf{x} \in \mathcal{S}, \end{aligned} \tag{1}$$

where $\mathcal{S} \subset \mathbf{R}^n$ denotes the set of feasible solutions. Problem (1) gives rise to a set of *Pareto-optimal* solutions or a Pareto-optimal front (P^*), providing a trade-off among the objectives. In the sense of minimization of objectives, Pareto-optimal solutions can be defined as follows [15]:

Definition 1. *A decision vector* $\mathbf{x}^* \in \mathcal{S}$ *and the corresponding objective vector* $\mathbf{f}(\mathbf{x}^*)$ *are Pareto-optimal if there does not exist another decision vector* $\mathbf{x} \in S$ *such that* $f_i(\mathbf{x}) \leq f_i(\mathbf{x}^*)$ *for all* $i = 1, 2, \ldots, M$ *and* $f_j(\mathbf{x}) < f_j(\mathbf{x}^*)$ *for at least one index* j.

In what follows, we assume that the Pareto-optimal front is bounded. We now define a nadir objective vector, that is, a nadir point, as follows.

Definition 2. *An objective vector* $\mathbf{z}^{\rm nad} = (z_1^{\rm nad}, \ldots, z_M^{\rm nad})^T$ *constructed using the worst values of objective functions in the complete Pareto-optimal front* P^* *is called a nadir objective vector.*

Hence, for minimization problems we have $z_j^{\rm nad} = \max_{\mathbf{x} \in P^*} f_j(\mathbf{x})$. Estimation of the nadir objective vector is, in general, a difficult task. Unlike the *ideal objective vector* $\mathbf{z}^* = (z_1^*, \ldots, z_M^*)^T$, which can be found by minimizing each objective individually over the feasible set S (or, $z_j^* = \min_{\mathbf{x} \in \mathcal{S}} f_j(\mathbf{x})$), the nadir point cannot be formed by maximizing objectives individually over $\mathcal{S}$. To find the nadir point, Pareto-optimality of solutions used for constructing the nadir point must be first established. This makes the task of finding the nadir point a difficult one. To illustrate this aspect, let us consider a bi-objective minimization problem shown in Figure 1. If we maximize f_1 and f_2 individually, we obtain points A and B, respectively. These two points can be used to construct the so-called *worst objective vector*, $\mathbf{z}^w$. In many problems (even in bi-objective optimization problems), the nadir objective vector and the worst objective vector are not the same point, which can also be seen in Figure 1.

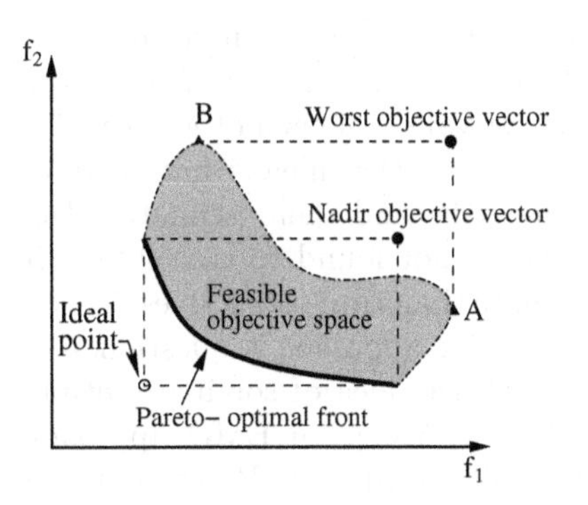

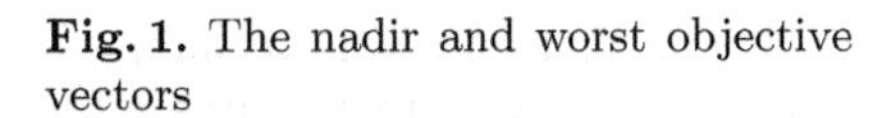
Fig. 1. The nadir and worst objective vectors

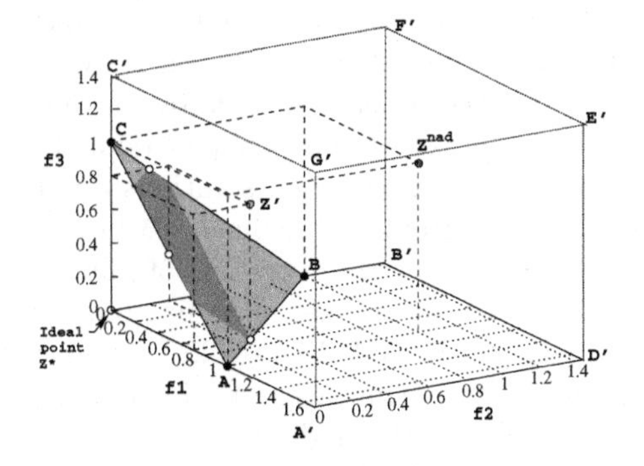

Fig. 2. Payoff table may not produce the true nadir point

3 Existing Methods

3.1 Payoff Table Method

Benayoun et al. [1] introduced the first interactive multi-objective optimization method for estimating the nadir point by using a *payoff table*. To be more specific, each objective function is first minimized individually and then a table is constructed where the i-th row of the table represents values of all objective functions calculated at the point where the i-th objective obtained its minimum value. Thereafter, the maximum value of the j-th column can be considered as an estimate of the upper bound of the j-th objective in the Pareto-optimal front and these maximum values together may be used to construct an approximation of the nadir objective vector. The main difficulty of such an approach is that solutions are not necessarily unique and thus corresponding to the minimum solution of an objective there may exist more than one solutions having different values of other objectives, in problems having more than two objectives. In these problems, the payoff table method may not result in an accurate estimation of the nadir objective vector. To illustrate, consider a three-objective problem shown in Figure 2. Minimization of the first objective will result in any solution on the trapezium CBB′F′C′C. If the point marked in a small circle on line CB is obtained by an optimization algorithm and similarly other two circles on lines CA and AB are obtained for minimizations of f_2 and f_3, respectively, a wrong estimate (z') of the nadir point ($z^{\rm nad}$) will be made.

3.2 Evolutionary Approaches

The nadir point is associated with Pareto-optimal solutions and, thus, determining a set of Pareto-optimal solutions will facilitate the estimation of the nadir point. Since an EMO algorithm is aimed at finding a set of Pareto-optimal solutions, it may be an ideal way to find the nadir objective vector. Several approaches are proposed recently.

In the naive approach, first a well-distributed set of Pareto-optimal solutions can be attempted to find by an EMO [4]. Thereafter, an estimate of the nadir objective vector can be made by picking the worst values of each objective [17]. In the context of the problem depicted in Figure 2, this means first finding a well-represented set of solutions on the plane ABC and then estimating the nadir point from them. Since EMO algorithms are not found to converge well and maintain a well-diverse set of solutions for more than three objectives [7], the accuracy of the estimated nadir point using the naive approach is questionable.

Szczepanski and Wierzbicki [17] have simulated the idea of solving multiple bi-objective optimization problems suggested in [8] using an EMO approach and construct the nadir point by accumulating all bi-objective Pareto-optimal fronts together. As discussed in our earlier study [5], such a technique is not generic and requires additional objective and variable-space niching techniques to correctly estimate the nadir point. Moreover, the procedure requires $\binom{M}{2}$ bi-objective optimizations, making it a daunting task particularly for problems having more than three objectives.

However, the idea of concentrating on a preferred region on the Pareto-optimal front, instead of finding the entire Pareto-optimal front, can be pushed further. An emphasis can be placed in an EMO approach to find only the critical extreme points of the Pareto-optimal front. Our earlier study [4] suggested two approaches in the crowding distance operator of the NSGA-II procedure and concluded in favor of the extremized crowding distance approach. In the extremized-crowded NSGA-II approach [4], we emphasized in concentrating on the best and worst solutions of each objective. In this approach, solutions on a particular non-dominated front are first sorted from minimum (with rank $R_i^{(m)} = 1$) to maximum (with rank $= N_f$) based on each objective. The rank of solution i for the m-th objective $R_i^{(m)}$ is assigned as $\max\{R_i^{(m)}, N_f - R_i^{(m)} + 1\}$. Two extreme solutions for every objective get a rank equal to N_f (number of solutions in the non-dominated front), the solutions next to these extreme solutions get a rank $(N_f - 1)$, and so on. After a rank is assigned to a solution by each objective, the maximum value of the assigned ranks is declared as the crowding distance.

Like other evolutionary optimization studies, the proposed extremized crowded NSGA-II approach did not ensure converging to the true extreme solutions exactly, as evolutionary algorithms are expected to find a near-optimal solution, rather than a true optimal solution in a finite number of solution evaluations. However, in the pursuit of estimating the nadir point for the purpose of normalizing objectives for executing different multi-objective optimization algorithms or for knowing the true range of Pareto-optimal solutions for decision-making, it is important to find the true extreme Pareto-optimal points, so that the nadir point can be estimated accurately.

In a recent study [6], the extremized crowded NSGA-II approach is ended with a bi-level local search operation on all extreme solutions to take them arbitrary closer to the true extreme solutions, so that the nadir point can be estimated more accurately. In this paper, we re-address the issue of the serial application of NSGA-II and the local search procedure and suggest a hybrid integrated approach for an accurate estimation of the nadir point.

4 Proposed Integrated Approach

Instead of applying the local search on the extreme solutions obtained by the extremized crowded hybrid NSGA-II procedure, we propose an integrated NSGA-II approach in which at certain generations the extreme solutions of the best non-dominated front are modified by the local search procedure to push them towards their true values. With such an integrated procedure, the attained accuracy is likely to be better and it would have a smaller chance of getting stuck to intermediate solutions, thereby leading to an accurate estimation of the nadir point. In the following, we outline an iteration of the proposed integrated NSGA-II procedure in which the population P_t is the current parent population of size N. Every member (i) of P_t is already ranked based on its non-domination level (ND_i) and its crowding distance (CD_i) within the population members of its own non-domination level.

Step 1: Population P_t is used to create an offspring population Q_t by using binary tournament selection, recombination and mutation operators. Two solutions are chosen at random from P_t and a hierarchical selection based on ND followed by CD is used to complete the tournament selection operation. Thereafter, two such selected solutions are recombined using the simulated binary crossover operator [3,2] to create two offspring solutions, each of which is then mutated by using the polynomial mutation operator [2]. These operators involve the following parameters: recombination probability p_c, SBX index η_c, mutation probability p_m, and mutation index η_m.

Step 2: Populations P_t and Q_t are combined together and ranked into different levels of non-domination: $P_t \cup Q_t = \{\mathcal{F}_1, \mathcal{F}_2, \ldots\}$. The set $\mathcal{F}_1$ contains non-dominated solutions of level one, and so on. Thereafter, the best N population members are chosen from the combined $2N$ population based on ranking and crowding distance criteria.

Step 3: Depending on a check on whether to perform the local search or not (which we describe a little later), in the set $\mathcal{F}_1$, we identify the worst solution ($\mathbf{x}^{(j)}$) with respect to each objective j, and modify it by using a local search procedure. The modified solution ($\mathbf{y}^{(j)}$) replaces the worst population member. For M objectives, there are M such local search operations performed in each iteration of the proposed procedure. The estimated nadir point ($\mathbf{z}^{\rm est}$) at generation t is then formed from the extreme solutions obtained by the local searches. Non-domination ranking and crowding distance computations are redone on the modified population, which we refer to as P_{t+1}.

This procedure is similar to the original NSGA-II procedure, except that the crowding distance computation is different suiting the need for emphasizing extreme solutions for the task of estimating the nadir point and that a local search procedure is used to update the extreme objective-wise solutions to make sure that the nadir point can be estimated with a desired accuracy.

We now describe the local search procedure here. The best ($f_j^{\min}$) and worst ($f_j^{\max}$) values of each objective j of the set $\mathcal{F}_1$ are first noted. We apply a bi-level local search procedure from each worst solution (solution $\mathbf{x}^{(j)}$ for which the j-th objective has the worst value in $\mathcal{F}_1$) to find the corresponding optimal solution $\mathbf{y}^{(j)}$ using the following bi-level optimization procedure. The upper-level optimization (described in (2)) uses an objective vector ($\mathbf{z}$, referred here as a reference point) as a variable vector and maximizes the j-th objective value of the optimal solution obtained by solving the corresponding augmented achievement scalarizing problem [15] (we refer to this task as the lower-level optimization task, described in (3)):

$$\begin{aligned} &\text{maximize}_{(\mathbf{z})}\ f_j^*(\mathbf{z}),\\ &\text{subject to}\ z_i \geq f_i^{(j)}{}_{\rm EA} - 0.5(f_i^{\max} - f_i^{\min}), \quad i = 1, 2, \ldots, M,\\ &\qquad\qquad\ \ z_i \leq f_i^{(j)}{}_{\rm EA} + 1.5(f_i^{\max} - f_i^{\min}), \quad i = 1, 2, \ldots, M. \end{aligned} \tag{2}$$

The term $f_j^*(\mathbf{z})$ is the optimal value of the j-th objective function of the optimal solution to the following lower-level optimization problem for which $\mathbf{z}$ is kept fixed [18]:

$$\begin{aligned} &\text{minimize}_{(\mathbf{y})} \ \max_{i=1}^{M} \left(\frac{f_i(\mathbf{y})-z_i}{f_i^{\max}-f_i^{\min}} \right) + \rho \sum_{k=1}^{M} \left(\frac{f_k(\mathbf{y})-z_k}{f_k^{\max}-f_k^{\min}} \right), \\ &\text{subject to} \quad \mathbf{y} \in \mathcal{S}, \end{aligned} \tag{3}$$

Figure 3 illustrates this local search procedure. In the lower-level optimization problem, the search is performed on the original decision variable space. The solution $\mathbf{y}{*}^{(j)}(\mathbf{z})$ to this lower-level optimization problem determines the optimal objective vector $\mathbf{f}^*$ from which we extract the j-th component and use it in the upper-level optimization problem. Thus, for every reference point $\mathbf{z}$ (a solution for the upper-level problem), the corresponding optimal augmented achievement scalarizing function is found in the lower-level loop. The upper-level optimization is initialized with the NSGA-II solution $\mathbf{z}^{(0)} = \mathbf{f}(\mathbf{x}^{(j)})$ and the lower-level optimization is initialized with the NSGA-II solution $\mathbf{y}^{(0)} = \mathbf{x}^{(j)}$.

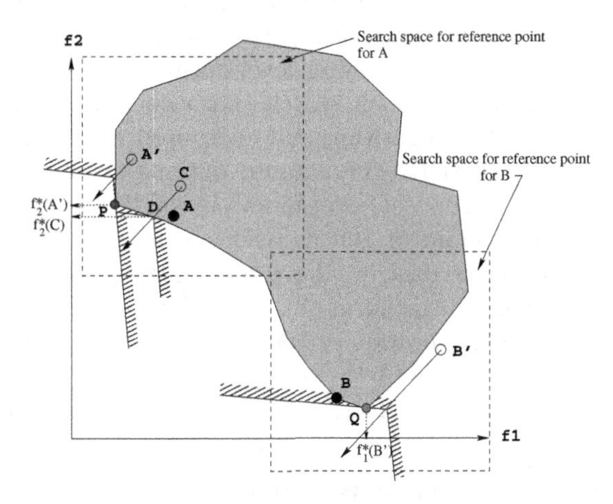

Fig. 3. Each arrow corresponds to a lower-level search for a specified reference point (C, A' or B'). The upper-level search finds a reference point having optimal worst objective (such as A' or B').

We now discuss the termination criterion of each optimization procedure. For terminating the overall NSGA-II procedure, we compute a *normalized distance* (ND) metric as follows:

$$\mathcal{D} = \sqrt{\frac{1}{M} \sum_{i=1}^{M} \left(\frac{z_i^{\text{est}} - z_i^*}{z_i^w - z_i^*} \right)^2}. \tag{4}$$

Here, the vectors $\mathbf{z}^*$ and $\mathbf{z}^w$ are the ideal and worst objective vectors of the optimization problem, respectively. These quantities can be computed once before the NSGA-II procedure by solving $2M$ different single-objective optimizations of minimizing and maximizing each objective at a time.

Since the exact final value of the $\mathcal{D}$ metric is not known a priori on an arbitrary problem, we record the change in $\mathcal{D}$ for the past τ (= 50 used here) generations. Say, $\mathcal{D}_{\max}$, $\mathcal{D}_{\min}$, and $\mathcal{D}_{\text{avg}}$, are the maximum, minimum, and average $\mathcal{D}$ values for the past consecutive τ generations. If the change $\Delta D = (\mathcal{D}_{\max} - \mathcal{D}_{\min})/\mathcal{D}_{\text{avg}}$ is smaller than a threshold Δ ($= 1(10^{-4})$ is used here), the NSGA-II procedure is terminated.

We use the same normalized distance metric to decide whether the local search needs to be performed in a particular generation of NSGA-II. At a generation, the change $\Delta_l D$ in normalized distance over the past τ_l (= 20 used here) generations is recorded. If $\Delta_l D \leq \delta$ (= 0.005 used here), the local search is performed. This reduces the number of local searches performed from not so good solutions. When the best non-dominated front has stabilized somewhat, the extreme solutions of the set are modified using the local search procedure.

Both upper and lower-level optimization tasks in the local search operation uses a point-by-point search approach which is terminated based on the chosen optimization algorithm and code used for the purpose. In all our simulations, we have used KNITRO (auto option) for the lower-level optimization task in which we have set a termination condition on the KKT error value ($\leq 10^{-6}$) or a maximum of 100 iterations whichever happens first. For the upper-level optimization task, we have used CFSQP solver [14]. The upper-level task is terminated if the norm of the Newton's direction is less than of equal to 10^{-8} or a maximum iteration of 100 is elapsed.

After the NSGA-II run is terminated, we construct the nadir point from the worst objective values of the final non-dominated set $\mathcal{F}_1$.

5 Simulation Results

In this section, we present simulation results on eight problems having three or more objectives. In most of these problems, the nadir point was difficult to obtain using the pay-off table. In all problems, we use a population of size $\max(60, 20n)$ (n is the number of variables), crossover and mutation probabilities of 0.9 and $1/n$, crossover and mutation indices of 10 and 50, respectively, and $\rho = 10^{-4}$. In each case, we make 10 different runs from different initial populations, but every time the procedure is found to converge near a particular set of extreme points, thereby leading to finding a similar nadir point every time.

5.1 Problem KM

The first problem KM, adapted from [12], is the following:

$$\begin{array}{rl} \text{minimize} & \left\{ \begin{array}{l} -x_1 - x_2 + 5 \\ \frac{1}{5}(x_1^2 - 10x_1 + x_2^2 - 4x_2 + 11) \\ (5 - x_1)(x_2 - 11) \end{array} \right\}, \\ \text{subject to} & 3x_1 + x_2 - 12 \leq 0, \quad 2x_1 + x_2 - 9 \leq 0, \quad x_1 + 2x_2 - 12 \leq 0, \\ & 0 \leq x_1 \leq 4, \quad 0 \leq x_2 \leq 6. \end{array} \tag{5}$$

The true nadir point of this problem is reported to be $\mathbf{z}^{\text{nad}} = (5, 4.6, -14.25)^T$ [9]. Table 1 shows the three extreme solutions ($\mathbf{x}^*$) found by our proposed approach. It is clear that when the worst objective values are collected together, we obtain an identical point (up to two decimal points) as that in the true nadir point. Figure 4 shows that the normalized distance value gets stabilized at around 40 generation and since $\Delta D = 50$ is used, it took another 50 generations to terminate the hybrid procedure. Interestingly, the $\mathcal{D}$ value reaches the final stabilized value very quickly, thereby indicating the efficiency of the proposed procedure.

Table 1. Extreme points found by the proposed approach on problem KM

$\mathbf{x}^*$		Estimated $\mathbf{z}^{\rm nad}$		
0.000	0.000	**5.000**	2.200	-55.000
0.000	6.000	-1.000	**4.600**	-25.001
3.500	1.501	0.000	-3.100	**-14.251**

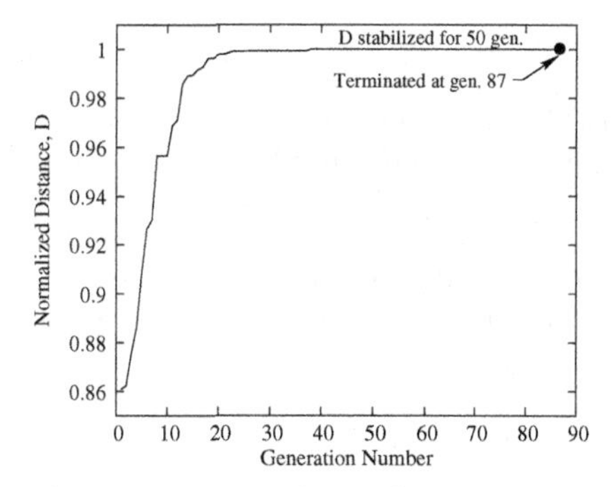

Fig. 4. Variation of $\mathcal{D}$ with generation on KM

5.2 Problem SW1

The second problem SW1 is as follows [17]:

$$\begin{aligned} &\text{minimize} \left\{ \begin{array}{l} f_1(\mathbf{x}) = -(100 - 7x_1 - 20x_2 - 9x_3) \\ f_2(\mathbf{x}) = -(4x_1 + 5x_2 + 3x_3) \\ f_3(\mathbf{x}) = -x_3 \end{array} \right\}, \\ &\text{subject to } 1\tfrac{1}{2}x_1 + x_2 + 1\tfrac{3}{5}x_3 \le 9, \quad x_1 + 2x_2 + x_3 \le 10, \\ &\qquad\qquad x_i \ge 0, \quad i = 1, 2, 3. \end{aligned} \tag{6}$$

The previous study [17] reported the true nadir point to be $\mathbf{z}^{\rm nad} = (-3.6364, 0, 0)^T$. Table 2 shows two extreme solutions ($\mathbf{x}^*$) (hence, the true nadir point) found by our proposed approach. Figure 5 shows the progress of the proposed approach.

5.3 Problem SW2

The third problem SW2 originates from [17]:

$$\begin{aligned} &\text{minimize} \left\{ \begin{array}{l} 9x_1 + 19.5x_2 + 7.5x_3 \\ 7x_1 + 20x_2 + 9x_3 \\ -(4x_1 + 5x_2 + 3x_3) \\ -(x_3) \end{array} \right\}, \\ &\text{subject to } 1.5x_1 - x_2 + 1.6x_3 \le 9, \quad x_1 + 2x_2 + x_3 \le 10, \\ &\qquad\qquad x_i \ge 0, \quad i = 1, 2, 3. \end{aligned} \tag{7}$$

The true nadir point for this problem is reported to be $\mathbf{z}^{\rm nad} = (94.5, 96.3636, 0, 0)^T$ [17]. The original study [17] found a close point $(94.4998, 95.8747, 0, 0)^T$ using multiple, bi-objective optimization simulation using an EMO procedure. The outcome is not identical to the true nadir point. Table 3 shows the three extreme solutions found by our proposed approach. We obtain the true nadir point. Due to an identical behavior of $\mathcal{D}$ variation with generation number on this and subsequent problems, we do not show the figures here.

Table 2. Extreme points found by the proposed approach on problem SW1

$\mathbf{x}^*$			Estimated $\mathbf{z}^{nad}$		
0.0000	3.1818	3.6364	**-3.6364**	-26.8182	-3.6364
0.0000	0.0000	0.0000	-100.0000	**0.0000**	**0.0000**

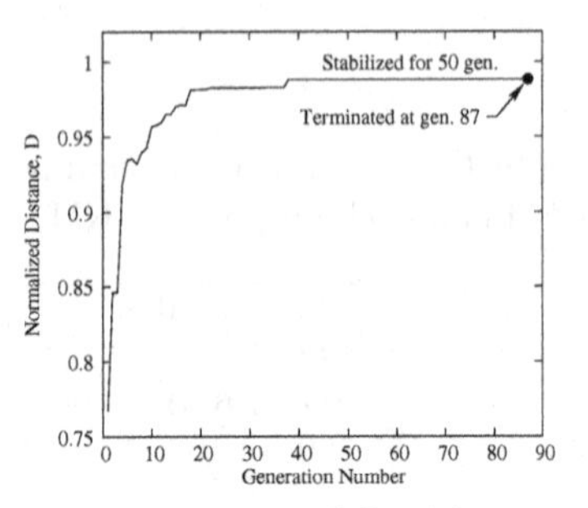

Fig. 5. Variation of $\mathcal{D}$ with generation on SW1

Table 3. Extreme points found by the proposed approach on problem SW2

$\mathbf{x}^*$			Estimated $\mathbf{z}^{nad}$			
4.0000	3.0000	0.0000	**94.5000**	88.0000	-31.0000	0.0000
0.0000	3.1818	3.6363	89.3182	**96.3636**	-26.8182	-3.6363
0.0000	0.0000	0.0000	0.0000	0.0000	**0.0000**	**0.0000**

5.4 Problem KSS1

The linear KSS1 problem [13] was found to be difficult for estimating the nadir point:

$$\begin{aligned} &\text{maximize} \left\{ \begin{array}{l} 11x_2 + 11x_3 + 12x_4 + 9x_5 + 9x_6 - 9x_7 \\ 11x_1 + 11x_3 + 9x_4 + 12x_5 + 9x_6 - 9x_7 \\ 11x_1 + 11x_2 + 9x_4 + 9x_5 + 12x_6 + 12x_7 \end{array} \right\}, \\ &\text{subject to } \textstyle\sum_{i=1}^{7} x_i = 1, \\ &\qquad\qquad x_i \geq 0, \quad i = 1, 2, \ldots, 7. \end{aligned} \tag{8}$$

The true nadir point is reported to be $\mathbf{z}^{\text{nadir}} = (0, 0, 0)^T$ [13]. Table 4 shows the three extreme solutions found by our proposed approach. Our approach finds a near nadir point with a slight error in the second objective value (as shown in Figure 6 the error is not visually detectable). This problem is a difficult one to solve for estimating the exact nadir point, because of the slow slope leading to each of the three extreme points, as shown by a set of representative solutions obtained through a *clustered* NSGA-II, in which NSGA-II's crowding distance method is replaced by the k-mean clustering method [2]. In this problem, it is easy to get stuck to a non-dominated point close to one or more extreme points. Our approach seems to have found the exact extreme values for first and third

Table 4. Extreme points found by the proposed approach on problem KSS1

$\mathbf{x}^*$							Estimated $\mathbf{z}^{nad}$		
1.000	0.000	0.000	0.000	0.000	0.000	0.000	**0.000**	11.000	11.000
0.000	0.994	0.000	0.000	0.000	0.001	0.004	10.910	**-0.026**	11.006
0.000	0.000	1.000	0.000	0.000	0.000	0.000	11.000	11.000	**0.000**

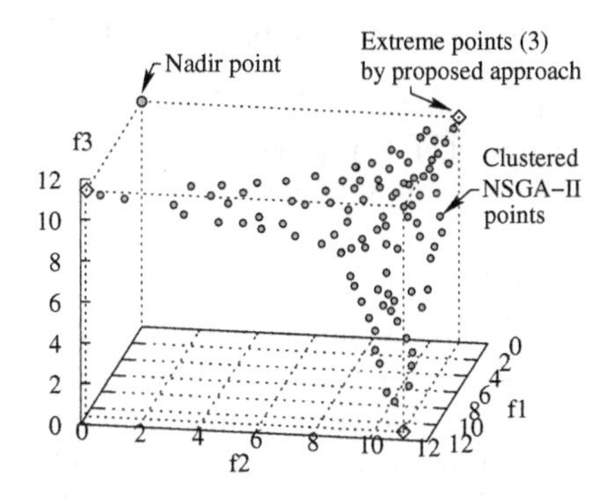

Fig. 6. Pareto-optimal front shows long narrow regions near extreme points in problem KSS1

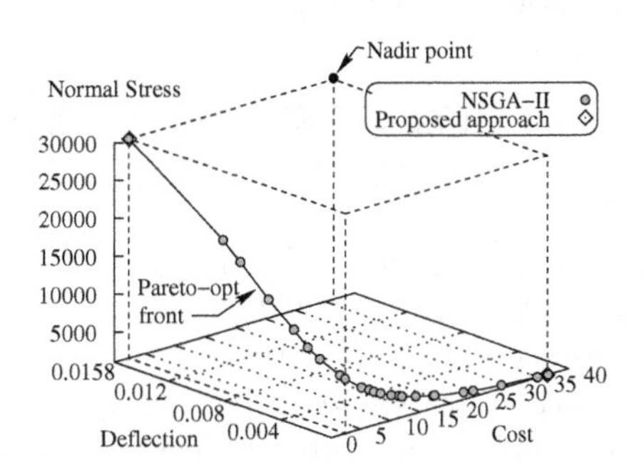

Fig. 7. Pareto-optimal front and two obtained points for problem WELD

objectives and managed to get to a near-by point around the extreme of the second objective.

5.5 Problem KSS2

Next, we consider another linear problem KSS2 [13]:

$$\begin{aligned} &\text{maximize } (x_1, x_2, x_3),\\ &\text{subject to } x_1 + 2x_2 + 2x_3 \leq 8, \quad 2x_1 + 2x_2 + x_3 \leq 8, \quad 3x_1 - 2x_2 + 4x_3 \leq 12,\\ &\qquad x_i \geq 0, \quad i = 1, 2, 3. \end{aligned} \tag{9}$$

The nadir point is reported to be $\mathbf{z}^{\text{nad}} = (0, 0, 0)^T$. Table 5 presents the extreme solutions obtained by our approach. The true nadir point is found by our approach in this problem.

Table 5. Extreme points found by the proposed approach on problem KSS2

$\mathbf{x}^*$			Estimated $\mathbf{z}^{\text{nad}}$		
0.000	3.818	0.166	**0.000**	3.818	0.166
3.344	0.000	0.432	3.344	**0.000**	0.433
3.253	0.628	0.000	3.253	0.628	**0.000**

Now we consider three more problems, borrowed from engineering fields. On each of these problems, the exact nadir point is not known, but wherever possible we explain the accuracy of the nadir point obtained by our approach.

5.6 Problem WELD

The WELD problem has four variables and three objectives, and is formulated in [6]. Our previous study [6] introduced the WELD problem which has four

Table 6. Extreme points found by the proposed approach on problem WELD

$\mathbf{x}^*$				Estimated $\mathbf{z}^{\text{nad}}$		
1.7356	0.4788	10.0000	5.0000	**36.4221**	0.000439	1008.0000
0.2444	6.2175	8.2915	0.2444	2.3810	**0.015759**	**30000.1284**

variables and three objectives. The nadir point was estimated to be $\mathbf{z}^{\text{nad}} = (36.4209, 0.0158, 30000)^T$. Table 6 presents two extreme points found by our proposed approach of this paper. The extreme points for the second and third objectives are found to be identical in this problem, indicating that although the problem has three objective functions, the Pareto-optimal front is two-dimensional, as is also confirmed by the original NSGA-II points in Figure 7. The nadir point estimated by our approach is $(36.4221, 0.0158, 30000.1284)^T$, which is close to that obtained by the earlier study [6].

5.7 Problem CAR

The seven-variable, three-objective CAR problem is described in [10]. No previous study exists on this problem for finding the nadir point. In Table 7, we present two extreme points obtained by our procedure. Thus, the nadir point estimated by our approach for this problem is $\mathbf{z}^{\text{nad}} = (42.767, 4.000, 12.521)^T$. Figure 8 shows the complete Pareto-optimal front with a set of representative clustered NSGA-II solutions. It is clear from the plot that the above two extreme points are adequate to cover the extreme objective values of the Pareto-optimal front and is able to locate the nadir point of the problem.

5.8 Problem WATER

Finally, we consider the WATER problem [16], which is also described in [2]. For this problem, the exact nadir point is not known. However, since there are

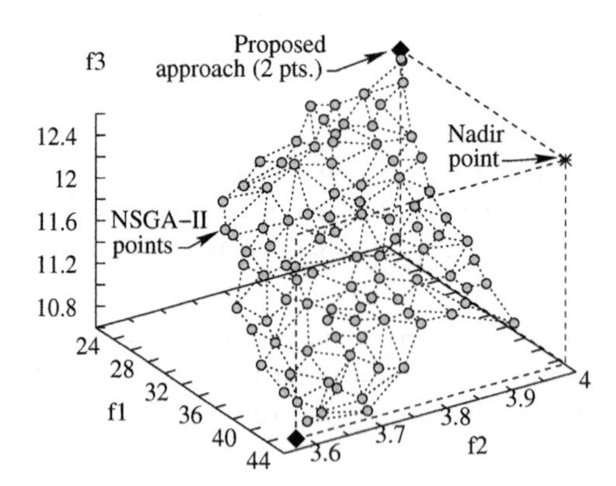

Fig. 8. Extreme objective vectors covers the entire Pareto-optimal front for problem CAR

Table 7. Extreme points found by the proposed approach on problem CAR

$\mathbf{x}^*$							Estimated $\mathbf{z}^{\text{nad}}$		
1.500	1.350	1.500	1.500	2.625	1.200	1.200	**42.767**	3.585	10.611
0.500	1.226	0.500	1.208	0.875	0.884	0.400	23.589	**4.000**	**12.521**

Table 8. Extreme points found by the proposed approach on problem WATER

$\mathbf{x}^*$			Estimated $\mathbf{z}^{\text{nad}}$				
0.010	0.100	0.100	**1.038**	0.020	0.949	0.075	5.649
0.450	0.098	0.010	0.916	**0.900**	0.936	0.033	0.002
0.114	0.100	0.010	0.918	0.228	**0.951**	0.031	0.285
0.098	0.010	0.100	0.918	0.197	0.095	**2.671**	**5.713**

three variables and five objectives, some redundancy in the objectives is expected for the Pareto-optimal solutions. An application of NSGA-II to this problem [2] (page 388) was found to indicate some correlations among the obtained representative solutions. Table 8 presents the extreme points obtained for this problem by our approach. We observe that the extreme points for objectives f_4 and f_5 come from an identical solution. The estimated nadir point using our procedure is $\mathbf{z}^{\text{nad}} = (1.038, 0.900, 0.951, 2.671, 5.713)^T$.

6 Conclusions

In this paper, we have extended our previous study on a serial implementation of an EMO procedure followed by an MCDM based local search approach to find extreme points accurately for estimating the nadir point of a multi-objective optimization problem. The nadir point in multi-objective optimization is used in normalizing objectives which is necessary for different multi-criterion optimization algorithms. Besides, the task of estimating the nadir point for three or more objectives is a open research task in multi-criterion optimization literature. Nadir points can only be estimated accurately if (i) objective-wise extremes and (ii) Pareto-optimal solutions are found. Due to this two-pronged requirements, we have suggested a bi-level local search task. The local search is employed with extreme non-dominated solutions only when the best non-dominated front has stabilized somewhat, thereby making the overall method computationally tractable. On a set of five test problems and three engineering design problems, the proposed integrated procedure has able to find the exact nadir point quite accurately.

This work is also important from another point of view. This work demonstrates how a local search approach can be integrated with an evolutionary population-based approach adaptively and used sparingly for a complex optimization to ensure accurate convergence.

Acknowledgments

Authors acknowledge the support from the Academy of Finland, Foundation of Helsinki School of Economics (grant 118319), and the Jenny and Antti Wihuri Foundation during the course of this study.

References

1. Benayoun, R., de Montgolfier, J., Tergny, J., Laritchev, P.: Linear programming with multiple objective functions: Step method (STEM). Mathematical Programming 1(3), 366–375 (1971)
2. Deb, K.: Multi-objective optimization using evolutionary algorithms. Wiley, Chichester (2001)
3. Deb, K., Agrawal, R.B.: Simulated binary crossover for continuous search space. Complex Systems 9(2), 115–148 (1995)
4. Deb, K., Chaudhuri, S., Miettinen, K.: Towards estimating nadir objective vector using evolutionary approaches. In: Proceedings of the Genetic and Evolutionary Computation Conference (GECCO 2006), pp. 643–650. The Association of Computing Machinery (ACM), New York (2006)
5. Deb, K., Miettinen, K.: A review of nadir point estimation procedures using evolutionary approaches: A tale of dimensionality reduction. Technical Report KanGAL Report Number 2008004, Kanpur Genetic Algorithms Laboratory (KanGAL), Indian Institute of Technology Kanpur, India (2008)
6. Deb, K., Miettinen, K., Chaudhuri, S.: Estimating nadir objective vector: Hybrid of evolutionary and local search. Technical Report Electronic Working Paper W-440, Helsinki School of Economics, Finland (2008)
7. Deb, K., Saxena, D.: Searching for Pareto-optimal solutions through dimensionality reduction for certain large-dimensional multi-objective optimization problems. In: Proceedings of the World Congress on Computational Intelligence (WCCI 2006), pp. 3352–3360 (2006)
8. Ehrgott, M., Tenfelde-Podehl, D.: Computation of ideal and nadir values and implications for their use in MCDM methods. European Journal of Operational Research 151, 119–139 (2003)
9. Eskelinen, P., Miettinen, K., Klamroth, K., Hakanen, J.: Pareto navigator for interactive nonlinear multiobjective optimization. OR Spectrum (in press)
10. Gu, L., Yang, R.J., Tho, C.H., Makowski, L., Faruque, O., Li, Y.: Optimization and robustness for crashworthiness of side impact. International Journal of Vehicle Design 26(4) (2001)
11. Isermann, H., Steuer, R.E.: Computational experience concerning payoff tables and minimum criterion values over the efficient set. European Journal of Operational Research 33(1), 91–97 (1988)
12. Klamroth, K., Miettinen, K.: Integrating approximation and interactibe decision making in multicrietria optimization. Operations Research 56, 222–234 (2008)
13. Korhonen, P., Salo, S., Steuer, R.: A heuristic for estimating nadir criterion values in multiple objective linear programming. Operations Research 45(5), 751–757 (1997)

14. Lawrence, C., Zhau, J.L., Tits, A.L.: User's guide for cfsqp version 2.5: A C code for solving (large scale) constrained nonlinear (minimax) optimization problems, generating iteration satisfying all inequality constraints. Technical Report Institute for Systems Research TR-94-16r1, Electrical Engineering Department and Institute for Systems Research, University of Maryland, College Park, MD 20742 (1997)
15. Miettinen, K.: Nonlinear Multiobjective Optimization. Kluwer, Boston (1999)
16. Ray, T., Tai, K., Seow, K.C.: An evolutionary algorithm for multiobjective optimization. Engineering Optimization 33(3), 399–424 (2001)
17. Szczepanski, M., Wierzbicki, A.P.: Application of multiple crieterion evolutionary algorithm to vector optimization, decision support and reference point approaches. Journal of Telecommunications and Information Technology 3, 16–33 (2003)
18. Wierzbicki, A.P.: The use of reference objectives in multiobjective optimization. In: Fandel, G., Gal, T. (eds.) Multiple Criteria Decision Making Theory and Applications, pp. 468–486. Springer, Berlin (1980)

Author Index